Residential Carpentry

Mortimer P. Reed

Residential Carpentry

John Wiley & Sons

New York

Chichester

Brisbane

Toronto

Designer: Judith Fletcher Getman

Cover Art: John Hite

Section Art: Joseph Gillians

Library of Congress Cataloging in Publication Data

Reed, Mortimer P
  Residential carpentry.

  Includes index,
  1. Carpentry. 2. House construction. I. Title.
TH5606.R43   694   79-18092
ISBN 0-471-08283-X

Printed in the United States of America

Preface

This book is not written exclusively for those studying carpentry in school. It is written for those interested in learning—or in learning more—about the procedures followed by carpenters in light construction.

As an industry, building is divided into heavy construction and light construction. Heavy construction includes roads and bridges, factories, high-rise office and apartment buildings, major shopping centers, schools, large churches, and most government buildings. The dominant structural materials are concrete and steel. Light construction includes homes, small apartment buildings, town houses, small churches, stores, farm buildings, and small office buildings. The dominant structural material is wood.

The content of this book is limited to the tools, materials, and building techniques used in light construction. Section 1 covers the materials, the basic tools, and the safety rules with which every carpenter must be thoroughly familiar.

Work in light construction is divided into structural, finishing, and mechanical phases. Structural workers prepare the site and build the skeleton of the building—its foundation, floor, wall, ceiling, and roof systems. They set heavy timbers to hold dirt banks in place. They build and place forms for concrete. They erect the wood structure of the building, and prepare it for application of finish materials. This work, called rough carpentry, is discussed in Sections 2 and 3.

Finishers cover the skeleton with skins—the visible surfaces of the building, both outside and inside. They apply roofing, siding, and trim to the outside. They install windows and hang doors. They apply finishing materials to walls, ceilings, and floors. They set cabinets and install hardware. They prepare surfaces for coatings. This work, called finish carpentry, is discussed in Sections 4 and 5.

Mechanical workers install the lines and equipment that make the building usable—heating, air conditioning, electrical, plumbing, and fuel supply systems. This phase must be carefully coordinated with the work of carpenters. Mechanical work is not covered in this text.

Within the sections on rough carpentry and finish carpentry the principles and techniques are presented in step-by-step procedures as fully illustrated as possible. Many of the basic steps followed in completing simple tasks are also part of a standard sequence in more complex tasks later in construction. As a result of this organization, the reader may steadily develop his skills and increase his knowledge, and find it easy to discover his area of greatest interest. In carpentry there is opportunity at all levels.

In small towns and rural areas, for example, carpenters are likely to be involved in all aspects of light construction. In larger communities with larger contractors, there is enough business volume so that carpenters usually do rough carpentry or finish carpentry, but not both. In major cities carpenters are often specialists. They may specialize in building and erecting forms for a concrete contractor. They may specialize in framing roofs, or in applying one type of finish material, such as wallboard or flooring. They may specialize in installing trim, or building cabinets or stairways in a small cabinet shop. Or they may be small businessmen who specialize in remodeling.

The field of carpentry offers many opportunities. Each requires a certain combination of skill, intelligence, training, education, and experience. Job opportunities fall into three general categories.

1 **Production workers** In this category are the rough carpenters, finish carpenters, helpers, and laborers who help to build on a site or in a factory. They work mostly with materials.

2 **Supervisors** In this category are the foremen and job superintendents who direct the activities of production workers. They work mostly with people. Many supervisors start their careers as production workers. Job superintendents more often hold degrees in such specialties as civil engineering.

3 **Managers** In this broad category are materials estimators, building supply dealers, materials salespersons, carpenter contractors, and builders. They work mostly with ideas and problems. Managers usually have a sound knowledge of carpentry, plus further education or experience in other specialized fields of knowledge.

This book will help you decide what phase of carpentry interests you the most, and where you best fit into the field of residential building.

Mortimer P. Reed

Acknowledgments

Many people, both known and unknown, contributed to the development of this book. Among the known are Professor Richard F. Bortz of Southern Illinois University, who proposed the format, and the many advertising and promotion people in the building industry who searched their files for photographs showing specific products and procedures. Unknown are the thousands of fine carpenters of the past who, over the years, developed and perfected the methods of quality construction presented here for future carpenters.

Thanks also go to the following individuals for their help in reviewing the manuscript.

J. G. Edgin, San Jose City College

Bill TeVogt, Hennepin Technical Centers

Rodney E. Gray, Southern Maine Vocational-Technical Institute

Gaspar Lewis, Pinellas Vocational-Technical Institute

E. R. McKoy, Mitchell Community College

G. A. Whitfield, College of the Redwoods

James Martin, Three Rivers Community College

E. L. Foster, Sr., San Antonio College

J. F. McKinley, San Antonio College

Philip O. Whitaker, Anson Technical Institute

Edgar Payne, Spokane Community College

M. P. R.

Contents

Section 3
Rough Carpentry   115

Section 4

Exterior Finish Carpentry 333

Section 5

Interior Finish Carpentry 495

1 Carpenters and Their Craft

By the broadest definition a carpenter is an artisan who builds with wood or wood substitutes. In ancient Rome men built two-wheeled carriages out of wood. The Latin word for this type of carriage was **carpentum,** and a man who built them was called **carpentarius.** He was the forerunner of today's versatile artisans, who are known in America as carpenters and in Great Britain as joiners.

Carpentry is now one of more than 25 skilled trades in the building industry. About one-third of all workers in construction are classified as carpenters. Some work only with wood. Some almost never work with wood. Some work with both wood and wood substitutes. Yet all are considered carpenters.

For many thousands of others carpentry is an avocation or hobby. They may not earn their living building things, but they get great enjoyment from it. Whether your interest in carpentry is as a future trade or hobby, this book will help you fulfill your need for knowledge of the materials, tools, terms, and procedures used in residential construction.

What does it take to be a good carpenter? First, you must have the natural ability to work with your hands, and the interest to convert those abilities into practical skills. You must know local building terms. Building terms are referred to in this book as they occur in construction procedures and are summarized in a glossary following the final chapter.

Next, you must learn everything you can about the materials with which you work. These materials and their characteristics are discussed in Chapter 1. You must know the tools of the trade, how to use them safely, and when and where to use them. The tools used in rough carpentry are discussed in Chapter 2. The additional basic tools used in finish carpentry are discussed in Chapter 14. Special tools used with special materials are included in the chapters that describe installation of those materials.

Perhaps most important, a good carpenter must know the proper procedures for achieving top quality. Whether you achieve top quality depends on your skills, knowledge, desire for quality, and the standards of quality set by your employer or immediate supervisor. If you know what constitutes quality construction, you can build to a lesser standard when costs or other factors require it. But if you don't learn how to do a job well, you won't be able to achieve high quality when it is required.

Finally, you must have the ability to use your mind, the education to stimulate this mental talent, and the experience to know how to solve the special construction problems that occur on every job. This book is organized to provide the education and the background for useful experience—from driving the first stake at the building site to hanging the last shutter on the finished house.

1

The Material You Use

In their normal work most carpenters use only three types of building materials. One is wood, which is grown and harvested like a crop, then processed for use in construction. Sheet materials—the second type —are manufactured in standard sizes for specific purposes. The third type is the fasteners that connect wood to wood, wood to sheet, and sheet to sheet.

Wood

Ever since we first learned to use tools, we have built with wood. In many ways it is an ideal building material. Wood is easily cut, shaped, fastened, and smoothed with simple tools. Compared to other structural materials, it is lightweight and very strong, yet flexible. It has good insulating value and is less affected by corrosion. Wood is durable and less expensive than most other materials. As a finished surface material, it is both attractive and easy to preserve.

Nevertheless, wood has its faults. It shrinks and expands. It has knots in it and, therefore, is not consistent in strength or hardness. It may split, splinter, warp, bend, or break. Wood is the favorite diet of some insects. Perhaps most important, it is not as plentiful as it once was.

People are largely responsible for the depletion of our resources of natural wood in this country. However, we are also responsible for reforestation, better management and care of forests, and prevention of destructive fires. We have found ways to overcome the faults in wood with chemical preservatives and have developed materials that replace wood in many of its original uses.

In spite of all these changes, nearly half of all wood today is processed for use in construction. This wood falls into six categories: dimensional lumber, plywood,

Fig. 1-0.1 This diagram of a tree shows how photosynthesis works.

fiberboard, siding, exterior trim, and millwork. To understand the properties of each of these woody building materials, you need to know how a tree grows.

How Wood Develops

In order to grow and remain healthy, a tree needs food, water, and sunlight. The trunk of the tree, from which all lumber is cut, is the conduit through which water travels from the earth to the leaves (Fig. 1-0.1). There the moisture combines with carbon dioxide from the air. Energy from sunlight triggers a process called **photosynthesis,** which transforms the water and carbon dioxide into the food that nourishes the entire tree.

A cross section through the trunk of a tree (Fig. 1-0.2) shows how this process works. At the center of the tree is the **heartwood**—dense dead wood that gives the tree its strength and rigidity against the forces of nature. Surrounding the heartwood is **sapwood,** a series of living layers that carries moisture, or **sap.**

The layers of sapwood show in a cross section as rings. A pair of rings indicates the amount of growth in a year; therefore they are called **annular rings.** One of the rings is light in color and the wood is rather porous. This is **springwood,** formed during a period of rapid growth early in the year. The other ring is darker and more dense, and usually thinner. This is **summerwood.** By studying the rings of sapwood you can determine the age of the tree. By studying the widths of the rings of springwood and summerwood, you can determine how favorable the climatic conditions were for growth in any year. The wider the ring, the better the growing year.

As trees grow older, the inner rings of sapwood die and become heartwood, and new layers of sapwood develop to form the conduit of life.

Surrounding the sapwood is a layer of reproductive cells called the **cambium.** The cambium consists primarily of long, thin tubes of wood fiber composed mostly of cellulose; they are held together with a similar but different substance called **lignin.** In chemical composition cellulose is similar to starch and sugar; because of this the sap of some trees, particularly the maple, has a sweet taste. The cambium layer produces new wood that turns into sapwood as the tree trunk thickens.

Protecting the tree from damage and disease are two layers of bark. The inner layer, called **bast,** is thin, fibrous, and flexible. The outer layer may also be thin and flexible, such as the bark on birch trees, or it may be thick and easily broken, such as the bark on cork trees.

Between the bast and the cambium are cells of wood fiber that extend at right angles through the growth rings. Known as **medullary rays,** they carry food between the two layers and, in the sapwood, act as

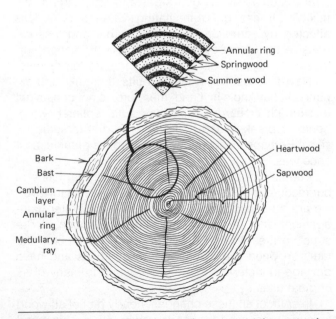

Fig. 1-0.2 A cross section through a tree and the parts of the trunk.

storage tanks for food. Although medullary rays are present in all trees, they are more noticeable in some species, such as oak, where the rays add beauty to the finished wood.

Although all trees are similar in their method of growth, the wood from these trees varies widely in its properties and uses. The most important division of woods is into hardwoods and softwoods.

Hardwoods and Softwoods

The terms hardwood and softwood for classifying trees are very misleading, because the wood of some hardwood trees is softer than the wood of some softwood trees. **Softwoods** (Table I) are the woods of trees with needlelike leaves that are not shed at the end of the growing season. In this category (called **conifers**) are

Table I Common softwoods and their characteristics

Species	Description and use
Douglas fir, hemlock, larch	A group of strong, dense softwoods with straight grain and a strong color contrast between rings in the sapwood. *Main use:* structural lumber.
Southern yellow pine	A strong, stiff, and fairly heavy wood with a slightly reddish tone. (The name comes from yellow pollen shed in the spring.) Has a tendency to warp. *Main use:* structural lumber.
Eastern hemlock, tamarack	A strong, dense, reddish brown wood. Has a tendency to splinter. *Main uses:* structural lumber and board sheathing.
Engelmann spruce, Sitka spruce, lodgepole pine	A group of lightweight softwoods with low shrinkage and small, tight knots. Not as strong as Douglas fir and Southern yellow pine. *Main uses:* general construction and scaffolding.
Eastern spruce	Similar to but heavier than western spruces. *Main uses:* framing and general construction.
White fir	A lightweight softwood with little color difference in the grain. Takes paint and stains well. *Main uses:* structural lumber and interior finishing.
Redwood	A strong softwood with excellent resistance to decay. Sapwood is almost white, while heartwood is reddish brown. If left unfinished, it weathers from a medium gray to black. *Main uses:* structural lumber where the wood is exposed, as on decks and patios, and on both exterior and interior surfaces of walls.
Western red cedar	A durable softwood with excellent resistance to damage from decay and insects. Reddish brown in color, but weathers to a soft gray. *Main uses:* exterior trim and shingles on walls and roofs.
Cypress	Native to swampy areas of the Southeast, it is highly resistant to decay from water. *Main uses:* siding, posts, windows, exterior trim, and other exposed members.
Ponderosa pine	A dimensionally stable and easily finished softwood with little variation in density or color between heartwood and sapwood. *Main uses:* trim, doors, windows, and paneling.
Eastern white pine	Once common but now scarce because of overuse. *Main uses:* interior trim, windows, and doors of high quality.
Idaho white pine, sugar pine	Soft, easily smoothed woods with light grain color and few defects. *Main uses:* cabinetry and paneling.
Eastern red cedar	Has dark red heartwood, white sapwood, a distinctive grain, and a pleasant aroma. *Main uses:* paneling and as a mothproofing liner for clothes closets.
Yellow poplar	Strong, soft, and lightweight with a high shrinkage factor. *Main uses:* paneling and as a core in plywood.

most of the woods used for structural framing—pines, firs, hemlock, spruces, and larch—and for exterior use —cypress, redwood, and cedar. The term softwood probably originated because foresters and carpenters found the woods of the most common conifers easier to saw and nail than other species.

Table II Common hardwoods and their characteristics

Species	Description and use
Oak	A very dense, durable wood with fairly uniform grain and a color ranging from white to light yellow. *Main uses:* flooring, stair parts, paneling.
Maple	Not as strong nor dense as oak, but as strong and durable. Cut to produce a wide variety of grain patterns. *Main uses:* flooring, stair parts, doors, paneling.
Beech	A stiff hard wood with a reddish brown tone. *Main uses:* flooring and interior trim.
Basswood	A soft, even-textured, and easily smoothed wood with very light brown heartwood and sapwood. *Main uses:* interior trim and cabinetry.
Birch	A dimensionally stable hardwood with an even texture and fine grain pattern. Takes a finish well. *Main uses:* doors and high-quality cabinets.
Mahogany	A very heavy, dark reddish brown wood with a smooth surface that takes a high gloss. Known for its beautiful grain, it grows in Central America and South America. *Main uses:* paneling and furniture.
Lauan	A general term for a group of woods similar to mahogany that grows in Southeast Asia and islands of the East Indies. More strongly grained and not as easily finished. *Main use:* paneling.
Walnut	A strong hard wood with dark brown heartwood and light brown sapwood. Cut to produce a wide range of beautiful grain patterns, which are often matched. *Main uses:* furniture and paneling.
Cherry	A hard, strong wood with reddish brown heartwood and dark brown annular rings. Grains range from straight to swirls. *Main uses:* furniture and paneling.
Gum	Softer than most hardwoods. Has a wide range of grain patterns and color variation from pink to medium brown. *Main use:* paneling.
Elm	A warm yellow brown wood with medium grain that takes finishes easily. *Main use:* paneling.
Pecan	A light tan wood with an attractive and uniform grain. *Main use:* paneling.

Hardwoods (Table II) are the woods of trees with broad leaves that fall off once a year. In this category (called **deciduous**) are many of the woods used for cabinets, trim, decorative panels, and furniture— birch, maple, oak, walnut, cherry, and mahogany.

A rough carpenter will use mostly softwoods, a finish carpenter will use both softwoods and hardwoods, and a cabinetmaker will use mostly hardwoods.

Logs to Lumber

From the time a tree is felled in the forest until lumber is delivered to a building site, wood passes through several important processes. First, loggers strip off all branches, and cut the trunk into logs 12 to 40′ long. Then the logs are transported on trucks, flatcars, or logging sleighs or floated on lakes and rivers to a sawmill.

At the sawmill, logs pass through a barker, a machine that strips off the bark and bast and leaves the logs clean for sawing. The wood expert who operates the saws—the **sawyer**—determines the best way to cut each log.

Just how the log is cut depends on the size of the log, current demand, and the power saws at the mill. The method of sawing affects both the strength and appearance of the lumber. One common method is to saw parallel slices tangent to the annular rings (Fig. 1-0.3). The result is called **plain-sawed** or **flat-sawed** lumber. The other method, called **quarter-sawing,** is to saw the log through the center into four pieces and then slice each piece at an angle of 45 to 90° to its surface (Fig. 1-0.4). The result is called **edge-grained** or **vertical-grained** lumber. Most logs are cut in a combination of flat-sawing and quarter-sawing.

Figure 1-0.5 shows a typical pattern for cutting a log. Note that most of the heartwood is cut into timbers and structural lumber. Heartwood warps less than wood cut from other parts of the log. Flooring is quarter-sawed to minimize the tendency of wood to warp, shrink, or absorb moisture. Plain-sawed pieces closest to the bark are the most likely to change their size and shape; heartwood is least likely to change (Fig. 1-0.6).

After a log has been cut into pieces of various sizes, each piece is edge-trimmed and cut to width and length. The pieces are dried to remove excess water.

Drying

At the time a tree is felled, 35 to 55% of its weight is water. Much of this water must be removed from the green wood before it can be used for construction. As wood loses water, it becomes lighter in weight, shrinks, gains strength, and becomes less susceptible to decay.

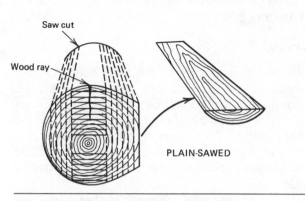

Fig. 1-0.3 How a log is cut in plain-sawing.

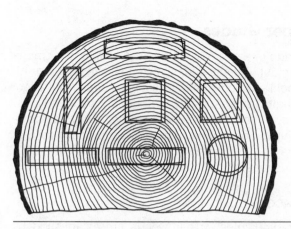

Fig. 1-0.6 Lumber shrinks and warps in different ways, depending on where it was cut from the log.

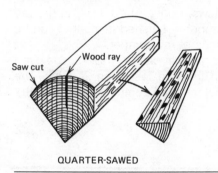

Fig. 1-0.4 How a log is cut in quarter-sawing.

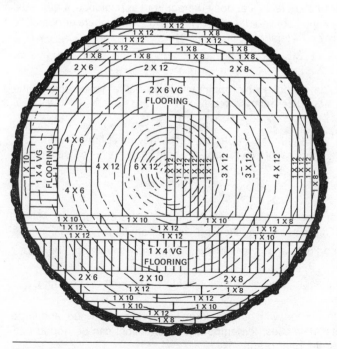

Fig. 1-0.5 Typical cutting plan of a large log.

Lumber is dried in one of two ways—in the open air or in a kiln (oven). In both methods the lumber is stacked in piles with thin wood strips between the layers so that air can circulate freely around each piece of lumber. In **air-drying** the lumber is stacked outdoors under cover and left alone until moisture content (abbreviated as M.C.) drops to 19%. By today's standards, 19% is the maximum allowable moisture content and 15% is the more normal figure. A 15% M.C. means that the weight of water remaining in the wood is 15% of the total weight of the wood.

The time needed to air-dry lumber varies with local weather conditions, but 4 months is about the maximum air-drying time. Because this process is so slow, most lumber today is **kiln-dried.** It is strip-piled on wheeled pallets, which are trundled into the ovens. Kiln-drying takes only about one-fourth the time of air-drying, and moisture contents as low as 5% can be reached, as is sometimes necessary for hardwoods.

Two kinds of water are driven out of wood by the drying process. Water in the lumber's cell cavities, called **free water,** is driven off first. When the cavities are empty, the wood has reached the **fiber saturation point**—25 to 30% M.C. Free water will not return to lumber unless it is submerged for a long time.

Lumber doesn't begin to shrink noticeably until it reaches the fiber saturation point. But, as the water that lies in wood fibers—the **absorbed water**—is driven off by drying, those fibers shrink. This shrinkage is not uniform, however. Wood shrinks very little in length, but quite a bit in width and thickness (see Fig. 1-0.6). When moisture content becomes the same as humidity in surrounding air, the wood has reached **equilibrium moisture content.** From that time on, wood will swell a little when the air is damp and shrink a little when the air is dry.

After drying, the final step is grading the lumber.

Lumber Grades

The grade of a piece of lumber is determined by moisture content, defects and, sometimes, stress rating. All wood contains some defects, however small, that affect its strength and beauty. Flaws in beauty are clearly visible. Flaws that affect strength are not always as obvious. The lumber industry grades its product so that builders and carpenters can be assured of the minimum qualities of the lumber they buy.

Hardwoods

The grading rules for hardwoods are established by the National Hardwood Lumber Association. Standards vary slightly from one species to the next, but the grades are the same. The grading is based on an inspection of the poorest side of the piece, since most hardwoods are used for finishing, where appearance is important.

The top grade of hardwood is **FAS,** an abbreviation for **first and seconds,** which are grouped together into a single grade. Next come **select** and **common.** All common-grade hardwood of some species is #1 common, while other species have common grades of #1, #2, #3A, and #3B.

Softwoods

The National Bureau of Standards, in its bulletin *Voluntary Product Standard PS 20-70,* sets down the basic rules for grading softwoods. Each major lumber association writes its own set of grading rules based on the national standard. These grading rules are similar, but they do vary from one growing region of the country to another because of variations in growing conditions for the trees.

For grading purposes softwood lumber is divided into three classifications: **factory and shop, yard,** and **structural.** Factory and shop lumber, often called by only one of the two names, is lumber of good quality that is further processed by manufacturers into components such as windows, doors, and trim. It is graded according to the number of finished items that can be milled from each piece of raw lumber.

Yard lumber is a general category for pieces used in light construction that are less than 2″ thick. The grades range from select to #5 common (see Table III),

Table III   Yard lumber grades and their uses

Grade	Description and use
Selects and finish	
B & better	Clear or with small defects that don't detract from appearance. *Main uses:* exterior and interior trim, moldings, and finish woodwork where appearance is the primary consideration.
C select	Not as good as B & better, but with only minor defects. *Main use:* interior woodwork.
D select	Small defects on the face which paint will hide, but larger defects on the hidden side. *Main use:* interior woodwork.
Boards	
No. 1 common [Western Wood Products Association (WWPA)] Select merchantable [West Coast Lumber Inspection Bureau (WCLIB)]	Has some small, tight knots in places where they don't limit usefulness. *Main uses:* exterior and interior purposes where knots are not objectionable.
No. 2 common (WWPA) Construction (WCLIB)	Tight knots and some small defects, including splits and checks. *Main uses:* sheathing, subflooring, shelving, knotty paneling.
No. 3 common (WWPA) Standard (WCLIB)	Medium-sized knots and other defects. *Main uses:* sheathing, shelving, fencing.
No. 4 common (WWPA) Utility (WCLIB)	Large knots, checks, and other major defects. *Main uses:* sheathing, braces, forms.
No. 5 common (WWPA) Economy (WCLIB)	Lowest grade of wood, with holes or large knots, diseased wood, and large pitch pockets. *Main uses:* for temporary construction; can be ripped into short lengths of higher grade.

Fig. 1-0.7 Common defects in lumber. Those shown on this page from top to bottom are knot, check, split, shake, and wormholes. From top to bottom on page 10 are mineral streak, pitch pocket, decay, and wane. (Knot, check, and worm hole photos courtesy Western Wood Products Association. All others courtesy U.S. Forest Products Laboratory.)

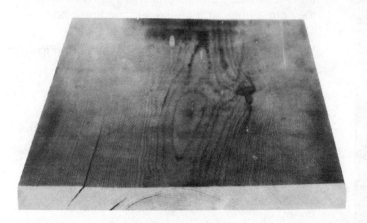

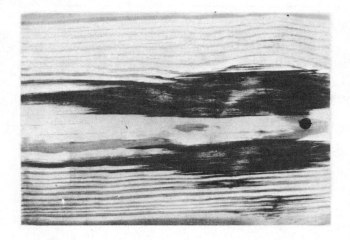

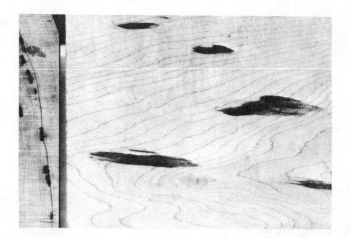

depending primarily on visual appearance. Yard lumber is graded on the best side, unlike hardwoods.

Structural lumber is a category for pieces at least 2″ thick. It is graded according to its ultimate use and its strength in resisting the stresses put on each piece in that use. Building codes specify the minimum grades that can be used for framing members such as floor and ceiling joists, wall studs, and roof rafters. Defects and the results of stress analysis are extremely important in grading structural lumber.

Common defects

Approximately two dozen defects in structural lumber may affect its grade. Among the most important defects, shown in Fig. 1-0.7, are:

Check. A lengthwise separation along the grain on one surface. The depth of the check affects grade.

Split. A lengthwise separation along the grain that extends from one surface of the wood to another.

Shake. A lengthwise separation along annular rings.

Pitch pocket. An opening between annular rings filled with tree sap.

Decay. Areas of destruction or discoloration of wood caused by fungi. Three diseases of wood are commonly caused by fungi: **mold,** which occurs mostly at the wood's surface; **sap stain,** which discolors wood from the surface inward; and **wood rot,** which destroys both the cellulose and lignin. Proper drying and chemical dips can kill the fungi and stop the spread of a disease.

Knot. A cross section through a tree's branch. Knots are classified by their size, frequency, and tightness. There are several types of knots, and they are all harder than the surrounding wood. Therefore knots are likely to increase warp.

Warp. Any distortion in wood from straight (Fig. 1-0.8). **Bow** is a distortion from flat along the grain. **Cup** is a distortion from flat across the grain. **Twist** is a bow in two directions. **Crook** is a lengthwise distortion of a straight edge. Warp may result from the way the tree grew, the way the lumber was dried, the way it was stored, or the way it was cut at the mill or building site. Once lumber warps, it cannot be straightened again.

Wane. Bark left on the edge of a piece of lumber during cutting.

Torn grain. A condition where the surface of the wood has been chipped, gouged, or torn during cutting.

Worm holes. Small holes drilled by wood-boring insects.

Stress analysis

All structural members in a building carry loads, and those loads produce stresses. There are two types of loads: dead load and live load. **Dead load** is the downward weight of the structural member itself, plus the weight of the structure that it carries. The dead load on a roof rafter, for example, includes the weight of the rafter and a proportionate part of the weight of roof sheathing and finish roofing materials.

Live load is the weight or force applied to the structural member or building component of which it is a part. A live load may be a downward load, such as the weight of snow or the weight of someone walking on a roof, or it may be sideways load, such as a gust of wind.

Most building codes specify the minimum dead loads and live loads that a building must be designed to withstand. The architect or designer then calculates the stress on each structural member. From tables published by lumber associations, the specifications writer determines what grade and species of lumber will meet the stress requirements. The purchasing agent then orders lumber to these specifications. When the carpenter builds the building, the grade stamp on the lumber being used should be checked against the requirements shown in drawings or specifications to make sure the correct material is being used.

Figure 1-0.9 shows allowable stresses for a typical species of structure lumber. The f_b value of a beam is the strength of extreme fibers in bending when a member is used horizontally, such as a floor joist. This means that the top fibers of a joist, which will be in compression, and the bottom fibers, which will be in tension, will resist a force or weight of so many pounds per square inch (psi) exerted downward at the center of the joist (Fig. 1-0.10). Note that members used by themselves, such as beams, can't resist as much stress as a series of members on a regular spacing, such as joists.

The f_b rating says that the member will not break under the load specified. How much it will bend—its deflection—can be determined by the **modulus of elasticity** (Fig. 1-0.9). The modulus of elasticity—the **E** value of a piece of lumber—is the ratio between the load on a member and the amount the member will deflect under that load. The higher its E value, the stiffer the lumber. Among woods commonly used for structural lumber, Douglas fir and Southern yellow pine have the highest f and E values.

The other fiber stress ratings in Fig. 1-0.9 are of equal importance to architects in determining the sizes of other members, such as plates and studs, as shown in Fig. 1-0.10.

Inspection

Lumber is graded at the sawmill by experts. The standards that the graders follow allow some leeway in the size, severity, and total number of defects. Actually, the difference between one grade of lumber and the grade above and below it is not great, and the grader must use good judgment and past experience.

At one time all lumber was graded by visual inspection. Today much of the grading is done electronically. Graders use a special meter to measure moisture content. They conduct stress tests on samples of the wood

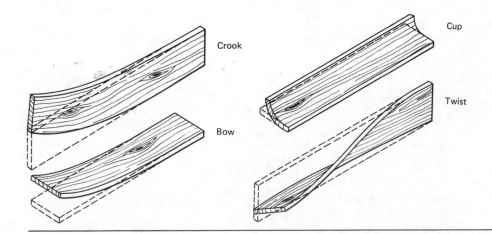

Crook

Cup

Bow

Twist

Fig. 1-0.8 The four types of warp.

to determine its resistance to bending, tension, and compression, and read the results on a computer. Yet graders may measure the density of a piece by counting the annular rings; generally, the more rings per inch, the stronger the wood. They study the **slope of the grain**—the angle the grain makes with the cut edge of the lumber. They measure the size of knots and note their location; a 1″ knot may meet grade standards if it is in the center of a piece of lumber, but it may not be acceptable at the edge. Graders look also at all other defects that would affect the grade. By comparing the results of their inspection with the rules of the grading agency for that particular species of wood, they accurately determine its grade.

The grades of structural lumber vary somewhat, depending on the rules of the grading agency and the species of wood. The top grade of most species is **select structural,** which is used only where high strength, stiffness, and good appearance are all required. **No. 1,** the next lower grade, may have tiny knots, but otherwise has almost the same qualities as select. **No. 2** lumber may have larger knots than No. 1, but they are tight knots, and the grade is excellent for floor and roof framing members. **No. 3** has still more and larger defects; it is adequate for sills and some plate members. The bottom grade is economy, which is used primarily for nonstructural purposes such as stakes and temporary bracing.

Species and commercial grade	Size classification	Allowable unit stresses in pounds per square inch							
		Extreme fiber in bending "F_b"		Tension parallel to grain "F_t"	Horizontal shear "F_v"	Compression perpendicular to grain "F_b"	Compression parallel to grain "F_c"	Modulus of elasticity "E"	Grading rules agency
		Single-member uses	Repetitive-member uses						
HEM-FIR (Surfaced dry or surfaced green. Used at 19% max. m.c.)									
Select Structural		1650	1900	975	75	245	1300	1,500,000	
No. 1	2″ to 4″	1400	1600	825	75	245	1000	1,500,000	West Coast
No. 2	thick	1150	1300	675	75	245	800	1,400,000	Lumber
No. 3	2″ to 4″	625	725	375	75	245	500	1,200,000	Inspection
Appearance	wide	1400	1600	825	75	245	1200	1,500,000	Bureau
Stud		625	725	375	75	245	500	1,200,000	and Western
Construction	2″ to 4″	825	975	475	75	245	925	1,200,000	Wood
Standard	thick	450	525	275	75	245	750	1,200,000	Products
Utility	4″ wide	225	250	125	75	245	500	1,200,000	Association
Select Structural	2″ to 4″	1400	1650	950	75	245	1150	1,500,000	
No. 1	thick	1200	1400	800	75	245	1000	1,500,000	
No. 2	6″ and	1000	1150	650	75	245	850	1,400,000	
No. 3	wider	575	675	375	75	245	550	1,200,000	
Appearance	wider	1200	1400	800	75	245	1200	1,500,000	
Select Structural	Beams and	1250	—	750	70	245	900	1,300,000	West Coast
No. 1	Stringers	1000	—	525	70	245	750	1,300,000	Lumber
Select Structural	Posts and	1200	—	800	70	245	950	1,300,000	Inspection Bureau
No. 1	Timbers	975	—	650	70	245	750	1,300,000	
Select Dex	Decking	1400	1600	—	—	245	—	1,500,000	
Commercial Dex		1150	1300	—	—	245	—	1,400,000	
Select Structural	Beams and	1250	—	850	70	245	900	1,300,000	
No. 1	Stringers	1050	—	700	70	245	775	1,300,000	Western
Select Structural	Posts and	1200	—	800	70	245	950	1,300,000	Wood
No. 1	Timbers	975	—	650	70	245	850	1,300,000	Products Association
Selected Decking	Decking	—	1600	—	—	—	—	1,500,000	
Commercial Decking		—	1300	—	—	—	—	1,400,000	
Selected Decking	Decking	—	1750	(Surfaced at 15% max. m.c. and				1,600,000	
Commercial Decking		—	1450	used at 15% max. m.c.)				1,500,000	

Fig. 1-0.9 This excerpt from the page of the *Supplement to National Design Specifications for Stress-Grade Lumber and Its Fastenings* lists the allowable unit stresses for hem fir, a common structural lumber having the combined characteristics of hemlock and certain firs.

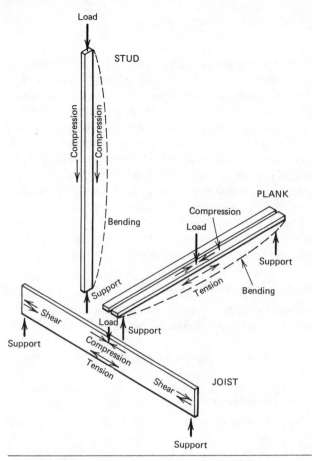

Fig. 1-0.10 The effect of a load on a beam or plank is to put top fibers in compression and bottom fibers in tension. The load on a stud puts bending stress on fibers.

Carpenters will often see four other grades of structural lumber. **Stud** grade is stiff, straight lumber with a high f_c stress rating; it is excellent for vertical wall members. **Construction** grade falls somewhere between Nos. 1 and 2. **Standard** and **utility** are still lower grades, but they are a bit better than economy.

Grade marks

The stamp that a grader uses to identify a piece of lumber provides all the information the carpenter needs (Fig. 1-0.11), although the information will vary. At the upper left is the number of the sawmill that processes the lumber. Below it is the logotype (identifying symbol) of the grading rules agency. In the lower right is the species. The remaining information, which appears in the middle and upper right of the stamp, states the grade or stress rating of the piece and its moisture content.

Lumber Sizes

Most of the lumber used in residential construction falls into the category of dimensional lumber (also called structural lumber). This lumber is further classified according to its width and thickness (Fig. 1-0.12).

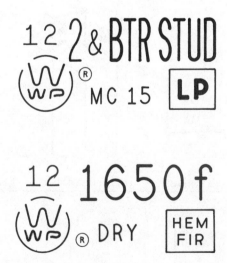

Fig. 1-0.11 Typical grade marks on lumber. (Courtesy Western Wood Products Association.)

Beams and Stringers. These are members at least 5″ thick and at least 2″ wider than they are thick. They are installed horizontally and are graded according to f_b value (bending strength) when loaded on the narrower of the two dimensions.

Posts and Timbers. These are members that are either square or nearly square in cross section. They are at least 5″ thick and no more than 2″ wider than they are thick. Normally, they are installed vertically and are graded according to their f_c value (strength in compression parallel to the grain). Loads are carried on the cross section.

Joists and Planks. These are members 2 to 4″ in nominal thickness and at least 6″ wide. They are graded according to their bending strength either when loaded on their narrow edge, such as joists or rafters, or on their wider side, such as planks and decking.

Lumber 1″ thick is called a **board.** Boards are cut in widths of 2 to 12″ and in lengths of 6 to 16′. They may be **rough sawn** (also known as common) or **surfaced** (also called dressed). Boards are produced in five grades. **Select** boards are used primarily when ap-

pearance is important. **Construction** grade is the standard where straightness and strength are more important, such as in building concrete forms. **Standard, utility,** and **economy** boards are adequate when neither fine appearance nor high strength is required.

Dressed boards are used almost exclusively in finish carpentry. They may be finished smooth on one side (SIS), on both sides (S2S), on one edge (S1E), on one side and one edge (S1S1E), on one side and both edges (S1S2E), or on all sides (S4S). A shelf in a bookcase would usually be S4S. Boards to finish an exposed ceiling beam would more likely be S1S1E on the sides and S1S2E on the bottom (Fig. 1-0.13).

Nominal and actual dimensions

Dimensional lumber and boards are known by their nominal size (nominal means by name). A 2 × 4, for example, is not 2″ thick by 4″ wide. When unseasoned or green, its actual dimensions are $1\frac{9}{16} \times 3\frac{9}{16}$″. Seasoning shrinks the lumber even further to $1\frac{1}{2} \times 3\frac{1}{2}$″. Figure 1-0.14 shows the nominal and actual dimensions of typical boards, joists, and timbers. The carpenter must be thoroughly familiar with the actual dimensions of all sizes of lumber.

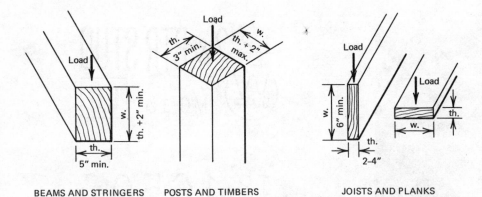

BEAMS AND STRINGERS   POSTS AND TIMBERS   JOISTS AND PLANKS

Fig. 1-0.12 Sizes and typical loading of structural lumber.

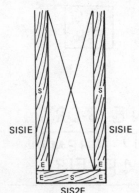

Fig. 1-0.13 Typical designations for dressed boards and their common usages.

Common name or nominal size	Actual size (S4S)	Common name or nominal size	Actual size (S4S)
1″ × 2″	$\frac{3}{4}″ × 1\frac{1}{2}″$	2″ × 10″	$1\frac{1}{2}″ × 9\frac{1}{4}″$
1″ × 3″	$\frac{3}{4}″ × 2\frac{1}{2}″$	2″ × 12″	$1\frac{1}{2}″ × 11\frac{1}{4}″$
1″ × 4″	$\frac{3}{4}″ × 3\frac{1}{2}″$	3″ × 3″	$2\frac{1}{2}″ × 2\frac{1}{2}″$
1″ × 5″	$\frac{3}{4}″ × 4\frac{1}{2}″$	3″ × 4″	$2\frac{1}{2}″ × 3\frac{1}{2}″$
1″ × 6″	$\frac{3}{4}″ × 5\frac{1}{2}″$	3″ × 6″	$2\frac{1}{2}″ × 5\frac{1}{2}″$
1″ × 8″	$\frac{3}{4}″ × 7\frac{1}{2}″$	3″ × 8″	$2\frac{1}{2}″ × 7\frac{1}{4}″$
1″ × 10″	$\frac{3}{4}″ × 9\frac{1}{2}″$	3″ × 10″	$2\frac{1}{2}″ × 9\frac{1}{4}″$
1″ × 12″	$\frac{3}{4}″ × 11\frac{1}{2}″$	3″ × 12″	$2\frac{1}{2}″ × 11\frac{1}{4}″$
2″ × 2″	$1\frac{1}{2}″ × 1\frac{1}{2}″$	4″ × 4″	$3\frac{1}{2}″ × 3\frac{1}{2}″$
2″ × 3″	$1\frac{1}{2}″ × 2\frac{1}{2}″$	4″ × 6″	$3\frac{1}{2}″ × 5\frac{1}{2}″$
2″ × 4″	$1\frac{1}{2}″ × 3\frac{1}{2}″$	4″ × 8″	$3\frac{1}{2}″ × 7\frac{1}{4}″$
2″ × 6″	$1\frac{1}{2}″ × 5\frac{1}{2}″$	4″ × 10″	$3\frac{1}{2}″ × 9\frac{1}{4}″$
2″ × 8″	$1\frac{1}{2}″ × 7\frac{1}{4}″$	4″ × 12″	$3\frac{1}{2}″ × 11\frac{1}{2}″$

Fig. 1-0.14 Nominal and actual sizes of lumber.

Fig. 1-0.15 Some glue-laminated beams are manufactured to exact specifications. They are carefully wrapped in waterproof paper for protection while in transit to the building site. Solid beams are often a stock item in large lumber yards. (Courtesy Western Wood Products Association.)

Glue laminating

The maximum dimensions of any one piece of lumber are limited by the available sizes of trees of a species. Furthermore, the greater its dimensions, the more likely the lumber is to have flaws that will make its use unsuitable in grade.

Modern technology has overcome these limitations by developing high-quality glues that have greater strength than the wood they fasten together. Now, boards up to 24″ wide are manufactured by laminating narrower boards edge to edge. Structural beams are manufactured to whatever depth is necessary by laminating pieces of lumber side to side (Fig. 1-0.15). Glue-laminated timbers are constructed in almost any size and shape to span broad openings decoratively in buildings such as large churches.

Tied in with the development of glues is progress in methods of jointing short lengths of wood. When end joints must be made, the ends are **scarfed** (lapped) in several ways. The most common method of scarfing is finger-jointing (Fig. 1-0.16). The glued surfaces must fit tightly together, and the joint is strongest when both surfaces are planed smooth.

In residential construction glue-laminated timbers or beams are most often purchased to span large openings, such as garage doors, and to support roofs in rooms with high ceilings. For carpenters, laminated timbers have one major disadvantage and two major advantages. They are heavy and awkward to set in place (see Fig. 7-4.1). Nevertheless, they are consistent in dimension and quality, and they do not deflect as much as individual pieces of lumber with the same strength.

This discussion of lumber has been limited primarily to softwood lumber used in framing a house or multifamily residence. Hardwoods used in millwork, exterior trim, and siding are discussed in Section IV.

Lumber Measure

Builders may order lumber in three ways. If the lumber is a stock size and they know the quantity needed, they order by the piece. Their order for ceiling joists, for example, might be for 32 2×8s 16′ long SPF No. 2 or better. If the item needed is not a stock size and the quantity needed is great enough, the builders might order the lumber **PET**, which is short for **precision end trimmed**. If they need studs 7′-10½″ long, for example, they might order 2×4s PET to that length instead of having the carpenters trim 8′ lengths on site. If builders need one size of lumber in a variety of lengths, they may order 200′ of 1×6 boards in **random lengths.** Individual lengths may vary from 6 to 12′, and carpenters select the closest length to meet their needs.

However it is ordered, lumber is priced by the board foot (Fig. 1-0.17). A **board foot** is a standard unit of lumber measure for a piece 12″ × 12″ × 1″ thick; **B.F.** or **b.f.** is the common abbreviation. To find the number of board feet in any piece of lumber, use this formula:

$$\frac{\text{thickness (in ″)} \times \text{width (in ″)} \times \text{length (in ′)}}{12}$$

For example, to find the board feet in a joist 2″ × 8″ × 16′ the formula becomes:

$$\frac{2 \times 8 \times 16}{12} = \frac{256}{12} = 21\tfrac{1}{3}\ \text{B.F.}$$

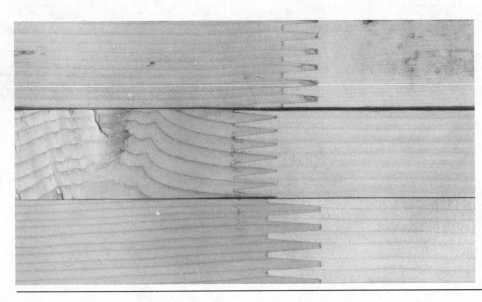

Fig. 1-0.16 Finger-jointing is a method of gluing short pieces of wood together to form a longer piece. Layers in glue-laminated beams are often finger-jointed. (Courtesy Western Wood Products Association.)

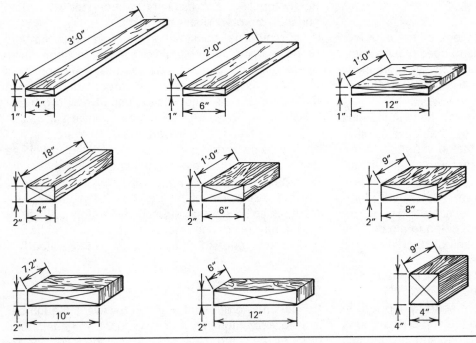

Fig. 1-0.17 Each of the pieces of lumber shown contains one board foot.

As a carpenter, you won't often need to calculate board feet. Nevertheless, you should know how to do so. Table IV will help. It shows the number of board feet for typical lengths of typical pieces of lumber.

If and when the United States converts from English to metric measure, the term board foot will disappear from use. Although the lumber sizes by metric measure to be used in the United States have not yet been agreed on, most foreign countries use 25 millimeters (mm) as a replacement measure for 1″. Therefore a 2 × 4 would be 50 × 100 mm. Metric lengths begin at 1.8 meters (m) (the equivalent of 6′) and increase in 300-mm increments up to 6.3 m (approximately 20′). Table V shows typical metric lumber sizes.

Lumber Storage

In a well-managed yard lumber is stored under either a roof or a waterproof cover to protect it from rain. The purpose is to keep the lumber from being soaked. It will not absorb enough moisture from the air, even in damp weather, to cause major problems with warping.

When the lumber is delivered to the site, it is often **banded** (tied together with steel strapping). As soon as

Table IV   Board feet in lumber of typical sizes and lengths

Nominal size (inches)	Actual length (feet)								
	8	10	12	14	16	18	20	22	24
1 × 2	—	$1\frac{2}{3}$	2	$2\frac{1}{3}$	$2\frac{2}{3}$	3	$3\frac{1}{3}$	$3\frac{2}{3}$	4
1 × 3	—	$2\frac{1}{2}$	3	$3\frac{1}{2}$	4	$4\frac{1}{2}$	5	$5\frac{1}{2}$	6
1 × 4	$2\frac{2}{3}$	$3\frac{1}{3}$	4	$4\frac{2}{3}$	$5\frac{1}{3}$	6	$6\frac{2}{3}$	$7\frac{1}{3}$	8
1 × 5	—	$4\frac{1}{6}$	5	$5\frac{5}{6}$	$6\frac{2}{3}$	$7\frac{1}{2}$	$8\frac{1}{3}$	$9\frac{1}{6}$	10
1 × 6	4	5	6	7	8	9	10	11	12
1 × 7	—	$5\frac{5}{6}$	7	$8\frac{1}{6}$	$9\frac{1}{3}$	$10\frac{1}{2}$	$11\frac{2}{3}$	$12\frac{5}{6}$	14
1 × 8	$5\frac{1}{3}$	$6\frac{2}{3}$	8	$9\frac{1}{3}$	$10\frac{2}{3}$	12	$13\frac{1}{3}$	$14\frac{2}{3}$	16
1 × 10	$6\frac{2}{3}$	$8\frac{1}{3}$	10	$11\frac{2}{3}$	$13\frac{1}{3}$	15	$16\frac{2}{3}$	$18\frac{1}{3}$	20
1 × 12	8	10	12	14	16	18	20	22	24
$1\frac{1}{4}$ × 4	—	$4\frac{1}{6}$	5	$5\frac{5}{6}$	$6\frac{2}{3}$	$7\frac{1}{2}$	$8\frac{1}{3}$	$9\frac{1}{6}$	10
$1\frac{1}{4}$ × 6	—	$6\frac{1}{4}$	$7\frac{1}{2}$	$8\frac{3}{4}$	10	$11\frac{1}{4}$	$12\frac{1}{2}$	$13\frac{3}{4}$	15
$1\frac{1}{4}$ × 8	—	$8\frac{1}{3}$	10	$11\frac{2}{3}$	$13\frac{1}{3}$	15	$16\frac{2}{3}$	$18\frac{1}{3}$	20
$1\frac{1}{4}$ × 10	—	$10\frac{5}{12}$	$12\frac{1}{2}$	$14\frac{7}{12}$	$16\frac{2}{3}$	$18\frac{3}{4}$	$20\frac{5}{6}$	$22\frac{11}{12}$	25
$1\frac{1}{4}$ × 12	—	$12\frac{1}{2}$	15	$17\frac{1}{2}$	20	$22\frac{1}{2}$	25	$27\frac{1}{2}$	30
$1\frac{1}{2}$ × 4	4	5	6	7	8	9	10	11	12
$1\frac{1}{2}$ × 6	6	$7\frac{1}{2}$	9	$10\frac{1}{2}$	12	$13\frac{1}{2}$	15	$16\frac{1}{2}$	18
$1\frac{1}{2}$ × 8	8	10	12	14	16	18	20	22	24
$1\frac{1}{2}$ × 10	10	$12\frac{1}{2}$	15	$17\frac{1}{2}$	20	$22\frac{1}{2}$	25	$27\frac{1}{2}$	30
$1\frac{1}{2}$ × 12	12	15	18	21	24	27	30	33	36
2 × 4	$5\frac{1}{3}$	$6\frac{2}{3}$	8	$9\frac{1}{3}$	$10\frac{1}{3}$	12	$13\frac{1}{3}$	$14\frac{2}{3}$	16
2 × 6	8	10	12	14	16	18	20	22	24
2 × 8	$10\frac{2}{3}$	$13\frac{1}{3}$	16	$18\frac{2}{3}$	$21\frac{1}{3}$	24	$26\frac{2}{3}$	$29\frac{1}{3}$	32
2 × 10	$13\frac{1}{3}$	$16\frac{2}{3}$	20	$23\frac{1}{3}$	$26\frac{2}{3}$	30	$33\frac{1}{3}$	$36\frac{2}{3}$	40
2 × 12	16	20	24	28	32	36	40	44	48
3 × 6	12	15	18	21	24	27	30	33	36
3 × 8	16	20	24	28	32	36	40	44	48
3 × 10	20	25	30	35	40	45	50	55	60
3 × 12	24	30	36	42	48	54	60	66	72
4 × 4	$10\frac{2}{3}$	$13\frac{1}{3}$	16	$18\frac{2}{3}$	$21\frac{1}{3}$	24	$26\frac{2}{3}$	$29\frac{1}{3}$	32
4 × 6	16	20	24	28	32	36	40	44	48
4 × 8	$21\frac{1}{3}$	$26\frac{2}{3}$	32	$37\frac{1}{3}$	$42\frac{2}{3}$	48	$53\frac{1}{3}$	$58\frac{2}{3}$	64
4 × 10	$26\frac{2}{3}$	$33\frac{1}{3}$	40	$46\frac{2}{3}$	$53\frac{1}{3}$	60	$66\frac{2}{3}$	$73\frac{1}{3}$	80
4 × 12	32	40	48	56	64	72	80	88	96

Table V Nominal metric dimensions of softwood lumber

Millimeters	75	100	125	150	175	200	225	250	300
16	x	x	x	x					
19	x	x	x	x					
22	x	x	x	x					
25	x	x	x	x	x	x	x	x	x
32	x	x	x	x	x	x	x	x	x
36	x	x	x	x					
38	x	x	x	x	x	x	x		
40	x	x	x	x	x	x	x		
44	x	x	x	x	x	x	x	x	x
50	x	x	x	x	x	x	x	x	x
63		x	x	x	x	x	x		
75		x	x	x	x	x	x	x	x
100		x		x		x		x	x
150				x		x			x
200						x			
250							x		
300									x

Actual dimensions are about 4 mm less. Thickness and width are given in millimeters, and length in meters.

it is delivered, all lumber should be stored off the ground so that it will not absorb moisture from that ground. Concrete blocks are commonly used for this purpose. Good builders leave the bands in place until the lumber is needed, and cover the stacks with a tarpaulin or canvas.

Lumber that isn't used within a few days of delivery should be piled on stickers similar to those used in the drying process at the mill. Thus each piece of lumber will pick up or lose airborne moisture on all sides and maintain its straightness and a uniform moisture content until the time of use.

Questions

1 Name four useful qualities and four faults of wood.
2 Which of the following is the densest wood in a tree—heartwood, cambium, cellulose, sapwood, or bast?
3 How many annular rings are formed in a normal year?
4 How can you tell the difference between springwood and summerwood?
5 Explain the difference between softwood and hardwood and their primary uses in building.
6 What are the two most common moisture content figures today?
7 Name and distinguish among the four kinds of warp.
8 Name and describe five other common lumber defects.
9 Name the two types of loads on a structure and give two common examples of each.
10 Which of these numbers is a typical value for the modulus of elasticity—90, 800, 1800, or 1,300,000?

11 For what can economy grade lumber be used?

12 Identify all the information in the grade stamp at the right.

13 Give the range of thickness, width, and length of boards.

14 What does S2S mean?

15 What are the actual dimensions of a 2 × 12 that is 20′ long?

16 How many board feet are there in the piece in question 15?

17 What is fingerjointing?

18 How should lumber be stored?

Plywood

Although few building materials are as strong as wood along the grain, wood is not strong across the grain. Plywood was developed to increase across-the-grain strength.

Plywood is made of thin plies (layers) of wood called **veneers**; they are bonded together with special glues under both heat and great pressure. Plywood has either three, five or seven plies, and alternate plies are glued at right angles to each other for maximum strength. The more plies, the stronger the plywood. The center ply is called the **core,** and the exposed veneers are called **faces.** Any other plies between the core and faces are called **crossbands.** The total thickness of plies with grain running in one direction is approximately the same as the thickness with grain running at right angles; this reduces the chances of warping.

Most veneers are produced by rotary cutting. Logs are first cut to length and softened with hot water or steam; then bark is removed. Each log is bolted at its center to a lathe. As the lathe turns the log, a knife slices off a thin veneer. A pressure bar just outside the long knife blade holds the veneer in place during cutting and helps to maintain a uniform thickness. The layer comes off the knife as a continuous sheet $\frac{1}{16}$″ to $\frac{5}{16}$″ thick and slightly wider than finished sheets of plywood.

The continuous sheet is trimmed into smaller sheets that are fed into a dryer that reduces moisture content to 3 to 8%. Cores and crossbands are now ready for use. Face veneers go through a patching machine that removes defects and matches face grains; they next go through a splicer that prepares them for the final manufacturing process.

Waterproof glue is applied by machine to the face plies, core, and crossbands, and the rough plywood sheet goes into a hot press. The pressure of the press squeezes the plies together to the approximate final thickness, and heat cures the glue. This process takes 2 to 20 minutes. The sheets are trimmed to dimension, squared, and sanded to finish thickness.

Sheet Sizes

The standard width of a sheet of plywood is 4′ and, in building construction, you will seldom find wider sheets used. The most common length is 8′, although lengths of 9′ and 10′ are manufactured for special uses, such as sheathing. A few mills manufacture plywood in sheets 6 to 8′ wide and up to 16′ long.

Finished thicknesses of plywood vary from 4 mm (which is about $\frac{5}{32}$″) up to a maximum of $1\frac{1}{4}$″. The most common thicknesses are $\frac{1}{4}$″ and $\frac{5}{16}$″ for finish paneling, and $\frac{3}{8}$″, $\frac{1}{2}$″, $\frac{5}{8}$″, and $\frac{3}{4}$″ for various structural purposes.

Plywood is made with three types of edges. Sheets may be cut square to fit at a *butt edge.* They may be cut to overlap at a *shiplapped edge.* Or they may be cut to interlock at a *tongue-and-groove edge.* Only opposite edges or all four may be cut to match.

Grades

After plywood is manufactured, the finished product is graded, similarly to lumber grading. The grade stamp on each sheet (Fig. 1-0.18) tells the story of that sheet.

Glues

The type of glue determines where plywood may be used. **Exterior glue** is 100% waterproof and is stronger than the wood itself. Neither the strength nor the shape of a sheet is affected by repeated wetting and drying. **Interior glue** is a lighter, thinner, less expensive glue that is adequate where moisture is controlled. The glue is moisture resistant, but it is not weatherproof. Most glues today are the exterior type, even though the plywood may be intended for interior use. The grade stamp indicates exterior glue when it is used and tells whether the panel is for exterior or interior application.

Some plywoods, designed as **structural panels,** are engineered and manufactured for special heavy-duty structural purposes. These panels are seldom used in residential construction. They may be either exterior or interior grade.

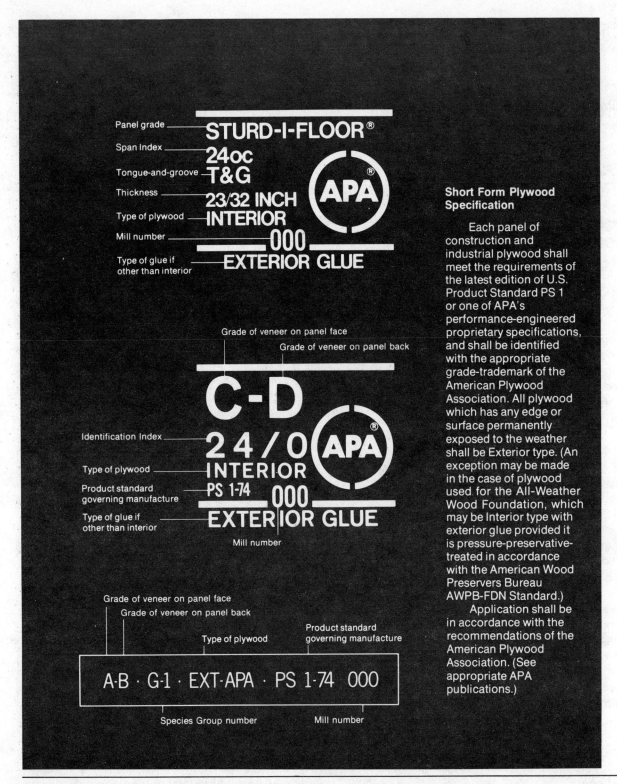

Panel grade — **STURD-I-FLOOR**®
Span Index — **24oc**
Tongue-and-groove — **T&G**
Thickness — **23/32 INCH**
Type of plywood — **INTERIOR**
Mill number — **000**
Type of glue if other than interior — **EXTERIOR GLUE**

Grade of veneer on panel face
Grade of veneer on panel back
C-D
Identification Index — **24/0**
Type of plywood — **INTERIOR**
Product standard governing manufacture — **PS 1-74**
000
Type of glue if other than interior — **EXTERIOR GLUE**
Mill number

Grade of veneer on panel face
Grade of veneer on panel back
Type of plywood
Product standard governing manufacture
A-B · G-1 · EXT-APA · PS 1-74 000
Species Group number   Mill number

Short Form Plywood Specification

Each panel of construction and industrial plywood shall meet the requirements of the latest edition of U.S. Product Standard PS 1 or one of APA's performance-engineered proprietary specifications, and shall be identified with the appropriate grade-trademark of the American Plywood Association. All plywood which has any edge or surface permanently exposed to the weather shall be Exterior type. (An exception may be made in the case of plywood used for the All-Weather Wood Foundation, which may be Interior type with exterior glue provided it is pressure-preservative-treated in accordance with the American Wood Preservers Bureau AWPB-FDN Standard.)

Application shall be in accordance with the recommendations of the American Plywood Association. (See appropriate APA publications.)

Fig. 1-0.18 Typical grade marks on plywood. (Courtesy American Plywood Association.)

Table VI Classification of species used for plywood veneers by group

Group 1	Group 2		Group 3	Group 4	Group 5
Apitong	Cedar, Port Orford	Maple, black	Alder, red	Aspen	Basswood
Beech, American	Cypress	Mengkulang	Birch, paper	Bigtooth	Fir, balsam
Birch	Douglas fir 2	Meranti, red	Cedar, Alaska	Quaking	Poplar, balsam
Sweet	Fir	Mersawa	Fir, subalpine	Cativo	
Yellow	California red	Pine	Hemlock, eastern	Cedar	
Douglas fir 1	Grand	Pond	Maple, bigleaf	Incense	
Kapur	Noble	Red	Pine	Western red	
Keruing	Pacific silver	Virginia	Jack	Cottonwood	
Larch, Western	White	Western white	Lodgepole	Eastern	
Maple, sugar	Hemlock, Western	Spruce	Ponderosa	Black (Western poplar)	
Pine	Lauan	Red	Spruce	Pine	
Caribbean	Almon	Sitka	Redwood	Eastern white	
Ocote	Bagtikan	Sweetgum	Spruce	Sugar	
Pine, Southern	Mayapis	Tamarack	Black		
Loblolly	Red lauan	Yellow poplar	Engelmann		
Longleaf	Tangile		White		
Shortleaf	White lauan				
Slash					
Tanoak					

Species

More than 70 species of wood are used to manufacture plywood. Most plywood today is cut from native softwoods, and this is likely to continue. Some hardwood is used, however; part of it is grown locally and part is imported from the Far East.

These species are placed in groups 1 to 5 (Table VI) according to their stiffness and other factors of durability. The strongest woods are in group 1. The group number appears on some grade stamps (Fig. 1-0.18) where group number is important in selection. Other stamps may have a class number (I or II), which indicates the strength of the plywood in withstanding stresses beyond simple bending, such as tension, compression, shear, and nail-holding ability. Still other stamps carry an index number, which consists of two numbers with a slash line between them. The first number specifies the maximum span in inches recommended between roof supports, and the second specifies the recommended span between floor supports.

Elsewhere on the grade stamp are the logo of the American Plywood Association, which establishes the grading standards for plywoods, the number of the standard that the plywood must meet in manufacturing, and the number of the manufacturing mill.

Cores

Plywoods for construction and industrial purposes are sometimes called **softwood plywoods.** They have a veneer core—a core made by the same process as other plies. This veneer core may be a single ply, or two plies with the grain running in the same direction glued together to form a single layer. A sheet of ½″ plywood, for example, may consist of four veneers, but only three plies (Fig. 1-0.19). Thus the number of plies in a sheet of plywood is always an odd number, even though the number of veneers may be an even number. Note that the grain of the two face veneers always runs in the same direction.

Plywoods used primarily for making finished components, such as furniture and doors, are called **hardwood plywoods.** They may have a veneer core, a lumber core, or a core of particle board (see page 25). The type of core is not indicated on the grade stamp, but you can easily distinguish one from the other by looking at the edge of the sheet.

Faces

Prominent across the top of most grade stamps are two letters separated by a hyphen. The only letters used are A, B, C, D, and N. They indicate veneer quality. The first letter tells the quality or grade of the face on the front of the panel, and the second letter tells the grade of the face on the back.

Top quality, **grade N,** is free of defects, selected to receive a natural finish. An N surface should be exposed to view and admired.

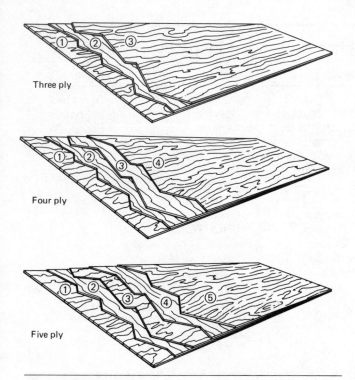

Fig. 1-0.19 Usually the number of plies and the number of veneers in a sheet of plywood is the same. Where the numbers are not the same, the grain of the two center veneers runs the same direction (center).

Three ply

Four ply

Five ply

Grade A is smooth, has no open defects, but may have some neat repairs. An A veneer will take paint easily.

Grade B veneer offers a solid surface, with no split wider than $\frac{1}{32}''$. Defects may be repaired with smooth plugs, and tight knots are allowed. Minor flaws in sanding are permitted. Like a grade A veneer, grade B may be painted, but the finished surface won't be as smooth.

Grade C veneer may have splits up to $\frac{1}{2}''$ and knotholes up to $1\frac{1}{2}''$ as long as they do not affect the re-quired strength of the panel. Some sanding defects are permitted.

Grade C plugged is an improved grade C with tighter limits on splits and knotholes, and a fully sanded surface.

Grade D is the poorest grade. It has less strength, rough appearance, and is usable only where it is not visible.

The veneer cores and crossbands of interior-type plywood are almost always grade D; the exception is panels with at least one grade N face. Veneer cores and crossbands of exterior-type plywood are either C or C plugged.

The so-called **appearance grades** for interior and exterior plywood are shown in Table VII. Note that most face veneers are either A or B, with some N and C grades for special conditions. The back veneer varies from N to D. depending on usage.

Engineered grades of plywood, which include sheathing, subflooring, concrete forms, and sheets for special structural purposes, bear slightly different grade markings. In general, the surfaces are unsanded or only lightly sanded, and face veneers are either C-C or C-D. Note in Table VIII, under both interior and exterior types, that Structural I and Structural II grades are lumped together (second grade from the top in each guide). All plies of Structural I must be group 1 species, while the plies of Structural II may be group 1, 2, or 3.

Storage

Plywood may be safely stored in two ways. The best way is flat and supported either under the entire surface or every 12" across the grain. Panels at least $\frac{1}{2}''$ thick may be stored on edge, provided they are maintained in a vertical position. Plywood stored at an angle will warp.

Questions

1 Name the three types of plies found in a five- or seven-ply sheet of plywood.
2 How does interior grade plywood differ from exterior grade plywood?
3 On the grade stamp shown below, identify all the information.

A-B · G-1 · EXT-APA · PS 1-74 000

4 Hardwood plywoods may have three types of cores. What are they?
5 How should plywood be stored?

Table VII Guide to appearance grades of plywood[1]

	Grade Designation[2]	Description and Most Common Uses	Typical[3] Grade-trademarks	Face	Inner Plies	Back	Most Common Thicknesses (inch)					
Interior Type	APA N-N, N-A, N-B INT	Cabinet quality. For natural finish furniture, cabinet doors, built-ins, etc. Special order items.	N-N G-1 INT-APA PS1-74 000 / N-A G-2 INT-APA PS1-74 000	N	C	N,A, or B						3/4
	APA N-D INT	For natural finish paneling. Special order item.	N-D G-2 INT-APA PS1-74 000	N	D	D	1/4					
	APA A-A INT	For applications with both sides on view, built-ins, cabinets, furniture, partitions. Smooth face; suitable for painting.	A-A G-1 EXT-APA PS1-74 000	A	D	A	1/4		3/8	1/2	5/8	3/4
	APA A-B INT	Use where appearance of one side is less important but where two solid surfaces are necessary.	A-B G-1 EXT-APA PS1-74 000	A	D	B	1/4		3/8	1/2	5/8	3/4
	APA A-D INT	Use where appearance of only one side is important. Paneling, built-ins, shelving, partitions, flow racks.	A-D GROUP 1 INTERIOR PS1 000 APA	A	D	D	1/4		3/8	1/2	5/8	3/4
	APA B-B INT	Utility panel with two solid sides. Permits circular plugs.	B-B G-2 INT-APA PS1-74 000	B	D	B	1/4		3/8	1/2	5/8	3/4
	APA B-D INT	Utility panel with one solid side. Good for backing, sides of built-ins, industry shelving, slip sheets, separator boards, bins.	B-D GROUP 2 INTERIOR PS1 000 APA	B	D	D	1/4		3/8	1/2	5/8	3/4
	APA DECORATIVE PANELS	Rough-sawn, brushed, grooved, or striated faces. For paneling, interior accent walls, built-ins, counter facing, displays, exhibits.	DECORATIVE B D GROUP 1 INTERIOR PS1 000 APA	C or btr.	D	D		5/16	3/8	1/2	5/8	
	APA PLYRON INT	Hardboard face on both sides. For countertops, shelving, cabinet doors, flooring. Faces tempered, untempered, smooth, or screened.	PLYRON -INT-APA 000		C & D					1/2	5/8	3/4
Exterior Type	APA A-A EXT	Use where appearance of both sides is important. Fences, built-ins, signs, boats, cabinets, commercial refrigerators, shipping containers, tote boxes, tanks, ducts. (4)	A-A G-1 INT-APA PS1-74 000	A	C	A	1/4		3/8	1/2	5/8	3/4
	APA A-B EXT	Use where the appearance of only one side is less important. (4)	A-B G-1 INT-APA PS1-74 000	A	C	B	1/4		3/8	1/2	5/8	3/4
	APA A-C EXT	Use where the appearance of only one side is important. Soffits, fences, structural uses, boxcar and truck lining, farm buildings. Tanks, trays, commercial refrigerators. (4)	A-C GROUP 1 EXTERIOR PS1 000 APA	A	C	C	1/4		3/8	1/2	5/8	3/4
	APA B-B EXT	Utility panel with solid faces. (4)	B-B G-2 EXT-APA PS1-74 000	B	C	B	1/4		3/8	1/2	5/8	3/4
	APA B-C EXT	Utility panel for farm service and work buildings, boxcar and truck lining, containers, tanks, agricultural equipment. Also as base for exterior coatings for walls, roofs. (4)	B-C GROUP 3 EXTERIOR PS1 000 APA	B	C	C	1/4		3/8	1/2	5/8	3/4
	APA HDO EXT	High Density Overlay. Has a hard, semi-opaque resin-fiber overlay both faces. Abrasion resistant. For concrete forms, cabinets, counter tops, signs, tanks. Also available with skid-resistant screen-grid surface. (4)	HDO A-A G-1 EXT-APA PS1-74 000	A or B	C or C plgd	A or B			3/8	1/2	5/8	3/4
	APA MDO EXT	Medium Density Overlay. Smooth, opaque, resin-fiber overlay one or both faces. Ideal base for paint, both indoors and outdoors. (4)(6)	MDO B-B G-2 EXT-APA PS1-74 000	B	C	B or C			3/8	1/2	5/8	3/4
	APA 303 SIDING EXT	Proprietary plywood products for exterior siding, fencing, etc. Special surface treatment such as V-groove, channel groove, striated, brushed, rough-sawn and texture-embossed MDO. Stud spacing (Span Rating) and face grade classification indicated on grade stamp.	303 SIDING 6-S GROUP 1 16 oc SPAN EXTERIOR PS1 N 000 APA FHA UM 64	(5)	C	C			11/32	15/32	19/32	
	APA T 1-11 EXT	Special 303 panel having grooves 1/4" deep, 3/8" wide, spaced 4" or 8" o.c. Other spacing optional. Edges shiplapped. Available unsanded, textured and MDO.	303 SIDING 6-S/W T1-11 19-32 INCH GROUP 1 16oc SPAN EXTERIOR PS1 N 000 APA FHA UM 64	C or btr.	C	C					19/32	
	APA PLYRON EXT	Hardboard faces both sides, tempered, smooth or screened.	PLYRON EXT-APA 000		C					1/2	5/8	3/4
	APA MARINE EXT	Ideal for boat hulls. Made only with Douglas fir or western larch. Special solid jointed core construction. Subject to special limitations on core gaps and number of face repairs. Also available with HDO or MDO faces.	MARINE A-A EXT-APA PS1-74 000	A or B	B	A or B	1/4		3/8	1/2	5/8	3/4

(1) Sanded both sides except where decorative or other surfaces specified.
(2) Can be manufactured in Group 1, 2, 3, 4 or 5.
(3) The species groups, Identification Indexes and Span Ratings shown in the typical grade-trademarks are examples only. See "Group," "Identification Index" and "Span Rating" for explanations and availability.
(4) Can also be manufactured in Structural I (all plies limited to Group 1 species) and Structural II (all plies limited to Group 1, 2, or 3 species).
(5) C or better for 5 plies. C Plugged or better for 3 and 4 plies.
(6) Also available as a 303 siding.

SOURCE: Courtesy American Plywood Association

Table VIII Guide to engineered grades of plywood

	Grade Designation	Description and Most Common Uses	Typical[1] Grade-trademarks	Veneer Grade			Most Common Thicknesses (inch)				
				Face	Inner Plies	Back					
Interior Type	APA C-D INT	For wall and roof sheathing, subflooring, industrial uses such as pallets. Most commonly available with exterior glue (CDX). Specify exterior glue where construction delays are anticipated and for treated-wood foundations. (7)	C-D 32/16 APA INTERIOR PS 1-74 000 / C-D 24/0 APA INTERIOR 000 EXTERIOR GLUE	C	D	D	5/16	3/8	1/2	5/8	3/4
	APA STRUCTURAL I C-D INT and APA STRUCTURAL II C-D INT	Unsanded structural grades where plywood strength properties are of maximum importance: structural diaphragms, box beams, gusset plates, stressed-skin panels, containers, pallet bins. Made only with exterior glue. See (6) for species group requirements. Structural I more commonly available. (7)	STRUCTURAL I C-D 24/0 APA INTERIOR PS 1-74 000 EXTERIOR GLUE	C[3]	D[3]	D[3]	5/16	3/8	1/2	5/8	3/4
	APA STURD-I-FLOOR INT	For combination subfloor-underlayment. Provides smooth surface for application of resilient floor covering. Possesses high concentrated- and impact-load resistance during construction and occupancy. Manufactured with exterior glue only. Touch-sanded. Available square edge or tongue-and-groove. (7)	STURD-I-FLOOR 24oc T&G 23 32 INCH APA INTERIOR 000 EXTERIOR GLUE NRB-108/FHA-UM-66	C Plugged	(4)	D				19/32 5/8	23/32 3/4
	APA STURD-I-FLOOR 48 O.C. (2·4·1) INT	For combination subfloor-underlayment on 32- and 48-inch spans. Provides smooth surface for application of resilient floor coverings. Possesses high concentrated- and impact-load resistance during construction and occupancy. Manufactured with exterior glue only. Unsanded or touch-sanded. Available square edge or tongue-and-groove. (7)	STURD-I-FLOOR 48oc 2·4·1 T&G 118 INCH INTERIOR 000 EXTERIOR GLUE NRB-108/FHA-UM-66	C Plugged	C[5] & D	D	1-1/8				
	APA UNDERLAYMENT INT	For application over structural subfloor. Provides smooth surface for application of resilient floor coverings. Touch-sanded. Also available with exterior glue. (2)(6)	UNDERLAYMENT GROUP 1 APA INTERIOR PS 1-74 000	C Plugged	C[5] & D	D		3/8	1/2	19/32 5/8	23/32 3/4
	APA C-D PLUGGED INT	For built-ins, wall and ceiling tile backing, cable reels, walkways, separator boards. Not a substitute for Underlayment or Sturd-I-Floor as it lacks their indentation resistance. Touch-sanded. Also made with exterior glue. (2)(6)	C-D PLUGGED GROUP 2 APA INTERIOR PS 1-74 000	C Plugged	D	D		3/8	1/2	19/32 5/8	23/32 3/4
Exterior Type	APA C-C EXT	Unsanded grade with waterproof bond for subflooring and roof decking, siding on service and farm buildings, crating, pallets, pallet bins, cable reels, treated-wood foundations. (7)	C-C 42/20 APA EXTERIOR PS 1-74 000	C	C	C	5/16	3/8	1/2	5/8	3/4
	APA STRUCTURAL I C-C EXT and APA STRUCTURAL II C-C EXT	For engineered applications in construction and industry where full Exterior type panels are required. Unsanded. See (6) for species group requirements. (7)	STRUCTURAL I C-C APA EXTERIOR PS 1-74 000	C	C	C	5/16	3/8	1/2	5/8	3/4
	APA STURD-I-FLOOR EXT	For combination subfloor-underlayment under resilient floor coverings where severe moisture conditions may be present, as in balcony decks. Possesses high concentrated- and impact-load resistance during construction and occupancy. Touch-sanded. Available square edge or tongue-and-groove. (7)	STURD-I-FLOOR 20oc 5 8 INCH APA EXTERIOR 000 NRB-108/FHA-UM-66	C Plugged	C[5]	C				19/32 5/8	23/32 3/4
	APA UNDERLAYMENT C-C PLUGGED EXT	For application over structural subfloor. Provides smooth surface for application of resilient floor coverings where severe moisture conditions may be present. Touch-sanded. (2)(6)	UNDERLAYMENT C-C PLUGGED GROUP 2 APA EXTERIOR PS 1-74 000	C Plugged	C[5]	C		3/8	1/2	19/32 5/8	23/32 3/4
	APA C-C PLUGGED EXT	For use as tile backing where severe moisture conditions exist. For refrigerated or controlled atmosphere rooms, pallet fruit bins, tanks, box car and truck floors and linings, open soffits. Touch-sanded. (2)(6)	C-C PLUGGED GROUP 2 APA EXTERIOR 000	C Plugged	C	C		3/8	1/2	19/32 5/8	23/32 3/4
	APA B-B PLYFORM CLASS I and CLASS II EXT	Concrete form grades with high reuse factor. Sanded both sides. Mill-oiled unless otherwise specified. Special restrictions on species. Available in HDO and Structural I. Class I most commonly available. (8)	B-B PLYFORM CLASS I APA EXTERIOR PS 1-74 000	B	C	B				5/8	3/4

(1) The species groups, Identification Indexes and Span Ratings shown in the typical grade-trademarks are examples only. See "Group," "Identification Index" and "Span Rating" for explanations and availability.
(2) Can be manufactured in Group 1, 2, 3, 4, or 5.
(3) Special improved grade for structural panels.
(4) Special veneer construction to resist indentation from concentrated loads, or other solid wood-base materials.

(5) Special construction to resist indentation from concentrated loads.
(6) Can also be manufactured in Structural I (all plies limited to Group 1 special species) and Structural II (all plies limited to Group 1, 2, or 3 species).
(7) Specify by Identification Index for sheathing and Span Rating for Sturd-I-Floor panels.
(8) Made only from certain wood species to conform to APA specifications.

SOURCE: Courtesy American Plywood Association

Building Boards

The development of plywood spurred manufacturers of building products to develop a whole new category of products in sheet form. The basic manufacturing process is similar for all of them. The main ingredient is vegetable or mineral fibers, which are mixed with a binder. The soft mixture passes through a press. The amount of heat and pressure put on the raw mixture determine the thickness and density of the finished board.

Vegetable Fiberboards

Much of the waste that once accumulated at sawmills, such as wood chips, sawdust, and cutoffs, now goes into building boards. The scrap wood is softened with heat and moisture before it is mixed with a binder and other ingredients. The finished wood products that come out of the presses are classified as softboards, hardboards, or particle boards.

Softboards

In general, softboards are fiberboards at least $\frac{1}{2}''$ thick. The most common thicknesses are $\frac{1}{2}''$, $\frac{5}{8}''$, $\frac{3}{4}''$, and $1''$, although boards up to $3''$ thick are manufactured for special purposes. Standard sizes for sheets are $4' \times 8'$ and $2' \times 8'$.

Because softboards have air cells trapped in them, their major uses are for insulating, sheathing, and controlling sound. Some types of softboard have unfinished surfaces. Others are either coated or impregnated with asphalt during manufacture so that they shed water. Still other types are finished on one side for decorative use on walls and ceilings.

The structural applications of softboards are discussed in Chapter 7 (Unit 18), Chapter 9 (Units 34 and 35), and Chapter 10 (Unit 5). Decorative application is discussed in Chapter 22.

Hardboards

Hardboards are fiberboards no more than $\frac{3}{8}''$ thick. Common thicknesses are $\frac{3}{16}''$, $\frac{1}{4}''$, and $\frac{5}{16}''$. Standard sheet size is $4 \times 8'$, but hardboards are made in widths up to $6'$ and lengths up to $16'$ for specialized usages.

Hardboards are extremely dense and much more water resistant than softboards. Although their surfaces are durable, they will break and chip at the edges unless properly supported. The product is manufactured with a smooth finish on both sides or with a smooth finish on one side and a screen pattern on the other.

Hardboard is manufactured in three grades. **Standard** hardboard is light brown in color, smooth, and accepts paint easily; it is for interior use, as in cabinets.

Service grade is not as dense, heavy, or strong as standard, and its surfaces are not as smooth. **Tempered** hardboard is standard board that has been coated with oils and resin and then baked to a dark brown. It may be used indoors or outdoors and is stronger, denser, and longer lasting than standard. It is also more brittle, however. Tempered hardboard is used for siding, wall paneling, and other decorative purposes. One type, called **perforated hardboard,** has holes punched in it $1''$ apart. Special hooks fit into the holes to support shelves and to serve as hangers for workshop tools, gardening equipment, or kitchen utensils.

The various applications of hardboard are discussed in Chapters 18, 22, and 24.

Particle boards

Particle boards have as a main ingredient wood chips or flakes, which are combined with resin binders and pressed under heat into panels. Therefore they are often called **chipboards** or **flakeboards.** Sheets range from $\frac{1}{4}$ to $1\frac{1}{2}''$ thick, from 3 to $8'$ wide, and up to $24'$ long.

Like plywood, particle board has equal strength in both directions. Because it has no grain, however, it has a smoother surface, is less likely to warp, and is more dimensionally stable (doesn't expand or contract). The product is used as a plywood core, for shelving, and as a base under finished flooring or under a countertop surface. Applications of particle board are discussed in Chapter 6 (Unit 14), and in Chapters 24 and 26.

Caneboards

One other vegetable product, the stalks of sugarcane, sometimes called **bagasse** or **megasse,** is used in manufacturing fiberboard. **Caneboard** is a fiberboard with usage similar to softboard. Some types are impregnated with asphalt for roof insulation. Others are left natural for insulating where moisture is not a problem. Still other types are factory finished for decorative purposes. The surface has an interesting texture.

Mineral Fiberboards

Boards made of mineral fibers serve the same general purposes as vegetable fiberboards. The main difference is that they will not burn or support combustion. The most commonly used minerals are glass, gypsum, rock, and asbestos.

Glass fibers mixed with a binder and sandwiched between two layers of asphalted paper form a rigid insulation. Various types of chemical foams, some mixed with glass fibers, also make good, rigid insulation. Mineral insulation will crush, however, and should never be used where it must support a heavy load. Its application is discussed in Chapter 5 (Unit 14) and Chapter 10 (Unit 5).

Gypsum is a rock that is heated until it turns to powder and is then mixed with water to form a **slurry**. The slurry is then pressed between two sheets of paper to form a board. Covered with kraft paper, the board makes a fireproof sheathing. Covered with lighter, more absorbent papers, gypsum board is used as a base for plaster or as a finished wall and ceiling surface. Its application as sheathing is discussed in Chapter 7 (Unit 18) and Chapter 12 (Unit 2). Finish application is discussed in Chapters 19 and 21.

Some types of mineral boards are made of molten rock that has been sprayed onto special forms. After the rock has cooled, the surfaces are trimmed to shape an acoustical material primarily for application to ceilings. The acoustical board may be in sheet form, but it is usually cut into smaller tiles. Its application is discussed in Chapter 23.

Asbestos is the fibrous form of various minerals. The fibers are mixed into a paste of portland cement; the mix is allowed to harden in a mold that controls both the thickness and shape of the finished product. The surface may be given a texture before it hardens, or a colorful, baked enamel finish may be·added. Asbestos-cement boards are strong but somewhat brittle and are most often used as an exterior finish material on roofs and walls. These applications are discussed in Chapters 15 and 18.

Questions

1 List five types of manufactured building boards and a common use of each type.
2 Which of the following boards is fire resistant—hardboard, gypsum board, caneboard, or chipboard?
3 With what is asbestos mixed to form a building board?
4 Which of these boards may be used for shelving—gypsum board, caneboard, hardboard, softboard, or particle board?
5 What is the advantage of mineral tiles over fiber tiles?
6 Name two types of cores for rigid insulation.

Thin Sheets

One of the most important responsibilities of carpenters is keeping moisture out of the structure that they build. The moisture barrier they use depends primarily on the source of the moisture. The choice is among sheet materials made of plastics, treated paper, treated felt, or metal.

Polyethylene

The ground on which a house is built always contains some moisture which tries to seep into a basement floor or crawl space beneath the house. A sheet material laid on the ground prevents this; the material is a vapor barrier called **polyethylene**. Polyethylene is a thin film of plastic, 2 to 4 mm thick, that comes in rolls as large as 20′ × 100′. It is also used as a vapor barrier in combination with loose fill insulation above a ceiling and with certain types of roof construction. These applications are shown in Chapter 10, Units 1, 3, and 5.

Building Paper

Building paper is a thin, soft kraft paper that has been soaked in asphalt or coated with paraffin. It comes in rolls 36″ wide and up to 144′ long. In most uses it is applied between two other materials to prevent the passage of air and to keep out water. The paper does "breathe," however, so that moisture isn't trapped in the house's structure. Application of building paper is discussed in Chapters 15 to 18 inclusive and 24.

Roofing Felt

Roofing felt is similar to building paper. It is made of vegetable or mineral fibers saturated with asphalt or pitch and formed into a long sheet that is cut into rolls. Felts are designated by weight, such as a 15-lb felt; 15 pounds is the weight of 100 square feet of the material.

Roofing felts are used as a vapor barrier beneath some types of shingle roofs and as a roofing material itself in a tarred roof. Application of roofing felt is discussed in Chapter 15.

Flashing

To protect a house against moisture at openings in the walls and the roof, such as around a chimney, windows, and exterior doors, the carpenter installs **flashing. Flashing** is a watertight material, usually metal or building paper, that is bent to form an angle that sheds water away from the opening it protects. The most common metals for flashing are aluminum, galvanized steel (called **sheet metal**), and copper. Application of flashing and the places where it is required are discussed in Chapters 15 to 19 inclusive.

Plastic Laminates

Inside a house the main sources of moisture problems are in the kitchen and bathroom. To protect working surfaces and the areas adjacent to them from moisture problems, countertops are surfaced with a plastic laminate. **Plastic laminate** is a very thin, stiff material manufactured in sheets. Laminates usually have a base layer of kraft paper impregnated with resin, a printed sheet saturated with resin that gives the material its color and pattern, and a protective transparent top sheet that is also saturated with resin. Yet total thickness is only about $\frac{1}{16}$". The various layers are bonded under pressure to form a surfacing material that is not only waterproof but is able to withstand heat, household cleaners and liquids, and impact. Application of plastic laminates is outlined in Chapter 26.

Questions

1 What is the common purpose of the thin sheets discussed in this unit?

2 Which thin sheet material "breathes?" What is the purpose of this feature?

3 Name three common places where flashing is required.

4 Which of the thin sheets is not affected by heat?

Fasteners

Over the centuries that wood has been used as a building material, carpenters have needed some means of fastening two pieces of wood together. Most fastening requirements still involve attaching one piece of wood to another. But carpenters also attach wood to other wood products, to masonry, and to metal. They often apply finishing materials that contain no wood at all. As the number of materials to be fastened has increased, so has the variety and types of fasteners. Of these, the nail is still the most important.

Nails

The wood dowel or wooden peg is the oldest known fastener—as old as the use of wood itself. Metal fasteners were first developed by the ancient Egyptians, and they were so effective that their invention was scarcely improved for more than 2000 years. The first major change in nail design occurred in France in the nineteenth century with the invention of a machine that could manufacture nails from wire. Today nails are still classified by the gauge of the wire from which they are cut.

Standard nails

A nail has three parts—a head, a shank, and a point (Fig. 1-0.20). The heads of standard nails are flat and circular in shape. The thickness of the head increases with the length of the nail. Shanks are smooth except for grooves cut near the head for greater holding power. Shanks are also shiny; standard nails are thus often called **bright nails.** Points are diamond shaped.

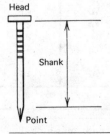

Fig. 1-0.20 The parts of a nail.

Standard nails are designated by the **penny**, which is a reference to length. Use of the term penny goes back many years, but its exact origin isn't known. Some historians think that the term may have referred to the cost in pennies of 100 nails of a particular length, because there is a fixed relationship between the length of a nail and its penny designation. As a further clue, the abbreviation of the word penny is the letter *d*; d was the British abbreviation for penny until 1970, when Great Britain shifted to the decimal system of currency.

Here is the penny-to-length relationship:

$$\text{nail length in inches} = \frac{\text{penny size}}{4} + \frac{1}{2}$$

Therefore, if you need a nail 3″ long, you need a 10-penny nail ($3 = \frac{10}{4} + \frac{1}{2}$). This length is measured from the underside of the head to the tip of the point.

Standard nails range in size from 2 to 60d (Fig. 1-0.21). Among standard nails, the smallest are called **box nails**, middle-sized nails are called **common**

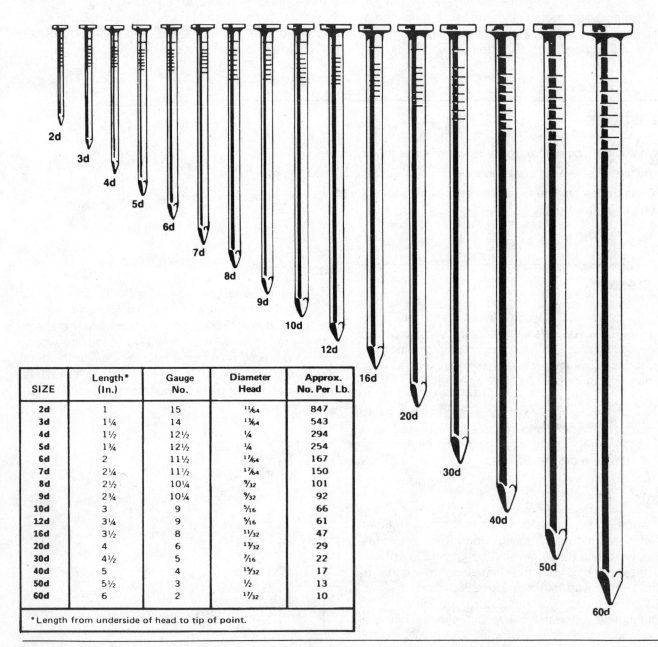

SIZE	Length* (In.)	Gauge No.	Diameter Head	Approx. No. Per Lb.
2d	1	15	$^{11}/_{64}$	847
3d	1¼	14	$^{13}/_{64}$	543
4d	1½	12½	¼	294
5d	1¾	12½	¼	254
6d	2	11½	$^{17}/_{64}$	167
7d	2¼	11½	$^{17}/_{64}$	150
8d	2½	10¼	$^{9}/_{32}$	101
9d	2¾	10¼	$^{9}/_{32}$	92
10d	3	9	$^{5}/_{16}$	66
12d	3¼	9	$^{5}/_{16}$	61
16d	3½	8	$^{11}/_{32}$	47
20d	4	6	$^{13}/_{32}$	29
30d	4½	5	$^{7}/_{16}$	22
40d	5	4	$^{15}/_{32}$	17
50d	5½	3	½	13
60d	6	2	$^{17}/_{32}$	10

*Length from underside of head to tip of point.

Fig. 1-0.21 Full-size drawing of standard nails and a chart of their characteristics. Nails of any given penny size may vary as much as ¼″ in length.

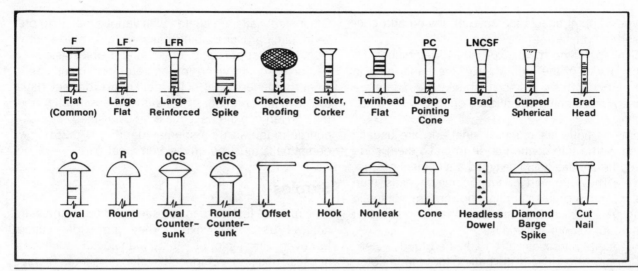

Fig. 1-0.22 Types of nail heads.

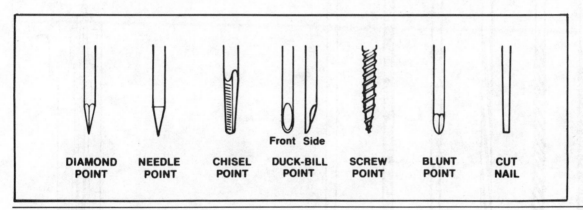

Fig. 1-0.23 Types of points.

nails, and the largest sizes are called **spikes.** All sizes are designed as all-purpose nails for easy nailing and good holding power.

Casing Nails

Close cousins to the common nail are **finishing nails** and **casing nails.** Like common nails, they have smooth shanks and diamond points. Their heads are different, however. Casing nails have a conical (cone-shaped) head, and finishing nails have a spherical head (see Fig. 1-0.22). Both types are designed to be set below the surface, or **countersunk.** Often the heads have a small depression in the top to accept the end of a nailset.

Casing nails range in size from 2 to 40d; 6 to 10d sizes are the most often used. Finishing nails range from 2 to 20d. Their lengths are measured from the top of the head to the tip of the point.

Other nails

For any purpose you can think of there is a nail manufactured to do the job; heads, shanks, and points vary to meet the need. Most heads (Fig. 1-0.22) are centered over the shanks, but a few are off center. Heads range in diameter from $\frac{1}{16}''$ for brads up to more than $\frac{1}{2}''$ for roofing nails. Most types are flat on top, but some are round or oval. One type, the **duplex** or **double-headed** nail, has two heads and is designed for temporary nailing. The lower head prevents you from driving the nail all the way in, and you pull on the upper head to remove it.

Nails may have one of seven types of points (Fig. 1-0.23). A **needle point** is round. A **diamond point,** which is the most common, has four sides. A **chisel point** has two sides, and a **duck-bill point** has one. A **cut nail** has no point at all; it is cut square at the end.

The type of point affects the ease of driving and controls surface damage to the wood.

The shape of the shank (Fig. 1-0.24) determines the holding power of the nail. Most of the nails used in rough carpentry have a smooth shank. The others are designed for fastening specific materials used in finish carpentry.

Common nails are made of steel and are usually covered with a thin coating of oil to make them slide into wood more easily. Some types are cement coated, however. They are thinner than common nails (see Table IX), have greater holding power, are lighter in weight, are $\frac{1}{8}$ to $\frac{1}{4}''$ shorter, and have sharper points that aren't as apt to split softwoods.

For special purposes some nails are **blued,** a sterilizing process that leaves the nails dark blue in color.

Other steel nails are coated with various metals to prevent staining. Nonstaining nails are also made out of aluminum, copper, zinc, brass, and stainless steel.

For fastening wood to concrete and sometimes to steel, hardened steel nails, called **powder-set nails,** are sometimes used. The nails are essentially steel pins driven into the base material by a small explosive charge triggered by a special tool (Fig. 1-0.25). Powderset nails must be driven with great care.

Staples

As structural fasteners for on-site building, staples are not yet used extensively. They are widely used, however, in factories that manufacture low-cost houses, housing components, and furniture. Staples

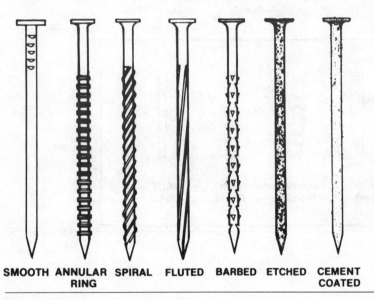

SMOOTH   ANNULAR   SPIRAL   FLUTED   BARBED   ETCHED   CEMENT
        RING   COATED

Fig. 1-0.24 Common shapes of nail shanks.

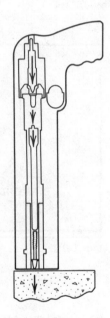

Table IX   Comparison of characteristics of common nails and coated nails

Common wire nails				Coated nails			
Size	Length (inches)	Wire gauge	Approximate number per pound	Size	Length (inches)	Wire gauge	Approximate number per pound
6d	2	$11\frac{1}{2}$	167	6d	$1\frac{7}{8}$	13	293
8d	$2\frac{1}{2}$	$10\frac{1}{4}$	101	8d	$2\frac{3}{8}$	$11\frac{1}{2}$	153
10d	3	9	66	10d	$2\frac{7}{8}$	11	111
16d	$3\frac{1}{2}$	8	47	16d	$3\frac{1}{4}$	9	64
20d	4	6	29	20d	$3\frac{3}{4}$	7	39

are driven into place with guns that operate under air pressure and require a source of compressed air. Hand-operated guns are satisfactory for application of light materials such as sheathing and ceiling tile.

Staples have three parts—legs, crown, and points—and are designated by all three (Fig. 1-0.26). The leg and crown dimensions are given first, in inches or fractions of an inch, and then the shape of the point is given. Staples may have one of three different points. With **standard points** the legs go straight into the material being fastened (Fig. 1-0.27). **Convergent points** are cut at opposite angles, so that the points cross to clinch the two pieces together. **Divergent points** spread front and back, like a gymnast doing a split.

Most staples are smooth, but some are cement coated for a better grip. Staples must be carefully se-

lected. Too small a staple won't have enough holding power, while too large a staple may damage a material's surface and also be weak in holding power. The main advantage of staples is time saved; an experienced carpenter can drive 200 staples per minute.

Screws

A rough carpenter uses few screws. A finish carpenter uses many. Most screws are made of brass or steel with various finishes on them.

Screws are designated first by length, then by wire gauge, then by head and, finally, by purpose. A typical screw is a 1" No. 8 RHWS. The 1" length is measured from the point to the bottom of the head if the screw will be exposed and to the top of the head if the screw

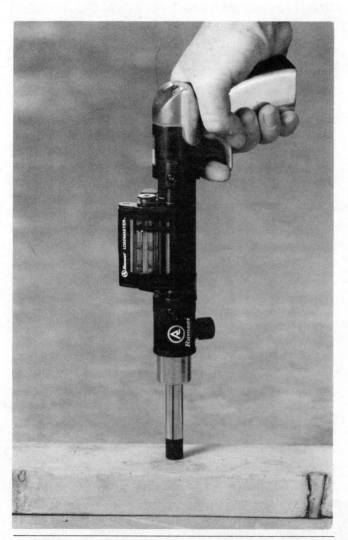

Fig. 1-0.25 Powder-set nail (left) and powder-actuated tool. Use of this tool requires an operator's license issued by the manufacturer after training. (Courtesy Ramset Fastening Systems.)

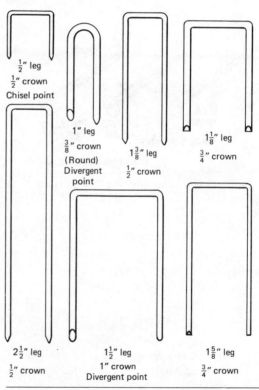

Fig. 1-0.26 Typical shapes of staples in commonly-used sizes.

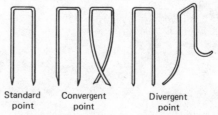

Fig. 1-0.27 The three types of points and how staples clinch when driven in.

will be countersunk. The shortest screw length is ¼″. Lengths increase every ⅛″ up to a 1″ length, then every ¼″ up to 3″, and then every ½″ up to a maximum of 5″. Gauge sizes range from No. 0 (about 1/16″) to No. 24 (about ¾″). The gauge size increases as shank diameter increases—just the opposite from nails.

A screw's designation always ends in four letters. The first two letters indicate which of the four head types the screw has—flat head (FH), oval head (OH), round head (RH), or Phillips head (PH) (Fig. 1-0.28). The last two letters indicate the purpose: WS for wood screws and MS for metal screws. The difference between wood and metal screws lies in the cut of the threads.

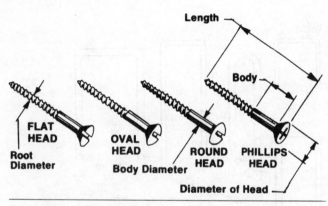

Fig. 1-0.28 The parts of a screw and the four common screw heads.

Fig. 1-0.29 Lag screws have screw points, threaded shanks, and heads that are tightened with a wrench.

Fig. 1-0.30 Anchor bolts have a bent shank or flat plate that holds firm in concrete after it has set.

With some types of screws for finishing you also indicate the finish, such as blued or chrome plated, or the material, such as brass or bronze.

When regular screws aren't long enough or strong enough for a special job, such as connecting heavy timbers, lag screws are often used. A **lag screw** (Fig. 1-0.29) is a heavy screw with a diameter of ¼″ to 1″ and a square or hexagonal head. Lengths range from 1 to 16″. Lag screws are tightened with a wrench instead of a screwdriver.

Bolts

A carpenter uses fewer bolts than any other type of fastener. Four types are used in building. **Anchor bolts**, also called **sill anchors** or **sill bolts**, are used to attach the first pieces of lumber to the top of a foundation wall. They have a square or hexagonal head (Fig. 1-0.30), a washer, and a hooked end or plate that holds them firmly in the concrete. **Expansion bolts** are a type of anchor bolt designed to expand inside a sleeve or inside the hole itself (Fig. 1-0.31) for a firm grip. **Toggle bolts** (Fig. 1-0.32) have a locking device at one end. The bolt and device fit into a hole that has space behind it, such as between two studs, then expand and lock against the surface material as the bolt is tightened.

One family of bolts has threaded shanks and nuts (Fig. 1-0.33). **Stove bolts** have slotted heads to accept the blade of a screwdriver, **machine bolts** have square or hexagonal heads for tightening with a wrench, and **carriage bolts** have round heads for positioning with a hammer. Nuts hold all types tight.

Special Fasteners

Many types and shapes of metal connectors have been developed for use in building construction to replace nails when good nailing practices are difficult to follow. Typical of these are **framing anchors** (Fig. 1-0.34) for connecting framing members that meet at right angles. **Splice plates** (Fig. 1-0.35) are flat metal plates with holes punched through them that are used to connect two members in line, such as two pieces of ridgeboard in a roof. **Split-ring connectors** are devices for fastening members in a roof truss (Fig. 1-0.36). **Corrugated fasteners** are small clips for fastening two boards together when other types of fasteners might split the wood. There are two designs (Fig. 1-0.37); both come in widths of ⅝ to 1⅛″ and penetrate into wood to a depth of ¼ to ¾″.

Instructions on how to select and install all special fasteners are given in the chapters where the fasteners are put to use.

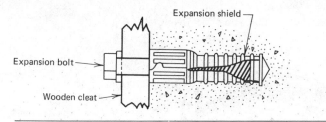

Fig. 1-0.31 Expansion bolts spread when driven into a sleeve.

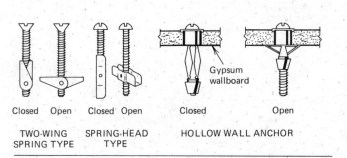

Closed Open Closed Open Closed Open

TWO-WING SPRING TYPE SPRING-HEAD TYPE HOLLOW WALL ANCHOR

Fig. 1-0.32 Toggle bolts have a locking device that opens after the bolts have been driven through a hole in sheet materials.

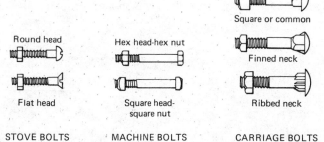

Round head

Flat head

STOVE BOLTS

Hex head-hex nut

Square head-square nut

MACHINE BOLTS

Square or common

Finned neck

Ribbed neck

CARRIAGE BOLTS

Fig. 1-0.33 Stove bolts, machine bolts, and carriage bolts all come with threaded shanks and nuts, but the heads are different.

Fig. 1-0.34 Typical framing anchors.

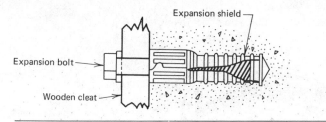

33

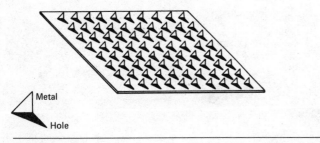

Fig. 1-0.35 Metal splice plates come in a variety of rectangular sizes.

Fig. 1-0.36 Split-ring connectors are used for fastening the chords of roof trusses. (Courtesy TECO Products Testing Corp.)

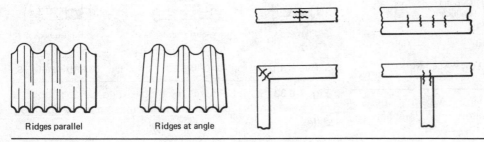

Fig. 1-0.37 Corrugates fasteners hold two boards together end to end or side to side.

Questions

1 Using the penny-to-length formula, determine the penny size of a common nail $4\frac{1}{2}''$ long.

2 What types of heads may casing nails have? How do they differ?

3 Name two advantages of cement-coated nails over common nails.

4 What is the chief advantage of staples?

5 What does PHMS stand for?

6 Name three types of bolts that do not have nuts.

7 What is the basic difference in usage between splice plates and corrugated fasteners?

2

The Tools You Use

Probably every tool was invented or developed in answer to the question, How can I do this job better? We are forever looking for ways to accomplish our ends more easily and with more effective results.

The tools you use in residential construction depend largely on the end to be accomplished. Every job, no matter how large, can be divided into a series of tasks. The building procedures in this book are organized by tasks, beginning with simple measuring up to the more complex task of assembling and installing cabinets. To accomplish each task efficiently, you must have the proper materials on hand and must know what you can do and should not do with those materials. To work on those materials you must have the right tools and know how to use them.

In school the materials and the tools are provided for you. On the job the materials and most, if not all, of the power tools will be provided. Carpenters are expected to bring their own hand tools, however, and to keep them in first-class working condition. Begin now to start your tool collection, and add to it as you find you need more tools.

Hand-Operated Tools

Before you buy any tool, make certain that

- You need it.

- It will do the job for which you need it.

- You know how to select the right tool from those on the market.

- You know how to use it.

- You know how to take care of it.

When you buy, go for quality. Select tools made by companies with a good reputation within the building industry for making tools that will help you produce accurate, quality work for a long time. You may see bargains occasionally, but most bargains turn out to be a waste of money after you have used them for awhile.

So start with quality and maintain that quality with good care. Keep small tools in a sturdy box with handles. You can build the box yourself (Fig. 2-0.1). Divide the space into compartments and keep similar tools in the same compartment. Store sharp tools so that their edges are protected against nicks and bumps. Store flexible tools so that they won't bend out of shape as you move the toolbox from place to place. The box should be strong, but it should not be so heavy that it is a chore to carry to and from the job.

During the workday, keep your tools out of the weather as much as possible. If they get wet or muddy, wipe them clean as soon and as often as possible. At the end of the day, clean and dry all tools so that they are ready for immediate use the next morning.

On days when you aren't working, wipe all metal surfaces with a clean rag soaked in fine oil; particularly oil working parts such as sawteeth, adjustable nuts, and jaws of wrenches. The oil prevents rust and helps tools work more easily. At least once a month, check all cutting edges for sharpness and sharpen all tools.

The beginning carpenter needs hand tools that serve these general purposes.

- Measuring.
- Locating.
- Marking.
- Checking square.
- Checking level and plum.
- Cutting.
- Shaping.
- Attaching.
- Boring.

Here are the woodworking tools available for these purposes, organized by the specific task at hand.

Fig. 2-0.2 Standard and metric bench rules. (Courtesy Stanley Tools.)

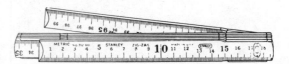

Fig. 2-0.3 Folding rule. (Courtesy Stanley Tools.)

Fig. 2-0.4 Steel tape (50′ and 100′). (Courtesy Lukkin/The Cooper Group.)

Fig. 2-0.1 To keep tools in tip-top shape, maintain them regularly and store them in a compartmented toolbox. (Design by John Gaynor.)

The task	The tool	Its description
To measure or check a measurement less than 12″	Bench rule (Fig. 2-0.2)	A metal ruler marked in eighths of an inch on one side and sixteenths of an inch on the other. Used flat
To measure or check a measurement less than 300 mm	Metric bench rule	A metal ruler marked in millimeters. Used flat
To measure lengths up to 6′ or 2 m	Folding rule (Fig. 2-0.3). Lengths of 6′, 8′, and 2 m.	A wood ruler with pivot points every 6″, or about 17 centimeters (cm), and with metal tips at the ends. Used on edge for short measurements and flat for lengths of more than 24″, or 60 cm
To measure distances up to 12′ and dimensions of materials	Steel tape. Lengths of 6′, 8′, 10′, 12′, and metric equivalents	Thin metal tape with a metal stop at the zero end. Retracts into a steel case. Some types have clips for attachment to a belt
To measure long distances (up to 100′)	Steel tape (Fig. 2-0.4). Lengths of 50′, 100′, and metric equivalents	Thin metal tape with hook on end. Retractable handle used to rewind tape into a metal case
To locate a building; *to check* level points on a building site	Builder's level, also called a dumpy level (Fig. 2-0.5), and leveling rod	An optical device that rests on a tripod. The carpenter looks through the telescopic instrument at a leveling rod to make horizontal sightings
To locate a building: *to check* level points on a building site; to mark square corners for a foundation	Transit level (Fig. 2-0.6) and leveling rod	An optical device similar to a builder's level, but more versatile. The telescope not only turns 360° but can be aimed 45° upward or downward from horizontal
To lay out an arc or full circle; *to step off* a short measurement	Dividers (Fig. 2-0.7)	A tool similar to a compass, having two metal legs with an adjustable distance between them. One leg may be replaced with a pencil for marking

Fig. 2-0.5 Builder's level and leveling rod. (Courtesy David White Instruments.)

Fig. 2-0.6 Transit level. (Courtesy David White Instruments.)

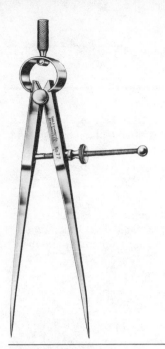

Fig. 2-0.7 Dividers. (Courtesy The L.S. Starrett Co.)

Fig. 2-0.8 Carpenters pencils. (Photo Drew Leviton, Atlanta.)

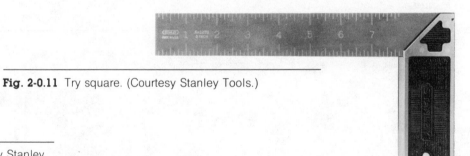

Fig. 2-0.10 Marking gauge. (Courtesy Stanley Tools.)

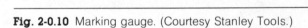

Fig. 2-0.11 Try square. (Courtesy Stanley Tools.)

Fig. 2-0.9 Awl. (Courtesy Stanley Tools.)

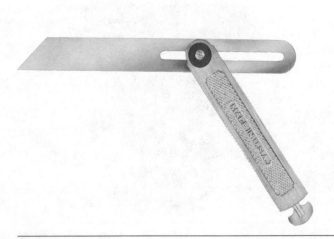

Fig. 2-0.12 Sliding T bevel. (Courtesy Stanley Tools.)

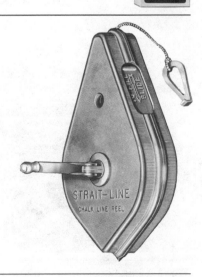

Fig. 2-0.13 Chalk line and chalk. (Courtesy The Irwin Auger Bit Company.)

The task	The tool	Its description
To draw a line	Carpenter's pencil (Fig. 2-0.8)	A flat wood pencil with flat lead that can be sharpened with a knife
To mark a point	Awl (Fig. 2-0.9)	A pointed metal tool with a bulbous handle
To mark a line parallel to an edge	Marking gauge (Fig. 2-0.10)	A tool, usually metal, consisting of a calibrated rod, adjustable head, and a marking point at one end
To mark a line perpendicular to an edge; *to check* the square of two adjacent surfaces	Try square (Fig. 2-0.11)	A tool with a short metal blade and a handle at right angles to it
To mark an angle	Sliding T bevel (Fig. 2-0.12)	Consists of a handle and a blade that can be set at any angle from 0 to 180°. The angle must be established with another tool, such as a protractor
To mark a long line temporarily on a surface	Chalk line and chalk (Fig. 2-0.13)	Consists of a long piece of cord, usually nylon, in a retracting case. The cord is rubbed with chalk, then snapped between two points
To check square; *to mark* and test a 45° angle of a miter	Combination square (Fig. 2-0.14)	A multipurpose tool with a long metal blade that slides in the handle. The handle contains a small awl and level
To check square; *to mark* a line at right angles to an edge	Framing square (Fig. 2-0.15)	A large angle of flat steel with a blade 24″ long and a tongue 16″ long
To lay out rafters and stairs; *to check* square and mark at right angles to an edge	Rafter square (Fig. 2-0.16)	Similar to a framing square except in edge markings and the information given on the blade and tongue. See Chapter 9 (Unit 1)

Fig. 2-0.14 Combination square. (Courtesy Stanley Tools.)

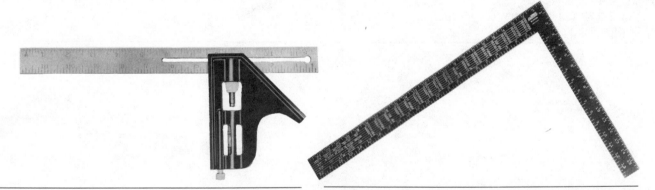

Fig. 2-0.15 Framing square. (Courtesy Stanley Tools.)

Fig. 2-0.17 Carpenter's level. (Courtesy Stanley Tools.)

Fig. 2-0.16 Rafter square. (Courtesy Great Neck Saw Manufacturers, Inc.)

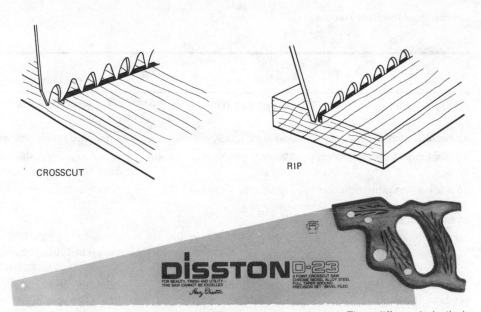

CROSSCUT

RIP

Fig. 2-0.18 Plumb bob and line. (Courtesy The L.S. Starrett Co.)

Fig. 2-0.19 Crosscut saw or ripsaw. They differ only in their teeth. (Courtesy Disston, Inc.)

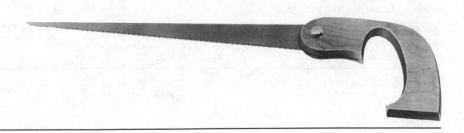

Fig. 2-0.20 Compass saw. (Courtesy Disston, Inc.)

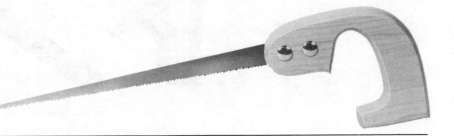

Fig. 2-0.21 Keyhole saw. (Courtesy Disston, Inc.)

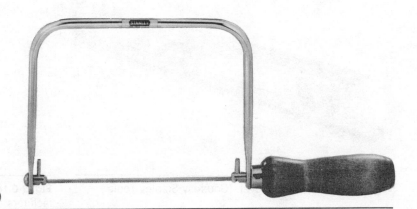

Fig. 2-0.22 Coping saw. (Courtesy Stanley Tools.)

The task	The tool	Its description
To check the plumb or level of a surface	Carpenter's level (Fig. 2-0.17)	Two or three leveling glasses set into a metal or wood frame
To transfer a point vertically from one level to another	Plumb bob (Fig. 2-0.18)	A pointed metal weight attached to a cord. At the upper end of the cord is a hook for hanging
To cut across the grain of wood; can be used to cut with the grain	Crosscut saw (Fig. 2-0.19)	A handsaw with 4 to 12 points (teeth) per inch. The most common general purpose crosscut saw has 10 points per inch and a 22″ blade. Teeth vary in length
To cut with the grain of wood	Ripsaw (similar to Fig. 2-0.19)	A handsaw with teeth farther apart and all the same length. Common size: $5\frac{1}{2}$ points per inch and a 26″ blade
To cut inside curves; *to cut* gentle curves	Compass saw (Fig. 2-0.20)	A shorter saw (12 to 14″) with a thin, tapered blade
To cut sharp curves and small holes	Keyhole saw (Fig. 2-0.21)	Slightly shorter and thinner than a compass saw and with finer teeth
To cut irregular shapes	Coping saw (Fig. 2-0.22)	Consists of a thin, fine-toothed blade in a U-shaped metal frame. The blade can be rotated 360°
To cut finely and accurately, as in a miter box	Backsaw (Fig. 2-0.23)	A crosscut saw with fine teeth and a metal stiffening band to prevent the blade from bowing
To cut accurately at 30°, 45°, 60°, and 90° angles	Miter box (Fig. 2-0.24, shown with a backsaw)	A manufactured or carpenter-made frame of wood or metal that acts as a holder for wood to be cut square or at standard angles
To cut and trim thin building materials	Utility knife (Fig. 2-0.25)	An all-purpose knife with a sturdy retractable blade that fits into a metal handle
To notch wood	Wood chisel (Fig. 2-0.26)	A straight tool, usually having a wood handle, with a thick blade that tapers to a knifelike edge at the end

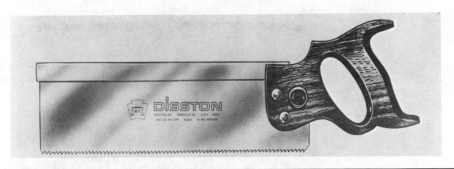

Fig. 2-0.23 Backsaw. (Courtesy Disston, Inc.)

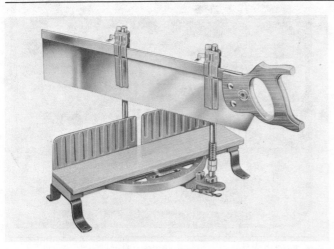

Fig. 2-0.24 Miter box saw and miter box. (Courtesy Stanley Tools.)

The task	The tool	Its description
To attach straight nails and to remove bent ones	Claw hammer (Fig. 2-0.27)	A tool with a flat surface at one end of the head for pounding and a curved claw at the other for pulling poorly driven or double-headed nails
To drive nails; *to rough-shape* structural lumber to fit	Hatchet (Fig. 2-0.28)	A two-faced tool with a hammer head at one end and a sharp cutting blade on the other. The notch in the blade is for pulling nails
To remove nails	Nail claw (also known as a cat's-paw)	A metal bar with one end bent like a hook that is slotted and dished for fitting under nails. The other end may be cut square or shaped like a chisel
To drive stakes	Sledge (Fig. 2-0.29)	A long-handled tool with a heavy metal head. Swung with both hands
To bore a hole up to 1″ in diameter	Brace (Fig. 2-0.30)	Two types. Standard brace (left) has two handles, and the circular drilling motion is around the shank. Ratchet brace (right) is for boring in close quarters, and the motion is parallel to the shank. Chuck opens and closes to accept bits of various sizes
To bore a hole less than $\frac{1}{4}$″ in diameter	Twist drill (top) or bit stock drill (bottom).(Fig. 2-0.31)	A short bit with a fluted shank. Ends may be round or square to fit various braces
To bore a hole of $\frac{1}{4}$ to 1″	Auger bit (Fig. 2-0.32)	A longer bit with a screw tip and twisted shank. Bit may be single-twist (top) or double-twist (bottom).
To bore a 1 to 2″ hole	Expansion bit (Fig. 2-0.33)	A bit with a screw tip and a slot for holding cutting blades of various sizes
To bore a smooth hole to a fixed depth	Multi-Spur® bit (Fig. 2-0.34)	A smooth-shanked bit with a jagged, circular cutting blade

Fig. 2-0.25 Utility knife. (Courtesy Stanley Tools.)

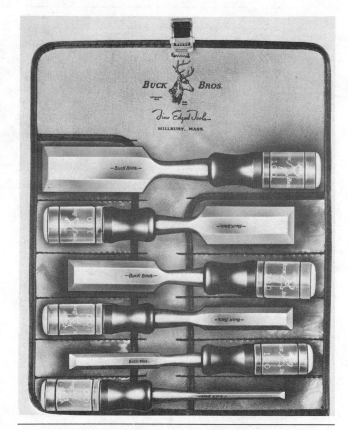

Fig. 2-0.26 Wood chisels. (Courtesy Buck Bros., Inc.)

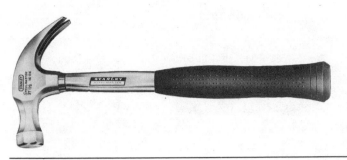

Fig. 2-0.27 Claw hammer. (Courtesy Stanley Tools.)

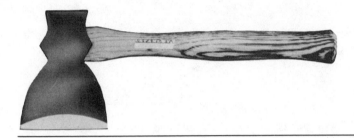

Fig. 2-0.28 Hatchet. (Courtesy Stanley Tools.)

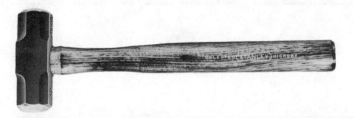

Fig. 2-0.29 Sledge.(Courtesy Stanley Tools.)

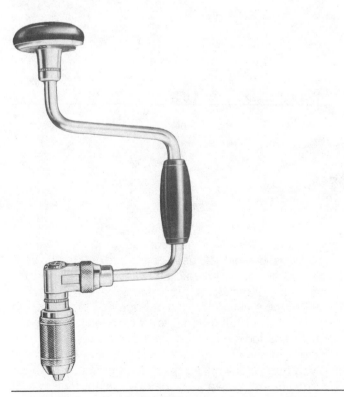

Fig. 2-0.30 Standard brace (left) and ratchet brace (right).   (Courtesy Stanley Tools.)

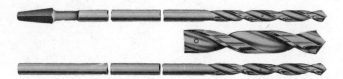

Fig. 2-0.31 Twist drill (top) and bit stock drill (bottom). (Courtesy The Irwin Auger Bit Company.)

Fig. 2-0.32 Auger bits, single twist (top) and double twist (bottom). [Courtesy The Irwin Auger Bit Company (top), Stanley Tools (bottom).]

Fig. 2-0.33 Expansion bit. (Courtesy The Irwin Auger Bit Company.)

Not all the tools a carpenter needs are used on wood. Some metalworking tools are required to set up and adjust power tools used on the site and removed at the end of each workday. Still others are needed for use on metal fasteners, pipe, and sheets. They are:

The tools listed are general-purpose tools used in many different building tasks. A carpenter uses many other hand tools designed for specific uses, such as a roofer's hatchet, a glass cutter, and a broadknife. These and woodworking tools used primarily in finish carpentry are shown in Section IV, where their use is discussed.

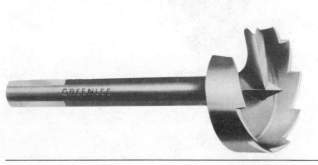

Fig. 2-0.34 Multi-spur bit. (Courtesy Greenlee Tool Co.)

The task	The tool	Its description
To cut thin sheets of metal	Metal shears (Fig. 2-0.35)	Heavy shears with a scissors action
To cut heavier metal	Hacksaw (Fig. 2-0.36)	Replaceable blades fit into a C-shaped frame. Thumbscrew adjusts tension on the blade
To cut the head off a nail or screw	Cold chisel (Fig. 2-0.37)	Similar to a wood chisel except that the tool is made of tempered steel for hardness
To hold, bend, or remove wire and small fasteners	Needle-nosed pliers (Fig. 2-0.38)	Nonadjustable pliers with long, thin, flat jaws
To hold or turn round pieces, such as pipes	Combination pliers (Fig. 2-0.39)	Pliers with jaws that may be set in two positions
To loosen or tighten nuts and bolts	Crescent wrench	A wrench with one fixed jaw and one adjustable jaw

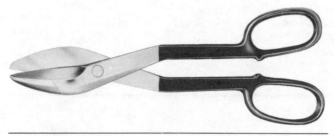

Fig. 2-0.35 Metal shears. (Courtesy Stanley Tools.)

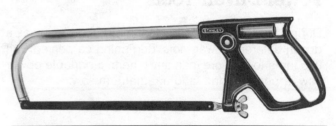

Fig. 2-0.36 Hacksaw. (Courtesy Stanley Tools.)

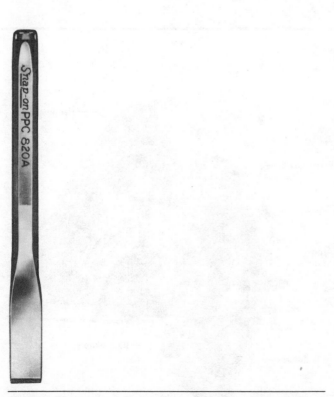

Fig. 2-0.37 Cold chisel. Compare with wood chisels (Fig. 2-0.26). (Courtesy Snap-on Tools Corp.)

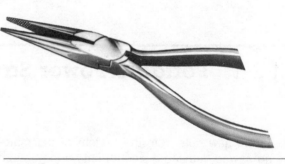

Fig. 2-0.38 Needle-nosed pliers. (Courtesy Snap-on Tools Corp.)

Fig. 2-0.39 Combination pliers. (Courtesy Snap-on Tools Corp.)

Questions

1 How does a bit differ from a drill?

2 Name two important differences between a 6' and a 50' steel tape, aside from their length.

3 What would you use to mark a 20' line on a floor?

4 Name four tools that can be used for marking a right angle.

5 What is the chief difference between a crosscut saw and a ripsaw?

6 Name the chief differences in purpose of a compass saw, a coping saw, and a keyhole saw.

Power-Driven Tools

During their careers, carpenters may use as many as a dozen different power tools. Beginning carpenters will use three tools more than any others: a portable power saw, a portable drill, and a radial arm saw.

Unit 2-1   Portable Power Saw

The portable saw cuts from the bottom of the material being sawed, leaving a smooth cut on the bottom and a rough cut on the top. Therefore the saw should be used with the good or finished side of the material face downward. The material must be fully supported on a table or sawhorses.

The diameters of saw blades vary from $4\frac{1}{2}$ to 12". A typical saw has a blade about 8" in diameter powered by a 2-horsepower (hp) motor. The base plate rests on the material to be cut, and the blade should be adjusted to cut about $\frac{1}{8}$" deeper than the thickness of the material. The angle of the blade should be adjusted before the power is turned on.

Fig. 2-1.1 Portable power saw and its key parts. (All photos courtesy Milwaukee Electric Tool Corporation.)

The task	The tool	Its description
To make a straight cut in lumber or sheet materials	Portable power saw; also called a cutoff saw, circular handsaw, and by various trademarks (Fig. 2-1.1)	A lightweight, electrically operated handsaw with a disc blade, base plate, fixed guard, and telescoping guard. Angle of the blade is adjustable up to 45°

Special Features

All portable power saws have the features shown in Fig. 2-1.1. Some types have blades that can cut either forward or backward, and the motor is reversible. The shape and size of the arbor hole varies from manufacturer to manufacturer (Fig. 2-1.2), and the blade selected must lock tight on the arbor.

General Safety Precautions

The following precautions should be observed whenever you use a portable power saw.

- Make all blade adjustments with the power *off.*
- Use blades with sharp teeth properly set.
- Use the proper size of blade for the cut to be made.
- Make sure that the arbor nut is tight at all times.
- Make sure that the telescoping guard is in place.
- Keep your hands and body out of the line of cut.

Operating precautions are listed as they occur in the following procedures.

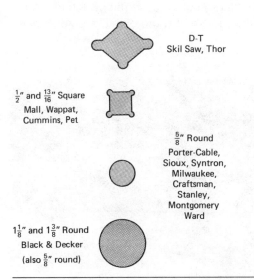

D-T
Skil Saw, Thor

$\frac{1}{2}''$ and $\frac{13}{16}''$ Square
Mall, Wappat, Cummins, Pet

$\frac{5}{8}''$ Round
Porter-Cable, Sioux, Syntron, Milwaukee, Craftsman, Stanley, Montgomery Ward

$1\frac{1}{8}''$ and $1\frac{3}{8}''$ Round
Black & Decker
(also $\frac{5}{8}''$ round)

Fig. 2-1.2 Shapes and sizes of arbor holes for portable power saw blades.

Procedure for Making a Straight Cut

1 Mark the line of cut on the material.
2 Adjust the depth of cut by loosening the nut or clamp, positioning the blade, and retightening the nut or clamp (Fig. 2-1.3).
3 Start the motor and let it reach full speed before beginning the cut. **Precaution.** Make sure that the teeth are pointing in the direction of the cut.
4 Place the base plate on the material and line up the blade with the cutting line.

Fig. 2-1.3

5 When making a long cut, use a protractor to assure a straight crosscut (Fig. 2-1.4) and use a fence for long, ripping cuts (Fig. 2-1.5).

6 Move the saw firmly but lightly along the cutting line (Fig. 2-1.6). **Precaution.** If the motor stalls, release the trigger at once and then carefully back the saw blade all the way out of the cut. Begin again at step 3.

7 After completing a cut, release the switch and set the saw aside. The guard protects the blade while it comes to a stop.

Procedure for Making a Bevel Cut

1 Loosen the nut that holds the base plate in place, adjust it to the required angle of bevel, and retighten the nut.

2 Follow steps 1 to 7 of the procedure for making a straight cut. **Note.** Bevel cuts are often made with a guide (Fig. 2-1.7).

Procedure for Making an Internal Cut (Plunge Cut)

1 Mark the lines of cut on the material.

2 Adjust the depth of the blade.

3 Slide the telescopic guard out of the way and hold it there.

4 Set the leading edge of the base on the material and line up the blade with the cutting line.

5 Start the motor and let the saw blade reach full speed.

Fig. 2-1.4

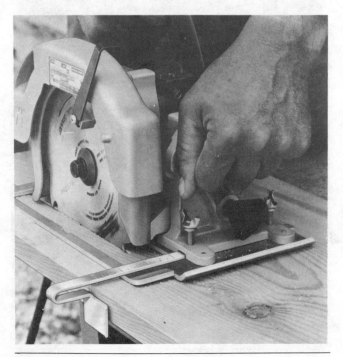

Fig. 2-1.5

Fig. 2-1.6

6 Slowly lower the blade onto the cutting line near, but not at, one end.

7 Complete the cut to a corner. If the saw does not have a reversible blade, shut off the motor, remove the blade from the cut, and repeat steps 5 to 7 to the opposite corner. If the saw has a reversible blade, cut slowly to both corners before removing the blade from the cut.

8 Repeat these procedures for the remaining cuts.

Unit 2-2   **Portable Electric Drill**

Power drills are shaped like small handguns. Some types have a pistol-grip handle with a trigger switch beneath the motor as in Fig. 2-2.1. Others have two handles, one with a trigger and one to guide the drill (Fig. 2-2.2). The better makes have a device in the end of the handle that keeps excessive strain from damaging the electrical cord.

Most chucks are geared for operation with a key (Fig. 2-2.3) and have three holes into which the key fits. Some types have a maximum capacity of $\frac{1}{4}''$, while others will open a full $\frac{1}{2}''$. Twist drills, auger bits with straight shanks, and spade bits for drilling large holes may all be inserted into the chuck. So may hole saws (Fig. 2-2.4). A combination drill and countersink for predrilling holes for wood screws is a handy attachment for finish carpenters.

The task	The tool	Its description
To drill a hole	Portable electric drill (Fig. 2-2.1)	A power drill driven by a $\frac{1}{2}$ hp motor that turns an adjustable chuck into which you insert drill bits

Fig. 2-1.7

Fig. 2-2.1 Portable power drill with pistol grip. (Courtesy Rockwell International.)

Fig. 2-2.2 Portable power drills with two handles. (Courtesy Rockwell International.)

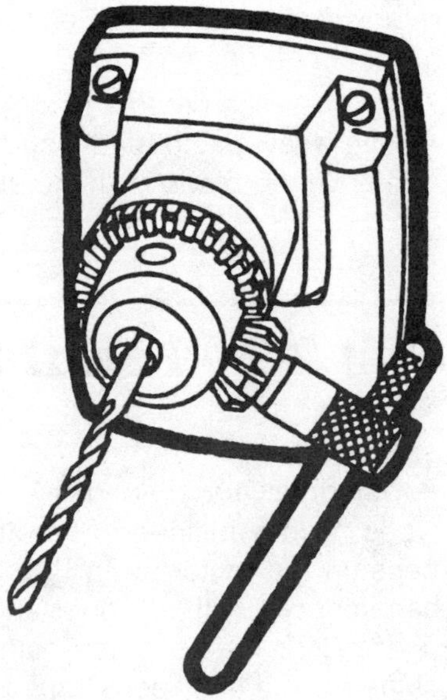

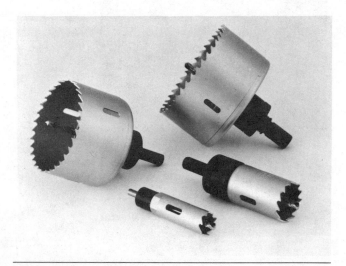

Fig. 2-2.4 Hole saw for use with power drills. (Courtesy Milwaukee Electric Tool Corporation.)

Fig. 2-2.3 Key for tightening drill chucks.

Special Features

The motor housing of an electric drill may be steel or plastic. Some types have a variable-speed motor; if so, the speed changes with the amount of finger pressure on the switch.

Screwdriver bits may be used in a chuck, thus converting a power drill into a power screwdriver. For such usage, however, the drill must be equipped with either a variable-speed motor or a special ratchet attachment.

General Safety Precautions

The following precautions should be observed whenever you use a portable power drill.

- Insert drill bits in the chuck with the power *off.*
- Make sure that the bit is properly seated and tightly clamped in position.
- Use only bits designed for use with a power drill. Bits with screw or tang ends should never be used.

Operating precautions are listed as they occur in the following procedures.

Procedure for Drilling a Hole

1 Determine the correct bit for the job to be done. Make sure that it is sharp and that the shank is straight.

2 Open the chuck by hand until its throat is wide enough to accept the bit.

3 Insert the bit and close the chuck by hand, making sure that the bit is fully seated.

4 Tighten the chuck by inserting the key in each of the three holes consecutively. By using all three holes, you reduce the possibility of the bit slipping and damaging both bit and chuck. **Precaution.** Be sure to remove the key.

5 Make sure the material to be drilled can't move. If necessary, hold it with a clamp or in a vise.

6 To prevent splintering on the back side, attach a piece of scrap wood to the back of the material.

7 With cross lines mark the center of the hole to be drilled.

8 Start the motor and let it reach full speed before starting to drill.

9 Holding the drill at right angles to the material's surface, begin drilling with just enough pressure to keep the bit cutting. **Precaution.** Do not push the bit into the material; too much pressure can break the bit or stall the motor. Too little pressure, on the other hand, can dull the point of the bit.

10 When the hole is drilled and while the motor is still running, slowly back the bit out of the hole at the same angle it entered.

11 Let the bit come to a complete stop before setting the drill down. **Precaution.** To prevent injury to you and damage to the bit, always lay the drill down with the point of the bit away from you.

Unit 2-3   **Radial Arm Saw**

The radial arm saw, unlike the portable power saw, cuts from the top of the lumber, leaving a smooth cut on the top and a rough cut on the bottom. Therefore the material should be set on the table with the finish side up.

Saw blades range in diameter from 10 to 16″. The blade may be replaced with a dado head for grooving and for other specialized work usually done in finish carpentry. The radial arm saw is basically an all-purpose tool for use during framing but, with proper maintenance, may also be used in finish carpentry.

Special Features

As you stand at the saw table, all controls are directly in front of you. The on-off switch is at the end of the arm (Fig. 2-3.1), and the handle for moving the saw is safely to one side of the blade. When the blade is turned for ripping or miter cuts, the guards and antikickback fingers turn with it. The height of the blade above the table is adjusted when you turn a crank that raises or lowers the column to which the over arm is attached. An adjustable fence serves as a stop or guide for the material to be cut.

General Safety Precautions

The following precautions should be observed whenever you use a radial arm saw.

• Make all blade adjustments with the power *off*.

• Cut only with sharp blades.

• Make sure the guard and antikickback fingers are in place.

• Make sure clamps and locking handles are tight.

• Keep the table clean of scraps and sawdust.

• Before turning on the saw, always grasp the

The task	The tool	Its description
To make a straight cut, dado, or groove in lumber	Radial arm saw (Fig. 2-3.1)	A versatile saw, supported on a fixed arm, that moves through the material to be crosscut. It is mounted above a worktable and can be turned 90° for ripping. It can also be set at an angle for bevel cuts

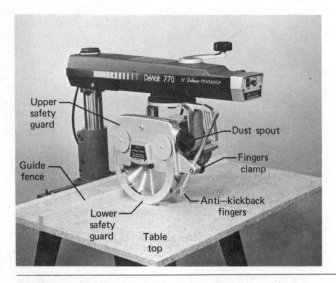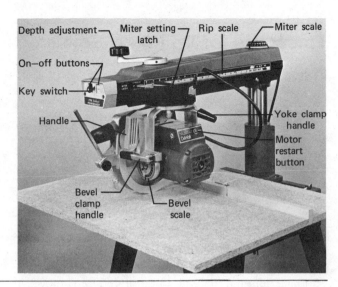

Fig. 2-3.1 The two sides of a radial arm saw and its key parts. (All photos courtesy DeWalt Division, Black & Decker Mfg. Co.)

handle to prevent the saw from moving forward until you are ready.

• Return the saw to its at-rest position at the back of the table after any cut.

Operating precautions are listed as they occur in the following procedures.

Procedure for Making a Crosscut

1 Insert a crosscutting blade on the arbor. **Precaution.** Make sure that the arbor nut is tight.

2 Set the radial arm at zero, which is at right angles to the fence. Lock the arm in place with the miter clamp handle.

3 Set the motor and blade assembly at right angles to the top of the table.

4 Turn the elevating crank or handle until the teeth of the saw blade fit into the **kerf** in the saw table about $\frac{1}{16}''$ below the table's surface (Fig. 2-3.2).

5 Adjust antikickback fingers about $\frac{1}{8}''$ above the table's surface.

6 Set the material to be cut on the table and line up the cutting line with the saw blade. Hold the material firmly against the fence with one hand.

7 With the other hand turn on the power and, with the handle, hold the saw behind the guide until the blade reaches full speed (Fig. 2-3.3). **Precaution.** The saw may tend to move by itself into the material to be cut.

8 With the handle pull the saw steadily but slowly through the material. **Precaution.** To prevent the motor from stalling, hold back on the handle. If the motor does stall, back the blade out of the cut and let it resume full speed before continuing the cut.

9 After you complete the cut, return the saw behind the fence and shut off the power.

Procedure for Making a Rip Cut

1 Insert a ripping or combination blade on the arbor. **Precaution.** Make sure that the arbor nut is tight and that the blade will rotate counterclockwise as you stand at the table.

2 Pull the saw assembly to the front of the radial arm.

3 Lift the yoke relocating pin and rotate the yoke 90° until the blade is parallel to the guide fence. For ripping lumber, turn the yoke clockwise so that the motor is toward the front of the worktable

Fig. 2-3.2

Fig. 2-3.3

Fig. 2-3.4

(Fig. 2-3.4). For ripping wide sheets of material, such as plywood, turn the yoke counterclockwise, with the motor toward the column (Fig. 2-3.5). This permits ripping a wider piece.

4 Loosen the rip lock and slide the motor assembly along the radial arm until the rip pointer points at the correct width of cut on the rip scale. Tighten the lock.

5 Lower the saw until the teeth of the blade just touch the worktable. **Precaution.** To test the correct height, slide a sheet of paper under the blade. The teeth should score the paper without tearing it.

6 Adjust the guard so that its lower end just touches the material to be cut (Fig. 2-3.6).

7 Adjust the antikickback fingers so that they lie about $\frac{1}{8}''$ below the surface of the material (Fig. 2-3.7).

8 Turn on the power. **Precaution.** Make sure the saw blade is spinning toward you as you feed the material toward it.

9 Hold the material against the guide fence and feed it slowly but steadily toward the blade. **Precaution.** Use a push stick (Fig. 2-3.8) to push the material.

10 After the cut is completed, pull the push stick straight back to you and shut off the motor.

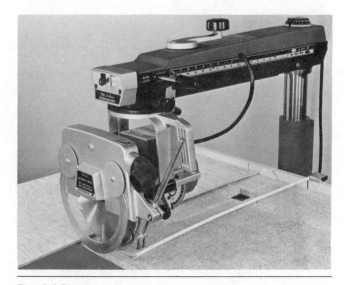

Fig. 2-3.5

Fig. 2-3.6

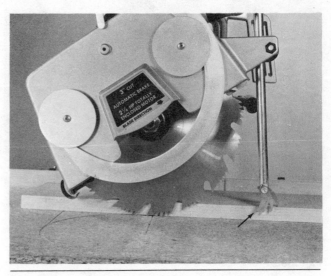

Fig. 2-3.7

Fig. 2-3.8

Fig. 2-3.9

Procedure for Making a Miter Cut

1 Insert a crosscutting or combination blade on the arbor. **Precaution.** Make sure that the arbor is tight.

2 Loosen the miter clamp handle and lift the miter latch.

3 Swing the radial arm to the correct angle of cut (Fig. 2-3.9). Lock the arm in place.

4 Lower the saw until the teeth just touch the work-table.

5 Follow steps 5 to 9 for making a crosscut.

Procedure for Making a Bevel Cut

1 Loosen the bevel clamp handle and pull out the bevel relocating pin.
2 Tilt the saw assembly to the correct angle of cut, as shown on the bevel scale (Fig. 2-3.10). Lock the assembly in position.
3 Lower the saw until the teeth just touch the worktable when the blade is tilted.
4 Follow steps 5 to 9 for making a crosscut.

Other power-driven tools and special attachments for the radial arm saw used in finish carpentry are shown in Section IV, where their use is discussed.

Fig. 2-3.10

Unit 2-4   Stapler and Nailer

The **hand-operated stapler** is swung like a hammer, but in a shorter arc. It is used primarily to attach thin materials on jobs where compressed air is not available. The staples come in cartridges that feed toward the head of the stapler. The impact of the head on the material drives a staple through it.

Much more common is the air-powered gun. **Heavy-duty nailers** weigh 7 to 11 pounds and can drive nails of various types up to $3\frac{1}{2}''$ long. They are used in structural framing and for application of thick sheet materials, such as sheathing and subflooring. **Finish nailers** are lighter in weight and are used to drive finishing nails into paneling and trim. **Staple guns** weigh 2 to 8 pounds and drive staples $\frac{3}{16}$ to $2\frac{1}{2}''$ long. They are used to apply thin sheet materials, attach roofing and siding, and build cabinets.

The chief advantage of these tools is saving time. Experienced operators, for example, can fasten down a sheet of subflooring as fast as they can walk along its edges, and can space fasteners with amazing accuracy.

The task	The tool	Its description
To attach structural parts to each other; to apply finishing materials	Stapler or nailer (Fig. 2-4.1)	A family of fastening tools operated by hand or compressed air. Most types have a trigger that releases fasteners with considerable force

Fig. 2-4.1 Air-operated stapler. (Courtesy Senco Producers, Inc.)

Special Features

Guns may fire in several ways. With one type, you place the tool where the materials are to be fastened and squeeze the trigger. With another type, you squeeze the trigger, and the gun fires as soon as the tool touches the material. With a third type, the gun may be adjusted to the operator's ability to fire up to a dozen fasteners per second with a single squeeze of the trigger. As a safety feature, the guns will not fire unless they are in contact with materials.

Fasteners

The fasteners for all types of air-powered staplers and nailers are packaged in cartridges by the manufacturer of the gun. Gun-driven fasteners are lighter and thinner than nails with comparable holding power, and are therefore less likely to split wood. Some types of top-quality fasteners have a plastic coating. This plastic melts from the friction of the fastener being driven at high speed, and fuses with the wood to provide good holding power.

General Safety Precautions

The following precautions should be observed whenever you use any pneumatic fastening tool.

- Wear safety glasses at all times.
- Use only clean, dry, compressed air at a pressure of about 100 pounds per square inch (psi). Never use oxygen or carbon dioxide or higher pressures.
- Disconnect the tool when you are not using it and when you clean it.
- Drain the air lines and wipe the tool clean of moisture once a day.
- Add two or three drops of nondetergent lubricating oil in the air inlet after each cleaning.Make sure all screws and caps are tight.
- When test-firing is required, always have a cartridge of fasteners in the tool and fire into scrap material.
- Practice on scrap material until you are used to the weight of the tool, the sensitivity of the trigger, and how to move the gun from one fastening point to the next.
- Keep your fingers away from the head of the gun. Stapling yourself can be a serious injury.
- Never point the tool at anyone. Like any gun, it can be a dangerous weapon.

Questions

1 What is the first safety rule with all power tools?

2 How does the cutting action of the portable saw differ from that of the radial arm saw?

3 What is the relationship between the thickness of material being cut with a portable power saw and the depth of cut?

4 What do you do if the motor of a portable power saw stalls?

5 At what point do you start a pocket cut with a portable power saw?

6 What is the purpose of tightening the chuck of a portable drill with the key in all three chuck holes?

7 When drilling a hole, how do you prevent splintering?

8 Why should the motors of all power tools be operating at full speed before you begin cutting or drilling?

9 How much pressure should you apply on the drill when you bore a hole?

10 Name four differences between crosscutting and ripping with a radial arm saw.

11 What is the purpose of a push stick? Describe your actions as you use it.

3

The Practices You Follow

As you study carpentry, you will learn how to select and use tools, how to work with construction materials, and how to complete correctly the many individual tasks that go into building a house. How well you learn will determine how good a carpenter you are when you go to work on building sites.

Knowledge is vital to your being a success. But your attitude is equally important—toward safety, toward procedures, and toward quality.

Safety

Accidents are expensive and can almost always be avoided. When you are hurt, you suffer the pain of the injury. Furthermore, if you are unable to work, your pay either stops or is sharply reduced while you are off the job. Your employer is required by law to carry workman's compensation insurance, so that you are paid part of your wages while you are injured. The amount you receive in benefits from workman's compensation varies from state to state, and so does the time when those payments begin. If you are hurt often, you earn a reputation for being an unsafe worker. Your employer then has to put in frequent claims for workman's compensation and, as a result, the insurance costs go up. Your employer may decide to let you go to keep costs under control. So it pays to be a safe worker.

Statistics indicate that most preventable accidents occur because of carelessness or poor safety practices. Carelessness stems from an "I-don't-care" attitude or inattention. You increase your chances of getting hurt when you don't concentrate fully on the job at hand. If, for example, you have an argument at home just before you leave for work and you keep thinking about the argument instead of your job, your poor mental attitude can get you in trouble.

To avoid poor safety practices, you must know and follow good safety practices. Safety rules are simple and logical. Learn them so well that you follow them as automatically as you breathe. Employers, if they are safety conscious, should provide you with a set of rules that they expect you to follow. If they do not, use the following list as a guide to becoming a safe worker.

Safety Rules

1 Wear the safety equipment recommended by the Occupational Safety and Health Act (OSHA). This equipment includes safety shoes, hard hats, gloves, and safety glasses. If you work for someone who doesn't furnish you with safety equipment, buy your own. And, when you have it, wear it!

2 Never wear rings, jewelry, or loose-fitting clothing on the job site. You can easily catch a sleeve in a power-driven tool and find your arm or hand slammed against the cutting edge. You can also lose your balance if a loose jacket catches on a nail or if long pants snag on scaffolding.

3 Make sure guards and other safety devices of power tools are in place before you turn on the motor. Guards do sometimes get in your way and make your job a little more difficult. But that is no excuse. A lost finger makes working even more difficult.

4 Never carry anything in your mouth. Use the pockets in your work apron for nails, screws, pencils, and small tools.

5 Never carry sharp tools in your pockets unless they have protective covers on them when they aren't in use. Carry tools such as chisels, utility knives, and screwdrivers in your toolbox.

6 Never leave your tools lying around. At the end of a work period, put all tools away. During a work period, when you must set a tool down momentarily, make sure you don't lay it in a normal walk path. *You* may know where you laid your saw, but a friend walking by with a keg of nails may not see it.

7 Do not indulge in horseplay. A job site is a place for work, not fooling around.

8 Use only standard ladders and check them regularly to make sure they are in safe condition. Falls are a frequent cause of serious injury.

9 Never stand on barrels, kegs, boxes, or crates to reach work overhead. Use ladders or scaffolding that meets OSHA requirements.

10 Keep work surfaces clean. Sweep up every night before you go home and, if necessary, just before

or after lunch. Tripping over scrap material, stepping on a nail, and filling an eye with windblown sawdust are common causes of minor but nevertheless lost-time accidents. Good housekeeping isn't easy, because you create a mess whenever you work with wood, plywood, and fasteners. One way to make the job of good housekeeping easier is to keep kegs or barrels handy, one for combustible materials such as scrap lumber, and one for bent nails and scrap metal.

11 Keep the job site picked up. Toss paper wrappings, pieces of roofing and insulation, and wallboard cutoffs into the scrap containers for later disposal. Instead of throwing bottles, cans, paper sacks, and drinking cups on the ground, get into the habit of using scrap barrels.

12 Drop all oily rags and other flammable waste in a metal container. When the container is about three-fourths full, burn or otherwise dispose of the contents.

13 Set yourself up as a good example for others to follow. If one of your co-workers fails to follow safety rules, you could be the one who is hurt as a result.

Procedures

The construction procedures presented in this book are standards used by good carpenters throughout the building industry. There are regional differences, of course. Because of differences in climate and the materials available locally, a house built in Minnesota is not quite the same as a house built in Florida or Connecticut or Oregon or Texas.

Regional differences in procedures don't cause problems. Where the carpenter may get into trouble is by deviating from standard procedures. Let's take two examples to illustrate the point.

Suppose, first, that your task is to set floor joists. One of the steps in the procedure (Unit 6-9) is to set all joists as plumb as possible. Another step later on in this same procedure is to look for cocked joists—joists set at a slight angle instead of plumb. Suppose further that you fail to follow these steps carefully and that one joist is set cocked.

This mistake may not seem important to you and may not even be easily noticed. But it can be costly. If that joist happens to support a seam in subflooring, your fellow carpenter won't be able to lay the subfloor level over that joist. Then one of three things must take place. The joist must be removed and reset straight, or the seam must be sanded level. Both of these corrective steps take time, and delay the job. The third possibility is that the second carpenter also fails to follow procedures and leaves the edges of subflooring panels at different levels. This failure makes it difficult for another worker laying floor tile to prevent that seam from showing in the finished floor. Such visible flaws make a house harder to sell.

Now suppose that your task is to cover roof sheathing with roofing paper. One of the steps in this procedure (Unit 9-38) is to lay each sheet so that it overlaps the sheet below it by half a width. Suppose that you allow only a few inches of overlap instead of the recommended 17".

The purpose of the overlap is to cover roof sheathing with two thicknesses of roofing paper as protection against leaks. If you fail to follow procedures and the roofer accidentally tears a hole in the paper while applying shingles, the roof may leak—not today, or tomorrow, but perhaps 6 months later. The builder is responsible for correcting the leak; this is time consuming and costly and cuts into the profits from which your wages are paid.

It is extremely important that you follow all procedures carefully. If you are forced by circumstances to vary from them, discuss the change with your supervisor to make sure you don't cause problems for other workers or your employer.

Quality

"Degree of excellence of workmanship" is one definition of the word quality. Every bit of work that anyone does can be judged against a standard for quality.

Who sets the standards? For the work you do while learning, your instructor sets the standard for you to try to attain. At the beginning, the quality of work you produce is not likely to be high, but the standard the instructor sets will allow for your lack of experience. As the year passes, however, and you gain in knowledge, ability, and experience, the standards will become higher and higher, and the quality of your work should improve to meet them.

When you are a carpenter working on the job, the standard of quality is initially determined by the builder or contractor. Total quality is the combined result of the quality of materials and the quality of work with which they are applied. The market for homes often enters into the picture here. If the local market for high-priced, good-quality homes is strong, smart builders build homes to suit that market. If the market is better in the low-priced field, builders may have to buy lower-priced materials. But good builders don't lower their standards of quality. They will build smaller houses instead.

One of the supervisor's responsibilities is to assure that the work of the crews is sufficiently high to meet these market requirements. The standards used will be similar to those of the builder. A supervisor who insists on higher quality than the builder can afford to sell, or tolerates a lower quality than the builder will accept, won't remain a supervisor for long.

In the long run, however, it is you who sets the standard for the quality of work that you turn out. And that standard should always be "to the best of my ability." It is not enough to be able to say to yourself, "I finished the job." If you intend to be successful in life, whether as a carpenter or salesperson or architect or whatever occupation you choose, you want to be able to say as often as possible, "I did that job well."

Questions

1 Why is the building contractor so interested in safety?
2 What are the two major causes of accidents?
3 Why should oily rags be discarded into a metal container?
4 Aside from knowledge and skill, what other personal characteristic is important to your success as a carpenter?

2 Site Work

Carpenters perform few tasks while a site is being prepared for construction. Most work in preparing a site for building is done by grading contractors and their crews. They clear the site of trees, underbrush, boulders, and existing buildings. They haul away the debris. They rough-grade the site—moving earth until the ground's surface throughout the building area is close to final level. They dig trenches for water lines, sewer lines, and drain lines. Later they excavate for the basement or foundation beneath the building.

The only part of this work that carpenters may be involved in is dismantling an existing building. But their other site work, although limited, is extremely important and must be completed with great care and accuracy. This other work occurs at three separate times.

Before grading crews attack the site, carpenters first help to establish the general location of the building to be built. They don't work alone. Builders or experienced supervisors may direct site work. They know the building code and local building ordinances that limit what can be built and where it can be built.

After the grading crew clears the site and completes rough grading, carpenters go back to set the exact location of the building, as shown on the plot plan. This is called **staking the building.**

Usually two carpenters, or a carpenter and an apprentice, work as a team.

Finally, after the grading crew has dug trenches for a foundation wall or excavated for a basement, a crew of carpenters erects the forms for the concrete work that supports the building. At this time they also shore up any banks to prevent them from caving in. All the other work done by carpenters on a building doesn't begin until the concrete finishers and masons who build the foundation have moved on to the next job site.

During site work the only materials carpenters use are lumber and nails of various sizes. They may build the necessary forms or erect special forms manufactured for the purpose.

For tools they use a marking pencil, framing square, folding rule, saws, ax, sledge, 100' steel tape, nylon cord, carpenter's level, hammer, plumb bob, and some sort of optical level. They must know how to use all of these tools except the last one and must know the principles of its operation.

4

Locating the Building

Many communities have laws that limit use of land. Land areas are divided into zones, and regulations, called **zoning laws,** state how land in each zone may be used. The most common zones are residential, multifamily, office, commercial, light industrial, heavy industrial, and farm.

In a residential zone only single-family houses may be built. In a multifamily zone apartments, townhouses, and condominiums are permitted in addition to single-family houses. A single-family house may be built in any zone, but most people prefer to live in a residential zone to protect their investment in their home.

The smaller the community, the less likely it is to have zoning laws. It may have building ordinances, however. A **building ordinance** may limit either the land or the house on it. It regulates how much of the site may be occupied by a building. It may limit the height or number of stories in the building or the type of architecture, or it may even specify a minimum floor area or lot size.

A common restriction found in city or county ordinances is the setback. A **setback** is the minimum distance permitted between a building and property lines around it. Some ordinances measure setbacks to the eaves of the house or garage. Usually, however, the measurement is to the foundation line.

Front setbacks (Fig. 4-0.1) are measured from the front property line, not from the curb or edge of the pavement. The distance from curb to property line is usually at least 20′. The front setback line is also called the **building line.**

Side and rear setbacks are most often measured in feet. Sometimes, however, they are stated as a percentage of the width of the lot. If an ordinance calls for a 10% side setback, for example, and the lot is 60′ wide, the house can be built no closer than 6′ to either side property line. Therefore the maximum width of a house that can be built on that lot is 48′.

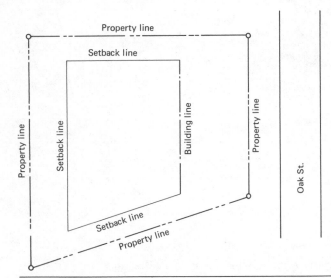

The carpenter's part in helping to locate a house on its site may include any or all of these tasks.

- Marking the outline of the foundation of the building at the exact location on the site.

- Establishing the locations of various working lines, such as the building line, footing lines, and excavation line, during the very early stages of construction.

- Building the wood frames to support cords marking these lines in the air.

Fig. 4-0.1 When a surveyor makes a survey of a piece of land, he marks the corners with special markers. These markers show the boundaries of the site or the property lines. But in most communities you cannot build right up to the property lines. You can only build within the setback lines, which are established by local ordinances.

Questions

Circle T if you think the statement is true and F if you think it is false.

1 A single-family house cannot be built in a multifamily zone.   T   F

2 A building line and side setback are the same thing.   T   F

3 A front setback is measured from either the curb or the centerline of the street.   T   F

4 A carpenter's work is more likely to relate to building ordinances than to zoning laws.   T   F

Unit 4-1   Set Up a Builder's Level or Transit Level

Two important tasks in site work are making right-angle corners and maintaining level when staking out a foundation. Both tasks can be completed without optical instruments. But they can be done better with a builder's level and even more quickly and accurately with a transit level.

Both the **builder's level** (Fig. 4-1.1) and the **transit level** (Fig. 4-1.2) are expensive instruments that must be treated with great care. Their manufacturers recommend the following steps in caring for their products.

- Keep the level in its case when not in use.
- Protect the lens with the lens cap when the level is in its case and when it is set up but temporarily not in use.

- Never put a level in its case when wet. Dry it with a soft cloth first.

- Protect the level from dust and dirt. To clean the eyepiece, use a soft brush or lens tissue; do not use a cloth.

- Handle the level only by its base when you set it up or take it down.

- To move the level a short distance, keep it upright. To move it over a long distance, whether carried on foot or in a vehicle, place the instrument in its carrying case.

- Do not use force on any part. All moving parts turn easily by hand.

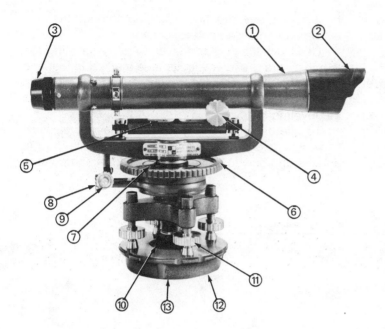

1. Telescope (26X)
2. Telescope sunshade
3. Eyepiece
4. Focusing Knobs
5. Instrument level vial
6. Horizontal graduated circle
7. Index vernier

8. Horizontal clamp screw
9. Horizontal tangent screw
10. Shifting center
11. Four leveling screws
12. 3½" x 8 thread base
13. Plumb bob hook and chain

Fig. 4-1.1 A builder's level is a useful tool in locating a house on its site, and in staking out the foundation. The telescope is fixed in a horizontal position. A leveling rod is what the carpenter looks at when he sights through the lens in the telescope of the builder's level. (Courtesy David White Instruments.)

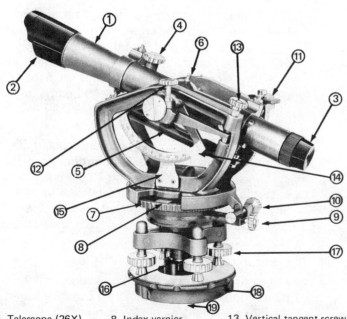

1. Telescope (26X)
2. Telescope sunshade
3. Eyepiece
4. Focusing Knob
5. Instrument level vial
6. Compass stud
7. Horizontal graduated circle

8. Index vernier
9. Horizontal clamp screw
10. Horizontal tangent screw
11. Telescope lock levers (two)
12. Vertical clamp screw

13. Vertical tangent screw
14. Vertical arc
15. Vernier scale
16. Shifting center
17. Four leveling screws
18. 3½" x 8 thread base
19. Plumb bob hook and chain

Fig. 4-1.2 A transit level is similar to and more expensive than a builder's level, but is more versatile. The telescope can be aimed not only horizontally but as much as 45° below and above horizontal. (Courtesy David White Instruments.)

- At least once a year have the level serviced at a qualified repair station or by the manufacturer. Do not try to make repairs or adjustments yourself.

Other precautions to take while setting up a level are included as part of the procedures that follow.

Both optical instruments rest on a tripod, and you adjust and fix the legs in position by tightening wing nuts. The level itself sets on a base, and you establish horizontal level by adjusting leveling screws. The telescope rests on a circle calibrated in degrees and can be aimed in any direction. You sight through the telescope and adjust a knob to focus it.

Used with the level is a leveling rod. A **leveling rod** is a thin, straight ruler with its face marked in feet and tenths of feet. In many operations during site work the object you look at through the telescope is a height on the leveling rod.

To use a builder's level or transit level, you must know how to set it up over a point, how to form right angles, and how to transfer a height from one point to another.

Questions

1 Identify the part of a level that does each of the following.

   a) Holds the legs of the tripod in position
   b) Holds the instrument on the tripod
   c) Shows the angle of direction
   d) Indicates that the telescope is level

2 The tool on which the telescope is often focused for readings is called a
   _____.

Procedure

1 Determine the point where the level is to be set up. This point is often called a **station mark. Note.** For most operations this point may be at any convenient location on the site from which you can see all corners of the future building.

2 Mark the point by driving a stake vertically into the ground and drive a nail halfway into the top of the stake. If the point is over a paved surface, mark the location with chalk.

3 Fit the plumb bob chain through the hole in the tripod and attach the plumb bob to the chain.

4 Set the tripod over the station mark, holding the base approximately level and centered over the point.

5 Loosen the wing nut on each leg in succession, extend the leg to the ground, and tighten the nut. The legs should be about $3\frac{1}{2}'$ apart at their tips. **Precautions.** When working on a slope, dig one leg of the tripod into the slope and place the other two legs downhill for maximum stability. When working in a wind, set legs as far apart as possible. When working on a slippery surface, block the legs until you have leveled the instrument.

6 Remove the level from its case. **Precaution.** Keep the level in its case when not in use and protect it from weather at all times.

7 Carry the instrument to the tripod. **Precaution.** Make sure the clamp screws (See Figs. 4-1.1 and 4-1.2) are loose so that the telescope is free to move if you bump it accidentally. Both telescope lock levers on a transit level should be in the closed position, however. To avoid bumps in close quarters, carry the instrument under your arm with the telescope in front of you.

8 Holding the base firmly in one hand, fasten the instrument onto the tripod. **Precaution.** Leveling screws should be tight enough so the instrument is stable, but not so tight that they scar the leveling base. Only firm contact between the screws and base is necessary.

9 Remove the dustcap from the eyepiece of the telescope and slip on the sunshade.

10 Adjust the position of the tripod's legs until the plumb bob hangs approximately over the nail in the stake. Within $\frac{1}{4}''$ is close enough.

11 Working with one leg of the tripod at a time, first loosen and then retighten the wing nuts. This action relieves any tension on the base leveling plate.

12 Line up the telescope so that it is directly over a pair of leveling screws.

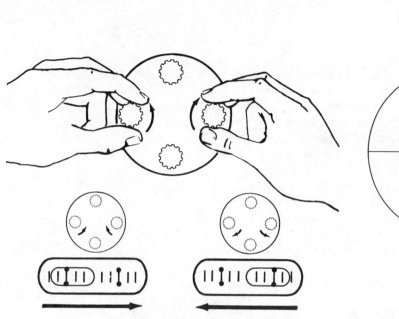

Fig. 4-1.3

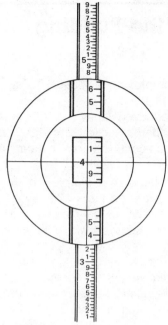

Fig. 4-1.4

13 Grasp these screws with the thumb and forefinger of each hand and turn them equally and simultaneously in opposite directions (Fig. 4-1.3) until the bubble in the spirit level is centered. **Note.** The bubble will move in the same direction as your left thumb.

14 When the bubble is centered, retighten the screws.

15 Turn the telescope 90° and repeat steps 13 and 14.

16 As a final check for level, rotate the telescope over each of the four leveling points to make sure the bubble remains centered.

17 To focus on an object, such as a leveling rod, rotate the telescope until it is aimed at the object as you sight along the top of the telescope.

18 Now look through the lens of the eyepiece at the object and turn the focusing knob until the object is sharp and clear (Fig. 4-1.4). For closer focusing, turn the knob forward or clockwise. For focusing on an object further away, turn the knob backward or counterclockwise.

19 If necessary, slowly rotate the telescope until the vertical cross hair is exactly centered on the object. You are now ready to start working with the level.

Questions

1 When you carry an optical instrument from one place to another, you should always _____ the clamp screws.

2 The hole in the leveling base of a tripod is for a _____.

3 Why do you loosen and then retighten wing nuts?

4 Leveling screws operate in _____ that turn in the _____ direction.

5 When you finish using the level, you protect the lens by replacing the _____.

Unit 4-2   **Stake the Building**

The stakes that mark the approximate location of a building are sometimes knocked out of position or lost while the site is being cleared. Even if they survive, the position of each stake must be rechecked and stakes must be reset before work can continue. This is a carpenter's job.

The process of staking a building is simple if both lot and building are rectangular and the site is level. When the land slopes and the lot is odd shaped, the process is more complex. You must work with great accuracy under all conditions.

The drawing from which you work is called a **plot plan** (Fig. 4-2.1). The plot plan shows the locations of corners of the property, any points between corners where property lines change direction, and lengths of all lines.

When surveyors establish the corners of a site, they mark those corners. Usually they drive a metal rod into the ground, but they may use a piece of iron pipe or even a stone marker. On the survey that they draw, they indicate the type of marker they used. This information also appears on the plot plan, from which the builder works to locate the building.

The plot plan also shows setback lines and where the house is to be built in relation to these setback lines. A house may be centered between side setbacks, or one side may lie on a setback line. Most houses, however, are located with the frontmost foundation on the building line—the front setback line. This line is therefore the starting point for staking.

While staking a building you use 2 × 4 stakes, a ball of nylon cord, a hammer and nails, 100′ steel tape, plumb bob, and an optical level. The job can be done with a builder's level and leveling rod, but it may be completed much more quickly and accurately with a transit level. The procedures that follow require a transit level.

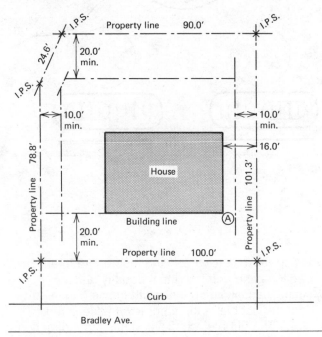

Fig. 4-2.1 A plot plan is a simple drawing of the site showing the exact location of the house on that site. Many city and county governments require a builder to provide a plot plan before a building permit can be issued.

Questions

1 Surveyors usually mark the corners of a site with a _____ or

_____ .

2 The starting point for staking a building is _____ .

3 On a plot plan you will find both _____ lines and _____ lines.

Procedure

1 At the site find the survey rods marking the front corners of the property. **Precaution.** If you have trouble finding the surveyor's markers, refer to the survey to see whether some other type of marker was used.

2 On the front property line if extended, and just outside the metal rods, drive 2 × 4 stakes vertically (Fig. 4-2.2). Leave at least 18″ exposed above ground.

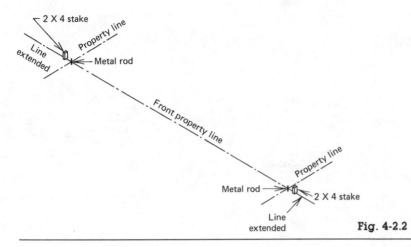

Fig. 4-2.2

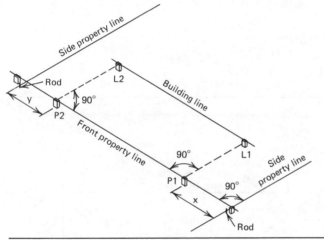

Fig. 4-2.3

3 Drive a 10-penny nail into the top of each stake exactly on the property line. Leave about an inch of nail exposed.

4 Stretch a cord between the nails and secure it around them. This cord marks the front property line.

5 At a distance x from one survey rod, drive a stake directly beneath the cord at point P1 in Fig. 4-2.3. **Note.** The measurement you use for x varies with the site. If the front and side property lines meet at exactly 90°, use for x the dimension from the side property line to the side of the building. If the property is irregular in shape, use the setback dimension.

6 At a distance y from the other survey rod, drive a stake directly beneath the cord at point P2. **Note.** The measure for y also varies with conditions. If the front and this side property line meet at 90°, use half the setback dimension for y. If the property is irregular in shape, use the full setback dimension.

7 Set the transit level directly over stake P1 and adjust it for level.

8 Sight through the lens at stake P2 and set the 360° scale at 0°.

9 Now turn the telescope 90° to the right and measure off a distance from stake P1 equal to the distance between the front property line and the building line.

10 Drive stake L1 at this point, sighting through the level to establish the exact location. Add a 10-penny nail in the top of the stake in line with the hairline in the telescope.

11 Now move the optical level over stake P2 and repeat steps 7 to 10 to locate stake L2.

12 Stretch a cord between L1 and L2 and secure it. This cord marks the front building line.

13 If stake L1 marks the corner of the building corresponding to point A in the plot plan, go to step 19.

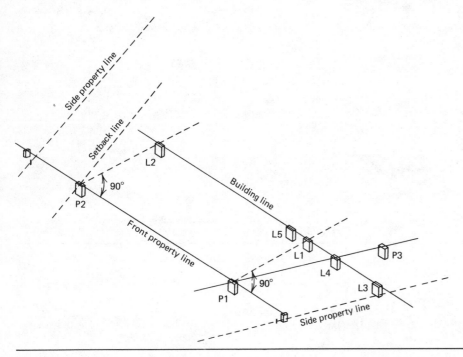

Fig. 4-2.4

14 If stake *L*1 does not mark the corner of the building, as in Fig. 4-2.4, first locate stake *L*3 near the property line by setting up the optical level at stake *L*2 and sighting on a line through stake *L*1.

15 Then locate stake *P*3 by setting up the level at stake *P*1 and sighting parallel to the side property line (the correct angle is indicated on the site plan or survey).

16 Stretch cords between stakes *P*1 and *P*3 and between *L*2 and *L*3.

17 Where the lines cross, drive stake *L*4. This stake marks where the building line and setback line cross, and most dimensions to the corner of the building are measured from this point.

18 If stake *L*4 does not mark the corner of the building, measure from it to the actual corner and mark that point with stake *L*5. This corner stake corresponds to point *A* in the plot plan.

19 From point *A*, using a steel tape, measure the width of the house, as shown on the foundation plan, along the building line. Drive stake *B* at this corner and mark the exact corner with a nail in the top of the stake. See Fig. 4-2.5.

20 To locate rear corner *D*, set the transit level at stake *A* directly over the nail and adjust for level.

21 Sight through the lens at stake *B* and set the 360° scale at 0°.

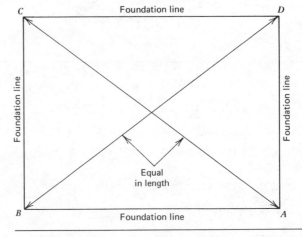

Fig. 4-2.5

22 Now turn the telescope 90° to the right and measure off a distance from stake *A* equal to the length of the side wall of the house.

23 Drive stake *D* at this point and add a nail.

24 Repeat steps 20 to 23, beginning at stake *B*, to locate corner *C*.

25 To check the accuracy of your work, measure with the steel tape the diagonal distance between *A* and *C* and between *B* and *D* (Fig. 4-2.5). These

measurements should be the same. Any difference of more than ⅛″ is not acceptable.

26 String a cord from stake *A* to stake *B* to stake *C* to stake *D* and back to stake *A*. You have now outlined the foundation for the main house.

27 Remove all other stakes and cords.

Questions

1 Why do you drive a nail into the tops of stakes?

2 On what drawing do you find the overall dimensions of the house?

3 What do you measure to make sure that corners of a building layout are square?

4 On what line do the front corners of a building usually lie?

Unit 4-3   Make Square Corners by Measuring

To make square corners when no builder's level is available, you follow the principle of the Pythagorean theorem. Pythagoras was a Greek mathematician who discovered over 2500 years ago that "the square of the hypotenuse of a right triangle is equal to the sum of the squares of the other two sides." The right triangles most often used in building to make right angles are 3-4-5 and 6-8-10 triangles. Here's how they work.

The hypotenuse of a triangle is its longest side. The hypotenuse of a 6-8-10 triangle, then, is 10′, and the lengths of the other two sides are 6′ and 8′ (Fig. 4-3.1). The square of any number is that number multiplied by itself. Therefore the "square of the hypotenuse" in a 6-8-10 triangle is 100 (10 × 10), and the "sum of the squares of the other two sides" is also 100 (6 × 6, which is 36, plus 8 × 8, which is 64).

A 3-4-5 triangle works the same way (3 × 3 + 4 × 4 = 5 × 5). But you can achieve greater accuracy in making right angles at a site by using the longer dimensions. You will have the opportunity to test the theorem in this procedure for staking a wing on a rectangular house (Fig. 4-3.2).

You will need a hammer, sledge, steel tape, 10 stakes, and 10 nails.

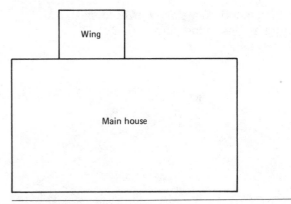

Fig. 4-3.2 The shape of the house used as an example in this unit.

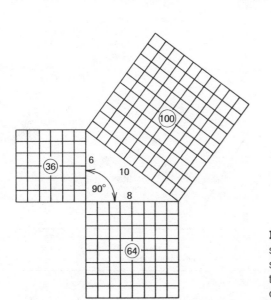

Fig. 4-3.1 In any triangle with a right (90°) angle in it, the square of the hypotenuse is equal to the sum of the squares of the other two sides. Known as the Pythagorean theorem, this statement is the basis for many measurements during site work.

Questions

1 If two sides of a right triangle measure 18′ and 24′, respectively, what is the length of the hypotenuse?

2 Why is a 6-8-10 triangle better for making right angles on a site than a 3- 4- 5 triangle?

Procedure

1 On the foundation plan find the length of the rear wall from corner to wing and the width of the wing.

2 From stake *D* measure off the length of the rear wall from corner to wing (*x*), along line *DC* (Fig. 4-3.3).

3 Again working from stake *D* along line *DC*, measure off the length in step 2 plus the width of the wing. Mark these two points with stakes *E* and *F*, as in Fig. 4-3.3.

4 From stake *F* measure back 6′ toward stake *D* and set stake *K*.

5 From stake *F* measure off 8′ at approximately right angles to line *DC* and scribe an arc in the dirt (Fig. 4-3.4).

6 From stake *K* measure off 10′ and scribe another arc that crosses the first arc (Fig. 4-3.5).

7 Where the two arcs intersect, drive stake *L*. Any cord stretched from stake *F* over to stake *L* will be at right angles to the cord from *D* to *C*.

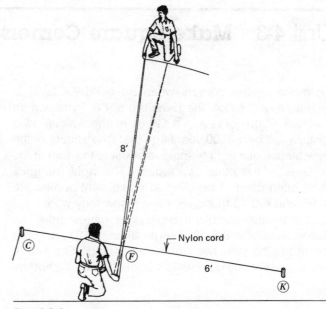

Fig. 4-3.4

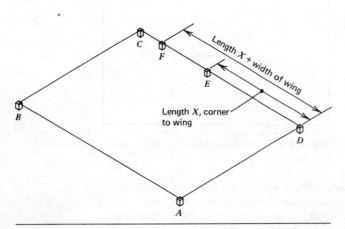

Fig. 4-3.3

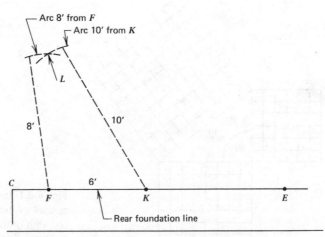

Fig. 4-3.5

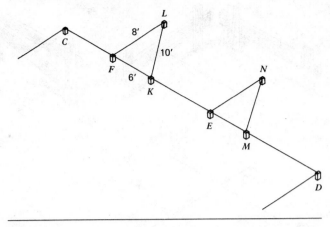

8 Working from stake *E*, repeat steps 4 to 7 to locate stake *N* (Fig. 4-3.6).

9 On the foundation plan find the length of the wing.

10 From stake *F* measure off this distance directly over stake *L* and set stake *G*.

11 Similarly, from stake *E* measure off the same distance directly over stake *N* and set stake *H* (Fig. 4-3.7).

12 Check the accuracy of your work by measuring diagonals *EG* and *FH*.

13 Run a cord from stake *F* to *G* to *H* to *E* to outline the foundation for the wing.

Fig. 4-3.6

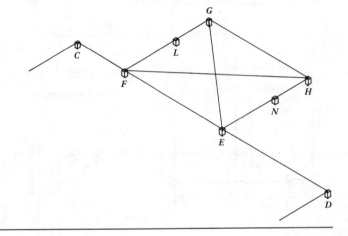

Fig. 4-3.7

Questions

1 When you know the locations of two corners of a right triangle, you find the third corner by _____.

2 Step 3 says to measure from stake *D* to locate stake *F*. Why do you think you should not measure from stake *E* if you knew the dimension?

Unit 4-4   Set Batter Boards

The stakes set to mark the corners of a foundation serve that purpose only temporarily. They will be in the way when excavation begins. To mark the excavation lines and foundation lines permanently, carpenters erect batter boards.

A **batter board** is a simple structure for supporting cords. There are several types; Fig. 4-4.1 shows a few of the more common shapes. A batter board consists of one or two cross-members (also referred to as batter

boards) attached to stakes set far enough away from the excavation so it does not interfere with earthmoving equipment. Usually this point is at least 4′ from corner stakes marking the foundation line. Batter boards must be placed at least 1′ beyond any excavation, and 3′ beyond is better.

The length of the cross-member depends on the size of the excavation. When soil at the site is firm and the excavation is no more than 36″ deep, the excava-

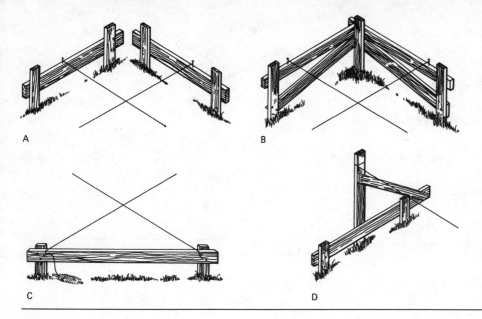

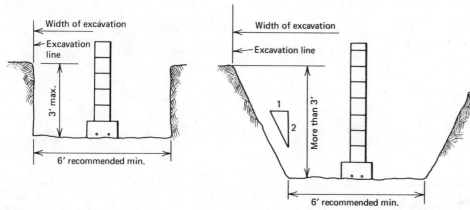

Fig. 4-4.1 Batter boards are made in several different shapes. These four are typical. Types A and B are most used at corners. Type C may be used at a corner or across a foundation from a wing to hold a single set of cords. Type D is for use on uneven ground or when a line must clear an obstruction.

Fig. 4-4.2 An excavation no more than 3′ deep (left) may have vertical sides. Any deeper excavation should have sloped sides (right) to prevent cave-ins. Any excavation must be wide enough at the bottom for carpenters and masons to work.

tion line normally lies 3′ beyond the foundation line (Fig. 4-4.2, left). For deeper excavations the sides should slope at a ratio no steeper than 2 to 1—2′ of depth for every 1′ of width (Fig. 4-4.2, right). These dimensions allow room outside the foundation line for carpenters to form for footings and foundation walls. Assume in the following procedure that the excavation line is 3′ outside the foundation line on all sides of the house. Cross-members must be at least 1′ longer than this dimension.

To build and set batter boards you need a sledge, hammer, carpenter's level, and an optical level. For materials you need a pair of 2 × 4 stakes and a length of 1 × 6 for each batter board, a pocketful of nails, and one 2 - 2 stake. You also need someone to help you with the optical level.

Questions

1 The two sizes of lumber used in batter boards are _____ and _____ .

2 The length of a cross-member of a batter board is determined by _____ .

3 Batter boards take the place of _____ .

4 If the bottom of an excavation is 7½′ below ground level and the excavation allows room to work beside footings, how far beyond the foundation line is the excavation at the floor of the excavation? At ground level?

Procedure

1 Assemble all stakes and cross-member stock for the job. You will need four stakes and two cross-members for each corner of the house.

2 Cut cross-members to a length of 7'.

3 Attach cross-members to stakes with a single nail into each stake. Place each stake near the end of the batter board and close to flush with its top (Fig. 4-4.3).

4 Four feet from the foundation line and parallel to that line set a pair of batter boards at each corner. Set one stake so that it lies 2' inside the foundation line extended and set the other 4' outside the line (Fig. 4-4.4). Always set the batter board with the stakes toward the foundation.

5 Drive stakes into the ground just far enough to keep them upright until you have determined the final level of each board.

6 To determine the level (if the builder has not already told you what it is), walk around the foundation line until you find the high point. Drive a 2 × 2 grade stake at this point.

7 From the sections in drawings, determine the height of the foundation above grade. The standard height is 8", which is the minimum.

8 Mark this height on the grade stake. The height is called a **benchmark**.

9 Set up the optical level at any point where the benchmark and all corner stakes are clearly visible. Adjust the telescope to horizontal.

10 While your helper sets the leveling rod beside the benchmark and rests it on the ground, focus the telescope on the rod and take a reading. For example, say the reading is 4.65'.

11 Next determine the distance between grade and the benchmark. Say this reading is 0.7'.

12 Subtract the distance in step 11 from the distance in step 10. The remainder is the distance from the benchmark to telescope level (Fig. 4-4.5). Write this figure down; you will use it again. In the example the figure is 3.95'.

13 Have your helper move to the first batter board and set the rod on the batter board at one stake. Take another reading, which should be less than the distance in step 12.

14 Slowly drive the stake into the ground until the reading on the rod is the same as in step 12.

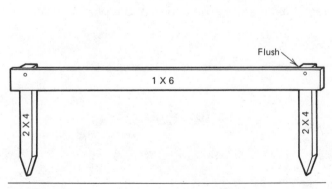

Fig. 4-4.3

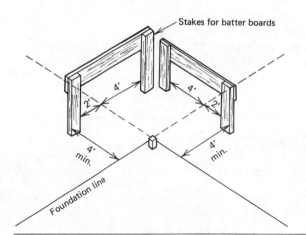

Fig. 4-4.4

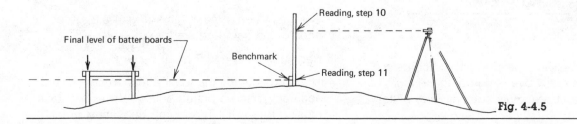

Fig. 4-4.5

15 Adjust the level of the adjacent stake by setting the carpenter's level across the batter board and driving in the stake until the top of the cross-member is level.

16 Then secure the cross-member to the stakes with another nail into each stake.

17 Repeat steps 13 to 16 at all corners until all batter boards are firmly in the ground, plumb, and level.

Questions

1 To assemble batter boards, you nail (boards to stakes, stakes to boards). (Circle one)

2 You set up the optical level at any point inside the foundation line.   T   F (circle one)

3 When you read the leveling rod at each corner, the final reading should be the same as the vertical distance from the benchmark to telescope level.   T   F (circle one)

4 Why do you think that cross-members should always be on the back sides of stakes when in their final position?

Unit 4-5   **Lay Out the Lines**

Batter boards are merely holders for the cords that serve as guidelines for future site work. Cords marking excavation lines guide the operator of the bulldozer, backhoe, trencher, front-end loader, or perhaps even the small power shovel.

Later on, carpenters work from footing lines to locate the forms they build for footings. Still later, the foundation lines are used either by carpenters, who set forms for poured concrete walls, or by masons who build a foundation wall of block, brick, or stone.

You need no materials to lay out the lines, but you do need a 2 × 2 stake to mark each corner of the excavation. For tools you need a pencil, crosscut saw, steel tape, folding rule, nylon cord, and a plumb bob.

Questions

1 The three types of lines marked on batter boards are:

a) _____ .
b) _____ .
c) _____ .

2 Why do you think nylon cord is better for marking lines than twine or heavy string?

Procedure

1 Stretch a cord between parallel batter boards so that it passes directly over a pair of foundation stakes (Fig. 4-5.1). Mark this point on both batter boards.

2 Cut a saw kerf at each mark (Fig. 4-5.2). Knot the ends of the cord and set it in the kerf.

3 Repeat steps 1 and 2 at all batter boards.

4 The cords outline the foundation. To check your accuracy, hang a plumb bob wherever two lines cross (Fig. 4-5.1). The plumb bobs must point

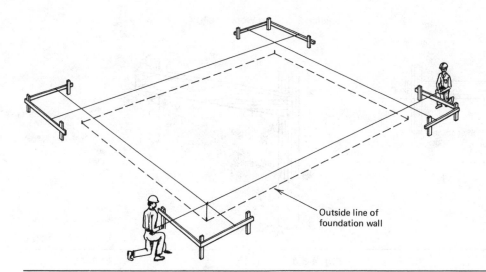

Fig. 4-5.1

Outside line of foundation wall

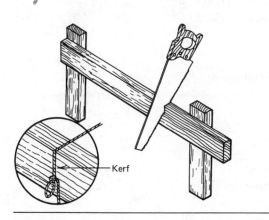

Kerf

Fig. 4-5.2

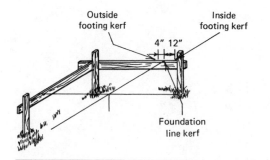

Outside footing kerf

Inside footing kerf

4" 12"

Foundation line kerf

Fig. 4-5.3

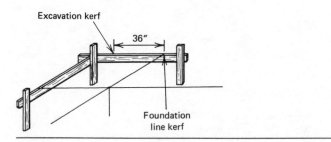

Excavation kerf

36"

Foundation line kerf

Fig. 4-5.4

directly at the nails in the tops of foundation stakes below.

5 As a second check, measure the diagonals between crossing points with your steel tape. The difference in the two dimensions should not exceed $\frac{1}{2}''$.

6 From the kerfs for the foundation line, measure 4" outside the foundation line and make a mark. Then measure 12" inside the foundation line and make another mark (Fig. 4-5.3).

7 Cut kerfs at these new points in each batter board. The points locate the outside and inside edges of the 16"-wide footing beneath the foundation wall.

8 Move the cords into the new kerfs and again measure the diagonals to check your accuracy.

9 Now measure 36" outside the foundation line kerf on each batter board and make another mark (Fig. 4-5.4).

10 Cut kerfs at these marks and move your cords into them. The cords now locate the excavation line.

11 Hang a plumb bob wherever two lines cross.

12 Directly below the plumb bob drive a 2 × 4 stake so that the plumb bob just touches the inside corner of the stake (Fig. 4-5.5). Leave at least 18" of the stake exposed. These stakes mark the corners of the excavation.

13 Paint the top 4" of the stake a bright color as a guide to the excavator in his work.

14 If the soil is soft, add a second stake, a short 2 × 2, about 2' from the corner stake. Connect the two stakes with a 1 × 2 (Fig. 4-5.6). The brace holds the corner stake in position.

15 Remove the original foundation stakes and their cords. You have now completed your work for the excavator.

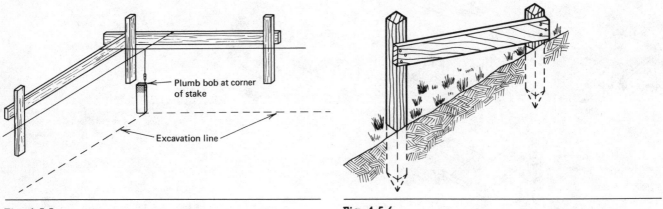

Fig. 4-5.5

Fig. 4-5.6

Questions

1 The standard distance between inside and outside footing lines on a batter board is _____.

2 The slot cut by a saw in a board is called a _____.

3 You check the accuracy of placement of lines by _____.

4 You paint the tops of excavation stakes to _____.

5 What is the difference in the length of a cord marking the location of a footing and the cord marking the location of the excavation line between the same two batter boards?

5

Shoring and Forming

After excavation is completed, and sometimes even before the last piece of equipment has left the site, carpenters begin the next phase of their work. The general purpose of this phase is to build and set in place forms that keep wet solids under control. Usually the wet solid is concrete, but sometimes it is nothing more than soft or muddy earth.

The carpentry during this time almost always involves building forms for concrete footings. Carpenters may build forms for concrete foundation walls or, instead, may erect manufactured forms. They build the forms necessary to shape concrete that is poured for piers, slabs, steps, and retaining walls. They may also have to build a form to prevent a bank from sliding and filling part of a trench or basement excavation.

Concrete forms are built and placed according to certain standard specifications. Forms for banks, however, vary according to the conditions to be overcome.

During the shoring and forming phases of rough carpentry it is easy for the new carpenter to develop sloppy work habits. Don't. The materials used for shoring are rough; pinpoint accuracy is not usually important, and physical strength is put to greater use than mental strength.

The process of building and placing forms, on the other hand, requires care, accuracy, and clear thinking. Much of the below-ground work done by carpenters is never seen by the customer who buys the house. But this owner can see the results of poor work a few years later when cracks appear in walls, floors begin to sag, and doors and windows stick. Footings and foundation walls are as much engineered parts of a house as floor, wall, and roof framing. Concrete work is done by other trades, but you, as a carpenter, establish the guidelines for their work with forms properly built and properly installed.

Unit 5-1 **Sheet a Bank**

Wet soil on a slope can be dangerous, especially if the soil is soft or the slope is newly cut and has not been exposed to weathering. It is easier to prevent a cave-in than to clean up after a cave-in occurs, and prevention may save a worker's life.

It isn't always possible to slope the sides of an excavation according to established standards (see Fig. 4-4.2). For example, the roots of a fine tree may not harm a foundation wall, but would have to be cut during excavation. The solution is to shore up, or **sheet**, the slope.

Sheeting consists of 2″ planks to hold the bank, 4 × 6 timbers to hold the planks in place, and 2 × 4s and 4 × 4s for bracing (Fig. 5-1.1). Planks may be any width. For most shoring work, planks with square edges are adequate. If the bank is very soft and likely to ooze, planks with tongue-and-groove edges are better. All planks should be at least 6″ longer than the bank's sloping dimension.

The heavy timbers, called **walers** or **whalers**, must be as long as the bank is wide. One waler is adequate for sheeting up to 3′ high. Two walers are needed for sheeting 3 to 6′, and three for sheeting more than 6′ high. Banks higher than 8′ are often shored with metal channels driven into place mechanically.

Braces are needed near the ends of walers and about every 8′ between. Each brace consists of a length of 2 × 4, which is wedged against the waler, and a 4 × 4 stake, called a **deadman**, to support the ends of the 2 × 4s. The length of each deadman is equal to about two-thirds the height of the sheeting.

The locations of deadmen vary with site conditions. The distance between sheeting and the point where each deadman enters the ground is equal to about one-third the height of the sheeting. All deadmen must be set outside the footing line or excavation line, depending on whether the bank to be shored is below or above finished grade at the site.

For tools you need a steel tape, folding rule, hand saw, sledge, and hammer. You can save time by using a power saw when temporary power is available at the site. For fasteners you need spikes and nails.

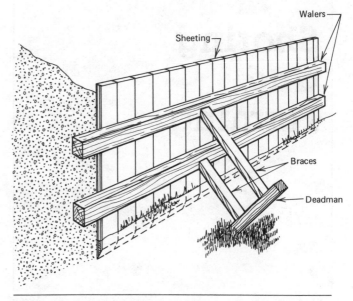

Fig. 5-1.1 To prevent cave-ins of loose soil, sheeting holds the soil in place, walers hold the sheeting in place, braces hold the walers in place, and deadmen holds the braces in place.

Questions

1 The length of a deadman is about _____ as long as the bank is high.

2 If a bank is 5′ high and 20′ long, how many of the following parts do you need for sheeting: Walers? _____ Braces? _____ Deadmen? _____ .

3 To sheet a 5′ bank, how long should planks be?

4 Which part of the shoring is most likely to interfere with construction?

Procedure

1 Determine the length of planks required by measuring along the slope of the bank at its highest point and adding 10″.

2 Mark all planks at this length, and bevel one end at a 45° angle.

3 Lay the planks against the bank, with the pointed edge of the bevel toward the bank (Fig. 5-1.2).

4 With the sledge, drive each plank 3 to 4″ into the ground. Fit tongue-and-groove planks tightly together. Gap square-edged planks about an inch to allow for drainage. Keep tops of planks as even as possible.

5 Cut walers to length, and attach them to planks with spikes. Space walers equally along the exposed length of the planks.

6 Cut deadmen to length, pointing one end.

7 At points dictated by the height of sheeting, drive each deadman into the ground at about a 45° angle to the planks (Fig. 5-1.3). At least half the length of each deadman should be in the ground.

8 Measure the right-angle distance between each deadman and each waler to determine the lengths of braces. **Precaution.** Measure separately at each deadman, because measurements will vary slightly from one location to the next.

9 Cut one end of each brace at about a 45° angle (Fig. 5-1.3). Saw so that the final length of each brace is $\frac{1}{4}$″ longer than the measurement.

10 Wedge each brace into position, and toenail at each end with four 10-penny nails.

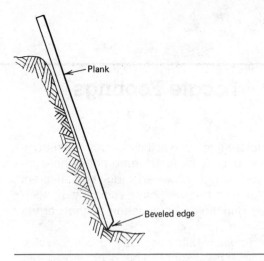

Fig. 5-1.2

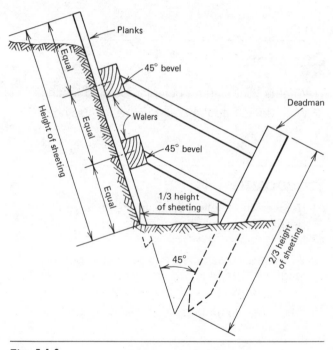

Fig. 5-1.3

Questions

1 Which pieces are cut with a 45° bevel?

2 Deadmen are driven into the ground at about a 45° angle to what?

3 Of its length, (about one-third, one-half, two-thirds) of a deadman should be below ground level. (Circle one.)

4 The length of each brace should be (a little less than, equal to, a little more than) the measured distance between the deadman and the waler. (Circle one.)

Unit 5-2 **Locate Footings**

Up to this point all line work at the site has been done at ground level, usually by transferring points horizontally. To locate footings, however, requires a different procedure. You must transfer points vertically after the excavator has dug holes or trenches to the correct depth.

The starting point is batter boards at ground level. If batter boards have been knocked out of position or destroyed, you must reestablish the foundation line (Unit 4-2) along one side of the house by building new batter boards (Unit 4-4). In order to locate accurately any

point at floor level of the excavation, you must first locate that point above ground level.

To locate footings, you need cords that run between batter boards, as well as some additional cord. You need a pair of 2 × 2 stakes for each outside or inside corner of the house. You need two 6″ lengths of metal rod for each corner—pieces of reinforcing rod work well in this case.

For tools, you need a plumb bob, hammer, sledge, steel tape, and folding rule.

Questions

1 What is your starting point for locating footings?
2 What do you do if a batter board is missing?

Procedure

1 Reinsert the knotted cords in the kerfs you previously cut into batter boards to mark outside footing lines.

2 Where two lines cross, hang a plumb bob on a long line. The plumb bob should swing freely 3 to 4″ above the floor of the excavation.

3 When the plumb bob stops swinging, drive a

piece of metal rod into the ground directly below the point (Fig. 5-2.1). Leave an inch or two exposed.

4 Repeat steps 2 and 3 at all corners of the house.

5 With your steel tape, measure the diagonals between rods to make sure the rods are accurately placed. Measure also between all adjacent rods to assure that the footings will be the correct length.

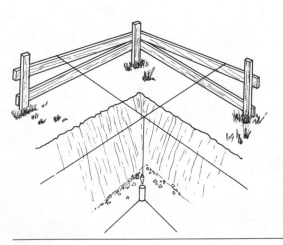

Fig. 5-2.1

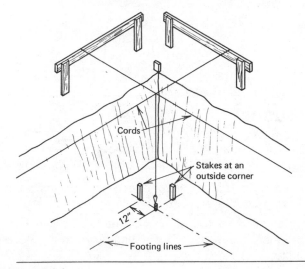

Fig. 5-2.2

6 About a foot away from the rods, drive a pair of stakes. The stakes should lie 12″ outside the outside footings lines when extended (Fig. 5-2.2). Leave about 12″ of stake exposed.

7 At any inside corner drive the stakes about 24″ from the rod, but inside the footing line (Fig. 5-2.3).

8 Mark on each stake the level of the top of the footing. The height varies with the method of forming (see Unit 5-3) and with the height of the footing. Footing height appears in working drawings on the main section. In this procedure assume that footings are formed on the floor of the excavation.

9 Stretch a nylon cord from stake to stake. The cords should cross directly above the metal rods. Form a triangle at corners, as in Fig. 4-3.6. Wrap the cord securely around each stake at the height mark. The cord marks the top outside edge of the footing.

10 Recheck your work by measuring the diagonals between opposite corners where the cords cross.

11 Now move the cords between batter boards at ground level into the kerfs marking the inside footing line.

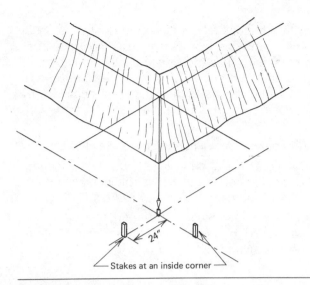

Fig. 5-2.3

12 Repeat steps 2, 3, 4, and 5. You have now located all inside corners of the footing. No stakes or cords are needed along inside footing lines.

13 Remove the metal rods.

Questions

1 While locating footings, how many times should you measure the diagonals?

2 The height of a footing appears on what drawing?

3 If footings are standard size, how far apart will the two pieces of rod at a corner be?

4 If the overall dimension of a wall is 48′-0″ and footings are 8″ × 16″, what is the distance between the two rods marking the corners of the footing for that wall?

Unit 5-3   **Form for Continuous Footings**

Footings that support walls are called **continuous** footings. They may be formed on the ground, in the ground, or half in the ground and half out (Fig. 5-14.1). The method used depends largely on soil conditions and on what rests on the footing.

With in-the-ground footings the sides of a trench cut into the earth act as forms. The trench may be from 8 to 18″ deep. The form is dug by a **trencher**, a machine that can dig a trench to the exact width needed. An in-the-ground footing is usually used when:

• The foundation wall is low, and the **frost line**—the point below the surface to which ground freezes—is close to grade level (ground level).

• A basement is called for in the plans, but footings at their normal depth would not rest on firm soil.

With a half-in-half-out footing the trench is usually only 4" deep. This type of footing is most often used when the house has a concrete basement floor that is poured on a bed of gravel 4" deep (see Fig. 5-14.1).

An on-the-ground form is the most common. It is used when:

- The foundation wall surrounds a crawl space.
- Drain tiles are to be installed outside the footing. (Drain tiles receive water that filters down through the soil and carry it safely away from footings.)
- Soil conditions require a level floor for the entire excavation, or a gravel bed thicker than 4" under the basement floor.

The carpenter prepares all types of forms. Procedures for building forms on top of the ground and half-in-half-out are identical. Only the forming materials differ.

For in-the-ground forming you need no lumber, but a pocketful of nails. For half-in-half-out forms you need stakes and 2×4s. For on-the-ground forms for standard footings you need 2×4 stakes, precut to length for secondary use as spacers. Form boards may be 2×8s, pairs of 2×4s placed edge to edge, $\frac{3}{4}''$ particle board or $\frac{3}{4}''$ exterior plywood, called **Plyform.** For forms greater than 8" deep you need wider forming materials.

It is common practice with on-the-ground forms to leave a gap between the bottom of the form board and the ground. A little concrete may seep through, but not enough to worry about. The gap serves three purposes. First, it leaves a place for sweeping out dirt that falls into the form. Second, it allows for variation in level of the excavation floor. And third, it lets rainwater run off if a shower falls after forms are built but before concrete is poured.

Tools needed for forming are a sledge, hammer, saw, carpenter's level, and folding rule.

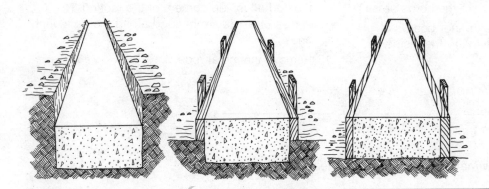

Fig. 5-3.1 Continuous footings for a foundation wall may be formed in the ground (left), on the ground (right), or half in and half out of the ground (center).

Questions

1 If a house has a crawl space, the usual type of footing used is the _____. If the house has a basement, the _____ type is more common.

2 Which of the following materials is not used for form boards (circle one): $\frac{1}{2}''$ plywood, 2×8, $\frac{3}{4}''$ plywood, 2×4.

3 Which of the materials in question 2 is used with half-in-half-out footings?

4 The three reasons for gapping below form boards are:
   a) _____.
   b) _____.
   c) _____.

Procedure for Forming on the Ground

1a. If forms are 2 × 8s, select from the stack the straightest pieces you can find. Lay them just outside the footing cords so that you know approximately where stakes are needed.

1b. If forms are Plyform, rip sheets into strips 7¾″ wide. Cut six strips from a sheet.

2 Distribute stakes where needed. You need two at every corner, two at every joint between boards, and intermediate stakes every 4 to 6′.

3 Begin at one corner by driving in a pair of stakes (Fig. 5-3.2) to support the ends of butting form boards. Place the stakes the thickness of a form board away from footing lines, and let the tops stick up about an inch above the lines.

4 Drive a second set of stakes 4′ from the corner pair as intermediate supports for corner boards.

5 Lay a form board against the two stakes. Hold the board on the toe of a safety shoe, and raise or lower it until it is flush with the top of the corner stake (Fig. 5-3.3). With a pair of 10-penny nails attach the board to the stake. Let the end of the board slightly overlap the end of the next corner board.

6 Nail the form board to the intermediate stake. With the carpenter's level, check both the form board and stakes for plumb, and make any adjustments.

7 Butt the second corner board against the first corner board, and nail it to its two stakes.

8 Check the stakes and form board for plumb. Then nail through one form board into the other to hold them together in a square corner.

Fig. 5-3.2 (Photo Drew Leviton, Atlanta.)

Fig. 5-3.3 (Photo Drew Leviton, Atlanta.)

9 Drive the four supporting stakes into the ground until their tops are level with the cords.

10 Repeat steps 3 to 9 at all other corners.

11 Next place stakes near the unsupported ends of all corner boards, and nail and plumb.

12 Set the remaining outside forms in this way, with the center form board in each wall last. **Precaution.** Plan the work so that no form board is less than 8' long.

13 Using a precut stake as a spacer, locate stakes to support inside form boards. Set the squared end of the stake against the inside of the completed form. The point of the stake is the location of the stake for inside forms. **Precaution.** With 2×8 form boards, stake length should be $17\frac{1}{2}''$. With Plyform, stake length should be only $16\frac{3}{4}''$.

14 Drive stakes and attach and plumb inside forms, beginning at inside corners.

15 To assure that tops of outside and inside forms are at the same level, lay your carpenter's level across the two forms (Fig. 5-3.4), and gently drive in inside form stakes until you reach level.

16 Brace forms at stakes with a deadman, and brace (A) as with shoring (Fig. 5-3.5).

17 If necessary to hold form boards straight and properly spaced, add spreaders (B) and form ties (C) (Fig. 5-3.5).

Fig. 5-3.4 (Photo Drew Leviton, Atlanta.)

Procedures for Forming in the Ground

1 Measure to the bottom of the trench about every 8' to be sure it is reasonably consistent in depth. Consider the depth of footing shown in the drawings as a minimum.

2 Check corners for square, and clean out any debris that has fallen into the bottom of the trench.

3 From the cords marking the top of the footing, measure into the trench the depth of the footing. If the trench is deeper than the depth of footing, mark the proper depth with nails that are pushed into the sides of the trench (Fig. 5-3.6).

4 Repeat step 3 in all trenches. **Note.** When nails protrude above the bottom of the trench, the footing is sometimes poured in two steps. The first pour fills the trench up to the nails. After this base has set, the regular footing is poured into the rest of the trench.

Fig. 5-3.5

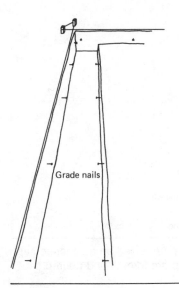

Grade nails

Fig. 5-3.6

Questions

1 You build (outside, inside) forms first. (Cross out one.)

2 When you first place stakes for forms, you drive them to a point (above, below, level with) the tops of footings. (Circle one.)

3 When you drive the stakes into their final location, you check them for

_____.

4 To locate inside forms, the most important tool you use is the _____.

5 To set forms, the place to begin is _____.

Unit 5-4   Form for Interior Footings

Footings that support foundation walls and interior basement partitions require continuous forms. When a house has a chimney, or its floor system is supported on piers, posts, or columns, each of these requires a separate footing. These footings must also extend below the frost line, rest on a good bearing soil, and be larger than the structure they support. Methods of forming are similar.

Piers, posts, and columns line up in rows, and therefore their footings are also in rows. The cords marking their locations form a checkerboard of lines above the excavation. The sizes of footings required are shown on the foundation plan and sections in the drawings.

Piers may be square, rectangular, or round. Their footings are square or rectangular and extend 4" beyond. In residential work concrete piers are usually 12" × 12", 8" × 16", or 14 to 16" in diameter, and require footings 20" × 20", 16" × 24", or 22 to 24" square, respectively. Piers of concrete block are one block long and one or two blocks wide—8" × 16" or 16" × 16".

When drawings call for piers, the majority of those piers are all the same size. To simplify forming, builders often cut a template to the size of the standard pier footing, and use it as a guide to forming.

A chimney is a type of pier, and requires a footing formed and poured completely separate from all other footings. Because of its tremendous weight, the chimney in a one-story house usually rests on a footing 8" thick that extends at least 6" beyond the chimney in all directions. The footing beneath the chimney of a two-story house is usually 12" thick and extends 8".

Footings for wood posts are formed and poured in two steps. The lower footing is square and formed like

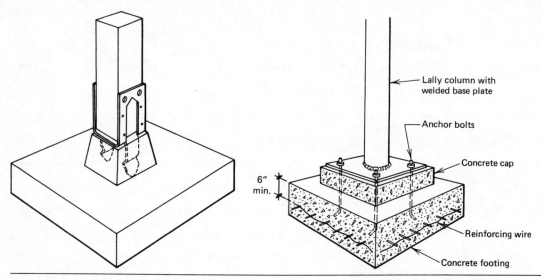

Fig. 5-4.1 Because of the weight they carry, posts and columns should have thicker support than a basement floor slab can provide. Wood post are therefore raised on a pedestal (left), and steel columns are raised on a concrete cap (right). Both pedestal and cap are formed and poured separately from the floor slab.

a pier footing. But the base of a wood post should never be in contact with either its footing or a concrete floor to avoid water damage. Therefore it rests on a separate pedestal (Fig. 5-4.1, left), which is formed after the footing has cured.

Steel columns—also called **lally columns**—rest on a steel base. Sometimes the base is bolted directly to the footing, and any basement floor poured around it. Otherwise, the base fits on a concrete cap (Fig. 5-4.1, right), which is formed like a pier footing.

Materials needed to form footings for piers, posts, and columns are plywood, form boards, and nails. Tools required are a hand saw, hammer, sledge, carpenter's level, framing square, and a chalk line.

Questions

1 Standard dimensions of a footing for a concrete block pier are _____ by _____ by _____.

2 Standard dimensions for the footing under the chimney of a one-story house are _____ inches greater than the dimensions of the chimney itself.

3 Which of these supports should never rest, either directly or indirectly, on a basement floor: lally column, chimney, or wood post?

4 Which of these is required, whether or not the house has a basement floor: pedestal for a wood post or concrete cap for a lally column?

Procedure for Forming a Pier Footing

1 Determine from the drawings the size of footing required.

2 From a piece of ¾″ plywood, cut a template to this size. A **template** is a pattern.

3 Select four lengths of forming material of the required width, and 4 to 12″ longer than the dimensions of the footing.

4 Lay these lengths around the template, lapping them at corners (Fig. 5-4.2). Nail them together. **Note.** Do not nail the template.

5 Check the form for square and level. Remove the template.

6 Lay the form directly below the cords that mark the four footing lines. Drive stakes outside the corners, and nail them firmly to the form extensions (Fig. 5-4.2). **Note.** Some carpenters use only two cords, and stretch them on the center lines of footings.

7 Recheck level, height, and alignment.

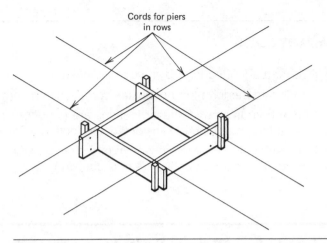

Fig. 5-4.2

Procedure for Forming a Pedestal

1 Determine the actual (not nominal) dimensions of the post to be supported.

2 From $\frac{3}{4}''$ plywood or straight boards, cut the sides for the form to dimension. Dimensions may vary, but Fig. 5-4.3 shows the basic shape of the boards and typical dimensions.

3 Nail the four form pieces together with duplex nails. Nail at top and center only.

4 Snap chalk lines diagonally across the base footing (Fig. 5-4.4), and set the form on the footing so that each corner lies on a chalk line.

5 Build a square collar of 1 × 4s (Fig. 5-4.5) that fits tight at a point about an inch above the bottom of the pedestal form.

6 Slip the collar on the form, set it level, and weight it with stones or concrete blocks. The weights prevent the force of fresh concrete from knocking the form out of position.

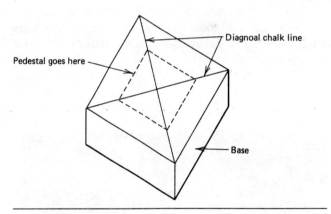

Fig. 5-4.4

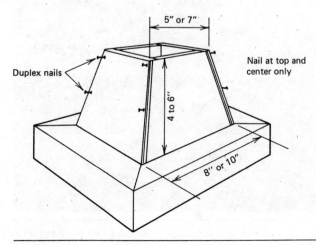

Fig. 5-4.3

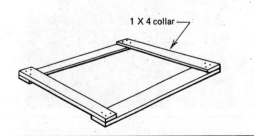

Fig. 5-4.5

Questions

1 To form a footing for a pier, you first make a _____.

2 To locate the form for a pedestal, you set its corners on _____.

3 Form boards around interior footings are (longer than, shorter than, the same length as) the dimensions of the footing. (Cross out two.)

4 The sides of a concrete pedestal (are straight, slope outward, slope inward) from top to bottom. (Cross out two.)

Unit 5-5 **Place Reinforcement**

Properly mixed and poured, concrete has great compressive strength and can support tremendous weight. But it has little tensile strength—that is, resistance to bending and twisting. In order to increase tensile strength, concrete must be reinforced.

Steel is the material used. In narrow sections of concrete, such as footings, special steel rods or bars are used (Fig. 5-5.1). Often called **rebars**, short for reinforcing bars, the rods come in sizes numbered from 2 to 8, and each number represents $\frac{1}{8}''$ in diameter.

If a drawing calls for No. 4 rebars, then their diameter is $\frac{4''}{8}$ or $\frac{1''}{2}$.

Reinforcing bars are most effective in increasing concrete's strength when placed in the lower third of concrete thickness, but not at the bottom. The rods must therefore be suspended in the form while concrete is poured, or they can be laid in the bottom of the trench and lifted into position in the wet concrete during the pour.

Broad areas of concrete, such as floor slabs and

Fig. 5-5.1 Steel bars are used for reinforcing footings and walls. Steel mesh is the standard reinforcement for large, flat slabs of concrete. (Courtesy Portland Cement Association.)

driveways, are reinforced with steel mesh. Strands of steel wire are welded at right angles to each other (Fig. 5-5.1), and the mesh is made in long rolls 6' wide. Meshes are specified by the spacing of the strands and by wire size. A 6 × 6 10/10 mesh, for example, has strands spaced 6″ on centers, with a wire size of 10 gauge in both directions.

Unlike rebars, mesh is laid in place directly on a gravel base or vapor barrier. It does not need to be suspended, but is more effective when it is.

To place reinforcing bars in footings, you need a bolt cutter or rebar cutter, lineman's pliers, piece of chalk, hacksaw, and hammer. To bend the bars at corners, a jig is quite useful (Fig. 5-5.2). You should have a jig for each diameter of rod used on the job.

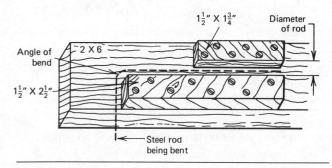

Fig. 5-5.2 This simple jig makes it easier to bend reinforcing bars for use at corners.

Questions

1 A No. 3 rod has a diameter of _____.
2 Specifications for wire mesh give first the _____ and next the _____.
3 Reinforcing rods are most effective when placed in the _____ of the concrete.

Procedure for Reinforcing Footings

1 Determine from drawings or specifications the size and number of rods required.
2 Lay out the rods in the forms, overlapping ends 18″. Cut to length where necessary.
3 Mark with chalk the points where bends are required, and make all bends using the jig.

4 Lay short pieces of 2 × 2 or 2 × 4 across the bottom of forms as temporary support for the rods. Space rods as specified. **Note.** If no spacing is mentioned, place rods about 2½″ from the sides of forms.
5 Cut two pieces of 16-gauge wire, each about 30″ long, for every 5' of perimeter of the footing.
6 Twist one end of the wires together, and nail to the top of the form (Fig. 5-5.3). Bend the nail over instead of driving it in.

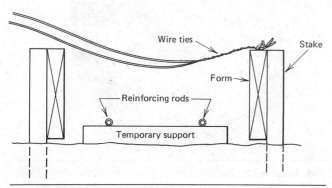

Fig. 5-5.3

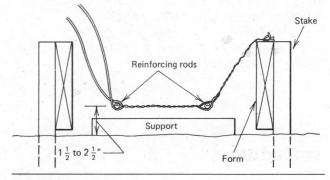

Fig. 5-5.4

7 Continue to twist the wires together, inserting the rods in the twists as you go (Fig. 5-5.4).

8 Nail loose ends of the twisted wires to the other form, and trim off excess wire. Remove temporary supports.

9 Repeat steps 6, 7, and 8 around the entire footing.

10 Where ends of rods overlap, tie them together with wire twisted with lineman's pliers (Fig. 5-5.5). Trim off any excess wire.

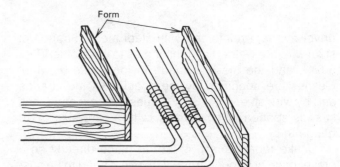

Fig. 5-5.5

Questions

1 Ends of rods should overlap _____ inches at joints.

2 To support rods temporarily, use _____. To support rods permanently, use _____.

3 Unless otherwise specified, space rods _____ apart under a concrete block foundation wall.

4 Why do you think it is better to hold the ends of wires by bending the nails over instead of driving them all the way into forms?

Unit 5-6   **Make a Keyway**

After footings are formed under concrete block walls, the final job before walls are built is to remove footing forms. But if walls are poured concrete, there is one more task to be accomplished while the concrete for footings is still **green**—that is, not fully set.

To assure a firm connection, or **key**, between cured concrete in the footing and fresh concrete in the wall, the footing may require a keyway. A **keyway** is a slot in the center of freshly poured concrete (Fig. 5-6.1). It runs the entire length of all footings.

To form a keyway you need a straight length of 2 × 4 or 2 × 6, some scrap lumber for a handle, and a saw, hammer, and chalk line.

Fig. 5-6.1 A keyway is a slot made in a wet concrete footing to provide a good mechanical tie with the concrete foundation wall that rests on the footing. A keyway is not required with walls of concrete block or other masonry.

Procedure

1 Study detail drawings of the footing to determine the size and shape of keyway required.

2 Cut a piece of lumber stock of the proper size, and shape it as required. Figure 5-6.2 shows a typical shape. Length may vary, but 6′ is a practical maximum.

3 Add a handle as shown in Fig. 5-6.2.

4 With a chalk line, mark the center of the keyway on the footing. Along the top of the key form, also mark its center.

5 After the concrete has been screeded (leveled), but while it is still plastic, press the key form into the concrete, lining up the centerlines. **Precaution.** Be careful not to disturb the screeded surface of the footing beside the keyway.

6 When one length of keyway is formed, lift the form and move it to the next position. Continue until the keyway is formed around the entire footing.

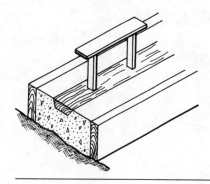

Fig. 5-6.2

Questions

1 Why do you think the sides of a keyway are tapered instead of straight?

2 Is a keyway required under a concrete block wall? Why?

Unit 5-7   **Build Wood Forms for Concrete Walls**

Large builders often buy manufactured wall forms made of metal, which they use over and over again. Smaller builders, for whom owning is not economical, may rent forms. They are made in shapes and sizes to fit almost any building condition. Yet some builders still prefer to build their own forms, either because they can't rent metal forms or because they like wood forms better.

The parts of any wall forming system (Fig. 5-7.1) include the forms themselves, some device for locking adjacent forms together, some means of maintaining opposite forms at a fixed spacing, horizontal supports to keep forms in alignment, and braces to keep forms upright and in position until the concrete has set.

Plyform plywood is used for the sides of forms. The plywood sheets are attached to a bottom plate and studs that form a frame. The plate is usually a 2 × 4. However, if forms for footing aren't removed, a 2 × 6

plate can be nailed to footing forms (Fig. 5-7.2). Forms may or may not have a top plate.

Studs are 2 × 4s, usually spaced 24″ on centers (o.c.) for walls up to 8′ high, and 16″ o.c. for taller walls. Forms may be built with the plywood running either vertically or horizontally. Typical sizes are 4 × 8′ and 8 × 4′. Special forms are often required at inside and outside corners, and in the centers of walls with odd dimensions.

The means of locking adjacent forms together varies. A typical device not only locks forms together, but secures form ties (Fig. 5-7.3). The simplest lock is a wire that is threaded through holes in adjacent studs near their ends and twisted tight.

To hold opposite forms in position there are a wide variety of form ties in use (Fig. 5-7.4). Most ties are metal rods or flat strips stiff enough to act as spacers prior to pouring, and strong enough to hold forms to-

Fig. 5-7.1 A system for shaping vertical concrete consists of vertical forms (either wood or metal), devices for securing each form to its neighbors and the form opposite, walers to keep forms in line, and braces to prevent forms from moving from the tremendous weight of wet concrete. (Courtesy Symons Corporation.)

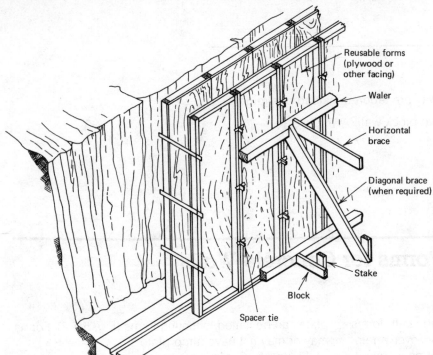

gether during pouring. At points that equal the wall's thickness, the metal is thinner to form a break point. After concrete sets and forms are dismantled, you break the tie at the break points and leave it in the wall.

Horizontal supports that keep forms in line are the same walers used in shoring (see Unit 5-1). Usually they are a pair of long 2 × 4s. One set of walers goes at the same level as the top row of ties and another at the same level as the bottom row of ties. In a strong forming system, walers are placed at all tie levels. Walers are held in place in many ways. Sometimes the ties have a clip at the end that fits over both walers. Sometimes they are tied with wires to walers behind the opposite wall form.

Walers must be frequently and securely braced to withstand the weight and pressure of wet concrete. The method of bracing is similar to that for bracing sheeting (Unit 5-1).

To build forms, the tools required are a crosscut saw or power saw, folding rule, hammer, drill, and a brush for coating the exposed faces of plywood with oil, if it does not come already oiled. Use either a special form coating or a good grade of oil. Do not use old motor oil, which discolors concrete.

To increase the life of plywood forms, repair large defects in the surface with waterproof putty, and seal all cuts, scuffs, and edges with aluminum primer, shellac, or lead and oil paint.

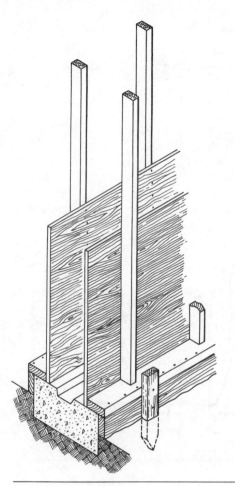

Fig. 5-7.2 If the bottom plate of wood forms is a 2 × 6, it may be attached to footing forms for added strength where it is needed the most.

Fig. 5-7.3 A typical locking device for manufactured forms includes metal tee that slips through slots in adjoining forms and is held fast by a metal wedge. Ties to opposite forms fit between adjoining forms, and the tee anchors them. (Courtesy Symons Corporation.)

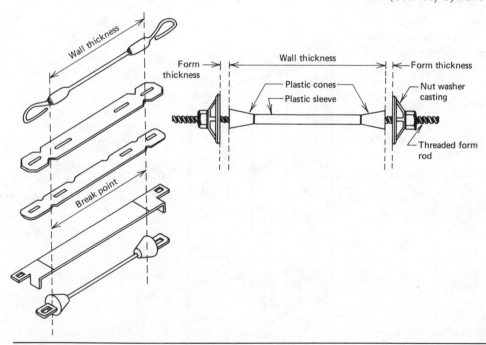

Fig. 5-7.4 There are many types of form ties. Most types have break points where the metal is thin. Part of the tie remains in the wall, and you break off the ends. With the tie shown at right, only the sleeves remain in the wall. All other parts are reusable.

Questions

1 Plywood for wall forms should be _____ grade made with _____.
2 When wall forms are to be nailed to footing forms, the bottom plate should be a _____.
3 The places in form ties when the metal is indented are called _____.
4 Defects in plywood surfaces may be corrected with _____.
5 Why do you think it is necessary to oil forms?

Procedure

1 If plywood is new, protect the surface with oil, fill defects, and seal edges. If plywood is used, clean it thoroughly before these preparations.
2 Cut the bottom plate to length. For standard forms, the length is 4' for a tall form and 8' for a low form.
3 Cut studs to length. Length is 7'-10$\frac{1}{2}$" for tall forms and 3'-10$\frac{1}{2}$" for low forms without a top plate. If the form has a top plate, cut off another 1$\frac{1}{2}$" (Fig. 5-7.5).
4 On the bottom plate, mark the locations of studs.
5 Nail studs to the bottom plate on the marks.
6 Lay plywood in position on the plate and studs, with the smoothest surface face up. Adjust the sheet so it is flush with the bottom plate and one side stud.
7 Nail through the plywood into the plate and side stud. **Precaution.** Use as few nails as possible, since their only purpose is to keep plywood flat. Every 2' is adequate.
8 Swing the loose ends of remaining studs into correct vertical position, and nail each in place.
9 Drill small holes in outside studs for locking devices.
10 Mark the locations of holes for ties, as shown in Fig. 5-7.6. Drill holes at these points just large enough for ties to fit easily through them.

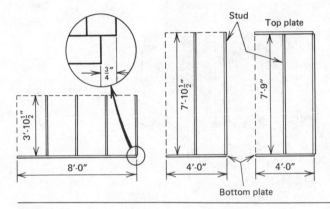

Fig. 5-7.5

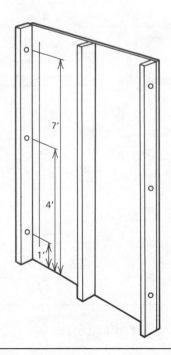

Fig. 5-7.6

Questions

1 How many holes for form ties do you need to drill in plywood 4′ wide and 8′ high?

2 How many nails are needed to attach plywood to its frame when the form is 4′ wide, 8′ high, has studs 2′ on centers, and no top plate?

3 At inside corners, plywood must overlap edge studs in forms either _____ or _____.

Unit 5-8 **Assemble Forms**

The procedures followed to assemble forms are almost the same whether the forms are bought, rented, or built for the job. For straight walls in residential construction, standard sizes of metal or plywood forms are 2′ × 8′, 2′ × 6′, and 2′ × 4′. Special forms are made for outside corners, inside corners, fillers, and junctions with interior walls.

Manufacturers of forms provide instructions for assembling them, since each type varies slightly in the method of locking together. With all types, however, the edges of individual sections in outside and inside walls must line up.

If the tops of footings are not level, you place wedges under forms to achieve a proper assembly. It is extremely important that the tops of forms be level, especially if the wall is the same height as the forms— and it often is.

To assemble any kind of wall forms, you need a plumb bob, chalk line, hammer, lineman's pliers, and carpenter's level.

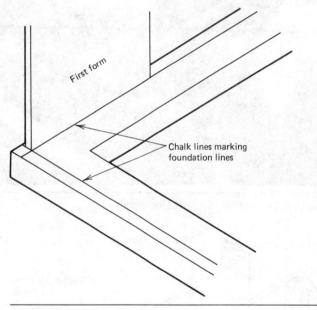

First form

Chalk lines marking foundation lines

Fig. 5-8.1

Procedure

1 Set cords between batter boards in kerfs marking the foundation line.

2 With plumb bobs hung where cords cross, bring the corners of the foundation down to the footing (see Fig. 5-2.2). Mark the corners with a nail or chalk.

3 Through the corner marks, snap chalk lines to locate the foundation line on the footing. These lines also mark the locations of outside wall forms.

4 Check the tops of footings for level. If not level, assemble wedges where needed.

5 Set the first wall form at an outside corner on the chalk line (Fig. 5-8.1).

6 Set the adjoining corner form next. If forms are wood, as shown in Fig. 5-8.2, hold them together with duplex nails. If forms are manufactured, add an outside corner piece, and lock the forms together (Fig. 5-8.3).

7 Complete outside forms at the other corners of the foundation.

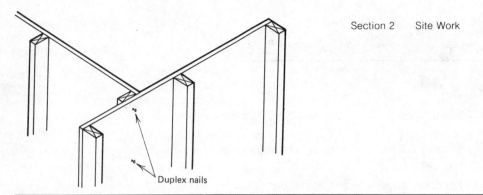

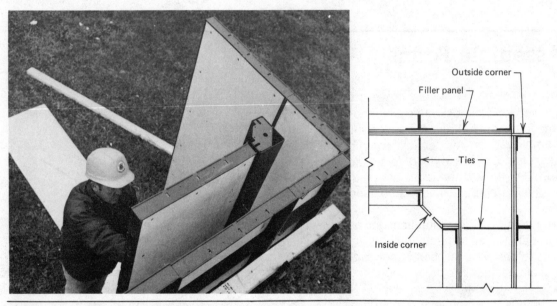

Duplex nails

Fig. 5-8.2

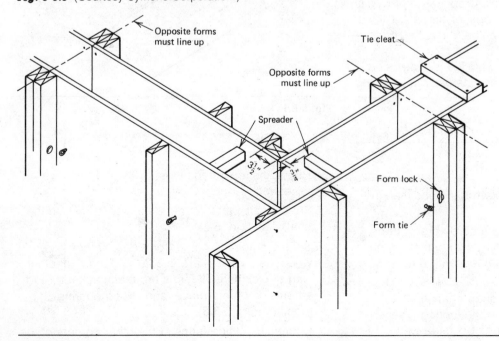

Fig. 5-8.3 (Courtesy Symons Corporation.)

Outside corner

Filler panel

Ties

Inside corner

Opposite forms must line up

Tie cleat

Opposite forms must line up

Spreader

$3\frac{1}{2}''$

$3\frac{1}{4}''$

Form lock

Form tie

Fig. 5-8.4

8 Working from corners toward the middle of each wall, continue setting standard outside forms. Lock each form to its neighbor, and insert ties through each hole in the forms as you set them.

9 Close the gap between the last two forms with a filler that fits tight.

10 Set the two forms at each inside corner in position (Fig. 5-8.4), using a spreader to assure the correct spacing. A **spreader** is a 2 × 2 equal in length to the thickness of the wall. Hold the tops of forms together with a cleat.

11 Insert ties through holes in inside forms, and check their positions. They should line up both vertically and horizontally with the matching holes in outside forms.

12 With corner forms positioned, set remaining inside forms. As you go, position any horizontal reinforcement required by specifications. Lay rods on ties about 2″ from forms. Bend rods around corners.

13 Check tops of forms for level. If uneven at any point, carefully drive wedges below the forms to make the adjustment. Set walers just above and just below ties, and lock them in place (Fig. 5-8.5). When placing walers, stagger the joints so that no two joints fall within 4′ of each other. Continue until all walers are locked in place around both inside and outside forms.

14 With 8-penny nails, tack cleats about every 8′ across the tops of forms (Fig.5-8.4) to hold them in position during bracing.

15 Line up and brace top and bottom walers. Set braces at the midpoint and near the ends of each waler. Brace corners in both directions.

16 Just before final staking of braces, recheck alignment and plumb of all forms.

17 If the height of the forms is greater than the height of the foundation walls, mark the exact height of the wall on the outsides of forms at each corner. Drive a nail, called a **grade nail**, into wood forms or between metal forms at each mark.

18 Snap chalk lines between nails around the perimeter of the wall.

19 Drive nails at approximately 4′ intervals along the chalk line (Fig. 5-8.6). The grade nails mark the height of the poured foundation wall.

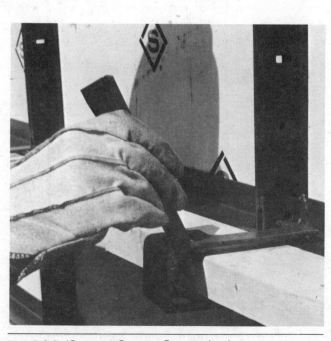

Fig. 5-8.5 (Courtesy Symons Corporation.)

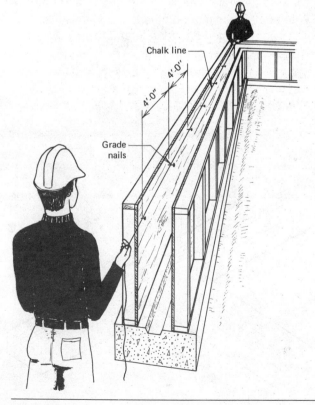

Fig. 5-8.6

Questions

1 To hold forms in place, you use _____ nails.
2 Under what conditions do you install wall pads?
3 The first wall form is placed at _____ of the wall along a chalk
  line marking _____.
4 Joints in pairs of walers should be staggered at least _____.

Unit 5-9   **Form a Tee-Wall**

Whenever an interior partition meets an outside foundation wall, the result is a **tee-wall** or **T-wall**. Forms for the T-wall must be set in position at the same time exterior wall forms are placed. Because the form in the exterior wall opposite the partition must be centered on that partition, T-wall forms are usually set right after corner forms. Forms are then set between the T-wall and corners.

Procedure

1 On the foundation plan find the dimensions from the corners of the foundation to the partition wall.
2 With chalk lines mark on the foundation wall's footing the centerline of the intersecting T-wall (Fig. 5-9.1).

3 Set a 24"-wide outside form in position, exactly centered on the centerline. Plumb and brace the form.

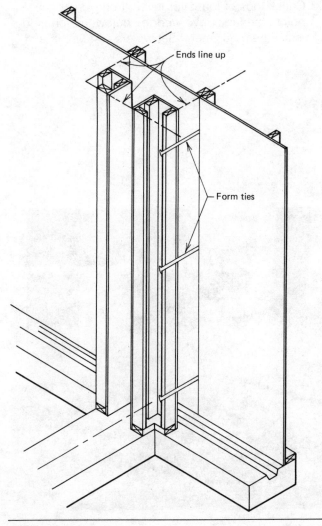

Fig. 5-9.2

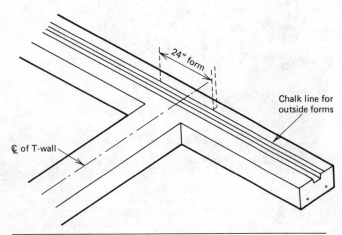

Fig. 5-9.1

4 Complete the outside form to the corners.

5 Build and set up inside corner forms at the T-wall so that the ends of the forms line up with the ends of the 24″ form (Fig. 5-9.2).

6 Insert form ties between inside and outside forms, and across the interior partition (Fig. 5-9.2).

7 Tie walers and lock them in position.

8 Where walers meet at inside corners (Fig. 5-9.3), connect them with metal plates for stiffness.

9 Stake and brace the assembly as outlined in Unit 5-8.

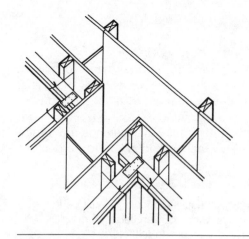

Fig. 5-9.3

Questions

Circle T if you think a statement is true and F if you think it is false.

1 T-wall forms are placed in position in their regular sequence between corners.   T   F

2 Edges of outside and inside forms at a T-wall must line up.   T   F

3 At any one level there are three form ties required between T-wall forms.   T   F

Unit 5-10   Form for Wall Openings

Poured foundation walls are solid concrete around most of their perimeter. But walls do have openings in them, and provision must be made for those openings during forming.

In some types of construction the floor framing system is supported not only on foundation walls but on a girder. **Girders** are horizontal beams of wood or steel. To provide support at the proper level, the tops of girders must be at the same level as the tops of foundation walls. Therefore the girders fit into the foundation wall, and pockets (Fig. 5-10.1) are formed in the wall in which to set girder ends.

Carpenters also must form around openings in walls for doors, windows, ventilators, and sleeves for pipes (Fig. 5-10.2). Manufacturers of metal windows and doors for use in basements also manufacture frames, called **bucks**, to fit various standard thicknesses of concrete wall. It is customary to tie metal bucks in place between forms, and leave them in the

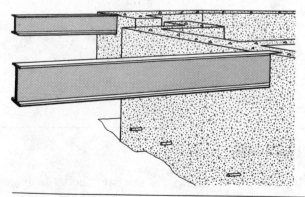

Fig. 5-10.1 In some types of construction, pockets must be formed in concrete walls to hold ends of wood or steel girders.

wall. Sometimes, however, you may need to build a buck out of wood—either a temporary buck for a metal door or window, or a permanent buck for a wood door or window.

Tools needed to form for wall openings are a folding rule, power saw, hammer, chalk, and carpenter's square.

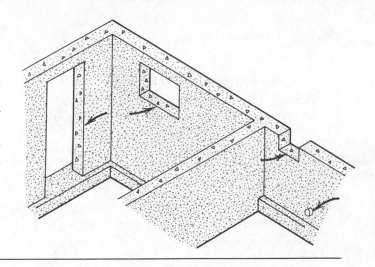

Fig. 5-10.2 In nearly all houses, openings must be formed for (left to right) doors, windows, ventilators in crawl spaces, and sleeves for incoming utilities.

Procedure for Forming a Girder Pocket

1 From working drawings determine the size of the girder and its location where it meets foundation walls.

2 From local codes determine the clearance required at the sides of the girder, if any. **Note.** Clearance is usually required around wood girders so that air can circulate and prevent rotting.

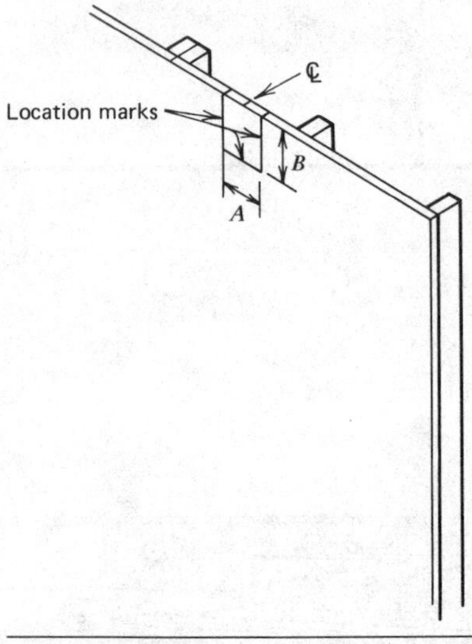

3 On wall forms mark the centerline of the girder and the locations of the bottom and sides of the girder pocket (Fig. 5-10.3).

4 Using A and B in Fig. 5-10.3 as two of the *outside* dimensions of the box that forms the girder pocket, build a box (Fig. 5-10.4) with three sides, a top, and a bottom. The third dimension is the depth of the pocket, taken from drawings.

5 Carefully position the box between vertical marks on the forms, and set the bottom exactly level on the horizontal mark.

6 Nail the box in position through outside wood forms, or wire-tie it to metal forms.

Fig. 5-10.3

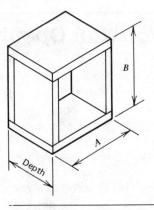

Fig. 5-10.4

Questions

1 The dimensions of a girder pocket match the (inside, outside) dimensions of the box form. (Circle one.)

2 Clearance is often required at the sides of girder pockets when girders are (wood, steel). (Circle one.)

3 The top of a girder is (slightly above, at, slightly below) the level of the top
  of a foundation wall. (Circle one.)

Procedure for Building and Setting a Buck

1 In the window or door schedule in the drawings, find the size of the rough opening.

2 Rip 2″ stock to the same thickness as the wall.

3 Cut side pieces to a length equal to the rough opening height.

4 Cut the top piece (plus the bottom piece for a window) to a length equal to the rough opening width less the thickness of the two side forms.

5 Cut 2 × 2 cleats to length, and fit them into the inside corners of the box (Fig. 5-10.5).

6 Nail cleats to form pieces as shown in Fig. 5-10.5. Drive duplex nails into side forms, 10-penny nails into top and bottom forms.

7 Square the frame, and brace it diagonally between cleats. This completes preparation of a temporary buck that will be replaced later by a steel buck.

8 To make a buck for wood doors or windows, add cleats on the sides. Rip a 2 × 6 in half, and bevel the sides. The length of these beveled cleats is the same as the rough opening height.

9 Nail cleats in position from inside the buck (Fig. 5-10.6). Determine the front-to-back position by studying details of wall sections in the drawings.

10 Mark on the wall form the location for the buck, and fix it into place in the same manner as for a girder box.

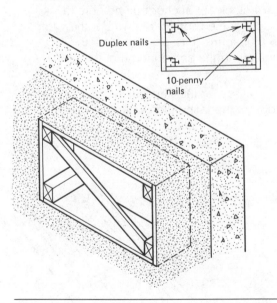

Fig. 5-10.5

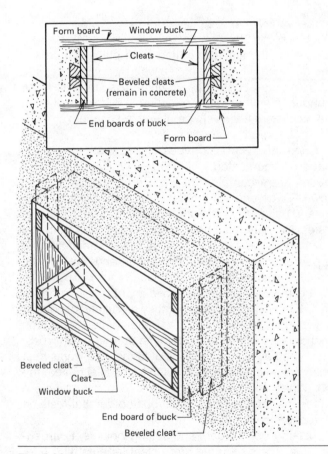

Fig. 5-10.6

Questions

1 Why do you think cleats are nailed as shown in Fig. 5-10.5?

2 What happens to the beveled cleats on the back of a wood buck (Fig. 5-10.6) when the buck is later removed?

Unit 5-11 **Set Sill Bolts**

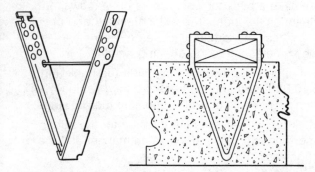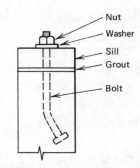

Fig. 5-11.1 V or J anchors hold the bottom wood member of a floor framing system to the foundation wall. The nut and washer come with the bolt at right.

The first piece of lumber to go into place atop a foundation wall is the **sill**, also called a **sill plate** or **mud sill**. The sill is held in place on the wall by anchors set into the concrete about 6' apart while it is still wet. If the foundation wall is made of concrete block, cavities in the blocks are filled with grout and the anchors embedded in the grout.

There are two basic types of anchors. One type is made of zinc-coated steel (Fig. 5-11.1, left). Its three pieces are assembled on the site to form an upsidedown A. The ends of the flat pieces are bent over the sill and nailed to it. The other type (Fig. 5-11.1, right) is a steel rod called a **J-bolt**, **sill bolt**, or **anchor bolt**. The lower end is bent to grip the concrete or grout, and the upper end is threaded. A washer and nut on the threaded end hold the sill in place.

It is possible to simply shove anchor bolts into concrete without any guides, but they are likely to move or sink slightly in the concrete. Better builders position anchor bolts with sill bolt holders; there are two types. One type lies across the tops of forms, and works best when the level of concrete in the forms is at or near the top. When concrete doesn't fill a form completely, a better answer is a **floating** sill bolt holder that rests on the concrete wall itself.

Making bolt holders and setting bolts is a carpenter's job. Tools required are a steel tape, folding rule, crosscut saw, and hammer.

Procedure

1 In the drawings find the size and spacing of sill bolts.

2 Mark these locations on the sides or tops of both outside and inside form boards.

3 From 2 × 4 stock cut anchor bolt holders to length—either the width of the wall or the width of the wall plus forms (Fig. 5-11.2). Mark the centerline along the entire length of the bolt holder.

4 Check drawings for the dimension between the outer edge of the foundation wall and the centerline of the sill plate.

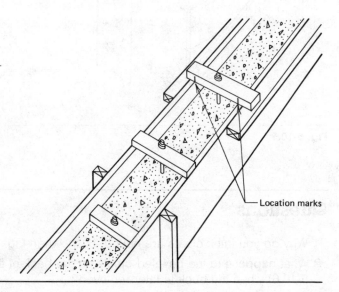

Fig. 5-11.2

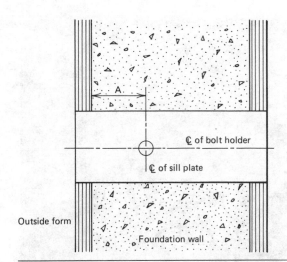

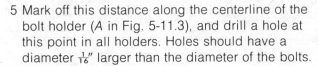

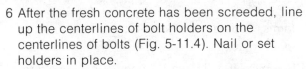

Fig. 5-11.3

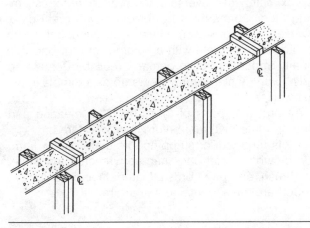

Fig. 5-11.4

5 Mark off this distance along the centerline of the bolt holder (*A* in Fig. 5-11.3), and drill a hole at this point in all holders. Holes should have a diameter $\frac{1}{16}$" larger than the diameter of the bolts.

6 After the fresh concrete has been screeded, line up the centerlines of bolt holders on the centerlines of bolts (Fig. 5-11.4). Nail or set holders in place.

7 Push sill bolts through holes into concrete. Check each bolt for plumb. Leave 1 to $1\frac{1}{2}$" of threaded end exposed above the holders.

8 Thread on a washer and nut to hold bolts at required height. When nuts touch washers, stop tightening.

9 After concrete has set, remove and store sill bolt holders, and loosely replace washers and nuts.

Questions

1 What is the length of a floating sill bolt holder for a standard residential foundation wall?

2 How does the thickness of a sill plate compare with the thickness of a sill bolt holder?

3 Are sill bolts set at the center of the tops of foundation walls? Why?

Unit 5-12   **Strip Forms**

Although concrete must cure for 28 days before it reaches full strength, forms can be removed much sooner. The longer forms remain in place, the less likely that concrete will be damaged. On the other hand, waiting delays other construction, so it is common practice to remove or **strip** wall forms following this schedule:

When midday air temperature is	Remove forms
Above 60 °F (16 °C)	After 3 days
40 to 60 °F (5 to 16 °C)	After 5 days
Below 40 °F (5 °C)	After 8 days

The procedure for stripping forms (Fig. 5-12.1) is approximately the reverse of the procedure for assembling them. Forms must be removed with great care, however, for two reasons. First, it is important not to damage the concrete with a hammer or crowbar. It is equally important not to damage reusable forms and parts. Every dent in a form shows up as a pimple in the next wall.

All reusable parts of forms must be cleaned and oiled or made rustproof before they are stored.

Tools required for stripping forms are a claw hammer, crowbar or flat bar, snap-tie wrench, broadknife, wire brush, and paint brush. Materials and equipment needed are metal containers and form oil.

Fig. 5-12.1 Forms are stripped in approximately the reverse order in which they are erected. (Courtesy Portland Cement Association.)

Procedure

1 Set containers around the excavation for storing reusable form parts.

2 Remove braces and stakes, and store them.

3 Remove tie cleats. Also remove exposed portions of wall ties if they prevent removal of walers. **Precaution.** Be careful not to put any lateral pressure on the wall, and keep all tools away from the fresh concrete.

4 Remove and store walers.

5 Remove insert panels near the centers of walls.

6 Work toward all corners to remove remaining forms. Separate and store reusable parts as you go.

7 Remove all nails and connectors between forms. Place them in a separate container for disposal.

8 With a broadknife or other tool with a wide, stiff blade, clean dried concrete off form boards.

9 Brush oil on the cleaned forms so they are ready for reuse. Store the forms carefully.

10 With a snap-tie wrench remove exposed portions of wall ties that were not removed in step 3.

11 Wire-brush and rustproof the reusable metal parts that you placed in containers. Cover the containers to keep out rain.

Questions

1 How long should forms remain in place when midday temperatures reach 50 °F?

2 Which of these parts—ties, tie cleats, nails, walers, stakes—are not reusable?

3 Name two reasons why you think it is better to oil forms just after stripping instead of just before assembling them.

a) _____.

b) _____.

Unit 5-13 **Form Concrete Piers**

Most concrete piers for residences are either 10″ square or 12″ in diameter. The tools used for forming concrete piers are the same as those you used before: hammer, saw, level, square, folding rule, chalk line, and wrench. But the procedures for forming them and the materials used for forms are quite different.

To form a square pier requires plywood, lumber, and mechanical ties, and the methods are similar to those for forming foundation walls. The form for a round pier, however, is a cardboard or plastic tube, and the carpenter builds a special support for the tube while concrete is being poured.

One of two devices may be used for holding forms together. One is a **collar** (Fig. 5-13.1, left) made to fit tight around the form. The other is a **yoke** (Fig. 5-13.1, right). A yoke consists of two slotted boards that are held together by a pair of threaded rods with nuts.

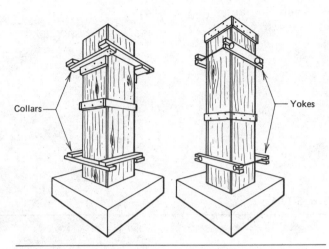

Fig. 5-13.1 Forms may be held together to form concrete pier by means of collars (left) or yokes (right).

Procedure for Forming a Square Pier

1 From the drawings determine the height and dimensions of the pier.

2 Rip two pieces of plywood to the width of the pier, and two more pieces to the width of the pier plus two thicknesses of plywood.

3 Cut all four pieces to length.

4 Form an open-ended box with the plywood. Nail the pieces together while carefully maintaining square.

5 Add cleats on all four sides, and cross-nail them at corners (Fig. 5-13.2). Around a pier less than 4′ high, locate the cleats at about the midpoint of the form. Around taller piers, place one set of cleats at the midpoint and another at the top.

6 Cut yokes from lengths of 2 × 4, and slot them as in Fig. 5-13.3.

7 Connect a pair of yokes with bolts near the bottom of the form, and another pair near the top of the form. Tighten nuts to hold form boards securely together.

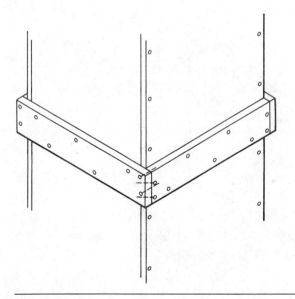

Fig. 5-13.2

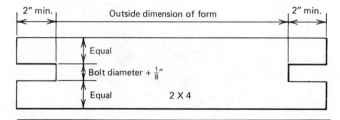

Fig. 5-13.3

8 Across the footing snap diagonal chalk lines.

9 Set the form so that each corner lies on a chalk line.

10 Hold the form in place with a pair of cleats nailed to footing forms (Fig. 5-13.4).

11 Secure the form with diagonal braces and stakes.

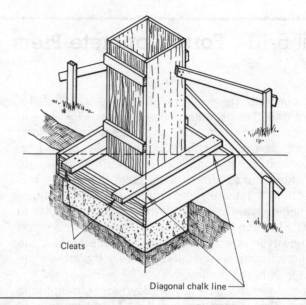

Cleats

Diagonal chalk line

Fig. 5-13.4

Procedure for Forming a Round Pier

1 From the drawings determine the height and diameter of the pier.

2 Snap four chalk lines on the footing, two in each direction (Fig. 5-13.5). The dimension between chalk lines is the diameter of the pier. The dimensions from chalk lines to all edges of the footing must be equal.

3 Cut to length a tube of the correct diameter. Set it on the footing so that its sides are tangent to all four chalk lines.

4 Make a collar with long sides (Fig. 5-13.6) that just fits over the tube. Sides must be long enough to extend beyond the excavation for the pier.

5 Stake the collar to the ground, and nail it securely.

6 Carefully fill the excavation with about 12" of earth to hold the base of the tube in position.

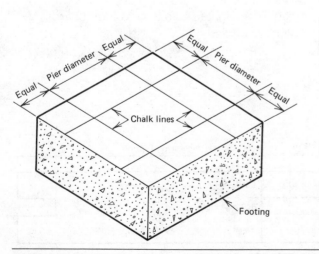

Equal   Pier diameter   Equal   Equal   Pier diameter   Equal

Equal   Equal

Chalk lines

Footing

Fig. 5-13.5

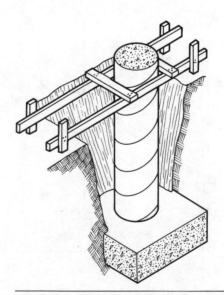

Fig. 5-13.6

Questions

1 You set _____ of forms for square piers on diagonal chalk lines.

2 If a round pier is 12″ in diameter, and the tube is $\frac{1}{8}$″ thick, how far should chalk lines be from edges of the footing?

3 Collars may be used in place of yokes around some pier forms. Why can't collars be used around the form in Fig. 5-13.4?

Unit 5-14   Form for Concrete Slabs

Many types of horizontal concrete, often called concrete **flatwork**, do not require forms. A basement floor, for example, is formed by foundation walls. Foundation walls also serve as forms for some first floor slabs. Therefore the only work for the carpenter is to place an isolation joint (Fig. 5-14.1) around the perimeter of a basement floor after the gravel base and vapor barrier are in place. An **isolation joint** is a strip of plastic material that allows walls and floor to move separately and prevents cracking. The carpenter also places rigid insulation (Fig. 5-14.2) at the edges of concrete floor slabs above grade.

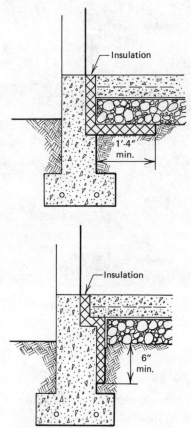

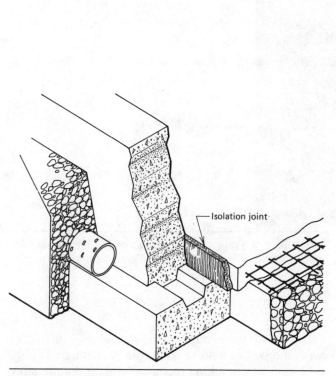

Fig. 5-14.1 An isolation joint separates two sections of concrete so that they can move independently without cracking.

Fig. 5-14.2 When a concrete slab lies above ground level, its edges must be insulated. Rigid insulation is used for this purpose. It lies between the foundation wall and the slab and extends either under the slab or down the wall.

Fig. 5-14.3 The form for a thickened-edge slab has one straight side and one sloped side (left). After a gravel base has been spread and tamped down, the inner form is removed (right), and the concrete may then be poured. (Courtesy Portland Cement Association.)

Fig. 5-14.4 When a concrete slab cannot logically be poured all at one time, temporary forms are used to break the slab area into more manageable sections. The temporary forms support one end of a screed. After the section of concrete has set, its edge serves as a form for the next pour.

Some types of slab floors do not rest on a foundation wall, but have a footing poured as an integral part of the slab (Fig. 5-14.3). The most common type is called a **thickened-edge** or **turn-down** slab, and it is often used in climates where the ground never freezes.

The form material for a thickened-edge slab may be metal, wood, or rigid insulation board. The excavator cuts a trench to the depth of the slab's edge, then cuts away the soil at a 45 to 60° angle on both sides (Fig. 5-14.3). This cut allows room for forming on one side, and shapes the thickened edge on the other.

The tool used to level flat concrete is a long 2 × 4 or 2 × 6 called a **screed**. To level and solidify the surface, the concrete finisher draws the screed back and forth across the fresh concrete. To do this properly, the finisher rests the screed on the tops of forms. If, as is often the case, the slab is wider than the screed, then the carpenter must provide temporary forms in the middle of the slab area (Fig. 5-14.4). The tops of these forms must be level with the tops of perimeter forms, and supporting stakes must lie below the tops of forms. After one section of concrete is screeded, the temporary forms are removed, and the edge of the new slab becomes the form for the next section of concrete.

The procedure for temporary forming is the same as forming for continuous footings in Unit 5-3. Tools required for building forms for concrete flatwork are a hammer, saw, folding rule, and level.

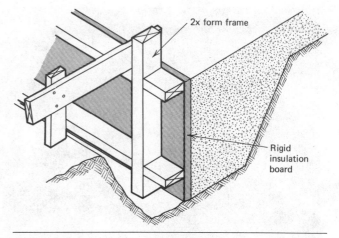

Fig. 5-14.5

Procedure for Forming a Thickened-Edge Slab

1 Cut exterior form material to the height of the edge of the slab, as shown in the drawings.

2 Build a framework of 2 × 4s (Fig. 5-14.5) around the entire perimeter of the building. Place the framework so that its inside edge is spaced the thickness of the form outside the foundation line.

3 Plumb the framework.

4 Add braces, and secure the braces to stakes outside the excavation.

5 Nail the forms to the framework, but only often enough to hold them in place. The tops of the forms must be at exactly the same level above grade as the finished surface of the slab. **Note.** The form acts as a support for a screed and remains in place after the concrete has set. Only the framework is stripped.

Questions

1 (All, Some, No) concrete flatwork must be insulated at its edges. (Circle one.)

2 The purpose of an isolation joint is to _____.

3 A temporary form in the middle of a slab is built like a form for a _____. The usefulness of a temporary form is as a _____.

Unit 5-15   Form for Outdoor Flatwork

In general, the procedures for staking and forming for driveways, walks, and patios are similar to those for footings. However, there are three differences.

First, you have no benchmark as a guide to the relationship of the slab to grade. You must stake the slab area to suit the contours of the site and the direction of drainage.

Second, the slab itself is not level, but slopes slightly for drainage. A slope of 1″ in 12′ is adequate for a patio. But for any walking or driving surface, the slope should be increased to a minimum of 1½″ per 12′, or ⅛″ per foot, to let water run off in winter before it turns to ice.

Third, outdoor slabs are thinner than footings. Stan-

Figure. 5-15.1 Expansion joints prevent large expanses of concrete from cracking. (Courtesy Portland Cement Association.)

dard slab thickness for a driveway is 6″, and the slab must be reinforced with steel mesh. Standard thickness for walks and patios is 4″. Walks don't require reinforcement, but patios do.

Outdoor concrete also needs expansion joints (Fig. 5-15.1). An **expansion joint** may be a length of treated lumber that is used as a form and left in place. Or it may be a strip of molded asphalt, $\frac{3}{4}″$ thick, set into place after a temporary form is removed. In a sidewalk or driveway an expansion joint is needed every 30′ of length. An expansion joint is also required where:

- A driveway slab meets the garage floor or any walkway.
- A walkway meets a patio, concrete steps, or another walkway.
- Any exterior concrete adjoins the house.

The tools required for forming outdoor flatwork are a steel tape, folding rule, ball of cord, sledge, hammer, and level. In addition to form boards and stakes, you need a long, straight 2×4.

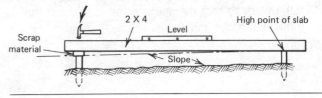

Fig. 5-15.2

2 Establish the high point of the slab. Drive in the stake at this point to the established level.

3 To maintain slope, set on top of an adjacent stake a small piece of material of a thickness equal to the amount of slope between stakes (Fig. 5-15.2).

4 Set the long, straight 2×4 across the two stakes, and place a level on the 2×4. Gently drive in the second stake until the 2×4 is level.

5 Repeat steps 3 and 4 at all other stakes.

6 Place 1×4 (for 4″ slabs) or 1×6 (for 6″ slabs) forms directly beneath the cords, and attach them to stakes every 4′. The tops of stakes and forms should be flush.

7 Overlap form boards at corners, as explained in Unit 5-4.

8 Brace the stakes, and remove cords.

9 Place and nail wood expansion joints.

Procedure

1 Mark with stakes the area to be paved, and outline the area with cords. Drive stakes in only far enough so that they remain upright.

Questions

1 A driveway slab is _____ thick, reinforced with _____, and has a minimum slope of _____ per foot of length.

2 A driveway 50′ long between a garage and a city sidewalk needs _____ expansion joints.

3 A sidewalk slab is _____ thick, with _____ reinforcement, and slopes _____ every 12′ of length.

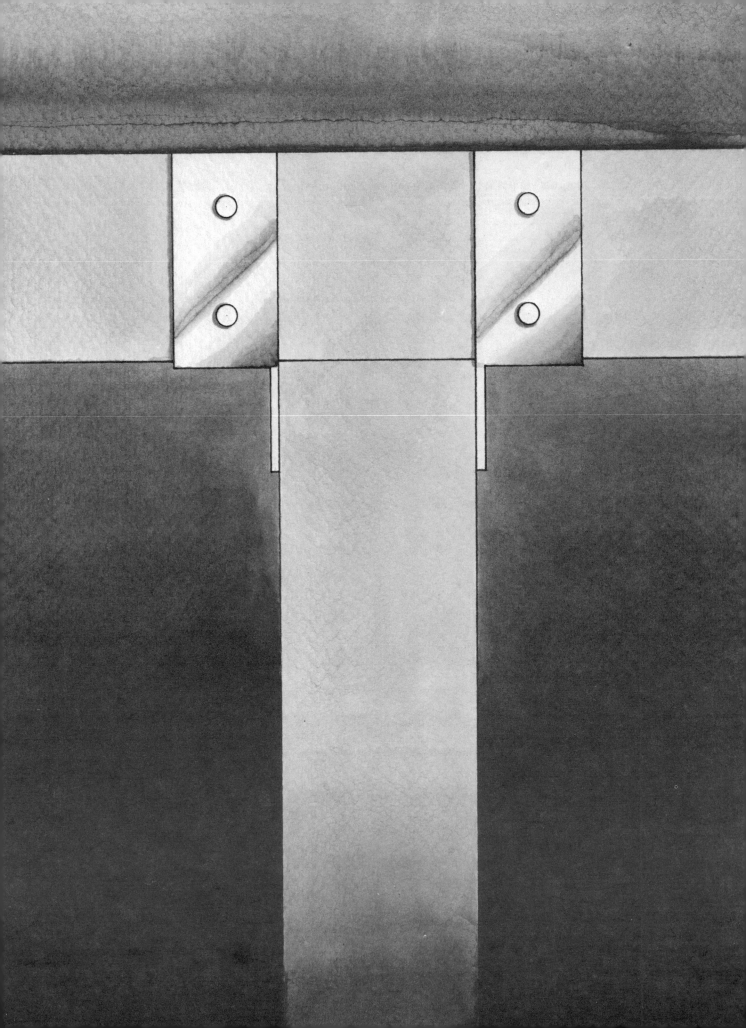

3 Rough Carpentry

The term **rough carpentry** applies to all the work a carpenter does to erect the structural wood framework of a house. Rough carpentry begins as soon as the foundation is ready to accept wood framing. It ends as soon as the skeleton of the house is completed and "skinned" with sheathing, ready for application of exterior and interior surface materials. The phase of work following rough carpentry is called **finish carpentry**.

The structural wood skeleton of a house consists of floor framing, wall framing, and roof framing. These elements work together to form a strong framework that serves two basic purposes: to withstand the effects of wind, water, and temperature, and to provide a base to which enclosing exterior materials and attractive interior materials are attached.

In floor framing the main horizontal structural members are **joists.** In wall framing, which includes not only exterior walls but also interior partitions between rooms, **studs** are the main vertical structural members. In roof framing, which includes ceiling joists above the uppermost floor level, the main structural members are **rafters.** Rafters run at an angle to all other framing members.

The structural relationship of floor to walls to roof is extremely important, because they work together to achieve their final strength. Wood is a very versatile building material, but it also has its weaknesses. It is strong in **compression**—that is, it can carry great weights when those weights are applied *with* the grain, as on a wall stud. Wood is not nearly as strong in resisting **deflection**—that is, it will bend when weights are applied *across* the grain, as on a floor joist or roof rafter. Some species of wood are stronger than others in withstanding compression, while others are strong in withstanding deflection. To increase the strength of wood frames, they are covered with sheet materials, of which the strongest is plywood. The sheet material applied across floor joists is called a **subfloor.** The sheet materials applied to wall studs and roof rafters is called **sheathing.**

Because the strength and durability of any building depends on the joists, studs, and rafters in the structural framework, these members must be carefully specified by the designer or architect according to the requirements of building codes. It is the responsibility of the designer to specify the size and grade of lumber for all major wood framing members. These members must be able to support the loads and stresses that are likely to be put on them during the life of the building. It is the responsibility of the builder to order framing materials that meet the designer's specifications.

When framing carpenters arrive at the site of a

new house to begin work, they should find all foundation work completed and ready to accept floor framing members. They should find the raw materials with which to begin work, stored and protected at the site. And they should have with them a set of working drawings for the building to be built.

What do rough carpenters need to know to do their job? First, they must be able to read the drawings, so that they use the specified materials where and how they are required. They must be able to distinguish between lumber sizes and grades, and know which size and grade to use for what structural purposes. All lumber and sheet materials, such as plywood, are stamped with their grade. Rough carpenters must be able to recognize defects in lumber and know how to make maximum use of the materials on hand without weakening the structure. They must know how to use the tools to cut, trim, measure, mark, and install framing members.

Perhaps most important of all, framing carpenters must know the proper methods for assembling and attaching the various pieces that make up the structural framework. You can be an adequate carpenter if you know and follow sound framing practices. You become a good carpenter when you use the correct methods and understand the reasons why these methods are correct. You become an excellent carpenter when you know not only the methods and the reasons behind them, but also can resolve a carpentry problem without creating problems for others later in construction. This ability to solve a construction problem correctly comes with experience, but the experience is useless unless you have the necessary knowledge of procedures and the reasons behind them.

6

Floor Framing Systems

Floor, wall, and roof framing systems may be combined in several different ways. By far the most common structural framing system is **platform framing,** also called **western framing.** In some parts of the country **balloon framing** is preferred, especially for two-story houses. Sometimes platform and balloon framing are combined to achieve the advantages and minimize the drawbacks of both systems, as in split-level houses. **Braced framing,** once almost as common as balloon framing, has virtually passed out of existence and is mentioned only briefly here. The other framing system—called **post-and-beam** or **plank-and-beam**—is different enough from platform and balloon framing that it is covered in Chapter 11.

Platform Framing

Of all structural systems, platform framing is the easiest and simplest to build. The entire floor frame is built first, including the subflooring. Thus the floor surface serves as a platform for building walls. All structural framing members lie above the level of the top of the foundation wall.

A box sill rests on the foundation wall. A **box sill** (Fig. 6-0.1) consists of a single 2 × 6 sill plate attached to the foundation wall, plus header joists and edge joists set upright on the sill plate. Header joists, edge joists, and regular joists are all the same size lumber. Where joists meet at a wood beam or built-up girder, they are in the same line across the house. This makes application of subflooring easy. Subflooring extends to the edges of the floor framing structure.

Because the floor frame and wall frames do not interlock, platform framing is weak in resistance to high winds. Yet because of its simplicity, platform framing has become almost a national standard for one-story

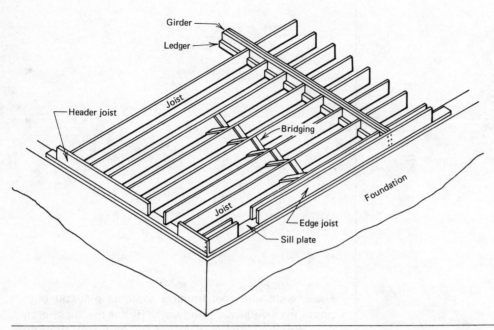

Fig. 6-0.1 In platform framing the sill plate is a single-thickness member. Header joists and edge joists form the perimeter of the box sill, and the tops of girders and all floor joists are at the same level. Joists line up at girders. Subflooring covers the floor framing members, and walls rest on the subfloor.

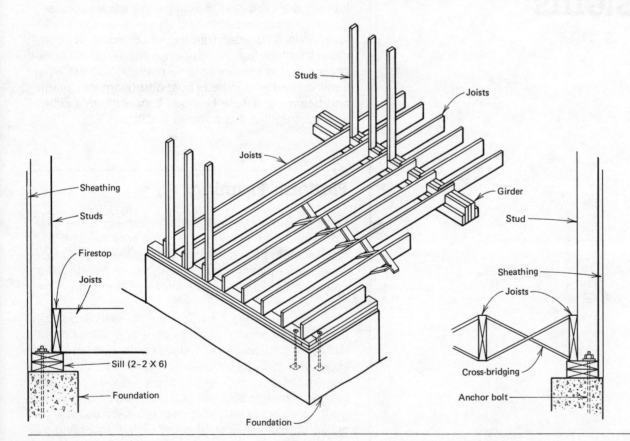

Fig. 6-0.2 In balloon framing the sill plate is doubled. Both floor joists and exterior wall studs rest on the sill plate. Steel girders fit into girder pockets, and their tops are flush with the top of the foundation wall. Joists overlap at the girder. Built-up beams may also be set in this way. More often they rest on the foundation wall, and are notched to fit over sill plates. Joists are also notched to fit over ledgers attached to the beam, as shown. Note that joists do not line up, and studs of center partitions fit against one joist.

houses. It is less desirable in two-story houses, not only because of weak wind resistance but also because it is the least desirable of all systems from the standpoint of wood shrinkage.

All lumber shrinks after it leaves the mill. It shrinks the least with the grain—that is, in length—and the most in cross section—that is, in thickness and width. Shrinkage across the width of flat members, such as header joists and edge joists, can cause the biggest problems by eventually forcing interior and exterior wall materials to crack from the stress.

Balloon Framing

Balloon framing has few members that are seriously affected by shrinkage. Joists rest on a doubled sill plate (Fig. 6-0.2). Studs in exterior walls rise from the sill plate and lie outside of edge joists along side walls. At front and rear walls, joists end not at a header but at the edge of the sill. Joists overlap at girders and

help to support studs of center partitions. Subflooring stops at the edges of studs in front and side walls. To prevent the spread of fire in a basement, fire-stops are required between joists on the inside edges of all sill members.

Braced Framing

Modern braced framing is similar to balloon framing as far as floor structure is concerned. The chief difference is at girders. Note in Fig. 6-0.3 that studs fit between overlapping joists, not beside them as in the balloon framing a system.

To build floor framing systems, you need only a few tools, but you will use all of them many times. You need a chalk line, steel tape, folding rule, try square, framing square, power drill, power saw, crosscut saw, carpenter's level, plumb bob, hammer, and of course a carpenter's pencil. Special tools for special jobs are listed in the individual learning units that follow.

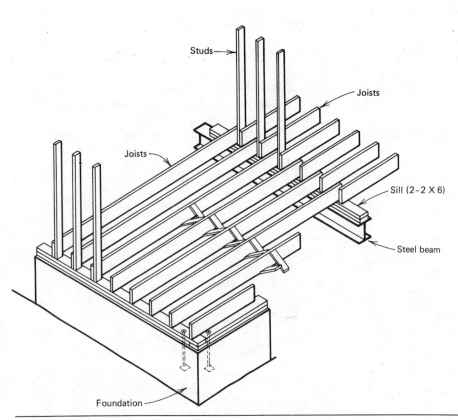

Fig. 6-0.3 Braced floor framing is the same as balloon floor framing except that studs of center partitions fit between floor joists over the girder.

Questions

1 The ends of wood girders equipped with ledgers must be notched in the (balloon, platform) framing system. (Circle one.)
2 The tops of girders and joists are at the same level in the (balloon, platform) framing system. (Circle one.)
3 Name two advantages of the platform framing system.
4 Name two disadvantages.

Unit 6-1   Cut Sill Plates

Of all the framing members that a carpenter installs, none is more important than the sill plate. Why? Because the plumb and level of all other framing members depend on the sill plate being installed with all pieces level, all corners square, and all set equidistant from the edge of the foundation.

In conventional construction sill plates are 2 × 6s. In special cases, however, they may be 4 × 6s, or even 2 × 10s or 2 × 12s set on edge and topped with a 2 × 4 laid flat. Sill plates may lie flush with the outside edge of the foundation wall, flush with the inside edge, or be inset from either edge (Fig. 6-1.1). The exact location is important in determining the lengths of individual sill members and the locations of holes for sill bolts that connect plates to the foundation wall.

When sill plates are a single thickness, it is customary to let the plates on the longest wall extend from side to side of the foundation and to fit shorter sill plates between them (Fig. 6-1.2, left). When sill plates are doubled, however, it is preferable to let the bottom plates on the shorter walls extend to the edges of the foundation (Fig. 6-1.2, right). The top plates then overlap at the corners and are cut the same way as single thickness plates.

Some sills will be too long to cut from a single length of lumber. When you need to use more than one length, neither should be less than 8' long. Study the framing drawings to determine the positions of joists or studs that rest on the sill plate, and plan the locations of butt joints so that they do not fall under a joist or stud. The reason is to avoid hitting nails in sill plates when you attach later structural members to them.

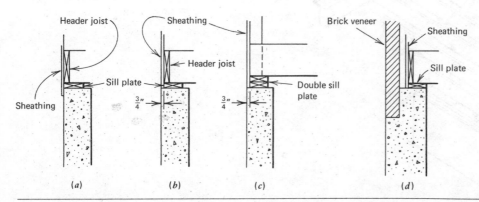

Fig. 6-1.1 Exterior wall materials and construction at the foundation wall determine the locations of sill plates in relation to the outside of the foundation wall. If sheathing overlaps the foundation, as in (a), sill plates are flush with the foundation. If sheathing is flush with the foundation, as (b) and (c), sill plates are set back $\frac{3}{4}''$. If the exterior wall is brick veneer, the foundation wall must be thicker, and sill plates are set flush or almost flush with the inside edge of the wall.

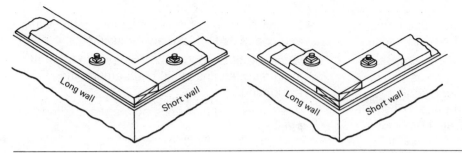

Fig. 6-1.2 When a single sill plate is required, the longest sill plate goes atop the longest wall (left). When a double sill plate is required, the top plate is cut this way, but the bottom plates are cut the opposite way (right) so that the plates overlap.

Questions

1 With what types of framing systems do you install a single 2 × 6 sill plate? When do you use a pair of 2 × 6s or their equivalent?

2 When the surface of sheathing is flush with the outside of the foundation wall, what is the common setback for sill members?

3 With a brick veneer wall, to which edge of the foundation wall do you set sill plates flush?

4 Where should any joints in long sill plates fall?

Procedure

1 From the drawings determine the size and setback of sill plates.

2 Mark the setback by snapping chalk lines on the top of the foundation wall (Fig. 6-1.3).

3 Check the locations of chalk lines where they cross at corners. Corners must be square, and diagonals equal from corner to corner.

4 Measure the length of the sill plate that extends the full length of the foundation wall.

5 If possible, cut a single piece of plate stock to fit. If more than one plate length is required, study framing plans to determine the best lengths. Cut all ends square for a tight butt fit between lengths and at corners.

6 Place plates upright along the chalk line, and mark the locations of sill bolts (A in Fig. 6-1.4).

Fig. 6-1.3 (Photos Drew Leviton, Atlanta.)

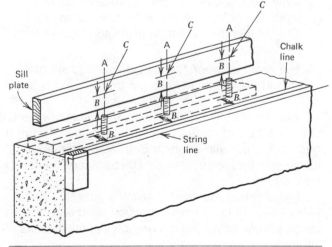

Fig. 6-1.4

7 Measure the distance from the chalk line to the center of each bolt (*B* in Fig. 6-1.4), and transfer these dimensions to sill plates.

8 Where two lines intersect (*C*), drill holes for sill bolts. Holes should be large enough so that you can adjust the position of the sill plate, but small enough so that washers with the sill bolts bear fully all the way around the hole. A hole $\frac{7}{8}''$ in diameter is standard.

9 Set the plate in position on the foundation wall, and check its fit. Make any adjustments.

10 When the fit is good, follow steps 4 to 9 for all remaining lengths of sill plate around the foundation.

11 Verify overall dimensions before proceeding further.

Questions

1 Which sill plate do you cut and drill first?

2 If sill plates are set back $\frac{3}{4}''$, and sill bolts are centered in an 8″ foundation wall, what should the dimension be from the outside edge of a 2 × 6 sill to the centerlines of bolt holes?

Unit 6-2 **Install Sill Plates**

No sill member should ever be in direct contact with the top of a foundation wall. A foundation seal under the sill plate serves three purposes:

1 It provides a level base for the sill. No matter how carefully a foundation wall is finished, it is never exactly level. Yet sill members must be.

2 If moisture somehow gets into exterior walls, it tends to pool on the top of the foundation wall. The sill members must be protected against this moisture.

3 The seam between foundation wall and sill plate should be airtight to keep out drafts.

Two types of foundation seals are commonly used. One is a bed of mortar; the other is a manufactured gasket treated with asphalt. Both seals are about $\frac{1}{2}''$ thick, and they compress as sill bolts are tightened (Fig. 6-2.1).

The only tool you need to install a gasket is a sharp utility knife. To lay a mortar bed you need a trowel. To install sill plates you need a wrench that fits the nuts of sill bolts.

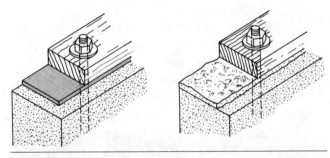

Fig. 6-2.1 The seal between foundation wall and sill plates may be an asphalt-treated gasket (left) or a thin bed of mortar.

Procedure for Installing a Gasket Seal

1 Read the manufacturer's instructions to determine the final thickness of the gasket when it is compressed. **Precaution.** You need to know this thickness in order to set girders at the correct

height. Final thickness is about ¼″ under the weight of a one-story house.

2 Cut gaskets for the longest walls to length. Cut other lengths about 1″ longer than necessary.

3 Lay the pieces in position along the chalk lines to determine where sill bolts will penetrate. Let longer side pieces overlap slightly at both ends.

4 With a sharp knife cut holes for sill bolts. Aim for a tight fit.

5 Fit gaskets over bolts and along chalk lines.

6 Trim off excess where ends overlap at corners. Again aim for a tight fit.

7 Set sill plates in place on the gasket.

8 Fit washers over sill bolts. Then tighten nuts until gaskets are compressed to the thickness recommended by the manufacturer.

9 With a carpenter's level, check the level of the sill around the entire foundation. Loosen or further tighten nuts to make the final adjustment.

10 Toenail sill plates at corners and joints, as shown in Fig. 6-2.2).

11 Add a second sill plate, if required. Nail as shown in Fig. 6-2.3.

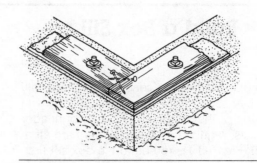

Fig. 6-2.2

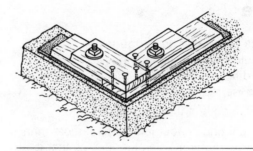

Fig. 6-2.3

Procedure for Installing a Mortar Seal

1 With a hose dampen the top of the wall. This dampening prevents the wall from absorbing moisture from the mortar too quickly.

2 Mix a thin, creamy paste of cement and water as a bonding coat. **Note.** Bonding coats are also manufactured, and sold in quarts and gallons.

3 Paint the top of the wall with the bonding coat.

4 Before this coat dries, trowel on a rich layer of mortar, to a depth of about ½″.

5 Set sill plates over sill bolts.

6 Add washers, and tighten nuts enough to squeeze out a little mortar on both sides.

7 Check the level of sill members around the entire foundation. Adjust level by further tightening nuts where necessary.

8 Trowel off excess mortar, and shape the edges of the mortar bed for drainage.

9 Toenail sill plates at corners and joints, as shown in Fig. 6-2.2.

10 If required, add a second sill plate that overlaps, as shown in Fig. 6-2.3.

Questions

1 What is the common thickness of a foundation seal before bolts are tightened?

2 From the standpoint of the carpenter, what is the most important reason for installing a foundation seal?

3 What do you think is the purpose of washers on sill bolts?

Unit 6-3   **Build a Box Sill**

The box sill in platform framing consists of a single sill plate, topped by header and edge joists set flush with the outside edge of the sill plate. The tops of header joists, edge joists, regular joists, and girders are all at the same level (Fig. 6-3.1). You build the box sill first, then set any girder in place.

Procedure

1 Select stock for header and edge joists. All pieces should be as straight as possible and equal in width. Use the longest pieces available that meet these requirements.

2 Measure the length of the longest sill plate.

3 If possible, cut one header joist to fit. If more than one member is required, plan the joint so that it falls between regular joists. Cut all ends square for a tight butt fit between members and at corners.

4 Determine the lengths of the two butting edge joists, and cut them to length.

5 At one corner place the header joist and edge joist flush with the outside edges of the sill plates (Fig. 6-3.2).

6 Toenail both joists to the sill, and nail the two joists together with three 16d nails.

7 Toenail the open end of the front header joist to the sill, then continue to toenail toward the corner.

Fig. 6-3.1 The tops of box sill members, built-up beams, and joists must all be at exactly the same level. (Photos Drew Leviton, Atlanta.)

Make sure the side of the header joist stays flush with the edge of the sill.

8 Where header or edge joists butt, toenail the ends together (Fig. 6-3.3). Add a temporary brace if necessary to hold the members in position.

Fig. 6-3.2

Fig. 6-3.3

9 Continue in this manner until the box sill is completed. **Note.** Although header and edge joists should be reasonably plumb, it is not necessary that they be exactly plumb at this time except at corners.

10 Recheck your work before marking joist locations.

Questions

1 What size nails should be used to connect header joists and edge joists?

2 Which are usually the longer, header joists or edge joists? Why?

3 Name an advantage to having splices in header joists fall between regular joists.

4 Name an advantage to having splices fall at the ends of regular joists.

Unit 6-4   Fabricate a Wood Girder

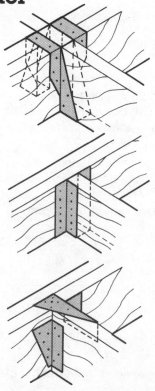

When the distance between sill plates is greater than can be spanned by single lengths of lumber, some intermediate support must be provided for joists. This support appears in framing plans as a girder.

Girders may be steel I-beams, solid timbers, or built-up beams. In many cases one type may be substituted for another, although their dimensions will be quite different. Ends of floor joists may be supported on the tops of girders, on **ledgers** attached to the sides of girders, or on metal hangers manufactured for this purpose (Fig. 6-4.1). The type of girder and the method of joist support must be determined before the foundation wall is completed.

A built-up beam (Fig. 6-4.2) is made of joist material. It is fabricated from two to four thicknesses, depending on the girder's span and support requirements. In most residential work, two or three planks form the girder. Splices between lengths must be staggered so that no two splices lie within 4' of each other. The planks are fastened together with 10d (two thickness) or 20d (three thickness) nails. Ledgers are added to the beams where necessary.

Drawings usually give the size of girder required, but they do not give its length. The length must be cal-

Fig. 6-4.1 Joist hangers are manufactured in shapes to fit almost any building condition.

culated from overall dimensions of the house, reduced by the thickness of wall construction at the ends of the girder. With platform framing, for example, the length of the girder is equal to the overall dimension, less the setback of the sill plate at each end, less the thickness of edge joists. With balloon framing the ends of a girder are notched to fit over a double sill, and lie flush with the outside edges of sill members. Therefore the length is the overall dimension less two setbacks.

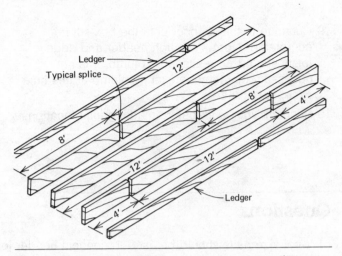

Fig. 6-4.2 Members of built-up beams are cut so that splices fall over supporting columns or posts.

Procedure

1 From the framing plan or sections, determine the size of girder required.

2 From floor plan and construction details, calculate the girder's overall length.

3 From dimensions between supporting posts or columns, determine the lengths of individual pieces. Plan splices so that they fall over grider supports.

4 Select girder stock carefully. Members must be straight, equal in depth, and as free as possible of defects such as large knots and splits.

5 Cut members to length, with all ends carefully squared.

6 Notch ends if required.

7 Lay out members with crowns up, ends butted tightly together, and edges flush and in a straight line (Fig. 6-4.3).

8 Beginning at one end, nail individual members together as required by local building codes, or in accordance with local practice. Drive the nails at a slight angle to increase holding power. Nail in a staggered pattern (Fig. 6-4.4).

9 If the girder is short enough and light enough to be positioned manually, completely fabricate the beam from end to end. If not, set supporting posts in place, and fabricate from support to support (see Unit 6-5).

Fig. 6-4.3 (Photos Drew Leviton, Atlanta.)

Questions

1 What is the minimum permissible spacing between splices?

2 What size nails are recommended for assembling a built-up beam $4\frac{1}{2}''$ thick? For a built-up beam 3" thick?

3 In what floor framing system must built-up beams be notched at their ends?

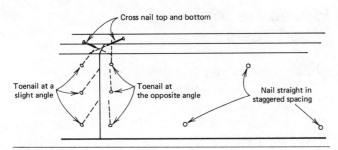

Fig. 6-4.4

Unit 6-5 **Install Posts and Columns**

Any long horizontal member, such as a girder, will deflect somewhat over a period of years. Under the weight of a floor system, wood fibers in a beam or built-up girder will stretch, and the beam will sag in the middle unless it has support. Even a steel I-beam will deflect a little.

It is possible to build and install a girder that would not deflect enough to notice—but it would be tremendous in size, be costly to build, and would take up too much headroom. It is much less expensive and better construction practice to use smaller girders and provide intermediate support between foundation walls.

The size of the girder and the spacing of supports is decided by the designer of the house. Details are shown on the foundation plan. However, these details tell only the locations and sizes of supports. The carpenter must know how to install them to meet structural requirements.

In Unit 5-4 you learned that vertical supports for girders rest on concrete caps or pedestals. These bases should be fully cured by the time you are ready to install posts and columns. At the time the pedestals were poured, some device should have been installed for anchoring the post. It may be an anchor bolt, a piece of pipe set vertically in the concrete, or a bracket of various types (see Fig. 5-4.1).

Similarly, anchor bolts are set into a concrete cap when it is poured. The base of the steel column is a flat plate with holes drilled in it for the anchor bolts. Columns may be round, H-shaped or I-shaped.

At their tops wood posts are usually attached to wood girders with plywood gussets or nails driven in at an angle. Lally columns have another flat plate at their tops, which is drilled for attachment to a steel or wood girder.

Girders fabricated to span from foundation wall to foundation wall must be supported as soon as the girder is positioned and before any deflection can take place. Girders built in sections to span only from support to support can be installed only after supports are in place.

No special tools are needed to install wood posts or lally columns. However you will need a length of cord, a chalk line, folding rule or steel tape, and a wrench to tighten fasteners.

Procedure for Installing Steel Columns

1 Stretch a taut cord from wall to wall along the centerline of the girder to mark the height of its bottom.

2 On concrete caps mark the location of the centerline of each column.

3 Measure the vertical distance from the cord to each cap. This is the length of column required, including top and bottom plates.

4 After columns have been fabricated, set the bottom plate of each column over bolts in the base.

5 Fit washers in place, and attach nuts to bolts. Fasten them finger-tight.

6 Check each column for vertical plumb.

7 Tighten bolts with a wrench.

Procedure for Installing Wood Posts

1 Stretch a taut cord from wall to wall along the centerline of the girder, to mark the height of its bottom (Fig. 6-5.1).

2 On pedestals mark the location of the centerline of each post.

3 Measure the vertical distance from the cord to each pedestal. This is the length of post required.

4 Cut posts to length.

5 Drill the post at its base, as required, to accept anchor bolts, pipes, or horizontal bolts.

6 Center each post on its pedestal, and attach it loosely at the base.

7 Check each post for plumb.

8 Complete the attachment.

Fig. 6-5.1 (Photos Drew Leviton, Atlanta.)

Questions

1 Why do you think it is necessary to measure from the cord to the top of each cap or pedestal?

Unit 6-6 Install Girders

Girders are not only heavy and awkward to work with; they must be installed very carefully and accurately. Their tops must be absolutely level and at the same height as points of joist support in foundation walls.

These points of support vary with the type of girder and the method of floor framing (Fig. 6-6.1).

• If the girder is an I-beam, it always fits into a girder pocket in foundation walls. Joists are supported on sill members attached to the top flange of the beam. This top flange must be at the same height as the top of the foundation wall, plus the thickness of a gasket or mortar seal.

• If the girder is wood—either a timber or a built-up beam—and joists rest on top, as in braced framing, then the *top* of the girder is at the same level as the *top* of the sill plate. A girder of this type must fit part way into a girder pocket.

• If the girder is wood and joists are supported by

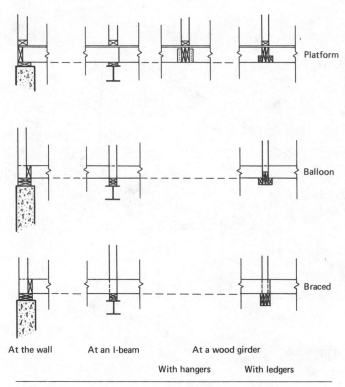

Fig. 6-6.1 Sections through foundation wall, beams, and girders show the level of the bottoms of floor joists in the various framing systems and methods of attachment.

Fig. 6-6.2 The simple temporary support used here consists of a 2 × 4 cut to length of the final support, plus another 2 × 4 nailed at right angles as a stiffener. (Photos Drew Leviton, Atlanta.)

joist hangers, then the *bottom* of the girder is at the same height as the *top* of the sill plate.

• If the girder is wood and joists are notched to fit over ledgers, then the *bottom* of the girder in platform framing is level with the top of the *sill plate*. In balloon framing the *bottom* of the girder is level with the top of the *wall*.

The construction sequence of girders and their supports varies with conditions, and sometimes with the builder's own preference. Steel beams are usually in-

stalled after supports are in place, but before sill members are attached.

Wood girders, on the other hand, are always installed after sill members are attached to the foundation wall, because the height of the sill establishes the height of the girder. If the girder is one continuous member, it is common practice to rest it on temporary supports (Fig. 6-6.2) until floor framing is completed.

If the girder consists of several timbers or a built-up beam to be spliced in place, then supports must be installed before the girder.

Questions

Underline the correct answer in the parentheses.

1 A steel I-beam used as a girder (always, usually, seldom) fits into a girder pocket in the foundation wall.

2 In balloon framing the (top, bottom) of a built-up girder with ledgers is level with the top of the (foundation wall, sill).

3 In platform framing the (top, bottom) of a built-up girder supporting joists on hangers is level with the (top, bottom) of the sill.

Procedure for Setting a Continuous I-Beam

1 On the foundation wall outside the girder pockets, place a scrap of material equal in thickness to the expected thickness of the mortar seal or gasket.

2 Swing the I-beam slowly and carefully until it is suspended directly above the girder pockets.

3 Gently lower the beam into position until it is in the pockets.

4 Set a carpenter's level across the end of the beam and the piece of scrap.

5 a. If the I-beam is too low, add a **shim,** made of steel plate no thinner than #10 gauge, under the beam in the pocket. Recheck the level. If the I-beam is now too high, follow step 5b.

5 b. If the beam is too high, measure the distance between the top of the beam and the top of the wall. Use this dimension as a guide to the final thickness of the seal.

6 When the proper height is established, check the level of the beam, both along and across the top flange, at the ends, and at all supports.

7 At each column align holes in the cap plate with holes in the flanges of the beam.

8 Place bolts through the holes, and add washers and nuts. Nuts should be finger-tight (Fig. 6-6.3).

9 Check each column for plumb.

10 Tighten all bolts with a wrench.

11 After sill plates have been installed on the foundation wall, attach wood sills to the girder with wire ties. Tie every 4′, with the twisted ends at the side of the beam between flanges.

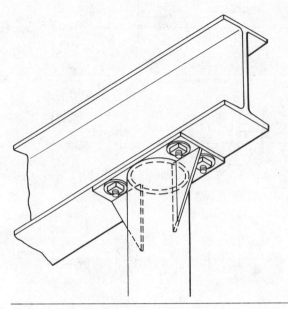

Fig. 6-6.3

Procedure for Setting a Wood Beam in a Girder Pocket

1 Remeasure the girder's length, and check the ends for square.

2 Gently lower the girder into the pocket. Center it so that you have the required $\frac{1}{2}''$ clearance on each side.

3 Place a carpenter's level across the girder and sill.

4 a. If the girder is low, raise the end and drive a shim between the girder and the floor of the pocket until level is correct. The shim must fully support the end of the girder.

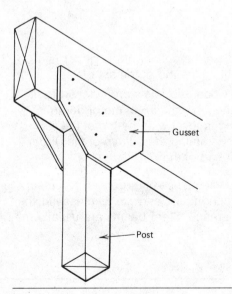

Fig. 6-6.4

4 b. If the girder is high, determine how much too high. Then remove the girder and notch the ends this amount.

5 When the proper height is established, check the beam for plumb and level, and for fit at center supports.

Fig. 6-6.5 (Photos Drew Leviton, Atlanta.)

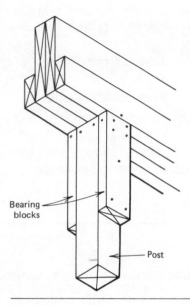

Bearing blocks

Post

Fig. 6-6.6

6 a. If the supports are steel columns, drill holes into the bottom of the girder through holes in the cap plate. Holes should be one size smaller than the lag screws used for attachment. After rechecking the plumb of columns, drive lag screws home.

6 b. If supports are wood posts, recheck each post for plumb. Complete the attachment with gussets (Fig. 6-6.4) or by toenailing.

Procedure for Setting a Built-up Beam on a Foundation Wall

1 If the foundation wall is built of concrete blocks, fill the cores with concrete directly below the ends of the girder all the way to the footing. With this step you are forming a concrete column to support the weight of the floor structure.

2 Fabricate the beam with ends notched to fit over side sills.

3 Determine the exact location of the girder where it rests on sill plates, and mark these locations.

4 Set the girder between the sets of marks.

5 Plumb the girder, and level it in both directions.

6 Attach the girder at foundation walls by toenailing through the sides and ends of girders into sill plates.

7 Check the fit at center supports.

8 Cut ledgers to length, and rip them to a width equal to the thickness of sill plates and the foundation seal.

9 Attach ledgers flush with the bottom of the girder (Fig. 6-6.5).

10 a. If supports are columns, attach them to girders with lag screws.

10 b. If supports are posts, toenail through posts into the girder. Then cut and attach bearing blocks, as shown in Fig. 6-6.6.

Procedure for Setting a Built-up Beam on a Sill

1 On side sills mark the exact location of the girder.

2 Verify that the girder will fit the space between edge joists.

134

3 Set the girder in position on the marks.

4 Set a level on the girder and edge joists.

5 a. If the girder is low, raise the end and level it with a pair of shims, one driven in from each side (Fig. 6-6.7).

5 b. If the girder is high, notch the ends slightly for a level fit.

6 Recheck level at both ends and at all supports.

7 a. If supports are columns, attach them to girders with lag screws.

7 b. If supports are posts, toenail through posts into the girder. Cut and attach bearing blocks (Fig. 6-6.6).

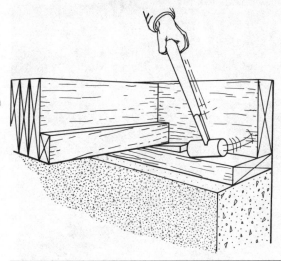

Fig. 6-6.7

Questions

1 When the ends of a girder rest directly on a concrete block wall, what is the first step you must take?

2 Why must a wood girder that is set into a pocket have clearance at the sides?

3 Why do you think ledgers are necessary with notched joists?

Unit 6-7 **Lay Out Joists**

With the platform framing system, the box sill and girders must be in place before you can mark joist locations. With the balloon framing system, girders and both sill members must be in place. Studs in the exterior walls may also be in place, however. Some builders prefer to assemble exterior walls on the ground and raise them into place before setting joists.

Whatever the floor framing system, the next step is to mark the locations of joists—either on the header joists and girders, or on the sills and girders.

In platform framing the subfloor runs under wall framing to the edges of the box sill. Therefore joist locations are measured from the corner of the box sill. In balloon and braced framing, the subfloor stops at wall framing. Therefore joist locations are measured from the inside edges of wall studs.

Procedure

1 Determine from the framing plan in which direction joists run. **Note.** They do not necessarily run the same way throughout the house.

2 a. From one corner of a box sill, measure off $47\frac{1}{4}''$ along the top of the header joist, and make a mark with your carpenter's pencil. This mark (Fig. 6-7.1) locates the third joist from the corner. **Note.** If you are right-handed, you will probably find it easier to begin at a right-hand corner and work to your left. If you are left-handed, it may be easier to work from a left-hand corner. The results should be the same.

2 b. From the corner of a doubled sill in balloon framing, measure off $3\frac{1}{2}''$, and make a mark. Then measure $47\frac{1}{4}''$ from this first mark to locate the third joist from the corner (Fig. 6-7.2).

3 From the mark for the third joist, measure every 16″ in each direction and mark, as long as joists are on a standard spacing. **Caution.** Make all measurements from your first mark. Do not measure 16″ and then move your tape. You increase the possibility of error.

4 On the floor framing plan look for any doubled joists under **parallel partitions**—partitions that run parallel to joists. They require extra support.

5 On the floor plan find the dimension to the

centerline of these partitions, and mark the centers on the sill.

6 On the floor framing plan, look for any joists not on standard spacing.

7 If the exact locations of these joists are indicated, mark the locations on the header joist or sill plate. If not, omit them until you can determine their exact location.

8 When all possible joist locations are marked, carry the marks down both sides of the header (Fig. 6-7.3) or sill plate (Fig. 6-7.4). Use your try square for this operation.

9 With an X, mark on which side of the lines the joists go.

10 Repeat steps 2 to 9 at the girder.

11 Before marking the opposite header joist or sill plate, study the framing drawing to determine how joists meet at the girder.

12 a. If joists are on the same line across the building, mark the second header joist as you did the first one. The Xs are on the same side of the lines (Fig. 6-7.5).

12 b. If joists overlap at the girder, mark the second sill plate as you did the first one. Then place Xs on the opposite side of the lines (Fig. 6-7.6).

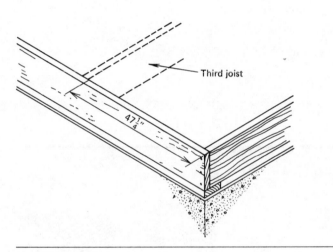

Fig. 6-7.1

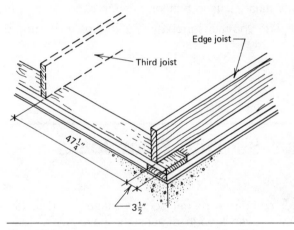

Fig. 6-7.2

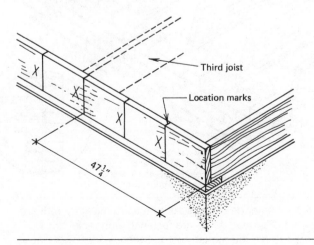

Fig. 6-7.3

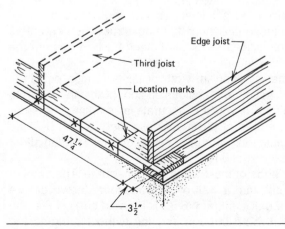

Fig. 6-7.4

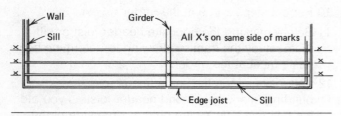

Fig. 6-7.5

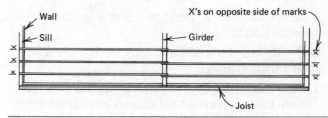

Fig. 6-7.6

Questions

1 In what framing system do floor joists line up from wall to wall?

2 Why must joists in balloon framing overlap at a girder?

3 Why do you measure only $47\frac{1}{4}''$ in Figs. 6-7.1 and 6-7.2 instead of 48''?

Unit 6-8   Trim Floor Joists

Lumber standards permit some variation in joist dimensions, even though dimensions have been standardized. These variations occur as a result of shrinkage. A 2×10 floor joists, for example, may vary in depth from $9\frac{1}{4}''$ (the minimum allowable) to $9\frac{5}{8}''$ (the maximum allowable).

Obviously if the joists used in a floor vary in depth from the maximum to the minimum, a level floor will be impossible. To set joists so that their tops are level across the entire floor framing structure, carpenters must select joist stock carefully, and trim or shim the ends of joists where necessary.

Some variation in width of joists is also possible. For this reason. dimensions on floor plans are shown to centerlines. If you are accurate in marking the locations of edges of joists, as described in Unit 6-7, the variation in joist width will not be great enough to affect future construction.

The ends of most joists are cut square. The ends of joists built into a wall of solid masonry, however, are cut on a diagonal (Fig. 6-8.1). Then, in case of fire, the joists will burn through and fall without bringing the masonry wall down at the same time.

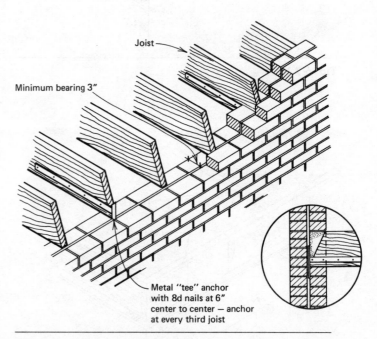

Fig. 6-8.1 The ends of joists set in a solid masonry wall are cut on the diagonal. Joists are fastened with metal tee anchors set into the mortar joint between wythes (vertical sections of masonry one unit thick).

Procedure for Leveling Joists

1 While joist stock is still bundled on the site, measure the ends of all joists.

2 Check every fifth or sixth joist to make sure both ends are the same size.

3 On the end of each joist mark the fraction of the width measurement. If, for example, the end measures $9\frac{1}{2}''$, write $\frac{1}{2}$ on the end.

4 After marking all joists, determine which fraction is the most common. Let it become your standard height on this job. For example, if you have 80 joists and 60 measure $9\frac{3}{8}''$, 15 measure $9\frac{1}{4}''$, and 5 are $9\frac{1}{2}''$, then you use $9\frac{3}{8}''$ as standard height.

5 Next sight down the length of all joist stock to look for any crown. On any joists with crown, mark the high edge with an arrow (Fig. 6-8.2).

6 Set aside all joists that are standard depth.

7 Mark "shim" near the ends of joists that are less than standard depth, and set them aside.

8 On the bottom side of over-depth joists, mark notches for trimming the excess where joists fit over sills and girders.

9 With a saw, cut to the depth of the notch (Fig. 6-8.3).

10 With a chisel or saw complete the notching (Fig. 6-8.4). The tops of all joists will now be at the same level.

11 If drawings call for ceiling material to be applied to the under sides of joists, mark on each joist the amount of shim required. In this case the joists with the maximum depth (step 4) become the standard.

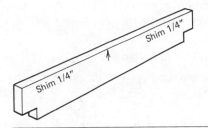

Fig. 6-8.2

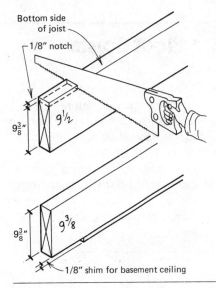

Fig. 6-8.3

Fig. 6-8.4

Procedure for Notching Joists at Ledgers

1 Determine the correct length of each joist.

2 Square both ends to this length.

3 Measure the distance from the top of the girder to the ledger.

4 Measure the thickness of the ledger.

5 Mark these measurements on the joist.

6 Make notches with a saw, being careful not to cut beyond the marks.

Procedure for Trimming Joists for Masonry Construction

1 Determine from drawings or local building codes the amount of bearing surface required for each joist on a masonry wall.

2 Mark off and square this distance across the joist (Fig. 6-8.5).

3 Draw a diagonal from the bottom end of the joist to the top of the joist where the line meets it (Fig. 6-8.6).

4 Cut along this line. This is called a **fire cut**.

5 Use the first joist cut as a pattern for other joists in the same floor.

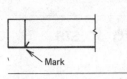

Fig. 6-8.5

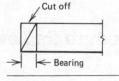

Fig. 6-8.6

Questions

1 Why is it wise to mark the crowned edge of joists with an arrow?

2 Why should cuts for notches not go beyond the cut marks?

Unit 6-9 **Set Floor Joists**

The tasks of marking joist locations and trimming and notching joists take as much time as setting joists in place. Although one person can set joists by working alone, the work goes faster, smoother, and more accurately when two people work together.

It is standard procedure to set standard joists on regular spacing first, then set standard joists on irregular spacing, and finally to set doubled joists under parallel partitions and around floor openings.

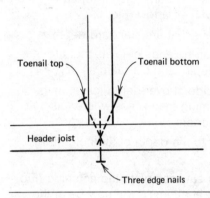

Fig. 6-9.1

Procedure

1 Position any shims for undersized joists.

2 Set each joist against its marks at both ends. Make sure that it is on the correct side of the line, and any crown is up.

3 Holding the joist as plumb as possible, drive one nail part way in at an angle near the bottom of the joist. With platform construction nail through the header joist into the end of each joist. Where there is no header joist, nail through the end of each joist into the sill.

4 Check the end of the joist for plumb and level. Make minor adjustments in placement with a hammer. Do not attempt to make a major adjustment with a hammer. Pull out the nail and start over.

5 a. With platform framing, complete the nailing by driving four additional nails, as shown in Fig. 6-9.1.

5 b. With braced or balloon framing, drive two additional nails at an angle through the joist and into the sill on the side away from future studs. Be sure not to knock the joist off its mark on the sill.

6 a. At griders with ledgers, set notched joists on spacing lines. Toenail at top and bottom through the end into the girder, and through the sides into the girder and ledger (Fig. 6-9.2).

6 b. When joists overlap at girders, toenail through the outer side of each joist and into the girder. Then nail the two joists together (Fig. 6-9.3).

6 c. When joists butt against girders, nail joist hangers to the girder through holes provided in flanges. Then drop the joist into the hanger, and nail into the joist through the hanger.

7 With platform framing, cut and fit wood splice plates between joists at splices in the header joist (Fig. 6-9.4).

8 Also with platform framing add a joist $\frac{1}{2}''$ from each edge joist to support framing for side walls (Fig. 6-9.5). When these joists are in the way of heating ducts, add spacers between edge joists and adjacent joists to provide the necessary support (Fig. 6-9.6).

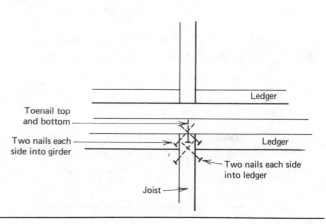

Fig. 6-9.2 (Photos Drew Leviton, Atlanta.)

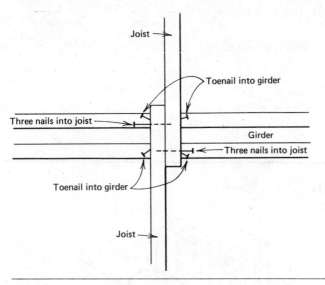

Fig. 6-9.3

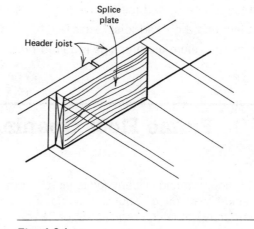

Fig. 6-9.4

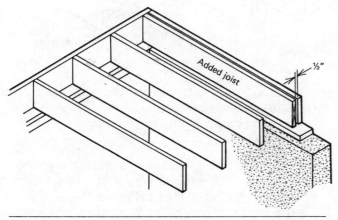

Fig. 6-9.5

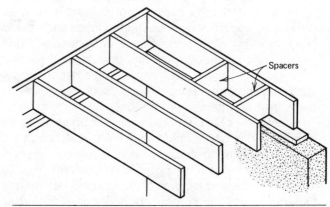

Fig. 6-9.6

9 After all joists are in place, recheck and adjust header joists and edge joists for straightness.

10 Sight across joist tops to check their level. Remove and reset any joists placed with crowns down.

11 Recheck levels of joists where they meet header joists, girders, or each other. Eliminate high spots.

12 Check all joists for plumb. A cocked joist prevents a level subfloor.

13 Look for bowed joists. Correct bow by nailing a 1 × 6 across half a dozen joists at standard spacing (Fig. 6-9.7).

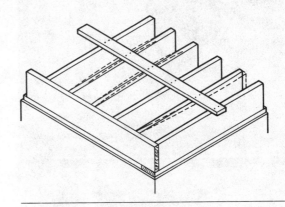

Fig. 6-9.7

14 When required, install blocking between joists over girders and firestops between joists at sills. A **firestop** is blocking that prevents the spread of fire.

Questions

1 What does the extra framing, such as that shown in Figs. 6-9.5 and 6-9.6, prevent?

2 Why does toenailing provide a stronger connection than straight nailing?

3 What problem can a bowed joist cause?

4 What problem can a cocked joist cause?

5 In what framing system are firestops required between joists at sills?

Unit 6-10 **Frame Floor Openings**

Over most of a floor framing system, joists march from side wall to side wall on a regular spacing. Some joists, however, may not fall on regular 16″ or 24″ centers; this occurs under two common conditions: (1) when joists support a heavy load, such as a parallel partition, and (2) when joists frame an opening in the floor, such as for pipes, ducts, stairways, or chimneys.

Parallel partitions that do not contain piping or ducts are supported by doubling the joists and nailing them to each other across their entire span. When parallel partitions do have ducts or pipes in them that enter from below, the doubled joists are separated by spacers or spreaders (Fig. 6-10.1). Floor framing plans normally state the space to allow between the joists. The centerline of the doubled-joist assembly should lie directly beneath the centerline of the parallel partition.

Around large openings, such as a chimney or stairway, all members must be at least doubled (Fig. 6-10.2). The size and location of the opening determines the amount of framing required. Joists used to narrow an opening are called **trimmers,** or sometimes **carry-**

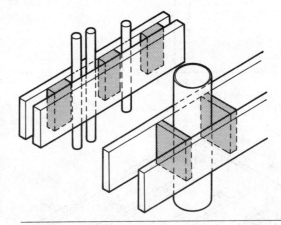

Fig. 6-10.1 Spreaders between doubled joists allow room for pipes and ducts but maintain the required stiffness of the floor structure.

ing joists. Members that run across the heads of openings are called **headers.** Short joists that butt against headers are called **tail joists.**

The method of attachment of trimmers and headers varies with local codes and building practices. Nailing

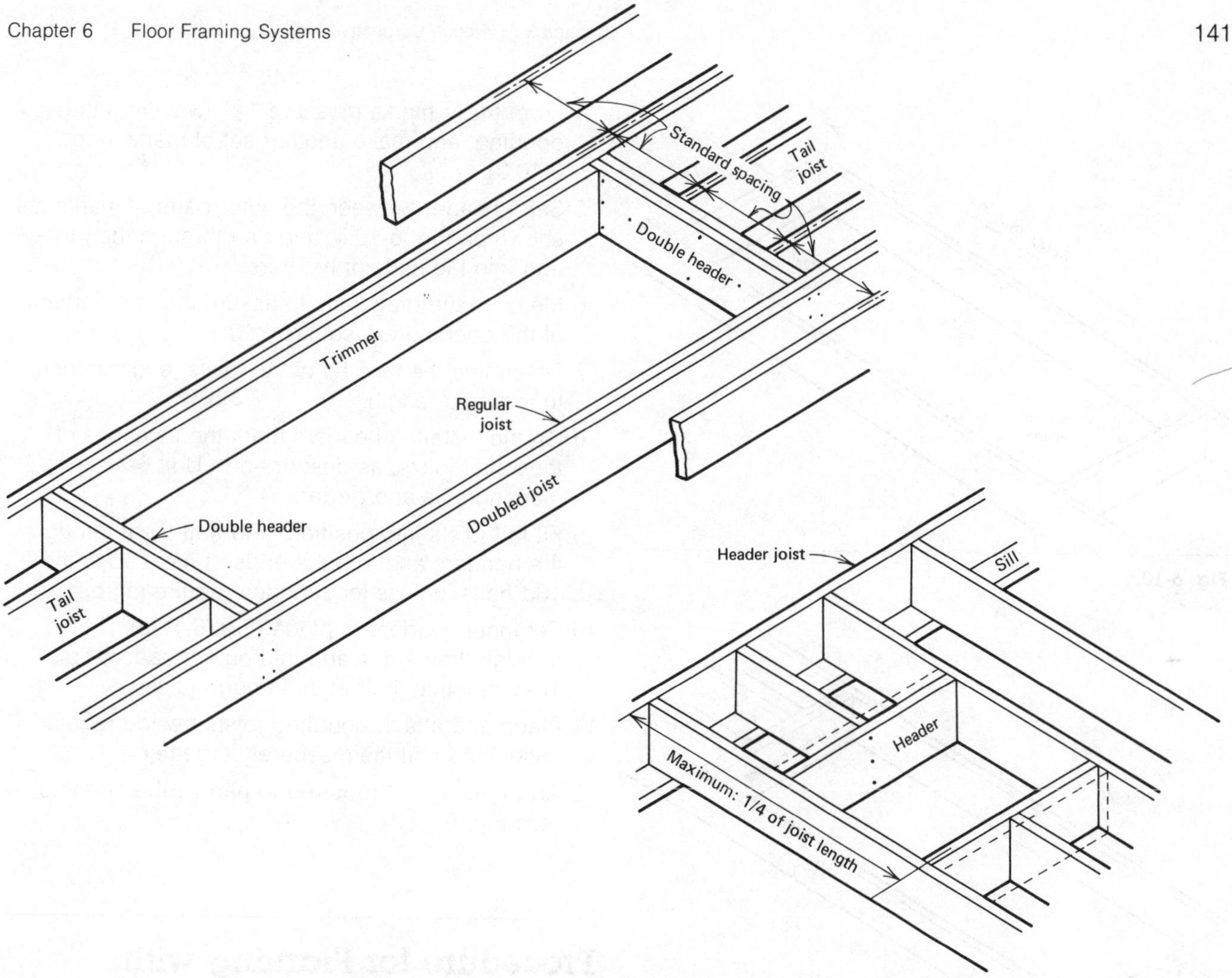

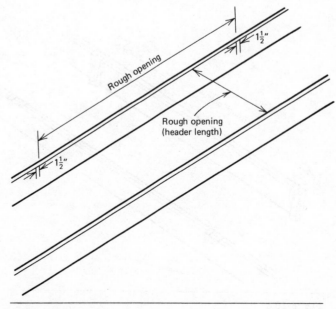

Fig. 6-10.2 Around large openings all floor framing members must be doubled. Small openings near joist supports may be framed with single headers and joists.

is usually permitted, but the number of nails and the angle of nailing vary. Joist hangers may be permitted or even required. If so, the framing operation becomes much simpler.

Procedure for Framing Without Joist Hangers

1 Fit and attach regular joists nearest to the opening, if this has not already been done.

2 Cut all header pieces to length. Their length is the distance between full-length joists and is equal to the width of the rough opening (Fig. 6-10.3).

3 On the joists mark the position of the length of the rough opening; follow the dimensions in the drawings.

Fig. 6-10.3

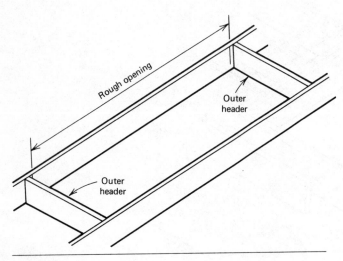

Fig. 6-10.4

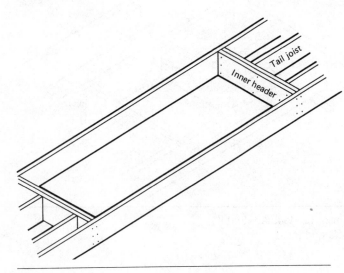

Fig. 6-10.5

4 From these marks measure $1\frac{1}{2}''$ away from the opening, and make another set of marks (Fig. 6-10.3).

5 Set a header between the outer pairs of marks, as shown on Fig. 6-10.4, and nail through the joists and into the ends of headers.

6 Measure the diagonals to assure that the corners of the opening are square.

7 Determine the lengths of tail joists, and cut them to required length.

8 On the installed headers mark the locations of these tail joists, as described in Unit 6-7 for marking sills and girders.

9 Fit tail joists into position, and nail them through the headers and into the ends of joists. Use three 16d nails, unless local codes require more.

10 Set inner headers in place (Fig. 6-10.5). Nail through their sides and into outer headers first. Then nail into their ends through joists.

11 Place and attach doubling joists beside regular joists, and nail them securely together.

12 Add trimmers, if required to narrow the opening (see Fig. 6-10.2).

Procedure for Framing with Joist Hangers

1 To regular joists nail the doubling joists beside the opening.

2 On the tops of the joists, mark the location and size of the rough opening.

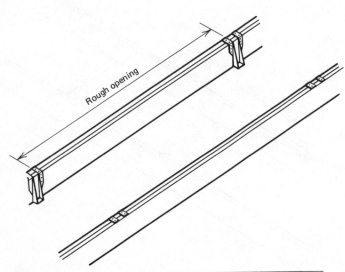

Fig. 6-10.6

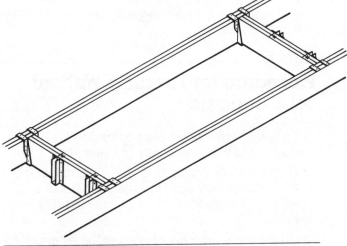

Fig. 6-10.7

3 Place hangers on the marks (Fig. 6-10.6), and nail them through holes in flanges to the joists.

4 Measure the width of the openings, and cut headers to length.

5 Measure the diagonals to make sure the opening is squared.

6 Fit doubled headers into the hangers. Then toenail headers to joists, and nail hangers to headers.

7 On the installed headers, mark the locations of tail joists. Make sure the marks line up with similar marks on sill and girder.

8 Nail single joist hangers on the marks (Fig. 6-10.7).

9 Fit tail joists into hangers, and nail securely.

10 Add trimmers as required.

Question

Below are the members used to frame an opening. In the space provided, list the order in which they are placed.

Member	Framing without hangers	Framing with hangers
Trimmers	_____	_____
Inner headers	_____	_____
Outer headers	_____	_____
Regular joists	_____	_____
Doubling joists	_____	_____
Tail joists	_____	_____

Unit 6-11   **Frame for Door Sills**

Exterior doors require a sill of their own. At one time almost all door sills were made of hardwood. Today many are made of extruded aluminum.

Most aluminum sills are one-piece units that include a threshold (Fig. 6-11.1, left). They come in several shapes, lengths, and widths—depending on the width of the door opening, the thickness of the exterior wall, and the thickness of finished flooring used in the entrance. Aluminum sills fit over subflooring, and no special cutting is required in the floor structure.

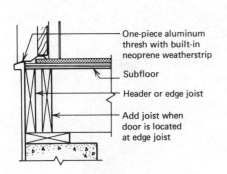
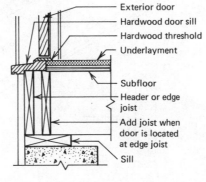
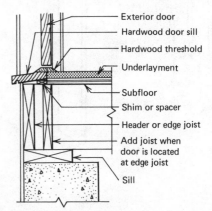

Fig. 6-11.1 Aluminum door sills (left) fit on top of subflooring, and require no preparation. Hardwood door sills butt against subflooring, and rest on floor framing. Their bottoms may be flat (center) or sloped (right).

Hardwood sills, on the other hand, require a separate threshold. Because of their thickness, builders seldom install them over subflooring. Instead, subflooring is cut so that the tops of hardwood sills lie flush with the finished flooring (Fig. 6-11.1, center and right). The bottoms of sills may be flat or sloped at the same angle as the top. The sill in the center of Fig. 6-11.1 is usually used when finished flooring is wood. The sill at the right in Fig. 6-11.1 is more common with thinner flooring.

The jambs of exterior door frames that accept wood sills extend below the subfloor and have a **dado** for the sill (Fig. 6-11.2). At the inside of the frame the distance from the top of the dado to the ends of jambs is $3\frac{1}{2}''$ if the sill is oak and $3\frac{1}{4}''$ if the sill is fir. The standard rough opening width for a door frame is the width of the door itself plus $3\frac{1}{2}''$. These dimensions and the thickness of flooring determine the size of cut required.

Procedure

1 From drawings determine the locations of centerlines for all exterior door openings.
2 Mark these centerlines on the existing floor structure.
3 From the door schedule in the drawings determine rough opening widths of all exterior doors.
4 Transfer these widths to the floor structure by measuring half the width on each side of the centerlines.
5 From sections in the drawings determine the distance from the outside of the floor framing to the inside edge of the door frame. **Note.** This

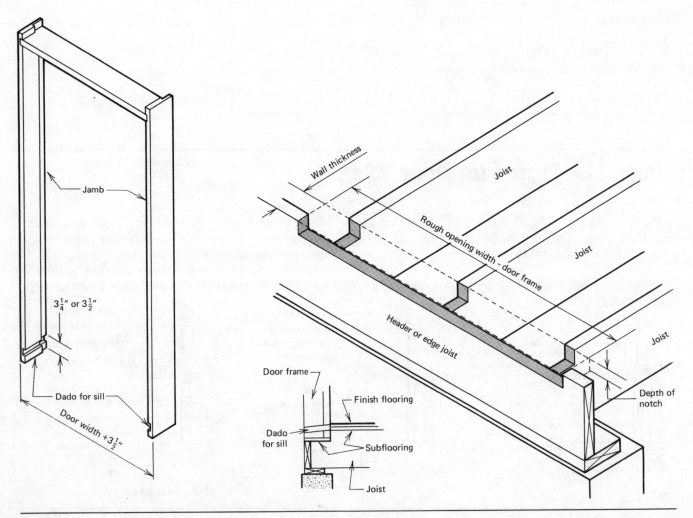

Fig. 6-11.2 Typical exterior door frame and dimensions.

Fig. 6-11.3

dimension is usually the thickness of a stud wall plus the thickness of the interior wall material.

6 Mark this depth across joists.

7 From drawings and specifications determine the total thickness of the finished floor, including any base (such as underlayment) and finished flooring material, but not including subflooring.

8 Subtract this thickness from the dimension measured from the top of the dado to the end of the jamb in the door frame (Fig. 6-11.3).

9 Mark this thickness on the side of the header joist or edge joist.

10 Notch the floor structure along the marks made in steps 4, 6, and 9.

11 You are now ready to lay subflooring. **Note.** In order for dimensions to work out correctly, the piece of subflooring cut out at each door opening is nailed to the cut joists as a base for installing door frames (see detail, Fig. 6-11.3).

Question

1 Name three differences between metal and hardwood door sills.

2 What is the standard rough opening width of a door frame?

3 Where do you look to find the dimension from the outside of floor framing to the inside edge of an exterior door frame?

4 How can you determine the dimension in Question 3 if you can't find it in the drawings?

Unit 6-12   **Cut Bridging**

Floor joists with long, unsupported spans are likely to twist, sway from side to side, bounce up and down, or vibrate under live loads on the floor they support. To stiffen the floor system, reduce movement of joists, and spread floor loads over more than one joist, the carpenter installs braces between joists called **bridging.**
  Three types of bridging are used (Fig. 6-12.1).

• **Wood cross-bridging,** also called **diagonal** or **herringbone bridging,** is set in pairs between

joists. Bridging stock may be 1 × 3s, 1 × 4s, or 2 × 2s, with 1 × 3s the most common and also the most economical.

• **Metal bridging** consists of manufactured steel straps. Metal bridging is also installed in pairs.

• **Solid bridging** is cut from joist stock. It is the least common, strongest, and most expensive. Because solid bridging gets in the way of wiring or piping that runs between joists, it is normally

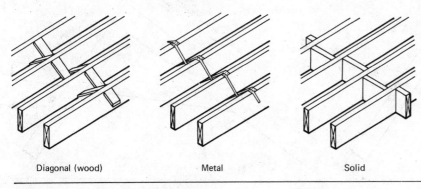

Diagonal (wood)   Metal   Solid

Fig. 6-12.1 The three types of bridging.

used only if much scrap is generated when joists are cut to length. Solid bridging is sometimes used to straighten cocked joists.

Diagonal wood bridging must be cut to length, with all ends beveled to fit tight against the sides of joists. Solid bridging must be cut square to fit the spacing between joists. Metal bridging requires no preparation.

Procedure for Cutting Diagonal Wood Bridging

1 Select a piece of bridging stock about 24″ long to mark as a template for other pieces.
2 Determine the space *between* joists. **Note.** This dimension is not the spacing on centers. When joist spacing is the standard 16″, for example, the space between joists is $14\frac{1}{2}$″.
3 Verify the depths of joists. **Note.** This is the *actual*, not nominal, depth. Typical depths are $7\frac{1}{2}$″ for an 8″ joist, $9\frac{1}{4}$″ for a 10″ joist, and $11\frac{1}{4}$″ for a 12″ joist.
4 Set the piece of bridging stock on edge.
5 On your framing square, carry the space between joists on the blade and the depth of joists on the tongue.
6 Lay the square on the stock, with the dimension on the tongue at the lower edge of the stock and the dimension on the blade at the upper edge of the stock (Fig. 6-12.2).
7 Along the outside of the tongue, mark the angle of cut.

8 Turn the piece of stock around so that the point of the cut line lies at the spacing mark on the blade, and the opposite edge of the sample meets the tongue at the depth mark (Fig. 6-12.3).
9 Again mark the angle of cut.
10 Saw along the two cut lines.
11 Test the sample for fit. If the fit is not good, cut a new sample.
12 When the fit is good, determine the angle of the two cuts, and set the blade on your power hand saw at this angle for cutting the remaining pieces of bridging.

Procedure for Cutting Solid Bridging

1 Assemble and stack the scrap pieces of joist to be used.
2 Check the end of each piece for square. If it is not square, square the end.
3 Along one edge measure off the space between joists, and make a mark.
4 Square a line across the width of the stock. Saw along this line.
5 Recheck your work to make sure that the two cuts are parallel and at right angles to edges.
6 Test the fit.
7 Cut the remaining bridging to length.

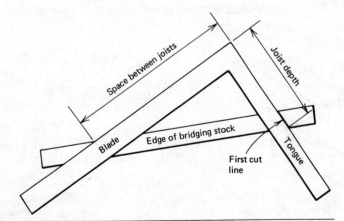

Fig. 6-12.2

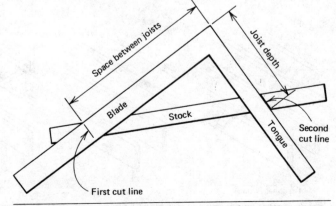

Fig. 6-12.3

Questions

1 If framing drawings call for 10″ joists 16″ on centers, what is the length of one side of a piece of diagonal bridging?

2 Can you cut bridging of this required length out of an 18″ piece of stock? If so, how much is left over? If not, how much short are you?

Unit 6-13   **Install Bridging**

Bridging is not required between joists less than 8′ long. If joists are more than 8′ but less than 16′ long, a single row of bridging at the midpoint of the span is adequate. If joists are more than 16′ long, a double row of bridging is required at the third points of the span.

Diagonal and metal bridging is usually put in place before subflooring is laid, because it is easier to nail from above. At this time, however, it should be fastened at only one end. The other end is not attached until after subflooring is installed. The delay allows joists to assume their final positions naturally and avoids setting up stresses that can cause floor squeaks.

Solid bridging is usually installed before subflooring. Because of its stiffness, it establishes the final positions of joists and helps to form a solid frame for subflooring.

Procedure for Attaching Diagonal Wood Bridging

1 After bridging is cut to fit, drive a pair of 6d nails into one end of each piece until the points just show through the bevels.

2 Snap a chalk line across the tops of joists at their midpoint or third points.

3 Between the first two joists at standard spacing near one exterior wall, nail a piece of bridging in place on one side of the chalk line (Fig. 6-13.1). Its top should be flush with the top of the joist.

4 Set the second piece of bridging on the other side of the chalk line and at the opposite angle, so that the two pieces form an X (Fig. 6-13.2).

5 Set the next pair of bridging in reverse, so that top meets top and bottom meets bottom on opposite sides of each joist (Fig. 6-13.3).

Fig. 6-13.1 (Photos Drew Leviton, Atlanta.)

6 Continue this procedure until all diagonal bridging is top-nailed.

Procedure for Attaching Metal Bridging

1 Snap chalk lines across joists as a guide to placement.

Fig. 6-13.2

Fig. 6-13.3

2 Nail pairs of metal bridging between joists, one on either side of the chalk line, so that they form an X. Nail only at the top through holes provided by the manufacturer. Use 1¼″ roofing nails, unless the manufacturer recommends otherwise.

Procedure for Attaching Solid Bridging

1 Snap chalk lines across joists as a guide to placement.
2 Between the first two joists at standard spacing near one exterior wall, fit a piece of solid bridging

between joists on one side of the chalk line. Make sure its top is flush with the tops of joists.
3 Drive three 12d nails through one joist and into the end of the bridging.
4 Check joists for plumb.
5 Nail through the second joist and into the other end of the bridging.
6 Between the next two joists, fit a piece of solid bridging on the other side of the chalk line.
7 Repeat steps 3 to 5.
8 Continue across the floor structure in this manner, nailing the bridging on alternate sides of the chalk line.

Questions

1 Which method of bridging do you think takes the least total time, including cutting the pieces? Why?
2 Why isn't solid bridging placed in a straight line?
3 Why is it important to place the tops of bridging flush with the tops of joists?

Unit 6-14 **Lay Subflooring**

For many years the material used for subflooring was 1 × 6 boards. Under finished hardwood flooring, boards were usually laid diagonally. This diagonal subfloor was strong and allowed flooring to be laid either at right angles or parallel to joists. There was quite a little waste, however, and extra time was needed to cut ends of boards at 45° to meet over the supporting joists.

As thinner flooring materials became more popular, board subfloors were more often laid at right angles to joists. This "straight" subfloor was not as strong as a diagonal subfloor, but could be laid more quickly and with less waste.

Because boards have a tendency to curl, sheet materials are now a standard for subflooring. Plywood is the most commonly used material, although some builders and home manufacturers use **particle board,** a thicker but light sheet material made of wood fibers.

Sheet flooring must be laid with the long dimensions across joists, and with ends of adjacent sheets staggered. The partial sheets used to start every other row may be one-half sheets, two-third sheets, or one-third sheets, depending on the overall dimension of the area to be floored.

Plywood for subflooring comes in sheets that are nominally 48″ X 96″, but often measure slightly less to allow for expansion. Thicknesses and edge treatment vary. Minimum thicknesses over typical joist spacings are:

$\frac{1}{2}$″ with joists spaced 16″ o.c.

$\frac{5}{8}$″ with joists spaced 20″ o.c.

$\frac{3}{4}$″ with joists spaced 24″ o.c.

Plywood is manufactured with its long edges square, shiplapped, or tongue-and-groove.

Edge treatment and edge support depend on the choice of finished flooring material. Wood strip flooring is a structural material itself, and requires no special support. Most other flooring materials are laid on **underlayment**—a thinner sheet material specifically designed for use under thin flooring materials. Underlayment is not applied until the house is completely enclosed. Subflooring with square edges provides adequate support for structural flooring and underlayment.

If a nonstructural flooring material, such as carpeting, is laid directly on the subfloor, then long edges of plywood must be either shiplapped or tongue-and-groove. If square-edged plywood is used, edges must be supported on 2 × 4 blocking between joists.

Particle board for subflooring comes in sheets 4′ × 8′ and 2′ × 8′, and it is 1$\frac{1}{2}$ to 2″ thick. It should not be used over joist spacings greater than 16″ o.c. Particle board does not warp, but it will absorb moisture. Some codes do not permit its use as subflooring.

Subflooring is attached with nails or staples. For a stronger floor less likely to squeak, high quality builders apply glue to the tops of joists and instruct carpenters to set panels carefully onto the beads of glue before nailing.

Procedure

1 From drawings or by measuring with a steel tape, determine the overall length of the subfloor from edge joist to edge joist or from side stud to side stud. This dimension determines the most economical cut for split sheets.

2 Check the stock of subflooring for severe warp; some warp is unavoidable. Set aside badly warped pieces for less important uses.

3 Begin at a corner with a full sheet (No. 1 in Fig. 6-14.1). Set the long edge flush with the header joist in platform framing, or with front studs in balloon framing. Center the short edge over a joist. If the floor structure is correctly built, the

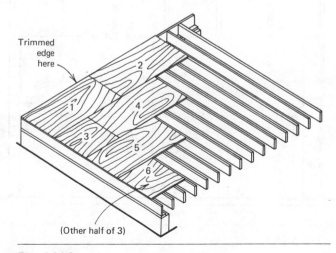

Trimmed edge here

(Other half of 3)

Fig. 6-14.1

outer short edge should be flush with the edge joist or side studs. **Note.** If plywood is tongue-and-groove, lay the grooved edge flush with the edge joist.

4 a. Begin attachment of plywood by nailing at the four corners with 8d nails. **Note.** With tongue-and-groove plywood, nail only at the two outer corners.

4 b. Begin attachment of particle board with a single 16d nail in the middle of the long side.

5 a. Complete attaching the first sheet of plywood with nails, every 6″ across short edges and every 10 to 12 ″ along long edges and into joists under the middle of the sheet, as shown in Fig. 6-14.2.

5 b. Complete attaching the first sheet of particle board by nailing on alternate sides of the first nail every 16″, then every 12″ across the sheet, working from the nailed edge as shown in Fig. 6-14.3

6 Fit and attach another full sheet along the header joist (No. 2 in Fig. 6-14.1). **Precaution.** Never pound on the edge of a sheet; pound instead on a piece of scrap 2 × 4 held against the edge.

7 Rip a full sheet to the length established in step 1.

8 Lay the partial sheet and another full sheet (No. 3 and No. 4 in Fig. 6-14.1).

9 Lay and attach full sheet No. 5 and partial sheet No. 6, which is the remainder of sheet No. 3.

10 Continue to lay subflooring in this diagonal pattern, with full sheets in each row, and alternating full sheets and partial sheets down the edge joist.

11 Add 2 × 4 blocking under unsupported edges between joists, if necessary, after you lay each sheet (Fig. 6-14.4).

12 At large openings carry subflooring into the opening, and cut off the excess with a power saw. **Precaution.** Cut immediately after the sheet is attached, to prevent someone from falling through the unsupported subfloor at the opening and to guide you in the placement of the next sheet.

13 At small openings, such as for pipes and ducts not yet installed, lay subflooring right over the opening.

14 Where pipes or ducts are already installed and rise above the level of joists, determine the location and size of holes required, and precut holes in the sheet prior to laying and attaching it.

15 At opposite edges of the floor, let subflooring overlap the sill. Then with a power saw trim all excess flush with sill members.

16 Complete attachment of bridging with staples (Fig. 6-14.5) or nails.

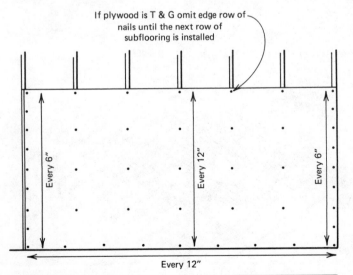

If plywood is T & G omit edge row of nails until the next row of subflooring is installed

Every 6″   Every 12″   Every 6″

Every 12″

Fig. 6-14.2

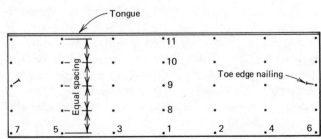

Tongue

Equal spacing

11
10
9
8

Toe edge nailing

7 5 3 1 2 4 6

Fig. 6-14.3

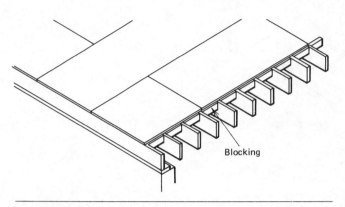

Fig. 6-14.4

Fig. 6-14.5 (Courtesy Senco Products Inc.)

Questions

1 How many full sheets of subflooring do you lay before you lay a partial sheet?
2 How many nails are needed to lay a sheet of plywood properly? How many nails are needed to lay a sheet of particle board? Can you think of a reason for the difference?
3 Under what circumstances is blocking between joists needed under edges of subflooring?

Unit 6-15   Frame for Masonry Floors

The plans for some houses call for brick, stone, or ceramic tile for finished flooring in areas where traffic will be heavy, such as entrances. Many of these materials are thicker than the more common flooring materials and require a different subfloor. Because the finished masonry surface should be at the same level as the surface of other flooring materials, the subfloor must be dropped below the level of subflooring elsewhere.

There are two accepted methods of framing. One is to use smaller joists than elsewhere, spacing them closer together and doubling them if necessary to meet load requirements (Fig. 6-15.1). The subfloor is then nailed to the tops of joists. The other method is to **chamfer** (bevel) the tops of regular joists to a V shape pointing upward (Fig. 6-15.2), then support pieces of subflooring on ledgers below the tops of the joists.

Subflooring must be $\frac{3}{4}''$ exterior grade plywood, nailed with noncorrosive nails. In top quality work the subflooring is covered with a layer of 15-lb felt, to keep moisture from the mortar bed out of the floor structure.

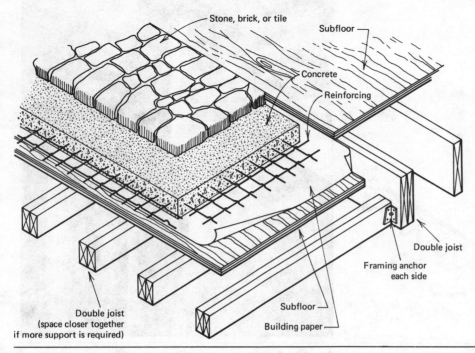

Fig. 6-15.1 Such masonry materials as slate, stone, brick, or tile need a specially built subfloor so that their surface is at the same level as other flooring adjacent to it.

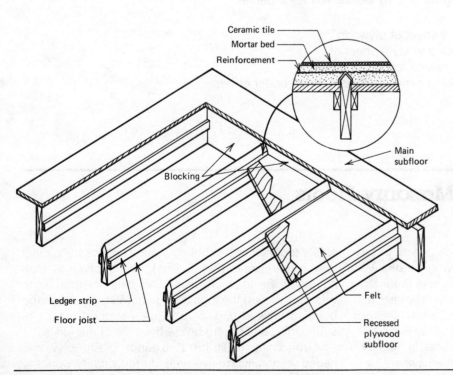

Fig. 6-15.2 The subfloor for a masonry floor may consist of several narrow pieces supported on ledgers attached to chamfered joists, or one large sheet as in Fig. 6-15.1.

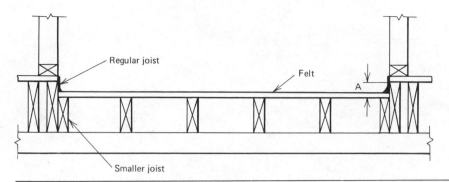

Fig. 6-15.3

Procedure for Framing with Smaller Joists

1 From the drawings determine the size of joists required and their spacing.

2 Determine the difference between the thickness of masonry flooring and adjoining flooring.

3 From the bottom of the regular subfloor, mark off this distance on headers and trimmers around the opening (*A* in Fig. 6-15.3).

4 Nail smaller joists to regular joists at the sides of the opening.

5 Space remaining joists as required.

6 Install bridging.

7 Cut subflooring to size, and install it with galvanized nails spaced 4″ o.c. around the edges and 6″ o.c. in the **field** (the remainder of the sheet).

8 Cover the subfloor with felt, overlapping 4″ at seams. **Precaution.** To prevent seepage of moisture or grout, carry the felt up the sides of the opening to the level of the underside of the masonry floor.

Procedure for Framing with Chamfered Joists

1 Determine the difference between the thickness of masonry flooring and adjoining flooring.

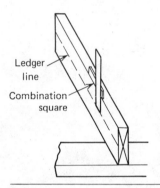

Fig. 6-15.4

2 Set this dimension on your combination square, and mark both sides of all joists that support the thicker floor (Fig. 6-15.4).

3 Nail 1 × 2 ledgers on the sides of joists with their tops on the marks.

4 Cut pieces of subflooring to length and width, and install them with galvanized nails spaced 4″ o.c.

5 Chamfer the top edges of joists to a point approximately level with the top of the subflooring.

6 Cut and install bridging.

7 Cover the subfloor with felt, as described in step 8 above.

Questions

1 Name one advantage of each of the two methods for framing under a masonry floor.

2 What tool would you use to chamfer the tops of joists?

7

Wall Framing Systems

As you learned in the previous chapter, wood has greater strength when used vertically than when used horizontally. Therefore the framing members that go into wall framing systems are smaller in size, lighter in weight, and easier to handle. But they must be just as carefully assembled.

In platform framing exterior walls and interior partitions have a single 2 × 4 plate that rests on the subfloor (Fig. 7-0.1). It is called a **bottom plate** or **sole plate.** The walls have a pair of plates at the top. They have a doubled **top plate** that supports ceiling joists and, in most cases, roof rafters. The lower of this pair of plates is called the **top plate;** the upper is called a **cap plate** or **doubler.** All wall frames are one story tall, and studs are usually less than 8' long. Walls and partitions are built flat on the subfloor, and then raised into position.

In balloon framing studs of exterior walls extend the full height of the house, from sill plate to doubled top plate (Fig. 7-0.2). These studs are notched for a ledger that supports second floor joists. Walls must be fully braced. Exterior studs are attached one by one to the sill. Because balloon framing is so open, firestops are required to prevent the spread of fire within walls.

Interior partitions that support ceiling joists in balloon framing rest on girders, are only one story tall, and end in a single cap plate that carries the joists. Supporting partitions on the second floor have a doubled top plate like exterior walls. All other partitions are shorter, and may be built on the subfloor as in platform framing.

In modern braced framing exterior studs rest on the sill plate as in balloon framing, but are only one story high as in platform framing (Fig. 7-0.3). Supporting partitions have a doubled top plate, and studs fit between joists above girders. All other partitions rest on the subfloor.

Installation of ceiling joists and sheathing is considered part of wall framing. Ceiling joists are installed

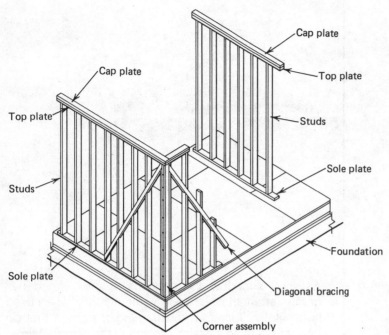

Fig. 7-0.1 Typical wall construction in the platform framing system. Walls have a sole plate and a doubled top plate.

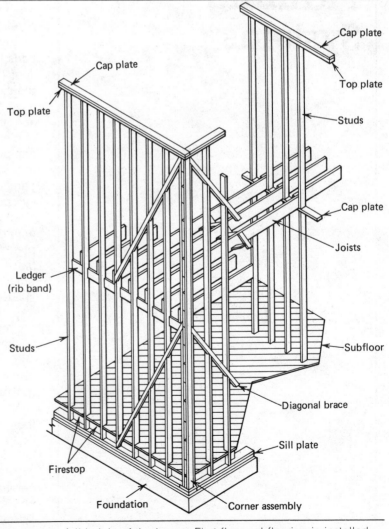

Fig. 7-0.2 Typical balloon framing in a two-story house. Studs rest on the sill, as do first-floor joists, and extend the full height of the house. First-floor subflooring is installed after exterior walls and supporting partitions are in place.

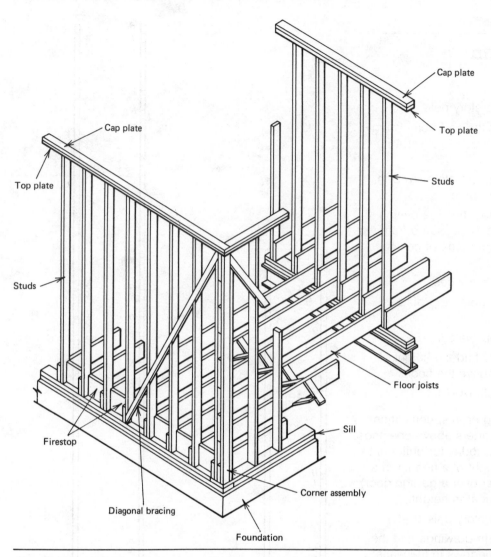

Fig. 7-0.3 Modern braced framing combines some of the features of balloon framing and platform framing.

after supporting walls are erected and firmly braced against the outward push of roof framing. Most builders do not sheathe walls until the roof itself is completely framed and sheathed.

The tools needed to build wall framing systems are the same as for floor framing systems, with the likely addition of a wood chisel. Special tools for special jobs are listed in the individual units that follow.

Unit 7-1   **Make a Story Pole**

Many of the parts that make up exterior walls and partitions are the same length or set at the same height. Building a wall is largely a matter of assembling precut parts and subassemblies. You can save some time in cutting these parts by making a story pole.

A **story pole** is a clean straight board, usually a 1 × 2, that is longer than the height of the wall to be built. On this board you mark the heights above the subfloor (in platform construction) or sill plate (in balloon and braced framing) of such parts as plates, window and door headers, and window sills. Then, when you need to know the exact length of any wall member, you have a place to measure that length accurately.

Procedure in Platform Construction

1 Select the board to use as a story pole.

2 Cut one end square. This is the end from which you begin all measurements, and is point *A* in Fig. 7-1.1.

3 Mark the thickness of the bottom plate (*A–B* in Fig. 7-1.1) on one side of the pole.

4 From sections in the drawings, find the ceiling height. **Note.** Ceiling height is measured from the top of the subfloor to the under side of ceiling joists.

5 Mark this height (*A–C*) on both sides.

6 Mark the thickness of the doubled top plate (*C–D*).

7 Mark the thickness of the top plate (*D–E*).

8 On the section or elevations, find the height of window and door headers above the floor.

9 Mark this location on the pole (*A–F*).

10 On the wall framing drawing or in specifications, determine the heights of headers above openings. **Note.** It is common practice today for builders to determine the maximum height of a header in a house (excluding the header over a garage door), and make all headers identical in height.

11 Mark header heights on the story pole (*F–G*).

12 From the window schedule in drawings find the rough opening heights of windows in the walls.

13 Mark these heights on the pole. (*G–H* and *G–I*).

14 Mark the thickness of window sills (*H–J* and *I–K*).

15 From specifications find the thickness of finished flooring and finished ceiling materials.

16 On the back side of the pole mark these thicknesses (*A–Z* and *C–Y*, respectively).

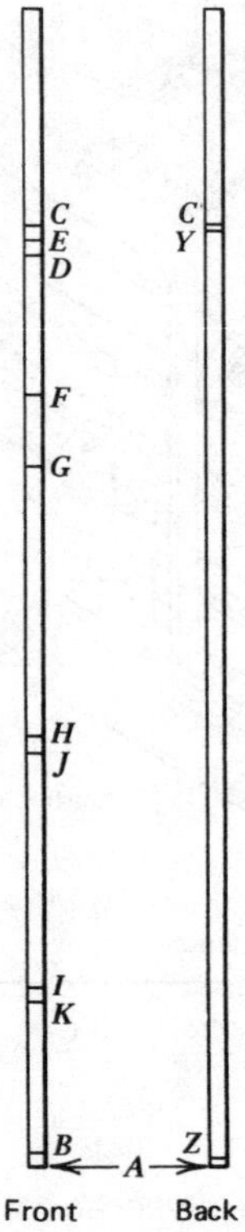

Fig. 7-1.1

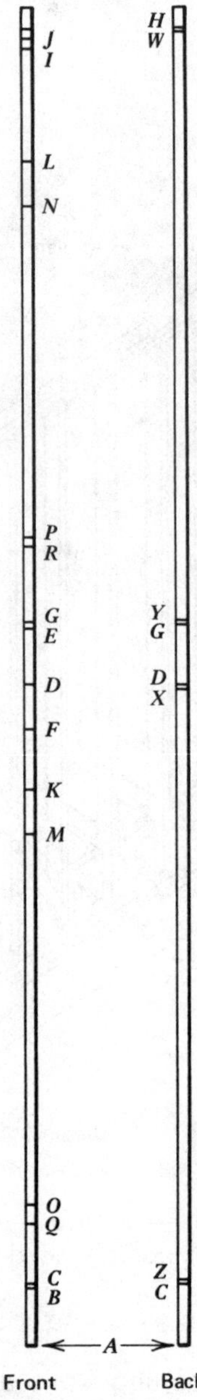

Fig. 7-1.2

Procedure in Balloon Framing

1 Select the board for use as a story pole.

2 Cut one end square. This is the end from which you begin all measurements, and is point *A* in Fig. 7-1.2.

3 Mark the depth of floor joists (*A–B*) on one side of the pole.

4 Mark the thickness of the subfloor (*B–C*) on both sides of the pole with a line at *C*.

5 From sections in the drawings, find the first floor ceiling height, and mark this height (*C–D*) on both sides of the pole.

6 From the drawings determine the height of the second floor joists and the height of the supporting ledger.

7 Mark these heights on the pole (*D–E* and *D–F*, respectively).

8 Mark the thickness of the upper subfloor (*E–G*) on both sides of the pole with a line at *G*.

9 In the drawings find the second floor ceiling height, and mark this height (*E–H*) on both sides with a line at *H*.

10 Mark the thickness of the doubled top plate (H–I).

11 Mark the thickness of the top plate (I–J).

12 On the section or elevations find the heights of window and door headers above the floors.

13 Mark these locations on the pole (C–K and G–L).

14 On the wall framing plan or in specifications determine the heights of the headers above openings. Mark these header heights on the pole (K–M and L–N).

15 From the window schedule find the rough opening

heights of windows in the walls, and mark these heights on the pole (M–O and N–P, for example).

16 Mark the thicknesses of window sills (O–Q and P–R).

17 From specifications, find the thicknesses of finished flooring, and mark these thicknesses on the back of the pole (C–Z and G–Y).

18 Also in specifications find the thicknesses of finished ceiling materials, and mark these thicknesses on the back of the pole (D–X and H–W).

Question

1 The front of a story pole helps you quickly find the lengths of wall components. Of what use is the information on the back of the pole?

Unit 7-2   Cut Studs to Length

As long as top and bottom plates in a wall are parallel, all full-length studs in a wall are the same length. Only when the ceiling of a house follows the slope of the roof is this not true, as in the side walls of a house with a shed roof, or a living room with a cathedral ceiling.

After the length of standard studs has been established on the story pole, studs can be cut to length in several ways. To save the carpenter's time on the site, some builders buy studs cut to length at the lumber yard or mill. Such studs are said to be **P.E.T.** (for precision end trimmed). Other builders with heavy-duty saws and feed tables cut studs to length five or six at one time. A stop is attached to the table at the proper position so that marking is not required.

When a number of studs of the same length must be cut by hand, the next step is to build a stud template. The template saves time in measuring, assures uniform stud length, and reduces the opportunity for errors.

Procedure

1 Determine stud length from the story pole.

2 Select for a template a piece of stud stock at least 6″ longer than stud length.

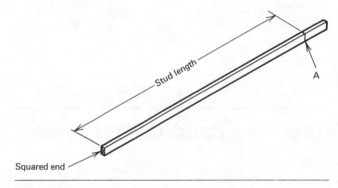

Fig. 7-2.1

3 Square one end.

4 From the squared end measure off and mark the length of a stud. Carry this mark across the face of the template. (A in Fig. 7-2.1).

5 Square the end of a 2 × 4 block 5 to 6″ long, and undercut the squared edge (as shown in Fig. 7-2.2).

6 Nail the block so that the squared end lines up exactly along line A (Fig. 7-2.3). Use duplex nails that can be removed.

7 Determine how many studs of standard length must be cut, and select the straightest available pieces from stud stock.

8 Square one end of each piece.

9 Set each piece against the block, and mark the other end at the squared end of the template.

10 Cut all studs along these marks.

11 Follow this same procedure for cutting **cripples**— studs that aren't full length. Find their lengths on the story pole, measure off and mark these lengths on the template, and move the block to the new lines (Fig. 7-2.3). **Note.** To keep the side of the template smooth and unaffected by nail holes, always cut the longest pieces first, then the next longest, and so on, with the shortest marked last.

12 If studs are to be used in a balloon frame, cut a notch $\frac{3}{4}''$ deep for a ledger. The length of the notch is $D-F$ in Fig. 7-1.2.

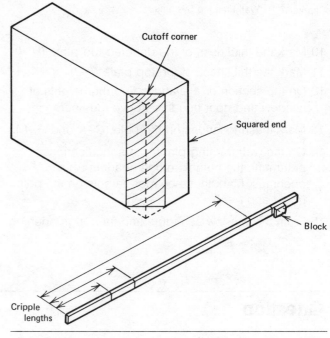

Fig. 7-2.2   **Fig. 7-2.3**

Question

1 What do you think is the reason for cutting off the corner of the block that acts as a measuring stop?

Unit 7-3 **Build Corner Posts**

As soon as studs are cut to length, the usual next step is to build exterior corner posts. The main members of posts are studs, and they may or may not include spacers. Posts are almost always set at the ends of bottom plates that extend to the corner of the house. A pair of 2 × 4s or a 4 × 4 have enough strength to act as corner posts. But strength is only one of two important considerations. The corner assembly also must provide a nailing surface for four surface materials— sheathing and an exterior material on two walls on the outside, and finish interior material on two walls on the inside.

Corner posts may be built in four different ways (Fig. 7-3.1). The post may be a single timber (A). It is the usual corner post in post-and-beam construction. A stud nailed to the timber during erection of walls provides a nailing surface for interior materials.

Corner B in Fig. 7-3.1 consists of two studs set at right angles to each other. The stud set back from the corner serves as a nailing surface for one wall, and a stud at the end of the other wall section provides a nailing surface for the second wall. This corner assembly uses the least materials and is the easiest to build.

In corner C in Fig. 7-3.1 the two corner studs lie parallel to each other. Between them is a strip of $\frac{1}{2}''$ plywood as a spacer. The stud with the spacer attached is offset about $1\frac{1}{2}''$ from the corner of the building as a nailing surface. As in corner B, the end stud of the adjoining wall provides the other nailing surface. This assembly is a little more expensive than corner B, but is a good way to use up an excess of scrap plywood.

Of the four corner assemblies, corner D is the most common. It consists of two studs parallel to each other and separated by a series of spacers made from scrap

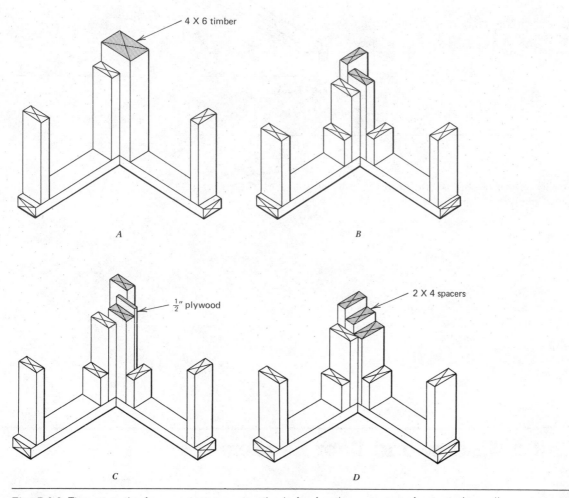

Fig. 7-3.1 These are the four most common methods for framing a corner of an exterior wall.

2 × 4s. The combined length of spacers should be about half the length of a stud, but no spacer should be less than 12″ long. The innermost stud of the assembly and the end stud of the adjoining wall provide the nailing surfaces.

Procedure for Building Corner D

1 Assemble spacer blocks. Their combined lengths should be about half the length of the stud.

2 To the sides of the corner stud, nail blocks at top and bottom. Make sure ends and sides of the blocks are flush with the ends and sides of the stud. Nail in the middle of each block with a pair of 8d common nails (Fig. 7-3.2).

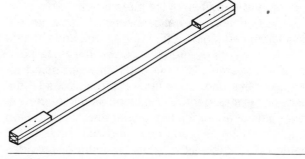

Fig. 7-3.2

3 Nail remaining blocks between the first two, maintaining approximately even spacing.

4 Lay the second stud on the spacers, with ends square and edges flush.

5 With 16d nails, attach the second stud, nailing as shown in Fig. 7-3.3.

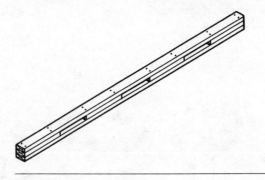

Fig. 7-3.3

Questions

1 If you had to build corner post B on the ground, draw on the sketch at right where the ground should be, and at which points you would nail.

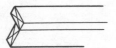

2 If you had to build corner post C, and all parts were cut ready for assembly, what would be:

a) Your first step?

b) Your second step?

c) Your third step?

Unit 7-4 **Build Window and Door Headers**

As you learned earlier, wood deflects or bends when used horizontally. The thinner the wood the more it bends. Also, the longer its span between supports the more it bends.

Headers over windows and doors must deflect as little as possible. Otherwise doors and windows will bind and not operate properly. During your life as a carpenter you are likely to install three types of headers, and build two of the three types.

Some builders use headers built of finger-jointed 2 × 4s laminated together (Fig. 7-4.1). The ends of lumber—usually 2 × 4s—are cut by saws in a zigzag pattern at the sawmill. Two cut ends are then glued together to make one long piece. If properly made, the glued joint is stronger than the wood as nature grew it. The lengths of finger-jointed lumber are then laid atop each other to the desired depth, and their sides glued together. This process of **glue laminating,** provides great strength against deflection.

More common over ordinary window and door openings are spaced headers (Fig. 7-4.2). Here two lengths of lumber rest on edge with a filler of $\frac{1}{2}''$ plywood between them. The filler is required to achieve the same thickness as a stud wall. The depth of the header depends on the span and location of the opening. For example:

Spans in a one-story house	Spans in a two-story house	Require a pair of
Up to 4'	Up to 2'	2 × 4s
From 4 to 6'	From 2 to 5'	2 × 6s
From 6 to 8'	From 5 to 7'	2 × 8s
From 8 to 10'	From 7 to 8'	2 × 10s
From 10 to 12'	From 8 to 9'	2 × 12s

Fig. 7-4.1 Laminated headers, such as this one over a garage door opening, are delivered to the site ready to install. They require a jamb stud on each side for support, and are capped with a doubled top plate. (Courtesy Drew Leviton, Atlanta.)

Fig. 7-4.2 Spaced headers are assembled on the site. They are made of two widths of 2X material with a sheet of plywood between to form a header 3½″ thick. Sometimes a 2 × 4 is added across the bottom to fill out the required header height.

To build spaced headers some builders use header stock of the full height required. Others, as in Fig. 7-4.2, use smaller header stock and add a 2 × 4 across the bottom for a better nailing surface. To save time and simplify ordering of materials, builders frequently determine the maximum header height needed in a house, then build all headers at that height.

Over very wide openings, such as a garage door, a truss is sometimes used. A **truss** is a braced framework; Fig. 7-4.3 shows typical designs. The main header member is built in the same way as a shorter header, and the bracing is added when the wall is assembled.

Headers are always longer than the rough width of the opening they span, because their ends are supported on vertical members variously known as **trimmers, jack studs,** and **jamb studs.**

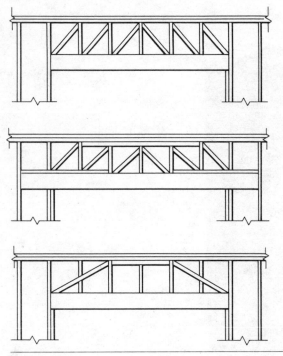

Fig. 7-4.3 Trusses are sometimes used as headers over broad openings in a wall. The truss construction permits use of a smaller spaced header across the opening.

Procedure for Building a Spaced Header

1 From drawings or local codes, determine the size of header material required.

2 On the drawings find the rough width of the opening.

3 Square header stock, and cut to this dimension.

4 Cut plywood filler to the same length and width as the header.

5 Coat both faces of the plywood liberally with glue.

6 Form a sandwich of the three pieces, and hold ends and sides flush with each other.

7 With 10d nails assemble the header. Use a pair of nails at each end and every 12″ between, driven from one side.

8 Turn the header over and nail in a similar pattern on the other side, avoiding nails already driven.

9 Add a 2 × 4 along one edge of the assembly if required.

Questions

1 The length of a header is usually _____ inches longer than the rough opening it spans.

2 To build a spaced header over a window 5′-8″ wide, what materials do you need, and what should their dimensions be?

Unit 7-5   Assemble Door and Window Frames

Frames around rough openings for windows and exterior doors consist of headers, sills (windows only), trimmers, and cripples (Fig. 7-5.1). In platform construction these pieces are often assembled into a complete component before the wall frame is put together, or while another crew is assembling the wall. Whether the components are set into the wall frame before it is raised into position or afterward depends on the total weight of the wall, the number of workers available to

raise it, and the builder's preferred method of assembly. In balloon framing window and door openings are usually cut into the wall after it is in position (see Unit 7-7).

The information on the heights of headers above floor level appears either in sections or elevations in the drawings. You must calculate the locations of window sills from rough opening dimensions in the window schedule, or use the story pole.

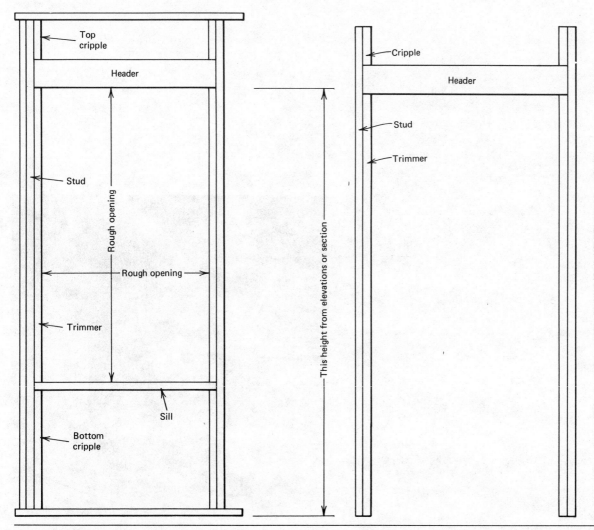

Fig. 7-5.1 These are the members that go into a window frame assembly (left) and a door frame assembly (right).

Some codes require the trimmers to be continuous, with the headers and sills between them.

Procedure

1 From a short length of 2 × 4 stock, cut window sills to the same length as window headers.

2 From the story pole, determine the length of trimmers. In Fig. 7-5.2 this dimension is *BG* for doors, and *GI* and perhaps *GH* for windows if there is more than one height.

3 Cut trimmers to length.

4 From the story pole determine the length of top cripples, if any. This dimension is *DF* in Fig. 7-5.2.

5 From the story pole determine the lengths of bottom cripples. This dimension is either *BK* or *BJ* in Fig. 7-5.2.

6 Cut all cripples to length.

7 On a pair of studs mark the locations of the under side of the header and the top of the sill, if any.

8 Place the header against the mark, and with four 12d nails, nail through the stud into the end of the header (Fig. 7-5.3). Make sure that the sides of the header and stud are flush, and that the angle between them is a right angle.

9 Repeat step 8 at the opposite stud.

10 Place the sill below its mark, and nail in position with a pair of 8d nails at each end.

11 Fit trimmer studs into position.

12 With a pair of 12d nails near each end and every 12″ between, nail through the trimmers into side studs.

13 Fit top and bottom cripples against side studs, and nail them into position.

14 Check the opening at all corners for square. Make any adjustments now. Then measure the diagonals before going any further.

15 To hold a window opening square, tack a length

Fig. 7-5.2

Fig. 7-5.3 (Courtesy Drew Leviton, Atlanta.)

of 1 × 4 diagonally across it (Fig. 7-5.4). Do not drive nails all the way in.

16 To hold a door opening square, tack a diagonal brace in place as for a window. Then add a

second brace across the opening near the bottom (Fig. 7-5.5).

17 Set aside the remaining cripples for installation after the window and door frames are in the wall.

Questions

1 Name the wall parts that each of the following members fit between.
   a) Window sill
   b) Top cripples
   c) Trimmers
   d) Headers
   e) Bottom cripples

2 What is the relationship between the length of a trimmer and the height of the rough opening?

3 Why do you set the header in position first?

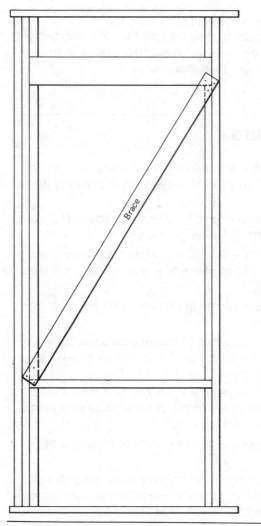

Fig. 7-5.4

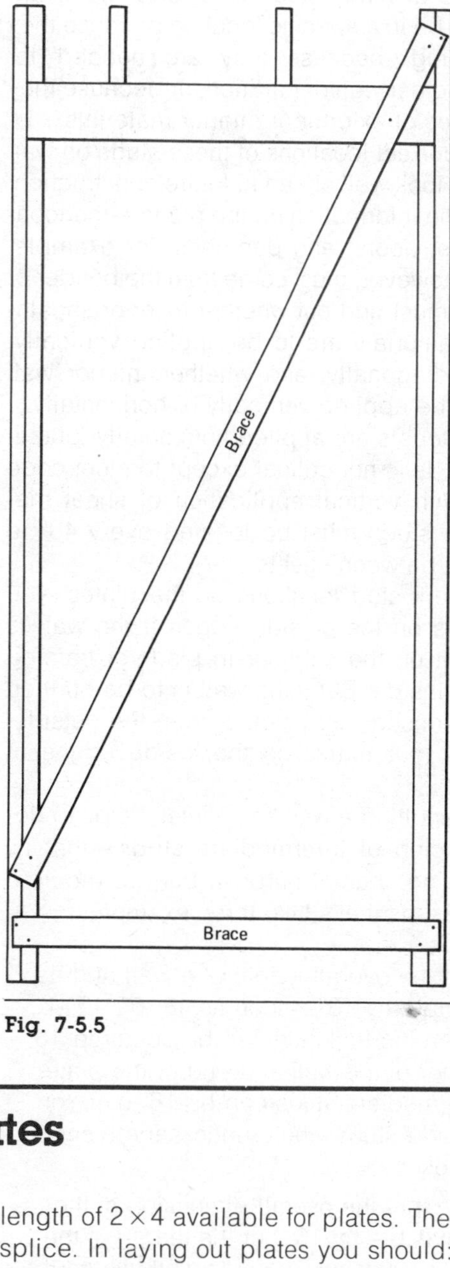

Fig. 7-5.5

Unit 7-6   Cut and Mark Wall Plates

The vertical framing members in a wall—corner posts, studs, and cripples at openings—must meet a wall's top and bottom plates at the same points. Therefore you save time and build walls more accurately if you mark the locations of vertical members on both plates at the same time.

In platform framing the top and bottom plates in any one wall are identical in all dimensions. Plates in the longest wall are the same length as sill plates. Plates in shorter walls are 4″ longer than sill plates (refer to Fig. 7-0.1).

In balloon framing a wall's bottom plate is the top sill plate, and is usually already in place. As in platform framing, the top plate in the longest wall is the same length as the top sill plate, and in shorter walls is 4″ longer (refer to Fig. 7-0.2).

Most exterior walls are longer than the longest

length of 2 × 4 available for plates. Therefore you must splice. In laying out plates you should:

- Use the straightest 2 × 4s available.
- Use long pieces to reduce the number of splices. number of splices.
- Save the best material for the top plate. Because bottom plates are fully supported by the subfloor in platform framing, knots are not a drawback, provided that they don't fall at points where you need to nail studs.
- Plan splices in the bottom plate so that they fall between studs or at door openings.
- Splice top and bottom plates at different points. Splices in top plates should fall on the center lines of studs.

The locations of many studs in exterior walls are fixed. They must be in a specific location because they frame an opening, because they are needed for attachment of another wall or partition, or because they support the edges of exterior or interior materials.

To mark the correct locations of these studs on wall plates, you must look well ahead to future construction. You will find some information on the plans—the locations of windows, doors, and partitions, for example. The remainder, however, must come from the builder or supervisor. You must find out whether exterior sheathing and finish materials are to be applied vertically, horizontally, or diagonally, and whether interior wall materials are to be applied vertically or horizontally.

If surface materials are applied horizontally, placement of other studs is not critical except to meet code requirements. With vertical application of sheet materials, however, studs must be located every 48″ to catch the seams between sheets.

When you mark stud locations on the plates, you make your marks on the outside edges if the wall is going to be built on the subfloor in platform framing and raised from inside. But if the wall is to be built on the ground and raised into position from the outside, you must make your marks on the inside edges of plates.

The overall length of a wall sometimes helps to determine the spacing of **intermediate studs**—that is, studs that serve no special purpose but are required for structural reasons. Let's take three examples:

1 Suppose that the overall length of a wall and its plate is divisible by 16″—such as 48′-0″, 49′-4″, or 50′-8″. Then the first stud can be centered 16″ from the corner of the wall and end of the plate. All other intermediate studs can be 16″ o.c. You then merely add studs where necessary to serve special purposes.

2 Now suppose that the overall dimension of the wall is not divisible by 16″ but the length of the plate is. This may occur in a side wall, where the side plates are set inside the front and rear plates. Then the first stud is centered 16″ from the end of the plate, and $19\frac{1}{2}$″ from the corner.

3 If neither dimension is divisible by 16″, it may be easier to start at some fixed stud locations in the middle of the wall, and work toward the corners. Stud spacing may be less than the standard for which the wall was designed—usually 16″—but never more than that standard.

Procedure

1 From the plans determine the overall lengths of plates. Verify this dimension by measuring the platform or sill plate.

2 Square one end of lengths to be spliced. Square both ends of full-length plates.

3 Determine splice points, mark these points on both members where they overlap, and cut ends square.

4 Lay bottom plates in position, with top plates on top of them.

5 Mark the locations of corner post assemblies on both plates (A in Fig. 7-6.1). Not all plates carry corner posts.

6 On the floor plan of the house find the dimensions from the corners to the centerlines of all window and door openings.

7 Mark these center lines on both plates (B in Fig. 7-6.1).

8 Measure from these centerlines to the sides of openings, and mark the locations of both trimmer studs and full-length studs (C).

9 On the floor plan find the dimensions from corners to the centerlines of all intersecting partitions.

10 Mark these centerlines on both plates, and the width of the intersecting partition plate (D).

11 Determine the locations of any studs required to support edges of exterior or interior materials.

12 If these locations are different from those of studs already marked, mark them also (E).

13 Mark the locations of remaining intermediate studs (F).

14 Place an X beside all marks so that you know on which side of the line to place studs.

Questions

1 Why should the top plate be cut of better lumber than the bottom plate?

2 Why do you think splices in bottom plates should fall between studs?

3 If the required stud spacing is 16″ o.c., can studs ever be spaced closer together? Can they be farther apart? Why do you think so?

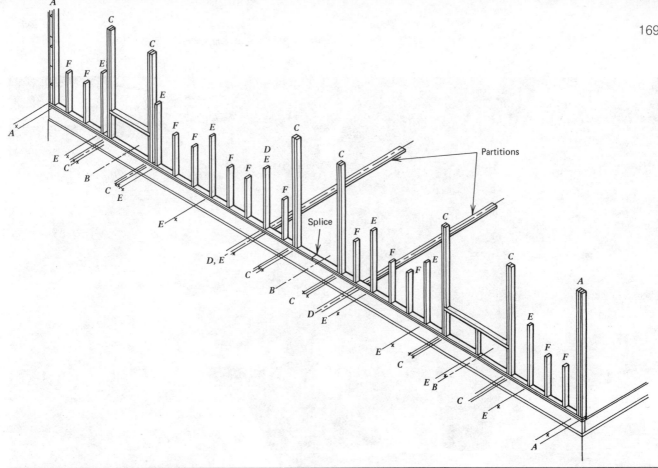

Partitions

Splice

Fig. 7-6.1

Unit 7-7 Assemble Wall Components

When all wall components are either premarked or preassembled, the next step is to build the walls. The order in which exterior walls are built depends somewhat on the floor plan. Usually the longest walls are built first, since they extend from one end of the subfloor to the other, and you need the entire length of floor to lay them out. Then come the shorter sidewalls, and last the walls around wings of the house.

With platform framing the easiest way to assemble the components is to lay them out flat on the subfloor, with the outside face of the wall up (Fig. 7-7.1).

With other types of framing, where bottom plates of walls support not only studs but floor joists, the subfloor is not in place when walls are erected. With braced framing, it is sometimes possible to assemble components for lower walls on the ground outside the house. In this process the components must be laid on the ground with the inside face of the wall up.

Long studs for a two-story wall built with balloon framing must be attached to the sole plate individually. Wall components are set in place after the wall in a vertical rather than a horizontal position.

Procedure for Assembling a Wall in Platform Framing

1 Lay the bottom plate flat on the subfloor, as close as possible to the point where it will be attached.

2 Set corner posts in position (Fig. 7-7.2).

3 Position the top plate.

4 Position all full length studs at their marks.

5 Set window and door frame assemblies into position.

Fig. 7-7.1 In platform framing, wall components are laid out and assembled with their outside faces up. This is a garage wall being assembled on a concrete slab. (Courtesy Drew Leviton, Atlanta.)

6 Before you do any nailing, recheck all studs for straightness. Replace any studs that are twisted or that have more than $\frac{1}{2}''$ bow. Studs with crooks should be set with the hollow of the crook toward the outside of the wall.

7 Drive a pair of 16d nails part way into the top plate above each stud.

8 Hold each stud in position on its mark and flush with the plate at both edges. Then drive the nails home.

9 Nail corner posts flush with the top plate on three sides.

10 Nail window and door frame assemblies if they won't make the wall too heavy to raise.

11 Place and nail top cripples above openings.

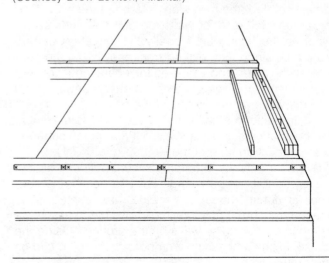

Fig. 7-7.2

12 Turn the bottom plate on edge, and nail it to studs through the bottom of the plate.

13 Place and nail cripples below window sills.

14 When all nailing is completed, measure the diagonals.

15 When the wall is square, run a pair of 1 × 6 braces diagonally across studs from the top plate to the bottom plate. **Caution.** Be sure that the braces do not extend beyond the edge of the bottom plate. The braces prevent the wall from being racked out of line when it is raised.

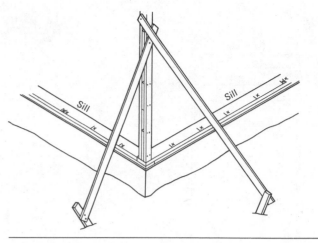

Fig. 7-7.3

Procedure for Assembling a Wall Vertically

1 Place the ledger or **ribband** for second floor joists against the marked plates, and mark stud locations on it.

2 Raise corner posts into position on their marks on the sill plate, and toenail them to the plate.

3 Place 1 × 6 diagonal braces between posts and ground on both sides (Fig. 7-7.3). Nail as high on the post as you can reach; leave the other end free.

4 Plumb corner posts. Then nail bottom ends of the diagonal braces to deadmen.

5 Next set on their marks any studs that flank joints in ribbands.

6 Toenail studs to the bottom plate. **Note.** If floor joists are already in position, plumb both joists and studs, and nail them together with three 8d or 10d nails. Then toenail.

7 Install the ribband with three 16d nails into studs and posts.

8 Install all other studs, and nail at plate and ribband.

9 On studs running beside and through window and door openings, mark the locations of the tops of headers and bottoms of sills (A and B in Fig. 7-7.4).

10 Outside these marks tack 1 × 4s horizontally with duplex nails.

11 On the 1 × 4s mark the width of the opening (C).

12 From studs that run through the opening, cut out the section between 1 × 4s (length AB in Fig. 7-7.4).

13 Insert precut header and sill members between regular studs across the opening, and inside lines A and B (Fig. 7-7.5). Nail them into position through studs.

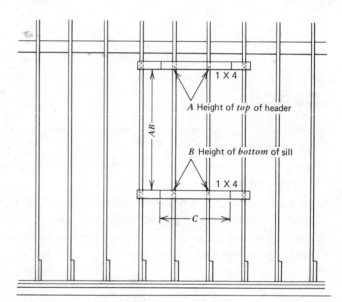

Fig. 7-7.4

14 Install trimmers, cripples, and jamb studs.

15 Nail cut studs to header and sill to complete the framing (Fig. 7-7.6).

16 Remove the 1 × 4s.

17 To plumb the wall, hang a plumb bob from the ledger 10 to 12′ from the corner of the wall.

18 Adjust the plumb of the wall until the plumb bob points directly at the sill plate. Renail braces at corners if necessary.

19 Drive a stake into the ground 10 to 12′ from the wall, and run a 2 × 4 brace between the stake and a nearby stud to hold the wall plumb.

20 Repeat steps 17 through 19 every 20′ across the wall.

21 Add firestops between joists at the sill.

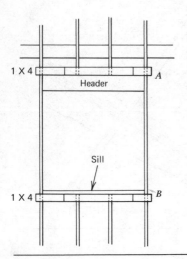

Fig. 7-7.5

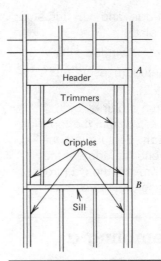

Fig. 7-7.6

Questions

1 In balloon framing, why do you think it is important to attach the ribband
   as soon as possible?

2 Why is the bottom plate laid flat during the early stages of wall assembly
   in platform framing?

3 In platform construction, what members are installed in the procedure
   above that could be installed later?

4 What other members could be installed before raising that wall that are not
   installed in the procedure above?

Unit 7-8   Raise and Fasten a Wall

In raising a wall frame built on the subfloor, several precautions are in order.

First, some means should be devised to prevent the wall from sliding off the floor when it is raised. Second, enough personnel must be on hand so that the wall can be raised at all points at the same time and at the same speed. Third, braces must be available in advance for holding the wall upright once it is raised into position.

The size of crew needed to raise a wall varies with the length of the wall. As a general rule it takes one person for every 8′ of wall length—one at each end and the others evenly spaced between. Then add one more person to nail the braces. If a wall is more than 40′ long, there are several alternate procedures:

- Get the help of another person to assist with the raising.

- Omit window and door frames until the wall is in its final position, plumbed, and braced.

- Build a long wall in sections, and raise it one section at a time. With this method top and bottom plates are not continuous, but are spliced between studs after the sections are in position. Splices should fall at the same point to simplify construction of the wall.

Procedure

1 To the edge of the platform nail 2 × 4 stops about 18″ long. Place the stops about 4′ from the ends of the wall, and let them stick up about 6″ above the subfloor (Fig. 7-8.1).

2 Lay a 1 × 6 or 2 × 4 about 12′ long within reach at each end of the wall, and one near the center of a long wall. These are for bracing.

3 Slowly but steadily raise the wall, letting it slide toward the stops (Fig. 7-8.2).

4 When the wall is upright, adjust its position if necessary by sliding it. **Precaution.** Do not shake or wiggle the wall. Even with diagonal braces on it, a wall can be easily knocked out of square at this time.

5 With the wall in position, and still held at its ends, attach the bottom plate to the box sill with 16d

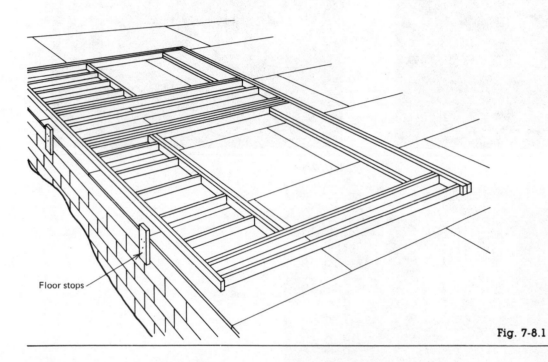

Floor stops

Fig. 7-8.1

Fig. 7-8.2
(Courtesy Drew Leviton, Atlanta.)

Fig. 7-8.3

Fig. 7-8.4

Fig. 7-8.5

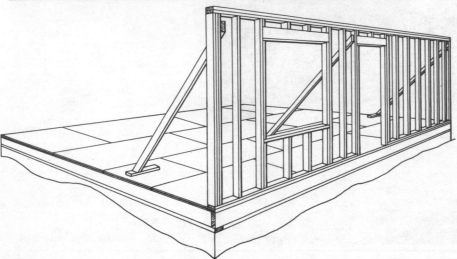

Fig. 7-8.6

nails between the corner post and first stud *only*. Make sure the face of the wall lines up with the edge of the floor. **Precaution.** If the edge of the floor is ragged, snap a chalk line on the subfloor $3\frac{1}{2}''$ from the edge, and set the inside face of the wall to this line.

6 Nail a brace high on the outside edge of each of the corner posts (Fig. 7-8.3).

7 Plumb the wall at one end. While the wall is held plumb, attach the other end of the brace to the box sill (Fig. 7-8.4) or to a stake firmly driven into the ground outside the wall (Fig. 7-8.5).

8 Repeat this step at the other corner.

9 To brace in the middle of the wall, or at ends of wall sections, nail wood blocks to the subfloor (Fig. 7-8.6), and run braces between the blocks and studs adjacent to splices. **Precaution.** Do not set braces where they will be in the way of building the remaining exterior walls on the subfloor.

10 Build and raise the remaining exterior walls, and straighten them before completing nailing to the sill.

Questions

1 Name three ways to brace a wall after it is raised.

a) _____

b) _____

c) _____

2 After a wall is raised, what must you do before you brace it?

Unit 7-9   **Straighten and Secure Walls**

Exterior walls that have been raised into position are far from complete. Walls that meet at a corner must be securely connected. All of them must be straightened over their entire length. Nailing must be completed. *Cap plates,* also called *doublers,* must be added to the top plates to support joists and rafters. And corners must be braced to withstand winds.

If firestops and wall bridging are required, they may be installed as the final step in completing exterior wall framing, or they may be installed after walls are sheathed.

In the procedures in this chapter, exterior wall frames are assembled and erected without cap plates and without sheathing. Many builders do otherwise, particularly volume builders who can temporarily increase or decrease the size of crew working at a particular site. The exact order of the steps in construction isn't as important as the result: sturdy walls that meet all code requirements and are both straight and plumb.

Cap plates should be cut from the straightest and longest pieces available. There is no strength in a series of short pieces. When you calculate the lengths of pieces you need, remember that the overall lengths of cap plates on exterior walls are either 7″ shorter or 7″ longer than top plates, because cap plates overlap at corners. This same overlapping is also required where partitions butt against walls; the cap plate on the partition extends to the outside face of the top plate.

Just as top plates often must be spliced, so must cap plates. The splices in a cap plate should fall at least 48″ from a splice in the top plate.

Corners may be braced in several ways. Sheets of plywood and boards laid diagonally as sheathing provide good bracing strength. Other types of sheathing do not. For maximum rigidity of a wall with nonstructural sheathing, let-in braces are used. A **let-in brace** is a 1 × 4 fitted at about a 45° angle into notches in studs near corners.

Procedure

1 Where two walls meet at a corner, plumb and square the intersection.

2 Nail through the end stud of the shorter wall into the corner post.

3 Cut cap plates to length and square their ends.

4 Fit cap plates at corners, overlapping the end of one cap plate onto the top plate of the other wall (Fig. 7-9.1).

5 Nail cap plates with 16d nails, working toward the center of the wall from the corners. Check before nailing to make sure cap plates and top plates are flush at the sides. Drive nails about $\frac{3}{4}''$ from the edges of the cap plate, and stagger them about 24″ apart (Fig. 7-9.1). **Precaution.** Avoid nailing directly above studs; you are likely to hit nails below.

6 Complete the task of nailing the bottom plate to the sill. Drive 16d nails as shown in Fig. 7-9.2. All nails should go into header joists, edge joists, or regular joists.

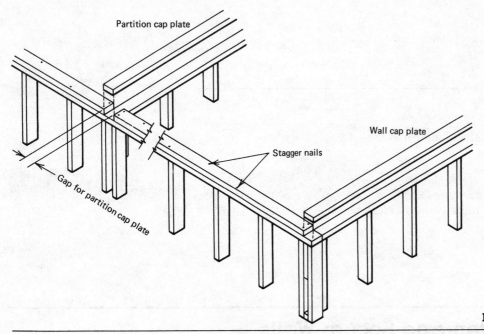

Fig. 7-9.1

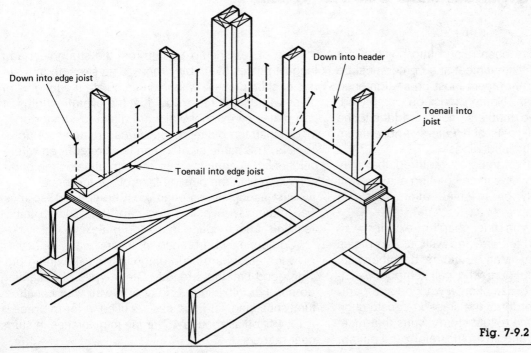

Fig. 7-9.2

7 To check your work, walk or crawl below floor joists and look up. If you see the points of more than a few nails, the connection between walls and floor is weak.

8 To check the straightness of a wall, cut three small blocks from a 1 × 2.

9 Tack a block on the front face of each corner post at the top plate. These are called **line blocks.**

10 String a line between the blocks.

11 At various points along the wall, slide the third block, called a **gauge block**, between the top plate and the line (Fig. 7-9.3).

12 If the line just touches the gauge block, the wall is straight. If the line does not touch the block, or bends over it, detach the lower ends of temporary braces in the middle of the wall.

13 Push on the wall until it is straight and plumb, then renail the brace.

14 If necessary, particularly where partitions intersect, add a brace to hold the wall straight.

15 To cut let-in bracing, lay a 1 × 4 12′ long diagonally across the faces of studs. The lower

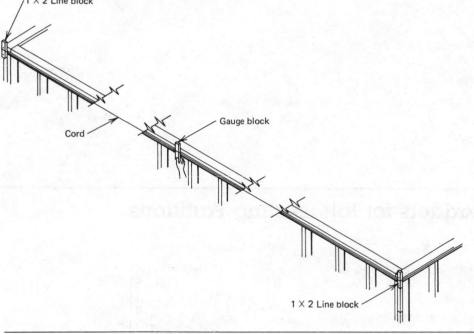

Fig. 7-9.3

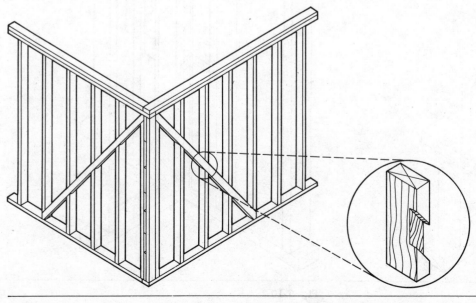

Fig. 7-9.4

end should rest on the top of the subfloor, and the upper end should reach to the upper outside corner (Fig. 7-9.4).

16 Mark the ends of the 1 × 4 for two angle cuts so that the brace fits between top and bottom plates.

17 Cut the ends along the marks.

18 Again lay the brace in position against the studs, and on the studs mark the locations of notches.

19 Cut notches in studs to a depth of $\frac{3}{4}''$ along the cut lines (inset, Fig. 7-9.4). **Note.** Cut for a tight fit.

20 With a chisel clean out the notches.

21 Fit the let-in brace into the notches, and nail with a pair of 8d nails into each stud.

22 At exterior door openings, cut out the bottom plate between trimmer studs.

23 Install firestops and wall bridging if required.

Questions

1 Can you name an advantage of attaching the cap plate before a wall is raised?

2 Can you name a disadvantage to doing so?

3 A splice in a cap plate should be at least how far from a splice in a top plate?

4 When is let-in bracing required?

Unit 7-10   **Build Ladders for Intersecting Partitions**

Where a partition meets an exterior wall, some means must be provided for attaching the two and for future attachment of interior wall materials in the two rooms separated by the partitions. Equally important, the fit between wall and partition should be as tight as possible to stop the passage of noise between the two rooms.

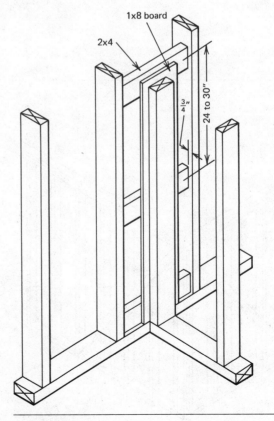

Fig. 7-10.2

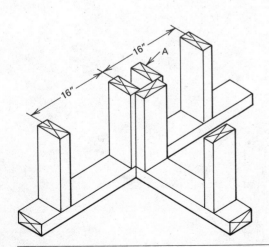

Fig. 7-10.1

When the end stud in a partition meets a stud in the wall, as in Fig. 7-10.1, all that is required to complete the intersection is a second stud in the wall (stud *A*). At least ¾″ of the edge of a stud must be exposed as a nailing surface.

When the partition falls between regular stud spacing, however, you have several choices. The simplest is to add studs in the wall where needed to attach the end stud of the partition and wall materials. Another method is to build a ladder for attachment (Fig. 7-10.2). The ladder consists of 2 × 4 crosspieces set on edge and a 1 × 8 vertical board. Crosspieces are spaced 24 to 30″ apart on centers. The end stud of the partition and the board provide the necessary surfaces for attaching interior wall materials.

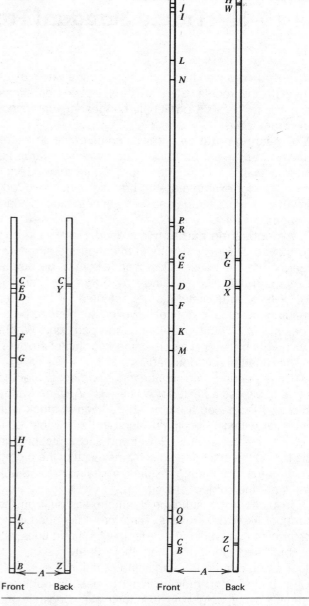

Fig. 7-10.3   Fig. 7-10.4

Procedure

1 Cut enough crosspieces 14½″ long out of 2 × 4 stock to form the rungs of the ladder. With a standard 8′ ceiling, you will need five pieces.

2 Cut a 1 × 8 board to length. Its length equals the distance between the top and bottom plates of the exterior wall. On the story pole in Fig. 7-10.3 this is dimension *BD*. In Fig. 7-10.4 it is *CF* on the first and *GI* on the second floor.

3 On the inside faces of studs flanking the partition, draw a vertical line ¾″ from the inside edges of studs (see Fig. 7-10.2).

4 Set crosspieces against the lines and also against bottom and top plates, and nail them into place through the studs.

5 Space the remaining crosspieces equally between the plates, and also against the lines, and nail them in place.

6 Place the 1 × 8 board upright with its center line on the center line of the partition. Make sure it is plumb from top to bottom.

7 Attach the board by nailing through it into the crosspieces.

Questions

1 Suppose that a stud in an exterior wall and a partition are on the same center line. How would you attach the partition?

2 If only ¾″ must be exposed as a nailing surface, why couldn't you use a 1 × 6 board in Fig. 7-10.2 instead of a 1 × 8?

Unit 7-11 Frame Standard Partitions

A **partition** is a wall that divides the space within exterior walls into rooms. There are three types of partitions in a house: bearing partitions, nonbearing partitions, and stack walls.

A **bearing partition** supports some of the structure above it, and must be built to the same standards as exterior walls. Full headers are required over door openings, for example. Let-in bracing is not required, however, and partitions end with a single stud instead of a post.

A **nonbearing partition** is a divider only, and can be moved or removed without affecting the structural integrity of the house. Door headers may be single 2 × 4s laid flat. Stud spacing may be greater than 16″ o.c. if finish wall materials are selected for the wider spacing. In actual practice, however, nonbearing partitions are often built just like bearing partitions. Where two partitions meet at a corner, the end stud of the overlapping partition is doubled.

Both bearing and nonbearing partitions are the same thickness as exterior walls—3½″. A stack wall is thicker. Also called a plumbing partition, a **stack wall** contains within it the main plumbing stack that connects to the sewer at the lower end and vents through the roof at the other. It also contains much of the piping to and from plumbing fixtures in bathrooms. Stack walls are the subject of Unit 7-12.

In platform construction partitions are built on the subfloor like exterior walls. Therefore it is important to determine in advance the order in which partitions will be built and erected, so that there is always enough floor room to build the remaining partitions.

In balloon and braced framing, bearing partitions are the same height as exterior walls, because the studs rest on the doubled sill over a girder. Bearing partitions are installed after subflooring has been laid up to the girder. They may be built flat and raised into place, or assembled in a vertical position. When bearing partitions are built on the ground, builders usually leave out one exterior wall temporarily so that they can move the partition and install it easily.

Nonbearing partitions rest on the subfloor, are therefore shorter, and have studs of a different length.

Procedure

1 Study the floor plan of the house to determine the type of each interior partition.

2 On the subfloor mark with chalk the center lines of partitions at their two ends.

3 Measuring from the chalk marks, locate the edges of all partitions between rooms, then snap chalk lines on the subfloor to mark the position of one edge of long partitions. For the time being ignore short partitions around closets.

4 Determine the order in which partitions should be built.

5 Cut and assemble the parts as described in Unit 7-7 for exterior walls.

6 Locate and nail ladders for intersecting partitions between studs of exterior walls.

7 Raise the partition into position against the wall. Shove it tight.

8 Drive one nail through the bottom plate into a joist.

9 Plumb the end of the partition, and attach it to wall studs or the ladder.

10 Position the other end of the partition on chalk lines, and attach it through the bottom plate.

11 Brace the intersection with a diagonal to the subfloor. Be careful not to set the brace where it interferes with partition building and setting.

12 When all partitions are in place and attached to each other and to walls at both ends, complete nailing the plate.

13 Add doublers, overlapping at ends.

14 Cut out the bottom plate at door openings between trimmers.

Questions

1 Assuming that the height of a wall from top of subfloor to bottom of ceiling joists is 8′-1½″, mark with an X which of the following walls and partitions would have studs 7′-9″ long:

	Exterior wall	Bearing partition	Nonbearing partition
Balloon framing	_____	_____	_____
Platform framing	_____	_____	_____
Braced framing	_____	_____	_____

Unit 7-12 **Frame a Stack Wall**

The plumbing system in most houses today is laid out so that a 3″ soil stack is adequate to vent the system. This 3″ dimension, however, is an inside pipe diameter. The outside diameter is $3\frac{1}{2}″$, and where two lengths of pipe join together, the connection is usually even thicker. Some builders use 3″ plastic vent piping, and build their stack walls out of 2 × 4 stock. Most builders, however, prefer to build their stack walls out of 2 × 6s. The extra space gives them a little leeway in construction, and the opportunity to insulate piping against water noise.

The main stack is usually located directly behind the toilet, also called a **water closet**. If you look carefully at floor plans, you will find that the wall behind the toilet, or at least a portion of it, is always thicker than the other walls in the bathroom.

Because pipes run both vertically and horizontally through a stack wall, the partition must be laid out carefully and accurately. Some sets of drawings include a diagram of water piping, and another of drain piping. These drawings are usually **schematics**—that is, they show where pipes run, but only in general terms without dimensions (Fig. 7-12.1).

While carpenters are setting partitions, plumbing, heating, and electrical contractors and their crews are also at work. By the time carpenters are ready to frame a stack wall, the chances are good that the drainage system has been completed in the basement or crawl space, and the main stack has been extended upward to a point just below the subfloor. It may even extend above the subfloor. The location of the stack is the main guide to positioning the stack wall.

Procedure

1 If not already done, snap chalk lines on the subfloor to locate the center line and edges of the stack wall.

2 a. If the stack extends above the subfloor, make sure that it fits between the side chalk lines and on the center chalk lines.

2 b. If the stack stops below the subfloor, hang a plumb bob from the under side of the subfloor so that it points directly to the center of the stack. Mark this point on the under side of the subfloor. Then, from underneath, drive a nail through the subfloor at this point. If all is well, this nail will lie on the center chalk line for the partition.

3 Lay out the top and bottom plates, using the chalk lines and position of the stack as guides to placement of studs.

4 Assemble the wall as described in Unit 7-7.

5 Mark the center of the stack on the bottom and top plates. Cut or notch the bottom plate if necessary to fit around an existing pipe.

6 Raise, plumb, and attach the partition.

7 From plumbing drawings or specifications determine the size of the vent pipe that will pass through the top plate.

8 Cut the top plate to allow the pipe to penetrate.

9 Reinforce the cut with pieces of steel strapping as shown in Fig. 7-12.2.

Question

1 Why isn't it necessary to reinforce the cut in the bottom plate?

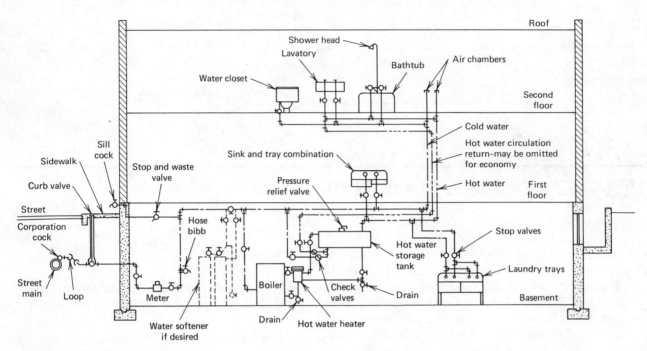

RESIDENTIAL WATER PIPING DIAGRAM

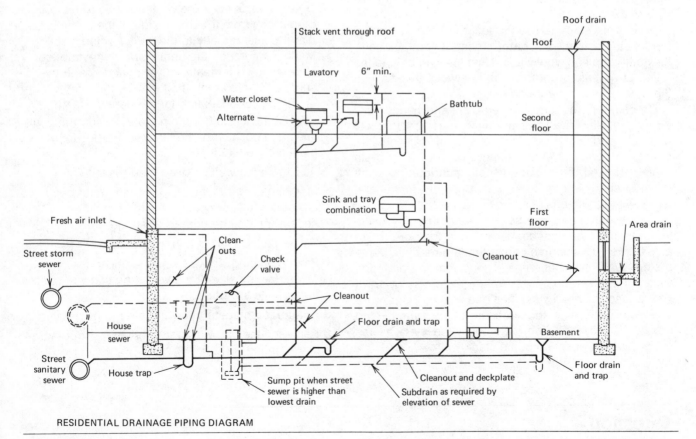

RESIDENTIAL DRAINAGE PIPING DIAGRAM

Fig. 7-12.1 A typical drainage piping diagram for a two-story house with a basement. Note that there are few dimensions. Schematic drawings such as this simply show the plumber how to route the piping. The plumber adapts the exact locations to the requirements of local codes and the structural conditions that are faced.

Fig. 7-12.2 (Courtesy Drew Leviton, Atlanta.)

Unit 7-13 **Notch Joists and Studs for Piping**

Building codes are very strict on what you can do and cannot do to notch joists and studs for piping, ducts, and wiring. At the time joists and stud walls are installed, it is seldom possible to know the exact point at which a service line will run. Therefore cutting and notching takes place at various times during rough plumbing and rough wiring. The cuts are usually made by the worker installing the utility system, but it is important that you know what is permissible and what is not.

Most building codes permit a notch in the top of a joist no greater than $\frac{1}{6}$ the depth of that joist. If joists are 10″ stock, then, their actual depth is $9\frac{1}{4}$″, and a notch $1\frac{1}{2}$″ deep is allowable. No notch of any kind is permitted in the bottom edge of a joist without reinforcement. (See Fig. 7-13.1, top.)

A circular hole may be drilled in a joist, as long as the diameter of the hole is not more than $\frac{1}{6}$ the depth of the joist, and as long as the circumference of the hole is no closer to the edge of the joist than 2″. Some codes do not permit any hole in the lower half of the joist.

If a hole or notch must be larger than the maximum allowed by code, or falls in a part of the joist not permitted by code, the cut must be reinforced. When the notch is in the top of the joist, you reinforce with lengths of 2×4 on each side of the joist that extend at least 12″ and preferably 18″ beyond the cut on both sides (top center in Fig. 7-13.1). When the hole is in the upper half of the joist but below the edge, cut a filler for a very tight fit, and glue it in place (Fig. 7-13.1, bottom center). When the notch must be in the bottom edge, the cut must be reinforced with a steel strap attached with lag screws (Fig. 7-13.1, bottom). The strap should extend the depth of the joist on either side of the cut.

Codes almost uniformly permit the notching just outlined *only* within 2′ of the ends of joists. No notching

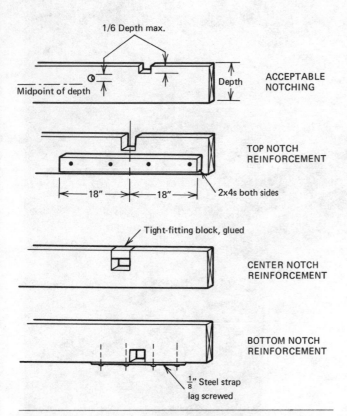

ACCEPTABLE
NOTCHING

1/6 Depth max.

Midpoint of depth

Depth

TOP NOTCH
REINFORCEMENT

18"   18"

2x4s both sides

Tight-fitting block, glued

CENTER NOTCH
REINFORCEMENT

BOTTOM NOTCH
REINFORCEMENT

$\frac{1}{8}$" Steel strap
lag screwed

Fig. 7-13.1 Not all building codes have the same provisions for notching joists. The acceptable notching and types of reinforcement shown here, however, meet the requirements of all major building codes.

of any kind, even with reinforcement, is allowed at any other point. Notching of a girder is not permitted at any point or under any circumstances.

Separate restrictions apply to notches in studs. Avoid making any notches in the middle third of any stud. In the lower or upper third, most building codes allow a notch up to one-third the width of a stud in an exterior wall or bearing partition, and up to one-half the width of a stud in a nonbearing partition. Thus you can cut a notch in a stack wall, which is usually nonbearing, as deep as $2\frac{3}{4}$"—more than enough for almost any horizontal pipe (Fig. 7-13.2, left).

Although notches of the sizes above are allowable, no two adjacent studs may be notched without reinforcement in all of them. The method of reinforcement shown at right in Fig. 7-13.2 is the most common, and is similar to joist reinforcement shown in Fig. 7-13.1, bottom center.

In addition to regular carpenter's tools, you may need a power drill with carbide-tipped bits, and a wrench.

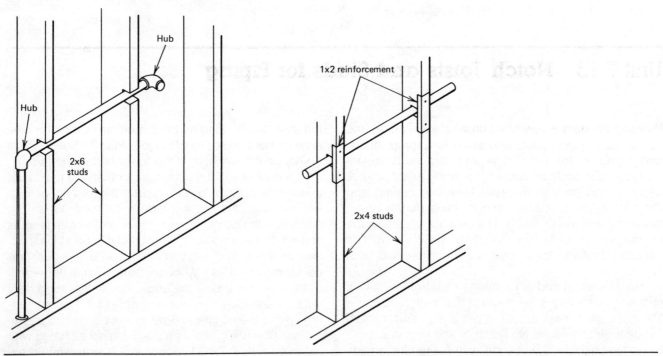

Hub

Hub

2x6
studs

1x2 reinforcement

2x4 studs

Fig. 7-13.2 Restrictions on notching studs are not as great as on joists. Most major codes permit an unreinforced notch up to one third the width of a stud, as long as the notch is not in the middle third of the length of the stud.

Procedure for Reinforcing a Notch with Wood

1 Measure the width of the notch in the joist or stud.

2 Measure the distance from the edge of the joist or stud to the pipe, and subtract $\frac{1}{16}''$.

3 Out of clear 2x stock, cut a block to fit the two measurements. Cut for a tight fit.

4 Coat the inside of the notch and the ends of the reinforcing block with glue.

5 Wedge the block in place flush with the edge of the framing member.

6 If necessary, hold the block in place with two finishing nails until the glue dries. With a proper fit, nailing should not be necessary.

Procedure for Reinforcing a Notch in a Joist with Steel

1 Measure the width of the notch.

2 Add 36″ to this dimension.

3 With a hacksaw cut a piece of steel plate $1\frac{1}{2}''$ wide and of the length and gauge required by code.

4 Drill holes in the strap for lag screws. Minimum screw size is $\frac{1}{8}'' \times 2''$.

5 Lay the strap in position across the notch, and mark the locations of holes.

6 Drill small starter holes in the joist.

7 Attach the plate with the lag screws.

Questions

1 What is the deepest notch permitted in the bottom of a 2×12 joist without reinforcement?

2 Why do you think notching is usually permitted only within 2′ of the supported ends of joists?

3 What part of a stud should not be notched? Why?

Unit 7-14   Install Second Floor Joists

Second floor joists are almost always the same size as first floor joists, and the method of installing them is quite similar.

With platform framing in a two-story house, the cap plate atop first floor walls and partitions becomes the sill plate of another box sill. The procedure for installing joists is the same as in Unit 6-9. However, joists overlap at bearing partitions, are not in line as at a girder, and are separated by blocking (Fig. 7-14.1).

With modern braced framing the doubled top plate of walls and partitions also becomes the doubled sill for second floor joists. The joists fit adjacent to studs in exterior walls, and on opposite sides of studs in bearing partitions over girders (Fig. 7-14.2).

With balloon framing, however, second floor joists rest on ledgers notched into studs in exterior walls. At bearing partitions they overlap on a single top plate (Fig. 7-14.3). Studs in the second floor partition rise directly over studs in the first floor partition, and joists overlap beside these studs. Standard procedure is to install joists in half the house, then complete installation in the other half.

In many houses first and second floor joists are identical in length, and can be cut at the same time. If first floor joists rest on a built-up wood girder, second floor joists may be 1″ shorter. This 1″ difference isn't worth bothering about unless second floor joists can be cut from shorter stock with less waste.

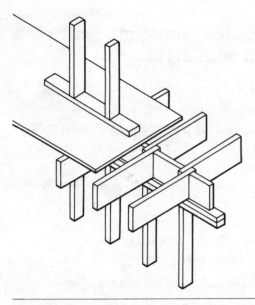

Fig. 7-14.1 Usually one bearing partition lies above another in a multi-story house. In platform framing the second floor is framed with a box sill just like the first floor. Joists do not line up at the partition, however. One joist may lie directly above the stud, or the two joists may touch on the center lines of studs.

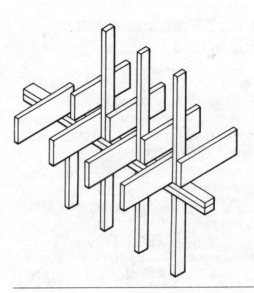

Fig. 7-14.2 In modern braced framing, second floor studs line up with first floor studs. Joists do not touch each other, but hold studs stiffly erect.

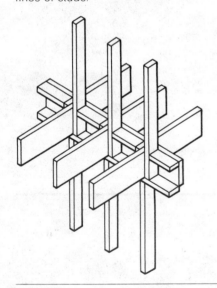

Fig. 7-14.3 With balloon framing, studs in the second floor bearing partition line up with first floor studs. Both joists are at one side of studs. Firestops and blocking add stiffness.

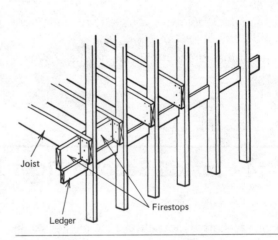

Fig. 7-14.4

Procedure with Balloon Framing

1 From framing plans determine the length of joists.

2 Cut joists to length and square the ends.

3 Beginning at a corner, set joists in position with their crowns up. Ends of joists must bear fully on the ledger and cap plate of the bearing partition.

4 With three 10d nails, attach joists to studs at the ledger. Nail the outer ends of all joists before attaching the other end.

5 Cut joist stock into 14½″ lengths for firestops.

6 Install the firestops as shown in Fig. 7-14.4.

7 Set inside ends of joists so that their sides line up with the edges of studs below the cap plate, as in Fig. 7-14.5.

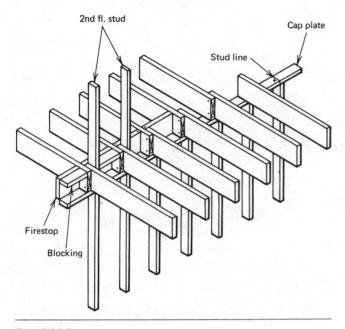

2nd fl. stud

Stud line

Cap plate

Firestop

Blocking

Fig. 7-14.5

8 Toenail into the end of the joist and down into the cap plate. Hold the joist on the line.

9 Repeat steps 3 to 6 for joists over the other half of the house.

10 Where joists overlap at the cap plate, toenail through the second set of joists into the cap plate, as in step 8.

11 Nail straight through second joists into first joists with a pair of 10d nails.

12 After studs in the second floor bearing partition are in position, nail through the studs into both joists with three 16d nails.

13 Install firestops as shown in Fig. 7-14.5.

14 Cut 2 × 4 stock into 13″ lengths for blocking.

15 Install blocking between studs (Fig. 7-14.5). Make sure the sides are flush with the tops of joists and firestops.

Questions

1 What is the main difference in platform framing between first floor construction and second floor construction?

2 Besides slowing the spread of fire, what other purpose does a firestop serve?

Unit 7-15 **Install Ceiling Joists**

The main difference between ceiling joists and floor joists is the loads they support. These loads determine joist size. Ceiling joists are lighter than floor joists, even when attic space is floored and used for storage. A span that requires 2 × 10 floor joists, for example, can be spanned with 2 × 8 ceiling joists. When a roof is carried on trusses, the bottom chords of the trusses replace ceiling joists, and are even lighter members. The procedures for installing trusses are discussed in detail in Unit 9-13.

In all three basic structural framing systems, the top horizontal member of walls and partitions is a doubled top plate. Ceiling joists usually bear directly over studs, although they don't have to as long as the top plate is doubled. To carpenters who frame walls, this top wall member is called a cap plate—the term used so far in this book. To carpenters who frame roofs, however, the top wall member is a **rafter plate.**

In balloon and braced framing, ceiling joists are the same length as floor joists from end to center line of center supports. In platform framing, however, they are each 1½″ longer. The outer ends of ceiling joists must be squared. It isn't necessary to trim the other ends unless enough can be trimmed off for blocking, or unless the attic is to receive a finished floor.

Roof rafters and ceiling joists rest side by side on the rafter plate. Because rafters are almost always smaller members than ceiling joists, the carpenter must trim the top corners of joists to the slope of the rafters. Some builders prefer to trim before ceiling joists are installed. Others prefer to trim after both ceiling joists and rafters are in place.

It is standard practice to leave out the end ceiling joists until roof framing begins, because the exact location of these joists varies with the type of roof and the amount of side overhang.

Procedure

1 From framing drawings determine the length of ceiling joists.

2 Select joist stock, square one end, and cut to length.

3 Borrow the sample common rafter (see Unit 9-3), and measure its thickness above the rafter plate (dimension *AB* in Fig. 7-15.1).

4 Transfer this dimension to the end of a ceiling joist.

5 From elevations or sections in the drawings, determine the pitch of the roof.

6 Beginning at point *B* in Fig. 7-15.1, mark this slope on the side of the joist.

7 Set the saw to cut at this angle, and trim the outer ends of all joists alike.

8 Study roof framing plans to determine the spacing and locations of joists, and the locations of ceiling openings.

9 Mark joist locations on the rafter plates and the cap plate of bearing partitions.

10 Beginning with the second joist from the corner, set joists in position with their crowns up.

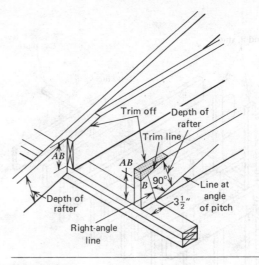

Fig. 7-15.1

11 Toenail through the sides of joists away from adjoining rafters into the rafter plate. Use a pair of 10d or 12d nails. Nail outer ends of all joists before attaching the other ends. **Note.** It is not necessary to plumb outer ends of rafters at this time. That can be done when rafters are installed.

12 At bearing partitions, line up joists on their marks, and toenail as at the rafter plate. Nail on the side opposite the overlapping joist.

13 Cut and install firestops or blocking between joists if required.

Questions

1 Can you name one advantage of trimming ends of ceiling joists before you install them?

2 Can you name an advantage to trimming ends after you install ceiling joists?

3 Which way do you prefer? Why?

4 Why are ceiling joists longer than floor joists in a house built with platform framing?

Unit 7-16   Attach Partitions to Joists

Where joists run above partitions, it is necessary to tie the two together. Corners where finish ceiling and wall materials meet must be square, and as tight as possible to prevent sound in one room from passing over the partition into the next room.

When attaching joists to partitions, the carpenter faces one of three conditions. If joists run across the partition, attachment is simply a matter of toenailing, although the task isn't easy if the joists have much crown. If a partition runs parallel to and between joists, you build a ladder similar to the ladder in Unit 7-10. If a partition lies directly under a joist, you must split the nailing strip.

Procedure When Joists Cross Partitions

1 Start an 8d nail into the side of each joist about an inch from the edge of the rafter plate (Fig. 7-16.1). You need a nail on each side of the joist, staggered on opposite edges of the plate.

2 Pull the joist down as tight as possible to the plate, and drive in one nail.

3 If there is still a gap between joist and rafter plate, drive in the second nail. Then return to the first nail and drive it in tighter.

4 When joists fit tight to rafter plates, drive all nails on one side of joists first, then go back and drive the second nails.

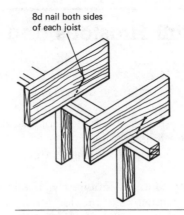

8d nail both sides of each joist

Fig. 7-16.1

Procedure When Joists Parallel Partitions

1 From a length of 1 × 6 cut a nailing base the same length as the exposed part of the rafter plate.

2 a. If the partition lies between joists, center the 1 × 6 over the plate and nail it every 16″ with 8d nails.

2 b. If the joist lies directly over the plate, rip the 1 × 6, and nail part of it to the plate on each side of the rafter (Fig. 7-16.2).

3 Cut cross blocks from 2 × 4 stock. Blocks should be 14½″ long, and spaced 16 to 24″ apart.

4 Set the blocks upright on the nailing base (Fig. 7-16.3), and nail through joists into the ends of the blocks.

5 Attach blocks to the nailing base by toenailing once on each side.

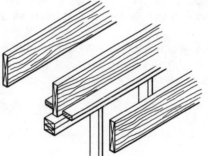

Fig. 7-16.2

Fig. 7-16.3

Questions

1 What is the purpose of the blocks above the nailing strip in Fig. 7-16.3?

2 Why aren't blocks necessary in Fig. 7-16.1?

Unit 7-17 Install Firestops and Blocking in Walls

The purpose behind placing firestops in walls is to prevent the spread of fire between studs by reducing the
length of the "flue" between top and bottom plates.
Most codes do not require firestops in platform construction, but make them mandatory in balloon construction.

Blocking, on the other hand, is required with all
types of construction. Its purpose is to provide a nailing base for various items that are attached to walls in
the house: kitchen cabinets, countertops, lavatories,
rims of bathtubs, towel bars and soap dishes, shower
rods and shower doors, shower heads, heat registers,
wainscoting, and built-ins. With proper placement
between studs, blocking can sometimes double as a
firestop.

Blocking is almost always set parallel with the floor.
Straight or staggered firestops are also set parallel.
Herringbone firestops, which are less common, are set
at an angle.

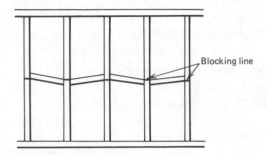

Fig. 7-17.1

Procedure

1 Determine requirements for firestops from
  drawings and local codes.

2 Determine requirements for blocking from plans
  and interior wall details.

3 Establish the horizontal center line of blocking
  and firestops. Draw this height across the faces of
  studs where necessary.

4 a. Cut herringbone firestops to length, following
  the same procedure but not the same angle used
  for cutting floor bridging (Unit 6-12). Aim for a
  snug fit.

4 b. Cut straight firestops and blocking to length with
  ends square.

5 a. To install herring bone firestops, set each piece
  with one top edge on the line, and one bottom
  edge on the line (Fig. 7-17.1). Alternate the
  pieces, and toenail carefully into studs. Be careful
  not to knock studs out of line.

5 b. To install staggered firestops, set every second
  piece above the line, and nail to studs. Set and
  nail alternate pieces below the line (Fig. 7-17.2).

5 c. To install straight single blocks, set the blocks
  on edge, with the center line of the block on the
  center line for nailing (Fig. 7-17.3). Nail through
  the ends.

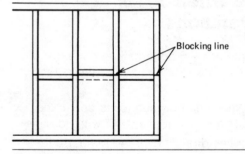

Fig. 7-17.2

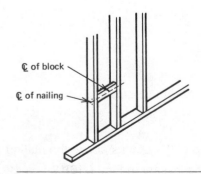

₵ of block

₵ of nailing

Fig. 7-17.3

5 d. To install staggered blocking, set every second piece with its bottom edge 1″ below the line, and alternate pieces with the top edge 1″ above the line (Fig. 7-17.4). Attach each piece with one nail driven straight in through studs, and another nail driven at an angle through blocking into studs.

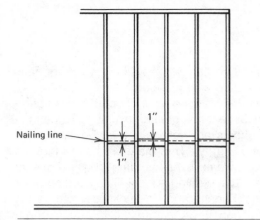

Fig. 7-17.4

Questions

1 Name one advantage and disadvantage of herringbone firestops.
2 Name one advantage and disadvantage of staggered firestops.
3 With staggered blocking, how much is one block offset from another when blocks are 2 × 4s on edge?

Unit 7-18 Install Wall Sheathing

In the normal sequence of construction, wall sheathing is not installed until the roof is framed, sheathed, and the finish roofing material applied. There are several good reasons for this delay. First, it is extremely important to get the house under cover, so that rainfall doesn't warp subflooring, and fill the basement or crawl space with water. Second, the location of the top edge of wall sheathing varies with the type of roof and the treatment at overhangs.

The subject of wall sheathing is covered here only because sheathing is a structural part of the wall, and logically belongs in this chapter.

Walls may be sheathed in one of five ways: with boards applied horizontally (1) or diagonally (2), and with sheets made of plywood (3), gypsum (4), or composition board (5).

For many years boards were the standard sheathing material, and they ranged in width from 1 × 6s to 1 × 12s. Horizontal application was the least expensive, but diagonal application was stronger in bracing strength. Development of sheet materials, however, has almost completely eliminated the use of boards.

Yet boards have two advantages: they can be cut to fit an uneven stud spacing, and there is often use for scrap. There is little other use for scrap sheet sheathing.

Today plywood is the most common sheathing for general use. It is durable, and has great strength, particularly when applied horizontally. Although plywood comes only in 4′ widths, it is available in 8′, 9′, 10′, and 12′ lengths, and in thicknesses from $\frac{5}{16}$″ up to $\frac{3}{4}$″ for use as sheathing. Most codes permit the use of $\frac{5}{16}$″ and $\frac{3}{8}$″ plywood for sheathing only when applied horizontally. The standard material is $\frac{1}{2}$″ CDX plywood, made with exterior glue that will stand quite a little wetting, and applied vertically.

Gypsum sheathing has three advantages. It does not burn, it won't absorb moisture, and it is low in cost. Its main disadvantages are its weight and limited insulating value. It comes in three sizes. Sheets 2′ × 8′ have V-shaped edges that lock together, and must be applied horizontally. Sheets 4′ × 8′ and 4′ × 9′ have square edges and should be applied vertically. Gypsum sheathing isn't as strong as plywood, and requires greater nailing if corners of the wall are not diagonally braced or sheathed with plywood.

Composition sheathing is also called insulating sheathing because it has greater insulating value than other types. Sheets are lightweight, and easy to handle, saw, and apply. The material has adequate strength, but is weak in nail-holding power. For that reason insulating sheathing is most often used on brick veneer houses. Sheets range in thickness from $\frac{1}{2}$ to 1″ for wall application, in width from 2 to 4′, and in length from 6 to 12′. The most common sizes are 4′×8′ with square edges for vertical use, and 2′×8′ with interlocking edges for horizontal use.

Plywood may be applied with 6d nails, or glue and nails. Gypsum sheathing may be applied with $1\frac{3}{4}$″ roofing nails or 16 ga. staples $1\frac{1}{2}$″ long and with a $\frac{1}{2}$″ crown. Insulating sheathing is applied with roofing nails only.

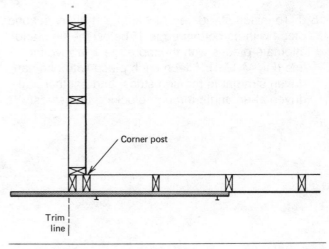

Fig. 7-18.1

Procedure for Applying Sheathing Vertically

1 Study stud spacing to determine how to apply the sheathing with the fewest vertical joints.

2 Mark the studs that will carry the seams.

3 Determine the length of sheet required, and trim sheets to length if necessary.

4 Wedge a pair of 16d nails between the sill and the foundation wall at points about 3″ and 45″ from one corner.

5 Set the first sheet at the corner with its bottom edge on the nails, one side edge on the center of the stud at the first seam, and any trimmed edge at the top (Fig. 7-18.1). The top edge should cover the top plate, and the other side edge should cover the corner post.

6 a. Attach plywood with 6d nails spaced every 6″ at edges and every 12″ into studs in the field.

6 b. Attach each sheet of gypsum wallboard with either roofing nails or staples spaced every 4″ around the edge and every 8″ in the field. If the wall does not have let-in bracing or will not be surfaced with siding or shingles, nail or staple every 4″ in the field. Otherwise every 8″ in the field is adequate.

6 c. Apply insulating sheathing with 2″ galvanized roofing nails, set $\frac{5}{8}$″ in from the edges. Space nails 3″ apart around the perimeter of each sheet, and 6″ apart in the field.

7 Continue down the wall by applying full sheets, even at door and window openings. Gap

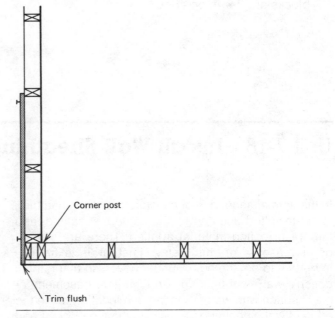

Fig. 7-18.2

insulating sheathing about $\frac{1}{8}$″ at studs; it is cut a little narrow to permit this.

8 Cut out excess material at openings.

9 Rout out smaller openings, as for a kitchen fan.

10 Where sheets meet at a corner, let one sheet overlap the other for a square corner (Fig. 7-18.2).

11 Remove nails driven in step 4.

Procedure for Applying Sheathing Horizontally

1 a. Set the first sheet of plywood at a corner, with its bottom edge on nails at the foundation wall, and one end centered on the stud at the first seam. The opposite end should cover the corner post.

1 b. Set gypsum sheathing with the V groove upside down so that the joint sheds water (Fig. 7-18.3).

1 c. Set insulating sheathing with the shiplapped or tongued edge up (Fig. 7-18.4).

2 a. Attach plywood with 6d nails spaced every 6″ at edges, and every 12″ to studs in the field.

2 b. Apply each sheet of gypsum wallboard with either roofing nails or staples spaced every 4″ around the edges and 8″ in the field. Drive staples parallel to framing members—vertically into studs, horizontally into sills, plates, and headers.

Crowns should not break the paper, but lie flush with the surface.

2 c. Apply insulating sheathing with galvanized roofing nails spaced 3″ apart around the perimeter and 4½″ apart into studs—6 nails per stud.

3 Trim the second sheet so that it ends on a stud at least 32″ from the stud supporting the first piece.

4 Apply the second sheet.

5 Continue down the wall, staggering seams at least 32″.

6 Cut out excess material at openings.

7 Rout out smaller openings.

8 Measure the distance between the top edge of sheathing and the top plate.

9 Rip sheathing to this dimension.

10 Apply the final course as in step 2.

11 Where sheets meet at a corner, let one sheet overlap the other for a square corner.

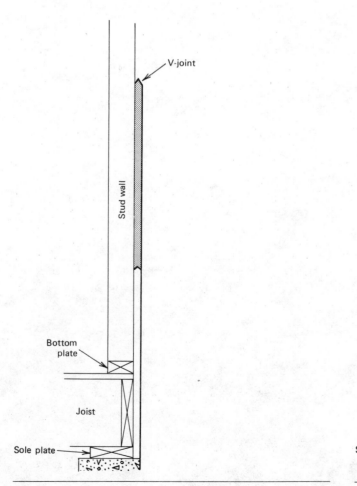

Fig. 7-18.3

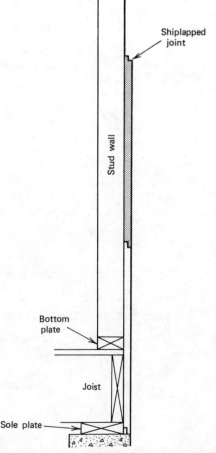

Fig. 7-18.4

Questions

1 Which of these materials are designed for horizontal application: $4' \times 8'$ gypsum sheathing, $\frac{3}{8}''$ plywood, $2' \times 16'$ insulating sheathing, 1×8 boards?

2 Which of the common sheathing materials may be applied with staples?

3 Is the space between fasteners always greater in the field of the sheet? Why?

8

Scaffolding

When carpenters work above the point where they can reach from ground level or a subfloor, they begin to climb ladders and use scaffolds. A **ladder** is a device for moving up or down from one working level to another. A **scaffold** is a temporary horizontal working platform at a fixed level. That level may be only a few feet off the ground or subfloor, or it may be 20' or more in the air.

The standards for safe design of scaffolding on building sites are spelled out in a thick book of requirements produced by the Occupational Safety and Health Administration (OSHA). Requirements vary with the height of the scaffold. For scaffolds less than 4' high there are no specific requirements for guard rails or width of floor. Scaffolds more than 4' but less than 10' high must have either a floor at least 45" wide, or guard rails on open sides and both ends. Scaffolds more than 10' high must have guard rails regardless of the width of the working platform.

The simplest low scaffold is a trestle. A **trestle** is a platform without rails that is supported at both ends. The supports themselves are also called a trestle. Sawhorses are a common support for a trestle that is used only briefly. When a trestle is used for a longer period, such as for application of finish ceiling material in a room, metal trestles and timbers may support the working platform (Fig. 8-0.1). Builders often build special scaffolds for special conditions, such as during roof framing (see Unit 8-1).

Higher scaffolds may be built of wood on the site, or assembled from steel parts manufactured for the purpose. Wood scaffolds may rest entirely on the ground, or be partly supported by the wall structure. Freestanding wood scaffolds are often called **double-pole scaffolds;** wall-supported types are called **single-pole scaffolds** (see Unit 8-2).

There are three basic types of manufactured scaffolding. One consists of a pair of jacks (Fig. 8-0.2) that

Fig. 8-0.1 You can build a temporary trestle scaffold with a couple of sawhorses and some planks. For a safer and more durable scaffold, use metal trestles and timbers underneath planking.

Fig. 8-0.2 The pump jack is widely used to provide a working platform. Vertical supports are pairs of 2 × 4s nailed together, and the jacks ride up and down on the 2 × 4s. The jacks are moved by pressure from a pedal. Scaffolds fit across the arms of the jacks. Jacks can be set up in about half an hour and taken down in about 10 minutes (Courtesy The Aluminum Association.)

rest on the ground and support platforms that can be raised or lowered as the work progresses. Another type consists of steel brackets attached to wall studs or roof rafters that support the working platform at a fixed height (see Unit 8-3). The third type is made up of tubular steel parts assembled on the site like the parts of an Erector set (see Unit 8-4). You adapt the assembly to the needs of the job.

The selection of the best scaffold depends somewhat on the job to be done. A narrow scaffold is often adequate if its sole purpose is as a walking or working surface. A broader scaffold may be required if it must also carry materials for installation and special tools not contained in the carpenter's tool bag.

To be safe, scaffolding must be plumb. It must be level. It must be strongly braced. It must be securely fastened. It must be strong enough for the load it will carry. It must have guard rails and toe boards when conditions warrant. It must be accessible by ladder; cross braces are not for climbing.

The planks that form the platform must be of a grade high enough to withstand the bending loads. Ends of planks should extend at least 6″ beyond the center of support, and overlap any other planks a full 12″. At the ends of scaffolds, planks should not cantilever more than 12″, and their ends should be cleated on the under sides to prevent them from sliding off their supports. If the fasteners do not get in the way of usage of scaffolding, planks should be nailed or clamped to the supports.

Questions

After each statement circle whether you think it is true (T) or false (F).

1 A trestle scaffold does not require rails.   T   F

2 Where two planks in a scaffold meet, they must overlap at least 6″.   T   F

3 A single-pole scaffold is self-supporting.   T   F

4 A scaffold more than 10′ above the ground or floor level must have guard rails.   T   F

5 Planks must not extend beyond the ends of supports.   T   F

Unit 8-1 **Build a Ridge Scaffold**

You can complete most carpentry work from forming for footings to raising walls without ever standing above the ground or subfloor. To assemble roof framing at the ridge, however, you need some sort of platform from which to work safely.

The level of the platform should be approximately 4' below the ridge. This height permits a carpenter to swing a hammer in natural motion. Beneath some low-pitched roofs the only platform needed is sheets of plywood laid across ceiling joists to form a catwalk directly below the ridge position.

When the ridge is more than 4' above the tops of ceiling joists, a low ridge scaffold is required. A good, safe scaffold must:

1 Be built to a comfortable working height.
2 Be wide enough to permit movement on both sides of the ridge.
3 Provide a safe, level working surface.
4 Prevent tools from falling on workers below.
5 Provide a safety rail to prevent falls.

Because the working conditions vary from job to job, most builders have their carpenters build a scaffold on site. The tools required are a steel tape, folding rule, saw, level, and hammer. The scaffolding must be sturdily built, yet it should be put together so that it may be disassembled and the parts reused.

Procedure

1 From the drawings, determine the distance from the top of ceiling joists to the top of the roof ridge. Subtract 24" from this dimension.

2 From 2 × 4 stock cut legs of this length for scaffolding. You need four legs for every 8 linear feet of scaffolding.

3 From 1 × 4 stock cut horizontal braces 57" long. You need as many braces as you have legs.

4 Lay a pair of legs flat on the subfloor or ground with $49\frac{1}{2}"$ between them. Attach brace *A* at right angles about 2" from one end (Fig. 8-1.1).

5 At a point 24" from the other end of the legs, attach another brace *B*.

6 Measure the diagonal distance between the ends of braces, and from 1 × 4 stock cut a diagonal

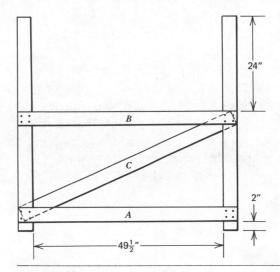

Fig. 8-1.1

brace to fit the space (*C* in Fig. 8-1.1). Attach the diagonal to the back side of the end frame.

7 Lay 1 × 10s or 1 × 12s across ceiling joists, centered 27" on each side of the future ridge (Fig. 8-1.2). Attach these boards to every second joist with duplex nails. Let boards overlap at their ends.

8 Place one end frame upright over a joist and toenail it to the 1 × 10 and joist.

9 At a point 7' away from the first frame, erect a second frame, with the diagonal brace running the opposite direction.

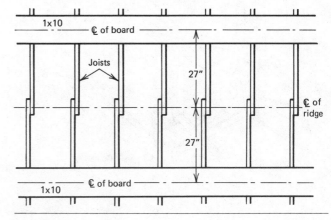

Fig. 8-1.2

10 Measure the diagonal distance between frames at braces (Fig. 8-1.3).

11 Cut four diagonals for each 8′ length of scaffold.

12 Attach the diagonals, first making sure that the frames are plumb.

13 Set another frame 1′ away from the completed section of scaffold, and nail it to the 1 × 10.

14 Repeat steps 9 through 13 until the scaffold is of the desired length.

15 Connect the end frames of adjoining sections with a length of 1 × 6 or 1 × 8 on each side (Fig. 8-1.4).

16 Between pairs of frames add a longitudinal 2 × 4 support about 85½″ long, and two 1 × 4 toe boards 8′ long (Fig. 8-1.5). The top of the 2 × 4

should be flush with the tops of horizontal braces, and the side rails no less than 48″ nor more than 48½″ apart.

17 Lay a deck of ½″ exterior plywood across the framework between toeboards. Begin with a half piece, so that full lengths bridge the gap between end frames.

18 Nail ends of plywood sheets to the 2 × 4 support. With duplex nails also nail 1 × 4 side rails into the edges of plywood every 24″.

19 Add a 2 × 4 safety rail on each side to complete the scaffolding (Fig. 8-1.6). **Note.** The location of these safety rails is critical. If they interfere with roof framing, they may be omitted.

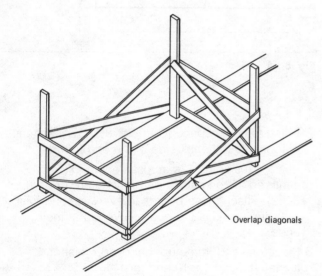

Fig. 8-1.3

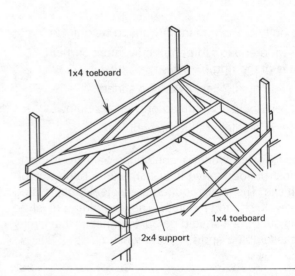

Fig. 8-1.5

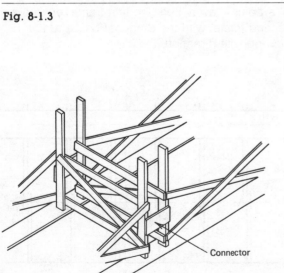

Fig. 8-1.4

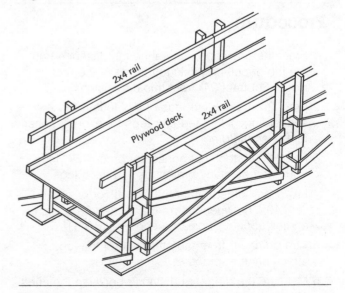

Fig. 8-1.6

Questions

1 Why are boards in step 7 located 27″ from the center line, but legs in step 4 are 49½″ apart?
2 What is the main value of a toeboard?
3 Name three ways in which the working platform is supported.

Unit 8-2 **Build a Wall Scaffold**

The work necessary to complete exterior walls and to treat the eaves of the roof almost always requires scaffolding. The height of the scaffolding depends on the distance from ground level to the eave. There are two types of wood scaffolding for these purposes: single-pole and double-pole.

A **single-pole scaffold** consists of one row of vertical supports and a series of horizontal supports attached to wall framing, or to sheathing if it is already installed (left, Fig. 8-2.1). The working platform rests on horizontal supports or ledgers.

A **double-pole scaffold** is a free-standing scaffold. It has two rows of vertical supports, is cross-braced, and is not attached to the house at any point (right, Fig. 8-2.1). The methods of construction are similar.

A single-pole scaffold is simpler and less expensive to build, provided the horizontal supports are not in the way of the work to be done. It is normal building practice, for example, to apply sheathing and siding from the bottom of the wall to the top. With a single-pole scaffold, both materials are applied at the top first, and the scaffolding is dismantled by levels as the work progresses downward. With a double-pole scaffold the entire wall and eave are available for work at any time.

When laying out a scaffold the carpenter must consider several conditions. The lengths of stud supports must be adapted to variations in ground levels. The spacing of stud supports must relate to spacing of studs in the wall. And the outer edge of the working platform must be far enough away from the building to permit working at the roof overhang. Ledgers may be 2 × 4s if their length is no more than 36″ and spacing of stud supports is no more than 48″. Otherwise use 2 × 6s.

To build either type of wood scaffolding the tools required are a steel tape, folding rule, saw, level, and hammer.

Procedure

1 Determine the width of working platform required. This width is normally the amount of roof overhang plus 2′.
2 To this dimension add 3″ if the wall is sheathed, and 7″ if it is not.
3 Determine the spacing between stud supports. This spacing is normally 48″.
4 From the spacing determine the number of stud supports required. Cut twice this many ledgers to the length established in step 2.
5 For a footing under each stud cut a square from a 2 × 8 or 2 × 10.
6 Cut a pair of 2 × 4 blocks 6″ long for ledger supports at each stud.
7 a. If the wall is not sheathed, cut another pair of blocks as in step 6 to support the ledgers at wall studs.
7 b. If the wall is sheathed, notch a pair of 2 × 6 blocks as shown in Fig. 8-2.2 to support ledgers at the sheathing.
8 Determine the required height of the working platform.
9 Attach one set of blocks to studs or sheathing 6″ below this height (Fig. 8-2.3), and another set at a

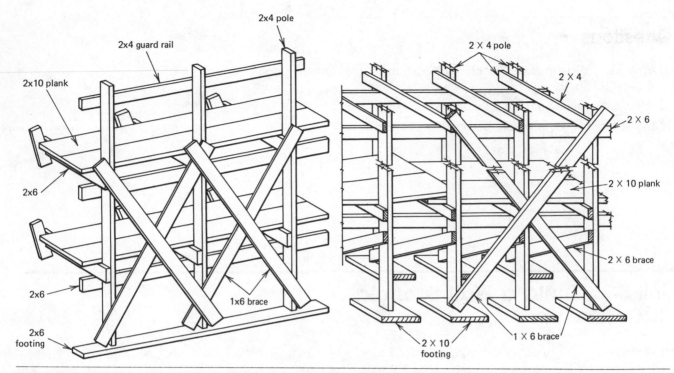

Fig. 8-2.1 In a single-pole scaffold (left) ledgers that carry the working platform are supported on the outer ends by a single row of studs, and at the inner ends by blocks attached to the wall. With a double-pole scaffold (right) ledgers are supported at both ends by rows of studs, which are cross braced. The wall provides no support.

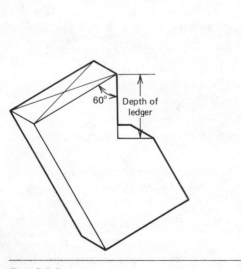

Fig. 8-2.2

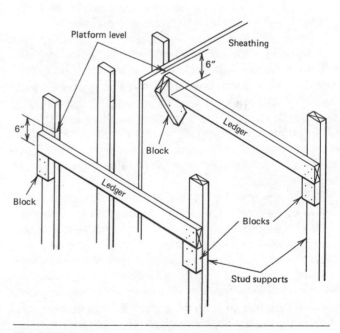

Fig. 8-2.3

lower point. **Note.** This lower point should be as low on the wall as possible, but no more than 4′ below the upper blocks. For tall scaffolds a third set of supports may be required.

10 Securely nail one end of ledgers to the wall support.

11 Set each stud on its footing in position.

12 Holding the stud plumb and the upper ledger level, securely nail the ledger, driving all nails home.

13 Add blocks below ledgers at stud supports.

14 Cross-brace the assembly with 1 × 6s every 48″.

15 Lay 2 × 10 planks across ledgers, overlapping at supports.

16 Add cleats on the under sides of end planks to keep them from sliding off ledgers.

Questions

1 If ledgers are 40″ long and 40″ apart, you would cut them from

_____.

2 A _____ -pole scaffold is not supported by wall framing.

3 How much longer must a ledger be if a wall is not sheathed than if it is sheathed in a single-pole scaffold? In a double-pole scaffold?

4 Is the angle of cut in the block in Fig. 8-2.2 important? Why?

Unit 8-3   **Build a Bracket Scaffold**

Several types of steel brackets are manufactured to support scaffolding for sidewall and roof installations, including special brackets for use at corners. The brackets may be attached with nails or bolts. Some types are designed to hook around studs in sidewalls. Nails are driven through slots in the brackets. When the need for scaffolding no longer exists, you remove the brackets, and drive the nails home. Nail pulling is not required.

The basic shape of wall brackets is a triangle (Fig. 8-3.1). Roof brackets, on the other hand, come in various shapes, depending on the steepness of the roof. Some types, for use on low pitches, hold a 2 × 4 or 2 × 6 flat against the roof (A and B in Fig. 8-3.2). Other types, for use on slightly steeper pitches, hold a 2 × 4 or 2 × 6 at right angles to the roof (C and D). Still others are adjustable, and hold planks as large as 2 × 10s level regardless of the pitch (E). The method of attachment is similar for all of them.

Tools required for attaching bracket scaffolds are a hammer, steel tape, chalk line, and saw.

Fig. 8-3.1 The typical wall bracket for scaffolding is a triangle. Here the leg attached to the wall and the hypotenuse leg are steel angles. The platform support and tension brace are flat steel. The platform here is lightweight metal that is thickest in the middle of the span. (Photo Jay Leviton, Atlanta.)

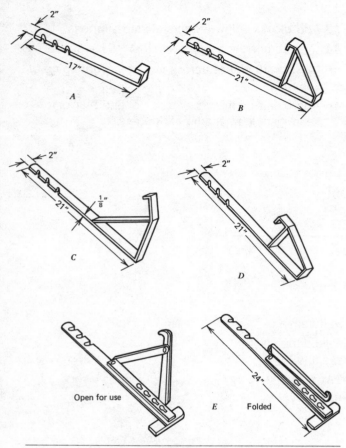

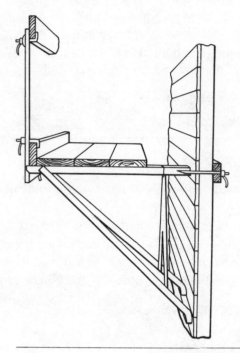

Fig. 8-3.2 The shape of roof brackets varies with roof pitch. Some types (a and b) are for low pitches, some for steep pitches (c and d), and some are adjustable to any pitch (e).

Fig. 8.3-3

Procedure for Attaching Wall Brackets with Nails

1 With a chalk line establish the height of the working platform.

2 Determine the spacing between brackets.

3 Attach brackets in position according to manufacturer's instructions. **Note.** Brackets usually fit between studs and through a brace bar that lies across the inside faces of studs. A nut on the threaded end of each bracket holds it tight against the studs.

4 Lay planks across the brackets. Lap ends over supports.

5 Add cleats to the undersides of planks.

6 Set the toeboard in position as shown in Fig. 8-3.3.

7 To the toeboard and each bracket, clamp a vertical support for a guard rail that is manufactured for this purpose.

8 Clamp the guard rail in position.

Procedure for Attaching Roof Brackets

1 Determine the required location for roof scaffolding.

2 Snap a chalk line across roof sheathing at this point.

3 Determine from manufacturer's instructions the spacing between brackets. The common spacing is 6 to 8', depending on the size of planking.

4 Measure the distance from final platform level to the center of the top nail hole.

5 Place brackets on the sheathing so that the first hole is above the chalk line by the distance measured in step 4.

6 Nail brackets with 8d nails. Make sure all nails go into rafters.

7 Lay planks across brackets.

8 If brackets are adjustable, set the support at the same angle as the pitch of the roof.

Questions

1 What size nails should be used for attaching roof brackets? Why do you think this size is recommended?

2 To what must brackets be nailed? Why?

Unit 8-4   **Assemble Manufactured Scaffolding**

For versatility and convenience, it is hard to beat manufactured scaffolding. You can build a scaffold high or low, short or long, fixed or movable—all from the same parts. The basic scaffold is ideally suited to the work of the framing carpenter, and there are special designs for other tradespeople, such as bricklayers.

All parts of manufactured scaffolding are made of tubular steel, and are designed to lock tightly together. The components consist of vertical frames, cross braces, flat plates, casters, rails and ladders (Fig. 8-4.1). You stack frames to the required height merely by inserting the bottom ends of one frame in the top ends of the one below. Ladder sections also fit into each other, and fit over vertical frames. Casters inserted into legs make the whole scaffolding movable if the surface is smooth enough to roll on. More commonly used are flat plates that serve as feet, and are adjustable to variations in ground slope.

The tools required for assembly vary with the manufacturer.

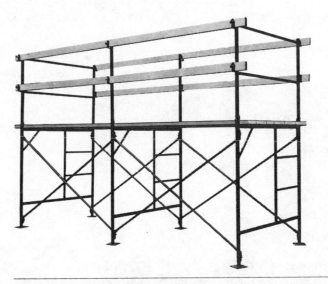

Fig. 8-4.1 Manufactured scaffolding is lightweight, easily adjustable to site conditions, and easy to erect and dismantle. Because all manufactured scaffolding must meet OSHA requirements, it is safer than most scaffolding built at the site.

Procedure

1 Determine the height of working platform required.

2 Insert feet or casters into the ends of lower frames.

3 Attach cross bracing.

4 Set the assembly in position, and plumb it. Locate the first assembly at the highest point of ground.

5 Repeat steps 2, 3, and 4 for all scaffolding on the ground.

6 Fit the next level of frames onto the lower frames, and cross brace.

7 Attach ladder sections after each level is completed.

8 When the level of the working platform is reached, insert end sections and guard rails.

9 Lay the platform, overlapping at tubular supports.

10 Cleat ends of planks.

9

Roof Framing

Up to the cap plate or rafter plate, the method of construction depends on the type of framing system used. From the rafter plate on up, the method of construction depends mainly on the style of roof used.

Two houses built from the identical floor plan can look quite different when only the roof style is changed. One house will usually look better than the other, however, because the type or style of roof is generally a matter of good architectural taste.

But roof styles change over the years, and at any given time in history some types are more popular than others. Two styles that have remained popular almost since the beginning of house building are the gable roof and the hip roof (Fig. 9-0.1). Other styles tend to come and go (Fig. 9-0.2).

Roof Styles

Of all roof styles and shapes, the **gable roof** is the most common. It has a high point at or near the center of the house or wing, called a **ridge**, that extends from one end wall to the other. The roof slopes downward from the ridge in both directions. This roof style gets its name from the **gable**, which is the triangular section of end wall between the rafter plate and the roof ridge. Usually the roof on one side of the ridge is the same size and slope as the roof on the other side, but this is not always true. The saltbox house built in early New England days had a gable roof with different slopes, and slopes of different lengths.

A **hip roof** also has a ridge, but it does not extend from one end of the roof to the other. The **eave**—the lower edge of a roof—is at a constant height, and the roof slopes downward to the eaves on all sides. The point where two roof surfaces meet at an outside corner is called a **hip.** Where two roof surfaces meet at an inside corner the junction is called a **valley.**

A **shed roof** slopes in only one direction, like half a gable roof. Walls supporting rafters are different heights, and the roof has no ridge. The shed roof has several variations. One is the **butterfly roof,** where two shed roofs slope toward a low point over the middle of the house. In another variation two shed roofs slope upward from the eaves, but do not meet at a ridge. The wall between the two roofs is called a **clerestory,** and is often filled with windows to let light into the interior of the house.

A **flat roof** may or may not be absolutely flat. It may slope just enough to let rain run off in one direction, but it appears flat. The roof frame for a flat roof must be built like a floor because of the live load (particularly snow) that it may have to support. In general, the less a roof slopes, the heavier the roof frame must be.

Two roof styles have double slopes—one pair of gentle slopes and one pair of steep slopes. A **gambrel roof,** like a gable roof, slopes in both directions from a center ridge. At a point about halfway between ridge and eave, however, the roof slope becomes much steeper. In effect, the lower slope replaces the upper exterior walls of a two-story house. It is common to add **dormers**—projections through the roof—for light and ventilation.

Just as a gambrel roof is like a gable roof with two different slopes, a **mansard roof** is like a hip roof. The roof drops from a shorter ridge in two distinct slopes to eaves that are the same height all the way around the house.

Fig. 9-0.1 The two most common roof styles are the gable roof (top) and the hip roof (bottom). In good residential design the two styles are never mixed. (Photos by Jay Leviton, Atlanta.)

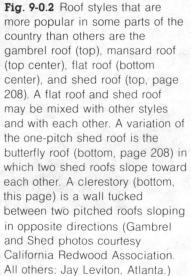

Fig. 9-0.2 Roof styles that are more popular in some parts of the country than others are the gambrel roof (top), mansard roof (top center), flat roof (bottom center), and shed roof (top, page 208). A flat roof and shed roof may be mixed with other styles and with each other. A variation of the one-pitch shed roof is the butterfly roof (bottom, page 208) in which two shed roofs slope toward each other. A clerestory (bottom, this page) is a wall tucked between two pitched roofs sloping in opposite directions (Gambrel and Shed photos courtesy California Redwood Association. All others: Jay Leviton, Atlanta.)

Roof Parts

On most houses, whatever the roof style, the roof extends beyond the line of the exterior walls. The horizontal distance that a roof extends beyond a wall is called an **overhang**. Overhangs serve two main purposes: they control the amount of sunlight that comes through windows, and they protect walls from heavy rains and snows. In hip, mansard, and flat roofs the amount of overhang is usually the same on all sides of the house. In gable, gambrel, and shed roofs, the overhang is usually greater at the front and rear—the eaves—than at the sides—**the rake.**

A roof has many parts, and carpenters must learn them all before they can cut the parts and assemble a roof. These parts all appear somewhere in Fig. 9-0.3.

A **ridgeboard** is the highest horizontal member in a roof with two or more slopes. There is no ridgeboard in a shed or flat roof. A ridgeboard has only one purpose: to make roof assembly easier. It has no structural value, and is not used with roof trusses. A ridgeboard is also known in various parts of the country as a **ridgepole, ridgebeam,** and **ridgeplate.**

A **purlin** is a horizontal framing member that runs at right angles to rafters and parallel to the ridgeboard. You will find purlins at the ends of upper slopes in gambrel and mansard roofs. They are also used as structural supports under long rafters, and as stiffeners in roofs built with trusses.

A **rafter** is the main framing member of a roof, just as a joist is the main framing member in a floor. There are many types of rafters:

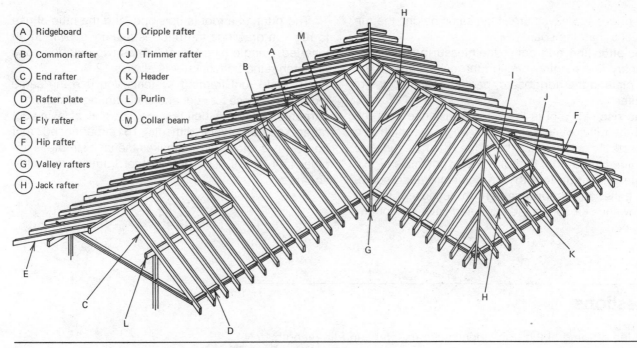

(A) Ridgeboard   (I) Cripple rafter

(B) Common rafter   (J) Trimmer rafter

(C) End rafter   (K) Header

(D) Rafter plate   (L) Purlin

(E) Fly rafter   (M) Collar beam

(F) Hip rafter

(G) Valley rafters

(H) Jack rafter

Fig. 9-0.3 This framing drawing, which combines a gable roof and hip roof, shows the main parts of any roof.

A **common rafter**, also called a **regular rafter**, is the framing member that runs from the ridge to the eave of a gable or hip roof, from ridge to purlin or purlin to eave of a gambrel or mansard roof, or from eave to eave of a shed or flat roof. A common rafter always runs at right angles to rafter plates and almost always to the ridge.

An **end rafter** is the outermost rafter in a gable roof with no overhang at the rake. It is identical to a common rafter, but is selected for its straightness. When a gable roof does overhang, the outermost rafter is called a **fly rafter**. It is the same length as a common rafter, but is cut differently.

A **hip rafter** runs diagonally from the end of a ridgeboard to the eave at the outside corner of the house. It is found only in hip and mansard roofs. A mansard roof has two hip rafters at each corner.

A **valley rafter** runs between the ridgeboard and the rafter plate at an inside corner of any house with a wing, regardless of the roof style.

A **jack rafter** is any short rafter that reaches either the eave or the ridge, but not both. A **hip-jack rafter** runs from hip rafter to eave. A **valley-jack rafter** runs from valley rafter to ridge. A **cripple-jack rafter** runs from a header to either ridge or eave. Like floors, roofs also have headers and trimmers, and they serve the same purposes.

A **collar beam** is a horizontal tie between pairs of rafters. It is a tension member that stiffens the roof structure.

Framing Terms

Whatever the style of the roof, certain framing terms are used to express the shape of the roof and to establish the lengths of its parts. These terms are span, run, pitch line, rise, and pitch (Fig. 9-0.4).

The **span** of a roof is the horizontal distance from the outside edge of the rafter plate on one wall to the outside edge of the rafter plate on the opposite wall. Every roof has a span.

The **run** of a roof is the horizontal distance from the outside edge of a rafter plate to the center line of the ridge. If the slope of the roof is the same on both sides

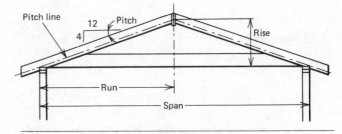

Fig. 9-0.4 Here are the five roof framing terms you will use over and over again.

of the ridge, and eaves are at the same height, the run is equal to half the span.

The **pitch line**, also called the **measuring line**, is an imaginary line running through the outer edge of the rafter plate to the ridgeboard, parallel to the edges of the rafter.

The **rise** of a roof is the vertical distance from the top of the rafter plate to the point where the pitch line intersects the center line of the ridge. The rise, run, and pitch line form a right triangle. The hypotenuse of this triangle is the **line length** of a common rafter, sometimes called the **theoretical length** because it makes no allowance for cuts at the ridge and eave, nor for any overhang.

The **pitch** of a roof is its slope, and the ratio of rise to run. On drawings the pitch is usually shown on an inverted triangle (see Fig. 9-0.4) as two numbers, one of which—the horizontal—is always 12.

To build roof framing systems you use the same tools that you use for floor and wall framing systems. To carpenters building their first roof, the task of laying out rafters appears very difficult. To experienced roof framers it is simple. What makes the task simple is another tool that is used only in roof framing: the rafter square.

Questions

1 In the drawing at the right, identify the rise, run, and line length.

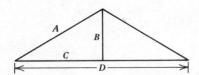

2 In the drawing with question 1, what is the formula for the pitch of the roof?

3 In which roof style or styles would you find a fly rafter? No ridgeboard? purlins?

Unit 9-1   Read a Rafter Square

To a carpenter framing floors and walls of a house, a framing square is a handy tool for marking and for checking squareness of openings. To a carpenter framing roofs, a rafter square is the most important tool he owns.

Like a framing square, a rafter square has a blade 24″ long and 2″ wide, and a tongue 16″ long and 1½″ wide. The two meet at the heel. But the edge markings and the information contained between rows of markings are somewhat different.

Markings

On one side of the square the markings on the outside edge of both blade and tongue are either in inches and sixteenths of an inch (Fig. 9-1.1) or in inches and eighths. On the inside edge, markings are in inches and eighths.

On the other side of the square, markings on the outside edge of both blade and tongue are in inches

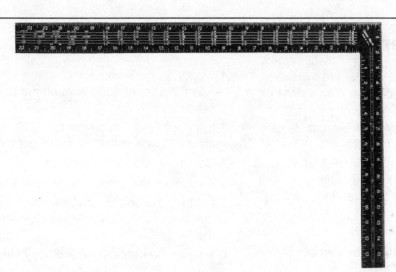

Fig. 9-1.1 A typical rafter square. The markings and the information provided on the tongue and blade vary with the manufacturer.

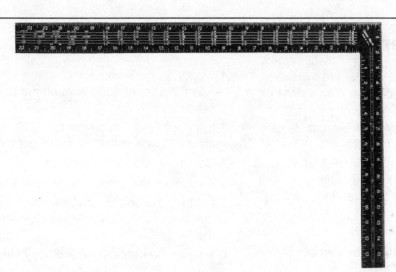

Fig. 9-1.2 Detail of a typical table on a rafter square.

and twelfths of an inch. On the inside edge of the tongue, they are in inches and tenths of an inch. Markings on the inside edge of the blade may be in tenths, twelfths, or sixteenths, depending on the manufacturer.

You may wonder why some markings are in tenths or twelfths of an inch. Information in rafter tables is sometimes given in tenths of an inch instead of in fractions. The twelfths marks give you a scale of 1″ = 1′-0″, which is useful in laying out roof rafters.

The information contained between markings varies with the manufacturer. All rafter squares contain at least one rafter table, however (see Fig. 9-1.2). From the tables you can calculate the lengths of main rafters, hip and valley rafters, jack rafters, and common rafters for roofs of various pitches. The tables are based on unit measurements—a form of scale.

Unit Measurements

Unit measurements used in roof framing are the unit run, unit rise, and unit length. **Unit run** is always 12″. **Unit rise** is the rise per foot of run; it is the other number beside the pitch triangle in drawings. **Unit length** is

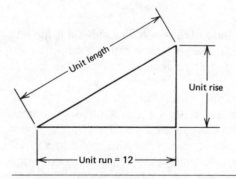

Fig. 9-1.3 Here are the unit measurements you use constantly to determine lengths of roof framing members.

the hypotenuse of a right triangle whose shorter sides are the unit run and unit rise (Fig. 9-1.3). In effect, by using unit measurements, you are reducing the actual run, rise, and rafter length to a scale of 1″ = 1′-0″.

To give you practice in using the rafter tables, solve these two problems:

1. Find the line length of a common rafter when the span of a gable roof is 30′-0″, and the pitch is 5 in 12.

2. Find the line length of a hip rafter in a roof with a span of 28′-0″ and a pitch of 6 in 12.

Procedure in Problem 1

1 Turn the rafter square to the side where the table on the blade gives "length of main rafters per foot of run."

2 Find the 5" mark (the unit rise) on the outside of the blade.

3 On the first line below the mark, read the figure. This is the unit length of a common rafter, per foot of run, stated in inches and hundredths of an inch.

4 Determine the run of the roof.

5 Since unit run is 12", determine the number of units in the run.

6 Multiply the unit length (from step 3) times the number of units (from step 5) to determine the length of the rafter.

7 Your answer should be 195", or 16'-3".

Procedure in Problem 2

1 Working on the same side of the rafter square, find the 6" mark (unit rise) on the outside of the blade.

2 In the second line below the mark, read the figure for the length of the hip rafter per foot of run.

3 Determine the run of the roof.

4 Determine the number of units in the run.

5 Multiply the unit length (from step 2) times the number of units (step 4) to determine the length of the hip rafter.

6 Your answer should be 252" or 21'-0".

Unit 9-2   **Mark a Plumb Cut**

Of all the cuts made in rafters, the plumb cut is the easiest to mark and make. A **plumb cut** is a cut in a framing member that is vertical when that member is in its final position in the roof structure.

When laying out any rafter, always lay the rafter stock flat across a working surface, such as sawhorses, with the crowned edge nearest you. Even a straight rafter will have a little crown—the high point in a crook. Make all measurements along the crowned edge, which will be the top edge of the finished rafter.

To make a plumb cut, you need a rafter square or framing square and a sharp pencil. You also need to know the unit rise and unit run of the roof.

You use the rafter square for information and also for marking. You measure the unit run on the blade and the unit rise on the tongue.

It doesn't matter whether you work from marks on the outside or inside of the blade and tongue, as long as you use only inside or only outside marks. It is usually easier to mark on the outside of the square, but on narrow stock the square is sometimes difficult to hold in place. To make marking easier, place a short piece of rafter stock under the ends of the square to help balance its weight.

Assume in the following procedure that the roof has a 4 in 12 pitch.

Questions

(Circle the correct answer.)

1 You measure the unit rise on the (tongue, blade).

2 You set the rafter stock with its (top, bottom) edge away from you.

3 You make all measurements on the (top, bottom) edge of rafter stock.

Procedure

1 Lay the rafter square on the piece of stock with the heel away from you, and the tongue at the right (Fig. 9-2.1).

2 Set the 4″ mark on the tongue against the top edge of the rafter.

3 Swing the square around until the 12″ mark on the blade is also against the top of the rafter.

4 Recheck to make sure both marks are on the top and same edge of the rafter stock.

5 Mark a line along the tongue. This is the line for a plumb cut.

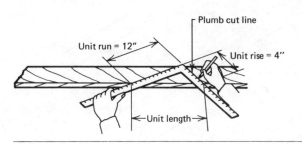

Fig. 9-2.1

Unit 9-3   **Mark a Sample Common Rafter**

In any section of roof, all common rafters are identical in size and length. Therefore many builders cut a pair of common rafters, test them for a fit, and use these rafters as templates for cutting the rest of the common rafters required.

A common rafter has at most two parts. The **body** is the part of the rafter between supports (Fig. 9-3.1). The **tail** is the part of the rafter that extends beyond the rafter plate to form an overhang. **Overall rafter length** is the combined length of body and tail as measured along the pitch line. Because of cuts at the ends of the rafter, the rafter stock required is somewhat longer than the overall rafter length.

Types of Cuts

A common rafter gets three cuts. Where it meets the ridgeboard it gets a **plumb cut** (see Unit 9-2). At the other end it gets a **tail cut**. The shape of the tail cut depends on the design of the roof and its overhang. It may be a plumb cut (vertical), a **square cut** (at right angles to the pitch), a **heel cut** (horizontal cut), or a **combination cut** (Fig. 9-3.2).

Where the rafter rests on the rafter plate, it must be notched to fit. The notch is called a **bird's mouth** or **rafter seat.** At least 3″ of the rafter must bear on the raf-

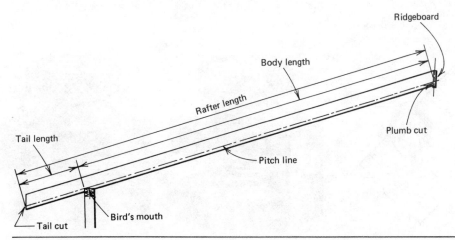

Fig. 9-3.1 A rafter has no more than two parts—a body and a tail. Note that line lengths of body and tail are measured along the pitch line. The actual overall length of the rafter is a little longer.

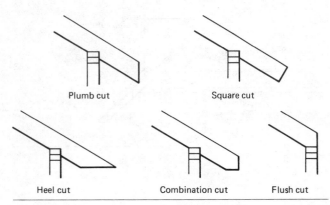

Fig. 9-3.2 The tail cut of a rafter may take many shapes, but there is a logical reason for all of them.

ter plate. If the rafter does not have an overhang, but ends flush with the rafter plate, it receives a **flush cut.**

Methods of Layout

Rafters may be laid out by two methods. In the **measurement method** the actual lengths of the rafters are used. In the **step-off method** unit measurements are used. The devices shown in Fig. 9-3.3 are useful in measuring speed and accuracy with the step-off method.

To become familiar with methods, assume that a house has a hip roof with equal pitches. The span is 28'-0", overhang is 2'-0", and the pitch is 4 in 12. The problem is to lay out a common rafter.

Procedure with the Measurement Method

1 Using the table on the rafter square, determine the approximate overall length of the rafter (see Unit 9-1).

2 From drawings determine the size of rafter stock, and select a straight piece at least 6" longer than the length determined in step 1.

3 Lay the piece across sawhorses with the crowned edge toward you. Mark this edge with an arrow.

4 Near the right-hand end of the stock, mark the plumb cut at the ridge (see Unit 9-2).

5 Determine the line length of the rafter.

6 With a steel tape measure this distance along the edge of the rafter nearest you, beginning at the plumb cut. Mark this point.

7 Through the mark draw a line parallel to the plumb cut (Fig. 9-3.4). This is the plumb cut of the bird's mouth.

8 Determine the length of the tail. The answer should be the answer in step 1 less the answer in step 5.

9 From the plumb cut of the bird's mouth measure off the distance in step 8 and make a mark.

10 Determine from drawings the type of tail cut required.

11 Mark the tail cut.

12 From the plumb cut of the bird's mouth measure off $3\frac{1}{2}''$ at right angles toward the upper end of the rafter (Fig. 9-3.5). **Precaution.** On roofs with a very steep pitch the measurement should be no more than half the width of the rafter.

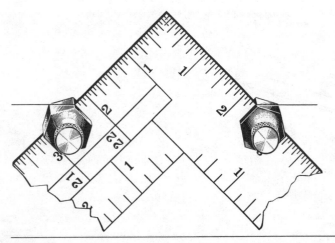

Fig. 9-3.3 Stair gauge fixtures come in pairs and are clipped to the square at the proper pitch line. (Courtesy

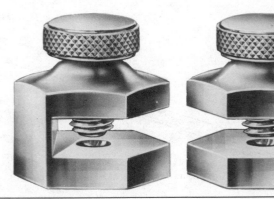

The L.S. Starrett Company.)

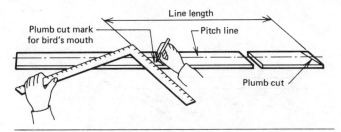

Plumb cut mark for bird's mouth — Line length — Pitch line — Plumb cut

Fig. 9-3.4

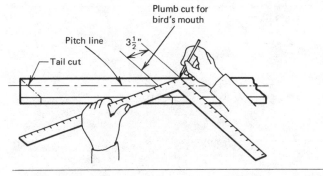

Plumb cut for bird's mouth — Pitch line — $3\frac{1}{2}''$ — Tail cut

Fig. 9-3.5

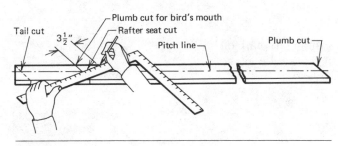

Plumb cut for bird's mouth — Rafter seat cut — Tail cut — $3\frac{1}{2}''$ — Pitch line — Plumb cut

Fig. 9-3.6

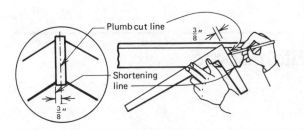

Plumb cut line — $\frac{3}{8}''$ — Shortening line — $\frac{3}{8}''$

Fig. 9-3.7

13 Through this point draw a line parallel to the plumb cut.

14 Beginning where this line meets the bottom edge of the rafter, draw a line at right angles between the parallel lines (Fig. 9-3.6). This line marks the seat cut of the bird's mouth.

15 Determine from drawings the thickness of any ridgeboard.

16 Measure off half this thickness at right angles to the plumb cut at the ridge (Fig. 9-3.7).

17 Draw a line through this point. This is the shortening line. The sample rafter is now ready to cut.

Procedure with the Step-Off Method

1 Repeat steps 1 to 4 inclusive of the measurement procedure.

2 Set the square on the plumb cut line with the unit run on the blade and the unit rise on the tongue.

3 Where the blade crosses the edge of the rafter nearest you, make a mark.

4 Repeat steps 2 and 3 as many times as there are feet in the line length of the rafter (Fig. 9-3.8).

5 At the last mark draw a line parallel to the plumb cut. This line marks the plumb cut of the bird's mouth.

6 Repeat steps 2 and 3 as many times as there are feet in the overhang. **Note.** If any measurement is not in even feet, such as 1'-9" for the overhang, set the tongue on the plumb cut line, and slide

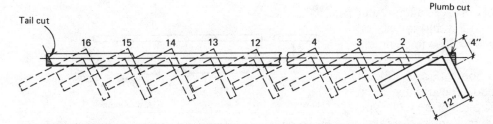

Tail cut — 16 15 14 13 12 4 3 2 1 — Plumb cut — 4" — 12"

Fig. 9-3.8

the square along the line until the blade crosses the edge of the rafter at the 9″ mark (Fig. 9-3.9). Then mark off the remaining distance in full feet as in step 3.

7 Repeat steps 12 through 17 of the previous procedure.

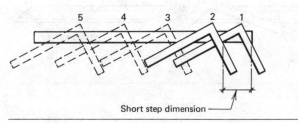

Fig. 9-3.9

Questions

1 A bird's mouth is formed by making two cuts. One is a _____ cut and the other is a _____ cut.

2 The correct answer in step 1 of the measurement procedure is 16′-10$\frac{3}{8}$″. Show mathematically how you arrive at this answer.

3 What is the length of the tail in step 8?

4 Which of the two methods of marking do you think is the quicker?

5 Which method do you think is the more accurate? Why?

6 Suppose that a roof has an 18″ overhang and a 6 in 12 pitch. To mark off the odd dimension, at what marks on the tongue and blade would you set your rafter square?

Unit 9-4   **Make a Layout Tee and Pitch Board**

Two simple devices can save you a lot of time in marking cuts in rafters.

One device is a **layout tee** (Fig. 9-4.1) cut from two pieces of board the same depth as a common rafter. You use the layout tee for marking repetitive cuts in common rafters, and similar cuts in jack rafters. After marking the location of the plumb cut and tail cut on the top edge of a rafter, use the layout tee to complete the markings.

The other device is a **pitch board** (Fig. 9-4.2) cut from a piece of exterior plywood. You use the pitch board whenever a cut follows the pitch of the roof. Typical examples are in stepping off the length of a rafter, and cutting studs in gables.

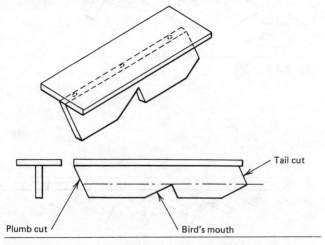

Fig. 9-4.1 A layout tee can save time in laying out rafters and, if carefully built, can increase accuracy in marking.

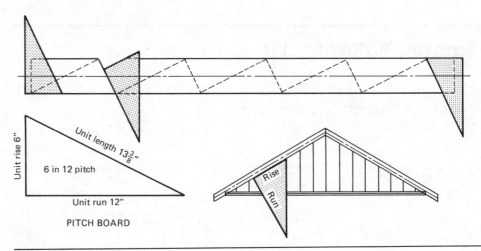

Fig. 9-4.2 A pitch board is a useful timesaver in marking cuts on rafters, studs in gables, and stairway supports.

You need a different pitch board for every roof pitch.

Procedure for Making a Layout Tee

1 Determine the length of the tail of a common rafter, then double it.

2 From a board of the same depth as the common rafter, cut two pieces to the length in step 1.

3 At one end of one board mark the tail cut.

4 From the plumb cut at the tail measure up and mark the plumb cut at the bird's mouth.

5 Mark the bird's mouth.

6 At the other end of the board mark the plumb cut.

7 Check all marks carefully, and cut accurately along the lines.

8 Square the ends of the second board.

9 Along the center line of the top of the tee, drill three holes at Intervals for screws (Fig. 9-4.1).

10 Center the second board over the first board.

11 Holding the two boards at right angles to each other, attach them with flat head wood screws.

Procedure for Making a Pitch Board

1 Along the straight edge of a piece of scrap plywood measure off 12". This is the unit run.

2 At right angles to the edge draw a line, and measure off the unit rise.

3 Draw a third line to form a right triangle.

4 Carefully and accurately cut the triangle out of the plywood.

5 Write the pitch on the triangle.

Questions

1 On the layout tee is the distance between the plumb cut at the bird's mouth and the tail cut fixed? Why?

2 On the layout tee is the distance between the plumb cut at the bird's mouth and the plumb cut at the ridge fixed? Why?

3 To what is the length of the hypotenuse of a pitch board triangle equal?

Unit 9-5 Test the Sample Rafter for Fit

Before the sample rafter is cut, it should be checked for accuracy of fit. The best time to test the sample is after the subfloor has been laid but before any walls have been raised. The subfloor makes an excellent worktable. Tools needed to test a rafter are a chalk like and a saw.

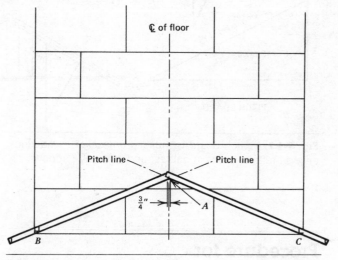

Fig. 9-5.1

Procedure

1 Down the center of the subfloor below the future ridge snap a chalk line (Fig. 9-5.1).

2 Along the chalk line from one edge of the floor, measure off the rise of the roof. Mark this point A.

3 Through this point to both corners of the floor snap two more chalk lines. These lines (AB and AC) represent the pitch lines.

4 Lay the two sample rafters on the floor. If cutting lines are on the same sides of both rafters, carry the lines on one rafter to the opposite side.

5 Fit rafters at the corners of the floor, and let them overlap at the center chalk line.

6 Check these points:

   a) Do the bird's mouths fit properly at the corners?
   b) Are the edges of rafters exactly parallel to the pitch lines?
   c) Do the plumb lines at the ridge fall exactly on the chalk line?

7 If the fit is not good, start over. Never try to force a good fit. If the fit is close, make adjustments at the ridge by relocating the plumb cuts.

8 Shorten both rafters.

9 Retest the fit, using a short piece of ridgeboard stock centered on the chalk line as a spacer.

10 When the fit is good, carefully make the cuts at bird's mouth and tail.

11 To mark additional rafters for cutting, use the sample rafters to locate the positions of plumb cuts. Use the layout tee for marking the cuts.

Questions

1 Do the bottom edges of sample rafters lie on the chalk lines?

2 If the fit is close but not exact, where is the best place to make adjustments?

Unit 9-6 Mark and Splice a Ridgeboard

A ridgeboard must be wider than a rafter, but may be thinner. If rafters are 2 × 6s, for example, the ridgeboard is usually a 1 × 8. A board (1″ lumber) is thick enough because a ridgeboard is nothing more than a means of attaching and aligning rafters during roof construction. But the board should be wider than raf-

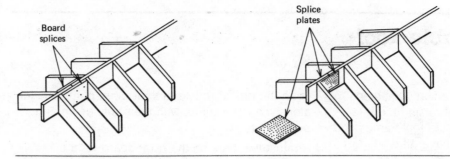

Fig. 9-6.1 Long ridgeboards must be spliced on both sides with splice plates made either of ridgeboard stock or perforated metal.

ters so that both top and bottom edges of rafters butt against the ridgeboard. If rafters and ridgeboard were the same width, the bottom of the plumb cut on rafters would lie below the bottom edge of the ridgeboard.

In most roofs it is not necessary to determine the exact length of the ridgeboard in advance. It is easier to let the ridgeboard overhang at its ends, and then cut the ends to fit after end rafters, fly rafters, or hip rafters are in place.

When the ridge is very long, the carpenter must splice the ridgeboard. Unless the top member is a heavy beam, the ridgeboard should be spliced before it is put in place. Splices, however, should fall between rafters. Therefore the ridgeboard must be laid out, cut, and spliced with care. For splices you can use lengths of ridgeboard stock 12 to 14″ long, or **splice plates**— metal sheets with holes punched in them (Fig. 9-6.1).

It is necessary to mark rafter locations on the ridgeboard, just as you marked joist locations on plates. There are three reasons:

1 You thus make sure that the ridgeboard is long enough for the job.
2 The marks assure proper placement of rafters.
3 The marks help you spot any twisted rafters that must be replaced.

Procedure

1 From the plans or the roof-framing drawing determine the approximate length of the ridgeboard.
2 Select two pieces of straight ridgeboard stock with a combined length no less than that required in step 1.
3 Square one end of each board, and butt the ends tightly together on a flat surface.
4 a. Cut two splice boards, and attach one to each side of the splice with 10 nails as shown in Fig. 9-6.1. Use 4d coated nails.
4 b. Across the joint place a metal splice plate, and hammer it flat into both boards. Then turn the ridgeboard over and repeat the process.
5 On the top edge and one side of the ridgeboard just outside the splice plate, mark the location of a rafter.
6 From this point measure off standard rafter spacings, and mark all rafter locations on the top and both sides of the ridgeboard. **Note.** The easiest way is to lay the ridgeboard along the cap plate on one wall, and transfer the spacings marked there.

Question

1 Why do you think the position of splices in the ridgeboard should be determined before rafter locations are marked?

Unit 9-7 **Frame a Gable Roof**

After the ridgeboard has been spliced and marked, and all common rafters cut, the task of raising a gable roof is ready to start. At the beginning the roof frame is very unstable, and it is important to have all working platforms, roof parts, and tools in place or at hand in advance of erection.

A gable roof may be framed in several ways, depending on conditions. Ordinarily two rafters are nailed to one side of the ridgeboard, then two to the other side. These four legs support the ridge while other rafters are added. If the ridge is a heavy beam, however, the ridgeboard is often supported temporarily on posts, and rafters are attached in pairs (see Unit 9-8). The finished result is the same (Fig. 9-7.1).

Procedure

1 Set and secure a working platform or scaffolding in place beneath the ridge.

Fig. 9-7.1 It is customary to frame a gable roof first, then to build the gables beneath its ends. (Courtesy Western Wood Products Association.)

2 Lean the precut rafters at their point of installation along the rafter plate, with their tails down and plumb cuts up.

3 Against the mark on the ridgeboard for the next-to-end rafter, set a common rafter (A in Fig. 9-7.2), and attach it by nailing through the ridgeboard with three 10d nails (Fig. 9-7.3, (left)).

4 Against the mark next to the splice, nail a second rafter (B in Fig. 9-7.2), on the same side in the same manner.

5 Slowly raise the assembly until the bird's mouths rest on the rafter plate against ceiling joists. Shove the rafters tight against the plate.

6 Toenail the rafters to the rafter plate with 16d nails, and nail straight into the ceiling joists with 10d nails (Fig. 9-7.3, middle).

7 Position the opposite pair of common rafters (C and D in Fig. 9-7.2), and nail them at both ends as with the two rafters (Fig. 9-7.3, (right)). **Note.** In some roofs opposite rafters will fit on the opposite sides of rafter marks on the ridgeboard.

8 Add temporary 1 × 4 braces between ridgeboard and ceiling joists. **Note.** There are many methods of bracing; Fig. 9-7.4 shows two of the more common methods.

9 Add a pair of rafters near the unsupported end of the ridgeboard. This pair matches rafters A and C in Fig. 9-7.2 at the other end of the roof.

10 Add a pair of rafters on the other side of the splice. This pair matches B and D.

11 Check the ridgeboard for level.

12 Add three more pairs of rafters at each end.

13 Check and adjust the plumb of the end pairs of rafters.

14 If necessary to achieve plumb, reset the temporary braces attached in step 8.

15 To maintain plumb, nail a temporary diagonal brace across the under sides of rafters from ridge to rafter plate (Fig. 9-7.5).

16 After you install each pair of rafters, sight along the top and bottom edges of the ridgeboard to assure that it is straight. At the same time recheck its level.

17 Install all remaining common rafters in pairs. Omit all other rafters until the entire roof frame is straightened.

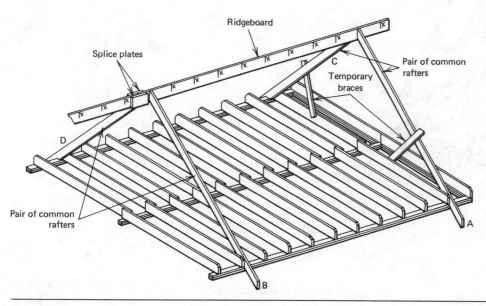

Fig. 9-7.2

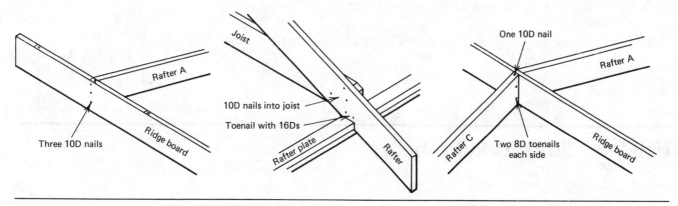

Fig. 9-7.3

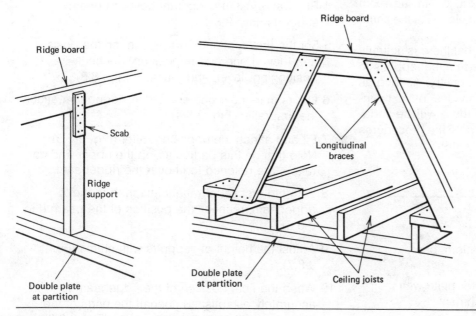

Fig. 9-7.4

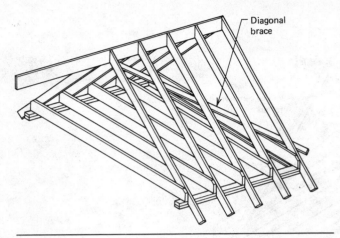

Diagonal brace

Fig. 9-7.5

Questions

1 Under what circumstances will rafters not meet on the same side of marks on the ridgeboard?

2 Why do you think the next-to-the-end rafters are attached before the end rafters?

Unit 9-8   **Raise a Ridgebeam**

When the spinal member of a gable roof is a heavy beam, many of the procedures in Unit 9-7 must be altered. Often this ridgebeam is exposed in the room below it, and the ceiling of the room slopes at the same pitch as the roof.

It was stated early in this book that there is actually no such thing as a standard procedure that applies to all conditions and all areas of the country. As long as individual houses are built from individual plans by individual builders and carpenters, there will be variations to standard procedures. Compare the procedures below, for example, with those in Unit 9-7.

Procedure

1 Build and secure a working platform or scaffolding beneath the ridge.

2 Lean cut rafters against the rafter plate with their tails up, near the point of attachment.

3 Set a pair of rafters in position on rafter plates to establish the heights of the top of the ridgebeam and the tops of supporting posts (Fig. 9-8.1).

4 Build and install temporary posts as beam supports (Fig. 9-8.2).

5 Set the first section of ridgebeam on the posts, and toenail one end in position. Recheck positioning, level, and plumb (Fig. 9-8.3).

6 On its mark near one end of the ridgebeam attach the first rafter (Fig. 9-8.4).

7 Fit and attach the opposing rafter (Fig. 9-8.5). Note that on this particular job the upper ends of rafters are notched to fit over the ridgebeam.

8 With the first set of rafters attached to the ridgebeam, recheck the position of the beam (Fig. 9-8.6).

9 Adjust the height of supports with shims (Fig. 9-8.7).

10 When the final position of the ridgebeam is accurately established, toenail the ends of the first set of rafters to the rafter plate (Fig. 9-8.8).

Fig. 9-8.1 (Photos Drew Leviton, Atlanta.)

Fig. 9-8.2

Fig. 9-8.3

Fig. 9-8.4

Fig. 9-8.5

Fig. 9-8.6 (Photos Drew Leviton, Atlanta.)

Fig. 9-8.7

Fig. 9-8.8

Fig. 9-8.9

Fig. 9-8.10

11 Continue to attach rafters in pairs (Fig. 9-8.9) up to the point of a splice in the ridgebeam.

12 Fit the next section of ridgebeam into position (Fig. 9-8.10). Add end supports and check the beam for position, level, and plumb.

13 Add a pair of rafters at the end of the next section of beam.

14 Install all remaining common rafters in pairs.

15 To maintain plumb, nail a temporary diagonal brace across the under sides of rafters from ridge to rafter plate.

Questions

1 Why are rafters laid against rafter plates with their tails up in step 2 of this procedure, but with their tails down in step 2 in Unit 9-7?

2 What single factor more than any other causes the differences in procedures between Units 9-7 and 9-8?

Unit 9-9   **Frame a Gable**

Many builders do not complete the sections of end walls that lie above the rafter plate until after they have framed most of the roof. By waiting, they can build the wall to fit the exact space available.

The method described in this unit is for framing a gable under a gable roof. The same general procedures apply, however, to gambrel walls and to end walls under a shed roof, whether those walls are built in one piece or two.

Gable walls may be framed in three ways (Fig. 9-9.1). If there is no opening in the gable for a window or

ventilating louver, upper studs may be placed directly above studs in the lower wall (A). This placement is not a structural requirement, however, and it is usually better to begin at the center line of the gable, and frame in both directions.

The second method is to center a stud directly under the ridgeboard (B in Fig. 9-9.1). This placement works well if a pair of windows or a standard 30″ window is centered in the gable.

The third method (C) is to place studs 8″ on each side of the center line. This placement works well if a

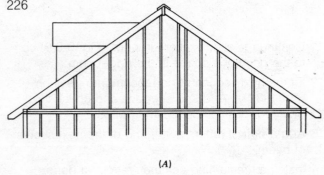

(A)

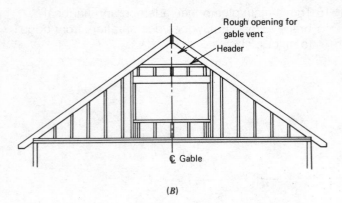

Rough opening for
gable vent

Header

℄ Gable

(B)

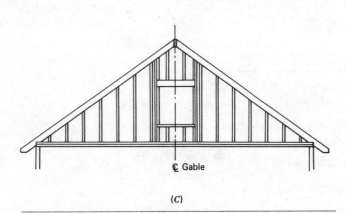

℄ Gable

(C)

Fig. 9-9.1 The three ways to frame a gable—stud above stud (a), with center line of gable on center line of middle stud (b), or with center line of gable between a pair of shorter studs (c).

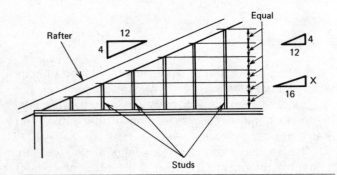

Rafter

12

4

Equal

4
12

X
16

Studs

Fig. 9-9.2 The common difference in length between adjacent studs in a gable wall is the unit rise of the roof times stud spacing in inches and divided by 12. A common difference also occurs between some jack rafters.

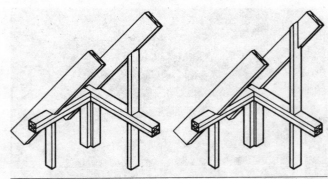

Fig. 9-9.3 When a roof has an end rafter, studs in the walls below end rafters may be cut to fit in two ways. Either method applies to any roof style that does not have a constant eave height.

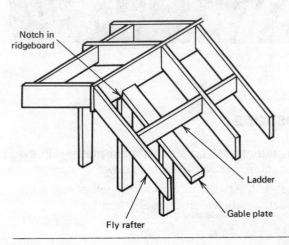

Notch in
ridgeboard

Ladder

Gable plate

Fly rafter

Fig. 9-9.4 When a roof has no end rafter, but an overhang, studs in the gable do not touch a rafter, as in Fig. 9-9.3, but end at a gable plate. The top surface of the gable plate and the under edge of rafters lie in the same plane.

standard 46″ window is centered in the gable. Because studs equidistant from the center line of the gable are the same length and are cut the same way, carpenters can save time by planning their work from a center line.

All the studs in half a gable are a different length. But the difference in lengths of adjacent studs is the same. This constant is known as the **common difference** (Fig. 9-9.2). To find the common difference between studs in inches, you multiply the unit rise of the roof times the spacing of studs stated in feet. Suppose that a roof has a 4 in 12 pitch and that studs are 16″ o.c. in the gable. Therefore if unit rise × stud spacing (in feet) = common difference (in inches), then $4 \times 1\frac{1}{3} = 5\frac{1}{3}$.

Stated algebraically from a pair of similar triangles, $\frac{x}{16} = \frac{4}{12}$. Then $x = \frac{64}{12}$ or $5\frac{1}{3}''$ (about $5\frac{5}{16}'$).

When there is an end rafter above end walls, studs may be cut to fit against the under sides of the rafter, or may be notched to fit around the rafter (Fig. 9-9.3). The second method is stronger, and the attachment is easier.

Most houses have an overhang at the rake. In these instances the top member of a gable wall is a **gable plate** (Fig. 9-9.4). Its top surface is in the same plane as the under sides of roof rafters. The ridgeboard is notched to fit over the gable plate. The fly rafter at the edge of the roof is supported on a ladder. The ladder lies across the gable plate and is attached to the first common rafter inside the gable wall.

Procedure for Framing a Gable Without Overhang

1 Set the two end rafters in position, and toenail them to the ridgeboard and rafter plate.
2 Trim the ridgeboard flush with the end rafter.
3 From drawings determine the position in the gable of any header to frame ventilating louvers. Mark this position on the end rafters.
4 Measure the horizontal distance between marks.
5 On a piece of stud stock mark this dimension. Use your pitch board to draw the cut lines (Fig. 9-9.5).
6 Cut along these lines.
7 Fit and level this header flush with the outer sides of end rafters. Toenail it to the under sides of the rafters.
8 Measuring vertically from the rafter plate to the top of the end rafter, determine the length of a gable stud. **Note.** It is customary to begin with either the shortest or longest stud in the gable.
9 Select a piece of stud stock longer than the measurement in step 8, and square one end.
10 Set the stock on its stud mark on the rafter plate, plumb it, and mark on it the position of the top and bottom edges of the rafter (*A* and *B* in Fig. 9-9.6).
11 Square these lines across one side of the stud.
12 Trim along the upper line (*C*).
13 Along the lower line make a cut $1\frac{1}{2}''$ deep (*D*).
14 Complete the notch with a vertical cut (*E*).
15 Determine the common difference between gable studs.
16 Mark, cut, and notch the remaining studs.
17 Cut all cripples below the header square and to length.
18 Install all studs on their marks.

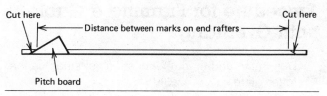

Fig. 9-9.5

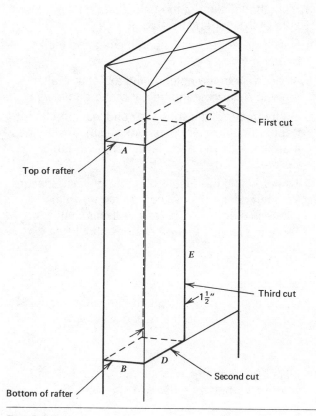

Fig. 9-9.6

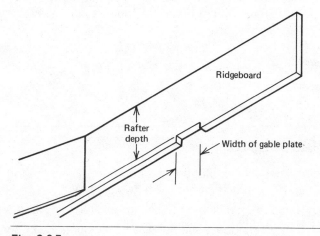

Fig. 9-9.7

Procedure for Framing a Gable with Overhang

1 Determine and mark the locations of studs on the rafter plate.

2 Measure the depth of the first rafter at the ridgeboard, and transfer this distance to the ridgeboard above the gable wall.

3 Mark the location of the gable wall on the under side of the ridgeboard (Fig. 9-9.7).

4 Notch the ridgeboard along these three marks made in steps 2 and 3.

5 Measure from the notch to the top of the rafter plate to determine the height of the gable wall.

6 On the rafter plate above the end wall, mark the depth of both rafter seats (Fig. 9-9.8). These lines mark the location of the lower ends of gable plates.

7 To determine the length of gable plates, measure from the center line of the ridgeboard to the marks in step 6. Measure in both directions to verify the length; the dimensions should be identical.

8 Mark the length from step 7 on two pieces of plate stock.

9 Using a pitch board, mark the lines for the plumb cut at the ridge and the heel cut at the plate (Fig. 9-9.9).

10 Cut gable plates.

11 Install gable plates on their marks, cross-nailing where they butt together beneath the ridgeboard.

12 From drawings determine the position in the gable of any header to frame louvers. Mark this position on the gable plates.

13 Measure the distance between marks.

14 Mark this dimension on a piece of stud stock, and use your pitch board to mark the cut lines (see Fig. 9-9.5).

15 Cut along these lines.

16 Fit and level this header. Toenail it to the under sides of gable plates.

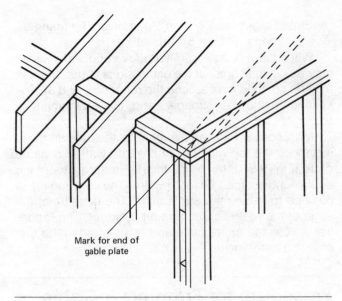

Mark for end of gable plate

Fig. 9-9.8

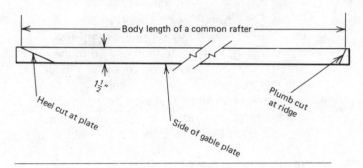

Body length of a common rafter

Heel cut at plate

$1\frac{1}{2}''$

Side of gable plate

Plumb cut at ridge

Fig. 9-9.9

17 Determine the length of the longest pair of gable studs, and cut them to fit.

18 Determine the common difference between gable studs.

19 Mark and cut the remaining studs.

20 Cut all cripples below the header square and to length.

21 Install all studs on their marks.

Questions

1 If studs in a gable are spaced 16" o.c., and the roof pitch is 5 in 12, what is the common difference between studs?

2 Name one advantage to laying out studs in a gable from the center line.

3 Can you suggest another way of handling the connection between a ridgeboard, two rafter plates, and a center stud in a gable?

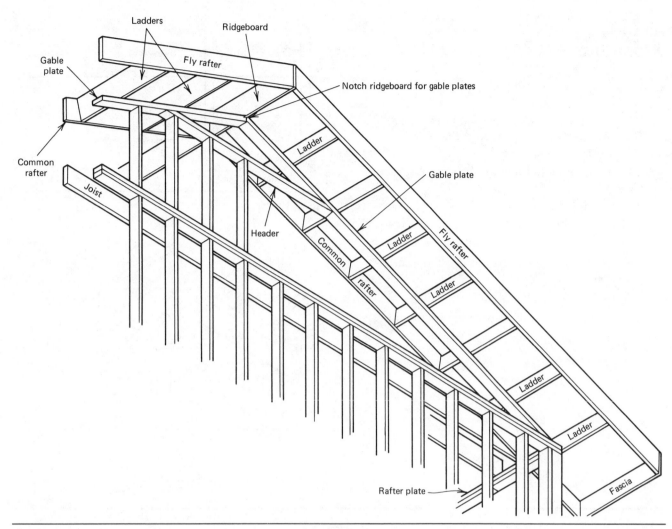

Fig. 9-10.1 The parts of a ladder supporting a fly rafter fit at right angles to rafters and flush with their tops.

Unit 9-10   **Build an Overhanging Rake**

When a gable roof overhangs the gable wall, the ends of the roof consist of a ladder laid at right angles to the gable plate, a fly rafter at the ends of the ladder, and a fascia at the lower ends of fly rafters. The underside of the roof is usually finished with a soffit.

Note in Fig. 9-10.1 that the ladder parts are not installed vertically, but at right angles to the roof pitch.

Spacing of ladder steps is usually either 16 or 24″ o.c. Every second or third step is positioned to support the seam between sheets of roof sheathing, which overhangs the fascia $\frac{3}{4}$″. The lowest ladder part may need to be notched to fit over the end ceiling joist.

A fly rafter is cut exactly like a common rafter except that it receives no bird's mouth.

Questions

1 Name one important factor to consider in determining spacing of ladder steps at a roof overhang.
2 What is the difference between a fly rafter and a common rafter?

Procedure

1 Determine from drawings the amount of side overhang.

2 From rafter stock cut ladder members to length. **Caution.** Remember to allow for the thickness of the fly rafter (Fig. 9-10.2).

3 On the gable plate mark the spacing of ladder steps.

4 If necessary, notch a ladder step to fit over the ceiling joist (Fig. 9-10.3).

5 Toenail steps to the gable plate on their marks.

6 Nail into ends of all steps through the endmost common rafter.

7 Cut the fly rafter to length.

8 Nail the fly rafter to all supporting members.

9 Trim the ridgeboard flush with the fly rafter.

10 Add fascia boards.

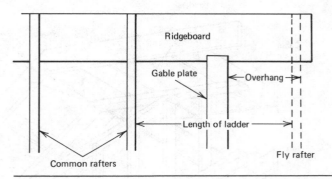

Fig. 9-10.2

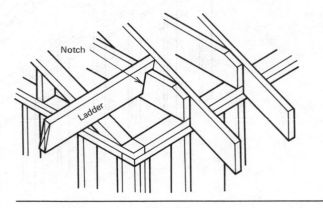

Fig. 9-10.3

Unit 9-11   **Plumb and Straighten a Roof**

The proper time to plumb a roof is after all common rafters are in place, including end rafters. Once the roof is plumb, bowed rafters should be straightened in preparation for attachment of sheathing.

When a roof has an end rafter, it is the end rafter that is plumbed. When the roof has a fly rafter, the last common rafter is plumbed. Plumbing a hip roof is not normally required, because hip rafters hold the main roof firmly in position.

Procedure for Plumbing a Roof

1 To the top of the ridgeboard tack a short length (about 18″) of 1 × 2 so that it overhangs the end of the roof. **Note.** This step is not necessary if the roof has an overhang.

2 Measure off an odd number of inches from the end of the ridgeboard (or the last common rafter

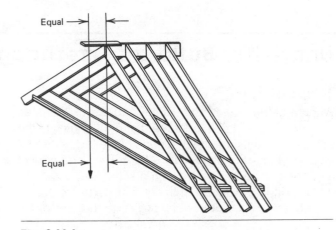

Fig. 9-11.1

of a roof with an overhang). Hang a plumb bob at this point (Fig. 9-11.1).

3 When the plumb bob stops swinging, measure to the rafter plate.

4 If the measurements in Fig. 9-11.1 are not equal, detach one end of the temporary braces. Then push or pull on the roof frame until the measurements are equal.

5 Toenail braces to hold the roof plumb until sheathing is completed.

Procedure for Straightening a Roof

1 Select a long 1 × 2 or 1 × 4, and on it mark standard rafter spacing.

2 Set the board on the end rafter at a mark and lightly nail it with an 8d nail driven part way in.

3 Work across the roof, nailing each rafter lightly on its mark.

4 To straighten a bowed rafter, cut two pieces of 1 × 2 or 1 × 4 long enough to span three rafters.

5 Mark rafter spacing on both boards.

6 To rafters flanking the bowed rafter, lightly nail the boards on their marks, one above and one below the bow (Fig. 9-11.2).

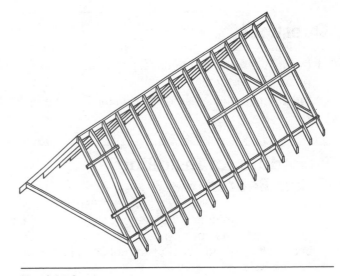

Fig. 9-11.2

7 Pull the bowed rafter up to the remaining set of marks, and nail tightly.

8 Leave all straightening boards in position until they interfere with sheathing application.

Unit 9-12 **Install Collar Beams**

Many building codes require collar beams every 48″ from one end of a roof to the other under a ridgeboard. Collar beams are tension members that prevent pairs of rafters from spreading and from pushing supporting walls out of plumb (Fig. 9-12.1).

Unlike other members in the roof structure, collar beams don't have to be cut for an exact fit. The ends should be cut to match the roof pitch, however, to provide as much surface as possible for nailing into the pairs of rafters.

Because collar beams tend to be in the way if attic space is to be used, many builders use 1″ × 18-ga steel straps instead. Straps are adequate if expected wind loads do not exceed 30 psf (pounds per square foot). Strapping is usually required on alternate rafters.

Fig. 9-12.1 Collar beams tie pairs of rafters together below the ridge. (Photos Drew Leviton, Atlanta.)

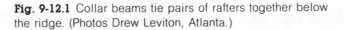

Questions

1 Of the two methods of connecting pairs of rafters, which method must take place before sheathing?

2 Which of the two methods do you think is stronger? Why?

Procedure for Installing Collar Beams

1 Determine from codes the size of collar beam required and the spacing called for between beams. If size and spacing are not specified, use 1 × 6s 32″ o.c.

2 Determine from drawings the height of collar beams. If no height is given, use 18″ below the bottom of the ridgeboard.

3 On a pair of rafters mark the height of the beam.

4 Measure the distance between marks to determine the lengths of beams.

5 With a pitch board mark end cuts.

6 Cut along the marks.

7 Use this collar beam as a template for marking and cutting all others required. After you have cut a sample collar beam, you can cut two more out of a single board with only three cuts (Fig. 9-12.2).

8 Nail collar beams at the required spacing, with their tops flush with or slightly below the tops of rafters. Install end beams first, then stretch cords between their bottom edges as a guide for locating the remaining beams.

Procedure for Installing Strapping

1 Determine from codes the length of strap required. If no length is given, use straps 24″ long.

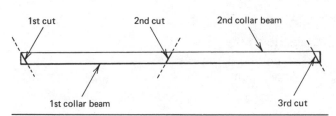

Fig. 9-12.2

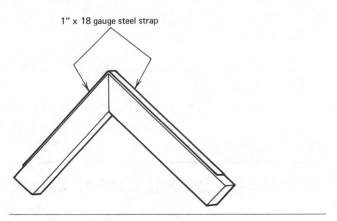

Fig. 9-12.3

2 Mark the midpoint of each length of strap.

3 Lay the strapping with the mark centered on the ridgeboard, and bend the strap to the shape of the roof peak (Fig. 9-12.3).

4 Begin nailing by fastening first on each side of the ridge, then working alternately down each rafter. Use 8d nails spaced every 3 to 4″.

Unit 9-13   Frame with Roof Trusses

A **truss** is a braced framework of triangular shapes, capable of supporting loads over long spans. Each truss has upper chords of structurally graded lumber that serve as rafters. It has lower chords of structurally graded lumber that act as ceiling joists. It has web members in various configurations that both separate and connect upper and lower chords. Chords are usually fastened to each other and to web members with **gussets**—plywood or metal stiffeners like the splice plates used on long ridgeboards.

Truss Designs

Of the many designs of roof trusses, three are commonly used in housing. The most common is the **Fink truss**, also called the **W truss**. It has four web members placed in the shape of a letter "W" (Fig. 9-13.1A). It is a practical shape for almost any residential roof span and pitch.

The **kingpost** or **K truss** (Fig. 9-13.1B) is the simplest to build. It has only a single vertical web member at the center. It is most often used when pitches are low and spans are short. It requires a better grade of lumber than a W truss.

The **scissors truss** (Fig. 9-13.1C) is designed for use when the ceiling of a room is not flat. Upper chords follow the pitch of the roof, as in W and K trusses. Lower chords, however, follow the pitch of the ceiling. The design may be similar to a kingpost truss with two shorter kingpost trusses added at the bottom, as in the drawing, or engineered for the job, as in the photograph.

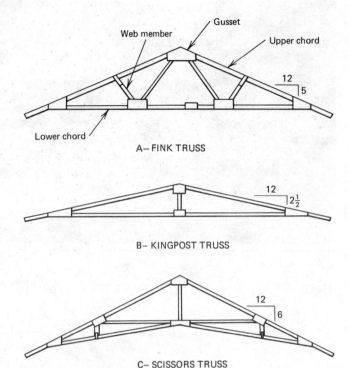

Fig. 9-13.1 The types of roof trusses commonly used in housing.

Fig. 9-13.1A A Fink truss consists of upper and lower chords braced by four web members in the shape of a W. (Courtesy Automated Building Components, Inc.)

Fig. 9-13.1B A kingpost truss has only one brace member—a short vertical web member at the center. (Courtesy American Plywood Association.)

Fig. 9-13.1C All chords of a scissors truss slope. Webs may take several shapes; compare this photograph with the drawings on page 233. (Courtesy Automated Building Components, Inc.)

Truss Fabrication

Trusses are built in three ways. In a **single-plane truss**, all members are the same thickness, and butt together in the same plane (Fig. 9-13.2, top). Where they meet, the joints are covered with gussets. Gussets are required on both sides of all joints except on end trusses, where wall sheathing provides the necessary stiffness. Excluding gussets, the single-plane truss is only 1½″ thick.

In a **layer-on-layer** truss the chords are in the same plane, but the webs overlap (Fig. 9-13.2, center). Usually long web members are attached on one side of the chords, and short web members are attached on the other. A single gusset at the ridge and pairs of gussets at the eaves complete the assembly. Web members are boards, and the total thickness of this type of truss is 3″.

In another type of layer-on-layer design all members overlap at the ridge (Fig. 9-13.2, bottom). Gussets are used only at two points. At other points the members are attached with **split-ring connectors** (Fig. 9-13.3). Each ring fits into a circular groove precut into a chord, and both members are drilled for a bolt. As the nut on the bolt is tightened, the wedge-shaped ring cuts into the other chord or web member for a strong, tight fit.

Large-volume builders often build their own trusses in a shop and truck them to the site. Smaller builders usually buy trusses from local fabricators or lumberyards. Trusses must be built to rigid specifications that have been established by national building codes. Designs are available from such sources as the American Plywood Association.

Trusses with split-ring connectors must be assembled before they can be erected. All other types are

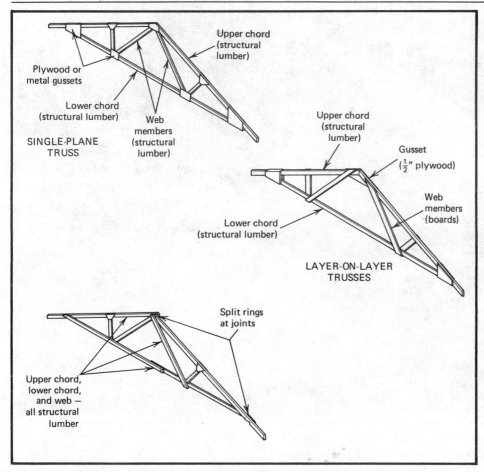

SINGLE-PLANE TRUSS

Plywood or metal gussets

Upper chord (structural lumber)

Lower chord (structural lumber)

Web members (structural lumber)

LAYER-ON-LAYER TRUSSES

Upper chord (structural lumber)

Gusset ($\frac{1}{2}$" plywood)

Web members (boards)

Lower chord (structural lumber)

Split rings at joints

Upper chord, lower chord, and web — all structural lumber

Fig. 9-13.2 Each type of truss in Fig. 9-13.1 may be built in several ways. In the single-plane truss (top) all members lie in the same plane and are held together with plywood gussets. The two lower trusses are the layer-on-layer type. For short spans the lightweight truss (center) is adequate. Note that it has fewer gussets than the single-plane truss. A layer-on-layer truss assembled with split-ring connectors (bottom) is the thickest of all trusses, but also the sturdiest.

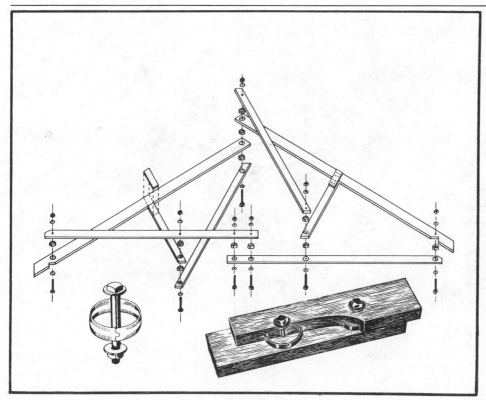

Fig. 9-13.3 How a truss is assembled with split-ring connectors. (Courtesy TECO Products and Testing Corp.)

Fig. 9-13.4 Trusses should be stored upright, braced and supported only at points where they will bear on walls. (Courtesy Automated Building Components, Inc.)

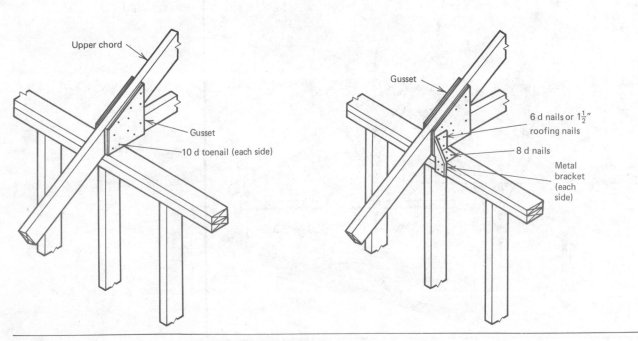

Upper chord

Gusset

10 d toenail (each side)

Gusset

6 d nails or 1½" roofing nails

8 d nails

Metal bracket (each side)

Fig. 9-13.5 Typical methods of attaching trusses at rafter plates.

ready to install as soon as they are unloaded at the site. Unload them in an upright position, and carry them at their normal points of support to the storage area. Always store trusses upright and supported only at their bearing points (Fig. 9-13.4).

All types of trusses may be fastened to rafter plates with nails driven at an angle into the rafter plate, or with metal brackets similar to joist hangers (Fig. 9-13.5). The brackets are made right-hand and left-hand, and one goes on each side of the truss at each plate.

Questions

1 Which is the best type of truss to use with:
   a low pitched roof
   a raised ceiling
   a long span
2 Why are gussets not required on end trusses?
3 Why should trusses be stored upright?
4 What is the best way to hold trusses in place at the rafter plate? at the ridge?

Procedure for Fastening Split-Ring Connectors

1 Through the hole predrilled in the end of the truss member insert a $\frac{1}{2}''$ machine bolt 1″ longer than the combined thickness of the members to be attached.
2 Position the ring around the bolt.
3 Fit the second member onto the bolt, until the ring slips into the circle precut to accept it. **Note.** To make this step easier, support the other end of the second member on a block of wood (Fig. 9-13.6).
4 When the connection is made, place a washer and nut on the bolt.
5 Align the edges of the two members for a flush fit.
6 Gradually tighten the bolt until the two members are tightly locked together. Check as you tighten to make sure edges remain flush.

Fig. 9-13.6 (Courtesy TECO Products and Testing Corp.)

3 Raise each truss onto the rafter plates, and rest it momentarily upside down (Fig. 9-13.7).
4 Place the end truss upright on its mark, and fasten it to rafter plates with metal brackets or by toenailing.
5 After five or six trusses are fastened, run a temporary brace across them near the ridge. Attach one end to the gable wall or plumbed end truss, and plumb each truss before nailing on the brace.
6 Raise, fasten, and brace the remaining trusses. **Precaution.** Trusses must be firmly tied together and well braced to keep them from collapsing until they are sheathed.

Procedure for Raising Trusses

1 When trusses arrive at the site, unload them in an upright position and carry them at their normal points of support to the storage area.
2 Store trusses upright and supported only at bearing points.

Fig. 9-13.7 (Courtesy American Plywood Association.)

Unit 9-14   **Mark Rafters for a Gambrel Roof**

In a true gambrel roof the rafters form the chords of a semicircle. When this is the case, the simplest way to determine the lengths of rafters and the shapes of cuts is to lay out the roof's shape on a flat surface, such as the subfloor.

If the roof is not a true gambrel, then the pitches of the two slopes are shown in the drawings. The procedure for marking the upper rafters is very similar to the procedure described in Unit 9-3. With lower rafters,

however, the unit rise is greater than the unit run, and measurements must be made with the rafter square held in reverse.

Gambrel rafters usually meet at a purlin (Fig. 9-14.1. left). The purlin is continuous and is the same size as the ridgeboard. Rafters are notched to enclose the purlin completely. Occasionally the rafters meet over a short wall. called a **knee wall,** and are notched to meet at the rafter plate (Fig. 9-14.1. right).

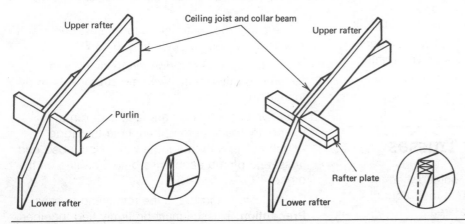

Fig. 9-14.1 Rafters in a gambrel roof usually meet at a purlin (left and left inset), but sometimes they meet at a rafter plate on a short wall.

Procedure Using a Layout

1 Along one side of the subfloor at right angles to the roof ridge. find the midpoint (A in Fig. 9-14.2).

2 Using point A as a center. and the run of the roof as a radius, scribe an arc on the floor (B) with chalk and string.

3 Through point A snap a chalk line (AC) at right angles to the edge of the floor. The point where the line intersects the circle (C) marks the top of the ridge.

4 From drawings determine the height of outer walls of the second story.

5 From the edge of the floor measure up the height in step 4. Snap a chalk line (D) through this point across the floor parallel to the edge. The intersections of this line with the circle (points E and F), mark the locations of the outer corners of purlins.

6 From the layout determine the lengths of upper rafters.

7 Mark on upper rafters the plumb cut at the ridge. the length of the rafter, and the shape of the notch at the purlin.

8 From the layout determine the length of the body of lower rafters, and the shape of the notch at the purlin.

9 From the drawings determine the amount of overhang, or the height of the overhang above floor level.

10 Using unit measurements, calculate the length of the tail.

11 Mark off the total length of lower rafters.

12 Mark the shape of the notch at the purlin, the bird's mouth, and the tail cut.

Procedure Using a Rafter Square

1 From data in the drawings determine the unit rise and unit length of upper rafters.

2 Determine the line length of an upper rafter. **Note.** The line length and actual length are identical here.

3 Mark the plumb cut at the ridge, and mark another plumb cut at the purlin (A in Fig. 9-14.3).

4 At right angles to the plumb cut at the purlin draw a level cut line (B).

5 Parallel to the plumb cut line at the purlin and $1\frac{1}{2}''$ from it toward the ridge, draw another line (C). Lines B and C mark the notch for the purlin.

6 From data in the drawings determine the unit rise and unit length of lower rafters.

7 Determine the line length of a lower rafter.

8 Mark the plumb cut at the tail, carrying the rise on the blade and the run on the tongue (Fig. 9-14.4).

9 Complete any further marking of the tail. Usually a combination cut is required (see Fig. 9-3.2).

10 Mark the bird's mouth.

11 Measure off the body length of the rafter.

12 Mark the plumb cut at the purlin (A in Fig. 9-14.5).

13 Mark the shortening line (B). **Note.** Since the line length of the rafter runs to the center line of the purlin, the amount of shortening is one-half the thickness of the purlin.

14 Measure down the shortening line the depth of the purlin, and make a mark.

15 At this point draw a line (C) at right angles to the shortening line. Lines B and C mark the notch for the purlin.

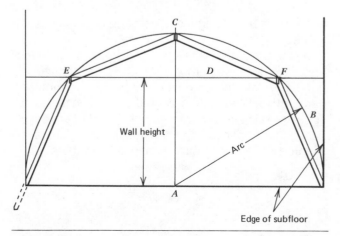

Fig. 9-14.2

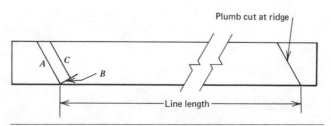

Fig. 9-14.3

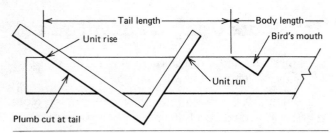

Fig. 9-14.4

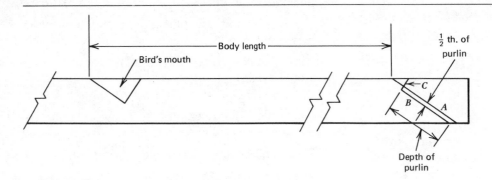

Fig. 9-14.5

Questions

1 Why are the line length and the actual length of an upper rafter the same?

2 Do you think that upper and lower rafters in a gambrel roof should be the same size? Why?

3 Why is it necessary to mark cuts on lower rafters differently than upper rafters?

Unit 9-15   **Frame a Gambrel Roof**

A gambrel roof must be built in stages, but the method of assembly varies from builder to builder and from carpenter to carpenter. The usual procedure is somewhat of a cross between balloon framing and framing a gable roof. The lower sections of the roof are framed as far up as the purlins, and the upper sections are installed like a gable roof. The lower ends of lower rafters are attached to studs with horizontal connectors called **lookouts**. End walls may be built before the roof is framed, after it is framed, or after each section is framed.

In the procedure in this unit, the lower roof is framed first, then the upper roof. End walls are treated like gables. Before the carpenter begins to assemble any gambrel roof, all rafters should be cut to fit as described in Unit 9-14. Purlins and the ridgeboard should be cut and spliced, and the locations of all rafters should be marked on both horizontal members.

Question

1 Name two purposes that lookouts serve.

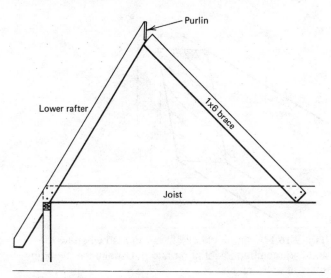

Fig. 9-15.1

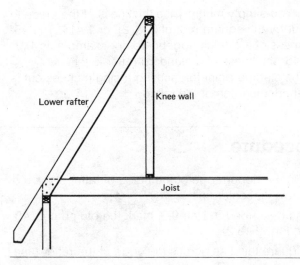

Fig. 9-15.2

Procedure

1 Set lower rafters in position on their marks against ceiling joists. Toenail rafters to the rafter plate with 16d nails, and nail straight into joists with three 10d nails.

2 Fit purlins into the notch in rafters and on rafter marks. With two 12d or 16d nails, nail through the purlin into rafters.

3 a   If purlins are installed upright, as at the left in Fig. 9-14.1, run a temporary 1 × 6 brace between the purlin and ceiling joists every 48″ (Fig. 9-15.1).

3 b. If rafters meet at a rafter plate, build the knee wall as a brace (Fig. 9-15.2).

4 Frame the other half of the lower roof as described in steps 1 through 3.

5 With the lower roof framed, frame the upper roof as described in unit 9-7. Align upper rafters with lower rafters, and attach with 10d nails driven vertically into the purlin and horizontally from lower rafter into upper rafter.

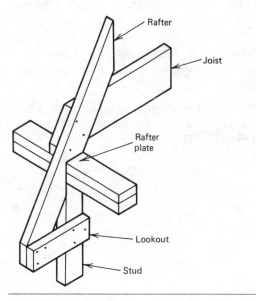

Fig. 9-15.3

6 Install collar beams as required.

7 At the eaves add 2 × 4 lookouts between rafter ends and studs (Fig. 9-15.3).

Unit 9-16   **Mark a Single Cheek Cut**

Frequently when one roof framing member meets another at an angle, one of the member's ends gets a single creek cut (Fig. 9-16.1). Typical examples are a hip rafter butting against the side of a ridgeboard, and a jack rafter butting against a hip or valley rafter.

When framing members meet at an angle, the amount of bearing surface is greater than when the same pieces meet at right angles. If both parts of the roof have the same pitch, then roof members meet at a 45° angle, and it is easy to determine the bearing surface. You simply multiply the thickness of the piece to be cut by the square root of 2 ($\sqrt{2}$), or 1.414. The 45° thickness of a $\frac{3}{4}$"-thick ridgeboard, for example, is $1\frac{1}{16}$". The 45° thickness of a standard rafter is $2\frac{1}{8}$".

The starting point for marking a single cheek cut is the shortening line of the rafter.

Procedure

1 On the upper edge of the piece of rafter stock draw its center line.

2 As described in Unit 9-3, mark the plumb cut (*AB* in Fig. 9-16.2).

3 Square this line across the edge of the rafter (line *BC*).

4 From the plumb cut line measure at right angles a distance equal to one-half the 45° thickness of the member to be butted against. Draw a parallel line (*DE*) at this point. This is the shortening line.

5 Square this line across the edge of the rafter (line *EF*).

6 At right angles to the shortening line measure a distance equal to one-half the thickness of the member being cut. Draw a parallel line (*GH*) at this point.

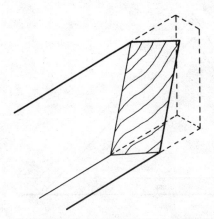

Fig. 9-16.1 A single check cut is required when two framing members meet at an angle. Usually the members meet at a 45° angle.

7 Square this line across the edge of the rafter (line *HI*).

8 a. For one cheek cut, draw a line from point *H* through point *J* to *K*. The single cheek cut is along lines *GH* and *HK*.

8 b. For an opposite cheek cut, draw a line from point *I* through point *J* to *L*. The opposite or reverse cheek cut is along lines *IL* and *LM*.

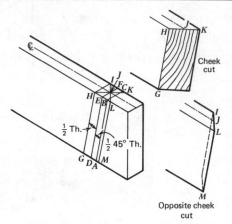

Fig. 9-16.2

Cheek cut

$\frac{1}{2}$ Th. $\frac{1}{2}$ 45° Th.

Opposite cheek cut

Questions

1 Assume in this procedure that the member being cut is a 2 × 6 hip rafter that butts against a 1 × 8 ridgeboard. What is the dimension in step 4?

2 What is the dimension in step 6?

3 Why is it necessary to find the center line of the member being cut in order to lay out a single cheek cut?

Unit 9-17 **Mark a Double Cheek Cut**

Whenever a roof framing member meets two other members that are at right angles to each other, the abutting end must receive a double cheek cut to fit into the corner (Fig. 9-17.1). Typical examples are a valley rafter butting against two ridgeboards, and the tail cut of a hip rafter.

If both parts of the roof have the same pitch, the diagonal members meet other members at a 45° angle. They bear equally on both members. The amount of bearing is not the 45° thickness of the member being cut, as with a single cheek cut, but half that thickness.

The starting point for marking is the shortening line of the rafter.

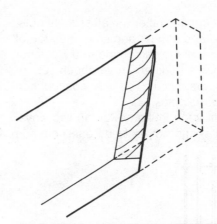

Fig. 9-17.1 A double cheek cut is required when a framing member meets two other framing members in a corner. Usually the meeting is at a 45° angle.

Procedure

1 On the upper edge of the piece of rafter stock draw its center line.

2 As in marking a single cheek cut (Unit 9-16), mark the plumb cut (line *AB* in Fig. 9-17.2) and square the line across the rafter (*BC*).

3 As in marking a single cheek cut, draw the shortening line (*DE*) and square it across the rafter (*EF*).

4 As in marking a single cheek cut, measure off half the thickness of the member being cut, draw a parallel line (*GH*) and square the line (*HI*).

5 Draw lines from *H* and *I* through point *J*. One cheek cut is along lines *GH* and *HJ,* and the other is along lines *JI* and *IK*.

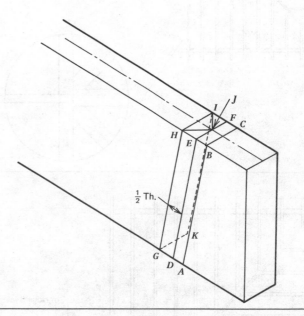

½ Th.

Fig. 9-17.2

Question

1 This procedure shows how to make a double cheek cut that points outward. Can you think of any condition where the cut would point inward?

Unit 9-18   **Mark a Hip Rafter**

The cuts made in a common rafter are all straight cuts, made with the saw blade at right angles to the side of the rafter. Almost none of the cuts made in a hip rafter are straight cuts. They are either single cheek cuts or double cheek cuts, depending on the construction called for in framing drawings.

Hip rafters may fit into roof framework in two ways. Sometimes the end section of roof has one common

rafter that butts against the end of the ridgeboard (Fig. 9-18.1). Here hip rafters fit into the corner made by two common rafters, and they get a double cheek cut at the ridge. At other times there is no common rafter in the end section, and the hip rafters are framed into the side of the ridgeboard (Fig. 9-18.2). Here they each get a single cheek cut at the ridge.

At the tail, hip rafters may get a plumb cut, square

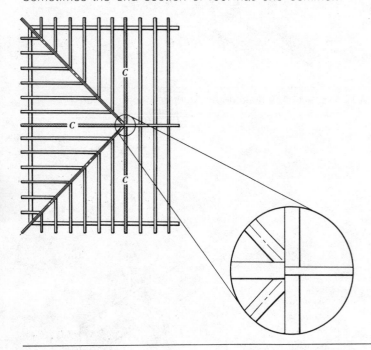

Fig. 9-18.1 When the end section of a hip roof has a common rafter in the center, hip rafters need a double cheek cut at the ridge.

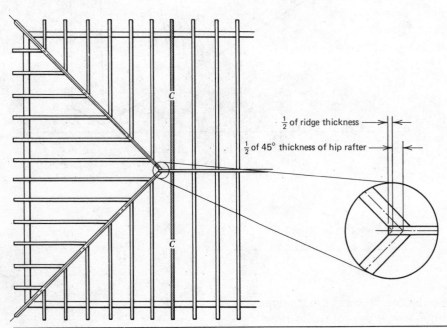

½ of ridge thickness

½ of 45° thickness of hip rafter

Fig. 9-18.2 When the end section of a hip roof has no common rafter, then hip rafters meet the ridgeboard in a single cheek cut.

cut, heel cut, or a double cheek cut if the ends of rafters are trimmed with a fascia.

The bird's mouth may be handled in two ways. The seat cut is a heel cut, as in a common rafter, but it is a longer cut. The plumb cut may be a double cheek cut to fit over the corner of the doubler (Fig. 9-18.3, left). Or the rafter plate may be trimmed, or **dubbed off**, to accept a plumb cut (right). The latter method is quicker.

A hip rafter acts as a type of ridgeboard. If you installed it with a flat top, the center line would be at the right height for roof sheathing, but the edges would be too high. Roof sheathing would buckle. To avoid this problem, hip rafters may be dropped or backed (Fig. 9-18.4).

Dropping a hip rafter—that is, lowering it slightly—takes less time. Sheathing meets with a slight gap above the rafter, and is more difficult to nail tight. Dropping a hip rafter alters all cuts except the tail cut, however. The seat cut is deeper and the plumb cut is shorter, although not always enough to require a change.

Backing a hip rafter means beveling the top edge. With this solution all roof members and sheathing fit well without adjustment.

If the roof pitch is the same on both sides of a hip rafter—and it usually is—then the hip rafter is the hypotenuse of a 45° right triangle in which the other two sides are common rafters. Applying the Pythagorean theorem, you know that the sides of a 45° right triangle are in the ratio of 1 to 1 to $\sqrt{2}$, or 1.414. Therefore, if you know the length of a common rafter, you can easily determine the length of a hip rafter in the same roof.

Another way to determine the length of a hip rafter is by unit measurement (see Unit 9-1). With a common rafter the unit run is always 12. With a hip rafter or valley rafter, when pitches are equal, the unit run is 17. Why? Because 12 (the standard unit run) times 1.414 is 16.97, which is rounded off to 17.

Your problem in this unit is to lay out a hip rafter for the roof in Unit 9-3. The house has a hip roof with equal pitches of 4 in 12. The span of the roof is 28'-0" and the overhang is 2'-0". Rafters are spaced 16" o.c.

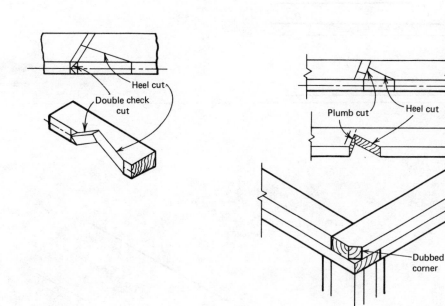

Heel cut

Double check cut

Plumb cut   Heel cut

Dubbed corner

Fig. 9-18.3 Where a hip rafter meets the rafter plate, either the hip rafter is shaped to fit over the plate (left), or the plate is trimmed to accept a plumb cut.

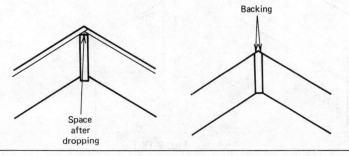

Backing

Space after dropping

Fig. 9-18.4 The upper edges of a hip rafter cannot protrude above the level of the tops of other rafters. Therefore hip rafters must either be lowered slightly (left) or have their edges beveled (right).

Procedure

1 From the table on the rafter square, determine the approximate overall length of the hip rafter.

2 Select a piece of straight stock that is the same size as the common rafter and at least 6″ longer than the length determined in step 1.

3 Lay the rafter stock across sawhorses with the crowned edge toward you, and mark this edge with an arrow.

4 Mark the plumb cut at the ridge.

5 Determine and lay out the line length of the body of the rafters. Mark the plumb cut at the bird's mouth.

6 Determine and lay out the length of the tail. Mark the tail cut.

7 Mark a single cheek cut at the ridge (see Unit 9-16).

8 Mark a double cheek cut at the tail (see Unit 9-17).

9 At right angles to the plumb cut at the rafter seat (line A in Fig. 9-18.5) measure away from the tail a distance equal to one-half the thickness of rafter stock.

10 Draw line B through this point parallel to the plumb line.

11 Square this line across the bottom of the rafter (C).

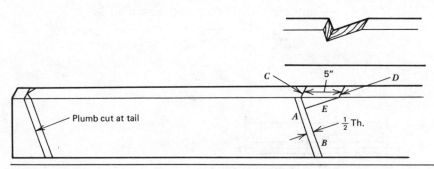

Fig. 9-18.5

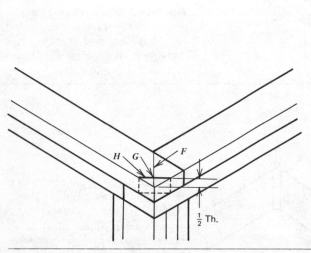

Fig. 9-18.6

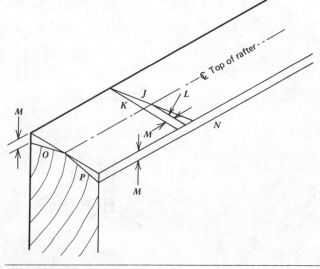

Fig. 9-18.7

12 Measure up the under side of the rafter a distance of 5″, and square a line (D) through this mark.

13 From this line draw a line E on the side of the rafter at right angles to line B. Lines C, D, E and part of B mark the notch for the bird's mouth.

14 Draw a line from the outside corner to the inside corner of the rafter plate (F in Fig. 9-18.6).

15 From the outside corner measure off horizontally half the thickness of the hip rafter. Mark this point (G).

16 Through point G and at right angles to line F draw line H. It marks the cutting line for dubbing off the corner of the plate to accept the bird's mouth of the hip rafter.

17 At any convenient point on the top of the rafter,

draw a plumb line (J in Fig. 9-18.7), using the unit rise and unit run for a hip rafter.

18 Square one end of this line across the top edge (K).

19 Lay out the center line of the top edge.

20 Where the plumb line and center line intersect, draw another squaring line (L) to one edge.

21 Measure the distance (M) between the two lines.

22 Mark off this distance down the sides of the rafter on both sides. Draw lines through these marks down the entire length of the rafter. These lines (N) are the backing lines.

23 At the ends of the rafter connect the backing lines with the center line. Lines (O and P) are the cutting lines for backing the hip rafter.

Questions

1 Name an advantage to dubbing the rafter plate compared to making a regular hip seat cut.

2 Name a disadvantage.

3 What is the length of the rafter in step 1?

4 In step 12 why is the measurement 5″?

Unit 9-19   Mark a Jack Rafter

Jack rafters that meet at a hip rafter or valley rafter are often identical in length (see Fig. 9-19.2, for example). At the ends where they meet they receive opposite cheek cuts. At the other end they are marked and cut like a common rafter, either with a shortening cut at the ridge if they are valley-jack rafters, or with a tail cut and bird's mouth if they are hip-jack rafters. Cripple-jack rafters receive a square cut or plumb cut where they meet a header, and receive cuts similar to adjoining rafters at their opposite ends.

Adjacent jack rafters are different in length, but there is a common difference. This common difference is usually shown on the blade of a rafter square for spacings of 16″ and 24″ (Fig. 9-19.1). If, for example, a roof has a pitch of 5 in 12, and rafters are spaced 16″ o.c., then each jack rafter is either 17.33″ longer or

shorter than its neighbor. Jack rafters have the same unit run and unit rise as common rafters in the same section of roof.

To verify that figure of 17.33″, follow this series of calculations. The first jack rafter lies 16″ from the corner of a roof. When that roof's pitch is 5 in 12, the rise of the jack rafter is 6.67″ (5 × 1⅓). The common difference (Fig. 9-19.2) is the hypotenuse of a triangle whose other two sides are 16″ and 6.67″. The square of 16 is 256. The square of 6.67 is 44.44. Add the two, and the sum is 300.44. The square root of 300.44 is 17.33 (see appendix)—the same common differences shown on the rafter square.

As an exercise in marking jack rafters, lay out hip-jack rafter A in Fig. 9-19.3. The roof has rafters 16″ o.c., a 4 in 12 pitch, and a 12″ overhang.

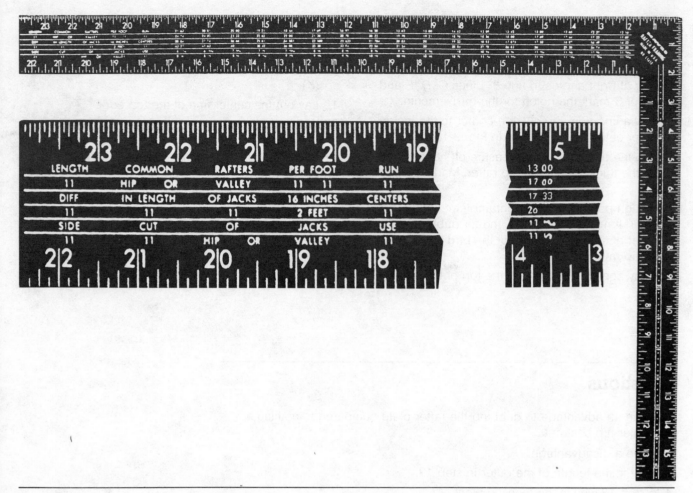

Fig. 9-19.1 The common difference between jack rafters appears in tables on most rafter squares. (Courtesy Great Neck Saw Manufacturers, Inc.)

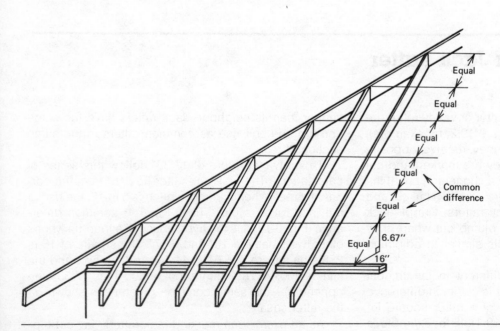

Fig. 9-19.2 The framing plan for the roof used as an example.

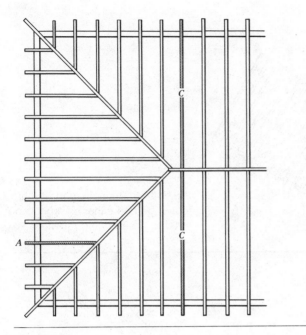

Fig. 9-19.3

Procedure

1 Determine the line length of the tail.

2 Counting from the corner, determine the number of the rafter to be cut. In Fig. 9-19.3 it is the third rafter.

3 On the rafter square find the common difference between jack rafters.

4 Multiply this number by 3 (the third rafter) to find the body length of the rafter.

5 Add together the lengths from steps 1 and 4 and select a piece of stock long enough for the job.

6 Mark the plumb cuts at the hip, for the seat, and for the tail as in Unit 9-2.

7 Complete marking the tail cut as required. See Unit 9-3.

8 Mark the heel cut for the rafter seat, as in Unit 9-3.

9 Mark the single cheek cut at the hip, as in Unit 9-16.

10 Recheck all marks before cutting.

Questions

1 What is the length of the tail in step 1?

2 What is the common difference in step 3?

3 What is the body length in step 4?

Unit 9-20   **Frame a Hip Roof**

The main section of a hip roof is assembled in the same way as a gable roof, with one exception. If the end hip has a common rafter at its center, this rafter serves as a brace. The actual length of the ridgeboard is its theoretical length (the length of the house less the roof's span) plus half the thickness of an end rafter at each end.

If the end hip does not have a common rafter, the two hip rafters serve as braces (Fig. 9-20.1). In this case the length of the ridgeboard is its theoretical length plus, at each end, half the thickness of the ridgeboard plus half the 45° thickness of a hip rafter. See Fig. 9-18.2.

In a hip roof—and occasionally in other styles, too—some rafters lie at right angles to ceiling joists. There are two ways to tie such rafters into ceiling framing. The simplest and most economical way is to strap them with metal ties to nailing blocks between the last two ceiling joists (Fig. 9-20.2, left). This method provides excellent resistance to wind uplift—a primary cause of damage to roofs with wide overhangs. The other method (right and photo) is with stub joists extending from the end ceiling joist to the rafter plate. These joists are in turn strapped to two or three ceiling joists beneath any attic subfloor.

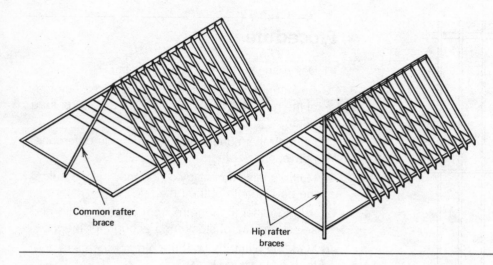

Fig. 9-20.1 The main part of a hip roof is braced either by a common rafter or a pair of hip rafters at each end.

Common rafter brace

Hip rafter braces

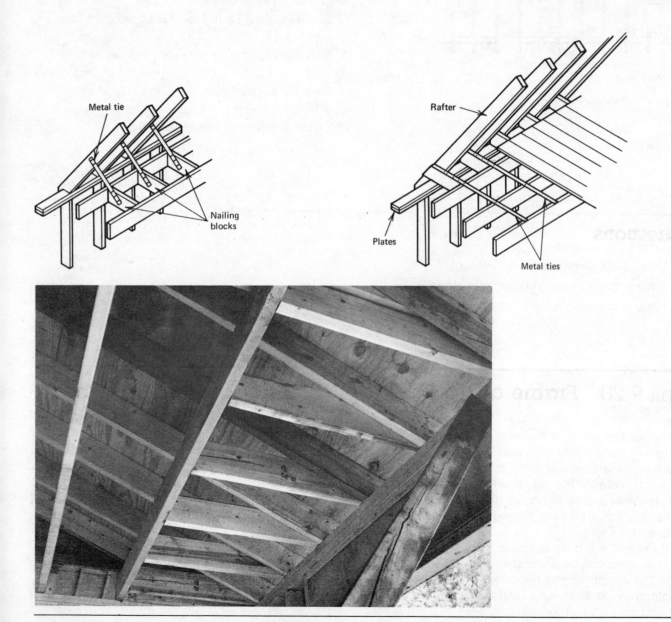

Metal tie

Nailing blocks

Rafter

Plates

Metal ties

Fig. 9-20.2 Two ways to strap down rafters that run at right angles to ceiling joists. (Courtesy Drew Leviton, Atlanta.)

Question

1 Of the two methods of strapping shown in Fig. 9-20.2, which do you think is the stronger? Why?

Procedure

1 Assemble the roof attached to the ridgeboard as described in steps 1 to 13 of Unit 9-7. Begin with the rafter at the end of the ridgeboard.

2 Trim the ridgeboard to length, with both ends receiving a plumb cut.

3 a. If the end hip has a common rafter at the center, install this rafter at each end of the roof. Then install hip rafters.

3 b. If the end hips have no common rafter, install both hip rafters at each end. Use the same nailing procedure as for common rafters at the ridge and bird's mouth.

4 Install jack rafters in the main roof and end roof in butting pairs. Nail each rafter on its mark on the rafter plate first, then nail through the hip rafter into the ends of jack rafters.

5 Cut and install blocks below (Fig. 9-20.2, left) or stud joists beside (Fig. 9-20.2, right) each end rafter.

6 Nail metal ties in position.

7 Install collar beams as required.

Unit 9-21   **Frame a Mansard Roof**

A mansard roof has some of the characteristics of a hip roof, and some of a gambrel roof. Like a hip roof it has a short ridge, and the roof slopes to a constant eave level on all sides of the house. Like a gambrel, a mansard has two pitches.

In a mansard roof, however, the break between pitches occurs, not at a purlin, but at the rafter plate of a wall two stories high (Fig. 9-21.1). The result, in effect, is a hip roof above the rafter plate with an overhang equal to the width of lower rafters. The pitch of the upper roof is close to flat. Therefore upper rafters are much heavier than lower rafters, and strapping is not required in end sections of the roof.

The lower roof is decorative rather than structural. Although it is almost as vertical as a wall, it is finished with roofing materials, not wall materials. Rafters may end on all sides at lookouts. More likely, however, they rest on lookouts on only two sides, like a gambrel roof. On the other two sides they rest on second floor joists that extend beyond the wall to form an overhang. The edges of lower rafters lie flush with the ends of rafters and lookouts (Fig. 9-21.2). The lower rafters line up with the ends of joists at both joist levels.

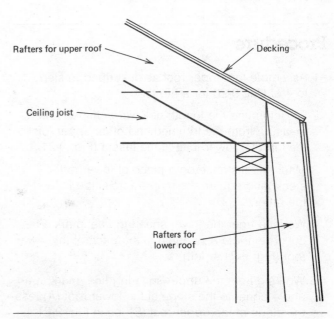

Fig. 9-21.1 The upper half of a mansard roof is built like a hip roof. Framing for the lower roof functions independently of the upper roof.

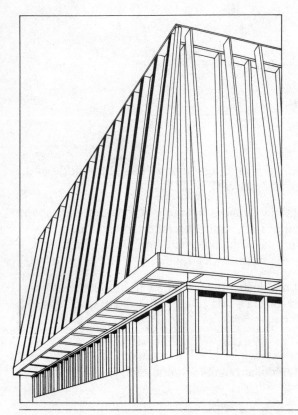

Fig. 9-21.2 Rafters in the lower roof of a mansard are attached at the upper end either to ceiling joist or rafters of the upper roof. At the lower end they rest either on floor joists or lookouts.

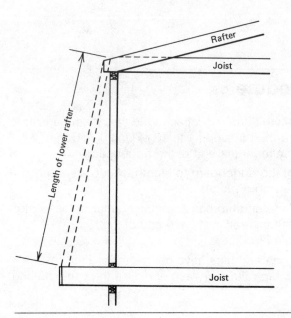

Fig. 9-21.3

Procedure

1 Assemble the upper roof as described in steps 1 to 4 of Unit 9-20 for a hip roof.

2 To determine the length of lower common rafters, measure from the trimmed end of an upper joist to the top of a lower joist at the fascia (Fig. 9-21.3).

3 Mark this distance on a piece of lower rafter stock, and square the marks across the face of the rafter.

4 Working from the lower squaring line, mark a heel cut. The angle A is equal to the pitch of the lower roof (Fig. 9-21.4, left).

5 Working from the upper squaring line, mark at an angle equal to the slope of the lower roof (A) less the slope of the upper roof (B), as in Fig. 9-21.4, right.

6 Cut the rafter and test the fit.

7 When the fit is good, cut the remaining common rafters.

8 Set common rafters in position, flush at both ends with joists.

9 Nail upper ends with 10d nails into upper rafters. Toenail lower ends into joists. Make sure the outer edges of rafters are flush with the ends of joists.

10 Determine the length of lower hip rafters by measuring as in step 2.

11 Mark and cut ends as described in steps 3 through 6.

12 When the fit is good, cut the remaining hip rafters.

13 Mark the backing for each hip rafter, as described in Unit 9-18, steps 17 to 23, and make the cuts.

14 Set hip rafters in position in line with the upper hip rafter and the diagonal lookout member.

15 Toenail or strap the two hip rafters together at each corner.

16 Toenail the lower end into the diagonal lookout.

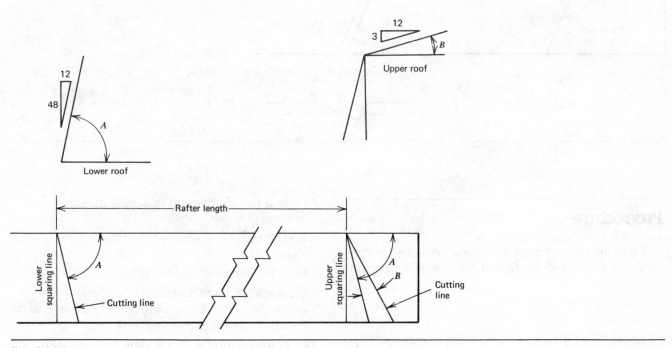

Fig. 9-21.4

Question

1 Match the type of cut with the correct rafter end in a mansard roof.
  Connect the two with a line.

upper end of upper rafter   heel cut

lower end of upper rafter   plumb cut

upper end of lower rafter   combination cut

lower end of lower rafter   difference-in-pitch cut

Unit 9-22   **Mark a Valley Rafter**

A valley rafter differs from a hip rafter in several ways, but in other ways is quite similar. At the ridge it may get a double cheek cut, a single cheek cut, or a plumb cut (see Unit 9-23). It does not need to be backed along its length. It may be notched at the bird's mouth like a hip rafter (see Unit 9-18), but for a better fit it should have a reverse double cheek cut.

When a roof overhang is great enough so that the common rafters next to the valley meet at the eave, builders often give the valley rafter a flush cut, and end it at the rafter plate. If the valley rafter extends to the eave, it receives a reverse double cheek cut at the tail.

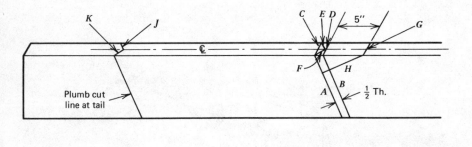

Fig. 9-22.1

Procedure

1 From the framing plan determine the length of the valley rafter, and the type of cut it receives at the ridge.

2 Select a piece of stock of the proper length for marking.

3 Using 17 as the unit run, mark the plumb cuts at the ridge, bird's mouth, and tail.

4 Mark the shortening cut at the ridge.

5 If required, mark the single cheek cut (see Unit 9-16) or the double cheek cut (see Unit 9-17).

6 From the plumb cut line at the seat (A in Fig. 9-22.1), measure at right angles a distance equal to one-half the thickness of rafter stock.

7 Draw a line (B) through this mark parallel to the plumb cut line.

8 Square both lines across the bottom of the rafter (C and D).

9 Mark the center line of the rafter on its under side.

10 Draw lines E and F between line C and the point where the center line crosses line D. Lines E and F mark the reverse double cheek cut.

11 From the plumb cut line squared (D) measure up the rafter 5″. Square this line (G) across the under side of the rafter.

12 From the end of line G draw line H at right angles to lines A and B. Line H marks the seat cut.

13 Repeat steps 4 through 8 at the tail, working from the plumb cut line. Lines J and K mark the tail cut.

Questions

1 Under what circumstances does the lower end of a valley rafter get a flush cut?

2 To cut the bird's mouth in Fig. 9-22.1, would you saw first along lines A and F or along lines G and H? Why?

Unit 9-23   Frame Intersecting Roofs of Equal Pitches

When two roofs of equal pitch meet at a valley, the carpenter faces one of several conditions. If the two ridgeboards are at the same height, the ridgeboard of the wing receives a vertical cut at the inner end. and butts against the main ridgeboard. The two valley rafters receive double cheek cuts (Fig. 9-23.1).

If the wing has a shorter span than the main roof, the intersection may be framed in two ways. In one

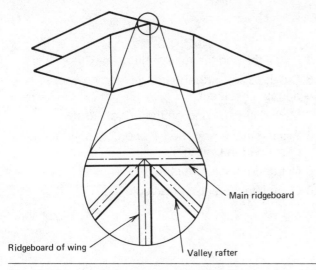

Fig. 9-23.1 Isometric and framing detail when two roofs' ridges meet at the same height and the roofs have the same pitch.

method, one valley rafter runs from rafter plate to main ridgeboard, and gets a single cheek cut at its upper end (Fig. 9-23.2). The other valley rafter is shorter, and butts against the longer valley rafter in a plumb cut. The ridgeboard meets the two valley rafters in a double cheek cut.

In the other way of framing, a hanger is attached to the main ridgeboard to support the end of the wing's ridgeboard (Fig. 9-23.3). Here valley rafters are of equal length, and meet the shorter ridgeboard in a single cheek cut.

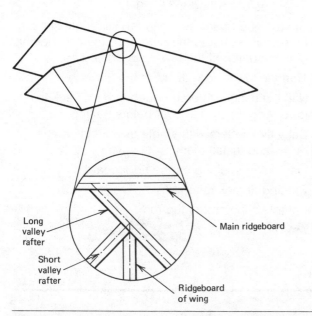

Fig. 9-23.2 Isometric and framing detail when roofs of equal pitch but different heights meet, and valley rafters are of different lengths.

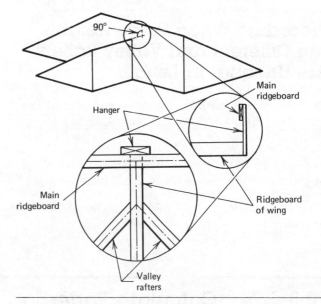

Fig. 9-23.3 Isometric and framing detail when roofs of equal pitch but different heights meet, and valley rafters are the same length.

Questions

1 Name an advantage of the construction shown in Fig. 9-23.2 over the construction shown in Fig. 9-23.3.

2 Name an advantage of the construction shown in Fig. 9-23.3 over the construction shown in Fig. 9-23.2.

Procedure When Heights and Pitches Are the Same

1 Square one end of the ridgeboard of the wing.

2 Mark spacing of jack and common rafters. **Precaution.** In working out the spacing, remember to allow for one-half the thickness of the main ridgeboard.

3 Assemble the ridgeboard and common rafters as described in Unit 9-7.

4 Attach the two sections of ridgeboard to each other.

5 Cut valley rafters with double cheek cuts at the ridge.

6 Cut jack rafters.

7 Install valley rafters, and then jack rafters.

8 Complete the remainder of the roof of the wing.

Procedure When Ridge Heights Are Different and Valley Rafters Are Unequal in Length

1 Measure and mark the longer valley rafter.

2 Cut the rafter to fit with a single cheek cut at the ridge.

3 Install the longer valley rafter.

4 Cut the shorter valley rafter with a plumb cut at the ridge.

5 Install the shorter valley rafter.

6 Cut the lower ridgeboard, giving one end a double cheek cut.

7 Mark rafter spacing on the ridgeboards.

8 Assemble the ridgeboard and common rafters as described in Unit 9-7.

9 Cut and fit jack rafters.

10 Complete the end of the roof.

Procedure When Ridge Heights Are Different and Valley Rafters Are Equal in Length

1 Assemble the lower ridgeboard and common rafters as described in Unit 9-7.

2 Cut a 1 × 6 board as a hanger. Its length should be at least equal to the difference in ridge heights plus the depth of the lower ridgeboard.

3 Trim the lower ridgeboard to length.

4 Nail the hanger to both ridgeboards. Check the lower ridgeboard for level before nailing.

5 Cut valley rafters with single cheek cuts at the ridge, and install them.

6 Mark rafter spacing on the ridgeboard.

7 Cut and fit jack rafters.

8 Complete the end of the roof.

Unit 9-24   Calculate Rafter Lengths When Roofs Have Unequal Pitches

Occasionally a house with a wing has a constant ridge height and a constant eave height, but the spans of the main roof and wing roof are different. Here all rafters—common, hip, and valley—have the same rise. But the run of hip and valley rafters is no longer 1.414 times the run of the common rafter because the roof has two common rafters.

There are two ways to find the length of a hip or valley rafter when the pitches of intersecting roofs are unequal. One way is to use the Pythagorean theorem, taking two steps. You first find the hypotenuse of a right triangle (Fig. 9-24.1) whose sides are the runs of the two roofs. Then you find the hypotenuse of a second triangle whose sides are the rise of the roof and the hypotenuse of the first triangle.

As an example, suppose that the major roof has a run of 18′ and a 6 in 12 pitch. The minor roof has a run of 12′ and a 9 in 12 pitch. The rise of both roofs is 9′.

To find the run of the valley or hip rafter (run V in Fig. 9-24.1), you square the other two sides (18 × 18 = 324) and (12 × 12 = 144), and add the two squares (324 + 144 = 468). If you really want to

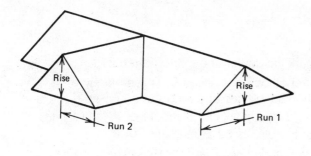

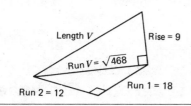

Fig. 9-24.1 The length of a valley rafter or hip rafter in a roof with unequal pitches is the hypotenuse of a triangle laid out on the hypotenuse of another triangle.

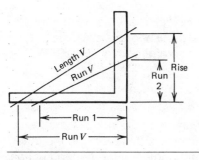

Fig. 9-24.2

know the run, you find the square root of 468 in your square root table.

But you don't need the run to determine the length of the valley rafter (length V). You reuse the square of one side (468), add to it the square of the rise ($9 \times 9 = 81$), to get 549. The length is the square root of that number: 23.431 or $23'\text{-}5\frac{3}{16}''$.

The other method for finding the length of a hip or valley rafter is to use your rafter square.

Procedure

1 On a large flat surface, such as the subfloor, lay your rafter square. On its blade mark the run of the major common rafter (run 1 in Fig. 9-24.2), and on its tongue mark the run of the minor common rafter (run 2).

2 Connect these two points, and measure the length of line (run V). This is the run of a hip or valley rafter.

3 On the blade of the square, mark the length measured in step 2. On the tongue mark the rise.

4 Measure the diagonal between points, using the $\frac{1}{12}$ scale on your rafter square. The answer is the length of a hip or valley rafter.

Questions

1 Why do you measure off the run of the shorter roof and the rise of both roofs on the blade, rather than the tongue?

2 Which method of calculating the length of a hip or valley rafter do you find easier? Why?

Unit 9-25 Mark Cuts on Rafters in Roofs of Unequal Pitches

Hip, valley, and jack rafters in roofs of unequal pitches get the same types of cuts as they do in a roof with equal pitches. But the cuts themselves are different. Single cheek cuts are not the same. Double cheek cuts still meet at right angles, but the lengths of the cuts are unequal. Even shortening cuts have to be determined differently.

Until you have had considerable experience in cutting rafters, the safest way to determine proper cuts is to draw on a sheet of graph paper the plan of a roof intersection. For this purpose use the twelfth scale on your rafter square. Once the cuts are correctly laid out and made, the method of framing is the same as for other roofs of similar types already described.

The procedures given here are for hip and hip-jack rafters. For valley and valley-jack rafters the procedures are similar.

Procedure for Laying Out Cuts

1 On the graph paper draw two sides of a rectangle. The length of one side, drawn to scale, is equal to the run of the major roof (Fig. 9-25.1). The length of the other side is equal to the run of the minor roof. These two lines represent the outside edges of rafter plates.

2 Draw in the other two sides of the rectangle as dot-dash lines. These lines represent the center lines of the two common rafters.

3 Draw a diagonal dot-dash line across the rectangle. This line represents the center line of the hip or valley rafter.

4 At full width ($1\frac{1}{2}''$) draw in the hip rafter and the two common rafters.

5 Where remaining space permits, draw in the center lines and full widths of two jack rafters. **Note.** Although the center lines of these two jack rafters would actually meet at the center line of the hip rafter, do not draw them this way. The lines yet to be drawn would overlap. Let them meet at least $1\frac{1}{2}''$ apart.

6 Mark the center line of the ridgeboard.

7 Strengthen lines A and B. These mark the shortening lines at the ridge.

8 Strengthen lines C and D. These mark the lines for the plumb cut at the seat.

9 Strengthen lines E and F. These mark the lines for cheek cuts for jack rafters.

Procedure for Marking Shortening Lines on a Hip Rafter

1 Determine the proper angle of the plumb cut (see Unit 9-24).

2 Draw the plumb cut line at the ridge on the scale drawing (Fig. 9-25.2).

3 On the side of rafter stock draw the plumb cut lines at the angle established in step 1. Square this line across the top of the rafter (Fig. 9-25.3).

4 On the drawing measure off the run of the cheek cut (Z). Transfer this dimension to the rafter by measuring at right angles to the plumb cut line.

5 On the drawing measure off the run of the other cheek cut (Y). Transfer this dimension to the rafter by measuring at right angles to the plumb cut line down the other side of the rafter.

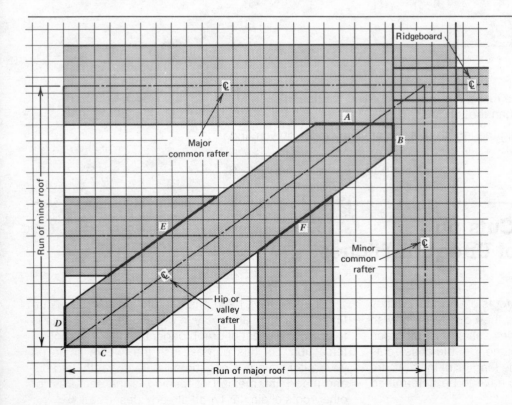

Fig. 9-25.1

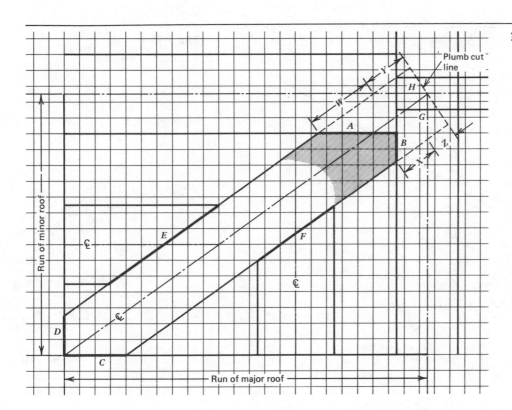

Fig. 9-25.2

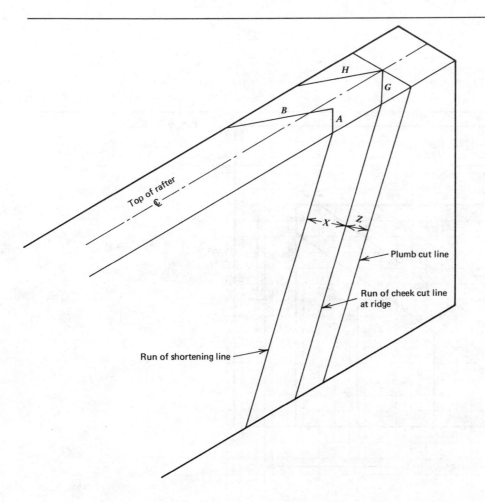

Fig. 9-25.3

Fig. 9-25.4

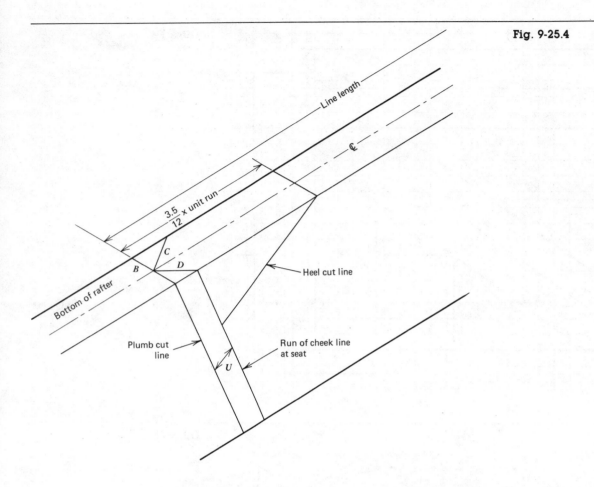

Line length

$\frac{3.5}{12}$ x unit run

Bottom of rafter

Heel cut line

Plumb cut line

Run of cheek line at seat

Fig. 9-25.5

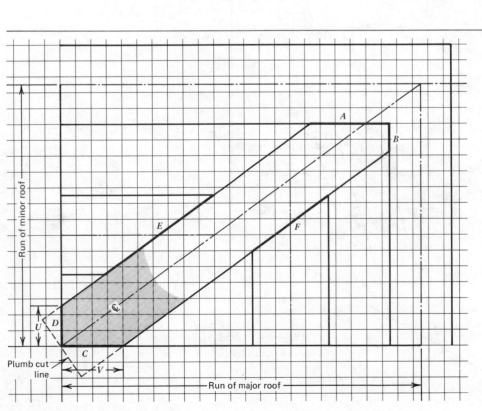

Run of minor roof

Plumb cut line

Run of major roof

6 Draw the center line of the rafter down its top edge.

7 Draw lines G and H in Fig. 9-25.3 from the run of cheek cut lines to the center line.

8 On the drawing measure off the runs of shortening (X) and (W), and transfer these dimensions to the rafter. Draw the two run of shortening lines parallel to the plumb cut lines.

9 Parallel to lines G and H draw lines A and B. The run of shortening lines and lines A and B mark the actual cut at the ridge.

Procedure for Marking the Seat Cut on a Hip Rafter

1 From the plumb cut line at the ridge measure off the line length of the rafter.

2 At this point draw the plumb cut line on the side of the rafter (Fig. 9-25.4). Square this line across the bottom of the rafter (B).

3 Draw the plumb cut line on the scale drawing (Fig. 9-25.5).

4 On the drawing measure off the run of one cheek cut (U), and transfer this dimension to the rafter by measuring at right angles to the plumb line. **Precaution.** In Fig. 9-25.5 you see the top of the hip rafter, whereas in Fig. 9-25.4 you see the bottom. Therefore the relationship of U to V is reversed.

5 Measure off the run of the other cheek cut (V), and transfer this dimension to the rafter.

6 Connect these two points to the center line as lines C and D. The run of cheek lines and lines C and D mark the actual plumb cut at the seat.

7 Mark the heel cut at the seat as with any hip rafter (see Unit 9-18), using the unit run of the rafter to determine the measurement from the plumb cut line squared to the heel cut line squared.

Procedure for Marking Cheek Cuts on a Hip-Jack Rafter

1 Determine the line length of a jack rafter in each slope, and the common differences between rafters.

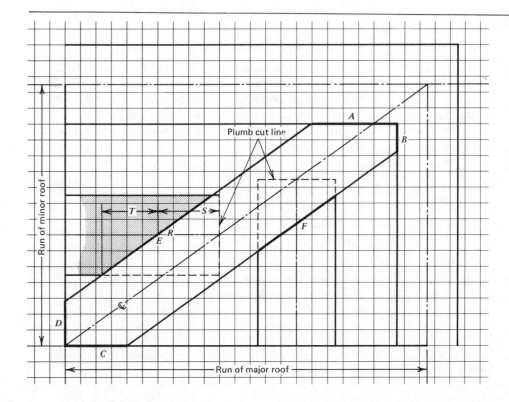

Fig. 9-25.6

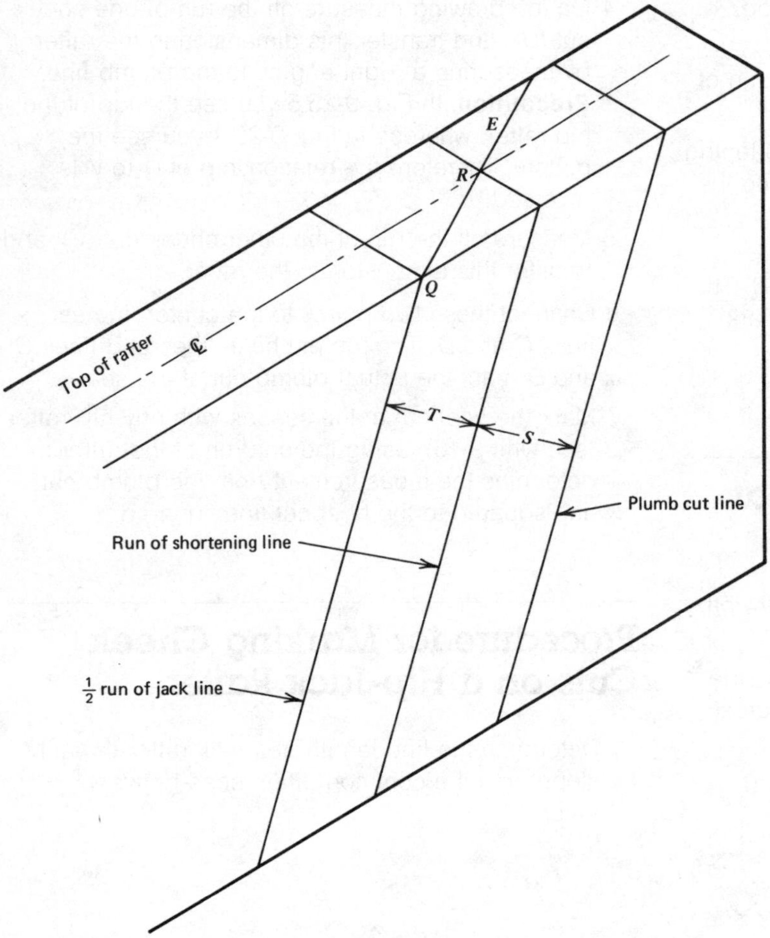

Fig. 9-25.7

2 Draw the plumb cut line at the hip on both the scale drawing (Fig. 9-25.6) and the rafter stock (Fig. 9-25.7). Square this line across the top of the rafter.

3 On the drawing measure off the run of shortening (S) and half the run of jack (T).

4 Transfer these dimensions to the rafter by measuring at right angles to the plumb cut line.

5 Square both lines across the top of the rafter.

6 Draw the center line of the rafter.

7 From point Q (where the half run of jack line touches the squaring line) through point R (where the shortening line squared crosses the center line), draw line E. The half run of jack line and line E mark the actual cheek cut.

Questions

1 Using the roof conditions set forth in Unit 9-24, determine the dimension between the plumb cut line and the heel cut line on the bottom of the rafter in Fig. 9-25.4.

2 To mark cheek cut F in Fig. 9-25.6, how would the procedure differ from the procedure given for cut E?

3 How would the procedure for the cheek cut differ if the jack rafter in Fig. 9-25.6 was a valley-jack rafter?

Unit 9-26 **Frame a Roof Opening**

Roofs, like floors, have openings through them, and the openings are framed in a similar way. Rafters beside the opening must be doubled, and doubled headers are required at both ends.

How the headers are set depends on what the opening is for. If a chimney rises through the opening, then headers must be set plumb, and their tops beveled to the slope of the roof. Cripple rafters receive a plumb cut at both ends. If the opening is for a dormer (see Unit 9-27), lower headers are set plumb, but are not beveled. Upper headers are set at right angles to the roof slope. If the opening is for a skylight, both headers are at right angles to the roof slope.

It is customary building practice to leave a hole in the roof structure until the roof has been plumbed and straightened. Then framing around openings, including installation of doubled rafters, follows as the final step in framing the roof.

In determining the size of the roof opening, remember that the frame must clear chimney masonry by at least 2″ on all sides. All other openings should be planned for a tight fit.

Procedure

1 From framing drawings determine the size of opening required. If the size is not given, and the opening is for a chimney, consult with the masonry contractor or your foreman.

2 On the rafter plate and ridgeboard lay out locations of doubled and cripple rafters.

3 Cut and attach trimmers (A and B in Fig. 9-26.1). Note that trimmers may be common rafters with tails (A) or without (B), depending on whether they fall on standard rafter spacing.

4 On rafters A and B mark the locations of the tops of doubled headers.

Fig. 9-26.1

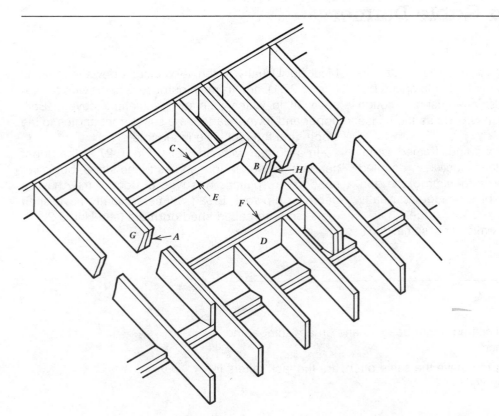

5 Cut headers to length, and bevel the tops if required.

6 Fasten headers (C and D) in position. Nail into ends through rafters. Determine from drawings whether headers should be set plumb or at right angles to rafters.

7 Measure, cut, and fit cripple rafters between headers and ridge or rafter plate.

8 Fasten doubled headers (E and F). Determine from drawings whether tops of plumb headers should be flush with each other or be offset for beveling.

9 Attach doubled rafters (G and H) to trimmer rafters.

Questions

1 Under what circumstances does a trimmer rafter have a tail?

2 To frame the openings listed below, indicate in the blanks what sort of cut you would give cripples where they meet headers.

Opening for a	Upper cripples	Lower cripples
Skylight	_____	_____
Chimney	_____	_____
Dormer with sidewalls	_____	_____

Unit 9-27   **Build a Gable Dormer**

A **dormer** is a vertical projection through a roof. Its purpose is to let light or ventilation or both into space beneath a sloping roof. It may be small—just big enough for a single window—or almost as wide as the house itself.

Dormers are built over roof openings framed as described in Unit 9-26. The distance between doubled rafters is determined by the outside width of the dormer. The distance between doubled headers is usually determined by the length of the dormer ridge.

Some dormers have side walls, and some don't.

Most do. Usually a dormer without sidewalls is small (Fig. 9-27.1), and its main purpose is to provide a vertical wall in which to place ventilation louvers. Both upper and lower headers are set at right angles to the slope, and the roof style is a gable.

In a dormer with sidewalls (Fig. 9-27.2) the upper headers are set at right angles to the slope, but lower headers are plumb. This type of dormer may have a gable or hip roof. Broad dormers usually have shed roofs, and are called **shed dormers** (see Unit 9-28).

Questions

1 Name two different ways of setting upper headers and an advantage and disadvantage of each method.

2 If the main roof and dormer roof have the same pitch, are the jack rafters in the two roofs the same length?

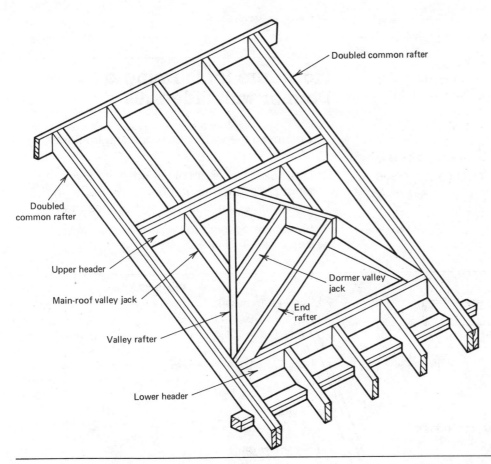

Doubled common rafter

Doubled common rafter

Upper header

Main-roof valley jack

Valley rafter

Lower header

Dormer valley jack

End rafter

Fig. 9-27.1 A dormer without side walls has no wall parts at all. End dormer rafters rest on the lower header. The

gable usually contains ventilating louvers.

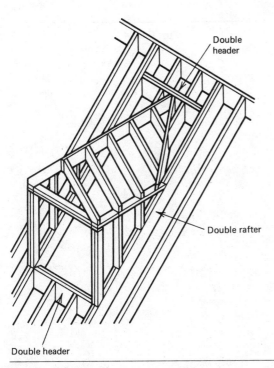

Double header

Double rafter

Double header

Fig. 9-27.2 A dormer with side walls is more conventional in construction. Side rafter plates are single members, and dormer rafters and studs must line up vertically.

Procedure for Framing a Dormer without Sidewalls

1 Determine the width of the dormer opening from framing drawings.

2 Determine the length of the dormer opening, using the length of the dormer ridge and the rise of the main roof as the other two sides of the triangle.

3 Frame the opening to these dimensions as described in Unit 9-26.

4 On the upper header mark the center line of the dormer ridge, and the width of the ridgeboard.

5 Give one end of the ridgeboard a reverse plumb cut at the header (Fig. 9-27.3). Trim the opposite end to fit.

6 Mark the locations of rafters on the ridgeboard.

7 Determine the length of end rafters. They have the same unit rise and unit run as common rafters in the main roof.

8 Cut end rafters. They receive a plumb cut at the ridge, and a beveled heel cut at the header (Fig. 9-27.3).

9 Nail end rafters to the ridgeboard.

10 Fit the assembly in position, and toenail it to headers. The lower ends of rafters rest on the inner header.

11 Determine the line length of valley rafters (Fig. 9-27.4).

12 Shorten the valley rafters as in Fig. 9-27.4. They receive a square double cheek cut at each end (Fig. 9-27.3).

13 Install valley rafters.

14 Install all dormer valley-jack rafters in pairs.

15 Install main roof valley jack rafters.

Procedure for Framing a Dormer with Sidewalls

1 Determine the width of the opening from framing drawings.

2 Determine the length of the opening, using the length of the dormer ridge and the rise of the main roof as the other two sides of the triangle.

3 Frame the opening to these dimensions as described in Unit 9-26.

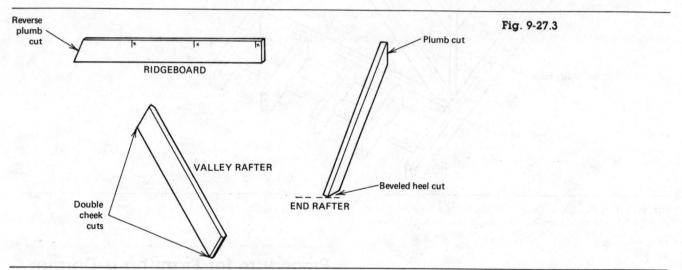

Fig. 9-27.3

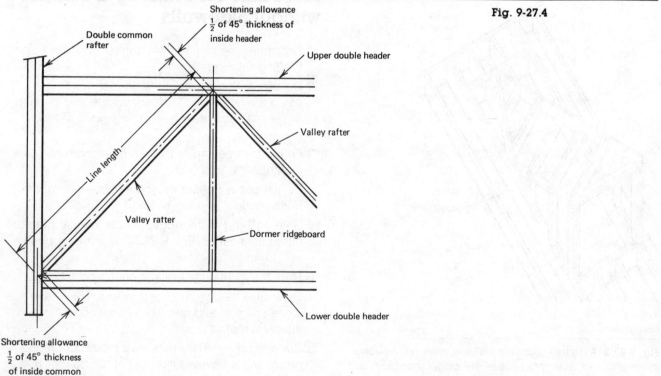

Fig. 9-27.4

4 From drawings determine the length of corner posts.

5 Cut posts to length. Bevel lower ends to the slope of the main roof (see Fig. 9-27.2).

6 Install posts plumb.

7 Install a top plate across posts.

8 On doubled rafters, and on the top plate and doubled header if required, mark the locations of dormer studs.

9 Cut studs to length as for a gable wall (see Unit 9-9).

10 Cut side rafter plates to length, and bevel one end to the slope of the main roof. (Fig. 9-27.5).

11 Install side rafter plates and studs.

12 Cut and install the cap plate.

13 Measure, cut, and assemble the ridgeboard and two end rafters as described in steps 4 through 9 for a dormer without sidewalls. **Note.** Rafters may get a heel cut at the seat, as shown in Fig. 9-27.3, or a combination cut as shown in Fig. 9-27.5.

14 Install this dormer roof assembly.

15 Determine the line length of valley rafters, and the shape of the cut at their lower ends. **Note.** The line length and cut of valley rafters varies with conditions. Usually the length is the diagonal distance from the intersections of center lines of the dormer ridge and header (*A* in Fig. 9-27.6) to the intersection of the rafter plate and doubled rafters (*B*). The cut is a simple cheek cut.

16 Cut and install valley rafters.

17 Install the remaining dormer common rafters in pairs.

18 Cut and install all dormer-valley-jack rafters in pairs.

19 Install main roof valley-jack rafters.

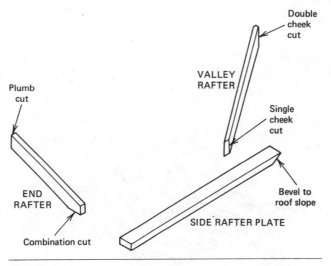

Fig. 9-27.5

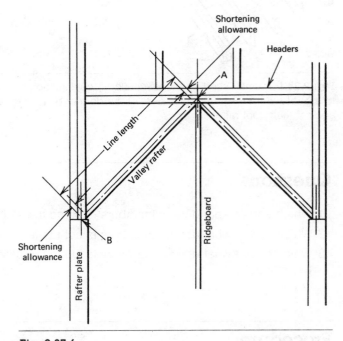

Fig. 9-27.6

Unit 9-28 **Build a Shed Dormer**

One major difference between a shed dormer and a gable dormer with sidewalls is in the studs between roofs. In a shed dormer, one end of sidewall studs is cut to the pitch of the main roof. The other end is cut to the pitch of the shed roof on the dormer. The common difference between dormer studs is the difference between roof pitches times the stud spacing stated in feet.

For example, assume that the pitch of the main roof is 6 in 12, and the gentler pitch of the shed is $1\frac{1}{2}$ in 12.

The difference between the pitches, then, is $4\frac{1}{2}$ in 12. If studs are spaced 16″ o.c., then the common difference is $4\frac{1}{2} \times 1\frac{1}{3}$, or $\frac{9}{2} \times \frac{4}{3}$, or an even 6″.

When a shed dormer is small (Fig. 9-28.1), the upper ends of dormer rafters butt against rafters in the main roof. The angle of cut of dormer rafters must be carefully worked out. In larger dormers (Fig. 9-28.2) rafters may extend all the way to the ridge. Only end dormer rafters receive a special cut. Ceiling joists in dormer rooms are installed after rafters are in place.

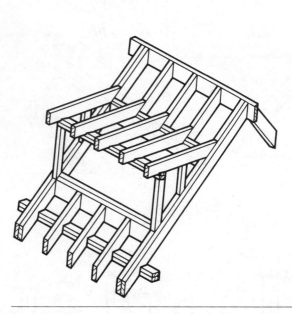

Fig. 9-28.1 A small shed dormer is similar in construction to a gable dormer except that rafters of the dormer butt against main roof rafters.

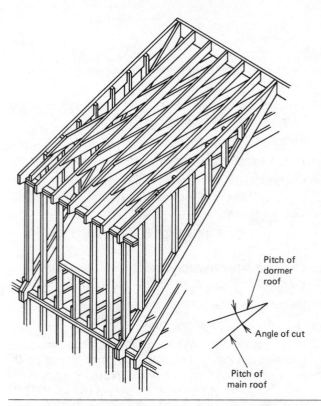

Fig. 9-28.2 The rafters of a broad shed dormer often extend all the way to the ridge of the house, and ceiling joists shape the upper reaches of the room.

Questions

1 Which end of what dormer members is cut to the slope of the main roof?

2 Which end of what dormer members is cut to the slope of the dormer roof?

3 Which end of what dormer members is cut to the difference between the two slopes?

Procedure

1 From drawings determine the size of the dormer opening, and frame it as described in Unit 9-26.

2 Build the front wall of the dormer, either as a standard exterior wall, or as described in Unit 9-27.

3 To determine the length of dormer rafters, use the run of the dormer roof and the dormer rise (Fig. 9-28.3).

4 Set the rafter square on a length of rafter stock as shown in Fig. 9-28.4, with the unit rise of the main roof (A) on the tongue, and the unit run (B) on the blade.

5 From the heel of the square measure down on the tongue a distance equal to the unit rise of the shed roof (C).

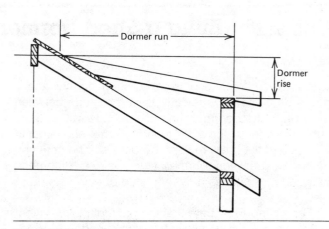

Fig. 9-28.3

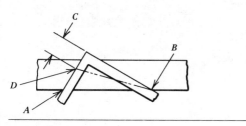

Fig. 9-28.4

6 Draw a line from point *D* through point *B*. This is the angle of cut of dormer rafters that butt against main roof rafters.

7 Cut rafters to length, and make required cuts at the ends and at the bird's mouth.

8 Mark positions of shed rafters on the rafter plate (and on the ridgeboard if required).

9 Set rafters in positions and toenail them to the rafter plate. Nail directly into rafters or ridgeboard.

10 On doubled rafters mark locations of sidewall studs.

11 Determine the common difference between studs.

12 Cut studs to length, and bevel the lower end. Notch the upper ends as described in Unit 9-9 for building a gable.

13 Install sidewall studs.

Unit 9-29   **Frame a Flat Roof**

For calculating purposes, a flat roof has zero rise, even though the roof may slope slightly to permit drainage. Therefore the length of a rafter is the same as the span plus two overhangs when the roof can be spanned with a single length. Usually, however, houses with flat roofs have a center bearing partition, and pairs of rafters must overlap 6″ at that point. Therefore the length of each rafter is its run plus one overhang plus at least 3″ (Fig. 9-29.1).

Ordinarily, a flat roof overhangs exterior walls the same distance all the way around the house. Because of this, common rafters are installed in only the center portion of the roof. The outer portions are framed with lookout rafters (Fig. 9-29.2).

Flat and low-pitched roofs require heavier rafters than steeper roofs. When rafters extend full depth from

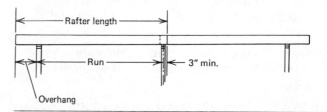

Fig. 9-29.1 The length of a common rafter in a flat roof is equal to the run plus overhang plus overlap at any bearing partition.

fascia to fascia, the roof looks heavy, and appears to squash the house. To overcome this appearance, designers often call for tapered ends to rafters (Fig. 9-29.2). Tapering permits a thinner fascia and a lighter looking roof.

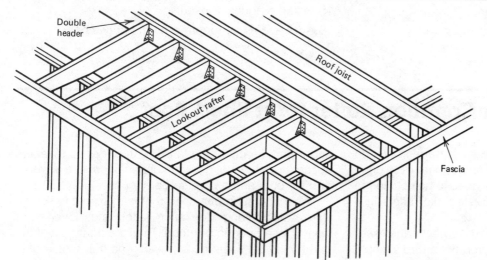

Fig. 9-29.2 On two sides of the house a flat roof is framed at the eaves with lookout rafters. Note that rafters taper from rafter plate to fascia.

Questions

1 Aside from common rafters, how many different lengths of rafters are there in a flat roof?

2 How many of these rafters receive double cheek cuts?

3 How much should rafters overlap at a bearing partition?

4 What is the unit rise of a flat roof?

5 How does the run compare with the span of a rafter?

Procedure

1 From the framing plan determine the quantity and lengths of common rafters required.

2 Mark rafters for trimming, allowing for overlap at bearing partitions and for any taper.

3 Cut rafters to shape and length.

4 Nail two pairs of rafters together to serve as headers (A in Fig. 9-29.3) near the ends of the roof.

5 On marks on the rafter plate 32″ from the ends of walls center the doubled rafters. Toenail them in place.

6 Install all single common rafters between the doubled rafters.

7 On doubled rafters mark the locations of lookout rafters.

8 Measure and cut the longest lookout rafters (B in Fig. 9-29.3). Their length should be $30\frac{1}{2}″$ plus the amount of overhang.

9 Beginning and ending 16″ from the corner of sidewalls, install long lookouts on their marks by toenailing at the rafter plate, and attaching with rafter hangers at the doubled rafters.

10 Cut the eight middle-length lookout rafters (C). Their length should be $15\frac{1}{2}″$ plus overhang.

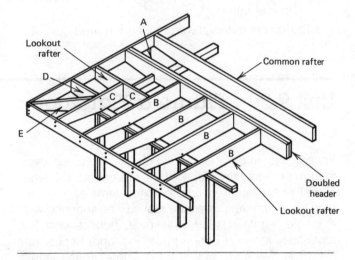

Fig. 9-29.3

11 Install these lookouts as shown in Fig. 9-29.3.

12 Cut and fit the four short lookouts (D) to complete the rectangular roof framework.

13 Determine the length of the four diagonal lookouts (E).

14 Give each a double cheek cut at both ends.

15 Install diagonal lookouts.

16 Add fascias as required by the drawings.

Unit 9-30   Mark a Common Rafter for a Shed Roof

The cuts made in rafters in shed roofs depend first on the pitch of the roof, and second on its span. If the pitch is less than 2 in 12, rafters do not need to be notched to fit over rafter plates. They are treated like rafters in a flat roof.

If the pitch is 2 in 12 or steeper, however, the lower ends of rafters at the lower eave require a bird's mouth and a parallel cut at the tail. The upper ends of rafters at the upper eave require identical cuts: a reverse bird's mouth and a reverse parallel cut (Fig. 9-30.1).

When the span of a shed roof is too great to bridge with a single length of rafter, rafters must overlap at least 6″ at a bearing partition. Then the treatment at the partition is the same as at a wall—no cut if the roof is

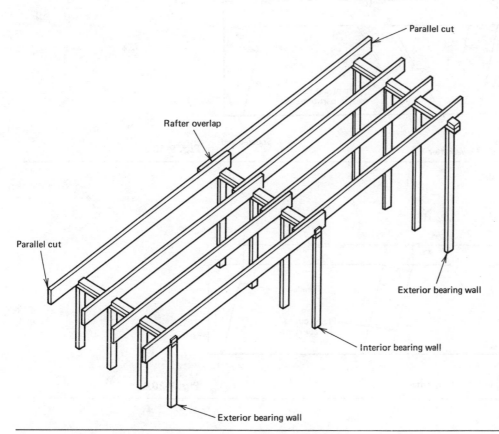

Fig. 9-30.1 The birds' mouths in the rafters of a shed roof are all identical in size and direction.

very low in pitch, and a bird's mouth if the roof is steeper. Note in Fig. 9-30.1 that all bird's mouths in a shed roof are notched in the same direction.

The run and rise of a shed roof are different from the run and rise of two-slope roofs. The total rise of a common rafter is the vertical distance between the tops of plates on which rafters bear (Fig. 9-30.2). The run of the rafter is the horizontal distance between matching edges of these rafter plates. To find the length of a rafter, you must then allow for the overhang and overlap. If a single rafter length is adequate, note that the run of the upper overhang is measured from the *inside* edge of the rafter plate.

In this unit assume that the house is 32'-0" wide, the overhang is 1'-6", the pitch is 2½ in 12, and the bearing partition is centered between exterior walls.

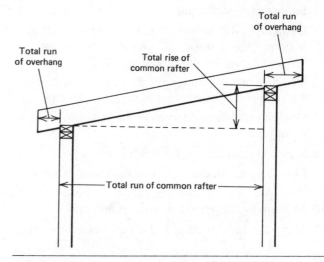

Fig. 9-30.2 In a shed roof the total rise is the vertical distance between the tops of rafter plates in opposite exterior walls. The total run is the horizontal distance between similar edges of these rafter plates. Therefore the run of the lower overhang and the run of the upper overhang are not the same.

Procedure

1 Determine the line length of the lower rafter, using the Pythagorean theorem. Write your answer here

2 Determine the line length of the tail. Write your answer here _____.

3 Estimate the minimum length of rafter stock needed.

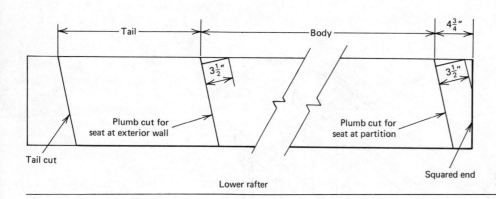

Fig. 9-30.3

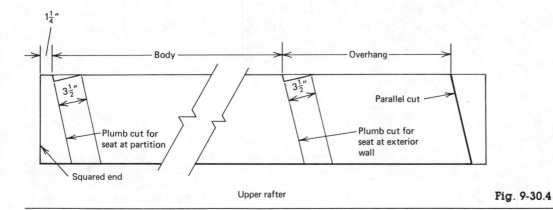

Fig. 9-30.4

4 Square one end of the rafter.

5 Measure in $4\frac{3}{4}''$ from the squared end along the bottom edge of the stock, and mark the plumb cut for the seat at the partition (Fig. 9-30.3).

6 Measure off the line length of the body, and mark the plumb cut for the seat at the exterior wall.

7 Measure off the line length of the tail, and mark a plumb cut.

8 Mark the heel cuts at both rafter plates.

9 Cut along the marks, and test the lower rafter for fit.

10 To mark the upper rafter, square one end.

11 Along the bottom edge of the rafter stock measure off $1\frac{1}{4}''$ (Fig. 9-30.4), and mark the plumb cut for the rafter seat at the partition.

12 Measure off the line length of the body, and mark the plumb cut for the rafter seat at the exterior wall.

13 Mark the heel cuts at both rafter plates.

14 From the heel cut at the upper rafter plate, measure off the line length of the tail and make a mark. **Caution.** Remember that the length of the overhang and the length of the tail are not the same in a shed roof.

15 Mark the parallel cut for the overhang.

16 Cut along the marks, and test the upper rafter to fit.

Questions

1 What is the difference in dimension between the run of the tail and the run of the overhang?

2 What do you think is the reason for the $4\frac{3}{4}''$ dimension in step 5 and the $1\frac{1}{4}''$ dimension in step 11?

3 In the example roof in this unit, what is the total rise? What is the total run?

Unit 9-31 **Frame a Shed Roof**

The procedure for framing a shed roof depends on what the roof covers. If the roof covers the entire house, and has overhangs at both ends, the method of framing is similar to that for a flat roof when the pitch is low, or a gable roof when the pitch is high. The dividing line between a low and a high pitch is 2 in 12.

There are three other common shed-roof conditions. The roof over a bay window (see Unit 9-33) is often a shed roof with little or no pitch. Its rafters butt against the header over the bay window openings. The shed roof over a **lean-to**—a one-story wing on a two-story house—rests on bearing walls at both ends (Fig. 9-31.1). Rafters are cut flush with the rafter plate at the higher end. The rafters in the roof of a shed dormer butt against the rafters of the main roof and the header at the dormer opening (see Unit 9-28). The method of cutting and installing rafters is the same as for a regular shed roof.

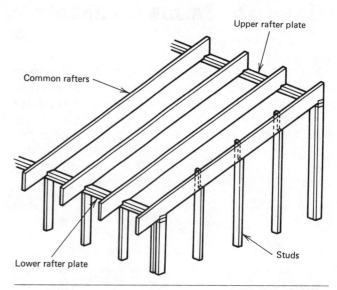

Fig. 9-31.1 Common rafters over a lean-to are cut flush with the rafter plate where they meet the upper wall of the house. At the tail they are cut like any other shed roof rafter.

Questions

1 Name two differences between a rafter in a shed roof over an entire house and one over only a lean-to.

2 The procedure for framing a low-pitch shed roof is similar to the framing procedure for a _____ roof. The procedure for framing a high-pitch shed roof is similar to that for a _____ roof.

Procedure for Framing a Low-Pitch Shed Roof

1 Cut all rafters to fit.

2 Beginning with the rafter 32″ from the corner of the wall, set lower rafters on their marks at the exterior wall, and toenail them in place.

3 Toenail the opposite ends of lower rafters to the rafter plate at the bearing partition. Make sure to hold rafters on their marks, and toenail from the side away from the adjoining upper rafter.

4 Install upper rafters as described in steps 2 and 3. Nail at the bearing partition first.

5 After all common rafters are installed, double end rafters as in Fig. 9-29.1.

6 Complete the roof by following steps 7 to 16 in Unit 9-29.

7 Install studs in end walls following steps 11 to 16 in Unit 9-9.

Procedure for Framing a High-Pitch Shed Roof

1 Cut all rafters to fit.

2 Beginning with the rafter 16″ from the corner of the wall, set lower rafters on their marks on the exterior wall, and toenail them in place.

3 Toenail the opposite ends of these rafters to the rafter plate.

4 Install upper rafters as described in steps 2 and 3. Nail at the bearing partition first.

5 Complete the roof and end walls as described in Units 9-7 and 9-9. The remaining rafters are normally installed when the side wall is completed.

Unit 9-32   Frame a Butterfly Roof

A butterfly roof consists of two shed roofs that slope toward each other. Rarely do the two roof sections have the same run. Nor do they always have the same pitch.

The outer ends of all rafters in a butterfly roof are cut like the upper ends of rafters in a shed roof—with a parallel cut and a reverse bird's mouth. The lower ends get a heel cut, and overlap 6″ at the bearing partition (Fig. 9-32.1). At rakes the roof is built like a gable roof.

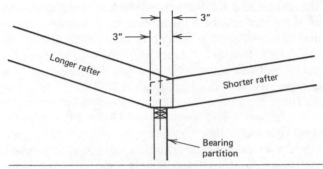

Fig. 9-32.1 Where rafters in a butterfly roof meet, they should overlap at least 6″ for maximum structural stability. If ceilings follow the slope of rafters, as they usually do in a house with a butterfly roof, finish wall and ceiling materials can be applied to cover the flat heel cuts.

Procedure

1 Cut all rafters to fit.

2 Beginning 16″ from the corner of the wall, set longer rafters on their marks at the bearing partition, and toenail them into place.

3 Set shorter rafters against longer rafters at the bearing partition, and toenail them into place.

4 Fit all rafters at exterior walls and toenail them in place.

5 Complete the roof as described in Unit 9-31.

Question

1 Why do you think you should attach all rafters at the bearing partition before you attach them at exterior walls?

Unit 9-33   Frame a Bay Window

A **bay** is a section of wall that projects beyond the main plane of the wall itself. There is almost always a window in the projection, and sometimes several windows. The side walls of the bay may be at right angles to the main wall, or at a lesser angle. When windows form a curve, the projection is called a **bow window** (Fig. 9-33.1).

Sometimes a bay has a full foundation, and is built by standard construction procedures up to subfloor level. More often the bay is cantilevered. The floor frame consists of doubled joists and a doubled header at the edges. Intermediate joists extend beyond the foundation wall, and blocking replaces the main header in this part of the floor (Fig. 9-33.2).

Wall framing is standard except when the sides of the bay are at an angle. In such a case (Fig. 9-33.3) the long wall of the bay is framed from corner of plate to corner of plate. End studs of angled walls are set in from the corners of the plate, and butt against other corner studs at the inside of the bends (points *A* and *B* in Fig. 9-33.3). The triangle between corner studs is strengthened with triangle blocking spaced 24 to 32″ o.c. Top members of bay walls may be a doubled rafter plate, or a standard window header.

A bay window must have a roof. The roof may be flat or a low-pitched shed, or pitched like other roofs in the house. The framing varies with the type of bay roof. It is similar to one-half of a standard roof frame.

Fig. 9-33.1 A bay window (left) has three walls and a roof, and may have a foundation. A bow window (right) has one wall shaped in a continuous curve, is usually tucked under an overhang, and almost never has a foundation.

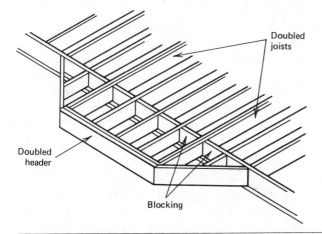

Fig. 9-33.2 Floor framing for a typical cantilevered bay window. Note that the main header joist stops at doubled floor joists. Blocking fills the gaps between floor joists that extend to the doubled header of the bay floor.

Procedure for Framing a Flat or Low-Pitched Bay Roof

1 Determine the length of common bay rafters (*A* in Fig. 9-33.4), and their size.
2 Give both ends plumb cuts.
3 On the header over the bay opening and on the rafter plate mark the spacing of rafters.
4 Set rafters in position on their marks, and toenail them. Check the plumb of all rafters. Also make sure that their under sides are flush with the bottom of the header in the main wall.

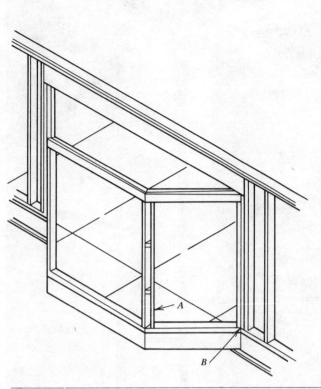

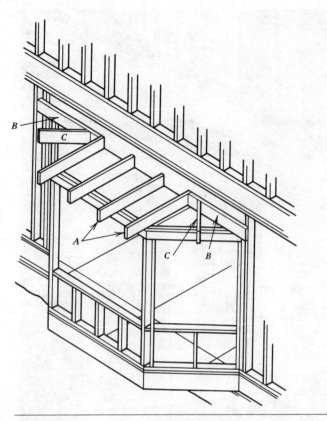

Fig. 9-33.3 Where studs meet other studs at an angle (*A* and *B*), studs butt at one edge, and are strengthened with triangles of stud stock.

Fig. 9-33.4

5 Cut right-angle rafters to length (*B* in Fig. 9-33.4). Give outer ends a single cheek cut, and nail them to the header.

6 Cut diagonal rafters (*C*) to length, and give inside ends a double cheek cut. Nail them into position.

Procedure for Framing a High-Pitched Bay Roof

1 On the rafter plate, the header over the bay opening, and the header joist above, mark the spacing of rafters.

2 Cut a piece of ridgeboard (*A* in Fig. 9-33.5) to the same length as the header in the long wall of the bay.

3 Mark rafter locations on the ridgeboard, then nail it in place on the header. One end should be flush with a rafter mark.

4 Cut bay ceiling joists to length (*B*).

5 Set joists on one side of the rafter marks, and toenail. Check their plumb. Make sure that their

under sides are flush with the header, and level over their entire length.

6 Determine the size and length of common rafters (*C*).

7 Cut rafters with plumb cuts at both ends and a rafter seat.

8 Set rafters on their marks, and toenail to rafter plate, ceiling joist, and ridgeboard.

9 Determine the length of right-angle rafters (*D* in Fig. 9-33.5), and mark the cuts. These two rafters get a plumb cut at the top and a single cheek cut at the tail. The cut at the bird's mouth is similar to that for a hip rafter.

10 To determine the line length of diagonal rafters (*E*), use the same unit rise as for a common rafter, but use 8.5 instead of 12 as the unit run.

11 Cut these two rafters to length, with a double cheek cut at one end, a tail cut at the other, and a bird's mouth at the rafter plate. **Caution.** These diagonal rafters have a steeper pitch than common rafters, and therefore the notch for the bird's mouth must be deeper.

12 Nail diagonal rafters in position.

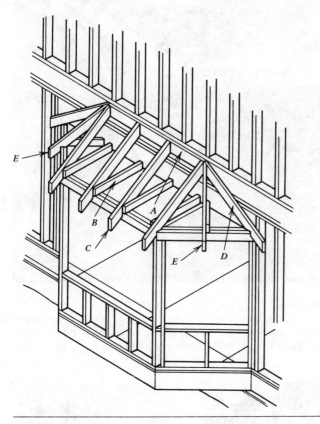

Fig. 9-33.5

Questions

1 In step 10, where does the figure 8.5 come from?

2 Under what circumstances would this not be the correct unit run?

3 Why is there a need for the ridgeboard in Fig. 9-33.5, and the right-angle rafters in Figs. 9-33.4 and 9-33.5?

Unit 9-34   Install Roof Sheathing

Sheathing on a roof has three important purposes. First, it adds great strength to a rather weak framework. Second, it protects the construction beneath it. And third, it serves as a base surface for application of finish roofing materials.

Of the sheet materials used for sheathing a wall, plywood is the most common for sheathing a roof. In a $\frac{1}{2}''$ thickness, it provides excellent bracing strength, and may be used under almost any roofing material except wood.

Under wood shingles or shakes, sheathing must be boards with spaces between them (Fig. 9-34.1). Unlike other roofing materials, wood shingles absorb some rain water, and air must be able to circulate underneath them to prevent uneven drying and subsequent splitting. The most commonly used board sizes are 1 × 3, 1 × 4, and 1 × 6. Their spacing depends on how much shingle is visible in any course—its **exposure**. If the sheathing you use is 1 × 3 or 1 × 4 boards, center line to center line spacing of boards should be

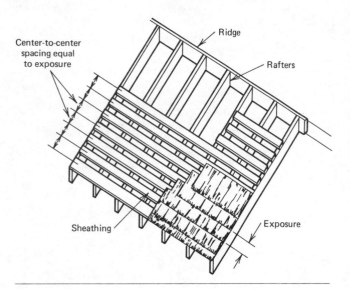

Fig. 9-34.1 When finish roofing is wood shingles, the sheathing beneath them must be boards spaced according to the amount of exposure of each course of shingle.

Fig. 9-34.2 A simple rack for sheet sheathing is a big time-saver. One man at ground level loads it, and the carpenter who is sheathing pulls from this stock as he needs it. (Courtesy American Plywood Association.)

Fig. 9-34.3 Long 2 × 4s nailed horizontally but temporarily across a roof make roofing work easier and safer. The 2 × 4s are called walking steps.

equal to the exposure. If sheathing is 1 × 6 boards, the spacing should be twice the exposure.

Boards as a backing for wood shingles should be painted with a coat of primer to seal them against penetration by moisture. Board sheathing may be placed directly on rafters in climates with no extremes in heat or cold. In better construction, however, sheet sheathing is applied first, and is covered with a layer of building paper. The boards are then applied over the paper.

When boards are used for sheathing in place of sheet sheathing, they are installed without spacing. This is called **closed** sheathing. If boards are installed with gaps between them, it is called **open** sheathing.

The procedure for applying insulating sheathing or gypsum sheathing is the same as for plywood except for spacing of fasteners. Follow manufacturer's instructions for proper spacing and size of fasteners. If compressed air is available at the site, the attachment process goes much more quickly with an automatic nailer. Good carpenters can nail with this tool almost as fast as they can walk.

Sheathing is heavy and awkward to handle. One way to raise it to roof level is with pump jacks. As an alternate, the job can be made much easier if plywood is stored for use in a lean-to rack that rests against the eave (Fig. 9-34.2). One man working from ground level keeps the rack full, while the sheather can lift each sheet from the rack without coming dangerously near the edge of the roof.

When roof slopes are gentle, carpenters have little difficulty working from the surface of sheathing already laid as they work toward the ridge. On steeper slopes, however, walking steps are needed. **Walking steps** (Fig. 9-34.3) are lengths of 2 × 4 laid across sheathing at convenient and comfortable intervals, usually about every 6′ up the slope. The 2 × 4s provide a good footing, and a handhold in an emergency.

Procedure for Sheathing with Plywood

1 Select a sheet of plywood, and check all four corners for square. The first piece applied must be square.

2 Begin work on the unbraced side of the roof by placing the first piece at the lower corner of the slope, with its inner edge centered on a rafter, and its lower edge either against the fascia or overlapping the tail of the rafters about ½″ (Fig. 9-34.4). Let the sheet overlap at the rake. **Note.** Right-handed carpenters usually find it easier to start at the lower left edge of a roof and work

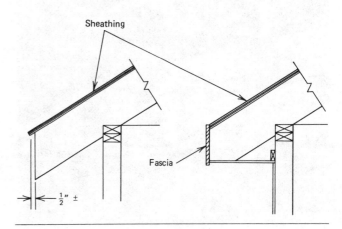

Fig. 9-34.4

toward the right. Left-handed carpenters usually prefer to start at the lower right edge.

3 When the first sheet is in position, attach with 8d common nails or 7d threaded nails driven into tops of rafters at the lower edge with a hammer or nailing gun (Fig. 9-34.5).

4 Continue to nail every 6″ along edges, and every 12″ into intermediate rafters.

5 Along the upper edge of each sheet place H-clips between rafters (Fig. 9-34.6).

6 Complete the first course of plywood to the opposite roof rake.

7 Begin the second course with a partial sheet. If possible, use the offall from the first course. Use no less than one-third of a sheet, however, and no more than two-thirds of a sheet.

8 Fit each sheet into the other half of the H-clips, and nail as in steps 3 and 4.

9 Continue the second course with full sheets. In succeeding courses, always stagger the seams between panels.

10 At the ridge let the final course overlap (Fig. 9-34.7, left). Trim it with a power saw flush with opposite rafters.

11 Begin sheathing the opposite slope as described in steps 1 to 9. Remove braces as you go.

12 At the ridge, again let the final course overlap (Fig. 9-34.7, right). Trim it flush with sheathing on the opposite slope.

13 Trim sheathing at rakes flush with the sides of end rafters or fly rafters.

Fig. 9-34.5

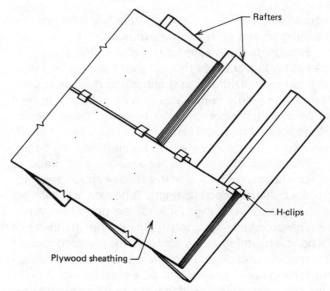

Fig. 9-34.6

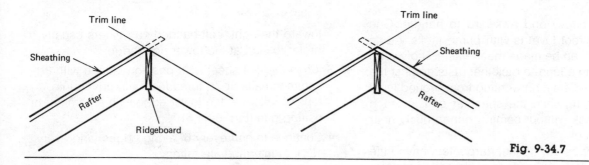

Fig. 9-34.7

Procedure for Sheathing with Boards

1 Select straight, good quality boards with squared ends for the first course.

2 Prime them on all four sides and at ends.

3 With 8d nails attach the first board, with one end centered on a rafter and the lower edge either against the fascia or overhanging the tails of rafters $\frac{1}{2}''$ (see Fig. 9-34.4, left). Use two nails per board for widths up to 1 × 6, and three nails in wider boards.

4 Cut other sheathing boards to length so that their ends fall on centers of rafters.

5 Complete the first course.

6 Apply the second course as closed sheathing. Joints between boards must be staggered between courses.

7 Continue closed sheathing to a point above the rafter plate (Fig. 9-34.8).

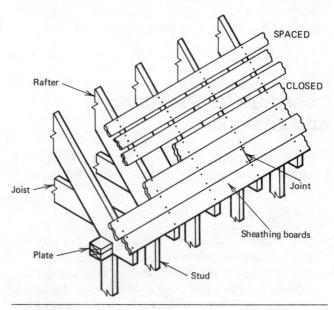

Fig. 9-34.8

8 Thereafter use either open or closed sheathing as required by finish roofing.

9 At the ridge, trim the final course flush with rafters on the opposite slope (see Fig. 9-34.7).

10 Trim the last course of sheathing in the other

slope at the ridge flush with sheathing on the first slope.

11 Trim sheathing at the rakes flush with rafters.

Questions

1 If the exposure for wood shingles is 5″, what should be the center-to-center spacing if sheathing boards are 1 × 3? 1 × 4? 1 × 6?

2 If a roof measures 38′-8″ wide, what should be the length of plywood used to start the second course of sheathing?

3 After how many courses would you use up all offall?

4 Why do you start work on the unbraced side of the roof?

Unit 9-35 Sheathe Hips and Valleys

Most pieces of roof sheathing installed by a carpenter have square edges and square corners. On a hip or mansard roof, however, or a roof of any style on a house with a wing, some sheathing must be cut to fit tightly at hips and valleys.

Most carpenters install one complete course of sheathing on one side of a house, including corners at hips and valleys before going on to the next. Often a piece trimmed in one course is usable at some other point farther up the roof.

The tools required for fitting sheathing at changes in slope are a saw, pencil, and your pitch board. The procedures are identical for both sheet and board sheathing.

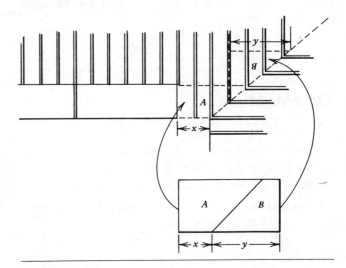

Fig. 9-35.1

Procedure

1 Install full-sized sheathing as close as possible to the turn in a roof.

2 Measure from the lower edge of the last full piece of sheathing to the center of the hip or valley (Fig. 9-35.1).

3 Transfer this measurement (x) to a piece of sheathing.

4 With the pitch board as a guide, draw a line from this point diagonally across the sheathing.

5 Make a plumb cut along this line. **Precaution.** Remember to set the blade of the saw to the proper roof angle.

6 Install the cut piece of sheathing (A in Fig. 9-35.1).

7 Check the fit of piece B at the valley or hip of the next course.

8 Trim if necessary. If the two slopes are at the same pitch, no trimming should be required to end at the center of a rafter (dimension y).

Unit 9-36   **Cut Holes in Sheathing for Pipes**

At large openings in the roof, such as for a chimney, it is customary to fit sheathing in place, then cut out the opening with a power saw or router. Codes require a 2" clearance between framing members and a chimney. Sheathing may extend to within $\frac{3}{4}$" of the masonry, however, to support roofing and sheathing (Fig. 9-36.1).

Where a roof must be cut for a vent pipe, plumbing work often isn't far enough along so that the exact location of the pipe can be determined. Usually, then, carpenters sheathe right over the future location of the vent, and come back later to cut the hole to fit.

Holes for pipes are cut about $\frac{1}{4}$" in diameter larger than the pipe itself. The sides of the hole are cut plumb, not at right angles to the sheathing. The procedure begins after the vent pipe has been completed to a height where its point of penetration through the roof can be accurately determined.

The tools required are a plumb bob, hammer, pitch board or rafter square, pencil, drill, and a saber saw.

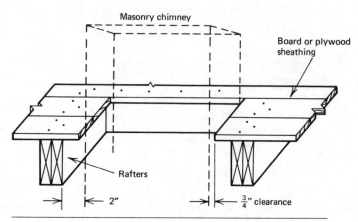

Fig. 9-36.1 Framing aroung a masonry chimney must clear it by at least 2". Sheathing, however, may extend to within $\frac{3}{4}$" to support roofing materials and flashing.

Questions
(Circle correct words in parentheses)

1 Where a pipe punctures a roof, the hole in the sheathing is cut (plumb, at right angles to sheathing).

2 The shape of a hole in sheathing for a vent pipe is (round, oval, square).

3 The diameter of a hole in sheathing for a pipe should be ($\frac{1}{8}$", $\frac{1}{4}$", $\frac{1}{2}$", $\frac{3}{4}$") greater than the pipe itself.

Procedure

1 From the under side of sheathing hang a plumb bob so that it points directly to the center of the vent pipe (Fig. 9-36.2).

2 Mark this point on the sheathing.

3 Drive a nail upward and vertically through the sheathing. Where it penetrates the roof, mark the center of the hole to be cut.

4 Determine the diameter of the pipe to be used, and add $\frac{1}{4}$".

5 On a piece of scrap board draw two parallel lines this distance apart (Fig. 9-36.3).

6 Using the pitch board or rafter square as a guide, draw a line between these two lines at an angle equal to the slope of the roof.

7 Measure the length of this line.

8 Move to the roof, and draw coordinates through the hole made by the nail (Fig. 9-36.4).

9 On the horizontal coordinate, measuring from the hole, mark off one-half the diameter from step 4 in each direction.

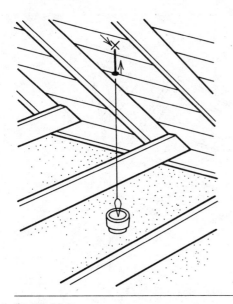

Fig. 9-36.2

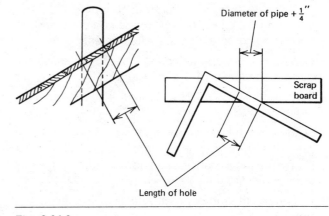

Fig. 9-36.3

10 On the vertical coordinate, again measuring from the hole, mark off one-half the length of the line from step 7.

11 Through these four points draw an oval.

12 Drill a starting hole at some point along the edge of the oval.

13 Make the cutout along the oval line with a saber saw. **Precaution.** Remember to make a plumb cut all the way around.

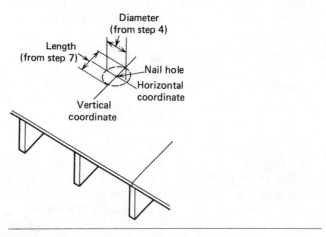

Fig. 9-36.4

Unit 9-37 **Build a Chimney Saddle**

When a chimney penetrates a roof at the ridge, rain can run off in all directions. As long as the joint between masonry and sheathing is properly sealed, there is little danger of a leak.

Sometimes, however, a chimney rises at a point where there is still roof above it. Then, some rain runs toward the chimney instead of away from it. To prevent leaks under this condition, the carpenter builds a saddle (Fig. 9-37.1).

A **saddle,** also called a **cricket,** is similar to a small gable dormer (Unit 9-27). But because the saddle carries little weight, it is built on the ground or subfloor, and then set in place on top of roof sheathing. Its main members are 1 × 4s and 2 × 3s. It usually has the same pitch as the main roof. It is sheathed like the main roof, and may be finished with the same material. If it is small, however, it may be finished only with flashing—either metal or carefully sealed building paper.

Procedure

1 From the masonry contractor determine the width of the chimney at the point where it passes through the roof (*D* in Fig. 9-37.2). If possible, wait until chimney construction is complete and you can take an accurate measurement.

2 By calculation, determine the vertical thickness of saddle sheathing (*A* in Fig. 9-37.2).

3 To this thickness add the vertical thickness of a valley strip (*B*).

4 Using the total dimension from step 3, and the ratio of unit rise to unit run, determine dimension *C*. This is the distance that saddle sheathing will extend beyond valley strips.

5 Determine the run of an end rafter, using the formula $\frac{1}{2}D - C$.

6 Determine the length of an end rafter.

7 Cut the two end rafters. At the ridge they get a shortened plumb cut. At the valley they get a cheek cut, shortened by the thickness of the valley strip (refer to Figs. 9-37.2 and 9-37.3).

8 Determine the length of the ridgeboard. Its length is equal to the line length of an end rafter less the dimension in step 2.

9 Cut the ridgeboard to length, with a flush cut at one end and the same as a seat cut for a common rafter at the other end (Fig. 9-37.3).

10 Assemble end rafters and ridgeboard.

11 Determine the length of valley strips. The procedure is the same as for valley rafters (Unit 9-22).

12 Set your rafter square on valley strip stock with the unit length of a common rafter on the tongue, and the unit run on the blade (Fig. 9-37.4). Mark the top cut along the tongue.

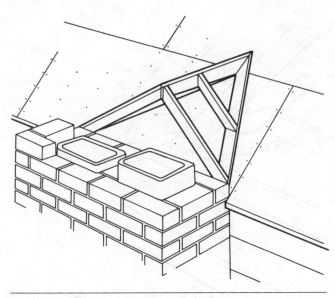

Fig. 9-37.1 A chimney saddle diverts rainwater flowing down a roof away from the chimney. It is similar to the roof on a gable dormer.

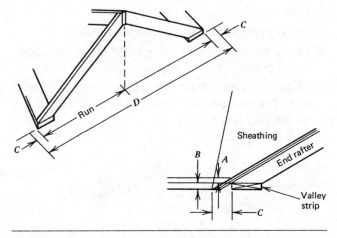

Fig. 9-37.2

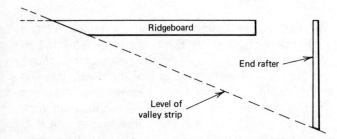

Fig. 9-37.3

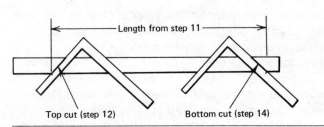

Fig. 9-37.4

13 Measure off the length from step 11 along the outer edge of the strip, and make a mark.

14 Set the rafter square with the blade on this mark and in the same position as in step 12. Mark the bottom cut along the blade.

15 Cut valley strips, and position them on the roof sheathing. The distance between ends should be equal to twice the run of a saddle end rafter (see Fig. 9-37.2).

16 Nail valley strips to the sheathing.

17 Nail the ridgeboard-rafter assembly to the valley strips.

18 Add valley jack rafters if the ridge is more than 4' long.

19 Sheathe the saddle as described in Units 9-34 and 9-35. Give lower edges of sheathing a horizontal cut for a tight fit.

Questions

1 What is the sole purpose of a saddle?

2 Why is the span of the saddle less than the chimney dimension where the two meet?

3 Why are the line lengths of the ridgeboard and end rafters usually the same?

4 Under what conditions would they not be the same?

Unit 9-38   Install Roofing Paper

As soon as roof framing is covered with sheet sheathing, the sheathing should be protected from weather, which can warp or watersoak it. Once this protection is in place, the task of applying finish roofing materials should follow as soon as weather and wind conditions are right.

The most commonly used protective layer is roofing felt. **Roofing felt**, also called **roofing paper**, is felt saturated with asphalt. It comes in rolls 36″ wide, and in 15-lb and 30-lb weights. The weight designation refers to the weight of 100 sq ft of felt. One hundred square feet of any roofing material is called a **square**.

Felt serves three important purposes:

• It protects the sheathing from weather.

• It prevents driving rains and a buildup of snow or ice in a gutter from working its way under roof shingles, causing serious structural damage to ceilings, roofs, and exterior walls.

• It acts as an **underlayment**—a base—for finish roofing material.

Roofing felt is best applied so that two layers cover the entire surface of the roof. Table X is a guide to usage. At the eave a **starter strip**, often called a sel-

vage edge, is laid first. The width of a starter strip varies from 9 to 18″, depending on local usage and weather conditions. Selvage edge and roofing felt may be applied with roofing nails or staples. The only purpose of the fasteners is to prevent the underlayment from blowing away.

To protect the exposed edges of sheathing, good building practice calls for a drip edge at eaves and rakes. A **drip edge** is a strip of corrosion-resistant metal 4 to 6″ wide. The metal may be 16 oz copper or 26 ga galvanized steel, but is usually thin aluminum. A drip edge is applied with nails of the same metal.

Procedure

1 Cut strips of metal for a drip edge.

2 Lay the first strip at a corner (Fig. 9-38.1). Overhang the eave and rake about $\frac{3}{4}$″.

3 Nail the drip edge in place, and bend outer edges downward at about 45°.

4 Add the remaining strips at the eave, overlapping at least 1″ at seams.

Table X   Summary of underlayment recommendations for shingle roofs

Type of roofing	Sheathing	Type of underlayment	Normal slope		Low slope	
Asbestos-cement shingles	Solid	15 lb. asphalt saturated asbestos (inorganic) felt, or 30 lb. asphalt saturated felt	$\frac{5}{12}$ and up	Single layer over entire roof	$\frac{3}{12}$ to $\frac{5}{12}$	Double layer over entire roof[a]
Asphalt shingles	Solid	15 lb. asphalt saturated felt	$\frac{4}{12}$ and up	Single layer over entire roof	$\frac{2}{12}$ to $\frac{4}{12}$	Double layer over entire roof[b]
Wood shakes	Spaced	30 lb. asphalt saturated felt (interlayment)	$\frac{4}{12}$ and up	Underlayment starter course; interlayment over entire roof	Shakes not recommended on slopes less than $\frac{4}{12}$ with spaced sheathing	
	Solid[c,e]	30 lb. asphalt saturated felt (interlayment)	$\frac{4}{12}$ and up	Underlayment starter course; interlayment over entire roof	$\frac{3}{12}$ to $\frac{4}{12}$[d]	Single layer underlayment over entire roof; interlayment over entire roof
Wood shingles	Spaced	None required	$\frac{5}{12}$ and up	None required	$\frac{3}{12}$ to $\frac{5}{12}$[d]	None required
	Solid[e]	15 lb. asphalt saturated felt	$\frac{5}{12}$ and up	None required[f]	$\frac{3}{12}$ to $\frac{5}{12}$[d]	None required[f]

[a] May be single layer on 4 in 12 slope in areas where outside design temperature is warmer than 0° F.
[b] Square-butt strip shingles only; requires wind resistant shingles or cemented tabs.
[c] Recommended in areas subject to wind driven snow.
[d] Requires reduced weather exposure.
[e] May be desirable for added insulation and to minimize air infiltration.
[f] May be desirable for protection of sheathing.

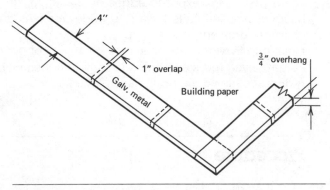

Fig. 9-38.1

5 Unroll a starter strip over the drip edge, overlapping sheathing slightly.

6 Nail or staple the starter strip in place, fastening every 6″ along lower edges and ends, and every 12″ along the upper edge (Fig. 9-38.2). Trim ends flush with sheathing at the rakes.

7 Complete the starter strip, overlapping 4″ at any seams.

8 Lay the first course of felt flush at all edges with the starter strip, and fasten in position as in step 6.

9 Start the second and all remaining courses with a 17″ exposure, so that all areas of the roof are covered with at least two thicknesses (Fig. 9-38.3).

10 At the ridge carry the final course at least 9″ over the ridge.

11 Repeat steps 1 to 10 inclusive on the opposite slope.

12 At rakes add a drip edge over the sheathing and roofing felt as in steps 3 and 4.

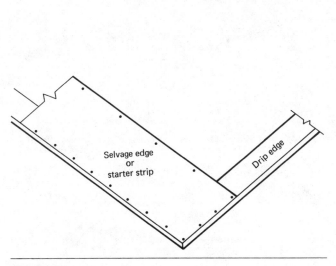

Fig. 9-38.2

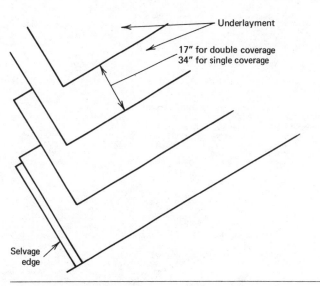

Fig. 9-38.3

10

Insulation

Insulation serves two very important purposes. In a single-family residence its main purpose is to protect against heat and cold, and to help maintain interior temperatures at a comfortable level as economically as possible. In multifamily buildings insulation is also a method of reducing the amount of noise between apartments. This purpose is discussed more fully in Chapter 12.

Heat Transfer

Heat always tries to move toward cold. This movement, called **heat transfer** by engineers and specialists in insulation, occurs in three ways. You have experienced all three ways many times in your lifetime.

Radiation is the transfer of heat directly from one surface to another. Heat transferred in this way is called **radiant heat.** The sun shining on any surface—you, a roof, a wall, a street—warms that surface even though the sun is millions of miles away. You may perspire from the sun's radiation, even though the temperature of the air around you is quite cool. The heat you feel from a camp fire or fireplace is also radiant heat.

Conduction is the transfer of heat from molecule to molecule through actual contact. The handle of an iron skillet on a fire or on a burner of a kitchen range gets hot even though the handle itself is not in direct contact with the source of heat. Some materials are better conductors of heat than others. Usually the more dense the material—that is, the closer the molecules are together—the better a conductor it is. Most metals, such as steel and copper, are good conductors. Most gases, such as air, are not. That is why radiant heat does not warm the air through which it passes.

Convection is the transfer of heat through a liquid

or gas. Warm air heating systems operate on the principle of convection. Air heated in a furnace and blown to various rooms heats those rooms. Similarly, hot water run into a cold bathtub heats the tub by convection.

The insulation installed in a house should protect it against all three types of heat transfer. To protect against radiation, insulation must reflect heat rather than absorb it. To protect against heat loss or gain by conduction, the insulation itself must be a poor conductor of heat. To protect against convection, the insulation must prevent the flow of air between the warm side and the cold side of a wall or roof.

Insulation must also protect against moisture problems.

Condensation

All air contains some moisture, even in the desert. The warmer the air's temperature, the more moisture it can hold. As warm air cools by moving toward a cold surface, it can no longer hold all its moisture. Therefore some moisture comes out of the air and forms as drops of water (condenses) on the cold surface. This process is called **condensation**. You have seen drops of water form in warm weather on the outside of a soft drink can or glass of lemonade. Soon the moisture runs down the side of the container, and leaves a puddle on the table below. Eventually the puddle dries up.

The same series of events can occur in a house in winter. Heated air in a room moves toward exterior walls and the ceiling. If the wall is not insulated, the air finds its way through the wall to the outside. But the moisture condenses when the air's temperature reaches the **dew point**—the temperature at which it contains more water than it can hold. This moisture remains in the wall until it eventually dries up. If the outdoor temperature is very cold, the moisture turns to ice, and stays in the wall until warmer weather.

As you can see, moisture is one of the worst enemies of houses. Builders must take three steps to prevent damage from condensation:

1 They must provide a vapor barrier on the warm side of exterior walls, ceilings, and exposed floors. The vapor barrier keeps moisture from working its way into the house's structural framework.

2 They must provide insulation, so that air temperature never reaches the dew point and therefore condensation can't occur.

3 They must provide ventilation, so that any moisture that does get into the structure will be

carried harmlessly away by air movement. No matter how carefully a house is built, it will not be completely moisture-tight. It is the builder's responsibility to select the best possible insulation. It is the carpenter's responsibility to provide the best possible installation.

Rating Insulation

The best insulation has a vapor barrier built into it to prevent condensation. It has a surface on one side that reflects radiant heat back into the room. It has air spaces in it to insulate against transfer of conducted heat. And the carpenter must install it in as airtight a manner as possible to control convected heat.

All insulations on the market do at least one job very well, but few do all jobs well. To help the builder know how good a job any one insulation will do, insulations are rated for their resistance to transmission (transfer) of heat. The scientific terms used to discuss heat and its control are all alphabetical. They are Btu, U, k, and R.

The measure of heat transfer is the **Btu**—also written B.T.U. and BTU. This is an abbreviation for British thermal unit. One Btu is the amount of heat required to raise the temperature of one pound of water one degree Fahrenheit. Btus are used by engineers to calculate heat loss in a house and to rate heating equipment. A building code may state that the maximum heat loss in a house shall not exceed 40 Btus per hour per square foot of floor space. By multiplying this maximum by the floor area of a house, you can calculate the minimum Btu output of the furnace or boiler. Heating plants are rated by Btu output, such as 110,000 Btus per hour.

The **k factor** is a decimal that expresses the amount of heat, measured in Btus, that can be transmitted in one hour through one square foot of material 1″ thick for every degree of difference between indoor and outdoor temperature. Every material, including air space, has a k factor. The lower its k factor, the better an insulator it is. An inch of concrete, for example, has a k factor 15 times as high as an inch of wood, and more than 11 times as high as an air space. Therefore wood is a better insulation than air, and both are better than concrete.

The **R factor** is a number that expresses the resistance of a material to transmission of heat. It is equal to $1 \div k$, and ranges from zero upward. Building codes frequently state requirements for insulation in terms of the R factor, and the R factor of packaged insulation is clearly stated on the bag.

The **U factor** is also known as the coefficient of heat transmission. It is a decimal that expresses the amount of heat in Btus transmitted in one hour through a square foot of wall, ceiling, or roof for every degree of difference between inside and outside temperature. To find the U factor for any given construction, the engineer adds up the R factors of the materials used in their particular thicknesses. The U factor is 1 divided by the sum of the R factors. The higher the R factor, then, the lower the U factor.

As you can see by reading the definitions of insulating terms, outdoor temperature is an important consideration in insulating a house. The colder the outdoor design temperature (Fig. 10-0.1), the greater the requirement for insulation, and the greater the R factor needed to meet the requirements.

Types of Insulation

All insulation is light in weight because its main ingredient is air. This air is trapped in hundreds of tiny cells. The materials forming the cells may be mineral—strands of glass or molten rock spun into fibers. They may be vegetable—wood or cotton fibers, wood chips or bark, stalks of corn or sugar cane. They may be animal—cattle hair, for example. Or they may be chemical—various types of foams.

There are five common types of insulation: loose fill, flexible, rigid, reflective, and foam.

Loose fill insulation consists of pellets of mineral wool or wood products of various types. It comes in bags, and may be poured or blown into place. It is effective only for insulating horizontal cavities. It requires a separate vapor barrier.

Flexible insulation is manufactured in two forms: blankets and batts. Blankets come in long rolls in widths to fit 16″, 24″, and sometimes 20″ spacings between framing members. The thicker the insulation, the shorter the rolls. Batts are blankets cut into short lengths, usually 4′ long. Batts in the most common thicknesses come seven lengths to a package.

Most blankets and batts consist of mineral wool, glass wool, cotton, or wood fiber insulation packaged in a sleeve. One side of the sleeve is thin paper with pinholes punched in it to allow it to breathe—that is, release any heat or moisture trapped in the insulation. The other side is a heavier asphalted paper that acts as a vapor barrier. The insulation is installed with the vapor barrier facing into the room.

Some blankets and batts have no covering at all, and require a separate vapor barrier. One type of batt made of glass fibers is a little wider than other types, and stays in place by friction. All other types are usually stapled in place.

Reflective insulation is a sheet material, usually aluminum. By itself it offers protection against radiant heat loss, and acts as a vapor barrier. More often, however, the reflective sheet is part of another insulating material. Some blankets and batts, for example, are made with a reflective foil surface instead of asphalted paper on the warm side. Gypsum wallboard, gypsum lath, and some types of rigid insulation also have reflective foil on one side.

Rigid insulation is made of various mineral and vegetable products. It is manufactured in sheets $\frac{1}{2}$ to 3″ thick and usually 2′ by 4′ or 4′ × 8′ in size.

Rigid insulation has many uses, some structural and some nonstructural. As a structural material, it is used primarily for sheathing, as described in Units 7-18 and 9-34. It has some insulating value, but its main purpose is structural. As a nonstructural material, it is used primarily for insulation, as described in Units 5-14 and 10-5. It has some structural value, but its main purpose is insulating.

Chemical foam is a comparatively new form of insulation that is growing in popularity. Foam is manufactured as a rigid board $\frac{1}{2}$ to 2″ thick, and is installed like sheathing, primarily on wall studs. Foam is also available in a liquid form that hardens quickly when exposed to air. It may be either poured or sprayed into place. For foam to be effective, however, the spaces to be insulated must be fully formed and relatively airtight. The foam completely fills the space, and conforms to its shape.

Foam is the most practical answer for insulating an existing house that is either uninsulated or underinsulated. The foam is mixed on the site, and is blown into each wall cavity through small holes. If the wall is masonry, the holes are drilled in mortar joints, then filled with fresh mortar after insulating. If the wall is wood, the practice is to remove a course of siding or shingles near the bottom and top of each wall, and drill holes in the sheathing behind. With thinner siding holes are drilled through both siding and sheathing. After the wall is insulated, the holes are filled with wood plugs, or the exterior wall material is put back in place.

If there is no practical way to insulate from the outside, the foam is sometimes sprayed through holes cut in the interior wall surface.

Where to Insulate

Drawings seldom show where to put insulation. Specifications usually state the type of insulation desired and the R factor required by heat loss calculations. The carpenter must know where to place insulation and vapor barriers, not only to make the house comfortable

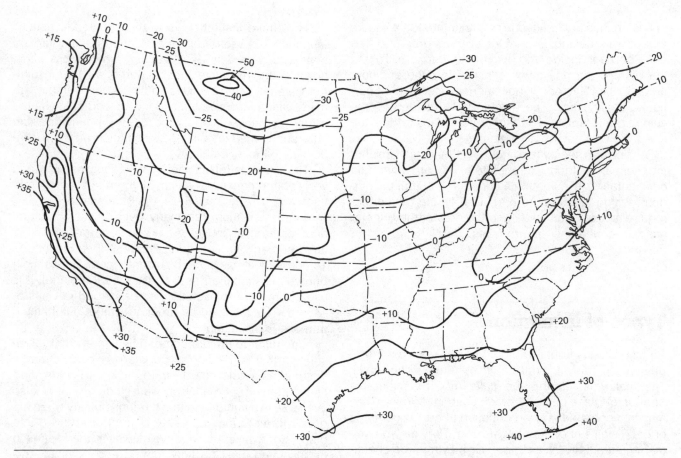

Fig. 10-0.1 This map shows the lowest temperatures during an average winter across the continental United States. These temperatures are one of the factors used in deter-mining outdoor design temperature and the R value of insulation required.

and economical to heat and cool, but also to prevent future damage. Most places are obvious; some are not (Fig. 10-0.2). Here is a checklist.

- On basement walls when spaces are used for living purposes (see Unit 10-6).
- Against header joists and edge joists between the sill and subfloor of houses with basements.
- Between floor joists over a crawl space (see Unit 10-3).
- Around the inside perimeter of concrete slab floors of heated rooms (see Unit 10-5).
- Between all studs in exterior walls.
- Between structural members and door and window frames on all four sides.
- In small spaces between window and door framing members.
- Behind outlet boxes and entrance panels in exterior walls.
- Against second floor header joists and edge joists—between the top plate and subfloor of two-story and split-level houses.

- Between ceiling joists of unheated garages or porches when rooms above are used for living purposes.
- Between ceiling joists below unheated attics.
- Below stairways to unheated attic space.
- In knee walls of heated attics.
- Between or over joists of decks above heated living space.
- Over ceiling joists of rooms with sloping ceilings.

Wherever insulation is needed, a vapor barrier is re-quired. Although wood is a good insulator, it is not a vapor barrier. Some builders therefore install a vapor barrier over all structural framing members exposed to outdoor temperatures, including plates and headers.

Fig. 10-0.2 Here is a guide to the areas of a house that should be fully insulated. ▶

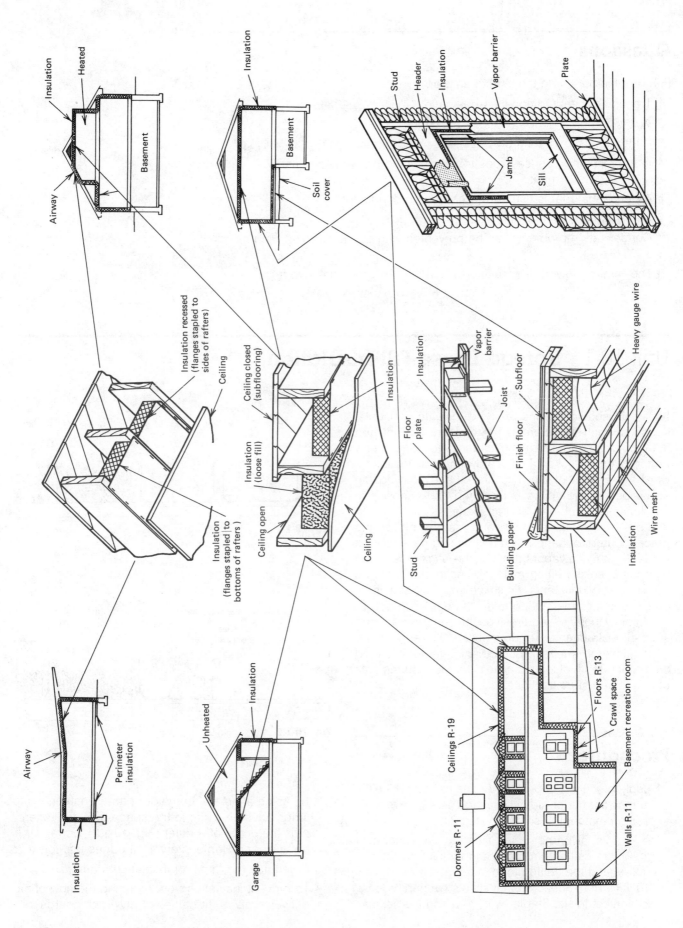

Questions

Fill the blank or circle the correct answer.

1 Transfer of heat through the air is called _____.

2 Air (is, is not) a good conductor of heat.

3 All materials have their own _____ factor.

4 In the summer a house needs to be protected primarily against _____ heat.

5 A dense material is usually a (good, poor) conductor.

6 All insulation is filled with _____.

7 Condensation problems can be prevented with _____ and a _____.

8 The (higher, lower) an insulation's R factor, the more effective it is.

Unit 10-1   Spread Loose Fill Insulation

Loose fill should be used only to insulate horizontal surfaces, such as ceilings. The material is light and fluffy when spread, but has a tendency to settle over a period of time. Loose fill used in walls tends to sink toward the bottom of the insulated space, leaving the upper part of the space uninsulated and exposed to condensation problems and air leaks.

Loose fill comes in bags, and may be blown into place, poured, or hand-packed. It should never be put in place without a vapor barrier on the warm side—that is, against finish ceiling material. Loose fill cannot be put in place until after finish ceiling material is in place, because the ceiling forms the bottom support.

To blow loose fill requires special equipment but no special instructions. To pour loose fill requires no tools, although a leveling board is useful for assuring the proper depth of insulation to meet the requirements of codes and specifications.

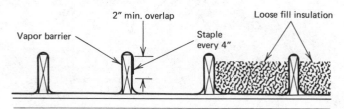

Fig. 10-1.1

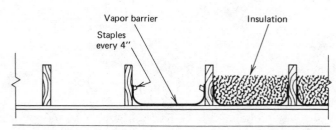

Fig. 10-1.2

Procedure

1 Look to see if a vapor barrier has been applied across the under sides of rafters, or is part of the ceiling material. If so, go to step 3.

2 a. If a vapor barrier is required, lay the sheet over the entire ceiling structure, and tuck it against rafters and the ceiling to form troughs (Fig. 10-1.1). Plan ahead so that sheets overlap at least 2″ against joists. Staple every 4″ along the seams.

2 b. As an alternate, cut vapor barrier materials into strips equal in width to the space between rafters plus the depth of a rafter (Fig. 10-1.2). Form troughs, and staple every 4″ at edges into joists.

3 Determine the depth of insulation required.

4 To prevent insulation from reducing the flow of air between rafters, fit baffles of plywood, boards, or

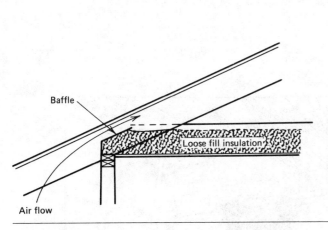

Fig. 10-1.3

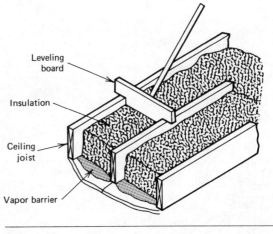

Fig. 10-1.4

screening between joists and rafters at the rafter plate. Baffles should follow the slope of the roof (Fig. 10-1.3), and clear sheathing by at least 2″.

5 Working in attic space and kneeling on a sheet of plywood as a working platform, pour loose fill into troughs to the tops of joists.

6 With a garden rake or similar tool push loose fill against baffles.

7 If the required thickness of insulation is less than the depth of joists, use a leveling board (Fig. 10-1.4) to spread loose fill to a constant thickness throughout the trough.

8 If the required thickness of insulation is greater than the depths of joists, tack 1 × 8 boards across joists for a catwalk as a safety measure. Insulate to full depth, but leave the walk visible.

Questions

1 Why not stop loose fill insulation at the top plate, and omit a baffle?

2 Why is air space needed between roof sheathing and attic insulation?

3 Which is the better way to install a vapor barrier—step 2a or 2b? Why?

Unit 10-2   Install Insulating Blankets and Batts

As you read in the previous unit, loose fill insulation can be installed only over finished ceilings. Flexible insulation, however, may be applied between joists either before or after ceiling material is applied. It must be installed between studs before finish wall materials are applied.

Batts are the form of insulation most commonly used above finished ceilings and in walls of standard heights. When ceilings are unfinished and wall heights vary, blankets are more practical.

Two simple devices can help speed the job of insulating. If pieces to be installed are consistent in length, the carpenter can save time by making a cutting board marked with common lengths (Fig. 10-2.1). The blanket

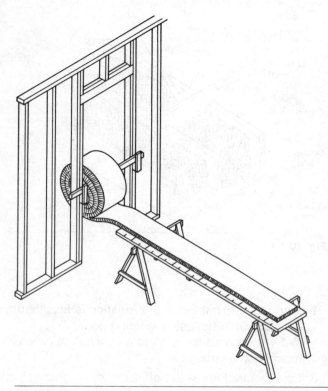

Fig. 10.2.1 A cutting board, marked with common lengths, saves time in cutting blankets. The roll of insulation is supported on a dowel.

Fig. 10-2.2 Insulating a ceiling is easier if one end of the roll of insulation is supported temporarily on a board.

roll rests on a board, dowel, or tool handle across a doorway. To offset the downward pull of gravity and to support the extra weight of blankets going between exposed joists, a short board laid across the tops of joists is helpful to the carpenter working alone (Fig. 10-2.2).

Whatever type of insulation you use, some cutting is necessary, and is most easily done on the back. Insulation may be cut to length with a saw, shears, or a sharp utility knife. If the insulation has an asphalted coating, a knife is best; it is easier to clean than saw teeth or shears blades. Although insulation may be nailed in place, the standard fasteners are staples with a $\frac{9}{16}''$ leg.

Flanged insulation may be applied in two ways. The flanges may be stapled to the edges of studs or joists, or to the sides, as shown in Fig. 10-2.3. Stapling to the edges provides a better job of controlling heat and vapor, but staples must be driven flush to prevent problems in applying finish wall and ceiling materials. Most manufacturers of insulation and wall finishing materials recommend recessing the insulation, and builders generally prefer this method too.

Insulating wool irritates some people's skin and lungs. To protect yourself when working with insulation of any kind, it is good practice to wear working gloves and a gauze filter mask.

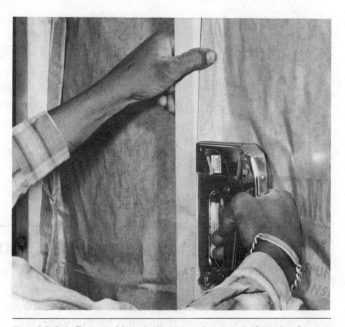

Fig. 10-2.3 Flanged insulation may be stapled to the faces or sides of framing members. Most manufacturers of insulation, however, recommend stapling to the sides as shown here. This leaves the faces of studs and joists free for installation of finish interior materials. (Courtesy Gypsum Association.)

Questions

1 The best tool for cutting insulation with a vapor barrier is a

_____ .

2 Give one reason for attaching the flanges of insulation to the sides of wood framing members.

3 Give one reason for attaching flanges of insulation to the faces of wood framing members.

4 Why do you think blankets are better than batts for insulating between ceiling joists?

Procedure for Insulating a Ceiling

1 Cut blankets to length. Use roof span as a guide to length.

2 If necessary to permit air flow between rafters, trim ends of blankets at eaves (see Fig. 10-1.3).

3 Begin stapling flanges to the sides of joists at one end (Fig. 10-2.4). The face of the insulation should be at least $\frac{3}{4}''$ above the under sides of ceiling joists.

4 As you work, stretch out the blanket so flanges are smooth and straight, and flat against framing members. Staple every 6" on both sides.

5 Where two lengths must butt, press ends tightly together to minimize the chance of air or moisture leaks.

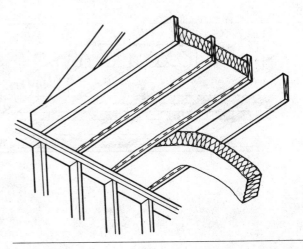

Fig. 10-2.4

Procedure for Insulating a Wall

1 Determine the length of cavities between horizontal framing members, and cut pieces about $\frac{1}{2}$ to $\frac{3}{4}''$ long to assure a tight fit.

2 Wedge blankets or batts between studs tight against the top framing members (top plate, window sill, etc.).

3 Staple from the top down, keeping tabs smooth and straight. Staple every 6 to 8". Staple approximately parallel with the edges of tabs.

Unit 10-3   Insulate Floors

For many years the need to insulate floors over crawl spaces was ignored. But with the increased interest in conserving energy, the advantages of insulating the floor are becoming more apparent. Although the great- est heat loss through the structure of an uninsulated house is through the roof, and next through walls, there is also some loss through the floor.

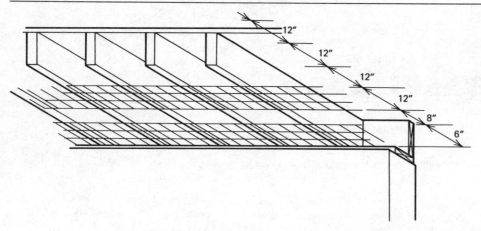

Fig. 10-3.1

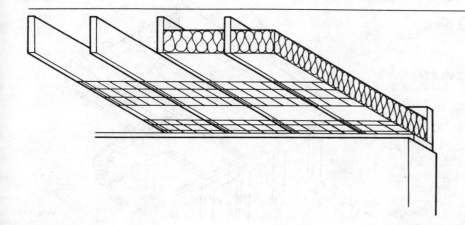

Fig. 10-3.2

Standard methods of insulating don't work very well here. If blankets are installed between joists from above before subflooring is laid, then the insulation interferes with bridging, heat ducts, and pipes. Loose fill can't be used because there is no way to support it. Use of foams isn't practical because there is no logical way to form a box for it.

Supported batts provide the best answer. The batts must have a vapor barrier on one side, unless a vapor barrier is laid over floor joists prior to subflooring. In this case friction-fit batts without a vapor barrier may be used.

Tools required to insulate beneath a floor are metal shears, a stapler, and a utility knife.

Procedure

1 Cut chicken wire or wire mesh into strips about 12″ wide, and as long as necessary to extend from edge joist to edge joist.

2 At the foundation wall attach a 6″ strip with staples across the under sides of joists (Fig. 10-3.1).

3 About 8″ from the first strip attach a 12″ strip.

4 On the two strips between each pair of joists lay a 48″-long batt, with the vapor barrier side *face up.* Slide each batt in from the open end until it is tight against the header joist (Fig. 10-3.2).

5 Add three additional 12″ supporting strips spaced 12″ apart. These strips will complete the support for the first row of batts, and half the support for the second set.

6 Continue to attach 12″ strips on a 12″ spacing, and to insert batts after attaching every set of three strips, until you are within 3′ of the opposite header joist.

7 Fit the last row of batts in place, supported only at their ends.

8 Add the final 12″ strip and a 6″ strip to complete support.

9 Cover the ground with a polyethylene vapor barrier to reduce dampness from ground moisture. Overlap pieces 12″ at seams.

Questions

1 Why can't blanket insulation be installed between floor joists from underneath?

2 Batts installed by this procedure could be laid on continuous rather than intermittent strip supports. Can you name an advantage to spaced strips?

Unit 10-4   Insulate Odd-Shaped Spaces

Most of the spaces to be insulated in houses are rectangular, and framing members are spaced to take standard insulation without cutting. There are always some spaces, however, that require special treatment. Typical examples are rectangles where the space between framing members is less than standard, and small cavities of various shapes.

Standard widths of blanket and batt insulation can be trimmed to fit narrower openings, without loss of insulating value or vapor barrier. Odd-shaped cavities can best be insulated in two steps: insulating first, then covering with a vapor barrier in a separate operation.

Tools required for insulating odd-shaped places are a utility knife, stapler, and possibly a hammer.

Procedure for Trimming Insulation

1 Slit the backing paper where it meets the flange (A in Fig. 10-4.1).

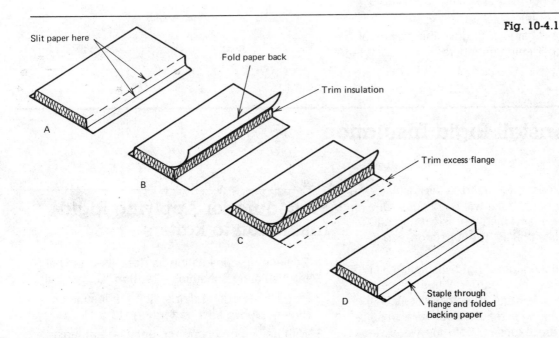

Fig. 10-4.1

Slit paper here

Fold paper back

Trim insulation

A

B

Trim excess flange

C

Staple through flange and folded backing paper

D

2 Carefully fold the paper back out of the way (*B*).

3 Determine the width of the opening to be filled.

4 With a utility knife trim off excess insulation (*B*).

5 Trim off the same amount of flange (*C*). If, for example, you cut away 2″ of insulation, trim off 2″ of flange.

6 Fold the backing paper to the new shape of insulation, and fold excess backing paper against the flange (*D*). **Precaution.** Do not trim off excess backing paper.

7 Install the narrower width of insulation by stapling through the flange and folded backing paper.

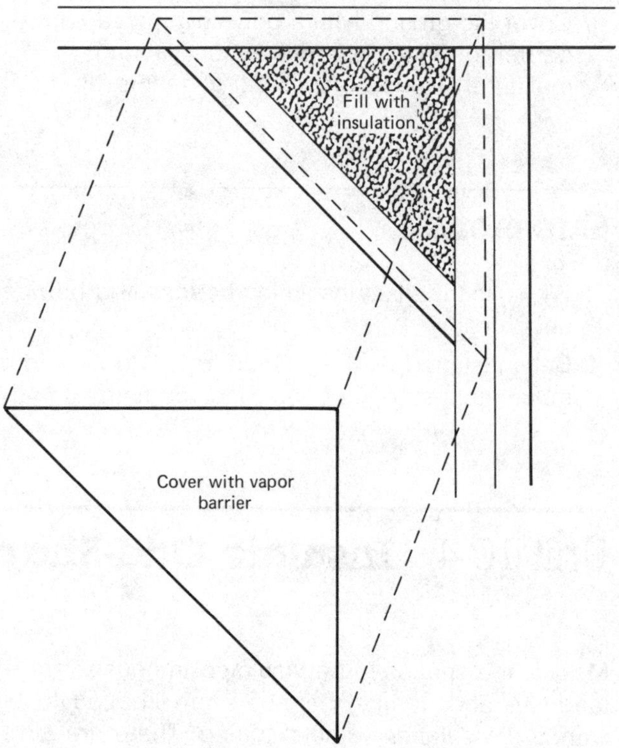

Fill with insulation

Cover with vapor barrier

Fig. 10-4.2

Procedure for Filling Cavities

1 From a piece of scrap insulating blanket or batt, slit or remove the backing paper.

2 Strip off, but save, the vapor barrier, leaving nothing but raw insulation.

3 Stuff the cavity full of insulation.

4 From the scrap of vapor barrier, cut a piece big enough to cover the cavity and leave flanges on all sides (Fig. 10-4.2).

5 Lay the vapor barrier over the cavity, and staple to framing on all sides.

6 If necessary, use a hammer to drive the crowns of staples flush with framing members.

Unit 10-5   **Install Rigid Insulation**

Rigid insulation is most often used as an insulator in place of blankets or batts under two conditions. One is with a sloping ceiling, and the other is with a ceiling, either sloping or flat, that is supported on exposed beams.

In neither installation does rigid insulation serve structurally or act as a vapor barrier. When the ceiling slopes, a vapor barrier is required across rafters on the warm side, and sheathing is required above the insulation. When beams are exposed, both the wood deck (which acts as sheathing) and the vapor barrier lie between the rigid insulation and supporting beams.

Procedure for Applying Rigid Insulation to Rafters

1 Position rigid insulation as described in Unit 9-34 for insulating sheathing. **Caution.** Do not nail.

2 Place 1 × 2 nailing strips atop the insulation directly above each rafter (Fig. 10-5.1).

3 With nails long enough to penetrate at least 2″ into rafters, nail through the 1 × 2s to attach the rigid insulation. Space nails every 4″. When strips

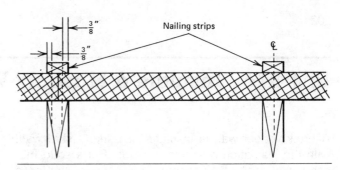

Fig. 10-5.2

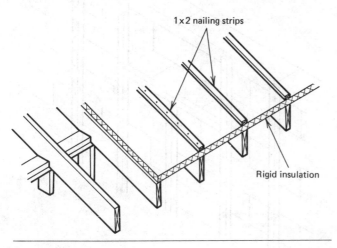

1 × 2 nailing strips

Rigid insulation

Fig. 10-5.1

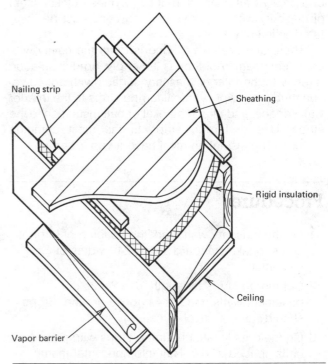

Fig. 10-5.3

cover seams in insulation, place nails $\frac{3}{8}''$ in from each edge of the 1 × 2s (Fig. 10-5.2). Otherwise nail down the center.

4 After all rigid insulation is in place, apply sheathing as described in Unit 9-34, nailing through the strips (Fig. 10-5.3).

Procedure for Applying Rigid Insulation over a Plank Ceiling

1 Lay a vapor barrier over the planks (Fig. 10-5.4). Overlap $1\frac{1}{2}''$ at edges, and 4″ at seams.

2 Set rigid insulation flush with the edges of planks, and nail through them into roof beams. Keep joints tight between sheets of insulation.

3 After all rigid insulation is in place, apply building paper as described in Unit 9-38.

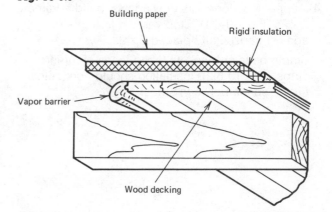

Fig. 10-5.4

Questions

1 If rigid insulation is 2″ thick, what is the minimum size of nail to use for attachment?

2 Why is building paper needed in Fig. 10-5.4 but not in Fig. 10-5.3?

3 Why do you think it is better to nail sheathing to nailing strips in Fig. 10-5.3 instead of nailing directly into the rigid insulation?

Unit 10-6   **Insulate a Masonry Wall**

A brick veneer wall is insulated just like a stud wall, with the insulation attached to studs. But a solid masonry wall, such as a basement wall of concrete blocks or a cavity wall consisting of two wythes (thicknesses) of masonry with an air space between, must be insulated differently.

There are three ways. One is to fill the cavity with chemical foam. Another is to apply rigid insulation directly to the interior masonry surface with adhesive. The third is to apply vertical furring strips to the interior wall surface, and attach blanket or batt insulation to the strips. Use of furring strips is the most common method. The insulation must have a vapor barrier.

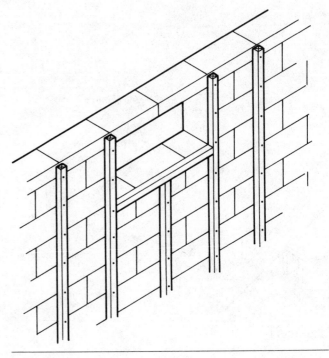

Fig. 10-6.1

Procedure

1 On the masonry wall mark the spacing of furring strips to take standard 16″ or 24″ widths of insulation.

2 Cut furring strips to length; use 2 × 2s on basement walls and 2 × 4s above ground. Strips should run from floor to ceiling.

3 On their marks attach furring strips with power nails spaced 8″ o.c. Nail into horizontal mortar joints.

4 At openings for doors or windows, build a frame of 2 × 2s (Fig. 10-6.1). Stop vertical furring strips against horizontally placed 2 × 2s.

5 Staple batts or blankets to the faces of furring strips (Fig. 10-6.2). Insulation should completely fill the space between strips.

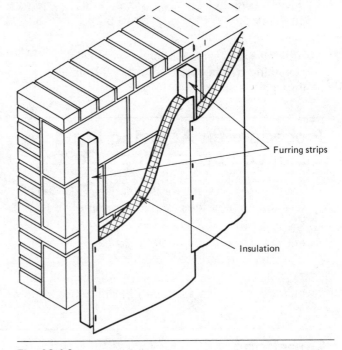

Furring strips

Insulation

Fig. 10-6.2

Question

1 Why is insulation stapled to the face of furring strips and not their sides?

11

Post-and-Beam Construction

Plank-and-beam or post-and-beam construction has many variations, but the underlying principle is the same for all. It is the use of heavier structural members placed farther apart than in standard construction (Fig. 11-0.1).

The floor frame usually begins with a box sill set on the foundation wall. Horizontal members, instead of being joists 2″ thick and 16″ on centers, are heavier beams spaced 48″, 72″, or 96″ o.c. Subflooring is replaced with 2″ planks matched at ends and edges. No bridging is needed to stiffen the floor.

Structural wall members are posts, usually 4 × 4s, that bear directly on floor beams beneath front and rear walls. The only structural member in a side wall is a post that supports the main roof beam. Wall studs serve no structural purpose, and may be placed wherever needed to frame openings or to support seams in interior and exterior materials. Walls are sheathed as in other framing systems.

Roof beams bear directly on posts in front and rear walls, and on a center ridgebeam supported by the posts in end walls and by intermediate posts under roof beams. Planks the same size as floor planks bridge between roof beams. They take the place of sheathing. Planks are often left exposed to living areas. They provide some insulating value, but usually not enough. Rigid insulation may be added above the roof deck, or below it if planks aren't exposed to living areas for decorative value.

In effect, the structural system in plank-and-beam framing is a series of five-sided frames spaced at wide intervals and connected and stiffened by 2″ planks.

The horizontal members-floor beams—are installed like regular floor joists. The two lines of vertical members—The posts—are nailed like studs and corner posts to top and bottom plates. The pairs of sloping members—the rafter beams—may be nailed but more often are fastened with metal hangers at both ends.

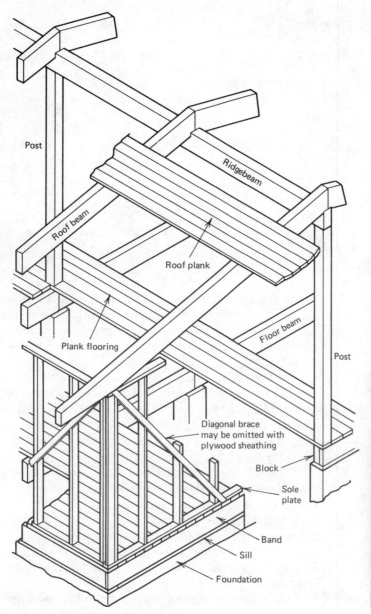

Fig. 11-0.1 Plank-and-beam construction has fewer structural members than conventional construction. Members are heavier and spaced farther apart, however, and they are connected by a skin of planks that adds great strength and stiffness. (Photo courtesy Western Wood Products Association.)

Question

1 Which of the following members are common to both standard construction and plank-and-beam construction?

Rafters	Studs
Sill plate	Doubler
Sole plate	Header joists
Ceiling joists	Floor joists
Wall sheathing	Roof sheathing
Bridging	Subflooring

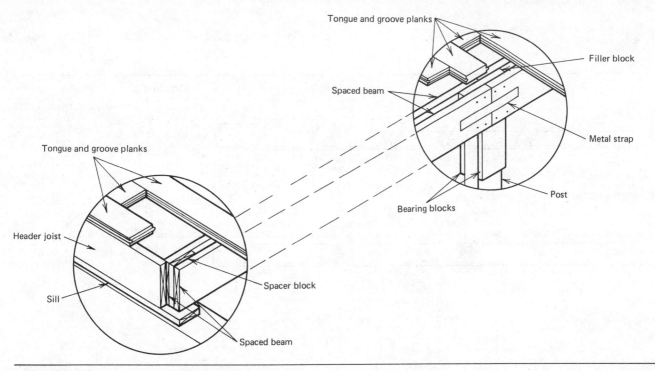

Fig. 11-1.1 A spaced beam is a site-built floor member used when timbers are not available or economical.

Unit 11-1 **Build Spaced Beams**

Floor framing members in plank-and-beam construction are usually solid timbers in parts of the country where such timbers are readily available. In much of the United States, however, the use of spaced beams is more practical and economical. A **spaced beam** consists of a pair of 2″ thick joists separated by blocks (Fig. 11-1.1). A single set of blocks forms a beam $4\frac{1}{2}″$ wide—adequate for short spans and light loads. Often, however, blocks are used in pairs to form a beam with a bearing surface 6″ wide for installation of floor planks.

Procedure

1 From drawings determine the length of individual beam members. Their length is equal to the distance from the center line of bearing posts to the edge of the foundation wall, less $1\frac{1}{2}″$ for the thickness of a header joist.

2 Cut beam members to length.

3 From beam stock cut a pair of filler blocks 24″ long, and nail them together side by side.

4 From 2 × 6 scrap cut spacer blocks equal in length to the depth of the beam. Cut one pair for every 24 to 30″ of beam length. Nail spacer blocks together in pairs.

5 Lay one beam member flat, and position spacers as shown in Fig. 11-1.2.

6 Position 24″ filler blocks so that their center line falls at the end of the beam (Fig. 11-1.3).

7 Add the remaining beam member, lining up all edges carefully.

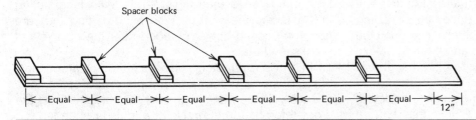

Fig. 11-1.2

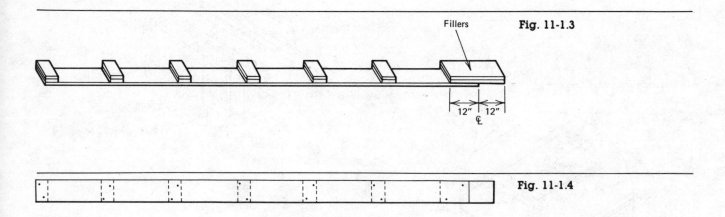

Fillers   **Fig. 11-1.3**

12" 12"
℄

Fig. 11-1.4

8 Attach all members with a pair of 20d spikes. Nail in a pattern such as that shown in Fig. 11-1.4.

9 Turn the assembly over, and nail with another pair of spikes in each location. **Caution.** To avoid

hitting the spikes you have already driven, use the same pattern as in step 8.

10 Build the butting floor beam in the same way, but omit the filler blocks.

Question

1 Why is it better to cut spacer blocks from 2 × 6 material than from joist stock?

Unit 11-2   **Frame the Floor**

The procedures for framing a plank-and-beam floor are not much different from those for building a floor system in platform framing. The box sill is the same. You use spaced beams or timbers in place of joists (Fig. 11-2.1). Members are heavier, but the method of attachment is the same.

There are two important differences, however. First, posts that support floor beams at their midpoints are usually 4 × 4 or 4 × 6 stock. Yet a spaced beam resting on a post will hang over it on both sides. To support the beam at the post, it is necessary to add bearing blocks of a thickness equal to the amount of overhang. It is also necessary to strap the beams.

Second, planks should be long enough to bridge at least two spans between beams. Planks may end on the center line of either member of a spaced beam, but

no two adjoining planks should end on the same beam member, and preferably not on the same beam.

Nailing methods vary somewhat with the type of plank used. Square-edged planks, for example, are attached with two spikes per beam, both driven before the next plank is laid. Tongue-and-groove planks, on the other hand, should be face-nailed with one spike, then blind-nailed (see Fig. 11-1.4). With face-nailing the fasteners are visible. With blind-nailing they are not. The second spike is added after the next row of planks is in place.

In previous chapters you have already learned most of the procedures for framing a plank-and-beam floor. In case you need to refresh your memory, the units are cross-referenced below. No new tools are needed.

Procedure

1 Cut and install sill plates as described in Units 6-1 and 6-2 respectively.

2 Build a box sill as described in Unit 6-3.

3 Install wood posts as described in Unit 6-5.

4 If necessary, fabricate spaced beams as described in Unit 11-1.

5 Install timbers or spaced beams on foundation walls and posts as described in Unit 6-6. Add bearing blocks where required.

6 Over each post connect spliced beams with metal straps (see Fig. 11-2.1). Use two straps at each post.

7 Cut floor planks to length.

8 Lay the first plank flush with the corner of the box sill. If planking is tongue-and-groove, set the first row of planks with the groove outward.

9 a. Attach square-edged planks with two spikes per beam, toenailing at a slight angle in opposite directions (Fig. 11-2.2).

9 b. Attach tongue-and-groove planks with one spike on the grooved side, and blind-nail at the tongue (Fig. 11-2.3) with 8d nails.

10 Complete the first row of planking.

11 Fit succeeding planks tight against planks already in place, working across the floor in a diagonal pattern, and staggering the ends of planks. **Caution.** To drive tongue-and-groove planks together without damage, pound against a grooved block (Fig. 11-2.4).

12 Nail each plank as in step 9. Drive the second spike into tongue-and-groove planks at each beam after the next plank is in position.

13 At a floor opening let planks extend beyond framing headers into the opening. Trim the first one or two planks at each side as you lay them. Trim remaining planks after the opening is surrounded by planks.

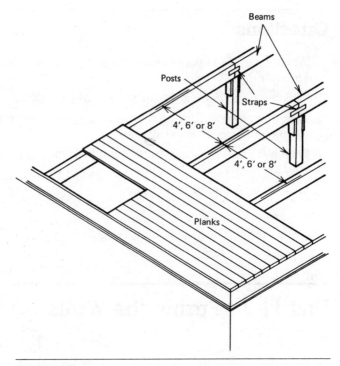

Fig. 11-2.1 Typical floor framing in plank-and-beam construction.

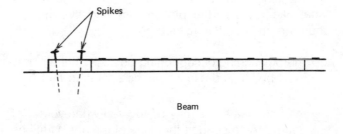

Fig. 11-2.2

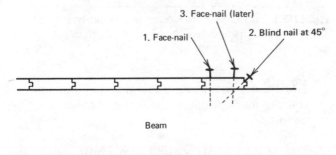

Fig. 11-2.3

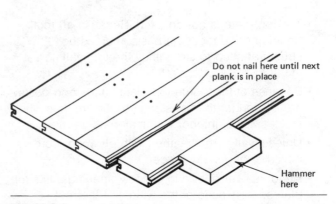

Fig. 11-2.4

Questions

1 Which type of planks—square-edged or tongue-and-groove—make the stronger deck?

2 Which type of deck is easier to lay? Why?

3 Why do you delay driving the second spike at each beam with tongue-and-groove planks?

4 Give two good reasons for trimming the first few planks at openings when they are laid.

5 What is the value of blind-nailing?

Unit 11-3 **Frame the Walls**

To frame the walls of a house built with plank-and-beam construction, you begin with supporting posts. The roof style is usually a gable roof. Roof beams bear on posts in the walls of the house parallel to the ridge (Fig. 11-3.1). The center ridgebeam bears on posts in the side walls.

The long bearing walls of a house are built somewhat like walls in the platform framing system, but there are these differences:

• The positioning of supporting posts is critical. They must be located exactly on the specified **module** (spacing).

• Intermediate studs may be located wherever necessary to support the ends of interior partitions or seams between lengths of finish wall materials.

• Corner posts may be built of 2 × 4s and spacers, as described in Unit 7-3. Or they may consist of a 4 × 4 post flanked on both sides by a 2 × 4 stud, as in Fig. 11-3.1.

• The top plate is not doubled. Because all roof beams bear directly on posts, there is no structural need to strengthen the top wall member.

• Because of the single top plate, studs and posts are 1½″ longer than in standard construction when stock sizes of interior wall materials are used.

• Unless walls are sheathed with plywood, corners must be diagonally braced.

• Headers over door and window openings, like top plates, carry none of the weight of the roof structure. Therefore spaced headers (see Unit

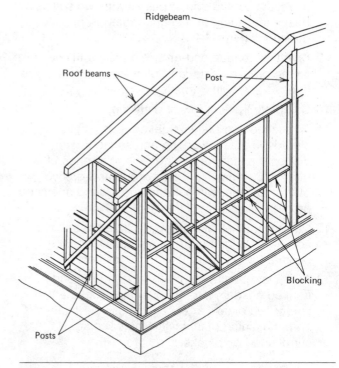

Fig. 11-3.1 Typical wall framing in plank-and-beam construction. How many differences from platform construction can you spot?

7-4) are not necessary, although many builders still use them. A pair of 2 × 4s laid flat is adequate.

Side walls in plank-and-beam construction are quite different from walls in other types of construction. The center post that supports the ridgebeam goes in

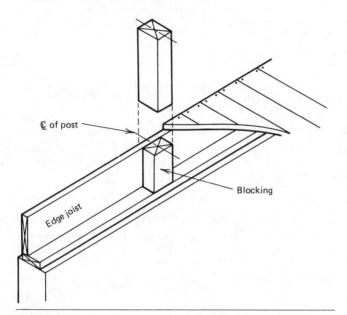

Fig. 11-3.2

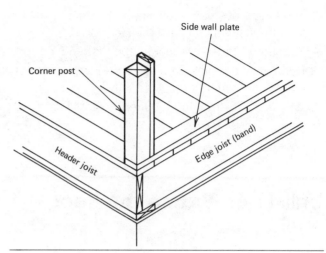

Fig. 11-3.3

place first. The stud walls between center posts and corner posts are built like a side wall under a shed roof, as described in Unit 9-9. They are normally built in two sections. The lower section ends with a top plate, as in Fig. 11-3.1, and is built flat and raised into position. In the upper section of the wall all studs are toenailed individually after rafter beams are in place, as in a gable wall. Note in Fig. 11-3.1 that bottom plates extend all the way to the corner of the floor structure, and support corner posts.

Procedure

1 From post stock cut blocks to support center posts in side walls. Install blocking as shown in Fig. 11-3.2.

2 From sections in drawings determine the length of center posts. **Caution.** Remember to allow for the depth of the ridgebeam.

3 Cut center posts to length, and toenail them with spikes driven through planking into the **band** (edge joist) and blocking.

4 Determine the length of corner posts and studs, and cut these members to length as in Unit 7-2.

5 Assemble window and door frames as in Unit 7-5.

6 Cut to length top and bottom plates for sections of side walls.

7 Mark the plates as in Unit 7-6. Note in Fig. 11-3.3 that the end stud is inset to allow for the corner post.

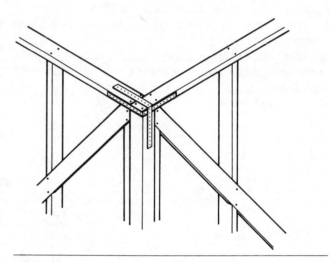

Fig. 11-3.4

8 Assemble side wall components as in Unit 7-7.

9 Raise and attach side wall sections. Nail through the bottom plate as described in Unit 7-8. Attach end studs to center posts with 10d nails spaced every 8".

10 Straighten and brace side walls as in Unit 7-9.

11 After lower sections of end walls are installed, assemble front and rear walls.

12 Raise and secure front and rear walls as described in Unit 7-9.

13 Install diagonal bracing if required.

14 Secure top plates together with steel strapping (Fig. 11-3.4).

15 Install blocking as described in Unit 7-17.

Questions

.1 The length of studs in plank-and-beam construction is _____ than in platform construction.

2 Additional under-the-floor support is required for _____.

3 Does any floor plate in plank-and-beam construction run from corner of box sill to corner of box sill? Why?

4 Why is steel strapping required in step 14?

Unit 11-4   **Frame the Roof**

The roof structure of a house built with plank-and-beam construction has few members, but all are heavy. The main member is a ridgebeam supported on center beams in end walls and usually overhanging those walls. Ends of the beams are cut square. Between end walls the ridgebeam is supported on posts wherever a splice is required. The lengths of beam are connected with metal gusset plates, and secured to posts with framing anchors.

Rafter beams may be supported in two ways. The most common method is to notch for a bird's mouth at the plate, and to notch also for a butt fit over the ridgebeam (Fig. 11-4.1). An alternate method at wall posts is to add a beveled doubler to the top plate (Fig. 11-4.2, left). This method is most common when the roof pitch is very low. An alternate method at the ridge is to butt rafter beams against the ridgebeam like rafters against a ridgeboard (Fig. 11-4.2, right). With this

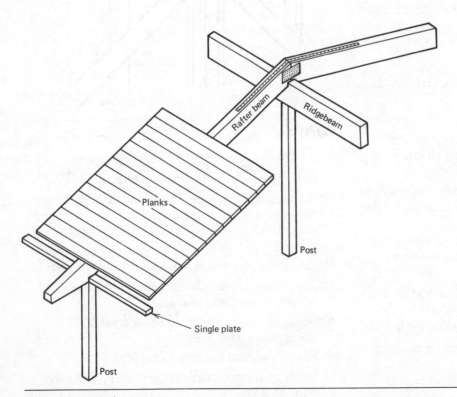

Fig. 11-4.1 Typical roof framing in plank-and-beam construction. Rafter beams are notched at both ends to fit over   lateral supports.

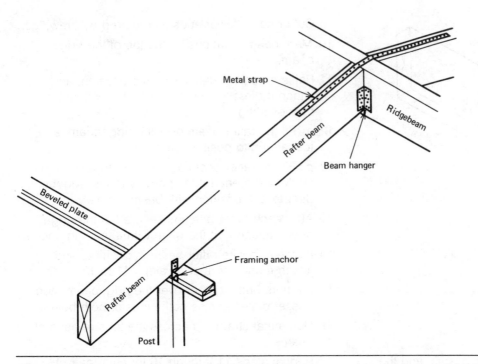

Fig. 11-4.2 In an alternate method of roof framing the rafter beams are not notched. Supporting plates are beveled to the pitch of the roof (left), and beams butt against the ridgebeam (right).

method framing members are connected with beam hangers instead of gussets.

In both methods of construction all roof beams receive a plumb cut at the ridge, and are connected with metal straps. At eaves the most common treatment is a heel cut to accept a fascia and soffit. Fascias are 2x members instead of boards, however, because of the spans between beams. Beams are attached to top plates with framing anchors. Planks are then laid across beams in the same manner as floor planks.

Procedure

1 Cut and install interior posts to support the ridgebeam on the spacing indicated in drawings.

2 Cut ridgebeam members to length.

3 Raise the first section of ridgebeam into position on posts.

4 Attach the beam to posts with framing anchors (Fig. 11-4.3).

5 Raise remaining sections of ridgebeam into position, and attach as in step 4.

6 Connect lengths of ridgebeam to each other with gusset plates (Fig. 11-4.4).

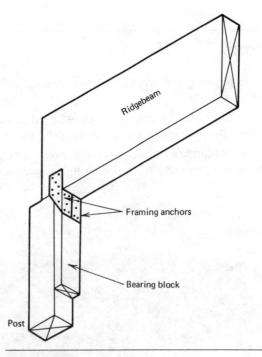

Fig. 11-4.3

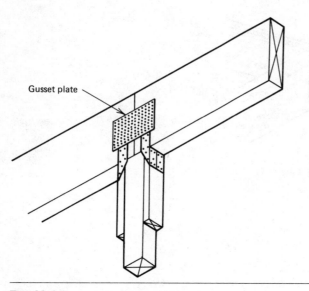

Gusset plate

Fig. 11-4.4

7 Determine the lengths of rafter beams, and the type of construction at the ridge and plate.

8 Either cut and notch (Fig. 11-4.1) or cut and shorten (Fig. 11-4.2) the upper ends of all rafter beams.

9 Cut or notch lower ends as required by drawings.

10 Mark beam locations on the top of the ridge-beam.

11 Raise a center rafter beam into position, and attach it directly over a post in the wall structure by toenailing.

12 a. Fit the rafter beam over the ridgebeam, and toenail it into position on its mark.

12 b. Fit the rafter beam against the ridgebeam on its mark, and toenail it temporarily to the ridgebeam. Secure it in position with beam hangers.

13 After each rafter beam is secured at the ridge beam, attach it at the plate with a framing anchor.

14 Fit the opposite rafter beam in position, and attach it as described in steps 11 to 13.

15 If rafters butt together over the ridgebeam, add gusset plates across both sides of both beams.

16 Nail metal strapping across the tops of pairs of beams.

17 Repeat steps 11 through 16 for each pair of beams, working alternately from the center pair toward side walls.

18 Add planks across rafter beams as described in Unit 11-2 for floors.

Questions

1 How does the length of interior posts compare with the length of center posts in side walls?

2 Whenever two rafter beams butt against a ridgebeam, how many attachment devices should you use to complete the connection?

3 Where two lengths of ridgebeam meet over a post, how many attachment devices should you use to complete the connection?

12

Multifamily Buildings

All the building codes and regulations that apply to one-family houses also apply to buildings that house more than one family—duplexes, townhouses, apartments, condominiums. In some areas of construction, however, the restrictions are even stiffer.

The differences lie in the walls between living units, and the floors and ceilings around them. In multifamily buildings these dividers must be **fire-rated**—that is, they must be able to slow the spread of fire from one living unit to the next for a minimum period of time. The minimum time varies from 10 minutes up to $1\frac{1}{2}$ hours. A fire resistance rating (**FRR**) applies not just to the surface materials, but to the entire wall, floor, or ceiling system.

Even under the intense heat of a fire, wood loses its strength very slowly. Under the test conditions used to determine fire resistance ratings, lumber graded #2 or better chars at the rate of about $\frac{1}{40}''$ per minute. Yet $\frac{1}{4}''$ inward from the char zone the temperature of the wood is so low that it still has most of its strength. Therefore solid wood beams and posts can withstand fire for at least 20 minutes. The main purpose of fire protection is to increase this time as much as possible.

Compare the requirements for various structural systems in Figs. 12-0.1, for detached single-family houses, 12-0.2 for detached two-family houses, and 12-0.3 for a townhouse. Note that in all three types of buildings, some parts of the structure may be **combustible**—that is, they have no specific fire rating and may burn. Roofs and nonbearing partitions, for example, have no specific rating since they do not affect **structural integrity**—the soundness of total construction. In other words, if a roof burns up or a nonbearing partition falls down, the rest of the structure is not affected.

When a structural system requires a rating such as $\frac{1}{6}$ hr. (structural integrity only), this means that the structure may burn, but must not collapse from fire for at least $\frac{1}{6}$ hour. A rating such as $\frac{3}{4}$ hr combustible means

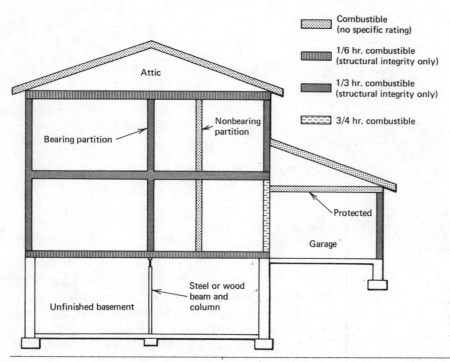

Fig. 12-0.1 The minimum construction permitted by fire codes is stated in terms of a fire resistance rating. This section through a detached single-family house shows the minimum ratings required for various floor, wall, and ceiling systems.

that the surface materials will prevent fire from attacking the structure for at least $\frac{3}{4}$ hr.

There is not too much danger of a fire starting in unused attic space. The danger is from fire getting into an attic and spreading across more than one dwelling unit. Therefore ceiling systems are fire-rated, and walls between dwelling units extend through attic space to the roof. Note in Figs. 12-0.1 and 12-0.2 that ceilings over attached garages aren't fire-rated, but that the garage ceiling joists must be finished with a fire-resistant material that protects the framing members. Note also that the ceiling structure over living spaces in one- and

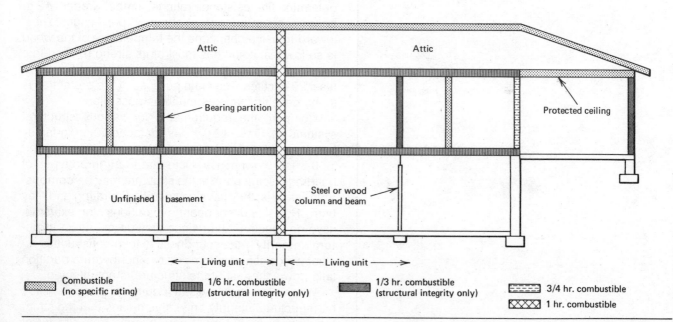

Fig. 12-0.2 In a detached two-family house fire ratings for structural systems are similar to those for a one-family house. The major exception is the party wall between living units. Note that it extends all the way to the roof framing.

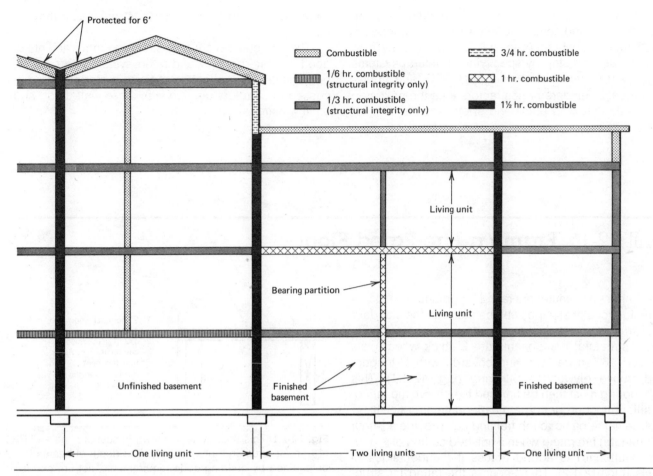

Fig. 12-0.3 The fire ratings for structural systems within living units of a townhouse are like those in all other type of dwelling unit. Between units, however, horizontal systems must be fire-rated for at least one hour, and vertical systems for at least 1½ hours.

two-family houses must carry a 10-minute FRR, and in townhouses must carry a 20-minute FRR.

Floor systems may require one of three ratings. The floor system between an unfinished basement (not used for living space) and first floor rooms must have a 10-minute FRR. Floors between living spaces in the same dwelling unit of a multifamily building must have a 20-minute FRR. Floors between different dwelling units (center in Fig. 12-0.3) must be fire-rated for a full hour.

Ratings required for wall systems vary greatly. In all types of residential buildings exterior walls must be designed for a 20-minute FRR. Bearing partitions within a dwelling unit also must carry a 20-minute rating. Bearing walls between an attached garage and living spaces must carry a 45-minute rating. The common wall in a two-family house and some bearing partitions in multifamily buildings must carry a one-hour rating. Common walls in multifamily buildings must carry a 1½-hour rating.

Thus, as the danger of loss of life from fire increases, so does the time period of the fire rating.

Sound Transmission Classes

Structural systems are fire-rated for the safety of residents of housing. Systems are also sound-rated for the comfort of residents in multifamily dwellings. There are two types of sound transmission: air-borne and structure-borne.

Air-borne sound, as the name implies, is sound carried through the air; conversation, electronic output of television, hi-fi, stereo and radio sets, and the hum of motors are typical examples. Each type of construction has a rating, called a **sound transmission class (STC),** that is a measure of transmission of air-borne sound. The higher the rating, the more soundproof that construction is.

Structure-borne sound, also called **impact sound,** is sound carried by the building's structure; the thud of shoes being dropped, the vibration of some kitchen and laundry appliances, and the noise of plumbing are examples. When measured, the rating is called an **impact insulation class (IIC).**

The architects or designers of any building take the first step in sound control by following the principles of good room arrangement in the floor plan. They also take the second step by specifying structural systems and finish materials that meet FRR, STC, and IIC requirements. Builders or contractors take the next step by rigidly following plans and specifications, and inspecting the work of their crews to make sure they do too.

This chapter covers only structural requirements of good fire-rated and sound-rated systems. Similar requirements for application of finish materials are discussed in chapters dealing with those wall, floor, and ceiling materials.

Unit 12-1   Frame a Fire-Rated Floor

The $\frac{1}{6}$-hour (10-minute) fire rating for ground level floors in residences of all types can be met with the standard floor framing procedures described in Chapter 6. The basic structural components are 2"-thick wood joists spaced 16" on centers, and covered with $\frac{1}{2}$" tongue-and-groove plywood subflooring (Fig. 12-1.1). This subflooring must then be finished to achieve the rating. Resilient flooring or carpet is adequate as a finish.

The same basic rough framing can provide a $\frac{1}{3}$-hour (20-minute) fire rating when a finished ceiling of $\frac{3}{8}$" gypsum wallboard is added across the under sides of joists (Fig. 12-1.2). This $\frac{1}{3}$-hour is the rating for structural integrity only required of floors *within* dwelling units.

Floors *between* dwelling units require a one-hour fire resistance rating (FRR). Now structural requirements change (Fig. 12-1.3). Joist spacing may be increased to 24" o.c. Plywood subflooring, however, must be $\frac{3}{4}$" thick instead of $\frac{1}{2}$", and applied with adhesive as well as nails. Gypsum board must be $\frac{5}{8}$" thick instead of $\frac{3}{8}$", and attached with screws.

These are the basic structural changes. Now look at Fig. 12-1.4 to see how various methods of finishing the structure affect the STC and IIC ratings.

In all four floor sections shown, the layer of $\frac{5}{8}$" gypsum wallboard is applied to resilient channels (see Unit 21-1). To block the passage of airborne sound between floor levels, joints in plywood subflooring must be closed off; adhesive isn't enough. You can achieve an STC rating of 40 with 6"-wide strips of $\frac{5}{8}$" gypsum wallboard stapled across the under sides of joints (*A* in Fig. 12-1.4). You can increase the STC rating to 45 by using R11 insulation in place of gypsum strips (*B*).

Airborne sound isn't affected by a change in floor covering, but impact sound is. Use of carpet instead of resilient flooring improves the IIC rating from 38 to 50 in one case (*C*), and from 42 to 60 in the other (*D*).

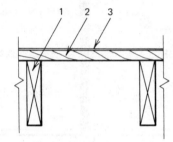

1. 2" nominal wood joist 16" oc.
2. 1/2" T&G plywood combination subfloor-underlayment.
3. Resilient flooring or carpet.

Fig. 12-1.1 This floor system, shown in section, has an FRR of $\frac{1}{6}$ hr. for structural integrity only. The depth of beams is determined by building code requirements, not fire codes.

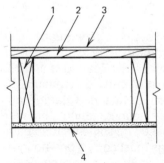

1. 2" nominal wood joist 16" oc.
2. 1/2" T&G plywood combination subfloor-underlayment.
3. Resilient flooring or carpet.
4. 3/8" gypsum wallboard attached with nails or screws.

Fig. 12-1.2 Addition of a layer of gypsum wallboard to the under sides of floor joists doubles the length of time the floor structure shown in Fig. 12-1.1 can withstand fire.

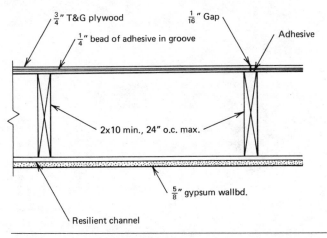

Fig. 12-1.3 The same basic floor structure can be upgraded in fire rating with the use of thicker wallboard and a change in the subflooring and its method of application.

Procedure for Building a 1-Hour Rated Floor

1 Set floor joists as described in Unit 6-9.

2 Just prior to applying adhesive, clean any sawdust, ice, snow, standing water, or other foreign material off the tops of all joists.

3 On all framing members, including blocking, lay a bead of approved adhesive under the first full sheet of plywood (Fig. 12-1.5).

4 Lay the first sheet in the adhesive, with the grooved edge flush with the header joist. Press firmly to assure a good bond.

5 Fasten the plywood with nails spaced 12" o.c. For plywood less than 1" thick use either 6d deformed

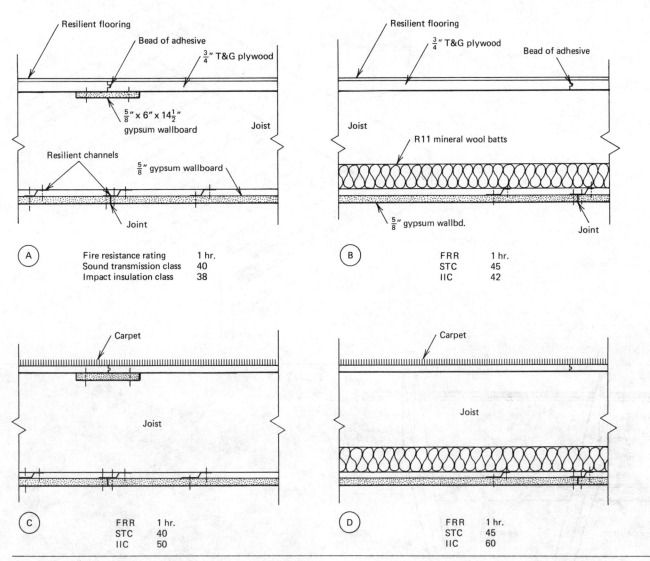

Fig. 12-1.4 The structure in these four sections—joist size, plywood and its method of application, wallboard and its method of application—is identical and, therefore, the fire resistance ratings are identical. But note the effect of insulation on the STC rating, and of carpet on the IIC rating.

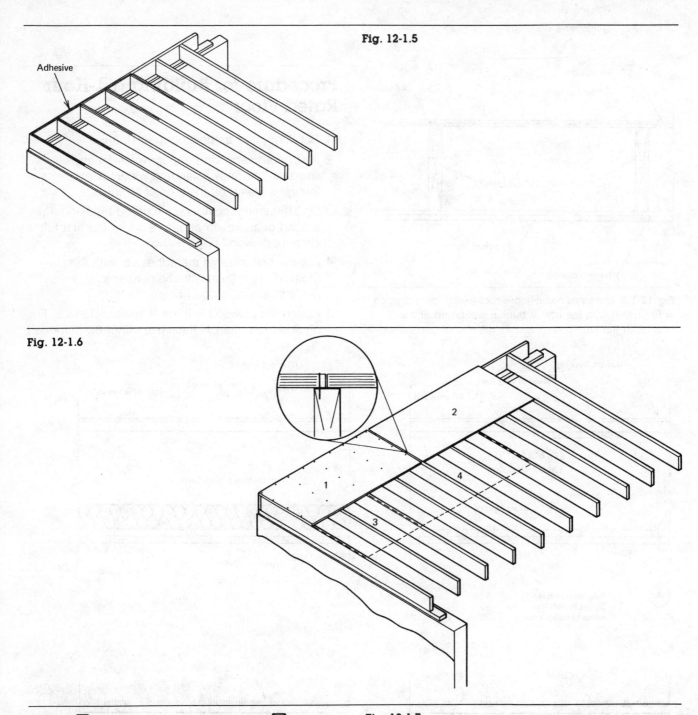

Fig. 12-1.5

Adhesive

Fig. 12-1.6

1

2

3

4

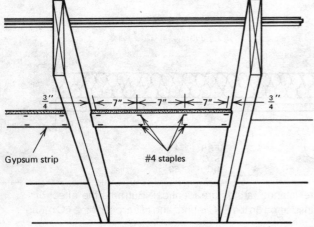

Fig. 12-1.7

$\frac{3}{4}''$ 7"   7"   7" $\frac{3}{4}''$

Gypsum strip

#4 staples

shank nails or 8d common nails. For plywood 1″ thick or thicker use 8d deformed shank or 10d common nails. **Note.** If the open time (setting time) of the glue permits, delay nailing at the tongued edge.

6 Repeat steps 3 and 4 for sheet 2. Where two sheets butt at a joist, spread a separate bead of adhesive under each panel. (Fig. 12-1.6, inset).

7 Nail sheet 2 as in step 5.

8 Cut sheet 3.

9 Before fitting panels 3 and 4 (Fig. 12-1.6) in place, completely fill the groove in them with a bead of adhesive $\frac{1}{4}$″ in diameter.

10 Carefully but nevertheless tightly fit the grooves of sheets 3 and 4 over the tongues of sheets 1 and 2. Nail immediately on both sides of the joint.

11 Remove any excess glue from the surface of the subflooring.

12 Continue steps 3 through 11 until subfloor is complete.

12 a. Cut strips of $\frac{5}{8}$″ gypsum wallboard $22\frac{1}{2}$″ long and 6″ wide.

12 b. Over every exposed joint between panels of subflooring attach a strip of wallboard from step 12a. Attach with eight #4 staples as shown in Fig. 12-1.7.

12 c. As an alternate, install R11 mineral wool insulation between joists, as described in Unit 10-2 and Fig. 10-2.4.

13 Apply gypsum wallboard to the under sides of joists as described in Unit 21-1.

14 Install finish flooring as described in Chapter 24.

Questions

1 What do you think is the primary purpose of the strip of gypsum wallboard or the insulating batts?

2 Why do you think a glued subfloor must also be nailed?

Unit 12-2   **Build a 1½-Hour Party Wall**

The basic structure for all bearing walls with fire ratings from 10 minutes up to one hour is 2 × 4 studs spaced 16″ on centers. The material applied between those studs and the surface materials across them is what makes the differences in FRR and STC ratings.

As a general rule, the more fire-resistant or the thicker the finish wall material applied to studs, the higher the fire rating and STC rating. Table XII gives a few typical ratings. Similarly, the use of mineral wool insulation between studs has a favorable effect. Insulation does not help either structural integrity or sound transmission, but it does improve the fire resistance rating.

When a fire resistance rating of more than one hour is required between living units, standard building practice is to erect a party wall. A **party wall** is a pair of similar walls with a space between them (Fig. 12-2.1). The two walls bear on the foundation, and extend all the way to roof sheathing. No horizontal structural member passes through the wall.

Note the construction details in Fig. 12-2.1. The pur-

pose of the sheathing is to wrap the wall's structural members in a cocoon of fire resistant material. Then, if a fire does start in one living unit, it cannot easily spread to adjacent living units. Each wall is at least $4\frac{5}{8}$″ thick, and the space between walls must be at least 1″. Therefore the foundation on which a party wall rests must be at least $10\frac{1}{4}$″ thick.

The general construction procedure for building a party wall is similar to that for ordinary walls except where the closeness of the two walls to each other forces changes. In attic space, however, the wall is built like a pair of gable walls. The bottom plates of attic walls serve as doublers on the wall below. Rafters are cut from standard rafter stock, but are assembled like spaced headers (see Unit 7-4), with a piece of $\frac{1}{2}$″ plywood between them to fill out the wall to a full $3\frac{1}{2}$″. Other roof construction is conventional. Roof sheathing may be continuous over the party wall without affecting the fire rating. If the roofing material or its color are to change from unit to unit, however, the gap in the party wall may be bridged with flashing (Fig. 12-2.1).

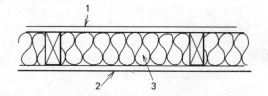

Table XII   Construction assembly

Material 1	Material 2	Material 3	FRR	STC
½" fiberboard weighing 0.7 lb./sq. ft.	Same as 1	None	⅙ hr. (⅓ hr. structural integrity only)	None
¼" hardwood plywood	Same as 1	None	⅙ hr. (⅓ hr. structural integrity only)	None
½" hardwood plywood	Same as 1	None	⅓ hr.	30
¼" hardwood plywood	Same as 1	R7 mineral wool insul.	⅓ hr.	None
¼" hardboard	Same as 1	R7 mineral wool insul.	⅓ hr.	None
¼" hardboard or hardwood plywood	⅜" gypsum wallboard	R7 mineral wool insul.	⅓ hr.	30
⅜" gypsum wallboard	Same as 1	None	⅓ hr.	32
½" type X wallboard	¼" hardboard or hardwood plywood nailed	R7 mineral wool insul.	¾ hr.	32
½" type X wallboard, nailed	Same as 1	None	¾ hr.	35
½" wallboard, nailed	Same as 1	R7 mineral wool insul.	¾ hr.	35
⅝" wallboard	¼" hardboard or hardwood plywood nailed	R7 mineral wool insul.	¾ hr.	35
⅝" wallboard nailed	Same as 1	None	¾ hr.	39
½" wallboard	¼" hardboard or hardwood plywood over ⅜" wallboard	R7 mineral wool insul.	¾ hr.	40

Procedure

1 On a foundation wall of adequate width, erect a pair of box sills as described in Unit 6-3. **Note.** The sill plate must be cut from treated lumber.

2 Apply subflooring as described in Unit 6-14. Stop flush with the edges of all header joists and edge joists.

3 From 2x stock, rip blocking 2" wide (A in Fig. 12-2.2) to fit along edge joists and between joists at the header joist.

4 Attach blocking.

5 Cut strips of ½" gypsum sheathing (B in Fig. 12-2.2) into widths equal to the distance between the top of the foundation wall and the top of subflooring.

6 Apply sheathing to edge and header joists, using adhesive at the sill and nails just below the subfloor.

7 Fill the cavity between sheathing with R7 insulation (C).

8 Assemble the framing for first floor party walls as described in Unit 7-7.

9 Apply ½" gypsum sheathing (see Unit 7-18) across studs from bottom plate to doubler.

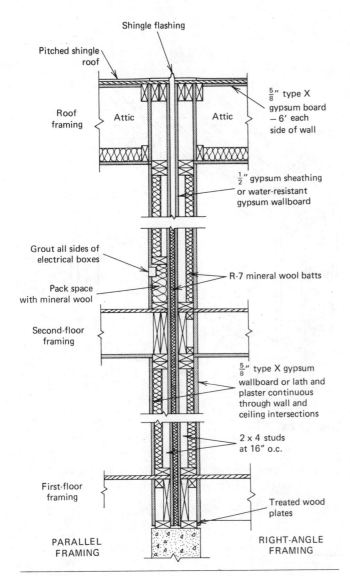

Shingle flashing

Pitched shingle roof

Roof framing

Attic   Attic

$\frac{5}{8}$" type X gypsum board — 6' each side of wall

$\frac{1}{2}$" gypsum sheathing or water-resistant gypsum wallboard

Grout all sides of electrical boxes

R-7 mineral wool batts

Pack space with mineral wool

Second-floor framing

$\frac{5}{8}$" type X gypsum wallboard or lath and plaster continuous through wall and ceiling intersections

2 x 4 studs at 16" o.c.

First-floor framing

Treated wood plates

PARALLEL FRAMING   RIGHT-ANGLE FRAMING

Fig. 12-2.1 A section through a typical party wall with a 1$\frac{1}{2}$ hour fire resistance rating. This wall has an STC rating of 45.

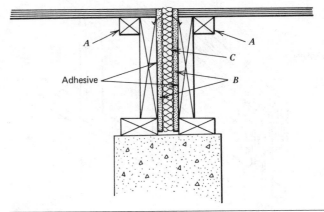

A   A

C

Adhesive   B

Fig. 12-2.2

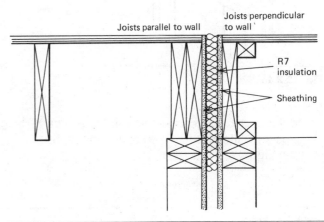

Joists parallel to wall   Joists perpendicular to wall

R7 insulation

Sheathing

Fig. 12-2.3

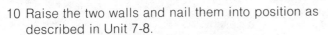

10 Raise the two walls and nail them into position as described in Unit 7-8.

11 Frame the second floor, repeating steps 1 to 6 as necessary to meet the construction shown in Fig. 12-2.3.

12 Fill the cavity between joists with R7 insulation.

13 Repeat steps 10 and 11 as many times as required by the drawings to reach roof framing. **Note.** Do not double the top plate of the wall in the topmost story (see Fig. 12-2.4).

14 Frame the wall in attic space in sections as shown in Fig. 12-2.4, to conform to the pitch of the roof.

15 Apply $\frac{1}{2}$" gypsum sheathing across studs, rafters, and cap plate.

16 Install the two sections of attic wall.

17 To the outer faces of the party wall in the attic, apply $\frac{5}{8}$" type X gypsum wallboard. Note in Fig. 12-2.5 that the wallboard does not cover the cap plate.

18 Lay out ceiling joists and roof rafters, placing the first set against the party wall.

19 Fill the cavity from cap plate to rafter with R7 insulation.

20 Sheathe the roof as described in Unit 9-34.

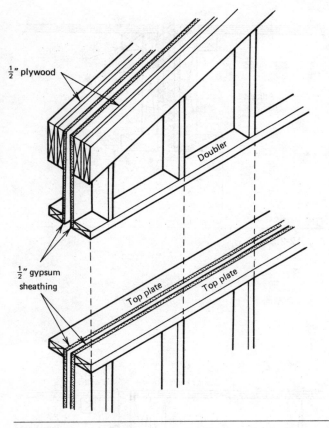

Fig. 12-2.4

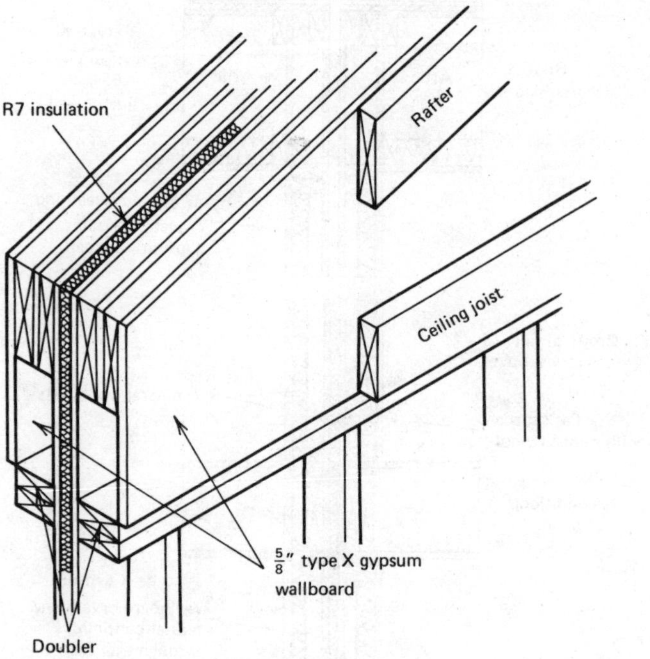

Fig. 12-2.5

21 To the under side of sheathing apply $\frac{5}{8}$" type X gypsum wallboard for 6' on each side of the wall (see Fig. 12-2.1).

22 Complete installation of insulation and surface materials as described in Chapters 10 and 21. See Fig. 12-2.1 for locations.

Questions

1 What areas of the cavity in a party wall are filled with R7 insulation?

2 Why are only these areas insulated?

3 Why doesn't the type X wallboard cover the rafter plate?

4 Note that the rafter and ceiling joists in Fig. 12-2.5 lie in the same vertical plane. Draw the construction where they meet at an outside wall.

Unit 12-3   **Build a Staggered Stud Wall**

One of the best ways to control the passage of sound between living spaces is with staggered stud construction. In this type of party wall the carpenter actually builds two thin walls close to each other but not touching (Fig. 12-3.1). The isolation of the two wall surfaces reduces impact sound transmission. STC ratings up to

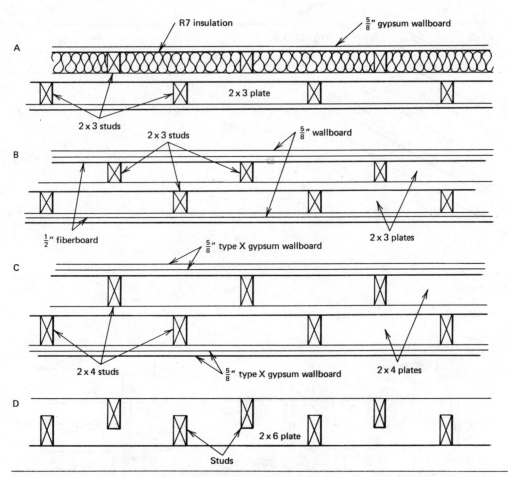

Fig. 12-3.1 Sections through four typical staggered stud walls. Type A is not fire-rated, but has an STC rating of 45. Type B has a one-hour FRR and an STC rating of 55. Type C has a two-hour FRR, and an STC rating of 50. The construction shown in D is an alternative that may be finished as A, B, or C.

55 and fire ratings up to two hours may be achieved with various surface materials and insulation.

When a fire-rated wall is not necessary, but high STC and IIC ratings are important, plates and studs may be 2 × 3s, and the finish material ⅝″ gypsum wallboard (A in Fig. 12-3.1). Between studs on one side goes a thickness of R7 mineral wool insulation.

For a fire-rated wall the insulation is not required. A one-hour rating can be achieved with 2 × 3 studs sur- faced with a ½″ layer of fiberboard plus a layer of ⅝″ gypsum wallboard (B). A two-hour rating may be reached with 2 × 4 studs surfaced with two layers of ⅝″ type X wallboard (C).

As an alternate, studs may be nailed to a single plate (D).

Question

1 To build a staggered stud wall with a one-hour fire rating, check the materials you would need from the list below:

2 × 3 studs _____   R7 insulation _____   ⅝″ gypsum wallboard _____

2 × 4 studs _____   ½″ fiberboard _____   ⅝″ wallboard, type X _____

Fig. 12-3.2

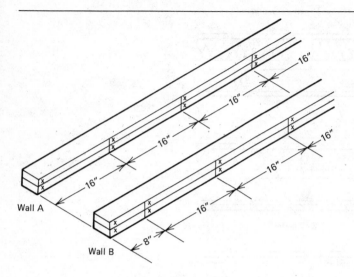

Wall A

Wall B

Fig. 12-3.3

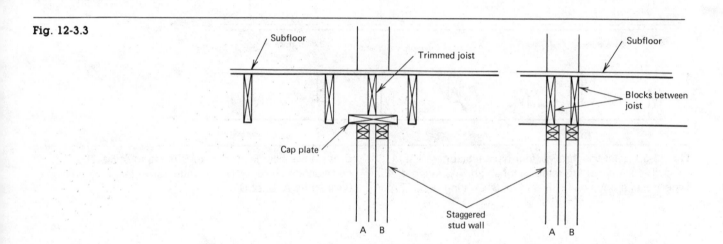

Subfloor

Trimmed joist

Subfloor

Blocks between joist

Cap plate

Staggered stud wall

A B

A B

Procedure

1 Determine from the drawings the size of stud and plate stock required.

2 Cut plates to length.

3 Mark stud locations on plates. Locations should be staggered 8″ between plates in walls *A* and *B* (Fig. 12-3.2).

4 Assemble the walls. Add cap plates.

5 On the subfloor mark the center line of the staggered stud wall.

6 One-half inch on either side of the center line snap a chalk line. The chalk lines mark the inside edges of wall plates.

7 Erect the two wall sections on the chalk lines, and attach them as described in Unit 7-8.

8 If the staggered stud wall runs the same direction as ceiling joists, fasten the wall at the top as shown at left in Fig. 12-3.3. If the wall runs at right angles, double-block between joists (right).

9 Add insulation and/or finish walls as required by specifications.

13

Pole-Frame Buildings

Most residential buildings rest on continuous foundations or piers, as described in Chapter 5. But some buildings rest instead on poles. Typical of these buildings are the barns and sheds on farms and ranches that shelter animals and protect machinery. Typical also are houses and small commercial buildings built on tidal land that floods easily and often.

Poles used in pole-frame construction range in diameter from 4″ for small sheds up to 8″ for larger barns, but 6″ poles are the most common. They are treated with wood preservative under pressure to prevent attack by rot and insects. Locust is a widely used species for poles.

When the soil is wet or sandy, poles are usually driven into the ground with a pile driver to a depth of at least 8′ below the surface. The pressure of the sand holds the poles in place. When the soil is dry and firm, poles are most often set in holes dug by a power auger. The poles may rest on concrete pads 6 to 8′ below the ground (Fig. 13-0.1, left). They may be embedded in a block of wet concrete (center); spikes driven into each pole before embedding hold it firmly in the block. Or they may be embedded in the ground and anchored with a reinforced concrete collar that is poured around the pole just below the ground's surface (right).

The carpenter may erect batter boards for pole-frame buildings. The batter boards are the same as for regular buildings, but the lines are at different points. The foundation line that guides the operator of a pile driver is tangent to the circumference of the poles at their outside edges (Fig. 13-0.2).

Pole-frame construction is similar in principle to post-and-beam construction. The main differences are at the plates supporting floor joists, and at roof rafters. Exterior joists do not rest on a sill plate, but are notched into the poles. Rafters may rest on a rafter plate as in standard construction, on a rafter plate set vertically, or on header joists in the ceiling.

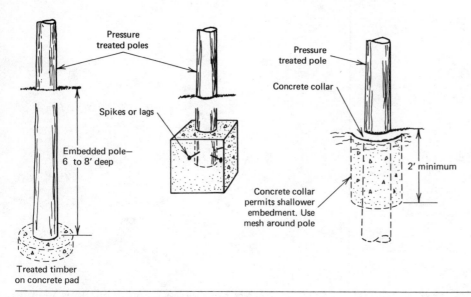

Fig. 13-0.1 In areas of the country subject to high winds, poles must be firmly anchored into the ground. They may be set on a concrete pad embedded at least 6' into the ground (left), anchored into a block of poured concrete below grade (center), or fiitted with a poured concrete collar just below ground level (right).

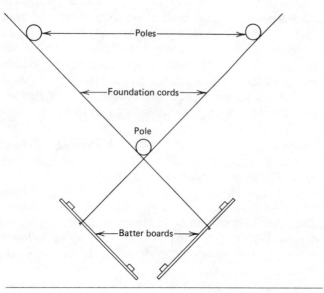

Fig. 13-0.2 The foundation line of a pole-frame building is tangent to the circumference of the poles.

Questions

(Circle T if you think the statement is true, and F if you think it is false.)

1 The foundation line of a pole-frame building passes through the center line of the poles.   T   F

2 Concrete is always used to hold poles in position in the ground.   T   F

3 Poles should be installed only when the soil is dry.   T   F

4 Poles must be coated with wood preservative.   T   F

Unit 13-1   **Frame the Barn Floor**

When the carpenter begins the work of framing a barn or other pole-frame building, the poles are likely to be already in place, and lined up along the cords marking the foundation line. The diameters of poles will be consistent, but their tops are likely to be at various levels. If the building is designed for conventional construction, poles are trimmed just below the floor line. If poles continue to the roof, they are usually covered on the outside of the building, but exposed on the inside.

The floor framing system in pole buildings is similar in many ways to a box sill. The chief differences are: (1) exterior joists are called **band joists** instead of header joists and edge joists, (2) band joists are attached to poles instead of a sill member, and (3) pairs of header joists serve as girders (Fig. 13-1.1). In the best construction, poles are notched so that joists fit tight. Total depth of notches should not exceed the diameter of a pole.

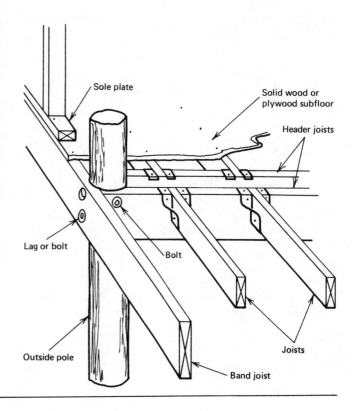

Fig. 13-1.1 Typical wood floor construction in a pole-frame building.

Question

1 What is the advantage of notching poles for joists?

Procedure

1 From drawings or other available data, determine the height above ground of the floor. **Note.** If batter boards are set up as described in Unit 4-5, this level is already established by foundation cords.

2 Determine the height of any notches by measuring band joist stock. Determine the depth of notches by using overall dimensions of the building (Fig. 13-1.2).

3 Notch poles for joists as required (Fig. 13-1.3). **Note.** Notch for a tight fit. No notch should be greater in depth than the thickness of the joist itself.

4 Determine the lengths of header joists. Headers butt against band joists, and any splices must fall at the centers of poles.

5 Cut and install header joists, bolting them in pairs (Fig. 13-1.4).

6 Cut and install band joists. Attach them to posts with bolts or lag screws.

7 If drawings do not require poles to extend above floor level, trim poles flush with the tops of joists.

8 Brace header joists at poles as shown in Fig. 13-1.5.

9 Install regular joists between header joists. They may be supported with joist hangers or ledgers.

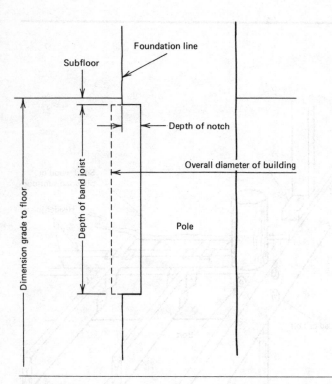

Fig. 13-1.2

Subfloor

Foundation line

Depth of notch

Overall diameter of building

Dimension grade to floor

Depth of band joist

Pole

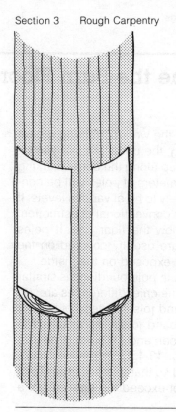

Fig. 13-1.3

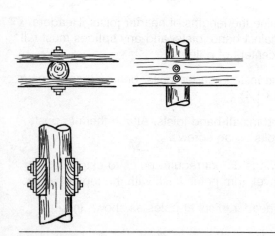

Fig. 13-1.4

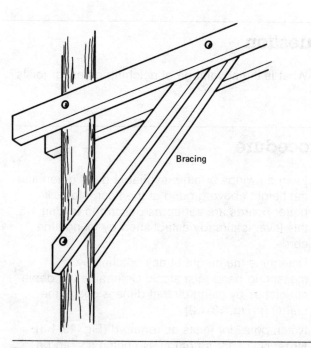

Bracing

Fig. 13-1.5

Unit 13-2 **Frame the Barn Roof**

When poles form only the foundation of a building, and are not continued above floor level, you may use conventional construction methods for framing walls and roofs. When poles extend all the way to the roof, however, that roof may be framed in two ways. The width of the building is one determining factor, and the roof load is the other.

Long, narrow buildings with light roof loads may be framed as in Fig. 13-2.1. Planks span from the ridgebeam to rafter plates. The ridgebeam consists of a pair of ridgeboards attached to center poles and running the length of the building. Rafter plates are attached vertically and offset slightly to allow for the pitch of the roof.

Roof framing in wider buildings begins with a pair of rafters flanking each exterior pole (Fig. 13-2.2). These rafters rest directly on header joists in the ceiling. Roof sheathing is supported, not by additional rafters as in conventional construction, but on purlins between header joists. **Outriggers** support sheathing

at the eaves, and ladders provide support at the rakes as in gable roof construction.

Procedure for Framing a Narrow Building

1 From drawings determine the height of the roof at the eaves.
2 Mark this height on all exterior poles.
3 To the poles bolt exterior rafter plates with their tops flush with the mark.
4 Using the known unit rise, unit run, and actual run from outside of pole to outside of pole (Fig. 13-2.3), calculate the rise for this section of roof.
5 Mark the rise on center poles.
6 At this height bolt ridgeboards on both sides of center poles.

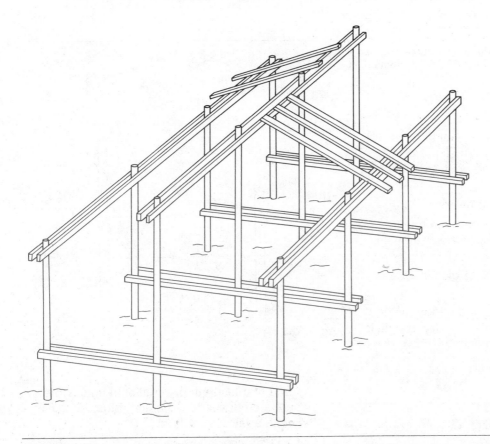

Fig. 13-2.1 In narrow pole-frame buildings, main roof framing members run at right angles to floor framing members. Planks serve as sheathing. This type of construction is most common in buildings without a horizontal ceiling.

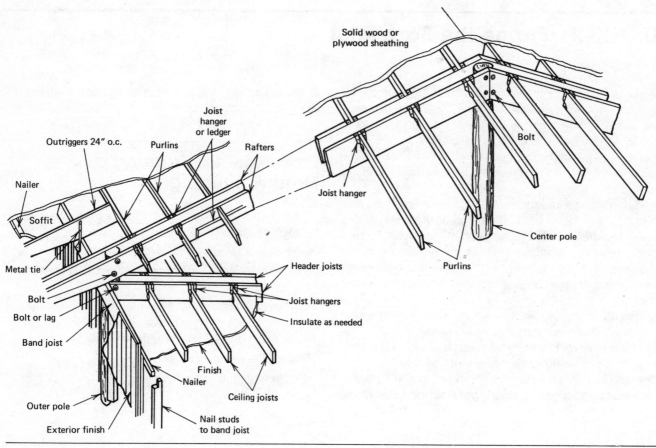

Fig. 13-2.2 In wide pole-frame buildings, and those with a flat ceiling, roof construction is similar to floor construction.

7 Determine the location of the remaining rafter plates, either by calculation as in step 4 or by setting a straight board on edge between rafter plate and ridgeboard.

8 Bolt inner rafter plates on their marks.

9 To maintain a constant space between framing members, add tieblocks centered between poles if pole spacing is greater than 8'. (Fig. 13-2.4).

10 Cut all roof planks to length, and give them a plumb cut at the ridge end.

11 Attach planks, beginning near the middle of the roof and working toward both ends. Use 10d nails at plates and ridgeboard, and toenail where ends of planks butt.

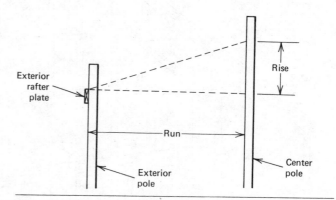

Fig. 13-2.3

Procedure for Framing a Wide Building

1 Determine the length of a common rafter.

2 Cut and fit a test pair of common rafters. **Note.** Rafters are not shortened at the ridge, because they butt against each other.

3 Cut all common rafters.

4 On all rafters mark the locations of purlins. Normal spacing is either 16 or 24".

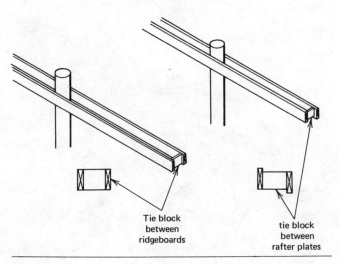

Tie block
between
ridgeboards

tie block
between
rafter plates

Fig. 13-2.4

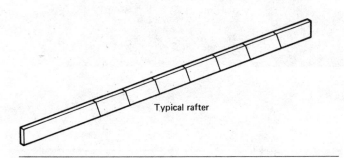

Typical rafter

Fig. 13-2.5

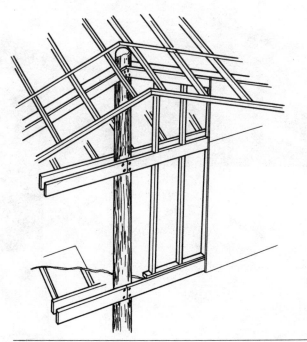

Fig. 13-2.6

5 Square the marks down the sides of rafters (Fig. 13-2.5).

6 Bolt rafters to poles.

7 Cut purlins to length.

8 Install purlins, supporting them either on ledgers or with joist hangers.

9 On band joists mark the locations of outriggers. Normal spacing is 24″ o.c.

10 Cut outriggers to length.

11 Attach outriggers, using metal ties at the band joist and 10d nails at the outer purlin (see Fig. 13-2.2).

12 If the roof overhangs at gables, follow the basic procedures described in Unit 9-9, but modified as shown in Fig. 13-2.6.

13 Apply sheathing and roofing paper as on a conventional roof.

Questions

1 When a pole-frame building has a ceiling, ceiling joists are attached to _____.

2 When a roof of a pole-frame building has an overhang on all four sides, sheathing is supported not only by rafters, but by _____, _____, and _____.

3 What is the difference in the line length and the actual length of a roof rafter in a pole-frame building?

4 Exterior Finish Carpentry

In a normal building sequence the first steps in enclosing a house take place at the same time as the final steps in framing. Bad weather is the enemy of builders, and they don't breathe a sigh of relief until the houses they are building are completely protected against all weather. This protection consists of applying roofing materials, surfacing exterior walls, installing windows and exterior doors, and applying trim around openings and wherever two materials or surfaces come together at a corner.

The finish carpenter uses many of the same tools as the rough carpenter, because many of the tasks call for tools that serve the general purposes listed in Chapter 2. But for certain tasks the carpenter now needs additional tools. Some are designed primarily for assisting in the application of one special product, such as roofing or wallboard. Others are used to put the finishing touches on wood to prepare it for paint or stain.

14

The Finishing Tools You Use

The steadily increasing popularity of factory-built components has reduced but not eliminated the use of wood finishing tools on the job site. There will always be a lot of fitting that must be done on the spot.

The tolerances and appearances permissible in rough carpentry are seldom acceptable in finish carpentry. With the use of millwork and higher grades of lumber comes a comparable need for hand operated and power driven tools of high precision that permit greater accuracy. The finish carpenter must know the capabilities and limitations of these tools, and how to use them safely and efficiently.

Hand Operated Tools

The carpenter needs hand tools that serve these general finishing purposes:

- Cutting
- Shaping
- Smoothing
- Concealing

On pages 336 to 338 are the woodworking tools available for these purposes, organized by the specific task the carpenter faces.

A number of other woodworking tools, such as skews and fluters, are also available. They are not included here, however, because their primary use is in making furniture and specialty items that require carving.

Other hand tools designed primarily for one specific purpose are described in the units of this section where they are used.

The task	The tool	Its description
To cut grooves; to shape small openings	Gouge (Fig. 14-0.1)	A type of wood chisel with a curved blade, usually but not always sharpened on the outside edge.
To round square edges; *to smooth* rough cuts	Wood file	A hard steel tool with ridged surfaces ranging from rough to very fine; blades may be flat or rounded.
To *rough-shape* an edge or end	Rasp	A very coarse wood file for removing excess wood quickly.
To remove surface material quickly	Draw knife	A beveled steel blade with handles at both ends.
To smooth long edges	Jointer plane	A long (22 to 24″) plane used extensively in hanging doors.
To smooth long edges and flat surfaces	Fore plane (Fig. 14-0.2)	Similar to a jointer plane, but shorter (18″) and wider.
To smooth rough surfaces	Jack plane (Fig. 14-0.3)	A heavy-duty plane of medium length (14 to 15″) for general use.
To smooth surfaces of small boards	Smooth plane (Fig. 14-0.4)	A smaller (7 to 9″), lighter plane similar to a jack plane.
To smooth surfaces and edges across the grain	Block plane (Fig. 14-0.5)	A small plane with the bevel of the blade up instead of down.
To smooth and *square* edges of boards	Edge plane (Fig. 14-0.6)	A small plane with the blade on the side instead of the bottom.
To smooth bottoms of grooves	Router (Fig. 14-0.7)	A two-handled tool with interchangeable blades in the center.
To smooth any surface as a final step before painting or staining	Sandpaper (Fig. 14-0.8)	Stiff paper with granules of sand or other minerals glued to it. Texture ranges from extra coarse to fine.
To *countersink* heads of nails	Nailset (Fig. 14-0.9)	A tapered steel cylinder with a blunt point that fits into the hollow in the head of a finishing nail.

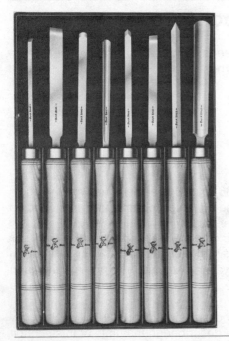

Fig. 14-0.1 A set of gouges. (Courtesy Buck Bros. Inc.)

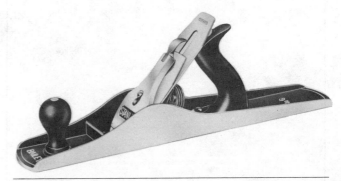

Fig. 14-0.2 Fore plane. A jointer plane is similar and a little larger. (All planes courtesy Stanley Tools.)

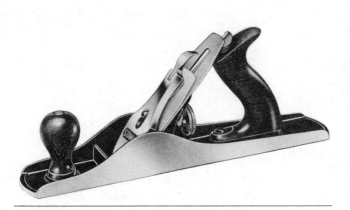

Fig. 14-0.3 Jack plane.

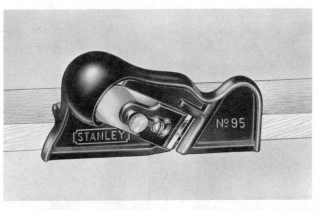

Fig. 14-0.6 Edge plane.

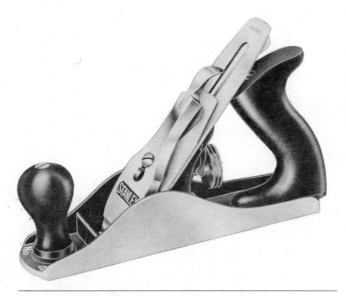

Fig. 14-0.4 Smooth plane.

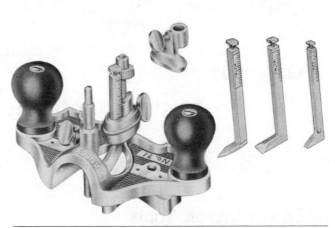

Fig. 14-0.7 Router. (Courtesy Stanley Tools.)

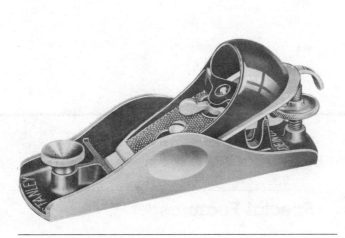

Fig. 14-0.5 Block plane.

Fig. 14-0.8 Sandpaper comes in both discs and sheets. Left to right: coarse, medium, fine. (Courtesy 3M Company.)

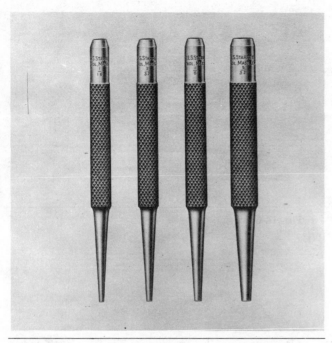

Fig. 14-0.9 Nailset. (Courtesy The L.S. Starrett Co.)

Questions

1 How does a gouge differ from a wood chisel?
2 What tools could you use to smooth the rough end of a board?
3 In what type of plane is the bevel of the blade reversed?

Power Driven Tools

Power driven tools do most jobs in finish carpentry better than equivalent hand tools, more quickly, and with greater accuracy. The finish carpenter will frequently use six such tools: a circular saw, saber saw, jointer, router, portable plane, and portable sander.

Unit 14-1   Circular Saw

The circular saw cuts from the top of the material, leaving a smooth cut on the top and a rough cut on the bottom. The blade spins in a narrow slot in a removable throat, and the throat supports the material to keep the bottom cut as smooth as possible.

Special Features

The angle and height of saw blades is adjusted by a pair of hand wheels, one on the front and one on the

The task	The tool	Its description
To cut lumber stock to size and shape	Circular saw, also called a table saw (Fig. 14-1.1). Comes in both floor and bench models.	A saw with interchangeable circular blades that rise out of a slot in a metal table. Angle of the blade is adjustable up to 45°. Common blade diameters are 8 and 10″.

side of the cabinet. The fence moves on a steel bar, and may be locked in place on either side of the blade, or removed entirely. The tabletop has a slot on each side of the blade for a **miter gauge,** a device for holding material at the proper angle as it moves against the blade. The gauge is adjustable for any degree of cut; many gauges have automatic stops at 30°, 45°, and 60°. Attached to the gauge is an adjustable **stop rod** that serves as a fence for some crosscutting operations. A guard fastened at the back or side of the table fits over the blade. This guard should be used whenever possible, although it can't be used with certain types of cutting. Most models also have a sheet of metal, called a **splitter,** directly behind the blade to keep the saw kerf open during cutting.

Types of Blades

Six types of flat blades may be used in a circular saw (Fig. 14-1.2):

- *The crosscut or cutoff blade*—like the crosscut saw, for cutting across the grain.
- *The ripping blade*—like the rip saw, for cutting with the grain.
- *Combination blade*—for general cutting, both with and across the grain, thus eliminating the need for frequent blade changes. Of the various arrangements of teeth on combination blades, the best for fine cutting is style S (shown in Fig. 14-1.2).
- *Easy-cut blade.* A combination blade with few teeth. Its advantages: reduces electrical usage by the motor and virtually eliminates kickbacks. Its disadvantages: cuts a wider and rougher kerf than other blades.
- *Planer blade.* The teeth on this blade are in line, instead of with some **set** (bent) to one side or the other. To prevent it from sticking in the kerf, the blade is not uniform in thickness; because of this feature, the planer blade is often called a **hollow-ground blade.** It cuts a smooth, narrow kerf.

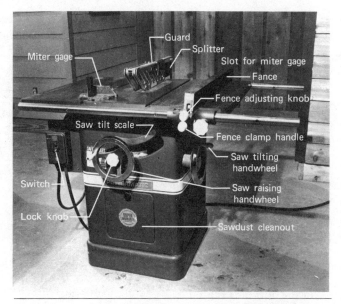

Fig. 14-1.1 Circular saw. (Photo Jay Leviton, Atlanta.)

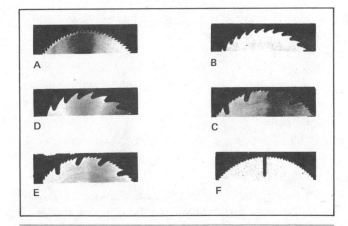

Fig. 14-1.2 Types of flat blades for a circular saw: (*a*) crosscut, (*b*) ripping, (*c*) combination, (*d*) easycut, (*e*) planer, (*f*) plywood. (Courtesy Simonds Cutting Tools.)

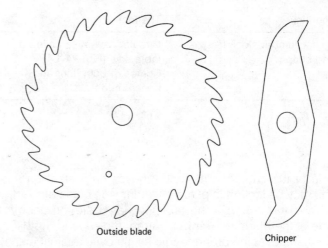

Outside blade   Chipper

Fig. 14-1.3 An assembled dado head consists of outside cutter blades and chipper blades. (Courtesy Simonds Cutting Tools.)

• *Plywood blade.* A special blade of tempered steel designed for cutting through veneers and the glue between them. It leaves a smoothly cut edge. The blade may also be used for crosscutting.

In addition to flat blades, the circular saw may be adapted for a dado head. A **dado head** has two outside blades, similar to combination blades, and chipper blades between (Fig. 14-1.3). Outside blades make a $\frac{1}{8}''$ cut, and chipper blades come in thicknesses of $\frac{1}{16}''$, $\frac{1}{8}''$, and $\frac{1}{4}''$. By using blades singly or in combination with chippers, you can cut dadoes or grooves from $\frac{1}{8}$ to $\frac{13}{16}''$ wide. A **groove** is cut with the grain; a **dado** is cut across the grain.

General Safety Precautions

The following precautions must be observed whenever you use a circular saw:

• Make all blade adjustments with the power off.
• Keep blades sharp, and use only sharp blades.
• Use the correct blade for the cutting operation to be done.
• Clamp the fence securely before starting any cut.
• Work with the guard and splitter in place if at all possible.
• Keep the table clean of sawdust and scraps with a brush or push stick. Do not use your hands.
• Wear nothing on your hands or arms that could catch in the revolving blade.
• Wear safety glasses.

• Do not saw warped lumber or plywood on a power saw.

Operating precautions are listed as they occur in the following procedures.

Procedure for Crosscutting Lumber

1 Mark the piece to be cut on the edge nearest the blade.
2 Remove the ripping fence.
3 Adjust the height of the blade. **Precaution.** The blade should be just high enough to cut through the material, and no higher.
4 Adjust the miter gauge to the correct angle, and set it in the slot on the same side of the blade as the longest part of the material to be cut.
5 Adjust the stop rod on the miter gauge to hold long stock in position.
6 Set the material against the miter gauge and stop rod about an inch from the blade, with the cutting mark in line with the blade.
7 Turn on the power, and let the blade reach full speed.
8 Standing to one side of the blade, move the material slowly and on a straight line toward the blade. **Precaution.** Never stand directly in line with the blade.
9 If the board being cut is unsupported over half its length, get someone to support the end.

Fig. 14-1.4 (Photo Jay Leviton, Atlanta.)

Precaution. Make sure the helper does not move the stock, but only supports it.

10 When the cut is complete, shut off the motor.
Precaution. If the piece cut off is small, remove it from the side after the blade has stopped moving. Never reach over a spinning blade.

Procedure for Cutting a Long Board into Repetitive Sizes

1 Clamp a small board or piece of plywood to the fence as a gauge block (Fig. 14-1.4). **Precaution.** The front edge of the block should not extend beyond the edge of the blade.

2 Adjust the fence so that the distance between the gauge block and the blade is equal to the length of piece desired. **Note.** Always measure to the end of a tooth set toward the fence (Fig. 14-1.5).

3 Follow steps 3 through 8 of procedures for crosscutting.

4 Repeat these procedures until all cuts are made.

5 Shut off the motor.

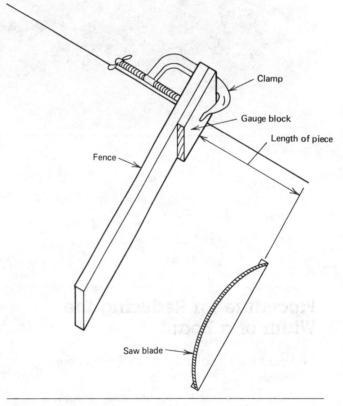

Fig. 14-1.5

Procedure for Trimming Lumber to Short Lengths

1 Adjust the height of the blade and the angle of the miter gauge.

2 Following procedures for crosscutting, square one end of each piece to be trimmed.

3 Set the stop rod on the miter gauge so that the distance from the end of the rod to the blade is equal to the length of material desired (Fig. 14-1.6).

Fig. 14-1.6

4 Repeat steps for crosscutting until all cuts are made.

5 Shut off the motor.

Procedure for Reducing the Width of a Board

1 Adjust the fence to the final width of board required.

2 Adjust the blade to proper cutting height.

3 Turn on the power, and let the blade reach full speed.

4 Set one end of the board on the table, and hold it against the fence with your left hand.

5 Feed the board into the blade at an even speed with your right hand. Keep that hand close to the fence.

6 a. If the board is long, have a helper support the two widths as they come off the table.

6 b. If the trimmed piece is narrow, use a push stick to feed the last foot of board past the blade.

7 Shut off the motor.

Procedure for Trimming the Thicknesses of Lumber

1 Adjust the fence to the thickness required.

2 Adjust the blade to cut a little more than halfway through the board.

3 Feed the lumber through the blade as outlined in steps 3 to 6 above.

4 Remove the partially cut material, turn it over, and feed it through again.

5 Shut off the motor.

Procedure for Cutting Plywood

1 Insert the plywood blade in the saw.

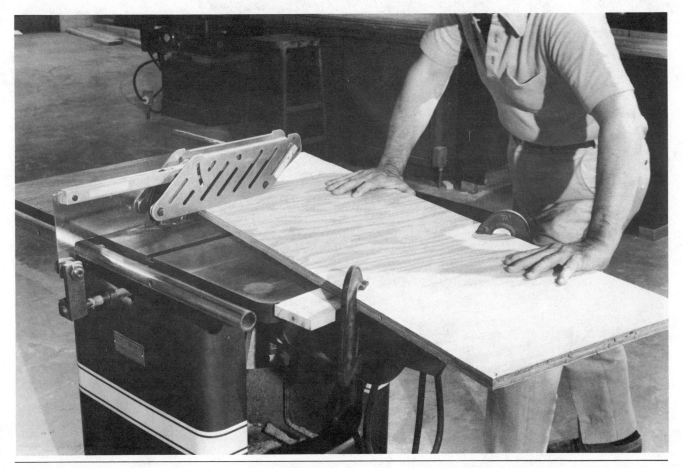

Fig. 14-1.7

2 Adjust the blade so that it will just barely cut through the material.

3 Mark the cutting line on the good side of the plywood.

4 Clamp a board to the under side of the plywood so that the distance between the board and cutting line is equal to the distance between the saw blade and the edge of the saw table (Fig. 14-1.7).

5 Feed the plywood through the blade at an even pace. **Precaution.** When cutting a full sheet, have helpers on both sides of the saw to support the weight and keep from jamming the blade.

6 Shut off the motor.

Procedure for Making an Angle Cut

1 Adjust the height of the blade.

2 Turn the miter gauge to the correct angle.

3 Set a small board upright against the miter gauge

(Fig. 14-1.8). This board keeps the saw from tearing the end of the board being cut.

4 Feed the board through the saw blade.

5 Shut off the motor.

Procedure for Making a Bevel Cut

1 Adjust the angle of the blade.

2 Adjust the height of the blade so that it just cuts through the material at the desired angle.

3 Insert the miter gauge in the slot away from the tilt of the blade.

4 Feed the board through the saw blade.

5 Shut off the motor.

Procedure for Cutting a Rabbet

1 On the end of the stock lay out the width and depth of the rabbet (Fig. 14-1.9).

2 Adjust the saw blade to the depth of the first cut (dimension *A* in Fig. 14-1.9).

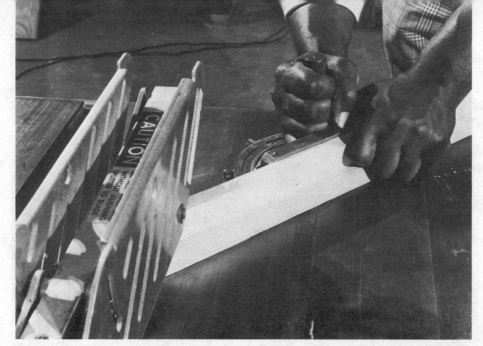

Fig. 14-1.8

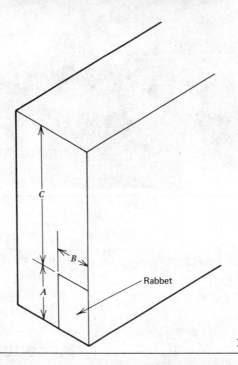

C

B

A

Rabbet

Fig. 14-1.9

Fig. 14-1.10

3 Adjust the fence to the width of the first cut (dimension *B*).

4 Place the stock face down on the table with the edge tightly against the fence.

5 Make the first cut.

6 If necessary, readjust the height of the saw blade to dimension *B*.

7 Adjust the fence, not to the width of the second cut, but to the thickness of the remaining lumber (dimension *C*).

8 Hold the stock firmly against the fence at the top. **Note.** When cut as shown in Fig. 14-1.10, the cutoff strip stays on the table; it will not bind between the blade and fence or kick back at you.

9 Make the second cut.

10 Shut off the motor, and recheck the dimensions of the rabbet.

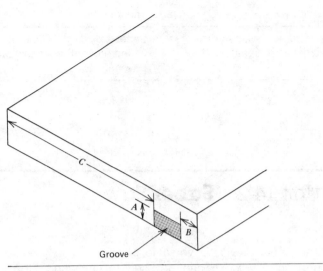

Fig. 14-1.11

Procedure for Cutting a Groove with a Standard Blade

1 On the end of the stock lay out the location, width, and depth of the groove (Fig. 14-1.11).

2 Adjust the blade to the depth of the groove (dimension *A*).

3 Adjust the fence so that the distance from fence to blade is equal to the dimension from one side of the groove to the edge of the material (dimension *B*).

4 Make the first cut.

5 Readjust the fence so that the distance from fence to blade is equal to the dimension from the other side of the groove to the other edge of the material (dimension *C*).

6 Turn the stock around and make a second cut.

7 Move the fence away from the blade and recut as often as is necessary to clean out the groove.

Procedure for Cutting a Groove or Dado with a Dado Head

1 Remove the throat and the single blade.

2 On the arbor insert enough outside cutters and chippers to give you the required width of groove or dado.

3 Mark the location, width, and depth of the groove on the end of the stock, or of the dado on the side of the stock.

4 Adjust the blade to the correct cutting height.

5 a. Adjust the fence if cutting a groove.

5 b. Adjust the stop rod if cutting a dado.

6 Turn on the motor.

7 Move the material across the dado head and out the back of the saw table. **Precaution.** Do not bring the material back through the dado head.

8 Turn off the motor.

Questions

1 Which type of blade for a circular saw gives the smoothest cut? The widest cut?

2 List the blades you would use in a dado head to cut a groove $\frac{7}{16}''$ wide.

3 What is the purpose of a miter gauge?

4 Name two uses of a stop rod.

5 What is the proper relationship between the height of the blade and the thickness of material to be cut on a circular saw?

6 After a board is cut, one side will be smoother than the other. Do you mark the cut on the smooth side or rough side?

7 What is the difference between a dado and a groove?

The task	The tool	Its description
To cut irregular shapes or holes in materials up to 2″ thick	Saber saw (Fig. 14-2.1); also called a portable jig saw and bayonet saw	A lightweight portable saw with a straight, pointed blade that cuts on the up stroke and bypasses the material being cut on the down stroke.

Unit 14-2 Saber Saw

The saber saw is a versatile cutting tool for on-site and shop work. It can be used to cut wood, fiberboards, gypsum boards, plastics, and metal. It may be used for making straight or bevel cuts, but is most valuable for making irregular cuts that can't easily be made with any other tool. All types have a motor, a handle on the top, a base plate, and a device for varying blade action. You operate the tool with one hand and, with the other, either steady the saw or hold the material being cut.

Special Features

Some models have a guide knob or grip on one side for better control of the cutting operation. They may have a device for reducing strain on the cord, and a base that can be tilted up to 45° for bevel cuts. One useful accessory is a guide for making circular cuts or as a fence for ripping.

Selection of the proper blade is important. For most work, blades with 7 or 10 teeth per inch are the best. For smoother cuts and for cutting sheet materials, a blade with 12 teeth per inch is better. For thick materials, blades with 6 teeth per inch are recommended. At least three teeth should be in contact with the material at all times.

General Safety Precautions

The following precautions should be observed whenever you use a saber saw:

- Change blades with the power disconnected.
- Use the correct blade for the cutting operation to be done.
- Make sure that the screw holding the blade is tight.
- Make sure that the material to be cut is clamped or otherwise held immobile.

Fig. 14-2.1 Saber saw. (Courtesy Rockwell International.)

Operating precautions are indicated where they occur in the following procedures.

Procedure for Making a Normal Cut

1 Mark the cutting line on the material.
2 Clamp or fasten the material so that it can't move.
3 Set the base on the material and line up the blade with the cutting line.
4 If the cut is to be straight, adjust the fence (Fig. 14-2.2).
5 Start the motor and let it reach full speed.
6 Begin the cut. **Precaution.** Hold the base flat and tight against the material.

Fig. 14.2.2 (Photo Jay Leviton, Atlanta.)

7 Move the saw slowly but steadily along the cutting line. **Precaution.** Use only enough forward pressure to keep the saw moving. Moving too rapidly results in a ragged cut, and may stall the motor or break the blade. Moving too slowly dulls the blade.

8 When the cut is complete, shut off the motor. **Precaution.** Let the blade come to a complete stop before setting the saw down on its side.

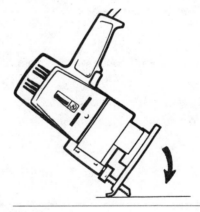

Fig. 14-2.3

Procedure for Cutting a Hole

1 Mark the size and shape of the hole to be cut.

2 Select a point somewhere within the cutting lines for starting the cut, and rest the tool on its base at an angle to the material, with the blade just above the starting point (Fig. 14-2.3).

3 Turn on the power and let the motor reach full speed.

4 Slowly lower the saw on its base until the blade has cut completely through the material and the saw is upright on its base.

5 Cut toward the mark for the hole, then along it.

6 If the required hole is rectangular, cut quarter circles at the corners until the hole is completed (Fig. 14-2.4).

7 Square the corners with additional short cuts.

8 Shut off the motor.

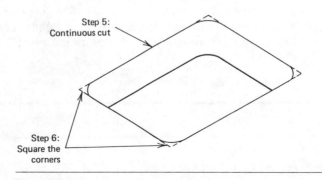

Fig. 14-2.4

Procedure for Cutting a Circle

1 Mark the center of the circle with a pencil.

2 Scribe a circle of the required radius around the center point.

3 At some point along the circle's circumference make a short cut as described in steps 2, 3, and 4 for cutting a hole.

4 Shut off the motor, and insert the fence upside down as a cutting guide.

5 Drive a small nail at the center of the circle.

6 Reinsert the blade in the cut, and adjust the guide so that it fits over the nail (Fig. 14-2.5).

7 With the blade free in the cut, restart the motor.

8 Complete the circular cut in the same manner as a straight cut.

9 Shut off the motor.

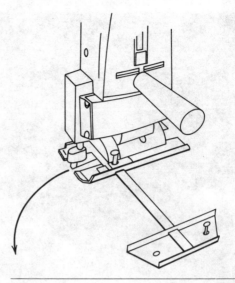

Fig. 14-2.5

Questions

1 Name three safety precautions concerning the blade of a saber saw.

2 Name four materials besides wood that can be cut with a saber saw.

3 What happens when you cut too quickly or too slowly?

4 Describe two uses of the fence attachment.

Unit 14-3   Jointer

A jointer is less often found on a residential building site than in a woodworking or cabinet shop. There it is used to cut rabbets, grooves, and bevels, and to taper wood stock, in addition to smoothing boards.

Common sizes for jointers are 6″ and 8″, dimensions that refer to the width of the cutting head. Jointers have two tables: an **infeed** table in front of the cutter and an **outfeed** table behind it. A guard over the cutting head either lifts up or swings to one side as material is fed through. The fence is adjustable to any angle from vertical to 45°. Most jointers are powered by a 1¼ h.p. motor that turns the cutting head at 4000 rpm (revolutions per minute).

The task	The tool	Its description
To smooth a surface or an edge	Jointer (Fig. 14-3.1), also called a planer and jointer	A power tool designed to replace or reduce the need for precision hand planing. A rotating cutting head, cylindrical in shape and with blades every 90° or 120°, does the planing.

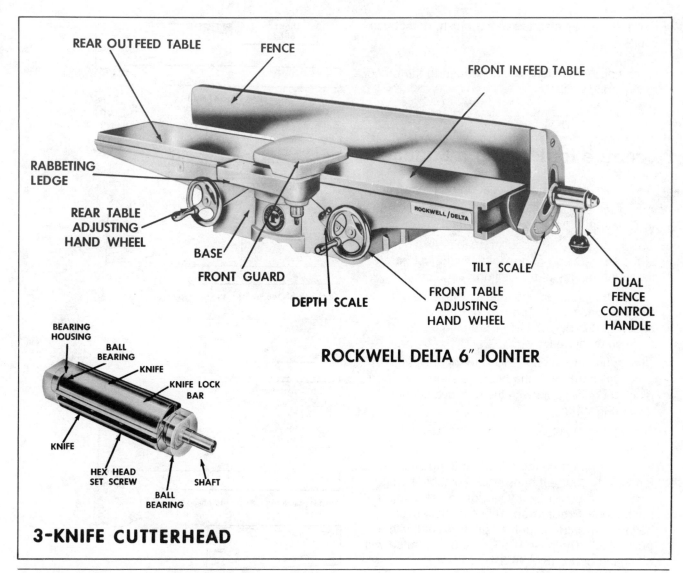

REAR OUTFEED TABLE   FENCE

FRONT INFEED TABLE

RABBETING
LEDGE

REAR TABLE
ADJUSTING
HAND WHEEL

BASE

FRONT GUARD

DEPTH SCALE

FRONT TABLE
ADJUSTING
HAND WHEEL

TILT SCALE

ROCKWELL/DELTA

DUAL
FENCE
CONTROL
HANDLE

ROCKWELL DELTA 6″ JOINTER

BEARING
HOUSING

BALL
BEARING

KNIFE

KNIFE LOCK
BAR

KNIFE

HEX HEAD
SET SCREW

BALL
BEARING

SHAFT

3-KNIFE CUTTERHEAD

Fig. 14-3.1 Jointer and its key parts. (Courtesy Rockwell International.)

Adjusting the Tables

In a few models the cutting tables are fixed in position, and you raise or lower the cutting head, as with many saws. On most models, however, the cutter is fixed, and the infeed table is raised or lowered by a hand wheel. The outfeed table may be fixed or adjustable.

To smooth a board, lower the infeed table just enough to provide a level finished surface without any high or low spots. To thin a board to an exact thickness, make a series of light cuts (about $\frac{1}{16}″$) if the material is a hardwood, and heavier cuts (up to $\frac{1}{8}″$) if the material is a softwood. When planing an end or surface, make a lighter cut than when planing an edge.

For most tasks performed on a jointer the outfeed table and the highest blade of the cutter should be at exactly the same height (Fig. 14-3.2). If outfeed table height is below cutter height, the end of the board will have a gouge in it. If the outfeed table is above cutter height, the board will have a taper.

General Safety Precautions

The following precautions must be observed whenever you use a jointer:

- Keep the blades of the cutter sharp for maximum safety and the best planing results.
- Check carefully the lumber to be planed. Do not try to work with pieces with defects, especially knots.
- Do not attempt to work on material less than $\frac{3}{8}″$ thick or less than 12″ long.

• Keep the tables cleaned with a brush; do not use your hands.

Operating precautions are described in the procedures that follow.

Procedure for Planing a Surface

1 Determine the species of wood to be planed.

2 Adjust the depth of the cut, either by lowering the infeed table or raising the cutter.

3 Check the level of the outfeed table. A straightedge laid on the table should just barely touch the blades of the cutting head.

4 Check the fence for squareness with a try square. **Precaution.** You can prevent uneven wear on the cutter blades by occasionally changing the position of the fence.

5 Check the stock to be planed for defects and warp. **Precaution.** If the board is slightly warped, place it on the table with the concave surface down (Fig. 14-3.3).

6 Turn on the motor, and let the cutter reach full speed.

7 Place the board on the infeed table against the fence, with your left hand back from the leading edge of the board, and your right hand guiding a push block. **Precaution.** Thin material should never be moved through a jointer except with a push block. The push block may be operated with one hand or two (Fig. 14-3.4).

8 Standing to one side of the jointer, move the material forward over the cutter.

9 Before your left hand reaches a point above the cutter, move it forward over the outfeed table. **Precaution.** Never have your hand directly above the cutting head.

10 Push the material all the way through with the push block.

11 Shut off the motor.

Procedure for Planing an Edge

1 Set the fence as close as possible to the left side of the table, and adjust it for square. **Precaution.** Never adjust the fence while the cutting head is turning.

2 Adjust the infeed table to the proper height.

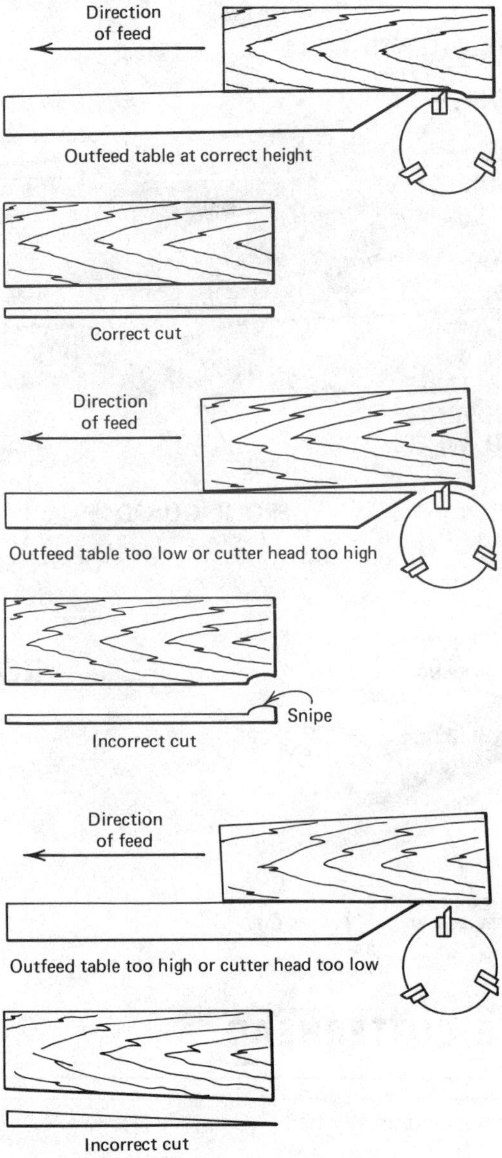

Direction of feed

Outfeed table at correct height

Correct cut

Direction of feed

Outfeed table too low or cutter head too high

Snipe

Incorrect cut

Direction of feed

Outfeed table too high or cutter head too low

Incorrect cut

Fig. 14-3.2 To achieve a straight cut, the outfeed and infeed tables must be at the same height. If not, the cut edge will either be notched or tapered.

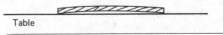

Table

Fig. 14-3.3

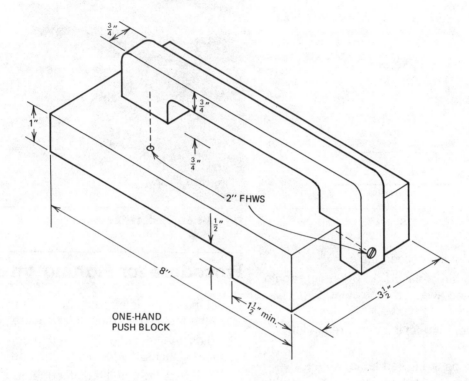

ONE-HAND
PUSH BLOCK

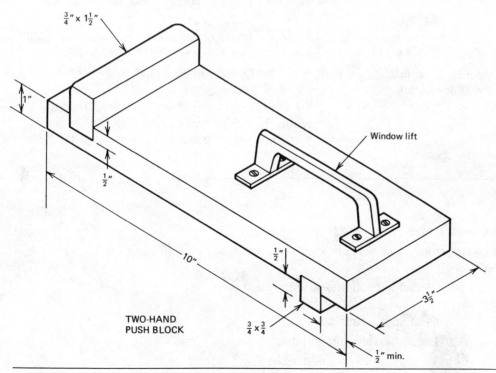

TWO-HAND
PUSH BLOCK

Fig. 14-3.4

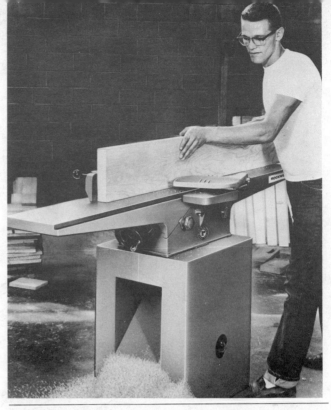

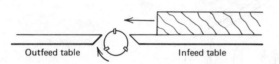

Outfeed table | Infeed table

Fig. 14-3.5

Fig. 14-3.6 (Courtesy Rockwell International.)

3 Check the level of the outfeed table.

4 Determine the grain of the stock. The grain should run downward and away from the cutter (Fig. 14-3.5).

5 Turn on the motor, and let the cutter reach full speed.

6 Place the board on the infeed table, with a smooth side firmly against the fence, and with the grain as described in step 4.

7 Move the material over the cutter. Use your left hand to hold the material against the fence and table, and your right hand to push (Fig. 14-3.6).

8 Move each hand over the outfeed table just before it reaches the location of the cutter.
Precaution. Move the board slowly and steadily for a smooth edge.

9 Shut off the motor.

Procedure for Planing an End

1 Set the fence as close as possible to the left side of the table, and adjust it for square.

2 Adjust the infeed table for a thin cut—no more than is necessary to smooth the end.

3 Check the level of the outfeed table.

4 Turn on the motor, and let the cutter reach full speed.

5 Place the board on edge on the infeed table, with a smooth side firmly against the fence.

6 Move the board into the cutter, and make a cut about 1″ long.

7 Turn the board around and complete the cut.
Note. By making two passes through the cutter, you avoid splitting the edge of the board.

Questions

1 How can you prevent uneven wear on the cutter blades of a jointer?

2 To plane $\frac{1}{4}$″ off the edge of an oak board, how many times should you run the board through the router?

3 Match up the safe position of hands with the location of a long board on a router:

At the beginning of the cut   One hand above each table

Halfway through the cut   Both hands over the infeed table

At the end of the cut   Both hands over the outfeed table

4 When you plane an end, about how long should the first cut be? Why?

The task	The tool	Its description
To shape an edge, *cut* a mortise, or *decorate* a surface	Router (Fig. 14-4.1)	A portable tool with a circular base plate and a chuck that accepts bits for cutting an infinite variety of shapes.

Unit 14-4   **Router**

The base plate of the router rests on the surface of the material to be shaped. In its center is a chuck that rotates clockwise, operated by a 1 or $1\frac{1}{4}$ h.p. motor. Bits locked into the chuck turn at up to 21,000 rpm when moving freely, and at a much reduced rate when into wood. Each router has a guide knob on one side. On the opposite side is either another knob or a handle, and the starting switch is located in conjunction with this handle. The better models have a disconnect switch that automatically shuts off the power if the router becomes overloaded.

Bits

Router bits have shanks machined to lock into the chuck. There are hundreds of head shapes (Fig. 14-4.2), and all cut on their sides. Some types have extended shanks that serve as a cutting guide. Among the most common uses of a router are to cut a mortise in doors for latches, trim plastic laminate, cut dovetail joints in drawer sides, or to add a decorative shape to the edge of a board. The depth of any cut can be changed by lowering or raising the motor above the base plate.

Router Control

Proper use of a router takes quite a bit of practice. It may be used for freehand work, such as cutting letters into a wood sign. Usually, however, some guide is necessary to keep the router moving in the right direction. Among those commonly used are a straight guide attachment, a circular guide attachment, a straight-edge clamped to the material being routed, or a template. The bits on some routers have pilot edges that act as guides.

The motor in a router turns the chuck and the bit clockwise. The direction you move the router should be against this clockwise action—that is, from left to right if you are making a straight cut, and counterclockwise if you are making a circular cut.

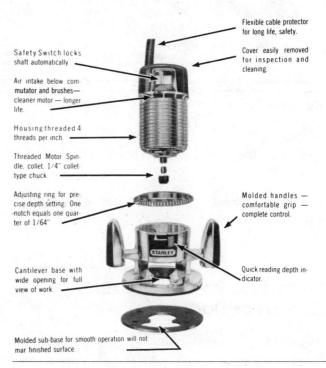

Safety Switch locks shaft automatically.

Flexible cable protector for long life, safety.

Cover easily removed for inspection and cleaning.

Air intake below commutator and brushes—cleaner motor — longer life.

Housing threaded 4 threads per inch.

Threaded Motor Spindle, collet, 1/4" collet-type chuck.

Adjusting ring for precise depth setting: One notch equals one quarter of 1/64"

Molded handles — comfortable grip — complete control.

Cantilever base with wide opening for full view of work.

Quick reading depth indicator.

Molded sub-base for smooth operation will not mar finished surface.

Fig. 14-4.1 Router and its main parts. (Courtesy Stanley Tools.)

General Safety Precautions

The following precautions must be observed whenever you use a router:

- Change bits only with the power off, and the router disconnected from its power source.
- Be sure the material to be routed is firmly clamped.

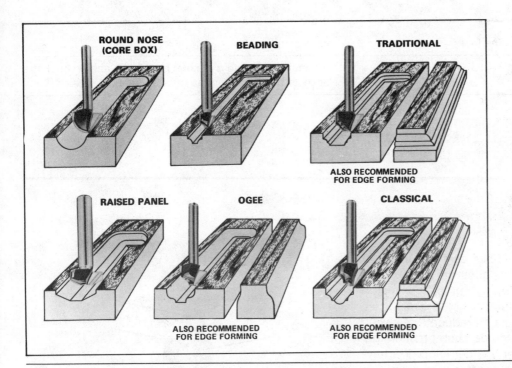

Fig. 14-4.2 Typical router bits and the shapes they cut.

• Be sure the routing guide is firmly clamped in position.

   Operating precautions are included with the following procedures.

Procedure for Routing a Groove or Dado

1 Clamp the material down.
2 Mark the location and width of the cut on the material.
3 Clamp the guide in its proper location.
4 Insert a straight router bit in the chuck, and lock it in place.
5 Adjust the depth of cut by screwing the motor into the base, using the adjustment scale as a guide to proper depth.
6 Plug in the router. **Precaution.** Make sure the power switch is off.
7 Just outside the beginning point of the cut, set the router guide against the edge of the material (Fig. 14-4.3).
8 Turn on the power. **Precaution.** The router will tend to move as the motor starts, so hold it firmly with both hands.

Fig. 14-4.3 (Courtesy Stanley Tools.)

9 Let the motor reach full speed.

10 Move the router through the material in the proper direction. **Precaution.** As quickly as possible become familiar with the sound of the motor when the router is cutting at maximum efficiency. Moving too slowly can damage the bit and burn the wood; moving too rapidly can overload the motor and leave a rough cut.

11 When the cut is completed, turn off the switch. **Precaution.** Lay the router down with the point of the bit away from you, and hold onto it until the bit has stopped turning.

Procedure for Finishing an Edge

1 Select and insert the bit that will provide the desired edge shape.

2 To the top of the work table clamp a scrap piece of material of the same thickness as the one you are edge-finishing. Make sure the scrap has one straight edge.

3 After taking the preparatory steps outlined in the previous procedure (1 to 3 and 5 to 9), make a test cut along the edge of the scrap piece.

4 If necessary, change bits or adjust the depth of cut.

5 Complete the edge cut, and shut off the motor.

Procedure for Cutting a Mortise

1 Follow procedures 1 to 6 inclusive for making a straight cut.

2 Set the router on the material with the base plate at an angle, so that the bit lies just above the area to be mortised.

3 Turn on the power, and let the motor reach full speed.

4 Lower the bit into the material, holding the router firmly.

5 Rout out the mortise.

6 Turn off the power, and set the router aside.

7 With a wood chisel cut the corners of the mortise square. **Note.** The bit will leave inside corners of the mortise slightly rounded, and they must be squared off.

Questions

1 What is the best way to determine if you are routing at the most efficient speed?

2 Name five types of cutting guides.

3 To make a straight cut, in what direction do you move the router?

4 What happens when you cut too quickly or too slowly?

Unit 14-5   Power Plane

A portable electric plane and a power block plane are similar in operation and purpose; the chief difference between them is size. Both types have a cutting blade between front and rear shoes (plates) that is protected by a guard. Both have a fence attachment for edge planing, which can be removed for surface or bevel planing. The front shoe of the larger plane is flat, and a thumb or forefinger here provides directional control and pressure (see Fig. 14-5.1). The smaller plane is controlled with one hand (Fig. 14-5.2). You establish the depth of the cut by turning an adjustment knob at the front or rear of the plane.

To operate either type, turn on the motor with the plane resting on the material. Then move the plane forward steadily and with uniform pressure. As with the jointer, the depth of cut and the speed of movement vary with the type of wood being planed.

The task	The tool	Its description
To smooth or *bevel* an edge	Portable electric plane (Fig. 14-5.1)	A motor-driven plane built for two-handed operation.
To plane an edge or surface	Power block plane (Fig. 14-5.2)	A small motor-driven plane light enough to operate with one hand.

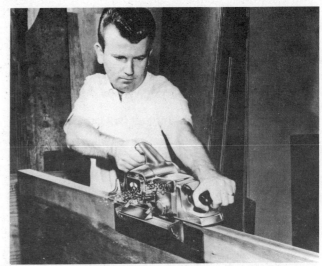

Fig. 14-5.1 Portable electric plane and how to hold it. (Courtesy Rockwell International.)

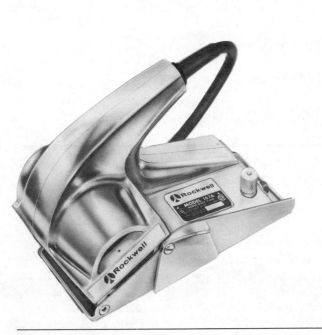

Fig. 14-5.2 Power block plane and how to hold it. (Courtesy Rockwell International.)

General Safety Precautions

The basic safety precautions to follow when using any power tool apply also to power-driven planes. In addition, observe these special precautions:

• Adjust the blade with the plane disconnected.
• Have a firm grip on the tool before beginning work.

• Do not put your hands below the shoes when the power is on.
• After use, turn off the power, and lay the tool on its side with the cutter away from you.

Unit 14-6 **Power Sander**

Sanding machines take much of the work out of smoothing surfaces, but their use requires extreme care to avoid oversanding. The sander does the work; the operator merely guides it. Too much pressure results in uneven sanding or a scratched surface.

There are a number of types of pad sanders, and

the difference between them is primarily the action of the pad. The sandpaper surface may move back and forth in a straight line, in an oval or orbital direction, or in a combination of the two (Fig. 14-6.4). For best results the sander must be flat against the material being smoothed, and moved lightly back and forth. Begin at

Fig. 14-6.1 Pad sander. (Courtesy Milwaukee Electric Tool Corporation.)

Fig. 14-6.2 Portable belt sander. (Courtesy Milwaukee Electric Tool Corporation.)

The task	The tool	Its description
To smooth flat wood surfaces ready for finish coating.	Pad sander (Fig. 14-6.1)	A small sander with a handle and knob for control and a sheet of sandpaper attached to a pad on the bottom. The motor activates the pad.
To smooth flat wood surfaces ready for finish coating.	Portable belt sander (Fig. 14-6.2)	Similar to a pad sander in design, but the sandpaper strip turns on a gear-driven belt.
To level uneven edges of sub-flooring; *to smooth* finish wood flooring.	Belt sander (Fig. 14-6.3)	A floor-type model that operates on the same principle as a belt sander.

Fig. 14-6.3 Floor-type belt sander. (Courtesy Rockwell International.)

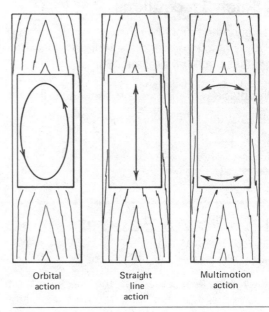

Orbital action | Straight line action | Multimotion action

Fig. 14-6.4 A pad sander may vibrate in one of three patterns.

one side of the material and work toward the other. Turn the motor on with the sander on the material.

The motor of a portable belt sander should be turned on before the sander touches the wood, and the cord should be out of the way. Slowly lower the back of the sander onto the material, being careful not to gouge the surface. Then, as with a pad sander, work back and forth from one side to the other (Fig. 14-6.5).

To smooth finish flooring, you move a floor sander in the same motion as a portable belt sander. To level a high edge where two pieces of material butt unevenly, sand first in the same direction as the seam until the high edge is rounded down. Then complete the leveling process by working across the seam.

General Safety Precautions

The basic safety precautions for use of any power tool also apply to sanders. In addition, observe the following precautions:

- Change sandpaper with the sander disconnected.
- Make sure the sandpaper is firmly clamped in the proper position and with adequate tension.
- Make sure the sandpaper has the proper coarseness or fineness for the job, and that it is in good condition.
- Do not plug in a sander with the power switch on.
- Do not put your hands anywhere on the tool except the handles when the power is on.
- Keep the sander unplugged when it is not in use.

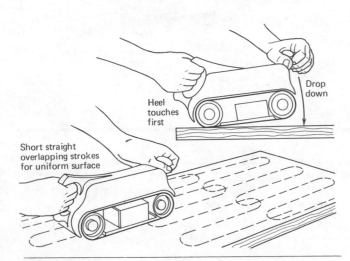

Fig. 14-6.5 The proper sanding on a surface is a series of pattern for smoothing a surface is a continuous series of Ws—up and down as you work across the material.

15

Roofing

Some parts of exterior finishing must be completed before others can begin, while other parts can be finished before, during, or after another stage. Window and door frames, for example, must be in place before siding or shingles can be applied over wall sheathing. Exterior wall materials must be in place before most exterior trim can be put in place. But, because the roof is the umbrella that protects the structure of the house, it is almost always the first area to be finished.

Every roof covering has one basic purpose—to provide a protective, water-shedding coating that keeps rain and snow away from the house's structure, its contents, and its occupants. The roofing material on a sloping roof may also serve two other purposes: to provide color and/or texture that enhance the exterior appearance of the house.

Most residential roofs slope, and the slopes are usually roofed with shingles. Asphalt shingles are the most common, the least expensive, and the easiest to apply. They are manufactured in several shapes and a wide variety of colors and textures. Wood shingles are popular because of their rich texture and natural color. They are most often specified in regions of the country where cedar, redwood, and cypress trees grow. Asbestos-cement shingles are manufactured in various shapes, colors, and textures, although the range is more limited than for asphalt shingles. Unlike other types, asbestos-cement shingles will not burn.

The choice of shingles depends partly on the secondary purposes the roof must serve, and partly on its slope. Asphalt shingles should never be applied to a slope of less than 2 in 12, and manufacturers recommend 3 in 12 as a minimum. Wood shingles may be used on slopes as low as 3 in 12 in dry climates, but are more economical on slopes of 5 in 12 or steeper. Asbestos shingles should not be applied to a slope of less than 5 in 12.

In parts of the country with warm climates and little

rainfall tile roofs are popular. The tiles are heavy, however, and the roof structure must be designed specifically to support the extra weight. Metal is sometimes used as a roofing material, although more frequently on commercial buildings than on houses. Aluminum, copper, and tin-plated steel in sheet form may be used on roofs of any pitch. For flat or nearly flat roofs, the usual material is roofing felt, laid in layers in hot tar or asphalt to form a **built-up roof.** Tile, metal, and built-up roofs require special procedures and equipment, and are rarely applied by carpenters.

For low-cost housing and such utility buildings as sheds, garages, and barns, roll roofing is sometimes used in place of asphalt shingles. **Roll roofing** is an asphalted material, manufactured and surfaced like shingles, that comes in long rolls.

The finished roofing material is established in the home's specifications. It is possible to use a different material, but the change may have a negative effect on the exterior appearance of the house.

Roofing materials are estimated and sold by the square. A **square** of roofing covers 100 sq ft of roof surface. **Coverage** is the term that describes the number of thicknesses of material applied. Shingles are applied in double or triple coverages in new construction. Single coverage is acceptable in remodeling over an existing roof.

Before roofing materials can be applied, the entire roof must be sheathed, and the sheathing covered with roofing felt (Units 9-34 to 9-38 inclusive). The final step in preparation is to provide extra protection at places where leaks are most likely to occur.

Questions

1 What is the minimum slope recommended for a roof finished with the following materials?
   a) Wood shingles
   b) Metal
   c) Asbestos-cement shingles
   d) Asphalt shingles
2 What is a built-up roof?
3 What is a square of roofing?

Unit 15-1 **Flash the Roof**

Whenever a roof turns a corner or butts against a vertical surface, flashing is required. **Flashing** is a construction process to prevent water from seeping under the roofing material at these junctures. Typical places for flashing a roof are at ridges, hips, and valleys, around pipes that protrude through the roof's surface, around chimneys and dormers, and wherever a roof butts against a wall (Fig. 15-1.1). Flashing is also recommended at eaves of low-pitched roofs covered with wood shingles in areas of heavy snow.

Several materials may be used for flashing. Roll roofing is the most widely used, and is generally acceptable at ridges, hips, and valleys. The best flashing for any purpose, however, is corrosion-resistant metal. The metal may be 0.019″-thick aluminum, 26-ga galvanized steel, or 16-oz copper. Valleys of wood shingle

roofs must be flashed with metal. Around pipes a sleeve of plastic works well. Vinyl flashing, a relatively new product, shows promise.

Order of Work

There are many acceptable ways to apply flashing, and they all serve the same purpose: to channel water away from the roof's structure. Some flashing must be in place before roofing is applied, such as at valleys and in some procedures for flashing walls and chimneys. At other locations, such as at hips and ridges, flashing is applied after roofing has been laid to the point to be flashed. Sometimes flashing and roofing in-

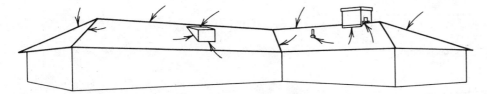

Fig. 15-1.1 Flashing is needed at ridges, hips, and valleys, and where chimneys, vents, and walls come through a roof.

terlock, and are applied at the same time. The flashing procedures in this unit all take place before roofing is applied. Other flashing procedures are included with the application procedures that follow in Units 15-2 to 15-5 inclusive.

Galvanic Action

Flashing is attached with nails. If the flashing is roll roofing, use regular roofing nails. If the flashing is metal, however, you must use nails of the same material as the flashing. Metals react to each other in the presence of water. This reaction, called **galvanic action,** causes one of the metals to corrode.

Metals are ranked in the order in which they react to other metals when they are in direct contact. This ranking is called the **electromotive series.** Here is the ranking of metals most often used in residential building:

1 Aluminum.	5 Lead.
2 Zinc.	6 Brass.
3 Steel.	7 Copper.
4 Tin.	8 Bronze.

Suppose that the flashing specified is aluminum, and it is applied by mistake with steel nails. Aluminum is higher than steel in the electromotive series, and it will therefore corrode when both metals are wet. If copper flashing is applied with steel nails, the steel will corrode. The farther apart the metals are in the series, the sooner and faster corrosion occurs.

The simplest way to avoid galvanic action is to attach metals with fasteners of the same metal. One alternative is to use fasteners with a nonmetallic washer, such as neoprene or asphalt, that keeps the metals from touching each other.

Tools

For applying asphalted flashing, the tools needed are a folding rule, chalk line, hammer, and a roofer's knife

Fig. 15-1.2 A roofer's knife has a curved blade that ends in a sharp point for trimming asphalt shingles and slitting building paper. (Courtesy Hyde Manufacturing Company.)

(Fig. 15-1.2). If the flashing is metal, you use metal shears in place of the roofer's knife.

Procedure for Flashing a Valley with Roll Roofing

1 Snap four chalk lines, one 6″ and one 12″ on each side of the valley.

2 Cut a piece of roll roofing into two strips as long as the valley, one strip 12″ wide and one 24″ wide. **Note.** The roofing should be the same color as the asphalt shingle to be applied.

3 Lay the 12″ strip face down in the valley between the chalk lines, and bend it to conform to the shape of the roof (Fig. 15-1.3). **Note.** If flashing must be spliced, allow a 12″ overlap at the seams.

4 Nail at the corners of the upper edges of the strip (or of each piece) to hold the flashing in place.

5 Lay the 24″ strip face up between the outer chalk lines.

6 Bend this strip to the shape of the roof, and nail only often enough to hold the flashing in place (Fig. 15-1.3).

7 With a folding rule measure 3″ on each side of the valley at its upper end, and mark the location with chalk. **Note.** With asbestos shingles measure 5″ instead of 3″.

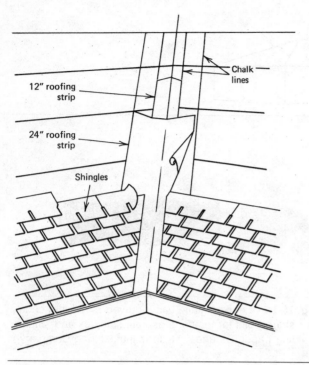

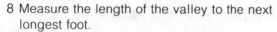

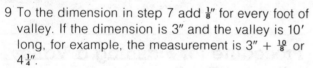

Fig. 15-1.3

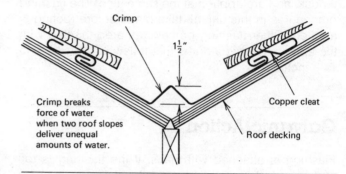

Fig. 15-1.4

8 Measure the length of the valley to the next longest foot.

9 To the dimension in step 7 add $\frac{1}{8}''$ for every foot of valley. If the dimension is 3″ and the valley is 10′ long, for example, the measurement is $3'' + \frac{10}{8}$ or $4\frac{1}{4}''$.

10 With a folding rule measure off the distance in step 9 on each side of the valley at its lower end, and again mark with chalk.

11 Now snap chalk lines between the pairs of marks. These lines mark the trim line for roofing shingles.

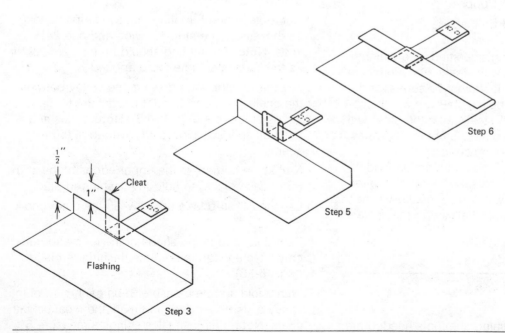

Fig. 15-1.5

Procedure for Flashing a Valley with Metal

1 Cut a strip of metal flashing at least 20″ wide and as long as the valley. **Note.** If more than one length is required, allow for at least a 3″ overlap at seams.

2 Crimp the metal to conform to the shape of the valley by marking the center with a chalk line, then bending the metal along the line over the edge of a 2 × 4. To complete the bend to the required angle, use a wood block and hammer. **Note.** If the roofs will shed unequal amounts of water into the valley, or if heavy rains are common in the community, crimp as shown in Fig. 15-1.4. The crimp breaks the force of the water, and prevents it from washing up the opposite slope.

3 Bend the outer edges of flashing 90° as shown at left in Fig. 15-1.5.

4 Fit cleats of the same metal against the flashing. Space the cleats about 24″ apart.

5 Bend each cleat 180° over the edge of the flashing (Fig. 15-1.5, center).

6 Bend both flashing and cleats another 90° to form a flat-lock seam (right).

7 Attach the cleats with nails of the same metal, and bend the end of each cleat over the nails.

8 Mark the trim line for shingles by following steps 7 through 11 for flashing a valley with roll roofing.

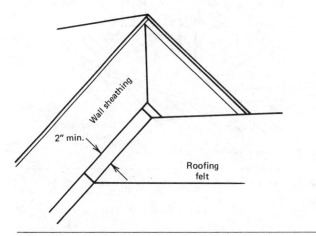

Fig. 15-1.6

Procedure for One-Piece Flashing at a Chimney or Brick Wall

1 Where the chimney and roof intersect, nail a small cant strip (Fig. 15-1.7).

2 Continue roofing felt from the roof over the cant strip and 10″ up the wall.

Procedure for One-Piece Flashing at a Wood Wall

1 Check to make sure that roofing felt on the main roof continues up the wall about 3″ (Fig. 15-1.6). If this does not happen, add another layer over the roof and up the wall as in flashing a valley with roll roofing.

2 Cut a strip of metal flashing as long as the intersection of roof and wall, and at least 9″ wide. A standard width is 14″.

3 Bend the strip to fit the intersection.

4 Attach the strip with cleats by following steps 3 to 7 inclusive for flashing a valley with metal.

5 Snap a chalk line ½″ from the wall along the flashing. This line marks the trim for roofing shingles.

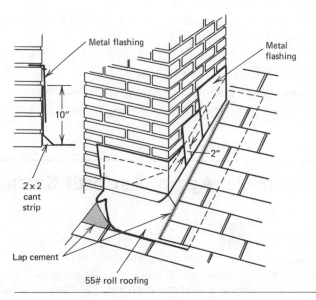

Fig. 15-1.7

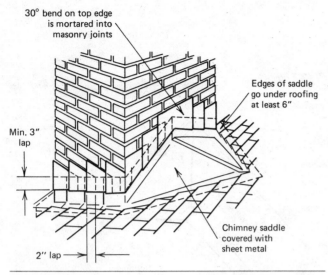

Fig. 15-1.8

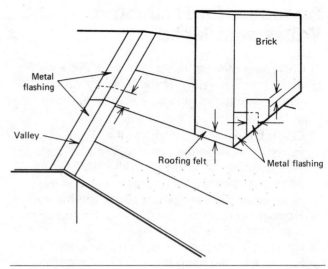

Fig. 15-1.9

3 Cut pieces of flashing to overlap roofing felt a minimum of 3", fit $1\frac{1}{2}$" into a mortar joint, and have a side lap over the lower piece of flashing of at least 2" (Fig. 15-1.8).

4 If necessary, clean out the mortar joints to a depth of $1\frac{1}{2}$" with a cold chisel.

5 Install the flashing, and refill the joint with fresh mortar or roofing cement. **Note.** In top quality construction metal flashing is mortared into the chimney or wall as it is built by the mason.

Questions

1 What is galvanic action?

2 If tin and copper are in contact with each other during frequent rains, which will corrode?

3 Why do you think the trim line for roof shingles is wider at the lower end of a valley than at the upper end?

4 Describe the steps for forming a flat-lock seam.

5 In Fig. 15-1.9, fill in the minimum dimensions for overlaps.

Unit 15-2 **Apply Asphalt Shingles**

Of all roofing materials, asphalt shingles are the most widely used. They are manufactured with a base of heavy felt or fiberglass that has been dipped in asphalt and then sprinkled with stone chips of various colors. Color plays an important part in the appearance of the house, and affects the temperature in the attic space beneath it. Light colors reflect heat, keep an attic cooler, and make a house look bigger and higher than it actually is. Dark colors absorb heat, and make a house look lower and smaller.

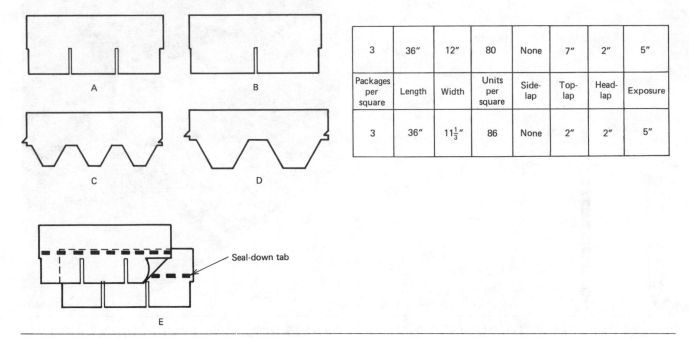

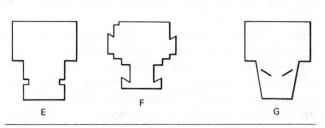

Packages per square	Length	Width	Units per square	Side-lap	Top-lap	Head-lap	Exposure
3	36″	12″	80	None	7″	2″	5″
3	36″	11⅓″	86	None	2″	2″	5″

Fig. 15-2.1 Square-butt strip shingles (A and B) are 36″ long, 12″ wide, and have a 5″ exposure. Hexagonal shingles (C and D) are 36″ by 11⅜″, and also have a 5″ exposure. Type E has a special adhesive strip that seals down the tabs to prevent damage by high winds.

The lightest-weight asphalt shingles weigh 215 pounds per square, but they are manufactured in weights up to 330 pounds. This weight is necessary if the shingles are to withstand the effects of the sun, which is the enemy of asphalt. The sun's heat can cause roofing to blister, buckle, or fade. If properly applied, however, a roof of light-colored shingles will last from 15 to 20 years.

Most asphalt shingles used today are strip shingles. Typically, **strip shingles** (Fig. 15-2.1) have two or three *tabs*—flaps separated by a notch. If the corners of the tabs are cut square, the roofing is called **square-butt shingles** (A and B). If the tabs are cut on a diagonal, the roofing is called **hexagonal shingles** (C and D). Strip shingles are also made with a serrated bottom edge and no notches. One type (E in Fig. 15-2.1) has a special adhesive strip that softens under the heat of the sun and seals down the tabs to prevent damage by high winds.

Three types of individual shingles are also manufactured (Fig. 15-2.2). Types A and B are designed for use on hips and ridges, although most roofers cut strip shingles to size for this purpose. Types C and D are so-called giant shingles, but they are seldom used in residential work today. Types E, F, and G have interlocking tabs to resist strong winds. The design of the shingles and the method of interlocking vary with the manufacturer.

Standard shingles should not be laid on a slope of less than 3 in 12. Interlocking shingles and strip shingles with seal-down tabs may be used on slopes as low as 2 in 12.

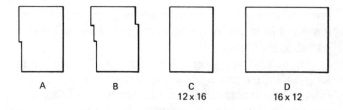

Fig. 15-2.2 Individual shingles for use on ridges and hips (A and B are typical) are 9″ long and 12″ wide and have a 5″ exposure. Giant shingles (C and D) are 16″ by 12″. Interlocking shingles come in many shapes; E, F, and G are typical of the variety.

Fasteners

Asphalt shingles may be attached with nails or staples. The standard roofing nail (Fig. 15-2.3) has a large head, at least ⅜″ in diameter, that is sometimes checkered for easier driving. Its shank, made of 10-gauge galvanized steel or 12-gauge aluminum, may be

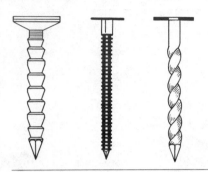

Fig. 15-2.3 Typical nails for applying asphalt shingles.

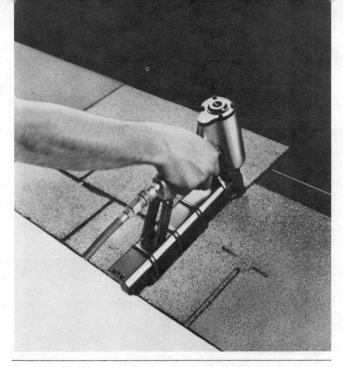

Fig. 15-2.4 Some stapling guns have a guide that acts as a spacer to assure uniform shingle exposure. One end of the guide fits against a shingle course already in place, and the other marks the location of the next course. (Courtesy Senco Products Inc.)

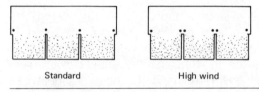

Standard High wind

Fig. 15-2.5 Standard (left) and high-wind (right) placement of fasteners.

barbed or threaded for maximum holding power. Roofing nails must be long enough to penetrate at least $\frac{3}{4}''$ into roof sheathing.

Staples for applying roofing have a 1″ crown, legs long enough to penetrate $\frac{3}{4}''$ into sheathing, and chisel points. Pneumatic guns and special hammers for staple application can be adjusted for staples with leg lengths up to $1\frac{1}{2}''$. Some types of guns have a guide (Fig. 15-2.4) that helps the operator apply shingles with uniform exposure.

Under normal weather conditions standard nailing practice calls for a fastener at the end of each notch between tabs (Fig. 15-2.5). Therefore, a three-tab shingle receives four nails. In areas with high winds, standard practice is a pair of nails at the end of each notch. The extra nails prevent the wind from tearing the shingle at the notches.

Staples may be used if the roof slope is 4 in 12 or greater. At least six staples per strip are required. At the eaves, rakes, hips, ridges, and where shingles touch flashing, nails should be used in place of staples because of their greater holding power.

Shingle Layout

Most asphalt shingle roofs are laid with double coverage. To provide two layers at the eave, a starter course is required. The **starter course** may be a row of standard shingles laid inverted—that is, with the tabs toward the roof ridge. Or it may be a strip of roll roofing at least 9″ wide, surfaced to match the color of the shingles.

Shingles are usually laid so that the notches between the tabs of one course fall over the center of the tabs in the course below (Fig. 15-2.6). Therefore every other course begins at the rake with a full tab, and alternate courses begin with a half tab. No strip shingle should be less than a full tab wide, however.

The most attractive roof has identical shingle patterns at both rakes. To achieve this result with hexagonal shingles, you must start from the center of the roof and work toward the rakes. Hexagonal shingles have a fixed width that can't be altered. With square-butt shingles, however, it is possible to adjust the spacing slightly as you work across the roof. For this reason many roofers start work at one corner. A similar vertical adjustment may be made so that exposure at the ridge is the same as exposure at the eaves. It is important to know what these adjustments are before you begin work.

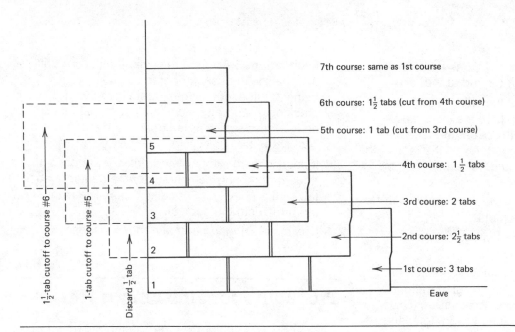

Fig. 15-2.6 In the most common shingle pattern the notches in one course fall at the midpoint of tabs in the course below.

Courses of shingles on the main roof are trimmed to fit at hips and ridges, at vertical walls and chimneys, and at open valleys. With **closed valleys** (Fig. 15-2.7) strip shingles continue through the valley and up the other slope, and are overlapped by similar courses on the adjoining slope. Hexagonal shingles may not be used in closed valleys.

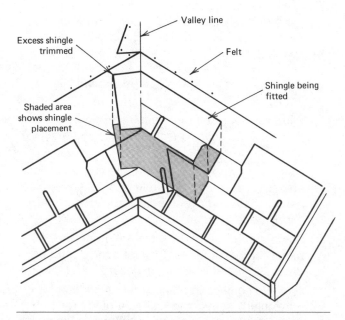

Fig. 15-2.7 Valleys are usually slashed as shown in Fig. 15-1.3. In a closed valley, however, shingles are laid across the valley line and overlapped on both sides of it.

Preparing to Roof

The ideal weather for shingling is a cool, cloudy day with little wind. In cold weather shingles are brittle and break easily. In hot weather they soften and tear easily when stepped on.

Most roofers bring as many bundles of shingles onto the roof as they expect to lay in one day, and scatter them where they are within easy reach. If the roof is a solid color, the shingles in one bundle will probably match those in another bundle. To guard against minor color variations, however, experienced roofers work from several bundles at once to avoid a blotchy appearance.

Tools Required

To roof with asphalt shingles you are likely to need a chalk line, steel tape, folding rule, try square, hammer, roofer's knife, and a putty knife.

Procedure for Shingling with Square-Butt Shingles

1 a. Roll out the starter strip, and position it so that the end is flush with sheathing at the rake, and the side is flush at the eave.

1 b. If inverted shingles are used as a starter strip, cut 3″ off the first strip to be laid at the rake.

2 Fasten the lower edge of the starter strip with nails placed about 3″ up from the eave, and spaced about 6″ apart. **Precaution.** Place the nails so that they will be hidden by the tabs of the first course of shingles, and not exposed in the notches.

3 Fasten the upper edge of the starter strip with nails spaced about 12″ apart.

4 Trim the rake edge of the first exposed shingle flush with the edge of the tab.

5 Nail the first shingle in position at the corner of the roof so that it overhangs the starter strip by $\frac{1}{2}$″ to $\frac{3}{4}$″ at both rake and eave. **Precaution.** Drive nails straight to avoid cutting the shingle, and leave the undersides of nail heads flush with the shingle's surface. Otherwise the tabs may curl.

6 Lightly tack the remaining shingles in the first course to determine if adjustments are required to come out even at the other rake.

7 After the proper spacing has been established, complete nailing the first course. Nail first beside the previously nailed shingle, and work toward the other end of each strip. **Note.** If the first course ends at a hip or valley, stop the course temporarily at least 24″ away from the turn. (See procedure for shingling at hips and valleys later in this unit.)

8 With a roofer's knife trim a full strip into a 2$\frac{1}{2}$-tab strip. Use a squaring tool to assure a straight cut at 90° to the edges.

9 Nail this down as the first shingle in the second course. Use your folding rule or staple gun to assure that you have the desired exposure. A 5″ exposure is standard.

10 Complete the second course, following the general procedures for the first course, and being careful to maintain equal exposure.

11 With two courses in place, snap chalk lines horizontally across the roof as a guide for even exposure. **Note.** Some manufacturers of roofing paper print guide lines on their product to help roofers. The number of chalk lines you snap depends on your experience. The distance between guidelines should be a multiple of 5″ except when adjustment is necessary.

12 Also snap chalk lines vertically, centered on notches, at about the quarterpoints of the roof. To check accuracy of the roofing job, sight along the butts of shingles and up the notches.

13 If you work from walking steps, lay about a dozen courses before moving the steps up the roof.

Precaution. Be sure to remove all nails, and fill the holes with asphalt-fibered or plastic roofing cement. Lift each shingle carefully to avoid cracking it, and apply cement underneath the holes. Press the shingle firmly into the cement to seal the hole.

14 If necessary to prevent wind damage, apply a 1″ square of roofing cement under the center of each tab with a putty knife or calking gun. Press the tab firmly into the cement.

15 Complete shingling on main roof surfaces.

Procedure for Finishing a Hip

1 Trim courses of strip shingles in the main roof to follow the angle of the hip.

2 Cut a strip of roll roofing 10″ wide for flashing, and crease it down the middle lengthwise.

3 Attach the strip, driving nails 4″ on each side of the hip and about every 36″ the length of the flashing. If necessary to splice, provide an end lap of 4″.

4 If you use strip shingles on the hip instead of individual shingles, cut them as shown in Fig. 15-2.8.

5 Bend a shingle lengthwise down its center. Measure with a folding rule to make sure the two leaves are equal.

6 Using the bent shingle as a guide, snap a chalk line down one side of the hip to assure uniform alignment.

7 Position the first hip shingle at the eave, and trim the edge as necessary to fit the corner.

8 Nail 5$\frac{1}{2}$″ from the lower end, and 1″ up from each edge (Fig. 15-2.9).

9 Nail all succeeding hip shingles in the same manner, maintaining a 5″ exposure.

10 a. At the top of a hip that meets another hip, overlap the shingles, and seal the top shingle with roofing cement at all edges.

10 b. At the top of a hip that meets a ridge, see the next procedure.

Procedure for Finishing a Ridge

1 Lay the top course of strip shingles in the main roof so that its upper edge lies at or just slightly below the ridge.

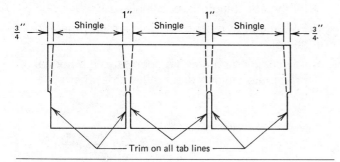

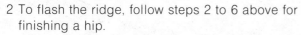

Fig. 15-2.8

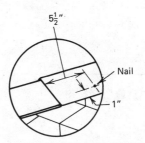

Fig. 15-2.9

2 To flash the ridge, follow steps 2 to 6 above for finishing a hip.

3 Determine the direction of the prevailing bad-weather wind, and set the first shingle in place at the end of the ridge or top of a hip *away* from the wind (Fig. 15-2.10). **Note.** The danger of leaks at a ridge is from blowing rain, and all ridge shingles should have their backs to the wind that brings rain.

4 Nail the first ridge shingle $5\frac{1}{2}$" back from the end of the ridge or hip (depending on wind direction), and 1" up from each edge.

5 Nail all succeeding shingles in the same manner. Maintain a uniform 5" exposure.

6 Cover the two nails in the last shingle with roofing cement, and seal all edges.

Procedure for Finishing an Open Valley

1 Set in position the strip shingle in the first course that fits at the side of the valley.

2 Using the chalk line on flashing as a guide, mark the cutting line on the shingle.

3 With a roofer's knife trim the shingle to fit.

4 Clip the top corner of the shingle (Fig. 15-2.11). **Note.** This step prevents the square corners of shingles from trapping water that runs down the valley.

5 Apply a ribbon of roofing cement on the back of the shingle along the cut edges.

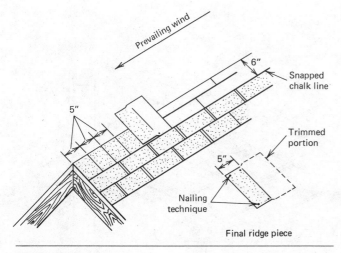

Fig. 15-2.10

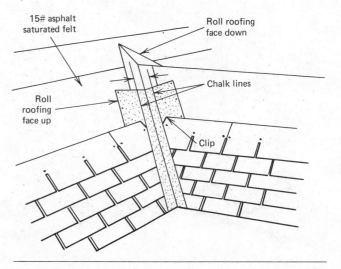

Fig. 15-2.11

6 Nail the shingle in position, following standard nailing procedures, and press down firmly to seal the cut edges.

7 Repeat this procedure for all remaining shingle courses.

Procedure for Finishing a Closed Valley

1 Continue the first course of strip shingles at least 12″ beyond the center of the valley (A in Fig. 15-2.12), bending the shingle to fit the shape of the valley.

2 Nail at the end of each notch, and add a nail at the upper corner of the strip. **Precaution.** If a notch ends within 6″ of the center of the valley, omit the nail to prevent leakage.

3 Overlap the first course of strip shingles from the adjoining roof (B in Fig. 15-2.12).

4 Repeat steps 1 to 3 inclusive to the top of the valley.

5 Trim the last shingles flush at the ridge.

Procedure for Shingling Around a Pipe

1 Lay shingles to the face of the pipe.

2 On a loose shingle mark the oval shape of the hole for the pipe at the proper location (Fig. 15-2.13). Use a roofing nail as a pencil.

3 With a roofer's knife carefully cut out the hole.

4 Fit the cut shingle in position and nail it.

5 Fit a metal or plastic sleeve over the pipe and shingle.

6 Lay succeeding shingle courses in the normal procedure, cutting around the pipe where necessary.

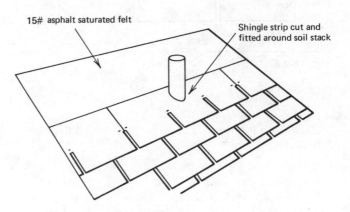

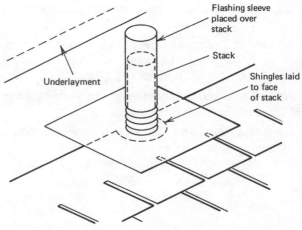

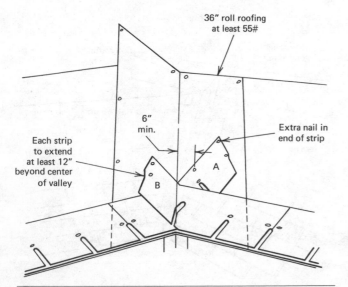

Fig. 15-2.12

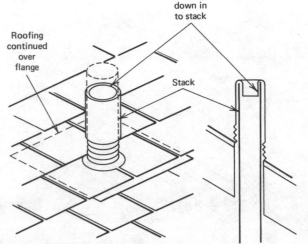

Fig. 15-2.13

Questions

1 Under normal wind conditions, how many nails are required in a three-tab strip shingle?

2 What is the minimum size of a strip shingle at a rake? At a hip or ridge?

3 What extra step is required to fit shingles in an open valley?

4 How many nails are needed in a ridge shingle, and where are they located?

5 Where do you apply roofing cement to seal holes in shingles left by walking steps?

6 Why is it necessary to cut 3″ off a starter strip made from inverted shingles?

Unit 15-3   **Apply Wood Shingles**

Wood shingles are cut from logs of cedar, redwood, or cypress, with Western red cedar by far the most common throughout the country. The logs are cut into sections of the desired length, then trimmed into blocks called **bolts.** Shingles are then sawed from the bolts.

Like lumber and plywood, wood shingles are graded, and the number of grades varies with the wood. Shingles of all species carrying a No. 1 grade stamp are cut from heartwood and are **edge-grained**—that is, the grain runs the long dimension of the shingle. Lower grades have some sapwood and areas of flatgrain—grain running at an angle across the shingle.

Shakes (Fig. 15-3.1) are wood shingles that have been split, rather than sawed, from bolts. There are three types—handsplit-and-resawn, tapersplit, and straightsplit. **Straightsplit** shakes are a uniform $\frac{3}{8}″$ thick. **Tapersplit** shakes, like standard wood shingles, are thicker at the exposed butt end than the concealed end. **Handsplit-and-resawn** shakes are split at maximum thickness, then tapered to final shape on a bandsaw.

15-3.1 A shake roof adds tremendous texture to a house. Because of their size, shakes are in proper scale only on large houses. (Courtesy Red Cedar Shingle and Handsplit Shake Bureau.)

Table XIII   Schedule of shingle types

		Exposure (at listed slopes)		
Grade[a]	Size	5 in 12 and up	4 in 12	3 in 12
1, 2 and 3	24″	$7\frac{1}{2}″$	$6\frac{3}{4}″$	$5\frac{3}{4}″$
1, 2 and 3	18″	$5\frac{1}{2}″$	5″	$4\frac{1}{4}″$
1, 2 and 3	16″	5″	$4\frac{1}{2}″$	$3\frac{3}{4}″$

[a] Grade description

No 1   Premium grade: 100% heartwood, 100% clear, and 100% edge grain.

No 2   Intermediate grade: not less than 10″ clear on 16″ shingles, 11″ clear on 16″ shingles, and 16″ clear on 24″ shingles. Flat grain and limited sap wood permitted.

No 3   Utility grade: 6″ clear on 16″ and 18″ shingles, 10″ clear on 24″ shingles (for economy applications and secondary buildings).

[b] The correct size of wood shingle and its exposure are determined by the slope of the roof. Coverage is triple on slopes, quadruple on the lowest possible slope.

Table XIV   Schedule of shake types

Type	Size (length & thickness)	Exposure[a] (4 in 12 slope)
Handsplit and resawn	$18″ \times \frac{1}{2}″$ to $\frac{3}{4}″$	$7\frac{1}{2}″$
	$18″ \times \frac{3}{4}″$ to $1\frac{1}{4}″$	$7\frac{1}{2}″$
	$24″ \times \frac{3}{4}″$ to $1\frac{1}{4}″$	10″
	$24″ \times \frac{1}{2}″$ to $\frac{3}{4}″$	10″
	$32″ \times \frac{3}{4}″$ to $1\frac{1}{4}″$	13″
Tapersplit	$24″ \times \frac{1}{2}″$ to $\frac{5}{8}″$	10″
Straight-split (barn)	$18″ \times \frac{3}{8}″$	$7\frac{1}{2}″$
	$24″ \times \frac{3}{8}″$	10″

[a] Shakes not recommended for roof slopes less than 4 in 12 without special construction.

[b] The three types of shakes are made in several sizes and thicknesses. The exposures shown are for a typical 4 in 12 slope.

Wood shingles come in 16″, 18″, and 24″ lengths, while shakes, being larger in scale, come in 18″, 24″, and 32″ lengths. The slope of the roof and the amount of exposure specified determine the length of shingle or shake required (Table XIII). The exposures shown provide triple coverage on roofs with a 5 in 12 or greater slope, and quadruple coverage on a 3 in 12 slope. Wood shingles should not be used on lower-pitched roofs.

Because shakes are thicker, they may be applied with double coverage on slopes of 4 in 12 or greater (Table XIV). On slopes as low as 3 in 12 shakes must be applied with reduced exposure that assures triple coverage. They must be laid over 30-lb felt applied to solid sheathing, with each course separated by interlayment of 30-lb felt.

Wood shingles are sold by the bundle, and a bundle laid with the exposure shown in Table XIII will cover 25 sq ft of roof surface.

Fasteners

Although widths of shingles in a bundle may vary from 3″ to as great as 14″, each shingle is fastened with only two nails, located as shown in Fig. 15-3.2. The penny size of the nail to be used increases with the thickness

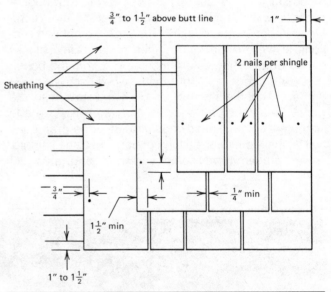

Size	Length	Shingles	Straight split	Taper split	Hand split
3d*	1¼″	16″ & 18″			
4d*	1½″	24″			
5d*	1¾″	16″ & 18″	18″ & 24″	24″	½″–¾″ butts
6d*	2″	24″			¾″–1¼″ butts

* Use nails 2-pennies large at hips and ridges.

Fig. 15-3.2 Recommended nail sizes for applying wood shingles and shakes, and the correct nailing pattern. Two nails per shingle, plus nails from courses above, provide ample attachment.

of the shingle or shake (Fig. 15-3.2). To attach shingles or shakes at hips and ridges, use nails at least 1″ longer than for main courses. Common nails are used on many roofing jobs, but threaded nails hold better in communities where the weather subjects the shingles to frequent wettings and dryings.

Shingle Layout

Wood shingles, unlike asphalt shingles, do not shed all water, but absorb some of it. It is therefore important to keep wood shingles and shakes from absorbing excessive water. For this reason it is standard practice to lay a double course of shingles at the eaves. To prevent dripping at rakes, it is good practice to lay bevel siding under the shingles at the rake to divert water away from the edge toward the middle of the roof.

The shingles in valleys must be raised at least $\frac{1}{2}$″ above the valley's surface, and are therefore supported on cleats (see Unit 15-1, Procedures for Flashing a Valley with Metal) or on small wood cant strips attached to sheathing above the cleat line (Fig. 15-3.3). At hips the common method of shingling is the Boston hip described in this unit. Either of the two methods shown in Fig. 15-3.4 may also be used. Ridges are finished like hips.

Shingles are laid with about $\frac{1}{4}$″ between them. This space allows for expansion, and permits quicker drying. The joints between shingles should be at least 1$\frac{1}{2}$″ away from joints in the course below. Because the shingles are three courses thick, joints should be staggered in three consecutive courses if possible.

Wood shingles can be laid to achieve a wide variety of patterns. In the simplest roof the butts of shingles form straight lines, as in an asphalt roof. For greater

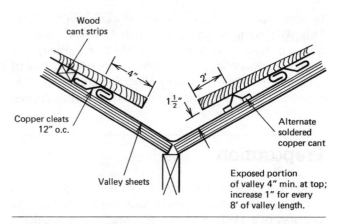

Fig. 15-3.3 If necessary to raise the butts of shingles $\frac{1}{2}$″ above valleys, wood cant strips may be added above cleats that hold flashing in place.

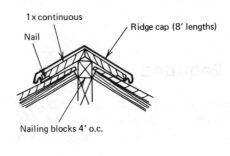

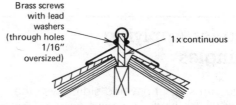

Fig. 15-3.4 These two methods of finishing a hip or ridge with wood shingles are alternates to the Boston hip.

Fig. 15-3.5 Interesting roof textures can be created by staggering shingle butts in courses (thatched) by adding extra shingles (Dutch weaved), or by doubling courses occasionally (striated). (Courtesy Red Cedar Shingle and Handsplit Shake Bureau.)

texture, however, butts can be laid in a staggered pattern, with some shingles on a butt line, and others $\frac{1}{2}''$ above or below that line (Fig. 15-3.5). For a strong horizontal shadow effect, every fifth or sixth course may be doubled. To achieve a wavy effect extra shingles may be added at random; the result is called **Dutch Weave**.

Preparation

Wood shingles and shakes may be applied as they come from the bundles, and allowed to weather to a silver gray or a medium brown. But weathering is seldom uniform, and shingles look better if they are stained before application. Each shingle is dipped individually in a container of stain to within 3″ of the thin end, then stood on the butt end in a trough until it is thoroughly dry. The trough funnels excess stain back into the container.

Tools Required

To roof with either wood shingles or shakes you will need a chalk line, steel tape, folding rule, roofer's knife, saw, and a shingling hatchet.

Procedure for Applying Wood Shingles

1 At rakes attach a strip of beveled siding with the thicker edge flush with the exposed edge of sheathing (Fig. 15-3.6).

2 a. On a straight roof lay a shingle at each rake so that it projects 1″ at the rake and 1″ to $1\frac{1}{2}''$ at the eave. Nail as shown in Fig. 15-3.4. **Precaution.** Drive nails so that the undersides of heads just touch the shingles. *Do not countersink.*

2 b. If the roof ends in a valley, saw a shingle to fit at the trim line.

3 Stretch a cord or chalk line between the two shingles as a guide for lining up the butts of other shingles in the doubled first course.

4 Complete the undercourse between the first two shingles. Be sure to gap $\frac{1}{4}''$ between all shingles.

5 Lay the exposed first course directly over the undercourse, but with the gaps staggered.

6 Lay additional courses in the same general manner. To maintain an even exposure, snap chalk lines or use a board or shingling hatchet as a gauge.

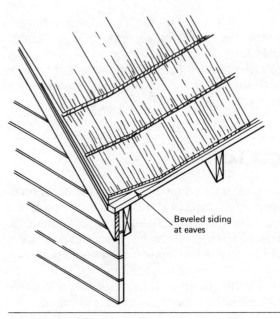

Beveled siding at eaves

Fig. 15-3.6

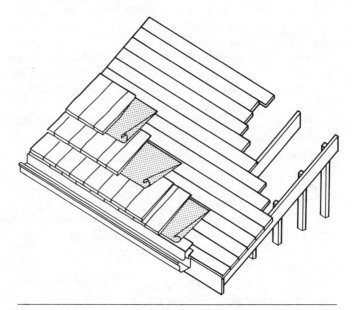

Fig. 15-3.7

7 With the hatchet split the last shingle in each course to the proper width.

8 Complete all straight courses on main roof surfaces up to the ridge. Continue to hips, but stop about 24″ short of valleys.

Procedure for Applying Wood Shakes

1 Follow steps 1 to 3 inclusive for applying wood shingles.

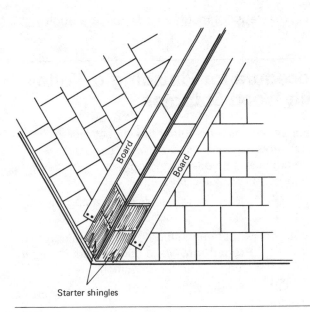

Starter shingles

Fig. 15-3.8

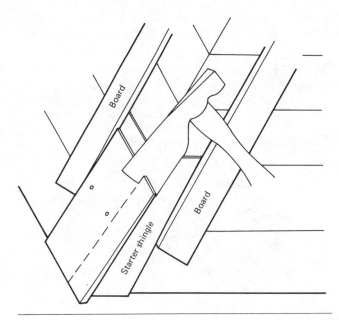

Fig. 15-3.9

2 Complete the undercourse with 15″, 18″ or 24″ shakes made specifically for this purpose. Gap $\frac{1}{4}$″ to $\frac{3}{8}$″ between shakes.

3 Lay the exposed first course directly over the undercourse, but with the joints staggered.

4 Over the top edge of each course lay a continuous strip of 30-lb roofing felt 18″ wide (Fig. 15-3.7). **Note.** The distance from the butt edge of the shake course to the lower edge of the felt should be twice the exposure.

5 Follow steps 6 to 8 inclusive for wood shingles to complete all straight courses on main roof surfaces.

Procedure for Finishing a Boston Hip with Wood Shingles

1 Trim the last shingle in each course of the main roof to gap $\frac{1}{4}$″ at the hip. Nail as before, but add one nail about an inch from the lower corner of the hip.

2 Lay a starter undercourse of two shingles 6″ wide, trimmed for a tight butt fit.

3 Using the starter shingles as a spacer, tack a straight board up each side of the hip as guide strips (Fig. 15-3.8).

4 Nail a shingle over the starter shingle, with one edge against the guide strip, and the other overlapping the hip. **Note.** Select hip shingles at least 6″ wide.

5 With the shingling hatchet trim off the overlap on

a bevel, so that the cut edge lies in the same plane as the top of the starter shingle (Fig. 15-3.9).

6 Nail the other hip shingle in place as in step 4.

7 Again trim the overlap, but now on the opposite bevel.

8 Apply the two hip shingles in the second course in the same manner, but reverse the order so that the bevels lie on opposite sides of the hip in alternate courses.

Procedure for Finishing a Boston Hip with Shakes

1 Select shakes with the least texture to complete the top courses of the main roof at a hip.

2 Trim the last shingle in each course to gap $\frac{1}{4}$″ at the hip. Nail as before, but add one nail about an inch above the lower corner at the hip.

3 Over the entire length of the hip lay a flashing strip of 30-lb felt 12″ wide, centered over the hip. Nail only often enough to keep the felt from blowing loose.

4 Follow steps 2 to 8 inclusive for finishing a hip with wood shingles to complete the procedure.

Procedure for Finishing a Ridge with Shingles

1 Lay the top course of shingles in the main roof so that they gap about $\frac{1}{4}$″ at the ridge.

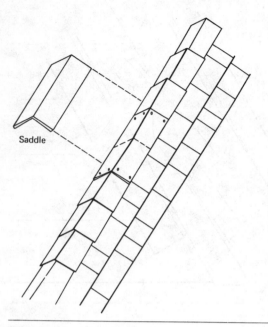

Fig. 15-3.10

2 Over the entire length of the ridge lay a flashing strip of 30-lb felt 12″ wide, centered over the ridge. Nail only often enough to keep the felt from blowing loose.

3 Lay a pair of undercourse shingles at each rake, using shingles 6″ wide and beveled for a close butt fit. **Note.** If the ridge ends in a hip, the top hip shingles serve as starter shingles.

4 Using the starter shingles as spacers, tack a pair of straight boards as guide strips on each side of the ridge.

5 Lay shingles as described in steps 4 to 8 for shingling a hip. Work from both ends toward the center of the ridge.

6 Where the center shingles meet, cover the joint with a saddle, face-nailed as shown in Fig. 15-3.10.

Procedure for Finishing a Ridge with Shakes

1 Select shakes with the least texture to complete the top course in the main roof.

2 Follow steps 1 through 6 for finishing a ridge with wood shingles.

Procedure for Finishing a Valley with Wood Shingles

1 For valley shingles select the widest pieces available, so that joints between exposed shingles at the valley drain onto shingles below, and not directly into the valley.

2 Lay each valley shingle in position along the cut line on the flashing, and mark the trim line.

3 With a saw, cut each shingle carefully along the trim line. **Precaution.** Make sure the cut is plumb when the shingle is in final position.

4 Nail each shingle with two nails as previously described.

5 Make sure that the cut edges are at least $\frac{1}{2}$″ above the valley after nailing, and that the valley is at least 6″ wide between shingles.

Procedure for Finishing a Valley with Shakes

1 Follow steps 1 to 4 inclusive of the procedure for applying wood shingles in a valley.

2 Cover each course with 30-lb felt, as described in step 4 of the procedure for applying wood shakes to the main roof.

3 Make sure that all cut edges are at least $\frac{1}{2}$″ above the valley after nailing.

Procedure for Shingling Around a Pipe

1 With either wood shingles or shakes, follow the same procedures at pipes as described for asphalt shingles.

Procedure for Shingling at a Wall

1 With either wood shingles or shakes, follow the same general procedures as at a valley.

Questions

1 How many bundles of wood shingles are needed to cover a square of roof surface?

2 What is the maximum exposure on a 4 in 12 roof slope with wood shingles? With wood shakes?

3 In what direction should the exposed edge of ridge shingles be laid?

4 What is the chief difference between laying wood shingles and laying shakes on a main roof?

5 What is the main feature of the Boston hip?

6 How far above a valley should the trimmed edges of shingles be?

7 What is the purpose of beveled siding at the rakes of wood shingle roofs?

8 How does the undercourse of wood shingles at an eave differ from an undercourse of shakes?

Unit 15-4   Apply Asbestos-Cement Shingles

Asbestos-cement shingles, often called simply as-bestos shingles, are manufactured from a mixture of about 30% asbestos fibers and 70% portland cement. They are formed under pressure in molds, and their surfaces are textured to give the appearance of natural roofing materials, such as wood and slate. Mineral coloring is added to the mix at the factory; the shingles may be painted, however.

Unlike asphalt and wood shingles, asbestos shin-gles won't rot, soften, or burn. But they are brittle, and can be easily broken by a blow from a poorly aimed hammer. They will also break under the weight of a roofer. Therefore most roofers hang a pair of ladders over the ridge of the roof and work from there. The lad-ders distribute the roofer's weight more evenly.

Valleys of asbestos shingle roofs are finished in the same way as asphalt shingle roofs, except that the width of the gap between shingles on adjoining roofs is 2″ greater. The procedure for finishing a hip is unique, and is described in this unit. Ridges may be finished in the same way as hips, covered with a metal ridge or topped with a ridge roll.

Fasteners

Asbestos shingles come in a variety of shapes and sizes (Fig. 15-4.1) and all are predrilled for nail appli-cation. Manufacturers generally recommend $1\frac{1}{4}″$ gal-vanized roofing nails.

Layout

Shingles are designed so that the nails are concealed under the next course up the roof. But because as-

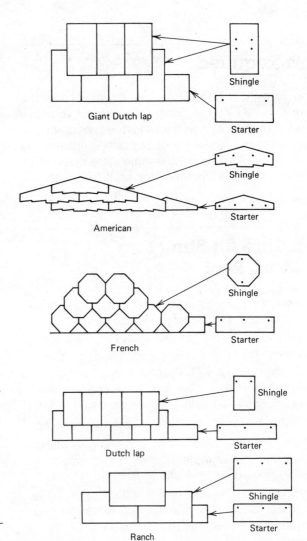

Fig. 15-4.1 Typical shapes of asbestos-cement shingles, the matching starter shingles, and the roof patterns they make.

bestos shingles are predrilled, there is no way to adjust the exposure, which is 5".

Preparation

The steps in applying roofing felt and a metal drip edge to roofs (Unit 9-38) must be followed in a different order if asbestos shingles are the finish roofing material. The 4"-wide metal drip edge is applied first, and nailed directly to sheathing at eaves and rakes. The roofing felt is then applied over the drip edge and exposed sheathing. The first course of felt is nailed at the eaves only often enough to prevent it from blowing loose in the wind, because the bottom edge must be raised during the shingling process.

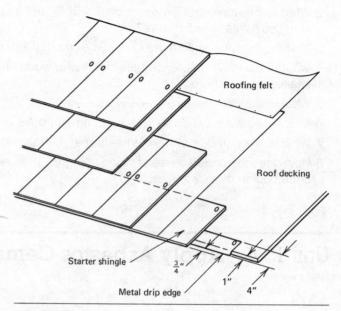

Fig. 15-4.2

Tools Required

To apply asbestos shingles you need a hammer and a knife or shingle cutter. A shingle may be cut vertically by scoring the trim line with a knife, then snapping the shingle along the scoring line. It is better, however, to make all cuts by inserting the shingle in the cutter. Cuts are cleaner and there is little danger of breakage.

Procedure for Shingling the Main Roof

1 Remove the nails from the bottom layer of roofing felt, and roll it back from the eave.

2 Lay a row of starter shingles along the eave so that they overhang the drip edge about $\frac{3}{4}''$ at eave and rake (Fig. 15-4.2). **Precaution.** Do not let shingles overhang more than 1", or they are likely to break under a heavy snow or ice dam.

3 Replace the felt, unrolling it over the starter strip.

4 Begin the first course of exposed shingles at the rake with a half shingle. Its bottom and side edges should line up with the starter shingle.

5 Complete the first course with full shingles to the opposite rake or a turn in the roof. **Note.** No gap is required between asbestos shingles.

6 Begin the second course with a full shingle. Follow the manufacturer's directions on the amount of exposure so that nails just miss the top edges of shingles in the course below.

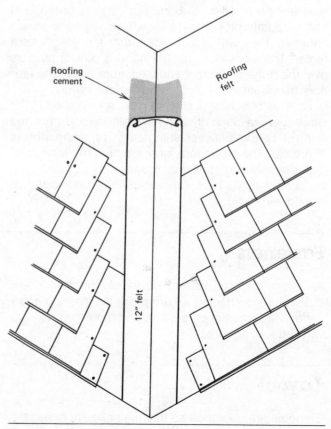

Fig. 15-4.3

7 Continue up the roof, alternating between half shingles and full shingles at the rake in order to stagger joints.

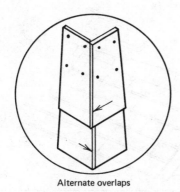

Alternate overlaps

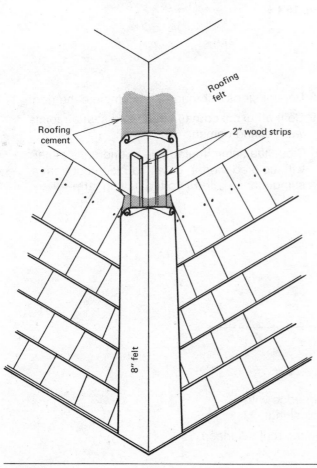

Fig. 15-4.4

Procedure for Finishing a Hip

1 Cover the roofing felt at the hip with another strip of felt 12″ wide. Center the strip over the hip, and embed it in a coat of roofing cement (Fig. 15-4.3).

2 Cut a pair of plywood strips 2″ wide and of the same thickness as the butts of shingles.

3 Nail one strip on each side of the ridge, following the shingle manufacturer's instructions for the exact location.

4 Trim shingles in the main roof to fit against the wood strips. Nail these shingles in position.

5 Over the wood strips lay another strip of felt in roofing cement, this one 8″ wide and centered on the hip (Fig. 15-4.4).

6 At the eave lay half a hip shingle over the felt so that it is flush with the hip. Nail it in position.

7 Nail another half hip shingle on the other side so that its edge is flush with the surface of the opposite shingle.

8 Seal the seam between the shingles and all nailheads with roofing cement.

9 Next nail a pair of hip shingles in position, but reverse the overlap (inset, Fig. 15-4.4). Again seal the seam and nail heads.

10 Continue up the hip in this manner, alternating the overlap from one side of the hip to the other.

Procedure for Finishing a Ridge with a Ridge Roll

1 Cut a wood nailing strip of the height specified by the manufacturer of the ridge roll. Nail this board upright atop the ridgeboard or the peaks of trusses.

2 Cover the strip with a double layer of roofing felt wide enough to overlap the upper edge of the next to top course of asbestos shingles in the main roof (Fig. 15-4.5).

3 Apply the top course of asbestos shingles, stopping just short of the felt flashing.

4 Set the first piece of ridge roll at the end of the roof *away* from the prevailing rain-bearing wind.

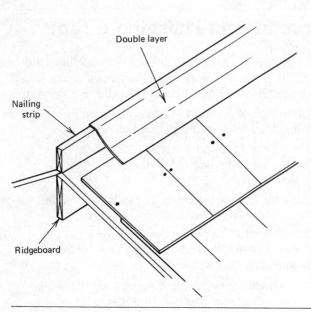

Fig. 15-4.5

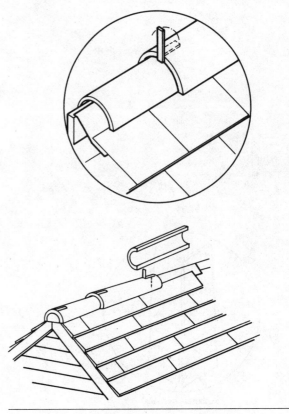

Fig. 15-4.6

Position it so that the large end overhangs the rake about $\frac{1}{4}$".

5 Secure the ridge roll by nailing into the nailing strip through L-shaped clips furnished with the roll (Fig. 15-4.6).

6 Set the second length of ridge roll so that it overlaps the first one by 2".

7 Bend the clip on the first piece over the exposed end of the second piece. Then nail the covered end as in step 5.

8 Repeat steps 5 through 7 to complete the ridge.

9 Coat all exposed nail heads and seal all joints with roofing cement.

10 Close the semicircular hole in end units either with colored mortar, or with a piece cut from a shingle and nailed into the nailing strip.

Questions

1 How do you adjust the exposure of asbestos shingles?

2 How does the application of roofing felt and a metal drip edge with asphalt shingles differ from the application with asbestos shingles?

3 What is the total number of layers of roofing felt between the roof's surface and sheathing at the hip of an asbestos shingle roof?

4 Name two methods for sealing the ends of ridge rolls.

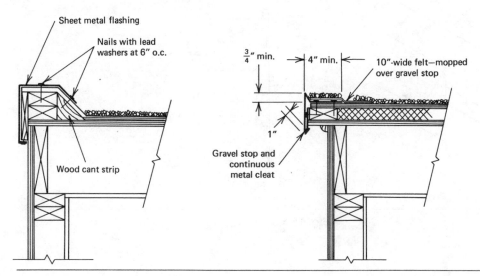

Fig. 15-5.1

Unit 15-5 **Build a Gravel Stop**

A carpenter is seldom involved in the application of a built-up roof. The durability of this type of roof is established by the number of layers of roofing felt, and the amount of **pitch** (a form of tar) that is used to make the roof waterproof. The preparation and application of the pitch requires special equipment—heaters, kettles, and mops—and is almost always done by a subcontractor who specializes in built-up roofs and who guarantees the length of its life. The roofing contractor also spreads gravel or crushed stone over the built-up roof as a finished surface. This type of roof is often called **tar and gravel.**

Built-up roofs may be laid over decks of metal, wood, gypsum, or, in commercial buildings, concrete. The roof is sheathed in the usual way (Unit 9-34), and the edge of the roof finished on all sides with a gravel stop. A **gravel stop** is a raised lip $\frac{3}{4}"$ to 3" high that prevents the gravel surface from being washed away by the force of water draining off the roof. The carpenter usually builds the gravel stop.

There are two types (Fig. 15-5.1). A **one-piece stop** is formed entirely of galvanized steel or copper bent to shape at the edges of the roof. A **two-piece stop** is higher, made of wood, and covered with metal. The two-piece stop permits water to collect to a shallow depth on the roof, where it acts as added insulation in hot weather.

Procedure for Building a Two-Piece Gravel Stop

1 Around the entire perimeter of the roof attach a 2×4 flat, flush with the edges of roof sheathing. Nail it securely into the edge member of the roof.

2 Add a doubler, staggering the joints at least 24".

3 Cut a continuous triangular cant strip (Fig. 15-5.2) from a 4×4.

4 Fit the cant strip against the two 2×4s, and nail it to them.

5 After the built-up roof is tarred but tar is still soft, add a fascia board around the roof.

6 To the bottom edge of the fascia nail a metal lock strip (Fig. 15-5.3).

7 Crimp lengths of metal flashing to fit the shape of the gravel stop. The flashing must cover at least half the cant strip, the fascia, and part of the lock strip.

8 Fit the flashing in place, overlapping the lengths 6" where necessary.

9 Attach the flashing either with metal screws with neoprene washers or nails with lead washers. Space fasteners 6" o.c.

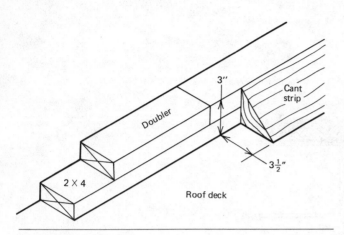

Fig. 15-5.2

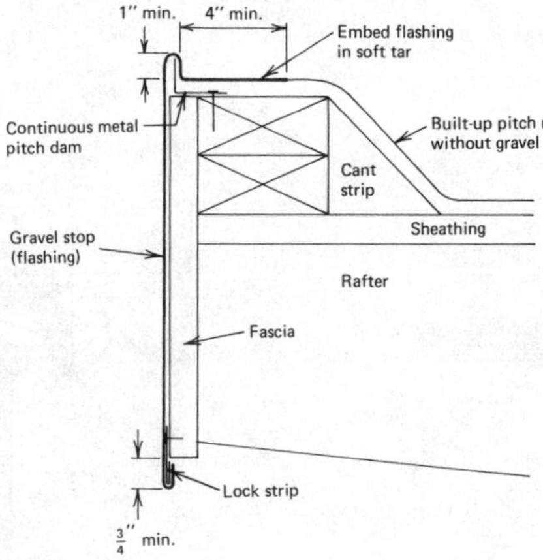

Fig. 15-5.3

Questions

1 What is another common name for a built-up roof?

2 What do you think is the purpose of the cant strip?

3 What is the purpose of the metal lock strip?

4 Why are washers recommended with fasteners into metal flashing?

16

Windows

When people first began to build dwellings, a window was an opening in a wall that provided ventilation. Today a window is an item of millwork that fills an opening. Most types of windows still provide ventilation, but they now serve two other important purposes: to let in light and to provide a view outward (Fig. 16-0.1).

Types of Windows

The architect or designer has a choice of eight ways to bring daylight into a home. They are:

- **Double-hung windows** have two **sashes** (a section of a window with glass in it) that slide up and down in their own tracks. Metal balances hold the sashes in any vertical position, and compressible weatherstripping provides a tight seal. Double-hung windows provide only limited ventilation—no more than 50% of the window opening.
- **Horizontal sliding windows** have two sashes that slide from side to side in top and bottom tracks. In one manufacturer's product the sashes are side by side when open, but in line when closed. Sliding windows also provide 50% ventilation.
- **Casement windows** have one sash that is hinged on the side, and provides 100% ventilation. They are sometimes used by themselves, but more often in groups of two or more. Most casements swing out, but they may be installed to swing into the house. Wood units are called left-hand casements when hinges are on the left as you look at the windows from *outdoors,* and right-hand casements when hinges are on the

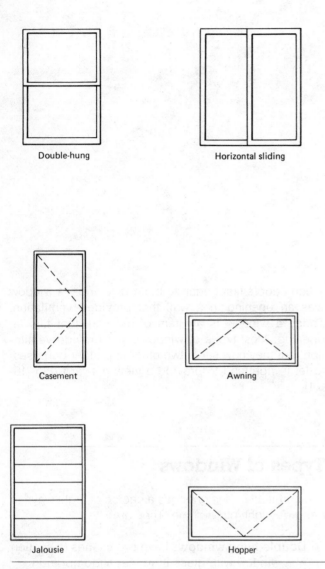

Double-hung

Horizontal sliding

Casement

Awning

Jalousie

Hopper

Fig. 16-0.1 The common types of residential windows as seen on elevations in drawings. The point of a dotted V points to the hinged side of casements, awning windows, and hopper vents.

right. With metal casements, the hands are reversed and called as you look at the sash from *inside* the house.

- □ **Awning windows** are one-sash units that are hinged at the top and swing outward. In essence a casement window turned 90°, the awning window also provides full ventilation, and good protection when open during all but a driving rain.

- □ **Hopper windows** are awning windows turned upside down and inside out. They are hinged at the bottom and swing inward. On floor plans and elevations, hopper and awning windows look

alike, except for a dotted V drawn on the windows in elevations. The V points toward the hinged side. With a hopper window the V is right side up. With an awning window the V is upside down. With a casement the V lies on its side.

- □ **Jalousie windows** are a series of glass slats set into metal clips in a frame. The slats pivot 90°, usually simultaneously, and do not project beyond the wall's surfaces when open.

- □ **Fixed windows** or **stationary windows** do not open. They are sometimes used by themselves in fully air-conditioned houses, but more often appear in combination with other types of windows.

- □ **Skylights** are a form of window installed in a roof. Some types are hinged to provide ventilation, but most are fixed. Skylights bring daylight into inside rooms, such as bathrooms, or to areas of large rooms too far from windows to receive good light.

In well-designed houses there is a definite relationship between the style of a house and the type of windows that look best. In traditional designs, such as Colonial and French Provincial, double-hung and casement windows with small panes of glass are the most appropriate. In more contemporary designs that feature larger glass areas, sliding, awning, and hopper windows are more in keeping with the style of living.

Parts of a Window

Each sash consists of a **top rail, bottom rail,** and **side rails** (Fig. 16-0.2). In double-hung and horizontal sliding windows the rails where two sashes meet are called **meeting rails** or **check rails.** If the sash has more than one pane of glass in it, it also has **muntins**—dividers between panes. Individual panes of glass are called **lights.**

Window sashes fit into a frame that holds them in place. The heads and jambs are usually identical in shape, and are therefore sometimes called **head jambs** and **side jambs.** The bottom member is called a **sill.** A strucutral member separating two windows in a group is called a **mullion.** A million may run either vertically or horizontally.

Frames of double-hung windows have two other parts not required with other types. A **blind stop** is a strip of wood or metal that forms a recess in which the upper sash slides. A **parting strip** is a similar piece that separates the upper sash and lower sash at the jambs, and forms a tight seal for the upper sash at the head. See Fig. 16-0.2.

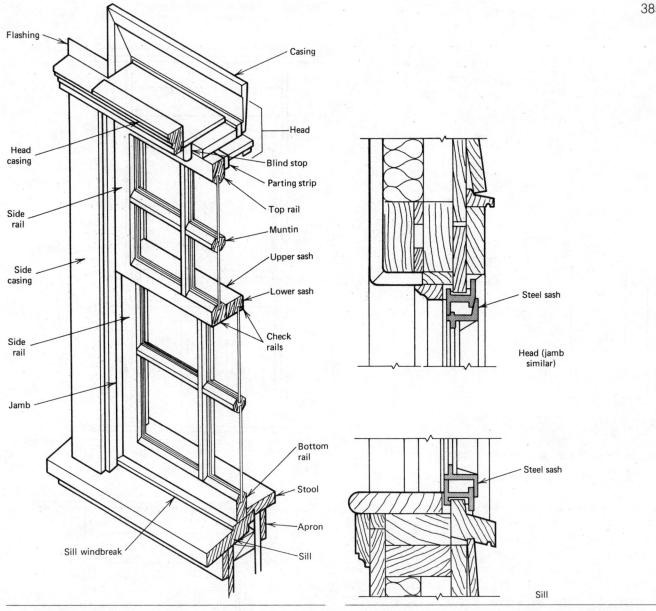

Fig. 16-0.2 The parts of a window.

Fig. 16-0.3 Section through a typical metal window.

Window frames come in several depths, because the jamb, head and sill members must be the same depth as the thickness of the finished wall in which they are installed. **Casings** for windows installed in frame walls are designed to fit over the exterior wall material. When walls are brick veneer, casings fit inside the masonry opening.

Materials and Accessories

Residential windows are made of wood, steel, and aluminum, and their frames are usually made of the same material as the windows. All parts of wood windows and frames are milled from heartwood of decay-resistant species such as Ponderosa pine, spruce, redwood, cedar, and cypress. Rails and sills are made from nominal 2″ stock, and heads and jambs of frames from 1″ stock.

Steel sashes (Fig. 16-0.3) are made from Z-shaped or T-shaped sections of rolled steel about $\frac{1}{8}$″ thick. Aluminum sashes, on the other hand, are usually extruded (forced through a die) from solid aluminum alloy.

Most windows used in houses today are fully assembled at the factory. They arrive at the site in their frames, with weatherstripping in place, and often with

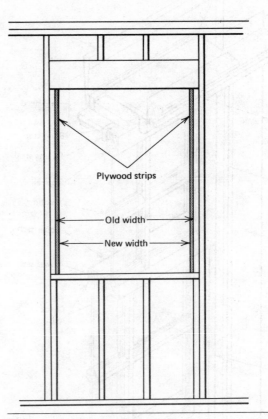

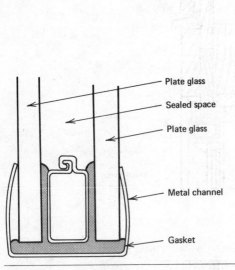

Fig. 16-0.4 Cross-section through a double–glazed window.

Fig. 16-0.5 How to narrow a window opening that is too wide.

lights already installed. Wood windows are treated with preservative to prevent deterioration from weather until they can be painted or stained. Steel windows receive a prime coat of paint at the factory. Aluminum windows are sprayed with a thin film that protects the finish until they are installed, then wears off in the weather.

Manufacturers of operable sashes also make screens that fit into or against window frames. Usually screens are installed on the outside of the window. The exceptions are on windows that open out—awning windows, jalousies, and most casements; here the screens fit on the inside.

A single thickness of glass offers little insulation in cold climates, and lights may become covered with condensation or ice on very cold days. The severity of the problem may be sharply reduced with storm windows. Storm windows fit in the same recesses as screens, and may be separate units to be changed by the owner in spring and fall. Combination screen-storm units are also manufactured that are installed by the carpenter and are never removed. Combination units have two disadvantages, however. In summer they limit ventilation to 50% of the opening, regardless of the type of window they protect, and in winter the windows are always screened. Screening blocks out up to 25% of the light coming through the window, and on dark

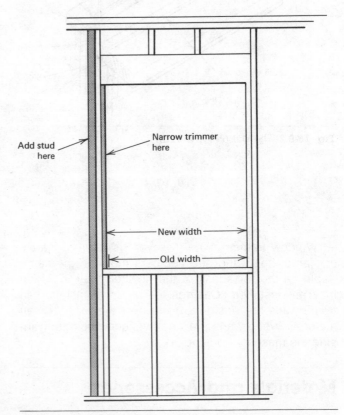

Fig. 16-0.6 How to enlarge a window opening that is too narrow.

winter days makes the interior of the house even darker.

To eliminate the need for storm windows, manufacturers produce windows double-glazed and triple-glazed (Fig. 16-0.4). **Double-glazing** consists of two sheets of insulated glass separated by a sealed air space that is a partial vacuum. Edges of the glass sheets fit against gaskets in a metal channel to provide the seal.

Hardware

All windows except fixed windows require hardware, which is usually supplied by the manufacturer. Casements, sliding windows, and jalousies are shipped with the hardware either attached or in plastic bags stapled to the frames. Awning and hopper windows may or may not come with hardware. Manufacturers of double-hung windows, on the other hand, provide and install only the balance mechanisms. Lifts and locking devices are provided by the builder.

When Windows Arrive at the Site

As soon as windows are delivered, they should be stored in a dry location and out of normal traffic so they won't be bumped and knocked out of square.

As they are stored, check the quantities and sizes of each type against the window schedule in the drawings. Then measure all rough openings to make sure that the windows on site will fit into the openings in the walls.

If an opening is too wide, narrow it by adding a stud or a strip of plywood. If possible, add the same thickness of strip on both sides of the opening to maintain the center line (Fig. 16-0.5).

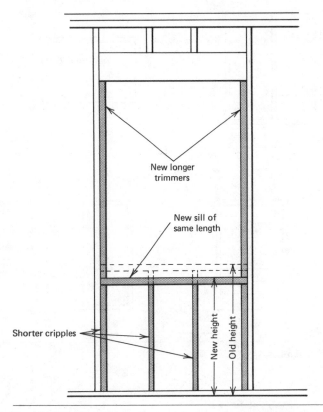

Fig. 16-0.7 How to lengthen a window opening that is too short.

If an opening is too narrow, add a stud on one side, then notch or cut out the trimmer that is in the way (Fig. 16-0.6). Do not weaken the structure by reducing the number of studs.

If an opening is too short, adjust at the sill, even though the window may be a little too low in the wall. It is far easier to remove the sill and trim cripples below it (Fig. 16-0.7) than to attempt to trim or raise a header.

Questions

1 Identify each of the wood windows in Fig. 16-0.8 on page 388, including left-hand or right-hand if necessary.

2 What is the difference between a muntin and a mullion?

3 Why are window frames made in various depths?

4 What is the purpose of double glazing, and how does it work?

5 What three steps should be taken as soon as windows arrive on the building site?

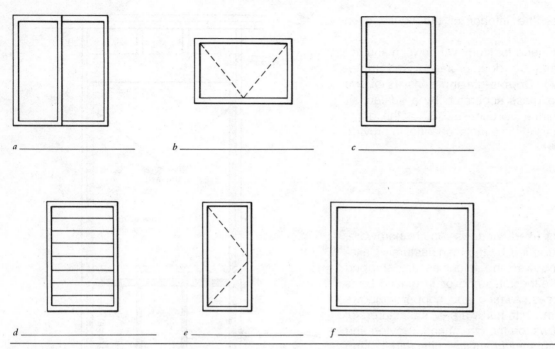

a _____

b _____

c _____

d _____

e _____

f _____

Fig. 16-0.8

Unit 16-1   **Install Skylights**

Most skylights used in residences today are stock items shaped from clear or translucent plastic. Where they meet the roof they may be square, circular, or rectangular in shape. Where they rise above the roof the shape may be a dome, a low vault, or a pyramid.

The smallest skylights have a minimum interior dimension of $14\frac{1}{4}''$, and fit over a standard opening between rafters. For larger sizes rafters must be cut and headers installed in the same manner as around any opening through a roof.

The simplest skylight has a horizontal flange on all sides that is attached directly to the roofing material at a properly flashed opening (Fig. 16-1.1, left). It has no curb, and is designed for sloping roofs only. Another type has a vertical flange that fits over a curb built on the job (center). The most complete type (right) has an integral curb and flashing. Curbs are required around skylights installed in flat roofs.

Curbs built on the job are usually made of lengths of 2×4 put together like a box sill (see Unit 6-3). The outside dimensions of the curb are about $\frac{3}{4}''$ smaller than the inside dimensions of the skylight. Manufacturer's instructions give the exact dimension.

When skylights open into an unoccupied space, such as an attic, the well is usually left unfinished, and the room below receives its light through a sheet of plastic set into a recess in ceiling joists below the skylight opening. When the skylight opens into finished space, such as an attic room whose ceiling follows the slope of the roof, the well is finished and the opening cased like any other window or door opening.

The tools you may need to install a skylight are a hammer, saw, roofing knife, metal shears, chalk, screwdriver, and putty knife.

Procedure for Installing a Site-Built Curb

1 Fit the curb accurately around the opening, and toenail it to the roof sheathing.

2 Flash the seam between curb and sheathing with roofing felt, and counterflash with sheet metal (Fig. 16-1.2).

3 Lay roofing material to the lower face of the curb.

4 Lay succeeding shingle courses as described in Chapter 15, cutting where necessary to fit around the curb.

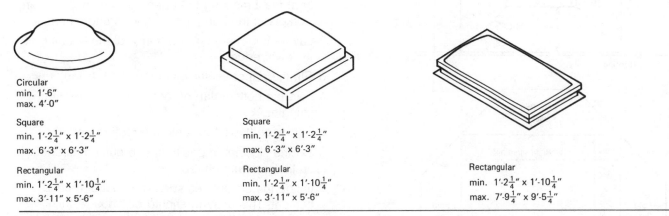

Fig. 16-1.1 Shapes of typical skylights.

Circular
min. 1'-6"
max. 4'-0"

Square
min. $1'\text{-}2\frac{1}{4}''$ x $1'\text{-}2\frac{1}{4}''$
max. 6'-3" x 6'-3"

Rectangular
min. $1'\text{-}2\frac{1}{4}''$ x $1'\text{-}10\frac{1}{4}''$
max. 3'-11" x 5'-6"

Square
min. $1'\text{-}2\frac{1}{4}''$ x $1'\text{-}2\frac{1}{4}''$
max. 6'-3" x 6'-3"

Rectangular
min. $1'\text{-}2\frac{1}{4}''$ x $1'\text{-}10\frac{1}{4}''$
max. 3'-11" x 5'-6"

Rectangular
min. $1'\text{-}2\frac{1}{4}''$ x $1'\text{-}10\frac{1}{4}''$
max. $7'\text{-}9\frac{1}{4}''$ x $9'\text{-}5\frac{1}{4}''$

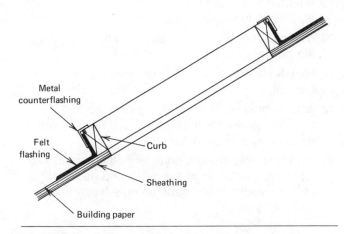

Fig. 16-1.2

Metal counterflashing

Felt flashing

Curb

Sheathing

Building paper

Procedure for Installing a Skylight on a Curb

1 a. If the skylight is a one-piece unit with vertical flanges, set it on the site-built curb, and attach it on all sides with roofing nails or screws spaced about 12" apart.

1 b. If the skylight comes with a separate metal curb frame, set the inner frame on the site-built curb, fit the skylight in position, and drop the outer frame over the horizontal flange (Fig. 16-1.3). Attach with screws.

2 Cover all heads of nails and screws with roofing cement.

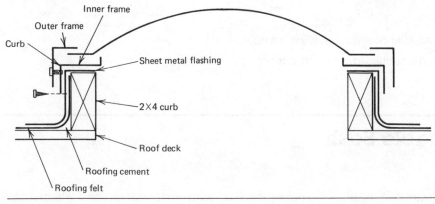

Inner frame

Outer frame

Curb

Sheet metal flashing

2×4 curb

Roof deck

Roofing cement

Roofing felt

Fig. 16-1.3

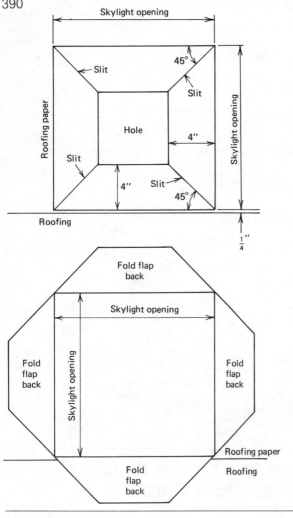

Fig. 16-1.4

Procedure for Installing a Skylight without a Curb

1 Carry roofing paper about 4″ over the opening on all sides. **Precaution.** Do not paper over the opening without cutting a hole in it immediately, or at least marking the opening with chalk to warn workers of the danger of falling through the hole.

2 Lay roofing material to within $\frac{1}{4}$″ of the opening at its lower edge.

3 Slit the paper as shown in Fig. 16-1.4 (top).

4 Fold the bottom flap of roofing paper over the shingles.

5 Fold side flaps 180° onto the roofing paper.

6 Set the skylight in position, and outline its edges on the roof with chalk.

7 Inside the chalk lines spread a coat of roofing cement. This coating should be about $\frac{1}{8}$″ thick, and should cover from the chalk line to within $\frac{1}{4}$″ of the opening.

8 Press the skylight firmly into the cement so that the excess oozes out around its perimeter.

9 Attach it with roofing nails every 6 to 8″ or according to manufacturer's instructions.

10 With a putty knife compress the cement around the perimeter to assure a tight seal.

11 Seal the heads of all fasteners with roofing cement.

12 Where shingles of succeeding courses butt against the raised dome or vault of the skylight, lay a bead of roofing cement.

13 Lay remaining shingle courses in the normal way, covering the flange at the sides and upper edge of the skylight. Press each shingle firmly into the cement.

Questions

1 To be safely installed in a flat roof, a skylight must have what feature?

2 Name three places where roofing cement should be used during installation of a skylight.

Unit 16-2   Install Operable Sash

The procedures for installing windows that open depend almost entirely on two conditions: whether the windows have wood or metal frames, and whether they are installed in a stud wall or masonry wall.

Wood Windows

It took manufacturers of wood windows many years to agree on a set of standards for the windows they produce. There are still slight differences between manufacturers' products, however, and between sections through heads, jambs, and sills of the types of windows described earlier in this chapter. None of these differences affect installation procedures.

Wood windows are normally installed in openings in stud walls after sheathing is in place but before the exterior wall material is applied. If the exterior is surfaced with a sturdy sheet material applied directly to studs without sheathing, however, the windows are installed after the exterior material is in place.

Wood windows for brick veneer walls are like those for stud walls, except that their frames are slightly smaller and must be caulked where they meet the masonry. Frames are usually installed as soon as the masonry has reached sill level.

Metal Windows

There is no comparable standardization in metal windows. There are likely to be differences in sections, in sizes, and in the method of installation. Rough opening sizes for metal windows are smaller than for wood windows, and the tighter fit permits attachment directly to framing around the opening without extensive shimming. Sometimes metal windows are held in place by wood bucks—strips into which screws are driven.

Basement sashes and their frames are usually made of steel or aluminum. If the basement wall is poured concrete, the frames are set in concrete forms before the wall is poured. Basement sashes are installed in a concrete block wall as the wall is being built. At the jambs they fit into sash blocks (Fig. 16-2.1); they are sealed at the head and sill after the basement wall is finished.

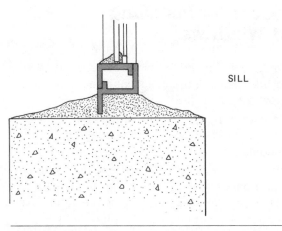

Fig. 16-2.1 Details of installation of a metal hopper window in a basement wall.

Preparation

Always study the manufacturer's literature before you begin work. Most windows arrive at the site already installed in preassembled frames. Manufacturers may recommend that their windows be installed as shipped, and that protective bracing and blocking be removed after installation. Or they may recommend removing the sash to protect them from damage and make installation easier. If a manufacturer's instructions are not followed, and the windows are damaged as a result, the warranty on the windows is no longer in effect, and the manufacturer is not obligated for the cost of damage or replacement—the builder is.

Some wood windows receive a coat of preservative at the factory, and require no further preparation on the site. If they are not primed, apply a coat of pentachlorophenol to the entire frame and all sashes. It is important to work this coat into corners and joints to protect them from moisture. It is equally important not to get preservative on the weatherstripping or on working parts, such as balances and hinges.

The heads of all windows in a house are usually at the same height above floor level—the same height as the head of the main entrance door. When head heights are constant, some carpenters cut a pair of 1 × 4s or 2 × 4s, called **story poles,** to support a window frame in its opening until they can level and plumb it. The length of these supporting poles is equal to the window height given in the drawings plus the thicknesses of finish flooring and any underlayment.

When window frames have a blind stop (Fig. 16-2.2, top) or are installed against a buck (bottom), the edges of sheathing must lie 1½" or 3" back from the rough opening.

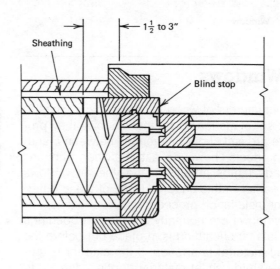

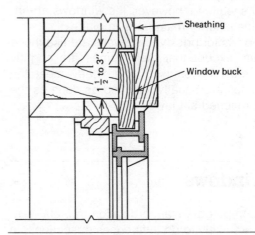

Fig. 16-2.2 The edge of sheathing must stop short of the rough opening if the window has a blind stop, or when it is installed against a window buck.

Tools

No special tools are needed for installing windows, but keep your carpenter's level within reach at all times. You will use it constantly.

Procedure for Installing Wood Windows

1 Upon delivery, check the dimensions of all windows against the dimensions specified in the window schedule in drawings.

2 If windows did not receive a prime coat at the factory, prime them now. Let the coat dry thoroughly.

3 Flash all four sides of each opening with strips of building paper 10 to 12" wide (Fig. 16-2.3).
Precaution. Make sure that flashing extends over the edges of sheathing far enough to form a weatherproof seal—about 6".

4 Working from outside the house, set the window frame in the opening, centering it as close as possible.

5 Lift the frame to the specified window height, and support it on shims 12 to 16" apart at the subsill.

6 Set your carpenter's level on the window sill, and gently tap on the shims until the frame is plumb.

7 Recheck the window height at the head against the height specified in drawings. Adjust shims if necessary.

8 In an upper corner of the outside casing, drive a casing nail part way into the studding (Fig. 16-2.4).

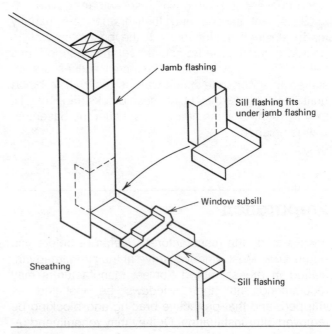

Fig. 16-2.3

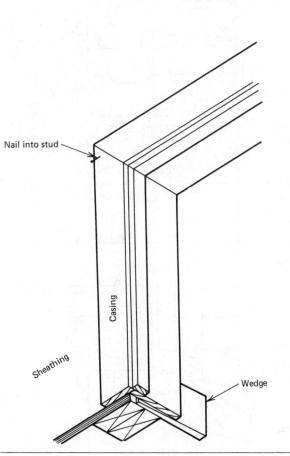

Fig. 16-2.4

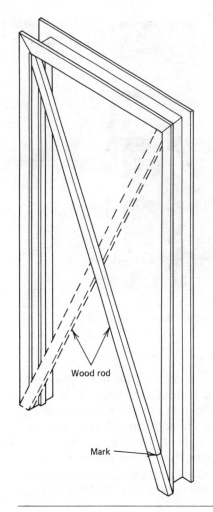

Fig. 16-2.5

9 In the opposite upper corner start another nail into the casing.

10 Check the head for level.

11 When the head is level, drive the second nail part way in.

12 Check the frame for square by measuring the diagonals (Fig. 16-2.5).

13 Wedge shims at the jambs to hold the frame securely in a plumb position.

14 Repeat steps 8 to 11 inclusive at lower corners of the frame.

15 Check window height, level at sill and head, plumb at jambs, and diagonal measurements once more. Also operate the sashes to make sure they move freely without binding. If they are not in their frames, measure the width between side jambs at several points to make sure it is exactly equal.

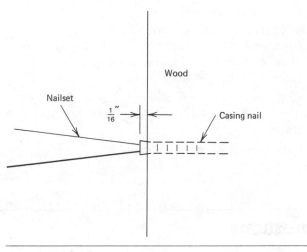

Fig. 16-2.6

Fig. 16-2.7 (Courtesy Senco Products Inc.)

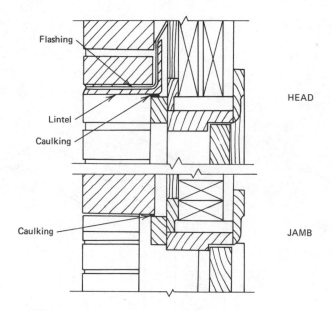

HEAD

JAMB

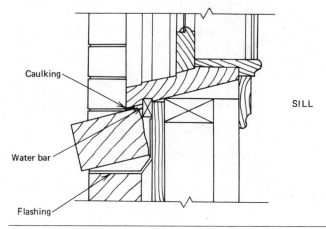

SILL

Fig. 16-2.8

16 When everything is in order, drive home the nails at the four corners. **Precaution.** When nailing any finished wood, such as a window frame or casing, stop hammering when the heads of nails are about $\frac{1}{16}''$ from the surface of the wood (Fig. 16-2.6). Then drive the heads just below the surface of the wood with a nailset. If wood begins to split, carefully pull out the nail, drill the hole, and renail.

17 With 16d galvanized casing nails or finishing nails spaced 8 to 12″ apart, nail through the window frame and shims into structural framing members. **Note.** If the window has blind stops, use 8d common nails.

18 a. Nail casings as described in step 17. **Precaution.** To hit studs, you may have to drive nails at an angle. Nailing into sheathing is not sufficient.

18 b. If windows have a nailing flange (Fig. 16-2.7), attach it to sheathing with $1\frac{3}{4}''$ galvanized nails or staples.

19 Trim shims flush with the frame.

20 Install the drip cap.

21 Fill nail holes with wood putty.

22 a. If the exterior material is already in place, caulk the joint around the frame on all four sides.

22 b. If the wall is masonry, wait until masonry work is completed around the window, then caulk at the lintel, both jambs, and the front edge of the sill (Fig. 16-2.8).

Questions

1 What extra step is required during installation of windows in a brick veneer wall that is not required in a stud wall?

2 Under what conditions are windows installed after exterior material is in place?

3 How are story poles used?

4 What should flashing cover?

5 Why are basement windows always the first to be installed?

Unit 16-3 **Apply Window Hardware**

The operating hardware for windows of all types usually comes already installed at the factory. The louvers of jalousies are already in their channels and the pivots attached to the frame. Only the cranks are shipped loose for installation on the job. The slides that support awning and hopper vents are already mortised into the frame, and only the latch comes separate. Casement windows arrive with one of two types of hinges, one of two types of operators, and one of two types of closers already installed (Fig. 16-3.1). Only the crank is shipped loose.

Before you install any loose hardware on these windows, read the instructions supplied either by the window manufacturer or the hardware manufacturer. Requirements vary considerably, and you can ruin a perfectly good window by cutting or boring in the wrong place.

Carpenters have a little more to do to install hardware on double-hung windows. Unlike carpenters of earlier generations, they don't have to adjust sash weights and balance the sashes; these steps are al-

ready taken at the factory. But they do have to install sash locks and lifts.

There are two basic types of sash locks. Both have two parts, one that fits on the top of the meeting rail of the lower sash, and the other on the top of the meeting rail of the upper sash. One part of the crescent type (Fig. 16-3.2, left) pivots into a groove in the other and draws the two sashes together. The handle of the triple-acting type (right) turns 180°. When the sash lock is open, a 90° turn of the handle swings the locking device over the latch piece, and another 90° turn pulls the two pieces together for a tight fit of the check rails.

There are also several designs of lifts, but all types serve the same purpose: to provide a grip for raising and lowering the bottom sash.

Hardware manufacturers provide a full-size template with their products showing where to drill holes for attachment. The sash lock is centered between jambs if the window has an odd number of horizontal lights. If it has an even number of lights, the center muntin interferes, and the lock must be placed just off

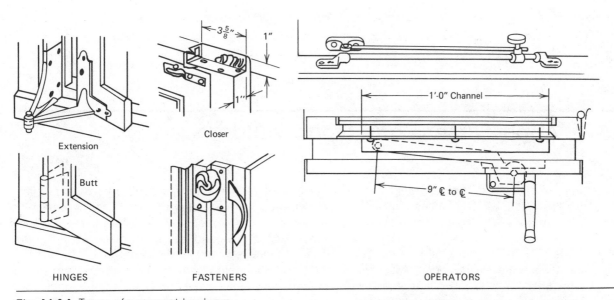

HINGES   FASTENERS   OPERATORS

Fig. 16-3.1 Types of casement hardware.

Fig. 16-3.2 The two types of sash locks for double-hung windows. (Courtesy Amerock Corporation.)

center. A window less than 2' wide gets a single lift centered on the bottom rail. Wider windows need two lifts. On three- and four-light-wide windows the lifts are centered under the lights nearest the jambs. On still wider windows place a lift 15" on each side of the center line of the window.

To install window hardware you need an awl for marking the locations of screws and a screwdriver for setting them.

Procedure for Installing a Sash Lock

1 Find the center line of the window, and mark it.
2 If the center line falls at a muntin, measure $2\frac{1}{2}$" to 3" to one side and make another center mark for hardware.
3 Remove the template from the package containing the sash lock.
4 Line up the template on the center line.
5 With the awl, punch holes for screws at the points marked on the template. **Precaution.** Be sure that the actual joint between check rails is directly below the check rail line on the template.
6 Attach half the lock with screws.
7 Fit the other half in place, and test the fit by locking the device. Make sure the holes in the lock and the holes made by the awl still line up.
8 When the fit is good, attach the other half of the lock.

Questions

1 Which of these windows is not operated with a crank?
  a) Casement
  b) Jalousie
  c) Double-hung
  d) Hopper

2 Which of these windows is usually not shipped complete with all hardware?
  a) Casement
  b) Awning
  c) Double-hung
  d) Hopper

Unit 16-4   Apply Screens and Storm Windows

The manufacturers of casement, awning, hopper, and jalousie windows design the window frames with a rabbet into which screen panels fit. This rabbet is on the inside of the frames of awning, jalousie, and most casement windows (Fig. 16-4.1, left), and on the outside of the frames when the sashes open in, as with hoppers and some casements (right). Clips or latches hold the screens in place, and they are easily removed.

If these windows are not double-glazed, heat loss may be reduced with storm panels. In most cases the panels fit into a rabbet on the outside of the operating sash (Fig. 16-4.2), and are left in place all year long. They may be removed for cleaning.

Good builders instruct their carpenters to fit screens and storm panels in place after windows are trimmed and painted. They sometimes mark them so

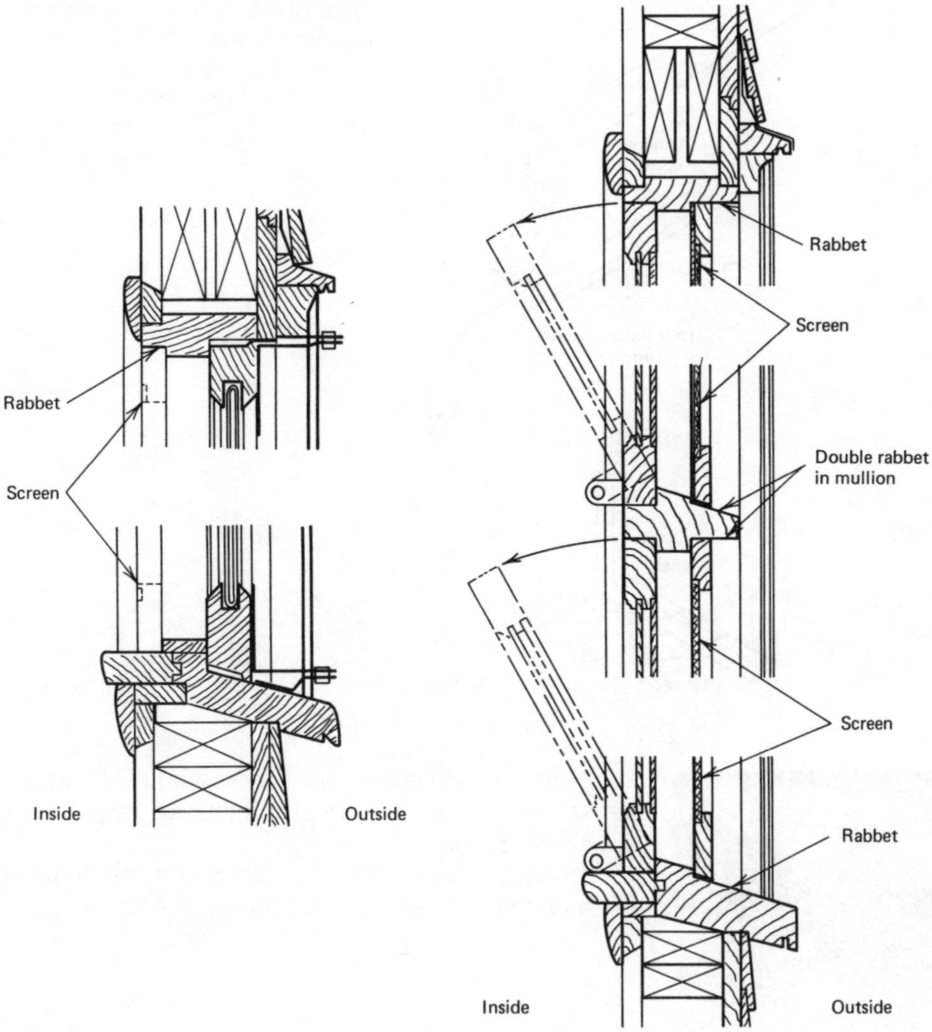

Fig. 16-4.1 Typical rabbets in window frames for screen panels.

Rabbet

Screen

Rabbet

Screen

Double rabbet in mullion

Screen

Inside   Outside

Rabbet

Inside   Outside

that the new homeowner knows which panels fit which windows.

Screens and storm windows fit on the outside of double-hung windows. There are two basic types—separate units and combination units. The separate units fit into the same rabbet in the frame, and the owner takes one unit down and puts up the other with the change in seasons. The units are supported on or in a pair of hangers at the top, and held in place' at the bottom by latches (Fig. 16-4.3).

The self-storing combination units are fastened permanently into window frames. Each unit has four parts (Fig. 16-4.4). The frame has three tracks in the jambs. The upper storm window fits at the top of the outer track and remains fixed there. In cold weather the lower storm window fits beneath the upper one, and the screen is stored on another track in the upper half of the window. In warm weather the screen moves down

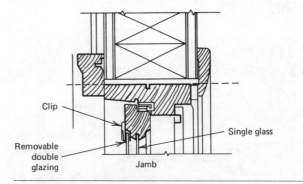

Clip

Removable double glazing

Single glass

Jamb

Fig. 16-4.2 Typical rabbet for storm panel.

into position to close the opening, and both storm sections are stored in the upper half of the frame.

Screen and storm units may be built of wood, metal, or plastic. Of these, wood units are the sturdiest but also the heaviest and the least likely to provide a tight

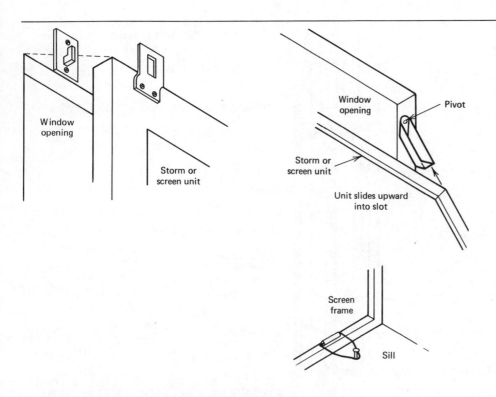

Fig. 16-4.3 Typical hanging and latching hardware for separate screen and storm units.

Fig. 16-4.4 A self-storing combination unit consists of a metal frame, two storm panels, and a screen panel. This side fits against the window opening, and shows the construction. All sections are removable. (Courtesy V.E. Anderson Mfg. Co.)

fit. Metal units are lighter, but may rust and attract condensation. Plastic units are also light, but are the most subject to damage.

The tools required to install storm and screen units are a pencil, awl, and a screwdriver.

Procedure for Hanging Separate Screen and Storm Units

1 Fit a screen unit into the rabbet in the window frame, with the hanging clips attached.
2 Mark on the frame the locations of holes for attaching the hangers.
3 Attach the hangers. **Precaution.** If the hangers pivot out to accept the screen unit, they should be attached loose enough to pivot but tight enough to hold their position.
4 Insert the screen unit to check the fit.
5 When the fit is good, mark on the sill the locations of the pins or eyes to which latches are attached.
6 Attach one pin or eye, and test the fit. The unit should fit tightly at the sill so that it cannot rattle.
7 Attach the other pin or eye.
8 Remove the screen and store it.

Questions

1 Why do you think storm panels are not clipped to the sash of double-hung windows, as they are in awning windows?

2 What is one advantage that you can think of to self-storing storm-and-screen units?

Unit 16-5 **Install Fixed Windows**

Fixed windows produced by manufacturers of other types of windows have the same frames and are installed in the same way as regular windows. Occasionally, however, the carpenter must build a frame for a large pane of insulating glass or for a network of smaller panes.

Frames may be built in several ways. The glass may be inserted into standard rail sections, and the fabricated window set into a standard window frame (Fig. 16-5.1). Or the glass may be fitted into a rabbet cut into the window frame, and held in place with a stop (Fig. 16-5.2). With this method the glass may be placed at either the outer or inner edge of the frame. Or glass may be installed between stops in an uncut frame (Fig. 16-5.3). Here the glass may rest at any point from front to back between stops.

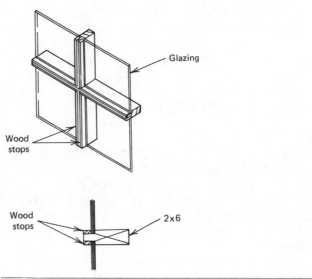

Fig. 16-5.2 Insulating glass set into a rabbet in the frame.

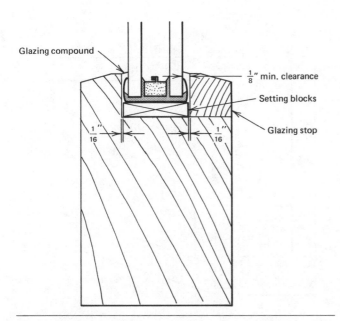

Fig. 16-5.1 Insulating glass inserted into a standard window frame.

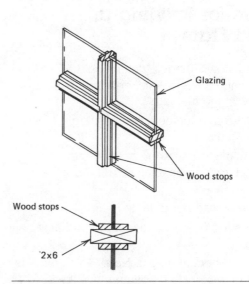

Fig. 16-5.3 Insulating glass set between two stops outside the frame.

Any new houses will settle a little, and settling may crack fixed glass unless it is free to move slightly. The gasket between panes of insulating glass and their metal frame (refer to Fig. 16-0.4) allows for some movement. To prevent settling from cracking the frame, however, each pane should be supported on a pair of wood blocks cut from exterior plywood or lumber treated to resist moisture. These blocks are 3 to 4″ long, about $\frac{1}{4}$″ thick, and $\frac{1}{8}$″ wider than the thickness of the metal frame (Fig. 16-5.4). They are glued at the quarter-points of each opening.

Except at these support points, insulated glass does not touch any part of a wood frame. There should be a clearance of at least $\frac{1}{8}$″ but no more than $\frac{1}{4}$″ at the head and jambs; this clearance should be the same on all three sides. There should also be at least $\frac{1}{16}$″ clearance between the metal frame around the glass and the stops on all four sides. These gaps are filled with **glazing compound**—a puttylike material that does not harden but remains plastic to absorb shock and any pressures caused by settling.

When fixed windows are grouped, the vertical and horizontal mullions between them may fit together in a halved joint or a notched joint (Fig. 16-5.5). Wood framing members must be precisely cut and rabbeted to assure a tight fit, and the clearances around the glass must be taken into account in determining the dimensions of the cuts.

Framing members are shaped with a jointer or router, and cut to length on a circular saw. The hand tools needed for assembly are a hammer, nailset, level, steel tape, framing square, and a putty knife.

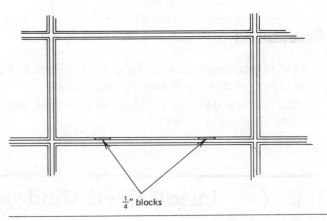

Fig. 16-5.4 Small blocks support double glazing at the bottom.

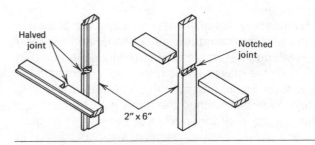

Fig. 16-5.5 Horizontal and vertical mullions may meet at a halved joint (left) or a notched joint.

Procedure for Building a Rabbeted Frame

1 From the window schedule determine the minimum lengths of stock needed for the window head, jambs, and mullions.

2 From select-grade lumber choose straight pieces of these lengths.

3 Cut rabbets along one edge of each piece of stock. Head and jambs receive one rabbet and mullions two.

4 Trim the head member to length.

5 Cut grooves in the ends to accept jamb frame members.

6 Cut jamb members to length. **Note.** This length is the height of the finished opening less the thickness of the sill member and the rabbeted thickness at the head (Fig. 16-5.6).

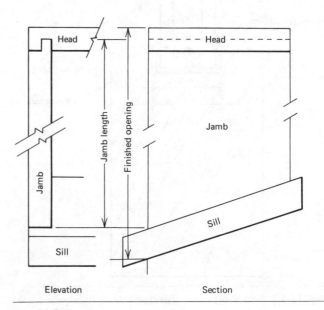

Fig. 16-5.6

7 Rabbet the upper ends of jambs, and bevel the lower ends at the angle of the sill.

8 On the head member mark the locations of any vertical mullions.

9 On jamb members mark the locations of any horizontal mullions.

10 Cut grooves for these mullions.

11 Cut vertical mullions to the same length as jambs.

12 a. If mullions meet at halved joints, cut horizontal mullions to the same length as the head, less the thickness of two jambs, plus the depths of two rabbets (Fig. 16-5.7).

12 b. If horizontal mullions are notched into vertical mullions, cut mullion pieces to length, allowing for the cuts in vertical framing members (Fig. 16-5.8).

13 Cut ends of mullions to fit into grooves.

14 Cut the sill to length out of sill stock.

15 Prime all framing members with an oil-rich coat of primer-sealer.

16 Assemble the outer frame, cross nailing at the upper corners. **Precaution.** Use your square and level, and measure the diagonals frequently to assure a square frame.

17 Insert and nail any vertical mullions. **Note.** For the best assembly, fill joints with an aliphatic glue, fit the mullions in position, and nail before the glue dries.

18 Recheck the frame for square.

19 Insert glue and nail horizontal mullions. If joints between mullions are notched, nail as shown in Fig. 16-5.9. If joints are halved, nail as shown in Fig. 16-5.10.

20 Again check the frame for square.

21 When the frame is square, add temporary cross bracing.

22 Install the frame in its opening as described in Unit 16-2.

Procedure for Installing Fixed Glass

1 Cut wood support blocks to size.

2 With waterproof glue install the blocks at the quarterpoints of each opening for fixed glass.

3 Cut all stops to size, and test their fit.

4 Fill the rabbets on all sides with glazing compound to the required depth.

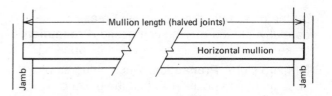

Fig. 16-5.7

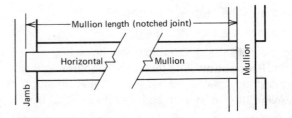

Fig. 16-5.8

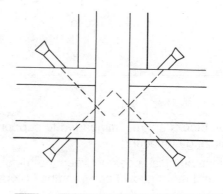

Fig. 16-5.9

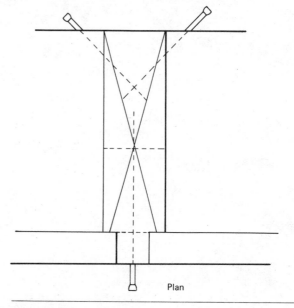

Plan

Fig. 16-5.10

5 Lift the glass onto the blocks, and press it carefully into the glazing compound. **Precaution.** Make sure the glass is centered in its opening.

6 Fit jamb stops. **Precaution.** When fitting stops, slide them into position so that the glazing compound is forced against the glass, and the excess oozes out between the glass and stop.

7 Tack the stops in place by driving finishing nails part way in.

8 Fit and tack the stop at the head.

9 Fit and tack the stop at the sill.

10 When the fit is good, complete nailing, using a nailset. **Precaution.** Into mullions, toenail slightly and alternate spacing of nails into adjacent stops to prevent the wood from splitting (Fig. 16-5.11).

11 With a putty knife compact the glazing compound around the perimeter of the glass on both sides. Fill any holes or pockets, and cut away any excess.

12 Use the excess compound to fill nail holes.

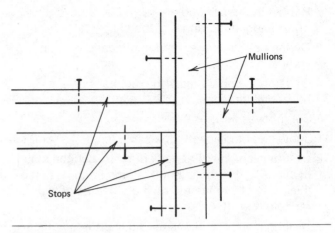

Fig. 16-5.11

Questions

1 Why do you think blocks are recommended to support fixed glass, rather than a bed of glazing compound?

2 What are the minimum clearances between the metal frame around insulating glass and the window head? Jambs? Stops?

3 Describe the two methods of fitting mullions together.

4 What do you think is the best way to join pieces of stop at a corner?

5 Why do you think it is better to rabbet frame pieces before they are cut to length?

17

Exterior Doors

Doors serve two opposite purposes: to permit passage from one area to another, and to prevent it. Exterior doors may also serve any or none of the same purposes as a window: to provide ventilation, to let in light, and to provide a view outward.

The architect or designer places exterior doors where they will best serve their various purposes. Most houses have three types, excluding a garage door:

- **A main entrance door.** The main entrance (Fig. 17-0.1) is the focal point of the front or street-side elevation of most houses. Its primary purpose is as an attractive passage for invited guests and a barrier against uninvited ones. It may provide ventilation, but seldom lets in much light or view. The entrance may have a single door hinged at one side, or a pair of doors that open inward. The door or doors may have a simple frame, or be part of a complete unit (Fig. 17-0.2) that includes a decorative frame and/or a **sidelight**—a panel of clear or translucent glass beside the door. Rough opening dimensions include any sidelight, but not the decorative trim. Entrance doors and their frames are usually wood. They are made in three widths and two heights, with 3'-0" × 6'-8" the common size.

- **The service door.** At one time years ago all deliveries of household supplies—milk, groceries, ice, pastry, etc.—came to the rear door of the house: thus its name of service door. Today the service door is primarily a family entrance, and opens toward or into a garage or carport (Fig. 17-0.3). As a secondary entrance, it tends to be functional rather than beautiful. If glazed, the glass is for convenience rather than light. Usually the service entrance has a lower cost, hinged wood door in a wood frame. The exception is a basement access door, which is often metal

Fig. 17-0.1 The main entrance door is the focal point of the front elevation of a well-designed house. (Photos courtesy Peachtree Doors Inc.)

Fig. 17-0.2 Some entrances are manufactured complete with sidelights.

Fig. 17-0.3 The service door is smaller, less conspicuous, and less expensive than the main entrance door.

because wood doors tend to warp when installed close to or below grade. Standard size is 2'-8" × 6'-8".

- **The patio door.** Known by a variety of names, the patio door provides access to outdoor living areas for the family. In most houses built today the patio door is a pair of sliding glass panels in a metal or wood frame that open onto a deck, patio, or garden (Fig. 17-0.4). In more formal homes casement doors, also called **French doors,** may open onto a screened porch. The patio door frequently fulfills all the purposes of a door, letting in light, air, and a view, and serving as a passageway between areas.

Fig. 17-0.4 Sliding glass doors provide ready access to outdoors, and let in more light than windows of the same width.

Parts of a Door

The names of door parts are quite similar to those for window parts (Fig. 17-0.5). The frame has a head, jambs, and a sill. If the door is glazed, there are muntins between the panes. If it has a sidelight, it has a mullion between the door and sidelight.

In a door, however, rails are horizontal members of wood or metal. Vertical members are called **stiles.** Every door has two stiles and two rails that form the outer frame. It may have one or two stiles and as many as six rails in the center, depending on the design. These are called **intermediate stiles** and **intermediate rails.** The areas between exposed stiles and rails are called **panels.**

The rail into which the door lock fits is, logically, called the **lock rail.** If a door has no lock rail, it must have a pair of **lock blocks** instead. Manufacturers provide two so that the door may be hinged on either side.

are usually plywood. There are dozens of different panel combinations. One variation of the one-piece panel door is the **Dutch door** (Fig. 17-0.8), which has two separately hinged sections that meet at center rails. The two sections may be locked together to form a single door.

- **Flush doors** have a wood framework surrounding a center core, and plywood veneer as a flush surface on both sides. Flush doors are designated by the type of core. In a **solid-core door** the

Door Construction

Doors are classified by their method of operation, by the material from which they are made, or by their construction. There are three types of construction:

- **Panel doors** (Fig. 17-0.6) have a top rail $4\frac{1}{2}$ to 6" wide, a bottom rail $9\frac{1}{2}$ to 12" wide, and intermediate rails from $4\frac{1}{2}$ to 12" wide. Stiles are $4\frac{1}{2}$ to $5\frac{1}{2}$" wide. These parts are heartwood $1\frac{3}{4}$" thick in exterior doors (interior doors are thinner). Stiles and rails are joined with wood dowels, and panels fit into mortise-and-tenon joints. The panels may be raised or flat (Fig. 17-0.7); raised panels are cut from heartwood, and flat panels

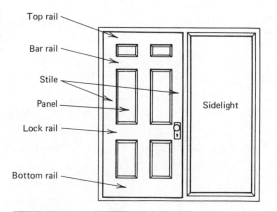

Top rail
Bar rail
Stile
Panel
Lock rail
Bottom rail
Sidelight

Fig. 17-0.5 The principal parts of a door.

Main entrance doors

Service doors

Fig. 17-0.6 Panel doors are manufactured in many designs, with and without glass.

center is made of narrow wood blocks glued together between stiles and rails. The blocks may all run vertically, or some may run horizontally like intermediate rails (Fig. 17-0.9, left). Usually the frame and core are $1\frac{1}{4}''$ thick, and the plywood is $\frac{1}{4}''$ thick on each side. A **hollow-core door** has a wider and slightly thicker frame, lock blocks, and a core of wood strips, stiff cardboard, or a paper honeycomb (Fig. 17-0.9, right). The plywood surfaces are thinner—either $\frac{3}{16}''$ or $\frac{1}{8}''$

thick. Total thickness of both types of exterior flush doors is $1\frac{3}{4}''$, but a hollow-core door is much lighter in weight and has less insulating value.

• **Framed glass doors** are just what they sound like: they consist of an exterior frame surrounding glass. The frames may be wood, aluminum, or steel. The glass may be a single pane of insulating glass, a number of small panes of standard glass between muntins, or a jalousie. The doors may swing on hinges, or slide in top and bottom tracks. Standard sizes of hinged French doors are 2'-6" × 6'-8" each, and they are usually installed in pairs. Sliding glass doors are a nominal 6'-8" high, and range in width in pairs from 5'-0" up to 8'-0", and up to 12'-0" in triples. The doors slide in separate tracks; both doors may slide, or one may be fixed. Metal doors are 1" thick around their perimeters, and $1\frac{1}{2}''$ thick where they meet. With their large glass area, sliding doors often take the place of windows in small to medium-sized rooms.

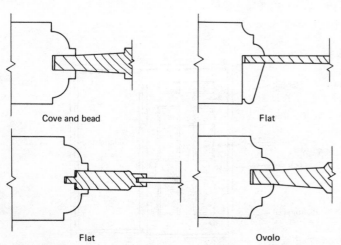

Cove and bead   Flat

Flat   Ovolo

Fig. 17-0.7 Door panels may be raised or flat, and there are many variations in the designs.

Accessories

Combination screen and storm door units are available to fit all standard size hinged doors. The frames may be wood or metal. The screen and storm inserts may fill

Fig. 17-0.8 A Dutch door has two separately hinged sections that meet at a shiplapped joint. The doors are equipped with a bolt so that the two sections can be locked together to open as a unit. (Courtesy Georgia-Pacific Corporation.)

half to two-thirds of the frame, with a solid panel below, or all of the frame. Some types have separate inserts, while others are self-storing, like combination window units. A storm door is also available with no screens;

it is one pane of insulating glass in a metal frame.

The glass in sliding doors is insulating glass, and storm units are not required. A door-sized screen section that slides in its own track comes with the door.

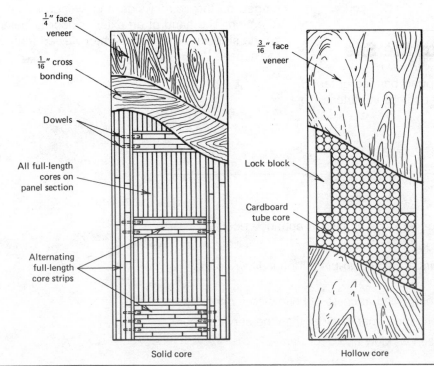

$\frac{1}{4}$" face veneer

$\frac{1}{16}$" cross bonding

Dowels

All full-length cores on panel section

Alternating full-length core strips

Solid core

$\frac{3}{16}$" face veneer

Lock block

Cardboard tube core

Hollow core

Fig. 17-0.9 Cutaway views of a typical solid-core (left) and hollow-core door.

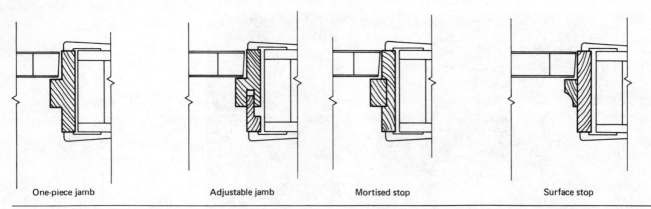

One-piece jamb   Adjustable jamb   Mortised stop   Surface stop

Fig. 17-0.10 Four types of door jambs, from left to right, are: one-piece jamb, adjustable jamb, jamb with mortised stop, jamb with surface stop.

Door Frame Construction

The frames around exterior wood doors are similar to window frames, but are built of heavier stock—a nominal $1\frac{1}{4}''$ instead of a nominal $1''$. Jambs and heads of good quality frames are rabbeted on the inside for the door and on the outside for a screen-storm combination (Fig. 17-0.10). Less expensive frames have either a mortised or surface door stop.

Door frames are available in three forms: **knocked down (K.D.)** for assembly at the site, preassembled without the door in place, or preassembled with the door already hung.

When Doors Arrive at the Site

Frames purchased K.D. are shipped banded or taped together in bundles, and the trim comes in cartons. To prevent warpage all parts should be primed at the earliest possible moment. Every bit of raw wood surface should be liberally coated, particularly in grooves and dados where moisture is most likely to settle. Be sure to wipe off any excess where pieces join.

Preassembled frames should be stored upright in a dry, well-ventilated area out of traffic. Even if frames are protected in shipment, carry instead of drag them from one location to another. Like all finished woodwork, doors and frames should be handled with clean gloves; fingerprints and dirt are hard to remove. Prime top and bottom edges of doors and frames promptly to prevent moisture penetration.

As you store each door that is already hung in its frame, check it against the drawings to make sure it is hinged on the correct side and swings in the proper direction. The hand of an exterior door is determined by the side with the hinges when viewed from outside the house.

Questions

1 What are the most common dimensions of a wood entrance door?

2 When are lock blocks required?

3 French doors and Dutch doors are both two-piece doors. How do they differ?

4 Name three materials commonly used in cores of hollow-core doors.

5 Why do you think the main entrance door is wider than the service door?

6 What is one difference between window and door frames?

Unit 17-1 **Install Frames for Hinged Doors**

The procedures for installing door frames are similar to those for window frames (Unit 16-2). If the heads of doors and windows are at the same height, you can reuse the story poles cut for windows. The main differences lie in the locations of shims behind jambs, and in treatment at the sill. Jamb members extend beyond the head and sill members, and sometimes these **horns** or **lugs** must be cut off.

At the time the floor of the house was framed, locations of exterior doors were marked on header or edge joists (Unit 6-11). At that time, too, joists were trimmed if necessary so that when the hardwood door sill is in place, its top would be level with finished flooring. Now that you are ready to install door frames, you must recheck this floor height and the size of all rough openings to make sure the frame will fit properly and at the correct head height. Standard rough openings for exterior hinged doors are the door width plus $3\frac{1}{2}''$, and 82″ if the sill is metal or $83\frac{1}{2}''$ if the sill is hardwood.

Door frames may be installed with or without the doors already hung in the frame. Doors are easier to hang when their frames are horizontal on saw horses, and they help to keep the frame square while it is being installed. On the other hand doors are more likely to be damaged if prehung, and should not be in place as long as bulk finishing materials are still being moved into the house.

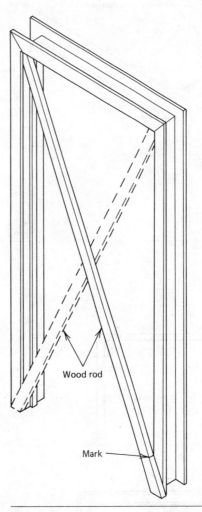

Fig. 17-1.1

Procedure for Assembling a K.D. Door Frame

1 Set the sill member in place to test the fit. Make any further adjustments to floor structure to accept the sill.

2 Check the length of jamb members and trim the horns as necessary.

3 Remove the sill and jambs, and liberally prime all framing members, including the head, on all surfaces. Let the primer dry thoroughly.

4 Fit the head member into the upper rabbet in one jamb. **Note.** If the door is to be prehung, set the door in position against one jamb, and assemble the frame around it. See Unit 17-2 for procedures for hanging a door.

5 Nail through the jamb into the head with three 12d nails. **Precaution.** Nail no closer than $\frac{1}{2}''$ to the edge of the jamb.

6 Repeat steps 4 and 5 at the opposite jamb.

7 Slide the sill member into its dadoes in the jambs.

8 Check the frame for square. If the door is not prehung, use a pointed rod (Fig. 17-1.1). A framing square isn't long enough for this job.

9 With the frame square, nail the sill in place like the head.

10 Cross nail at all four corners with a pair of 16-penny coated nails (Fig. 17-1.2).

11 Recheck the square.

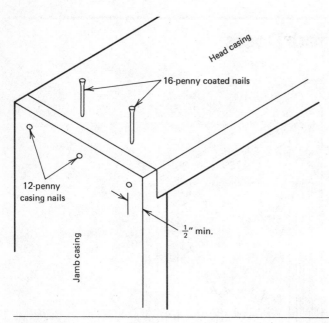

Fig. 17-1.2

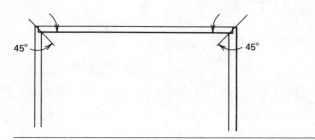

Fig. 17-1.3

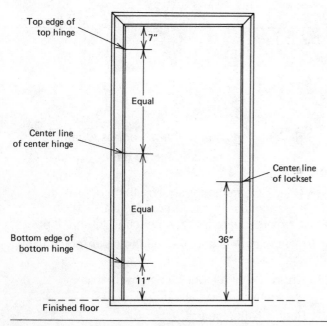

Fig. 17-1.4

12 On the frame draw lines at 45° upward from the corners where the head and jambs meet (Fig. 17-1.3).

13 Center the head casing between the lines, and tack it with 10-penny casing nails.

14 Fit and tack the jamb casings.

15 When the fit is good, complete the nailing.

Procedure for Installing an Exterior Door Frame

1 Flash all four sides of the opening with strips of building paper 10 to 12″ wide.

2 Set the door frame in its rough opening as close to centered as possible, and slide it into place.

3 Set your carpenter's level on the sill, and shim under the jambs if required to plumb the frame.

4 Between each jamb and trimmer stud drive a pair of wedges just above the sill line. **Note.** Work first on one side and then the other so that the space between jambs and studs is as identical as possible.

5 Drive a 16-penny casing nail through the jamb casing at shim level into the stud on each side of the door. Do not drive nails all the way in.

6 Repeat steps 4 and 5 just below the head, making sure that the jambs remain plumb.

7 Check the frame for square either by measuring the diagonals or by opening and closing the door to make sure it operates freely.

8 About $\frac{3}{4}$″ from the outside edges of the casings drive additional 16-penny casing nails into the header and jamb studs. Space these nails 12 to 16″ apart.

9 Drive all nails home.

10 If the door is not prehung, mark lightly on the door frame the locations of door hardware (Fig. 17-1.4). **Note.** Hinges are usually placed 7″ down from the head, 11″ up from the sill, and at the midpoint between. The lock is usually 36″ above the sill, and any deadbolt or other antiburglary hardware up 60″.

11 Add shims behind the frame at these points, and nail them in place with 6-penny casing nails.

12 Trim shims flush with the frame.

13 Countersink all nails, and fill nail holes with putty.

14 If exterior material is already in place, caulk the joint around the frame on all four sides.

15 Install a drip cap over the head casing.
16 Cover the sill with a piece of plywood to protect it against wear until the threshold is laid.

Questions

1 How do you measure the diagonals of a door frame?
2 Assuming that the entrance door is fully equipped with hardware, what is the minimum number of wedges used in the installation procedure?
3 Why do you think door frames assembled on the job should be cross nailed?

Unit 17-2 *Hang a Hinged Door*

The direction a hinged door swings is always shown on floor plans, and the hand of a door is shown in the door schedule. The hand is always stated as you face the door from outside. In the case of an exterior door "outside" means from outdoors. In the case of an interior door "outside" may mean outside a room, outside a closet, or, when the door is between two rooms, on the side where hinges are concealed.

A door may have one of four hands: (Fig. 17-2.1):

- A **right-hand door** (R.H. in a door schedule) swings away from you to the right.
- A **left-hand door** (L.H.) swings away from you to the left.

- A **right-hand-reverse door** (R.H.R.) opens toward you and to the right. Because the edge of the door must be beveled the opposite way from standard, this type of door swing is sometimes called a reverse bevel (R.H.R.B.).
- A **left-hand-reverse door** (L.H.R. or L.H.R.B.) opens toward you and to the left.

Reverse doors are not common except on closets and cabinets. Exterior doors are always either right-hand or left-hand doors. Carpenters must know which before they can prepare doors for hanging.

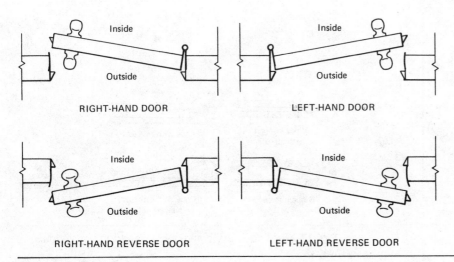

RIGHT-HAND DOOR   LEFT-HAND DOOR

RIGHT-HAND REVERSE DOOR   LEFT-HAND REVERSE DOOR

Fig. 17-2.1 The four hands of a door.

Hinges

Exterior doors swing on butt hinges (Fig. 17-2.2). Each hinge has three parts—a pin, a two-knuckle leaf, and a three-knuckle leaf. The three-knuckle leaf goes on the jamb, and the two-knuckle leaf on the door. The pin fits through the knuckles to hold the leaves together. Quality hinges have removable pins, and are called **loose-pin hinges.** On some types, such as security and hospital hinges, the pin cannot be removed. The tips of pins come in several designs.

Hinges on exterior doors are almost always mortised (recessed) into both the edge of the door and one jamb. This is called **full mortising.** Some hinges, such as decorative strap hinges, are surface mounted, however. Figure 17-2.3 shows in plan several ways of mounting hinges.

The leaves of most hinges are **swaged**—bent slightly at the knuckle for a tight fit (Fig. 17-2.4). When closed, the two leaves are about $\frac{1}{16}''$ apart. Hinges are also made with neither leaf swaged, and with only one leaf swaged.

Door hinges are made in three weights, four common sizes, and six different finishes. For exterior doors hinges are usually standard weight, finished in bronze or brass, and $3\frac{1}{2}''$ or 4″ in size. The dimension refers to both the length and the width of the hinge when open, which are generally the same.

The number of hinges required on any door is always expressed in pairs. Exterior doors of standard sizes require $1\frac{1}{2}$ pairs (three hinges). Doors that open 180° require heavier hinges than doors that open only

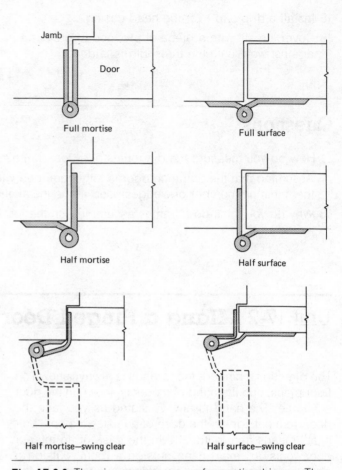

Fig. 17-2.3 The six common ways of mounting hinges. The hinges are especially designed for the method of mounting.

90°. The rule for hinge size on wide-swinging doors is:

$2 \times$ door thickness + thickness of
casing $- \frac{1}{2}$ = hinge width.

With standard exterior doors and casings, then, the hinge size is

$$2 \times 1\frac{3}{4}'' + 1\frac{1}{2}'' - \frac{1}{2}'' = 4\frac{1}{2}''$$

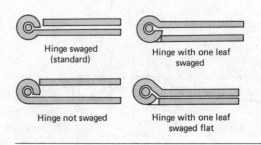

Fig. 17-2.4 Hinges are commonly swaged in one of four ways.

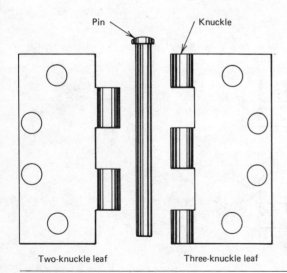

Fig. 17-2.2 Parts of a butt hinge.

Preparation

Doors that are not prehung in their frames are manufactured with stiles that extend beyond the top and bottom rails. The purpose of these horns is to protect the finished corners from damage. Horns must be removed before the door can be hung.

Clearances

If properly installed, door frames should be square. Even when they are not, the shape of the door must be adapted to the shape of the opening with minimum clearances. The width of the door should be $\frac{1}{8}''$ less than the width of the opening, allowing a $\frac{1}{16}''$ clearance on both sides. Clearance at the head should also be $\frac{1}{16}''$. The door should clear a sill without a threshold by $\frac{1}{16}''$, or clear a threshold by $\frac{1}{8}''$.

Tools Required

To fit and hang any door you will find a door jack, a butt gauge, and a butt template handy devices to have available. A **door jack** is a frame for holding a door so you can trim its edges. It is made commercially, but many carpenters build their own (Fig. 17-2.5). A **butt gauge** is a small tool for marking the exact locations of hinge leaves. A **butt template** is an adjustable tool that provides a guide for a router. Other tools needed are a steel tape, framing square, hammer, saw, jointer plane, block plane, router, and wood chisel.

Procedure for Fitting a Door

1 Set any threshold temporarily in position.

2 Measure the height of the opening at both jambs, and the width of the opening at the head, sill, and midpoint.

3 Check the square of all four corners of the frame.

4 From drawings determine how the door will swing, and mark the jamb that will take the hinges.

5 On the head and both jambs, mark the location of the door stop (exterior door frames do not have a separate stop). It is located the thickness of the door from the edge of the frame in the room into which the door swings.

6 Cut stop members to length, and miter the corners.

7 Tack the head stop in place on its marks, then set the jamb stops temporarily.

8 Now compare the two sides of the door to be hung, and lightly mark the best side to face outward.

9 Set the door on a pair of sawhorses, outside up.

10 Mark the hinged side of the door.

11 Working from the hinged side, measure off the final width of the door. If cutting is required, mark the cutting line on the face of the door near the latch edge.

12 Trim any lugs at the top of the door.

13 Working from the top, measure off the final height of the door, and mark the bottom cutting line.

14 Trim the bottom along the cut line, beveling slightly upward toward the outside.

15 Now set the door in a door jack with the hinge side up.

16 Plane the edge to conform to the shape of the jamb, again beveling slightly toward the outside. **Note.** The angle of bevel is the same for all doors, but the amount of bevel varies with the door's thickness (Fig. 17-2.6).

17 Now turn the door latch side up in the jack.

18 Trim about three-quarters of the way to the cutting mark.

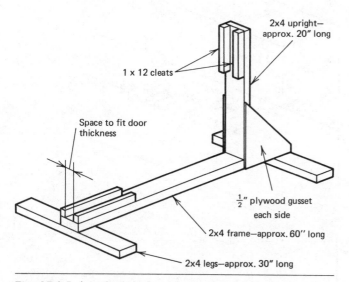

2x4 upright—approx. 20" long

1 x 12 cleats

Space to fit door thickness

$\frac{1}{2}''$ plywood gusset each side

2x4 frame—approx. 60" long

2x4 legs—approx. 30" long

Fig. 17-2.5 A typical design for a carpenter-made door jack. They are also manufactured commercially.

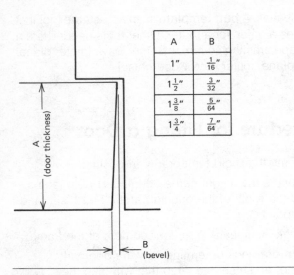

A	B
1"	$\frac{1}{16}$"
$1\frac{1}{2}$	$\frac{3}{32}$"
$1\frac{3}{8}$	$\frac{5}{64}$"
$1\frac{3}{4}$	$\frac{7}{64}$"

Fig. 17-2.6

19 Set the door in its frame and test the fit. Lay two 10-penny nails on the threshold to support the door; they are $\frac{1}{8}$" thick and establish the necessary bottom clearance.

20 Mark any points that need additional planing.

21 Put the door back in the jack. Finish trimming on the latch side and planing elsewhere to match the contours of the frame.

22 Recheck the fit.

23 When the fit is good, relieve the sharp edges of the door with a block plane or scraper.

Procedure for Hanging a Door

1 Remove the threshold, and replace it with a pair of wedges. The midpoint of the wedges should be about equal to the bottom clearance.

2 Set the door in its frame, tight against the jamb on the hinge edge.

3 Drive a wedge on the latch side near the center of the door to hold it in place.

4 Set a pair of 4-penny finishing nails atop the door to maintain the $\frac{1}{16}$" clearance.

5 Gently drive in the bottom wedges until the door has the proper top and bottom clearances.

6 On both the door stile and the jamb mark the locations of the top of the top hinge, the bottom of the bottom hinge, and the center point between these two marks (see Fig. 17-1.4). Use a sharp knife or very sharp pencil.

7 Before removing the door, recheck clearances on all sides.

8 Move the door back into the jack, and mark the length of each leaf on the edge of the stile (A in Fig. 17-2.7). Square the lines across the edge and down the back surface of the stile.

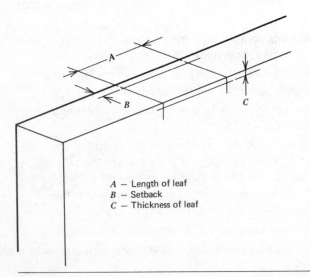

A — Length of leaf
B — Setback
C — Thickness of leaf

Fig. 17-2.7

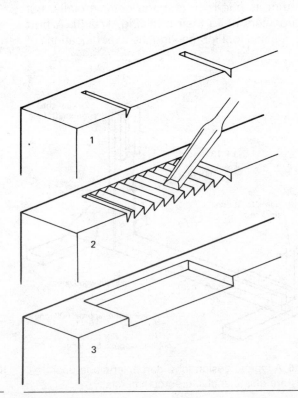

Fig. 17-2.8

9 Mark the locations of the backs of the hinges. **Note.** The backs of hinges are set $\frac{1}{4}$ to $\frac{3}{8}''$ away from the outside edge of exterior doors (B in Fig. 17-2.7). Set one scriber point of the butt gauge at this dimension to make the marks.

10 Set the other scriber point on the butt gauge to the thickness of a hinge leaf (C), and mark this thickness on the face of the stile. **Note.** The mortise for a door hinge is called a **gain**.

11 Cut the gains with a router, using the butt template as a guide. If a router is not available, pare the gain with a wood chisel in the three steps shown in Fig. 17-2.8. **Precaution.** Be careful to make the first two cuts at right angles to the edge of the door, and pare with the chisel at right angles to the face of the door.

12 Repeat steps 8 through 11 to mortise the jamb.

13 Set the hinges in the door mortises with the pins toward the top of the door.

14 Insert the screws toward the back of each hole so that the hinge moves against the back of the gain as you drive them home. **Precaution.** Use only two screws at this time.

15 Remove the pins, and screw the other leaves into the jamb mortises.

16 Test-hang the door, inserting the pins only far enough to hold the door in position. Make sure it swings without binding, and that you have the proper clearances all the way around.

17 a. If the clearance at the jamb is more than $\frac{1}{16}''$, deepen the gain in the jamb.

17 b. If the clearance is less than $\frac{1}{16}''$, remove the door and add cardboard shims behind the leaves to prevent the door from binding.

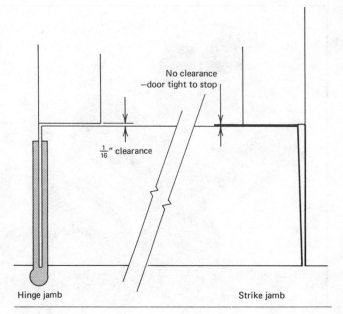

Fig. 17-2.9

18 When all clearances are perfect, insert the remaining hinge screws, and drive pins all the way into the knuckles.

19 With the door closed, check clearances against the stops. There should be about $\frac{1}{16}''$ clearance at the hinge jamb, and zero clearance at the latch jamb (Fig. 17-2.9).

20 Adjust the stop as necessary, and complete the temporary attachment (step 7 in the previous procedure) by driving the nails home, countersinking them, and filling the holes.

Questions

1 What is the correct hand of the door in Fig. 17-2.9?

2 What is full mortising?

3 What is a gain?

4 What are the standard clearances around an exterior door with a threshold?

5 Which edge or edges of an exterior door are beveled? Why?

6 How far are hinges normally set back from the outside face?

7 What two tools may be used to pare the gain?

8 How much clearance should there be between the door and the door stop?

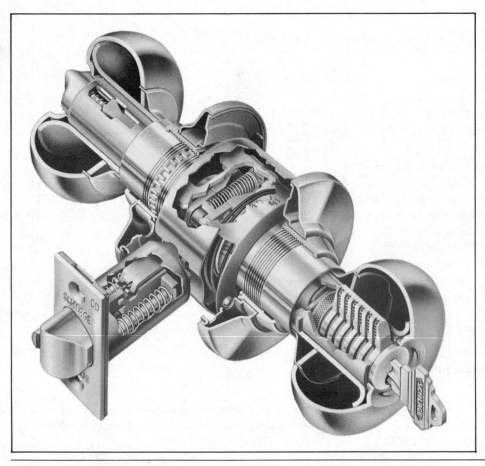

Fig. 17-3.1 The parts of a typical lockset. (Courtesy Schlage Lock Company.)

Unit 17-3   Apply Locking Hardware

The hardware still to be installed on a door is called a **lockset.** This basic set of parts to open, close, and lock a door includes a lock with keys, a latch bolt, strike plate, knobs, and roses (Fig. 17-3.1). A **strike plate** is the metal plate mortised into the jamb that receives the latch bolt and holds the door shut. A turn of the knob retracts the latch bolt and the door opens. A **rose** is the circular base trim around the knob, which holds the lockset in place on the door. An **escutcheon** is a decorative plate often added behind knobs or latches on the outside of exterior doors. There are many designs of escutcheons (Fig. 17-3.2).

Types of Locksets

Locksets are classified by the type of locking mechanism, by the locking function, or by the method of in-

stallation. Most common mechanism is the **cylinder lock.** It has at least five tumblers (Fig. 17-3.3), which are pushed upward against tiny springs by the notches in the top of a key. When the matching key is inserted, all tumblers are in the proper position, and the key can be turned. The wrong key won't set the tumblers properly, and can't be turned. The **bit key lock**—the second type of locking mechanism—is operated by the end of the key.

Of the six types of locking functions, four are found in locksets for exterior doors. (The other two are discussed in Chapter 25). The **exterior door lock** (A in Fig. 17-3.4) is locked from inside by pushing in and turning the knob, or by turning an insert in the knob. To unlock from the inside, you turn the knob or insert. The key turned in the lock on the outside releases the locking mechanism.

The **all-purpose lock** (B) is locked on the inside by pushing in a button. To unlock the door from the inside

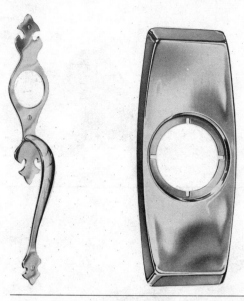

Fig. 17-3.2 Escutcheons are decorative plates that fit behind the exterior rose of a lockset. (Courtesy Schlage Lock Company.)

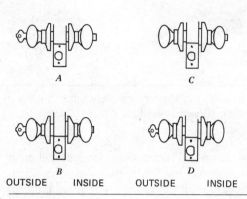

Fig. 17-3.4 The four types of locksets used with hinged exterior doors.

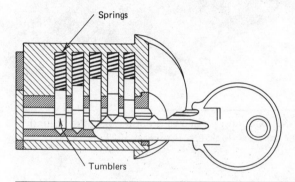

Fig. 17-3.3 Cross section through a cylinder lock showing how a key operates it.

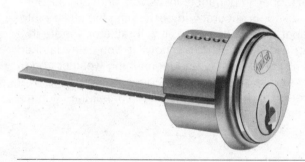

Fig. 17-3.5 Typical night latch. (Courtesy Kwikset.)

you either turn the knob or push the button again to release it. The key releases the button from the outside.

The **button lock** (C) operates like an all-purpose lock, but can be unlocked only from the inside. It has no slot for a key. Its use is limited to secondary exterior doors, such as onto a deck or out of a basement.

The **utility lock** (D) has no button and the knob always turns on the inside. A latch bolt automatically locks when the door is closed, and a key must be used to lock or unlock the door from the outside.

Method of Installation

The locksets installed in exterior doors of most homes today are mortised into the door. Years ago, when the keyhole was in a separate hardware unit located in the stile of the door, installing a mortised lock took great

precision. Development of the cylinder lock simplified the procedure.

Three types of locking devices are surface mounted. A **rim lock** is a type of bit key lock that is fastened to the inside surface of the door. It requires a deep jamb. A key turns the bolt that locks the door, and usually remains in the keyhole. Handles turn the latch separately. A **night latch** (Fig. 17-3.5) has no handles. It may or may not be operated with a key. A button on the inside can be set to hold the latch in a retracted position, or in the strike plate. A night latch is usually used in addition to a lockset. A **deadlock** (Fig. 17-3.6) is similar to a night latch, but has a heavy rectangular bolt in place of a latch. The latch plate of some types must be mortised. Others have a surface mounted plate, and pins in the deadlock fit into rings in the plate to lock the door. Some types of deadlocks are key operated, and some are not.

Manufacturers of mortised (also called **bored-in**)

Fig. 17-3.6 A deadlock is similar to a night latch, but the bolt is rectangular. (Courtesy Kwikset.)

locks provide complete instructions for installation of their locksets. Follow the instructions carefully, because they do vary from manufacturer to manufacturer. They usually provide paper templates that show the exact locations of holes. Some hardware manufacturers provide a special jig that permits extreme accuracy in fitting the lockset.

Of equal importance, however, and all too often overlooked, is accuracy in positioning the strike plate. If a door doesn't fit snugly, it will rattle or vibrate in a strong wind. If it fits too snugly, it won't stay latched when it warps slightly under changing weather conditions. Always allow, too, for a door to sag a little over the years, and set the strike plate so that the latch fits near the top of the hole.

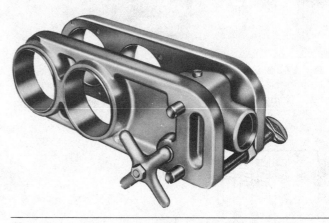

Fig. 17-3.7 A boring jig allows great precision in cutting holes for bored-in locksets. (Courtesy Kwikset.)

Tools

To install a lockset you will need a steel tape, awl, brace and expansion bit, wood chisel, and a screwdriver. A boring jig (Fig. 17-3.7) saves time, but is not a necessity.

Procedure for Installing a Lockset

1 Wedge the door open so that you can work conveniently on the edge and both sides.

2 Measure up from the floor to locate the center of the knob. **Note.** Normally this dimension is 36″. Sometimes, however, homeowners prefer the knobs lower so children can reach them, or higher so they can't.

3 Apply the lockset template to the door and, with an awl, mark the centers of holes in the face and edge of the stile (Fig. 17-3.8). **Note.** If a boring jig is available, a template is not required.

4 From outside the door bore a hole into the face of the stile of the size recommended by the manufacturer. Standard is $2\frac{1}{8}″$. **Precaution.** Be careful to bore at right angles to the stile.

5 When the tip of the expansion bit is just barely visible on the inside of the door, stop boring. Remove the brace and complete the hole by boring from the inside of the door.

6 In the edge of the stile drill a second hole of the size recommended by the manufacturer, again working at right angles.

7 Insert the latch mechanism into the jamb hole, and mark with a sharp pencil the shape of the faceplate (Fig. 17-3.9).

8 Remove the latch, and mortise for the faceplate. **Precaution.** The faceplate must be flush with the edge of the door.

9 Reinsert the latch mechanism, with the beveled side of the latch bolt toward the outdoors. Insert and drive home any screws. **Note.** Some types of latches stay in place with a friction fit, and have only a small rim without screw holes.

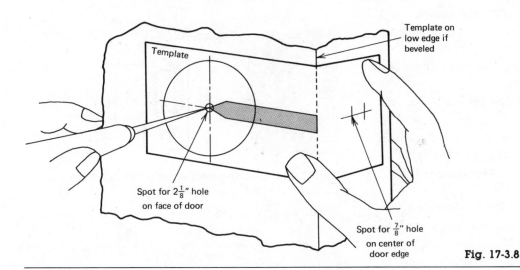

Template

Template on low edge if beveled

Spot for $2\frac{1}{8}$" hole on face of door

Spot for $\frac{7}{8}$" hole on center of door edge

Fig. 17-3.8

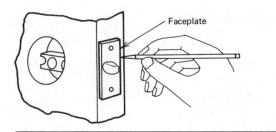

Faceplate

Fig. 17-3.9

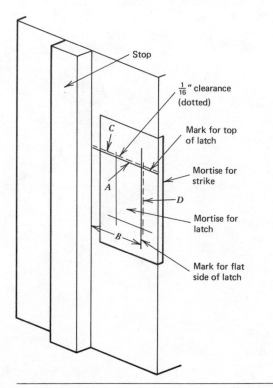

Stop

$\frac{1}{16}$" clearance (dotted)

C

A

B

D

Mark for top of latch

Mortise for strike

Mortise for latch

Mark for flat side of latch

Fig. 17-3.10

10 From outside the door fit the knob in place through the latch, and push it in until the rose is tight against the stile. **Precaution.** The keyway should be down so that the key is inserted as shown in Fig. 17-3.3. The lockset will operate if installed upside down, but all locksets that are keyed alike should be installed identically.

11 Fit the inside knob over the stem, and push it until the rose is tight against the stile.

12 Test the operation of the lockset.

13 If it works smoothly, screw through the inside rose to complete the assembly.

Procedure for Locating the Strike Plate

1 Remove the wedge holding the door open, and carefully close the door.

2 With a pencil mark the point where the top of the latch hits the jamb.

3 Square this line across the jamb from edge to stop (*A* in Fig. 17-3.10).

4 Measure the distance from the inside surface of the door to the flat surface of the latch.

5 Measuring from the stop, mark this distance (*B*) on the jamb.

6 Mark this edge with a vertical line.

7 Mark the other two sides of the latch hole.

8 Mortise the hole for the latch. **Precaution.** If the strike plate has only a hole for the latch (Fig. 17-3.11, left), cut $\frac{1}{16}$" above line *A* and $\frac{1}{16}$" outside

line *B* (dotted lines *C* and *D*). If the plate has a metal recess (Fig. 17-3.11, right), allow for the thickness of the recess equally on all sides.

9 Clean the gain, keeping the vertical cut closest to the stop as flat as possible.

10 Close the door to make sure the latch drops into the mortise.

11 When the fit is good, fit the strike plate carefully in position, and mark its shape on the jamb.

12 Mortise for the strike plate. **Precaution.** The mortise should be small enough so the strike plate will stay in place without screws.

13 Fit the strike plate in its mortise, and close the door again to retest the fit.

14 When the fit is right, install the strike plate with screws.

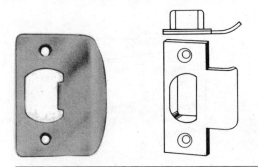

Fig. 17-3.11 [Courtesy Kwikset (left), Schlage Lock Corporation (right).]

Questions

1 Identify the parts of a lockset in the sketch.

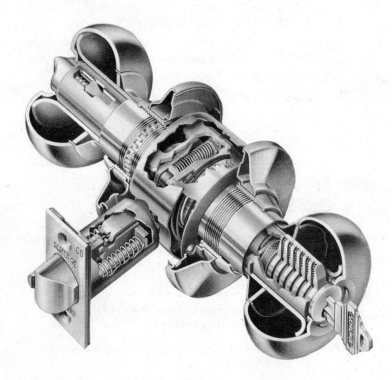

2 How do the exterior door lock and the all-purpose lock differ?

3 What is the standard height above the floor for a lockset?

4 How can you determine whether a lockset is installed right side up or upside down?

5 Why do you think the cuts in step 8, "Procedure for Locating the Strike Plate," are not on the pencil lines?

Unit 17-4 Install Sliding Glass Doors

Sliding glass doors, also called patio doors, may be made of wood or metal. Their frames are provided by the manufacturer, along with weatherstripping and hardware. Heads and jambs of frames are usually made of the same material as the doors, but the sill of both types is a heavy aluminum extrusion.

The frames are designed to fit into their rough openings with $\frac{1}{2}''$ clearance at the head and jambs. Some builders prefer to have a wood or stone sill or a rowlock course of bricks already in place before the door frame is installed. Others prefer to install and level the frame first, then fit the sill in later. Trim at the head and jambs always follows installation of the doors.

The doors move over tracks in the sill on nylon rollers. At the head they move either in channels or on guides that hold them in place. At jambs they slide into pockets made airtight with weatherstripping. Almost all patio doors are equipped with insulating glass, eliminating the need for storm doors. A screen section one door wide moves on its own track ; it may fit toward the inside or the outside of the frame.

In a two-door unit both doors may slide, or one may be fixed. In a three-door unit only the center door slides. Installation procedures are generally the same for all types of doors. Tools required are a carpenter's level, screwdriver and, with wood doors, a hammer.

Procedure

1 Carefully read the manufacturer's instructions for assembly of the frame. They may vary from these procedures.

2 Begin assembling the frame by nailing or screwing the head to the jambs.

3 Caulk the joint between the sill and jambs, and attach them together.

4 Caulk the subfloor thoroughly across the door opening.

5 Set the assembled frame in the caulking, and check the sill for level. If necessary, shim at the ends of the sill and behind screw holes.

6 With the sill level and the frame centered in the rough opening, add wedges at the jambs at the level of the locking mechanism.

7 Check the jambs for plumb, check the corners for square, and recheck the sill for level.

8 With the door frame in its proper position, attach it at the sill with screws provided by the manufacturer.

9 Attach the frame at the jambs with $1\frac{1}{2}''$ wood screws driven through the shims.

10 Attach the head with screws. **Precaution.** The head may be shimmed to maintain level. If not, be careful not to drive the screws far enough to bend the head out of square.

11 Install the drip cap as required by the manufacturer.

12 Fit the inner door in place by raising it to the top of the recess for it in the head, and lowering it onto the bottom track. **Note.** With some models it is necessary to remove the top screen track in order to have sufficient clearance.

13 If required, install closures at the head and sill, and replace the top screen track.

14 If the inner door is not fixed, test its operation to make sure it slides easily and does not bind. **Note.** To adjust the height of the door, turn the screws that raise or lower the rollers. These screws may be in the side or in the end of the bottom rail (Fig. 17-4.1).

15 Fit the outer door in place in the same manner as the inner door.

16 Test the door's operation and adjust the height of the rollers.

17 Attach the handles and locking assemblies. **Precaution.** The latch and locking device must not be installed until after the height of rollers has been adjusted.

18 Install the screen by setting bottom rollers on their track, and pushing down the top rollers until they snap onto the top track.

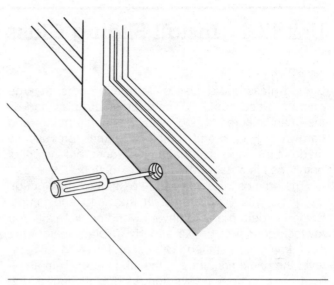

Fig. 17-4.1

Questions

1 What is the difference in width between the rough opening and the frame for sliding glass doors?

2 Why do you think the door frame should not be shimmed at the head?

3 How is the height of a sliding door adjusted?

Unit 17-5 Install Garage Doors

Two generations ago, when garages were first built to shelter the newly invented automobile, they were placed as far from the house as possible. Probably the reason for this was that they replaced the barn, and the barn smelled. Garages were just wide enough for one car, and the doors were hinged to swing outward.

Today a few garages are still built with hinged doors. But most of them have an overhead-type door. These doors are 6½′ or 7′ high, and come in widths of 8′, 9′, and 10′ for one-car garages and 15′, 16′, and 18′ for two-car garages. There are two types of overhead doors, and their operation is quite different.

Overhead-Type Doors

The swing-type overhead door is made in one piece, usually of aluminum or molded fiberglass. Its frame is aluminum. A pair of rollers mounted on the sides of the door at the top move in a horizontal track anchored to the ceiling structure just below head level. Mounted beside the frame at about the midpoint of the door's height are a pair of arms counterbalanced by a spring. As the door opens, the top moves inward on the tracks, and the bottom swings outward and upward. In its final open position the door lies horizontal with its bottom edge visible below the head of the door.

Sectional overhead doors are usually made with stiles and rails of 1⅜″ Douglas fir or spruce and panels of hardboard. The panels may be flush, raised or recessed. One row is often glazed. Each door is four or five panels high, hinged together on the inside.

Except for the bottom edge of the bottom panel and the top edge of the top panel, the edges where panels meet are shiplapped. Two types of hinges connect the panels (Fig. 17-5.1). Edge hinges have a sleeve into which the axles of the rollers fit. Center hinges are

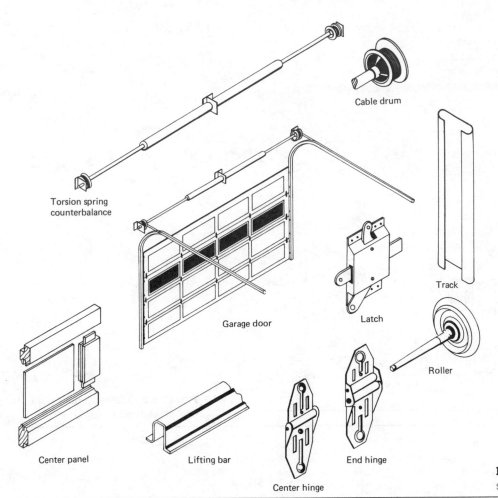

Torsion spring
counterbalance

Cable drum

Track

Garage door

Latch

Roller

Center panel

Lifting bar

End hinge

Center hinge

Fig. 17-5.1 The parts of a
sectional overhead-type door.

smaller and flatter, and are spaced about every 4'
across the door. Hinges are usually shipped separately, but may be attached to the panels at the factory.

At the ends of each panel are rollers. These run in a
track that begins just above the garage floor level,
rises at a slight angle to true vertical to just below door
head height, then curves gently 90° and continues horizontally above head height (Fig. 17-5.2).

The tracks come in two sections, a straight lower
section and an upper section with the curved track at
one end. Brackets attached to the door frame support
the vertical section, and hold the two together. Tracks
are punched and slotted for the brackets. Some manufacturers attach the brackets at the factory. Counterbalancing springs may be the torsion type mounted just
above the door when closed, or the stretch type, which
lie above the horizontal sections of track.

Brackets are attached to jamb framing with lag
screws and washers, and the center lines of the screws

fall about $2\frac{1}{2}$" from the edge of the framing. If their center lines hit the seam between framing members, or if
the garage wall is masonry, add a 2 × 4 at least 8'-3"
long flat at each jamb, and flush with the rough
opening to assure solid attachment.

Overhead-type doors are not airtight, but they are
effectively weathertight at the jambs and head. Where
they meet the garage floor, however, the concrete
apron must pitch away from the garage for drainage,
and the edge of the door must be shaped to the pitch
and to accommodate any unevenness in the floor. Wood
doors must be **scribed** (marked and cut to fit). On metal
and plastic doors a rubberized gasket will serve the
same purpose.

To install an overhead-type door you will use a
hammer, wrenches, folding rule or steel tape, level, pliers, and a scriber (Fig. 17-5.3). A **scriber** is similar to a
pencil compass.

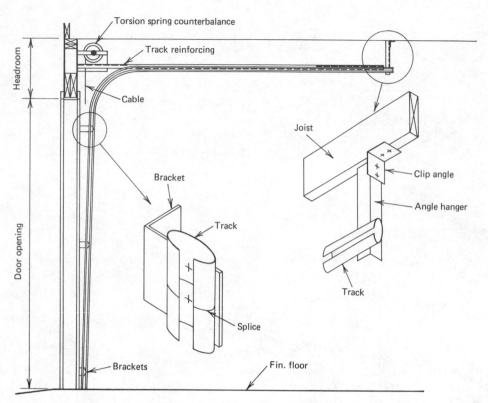

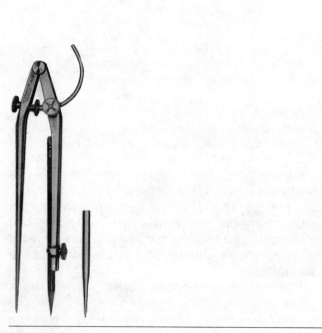

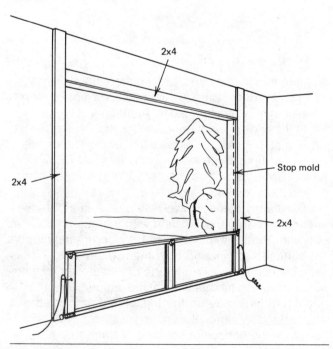

Fig. 17-5.2 Overhead door track and its method of support.

Fig. 17-5.3 A scriber for marking irregular cuts. (Courtesy The L.S. Starrett Co.)

Fig. 17-5.4

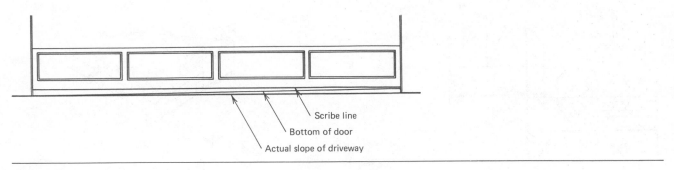

Fig. 17-5.5

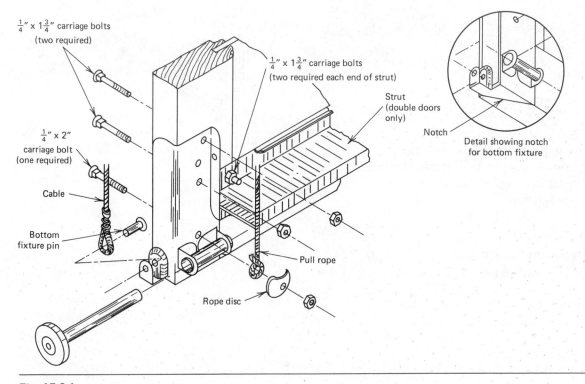

$\frac{1}{4}'' \times 1\frac{3}{4}''$ carriage bolts
(two required)

$\frac{1}{4}'' \times 1\frac{3}{4}''$ carriage bolts
(two required each end of strut)

$\frac{1}{4}'' \times 2''$
carriage bolt
(one required)

Cable

Bottom
fixture pin

Strut
(double doors
only)

Notch

Detail showing notch
for bottom fixture

Pull rope

Rope disc

Fig. 17-5.6

Procedure

1 Read the manufacturer's instructions for assembly carefully. They may vary from this procedure.

2 On each side of the door opening nail a 2 × 4 or 1 × 4 for attachment of track brackets (Fig. 17-5.4).

3 Cut and fit a stop molding around the opening, and tack it to the frame so that it extends $\frac{1}{8}''$ beyond the inside edges of the jambs (Fig. 17-5.7).

4 Set the bottom panel of the door upright on the floor and centered in the opening. **Precaution.** In this position the flush edge is downward and the flange of the shiplapped edge is on the inside.

5 Check the fit at the floor. If the floor is level across the opening, go to step 8.

6 With a scriber mark the contours of the floor across the bottom of the panel (Fig. 17-5.5).

7 Cut along the scribed line.

8 To the bottom panel install the bottom fixture, notching the bottom rail for flush attachment (Fig. 17-5.6).

9 Install the lift handle at the center of the bottom rail.

10 Set the bottom panel back in position, and hold it in position on each side with nails driven as shown in Fig. 17-5.7.

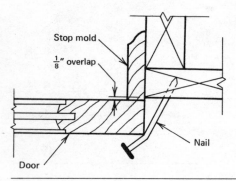

Fig. 17-5.7

11 Stack the second panel on the bottom panel, line up the end stiles, and tack it in position.

12 Connect the two panels with edge hinges and center hinges. **Note.** Spacing of center hinges is about every 4' o.c.

13 Repeat steps 11 and 12 for the remaining panels, and install the top fixture.

14 Using the template and instructions provided by the manufacturer, install the lock. **Note.** The lock goes on the second panel of four-panel doors, and on the third panel of five-panel doors.

15 Fit rollers into edge hinges on both sides.

16 To the straight sections of track attach brackets, if they were not attached at the factory.

17 Fit the vertical sections of track over the rollers, and fasten the brackets with $\frac{5}{16}'' \times 1\frac{3}{4}''$ lag screws. Gap $\frac{1}{2}''$ between track and door. **Precaution.** If the bottom panel was scribed, position the track above the floor on the low side a distance equal to the amount trimmed. The tops of the two vertical sections of track must be at exactly the same level.

18 Bolt the curved sections of track to the lower track at a hanger (Fig. 17-5.8). **Precaution.** Make sure that the spacing between supports is the same as the spacing between brackets.

19 Support the horizontal track temporarily so that the track rises about 1" toward the open end.

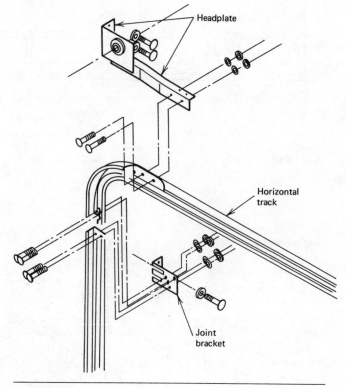

Fig. 17-5.8

20 Install permanent supports for the end of the horizontal track.

21 Add door stiffeners and tension braces at the locations recommended by the manufacturer.

22 Install the counterbalancing mechanism over the door, following the manufacturer's instructions closely.

23 Remove the nails driven in step 10 to hold panels in place.

24 Test the door for smoothness of operation, and adjust the positions of rollers and brackets if necessary.

25 Weatherstrip the bottom edge of the bottom panel if required.

Questions

1 What are the two types of counterbalancing springs?

2 Why do you think 2 × 4s may sometimes be needed for lag screws beside the garage door opening?

3 What two steps may be necessary to provide a watertight fit at the garage floor?

4 How can you identify the top and bottom panels of an overhead-type door?

18

Exterior Wall Finishing

As they decide what exterior material to use on the walls of a house, architects or designers keep in mind the effect of those materials on the appearance of that house. Exterior appearance, more often than the floor plan, attracts prospective homeowners to buy. For a strong horizontal effect with smooth texture, the material most commonly chosen is siding (Fig. 18-0.1). Because siding takes paint readily, the exterior can be any color. For a horizontal effect with more texture, the material may be shingles or shakes (Figs. 18-0.2 and 18-0.3). For a vertical effect with interesting shadow pattern, drop siding does the job. For a vertical effect with either smooth or rough texture, boards and sheet materials offer the best results (Fig. 18-0.4).

Siding, shingles, and sheet materials may be used together, or with masonry materials such as brick and stone, to achieve multiple effects. An important rule of good design, however, is that no more than two exterior materials should be used together at one time.

Siding

The term **siding** is often used to describe any nonmasonry material applied to the exterior of a house. Technically, however, siding is an exterior material applied in strips, as opposed to small pieces or sheets. In the early Colonial days in the United States wood siding was the material most often used on houses, and it still is. Today, however, siding is also made of such wood products as hardboard, plywood, and particle board, and of such nonwood products as aluminum, steel, and vinyl.

Most siding is applied horizontally in courses, with the bottom of one course overlapping the top of the course below. The front and back surfaces may taper toward each other or be parallel. The effect is the same

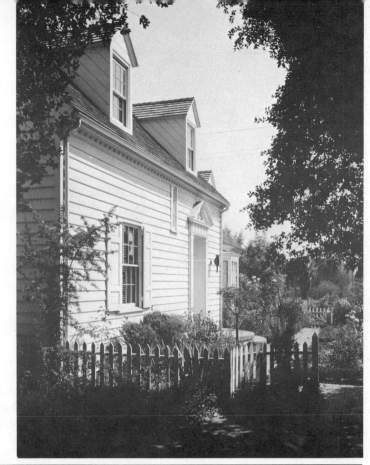

Fig. 18-0.1 Of all exterior materials wood siding is the most popular for achieving strong horizontal lines on a house. Note that the siding on the dormers is flush and laid with the slope of the roof. (Courtesy California Redwood Association.)

Fig. 18-0.2 Wood shingles combine strong horizontal lines, a pleasant texture, and a variation in wall color. (Courtesy Red Cedar Shingle & Handsplit Shake Bureau.)

—strong horizontal shadow lines created by the thick butts of the lengths of siding.

The method of application of siding varies with the raw material from which it is made, and the shape of a section through it. Plain beveled siding (see Unit 18-1) is the most popular because the amount of exposure of the courses can be adjusted to vertical dimensions of the wall. Rabbeted beveled siding (see Unit 18-2) does not permit this flexibility, but the courses are easier to align. Lap siding made of wood products (Unit 18-3) has parallel surfaces. Manufacturers of both metal siding (Unit 18-4) and vinyl siding (Unit 18-5) also make the necessary fittings and accessories required for the specialized method of attachment.

Shingles

The same wood shingles and shakes used to cover roofs (refer to Unit 15-3) are also used on walls. The method of application is different, however, and is described in Units 18-6 and 18-7. Allowable exposures are much greater on walls than on roofs.

One type of wood shingle exclusively for walls is manufactured with a backing of fiberboard. These oversized shingle units—$46\frac{3}{4}''$ wide and more than $16''$ high—are specified by designers where their scale is appropriate to the size and shape of the building.

The asbestos-cement shingles available for walls are larger than similar shingles for roofs (refer to Unit 15-4), and for that reason are often called **siding shingles.** The exposure must be carefully worked out, because there is little leeway for adjustment in application (Unit 18-8).

Fig. 18-0.3 Asbestos-cement siding shingles are used both on new houses and as a resurfacing material in exterior remodeling. (Courtesy Supradur Manufacturing Corp.)

Boards are a uniform $\frac{3}{4}''$ thick. **Drop siding** is $\frac{3}{4}''$ thick at its thickest point, and its surface is milled in a wide variety of shapes, some of them patented. For this reason drop siding is also known as **patent siding.** Its edges are rabbeted for a close fit. The effect is a patterned exterior that changes as outdoor light conditions change.

Boards

When narrow exterior materials are applied for vertical effect, such as plain boards with battens or drop siding, they must be nailed to a solid backing. If sheathing is plywood, these materials are nailed directly into the sheathing as described in Unit 18-9. If sheathing is a softer material, such as insulation board or gypsum, then the boards must be fastened to horizontal nailing strips. These strips may be laid over the sheathing, or fitted between sheets as described in Unit 18-10.

When applied horizontally, boards and drop siding are usually nailed over sheathing into studs. On such buildings as unheated garages and sheds, however, they are sometimes applied directly to studs without sheathing.

Sheet Materials

Two types of sheet materials have found their place in residential construction. One is plywood. Exterior plywood usually has shiplapped edges, and is applied vertically (see Unit 18-11). Its surface may be finished or unfinished, textured or plain. Sheets from $\frac{5}{16}''$ up to $\frac{1}{2}''$ thick must be applied over sheathing. Panels of greater thickness may be used as a combination sheathing and finishing material.

Asbestos-cement sheets are one of the few materials that provide a plain, smooth exterior surface. Sheets are $4 \times 8'$ and have square edges. As a result the joints must be protected against moisture (see Unit 18-12). Asbestos-cement may be painted or left its natural color, which is a pearl gray.

Fig. 18-0.4 Exterior plywood is manufactured in a variety of interesting textures and is usually applied vertically.

(Courtesy California Redwood Association.)

Other Exterior Materials

The carpenter is seldom involved with most other exterior materials, such as brick, stone, masonry blocks, and steel sheets. On occasion, however, he may be called on to apply three types of materials that imitate brick and stone. Their use is limited today, but they are likely to become more common as technology and appearance improve.

One of these materials is asphalt that comes in rolls like roll roofing. Asphalt siding is a low-cost, low-quality material seldom used on homes, but is adequate on farm and utility buildings. It protects the structure of the building it covers, but does little for its appearance.

Another is thin bricks and there are several types (Fig. 18-0.5). Some are high quality clay products man-

ufactured with a flange at the top edge. The flange comes predrilled, and the bricks are attached with nails. Another type comes in strips of several bricks, and the strips are attached to sheathing with metal clamps. There are special pieces for turning corners. A third type of imitation brick consists of lightweight vermiculite units that are applied individually with mastic. They can be cut with a hacksaw. All three types are approximately an inch thick, and the joints between bricks must be filled either with mortar applied with a trowel or with caulking compound applied with a gun. Unlike brick-patterned asphalt siding, thin bricks look like bricks when the wall is finished. Their chief advantages are lower cost and a thinner foundation wall. Because of the variety of methods of application, manufacturers' instructions must be carefully followed. No procedures are given here.

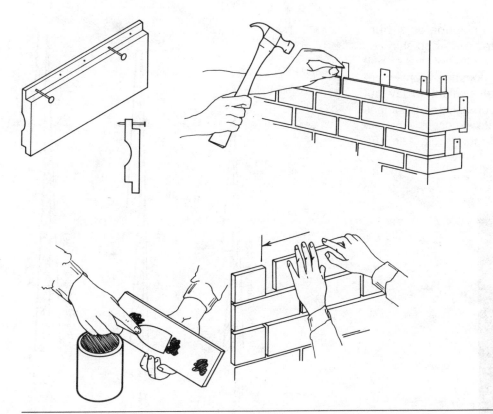

Fig. 18-0.5 Typical imitation masonry units and their method of application.

Wall Preparation

As long as any exterior material has joints in it, there is always the possibility of leakage during severe weather conditions. The best protection against moisture penetration is building paper, and its use is re-quired in many of the procedures in this chapter. Board sheathing must be covered with a layer of building paper applied with a 4″ lap at all seams. This paper is required over any type of sheathing if the exterior material is stucco. It is not required over sheathing grade plywood or over insulating or gypsum sheathing that has been factory-treated with a water-repellent finish.

Questions

1 Name seven materials from which siding is made.

2 What two exterior materials are available in sheet form?

3 What type of siding may be applied vertically?

4 The exterior materials used on any one house should be limited to how many?

5 Behind what exterior material is building paper required over any type of sheathing?

Unit 18-1 Apply Beveled Wood Siding

Beveled wood siding, also called **clapboards** or **bungalow siding,** has a tapered shape like wood shingles and must be applied horizontally with the thicker butt edge down. In colonial days the common widths were 4″, 5″, and 6″. These widths are still available today, but 8″ and 10″ widths are the most popular. The thickness of the butt edge and the allowable exposure vary with the width (Fig. 18-1.1). Lengths range from 4 to 16′, and the siding is supplied in random lengths.

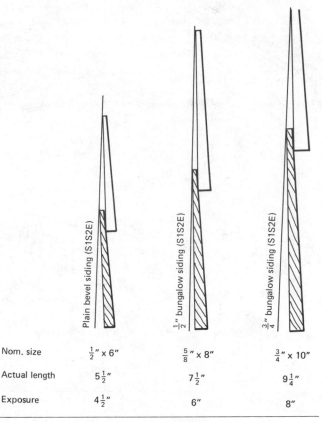

Nom. size	$\frac{1}{2}″ \times 6″$	$\frac{5}{8}″ \times 8″$	$\frac{3}{4}″ \times 10″$
Actual length	$5\frac{1}{2}″$	$7\frac{1}{2}″$	$9\frac{1}{4}″$
Exposure	$4\frac{1}{2}″$	6″	8″

Fig. 18-1.1 Sizes, exposures, and minimum overlaps of typical widths of beveled wood siding.

Siding is milled from clear, all-heart grades of softwoods—cedar, redwood, white pine, sugar pine, and cypress. The allowable moisture content is lower than in structural lumber—about 12% in most of the country and 9% in the drier climate of the southwestern states.

One surface of beveled siding is smooth and the other rough. If specifications call for painting, apply the siding with the smooth side exposed. If the siding is to be stained, either side may be exposed, and the amount of texture desired governs the decision.

For protection against the elements until it can be painted or stained, siding must be treated with a preservative. Some siding is pretreated at the mill. If not, each piece should be brushed with or dipped in a water repellent solution so that all surfaces, including ends and edges, are treated. While they are drying, lengths of siding should be stacked on strips of wood called **stickers** that permit air to circulate. The stickers should also be treated so that they don't draw repellent out of the siding.

Preparation

Before the first piece of siding can be applied, the carpenter must determine what happens at the foundation, at the eave, and at inside and outside corners of the house. Complete working drawings include details at these points. If no details are shown, begin planning at the foundation.

At the Foundation

Some sort of barrier must be provided to keep rain and snow from entering the wall and puddling on the top of the foundation wall. The primary purpose of this barrier is to carry water below the level of the top of the wall, where it can drip off and run away from the house. The treatment at the foundation establishes the bottom limit of the courses of siding.

There are two basic solutions, and many variations

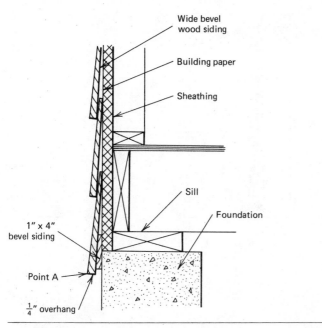

Fig. 18-1.2 Section at a foundation wall with a backer strip behind the bottom course of beveled siding or shingles.

of each. The simplest method is to nail a beveled board at about its midpoint into the sill plate (Fig. 18-1.2). This backer strip may be a piece of beveled siding, or a 1 × 2 or 1 × 3 strip beveled at the top. In regions with heavy rainfall, some builders use a beveled 2 × 3 to direct the water as far as possible from the foundation wall. The bottom course of siding must extend at least $\frac{1}{4}''$ below the bottom of the board as a drip edge. The butt edge of this first course (point A in Fig. 18-1.2) is the bottom limit for calculating exposure.

The other solution, used on better quality houses, is to provide a water table. A **water table** consists of a board and a drip cap or drip strip (Fig. 18-1.3). The board is $1\frac{1}{4} \times 6$ or $1\frac{1}{4} \times 8$ (these are nominal dimensions) that extends at least 1″ below the top of the foundation wall. The bottom edge may be squared or beveled. The top edge is square when covered with a drip cap—a stock shape at most lumber yards. The top edge is beveled when covered with a beveled drip strip, which is cut on the job. A backer strip establishes the angle of slope of the bottom course and provides a surface against which to nail. With a water table the bottom dimension point for calculating exposure (point A in Fig. 18-1.3) is the upper end of the slope of a drip cap or the bottom edge of the backer strip over a drip strip.

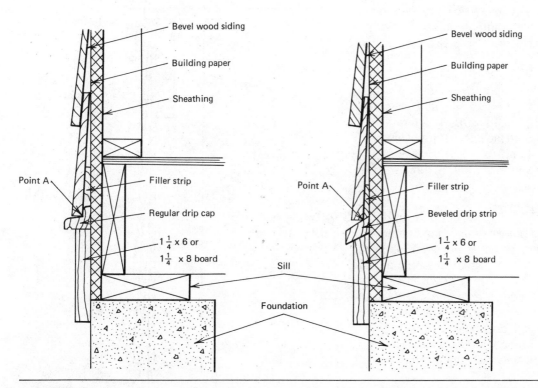

Fig. 18-1.3 Sections at a foundation wall with water tables below the bottom course of siding or shingles. The drip cap (left) and drip strip (right) serve the same purpose of directing water away from the foundation.

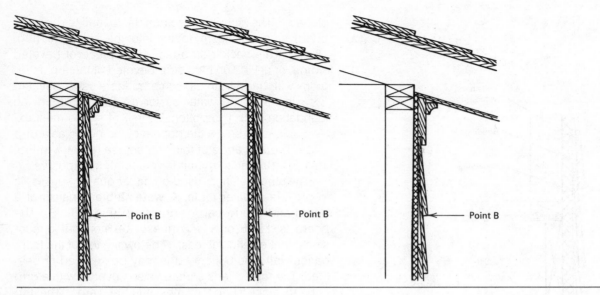

Fig. 18-1.4 Three ways to trim the top of a wall at an overhanging eave.

At the Eaves

When the roof has an overhang, the wall may be trimmed in three ways at eaves. Siding may continue to the soffit, and a crown or bed molding added to cover the seam (Fig. 18-1.4, left). The simplest but least attractive solution is to apply the top course of siding in reverse as a trim piece (center). A more decorative answer is to stop the upper siding course short of the soffit and trim with a rabbeted fascia board and cove molding (right).

At side walls where there is no overhang, the common treatment is to fill the gap between the top course of siding and roof sheathing with a filler strip, then trim with a flat fascia board (Fig. 18-1.5).

For calculating exposure the upper dimension point is point B in Figs. 18-1.4 and 18-1.5. This point can be adjusted slightly by varying the width of trim.

Planning the Exposure

Points A and B mark the bottom lines of the bottom and top courses of siding. It is possible but not good practice to use only these two points in determining the best exposure. The wall will not only look better but be less likely to leak at windows and doors if the butt edges of courses also line up with the bottoms of window sills and the tops of door and window frames. When the heads of doors and windows don't line up, plan the exposure to fit windows; there are more of them.

The ideal exposure is one that is uniform from top to bottom of the wall for best appearance, and is the max-

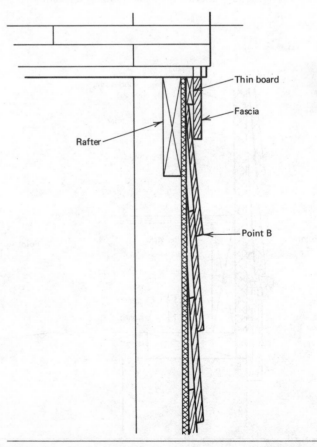

Fig. 18-1.5 Typical treatment at the rake of a roof or at an eave without an overhang.

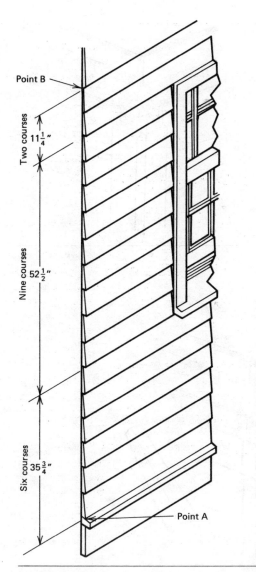

Fig. 18-1.6 Measuring points for calculating the best exposure and the number of courses of siding or shingles required.

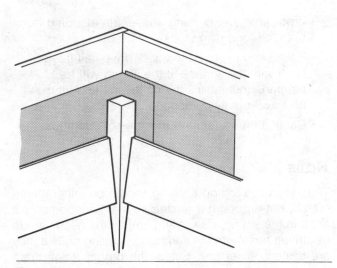

Fig. 18-1.7 Use a square board at an inside corner.

imum allowable for greatest economy. To come as close as possible to the ideal takes some calculations. To illustrate the method of establishing exposure, assume that 8″ siding is specified. Assume also that the dimension from point A to window sills is $35\frac{3}{4}″$, from sills to heads is $52\frac{1}{2}″$, and from heads to point B is $11\frac{1}{4}″$ (Fig. 18-1.6).

Siding up to 6″ in width must overlap 1″, and the minimum overlap for wider siding is $1\frac{1}{4}″$. Therefore the maximum exposure (see Fig. 18-1.1) is 6″. Divide 6 into $35\frac{3}{4}$, and you find you need six courses, each with an exposure of about $5\frac{15}{16}″$. Divide 6 into $52\frac{1}{2}$; and you find you need nine courses with an exposure of $5\frac{13}{16}″$. Divide 6 into $11\frac{1}{4}$; two courses are needed with an exposure of $5\frac{5}{8}″$ each.

These three exposures are all within the maximum allowable, and the difference between them will never be noticed. In practice, a variation of up to $\frac{1}{2}″$ is permissible.

In this example apply the first six courses with an exposure of $5\frac{15}{16}″$ and the remaining courses at $5\frac{13}{16}″$. Adjust the height of the trim at the eave; point B now moves upward $\frac{3}{8}″$.

At Inside Corners

Where walls come together to form an inside corner, the simplest and most common solution is to nail a board $1\frac{1}{4}″ \times 1\frac{1}{4}″$ into the corner (Fig. 18-1.7). The ends of siding courses must be cut square for a tight fit.

At Outside Corners

Outside corners may be finished either by butting or mitering, and there are several ways of accomplishing each type of corner. The simplest method of corner finishing is with a one-piece molding called a **corner post** or **corner bead** (Fig. 18-1.8, left). The molding measures $1\frac{1}{4}″$ on each side and is manufactured in 10′ lengths.

Another method that brings a similar result is to fit two boards of equal width (but neither less than 1″ wide) at the corner, and fill the gap between them with a length of quarter-round or cove molding (Fig. 18-1.8, center). A third method of butting is to rip a $1\frac{1}{4}″$ board into two pieces, one an inch wider than the other, and butt the two widths (Fig. 18-1.8, right). The width of these corner boards may vary from 2 up to 8″ on a side, depending on the size of the house and the exposure of the siding.

The corners of thicker siding may be mitered in a miter box. This is a slow and painstaking process, however, especially if the structural corner is not quite square and plumb. Procedures for mitering are shown in this unit. More common today is the use of metal cor-

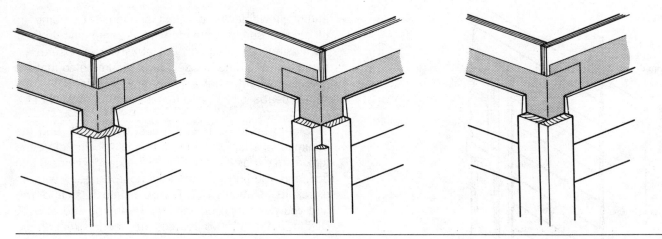

Fig. 18-1.8 Three types of trim for finishing an outside corner.

ners (Fig. 18-1.9). Courses of siding gap slightly at the corner, and the metal corners cover the gap. See procedures in Unit 14.4.

Installation

On one-story houses, siding is installed from the bottom up. On a two-story house, where the upper level has siding and the lower level is masonry, it is better to work from the top down. In this way you can work from bracket scaffolding, which remains below the level of the finished surface. Each course is nailed only at the top until the course below is slipped underneath it. You complete attachment by nailing through the bottom of the upper course.

Here are some basic rules to follow while installing wood siding.

- Always prime the ends of all pieces after they have been cut to length.
- Apply the siding with one nail per stud, driven at a point just above the upper edge of the course below (Fig. 18-1.10).
- Use a spacer block (Fig. 18-1.11), notched to the exposure you have decided upon, to assure uniform exposure throughout the wall.
- Where two lengths of siding butt together, cut them for as tight a fit as possible. Some carpenters cut siding about $\frac{1}{16}''$ long, then spring the piece into place by bowing it.
- The ends of all lengths of siding above the first course must fall on the center lines of studs.
- Predrill all nail holes in fir siding. Predrill nail holes near ends of pieces to prevent splitting.

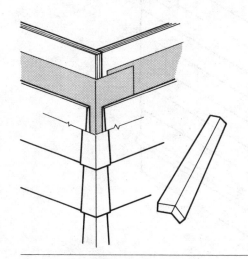

Fig. 18-1.9 Metal corners give the appearance of mitering. The ends of siding are cut square and gapped slightly.

- If possible, avoid using any length of siding less than 32″ long (two studs).
- If the siding is to be stained, drive nail heads flush with the surface. If the siding will be painted, countersink the nailheads and fill the holes with putty compound.
- Caulk all joints around all sides of openings.

Nails

The nails used to apply siding should be either aluminum or hot-dipped galvanized so that they do not rust or stain the siding. Use siding nails if the heads are to be driven flush with the surface, or casing nails if they are to be countersunk. Select the proper length from the table right.

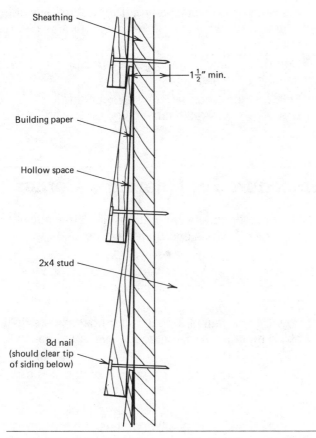

Fig. 18-1.10 Where and how to nail beveled wood siding.

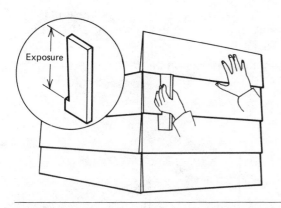

Fig. 18-1.11 A spacer block notched to the correct exposure assures proper alignment of siding courses.

Tools

For the typical siding job you will need a portable power saw, hammer, chalk line, carpenter's level, spacer block, and a paintbrush. You may also need a nail set, block plane, rasp, combination square, and a miter box.

Procedure With a Water Table

1 At the corners of the walls locate point A and mark it on the sheathing.

2 From point A measure down to the bottom edge of the water table and mark its location at each corner.

3 Snap a chalk line on the foundation wall between these corner points. Carefully check the line for level, preferably with an optical level.

4 Cut inside and outside corner boards to length, if required. They should extend from the soffit to the chalk line.

5 Prime the cut ends and all four sides with water-repellent preservative if corner boards are to be stained, or with a prime coat if they are to be painted.

6 Install corner boards with noncorrosive nails long enough to penetrate $1\frac{1}{2}''$ into studs.

7 Cut and prime the two pieces of the water table.

8 Nail the water table in place above the chalk line. **Note.** Space nails 16" o.c., and check the level of the water table as you nail. **Precaution.** Because the butt edges of all courses of siding will be parallel to this base line, it is very important that the starting point for the butt edge of the bottom course (point A) is absolutely level.

9 Tack the primed filler strip in place atop the drip cap or drip strip.

10 Set the first length of primed siding at a corner on the water table. **Note.** If the house has an inside corner, start there and work outward in both directions. Otherwise start at outside corners and

	Type of sheathing			
Siding thickness	Wood	Plywood	Gypsum	Fiberboard
$\frac{1}{2}''$	6-penny	6-penny threaded	8-penny	8-penny
$\frac{5}{8}''$ or $\frac{3}{4}''$	7-penny	8-penny threaded	9-penny	9-penny

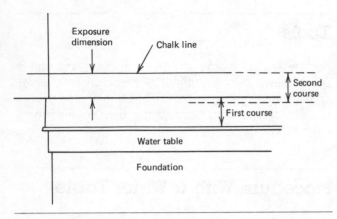

Fig. 18-1.12

work toward the center of the wall or a door opening.

11 Attach the corner length and adjoining lengths of siding, nailing 16″ o.c. just above the filler strip into the header joist, edge joist, or sill plate. **Precaution.** Be careful to avoid nailing into the subflooring, as it will not hold the nails well.

12 Cut, prime, and fit the last piece of siding to complete the first course.

13 Above the first course snap a chalk line to mark the height of the thin edge of the second course (Fig. 18-1.12).

14 Select and cut primed pieces of siding for succeeding courses so that they fall on stud lines, with the joints staggered from course to course. Use the lines of nails or staples in the sheathing as a guide to stud locations, or snap vertical chalk lines when building paper hides fasteners.

15 Continue to apply siding courses up to the level of the first window sill, using the spacer to assure uniform exposure and nailing as shown in Fig. 18-1.10.

Procedure Without a Water Table

1 At the corners of the wall locate point A.

2 Snap a chalk line along the foundation wall between these two points. Carefully check the line for level.

3 Cut, prime, and install any corner boards.

4 Nail the primed backer strip in place $\frac{1}{2}″$ above the chalk line.

5 Using the chalk line as a guide, attach the corner length and adjoining lengths of siding, and check

the level before each nailing. **Precaution.** Nail this course just above the backer strip into the floor structure.

6 Cut, prime, and fit the last piece of siding to complete the first course.

7 Follow steps 14 and 15 of the preceding procedure to install succeeding courses.

Procedure For Mitering a Corner

1 Hold the length of siding in position against the sheathing, with its end projecting beyond the corner.

2 On the back and top of the siding, mark the line of the corner to establish the cut length (Fig. 18-1.13).

3 To the back wall of a deep miter box tack a strip of siding equal in height to the overlap of courses (Fig. 18-1.14).

4 Set the siding to be cut against the strip and saw on the cutting line.

5 To cut the matching mitered corner, hold the length of siding in position against the sheathing, with its end projecting beyond the corner.

6 On the back, mark the shape of the adjoining length of siding (Fig. 18-1.15).

7 On the top and bottom edges carry these marks across, then away from the corner at a 45° angle (Fig. 18-1.16).

8 Connect these marks on the face of the siding.

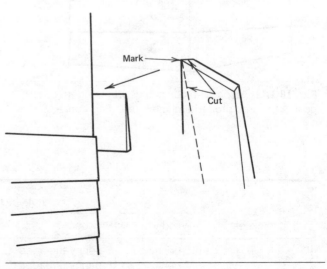

Fig. 18-1.13

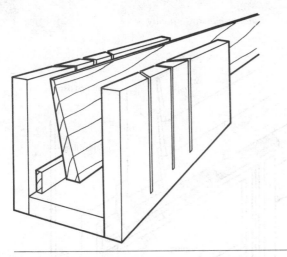

Fig. 18-1.14

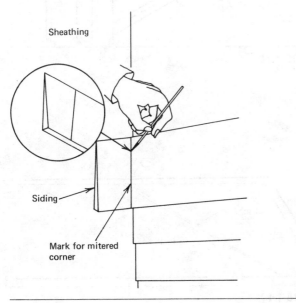

Fig. 18-1.15

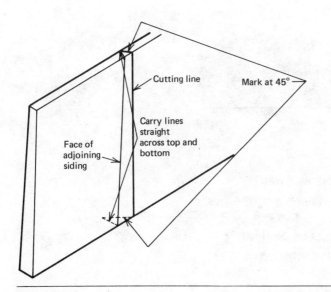

Fig. 18-1.16

9 Cut carefully along the line on the face. **Note.** Try to stay on the back line as well, but it is the face line that is important for a tight fit.

10 Test the fit.

11 If necessary, adjust the cut with a block plane or rasp.

12 Fill any open part of the miter with caulking compound, and nail the two lengths of mitered siding together.

Procedure for Notching at a Window Sill

1 Using the spacer block as a guide, mark on the last full course of siding below the sill the butt line of the length to be notched (Fig. 18-1.17).

2 Snap a chalk line at this level.

3 Measure the distance from the chalk line to the underside of the sill, and transfer this dimension to a piece of siding long enough to extend at least one stud beyond both jambs.

4 Mark the width of the sill on the siding.

5 Notch the siding by cutting carefully along the three lines.

6 Test the fit.

7 When the fit is good, spread caulking around the perimeter of the sill.

8 Press the siding into the caulking and nail in the usual manner.

Procedure for Fitting Siding at a Window or Door Head

1 Using the spacer block as a guide, mark on the last course of siding below the head the butt line

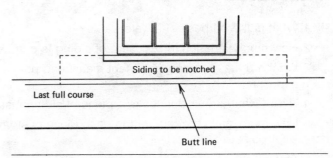

Fig. 18-1.17

of the next course. **Note.** With good planning of the exposure, this line should fall at the same level as the drip cap.

2 If necessary, mark and notch a length of siding as described in steps 3 through 6 above.

3 Fit a length of metal or vinyl flashing over the drip cap the full width of the opening and extending at least 2″ up the sheathing (Fig. 18-1.18). **Note.** The flashing may be face-nailed, but in quality construction it is blind-nailed.

4 Fit the notched length of siding over the opening, resting it on the drip cap, and nail it in place. **Precaution.** Do not nail through the flashing. Either nail above it or, if the width of siding does not completely cover the flashing, add caulking at the bend in the drip cap and press the siding into the caulking.

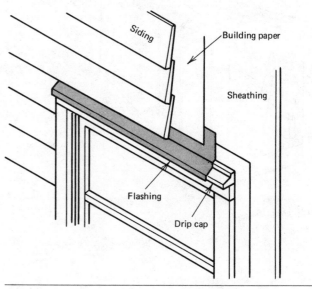

Fig. 18-1.18

Procedure for Finishing a Gable

1 At the rakes of roofs in gables, mark the pitch line on the face of each length of siding with a pitch board.

2 Along the thin edge of the siding, measure off ⅛″ outward from the pitch line and make a mark.

3 From this mark draw a new line to the end of the pitch line at the butt edge (Fig. 18-1.19).

4 Cut along this new line.

5 Test the fit.

6 If necessary, use a block plane to touch up the cut edge for a closer fit.

7 Because the fit will rarely be exact, caulk the joint tightly before nailing the siding in place.

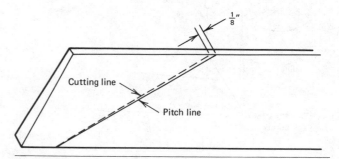

Fig. 18-1.19

Questions

1 What is the purpose of a water table?

2 How much variation in exposure of siding is permissible in a wall?

3 What are the four most important measuring points for determining the best exposure?

4 Under what circumstances is a piece of siding installed upside down?

5 How many nails are required in each piece of siding at each stud, and at what point are they driven?

6 At what points should joints be caulked?

Unit 18-2   **Apply Rabbeted Beveled Siding**

Plain beveled siding has square-cut edges. **Rabbeted beveled siding** and its thicker version, **Dolly Varden siding,** have a rabbeted bottom edge that fits over the square edge of the courses below (Fig. 18-2.1). Because of this rabbet, the back surface of the siding lies flat against sheathing.

Anzac siding is not a rabbeted siding but is included here because of similarity of application. Developed in New Zealand and usually cut from redwood, Anzac siding has a back shaped to lie flat against sheathing at the top and flat against the course below at the bottom (Fig. 18-2.2). The butt edge is thicker than on other types of siding and provides a heavy shadow line. Near the thin edge the face has two grooves. The upper groove prevents wind-driven rain from working its way upward, while the lower groove is a guide to alignment of the next course above.

All three of these types of siding have a much smaller air space behind them than plain beveled siding. They are a little more difficult to work with because the exposure can be adjusted very little. On the other hand it is easier to line up the courses. Installation procedures are similar to those for plain beveled siding, and the preparation, treatment at corners, foundation, and roof, and the basic rules for installation given in Unit 18-1 also apply here. Rabbeted beveled siding is applied with 6-penny nails; Dolly Varden and Anzac siding require 8-penny nails.

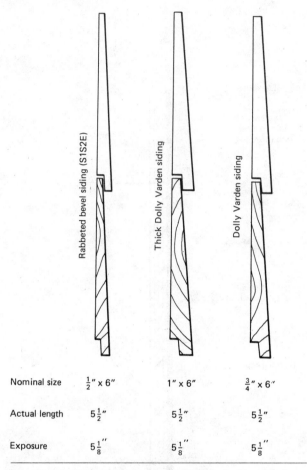

	Rabbeted bevel siding (S1S2E)	Thick Dolly Varden siding	Dolly Varden siding
Nominal size	$\frac{1}{2}'' \times 6''$	$1'' \times 6''$	$\frac{3}{4}'' \times 6''$
Actual length	$5\frac{1}{2}''$	$5\frac{1}{2}''$	$5\frac{1}{2}''$
Exposure	$5\frac{1}{8}''$	$5\frac{1}{8}''$	$5\frac{1}{8}''$

Fig. 18-2.1 Typical thicknesses of rabbeted beveled siding.

Procedure

1 Establish and mark the location of the bottom course.

2 Cut, prime, and install corner boards.

3 Cut, prime, and install the water table, if specified.

4 Apply the bottom course of siding with its bottom edge resting either on the drip cap or drip strip of the water table, or aligned with the chalk line marking point A. **Note.** No filler strip or backer strip is required with these types of siding.

5 Nail the first course. With rabbeted beveled siding or Dolly Varden siding nail 1″ above the butt edge and 16″ o.c. (Fig. 18-2.3). With Anzac siding nail 16″ o.c. just above the point where the siding touches sheathing (Fig. 18-2.4).

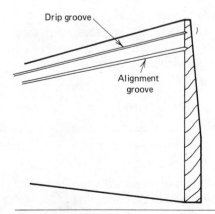

Drip groove

Alignment groove

Fig. 18-2.2 Anzac siding has a groove that establishes the exposure.

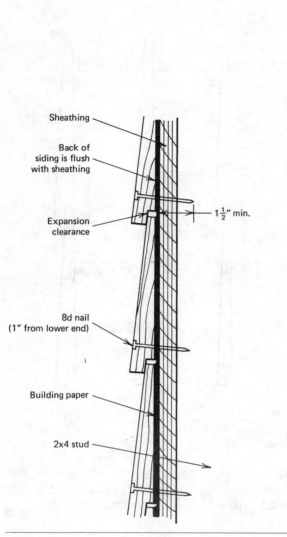

Sheathing

Back of
siding is flush
with sheathing

Expansion
clearance

$1\frac{1}{2}$" min.

8d nail
(1" from lower end)

Building paper

2x4 stud

Fig. 18-2.3

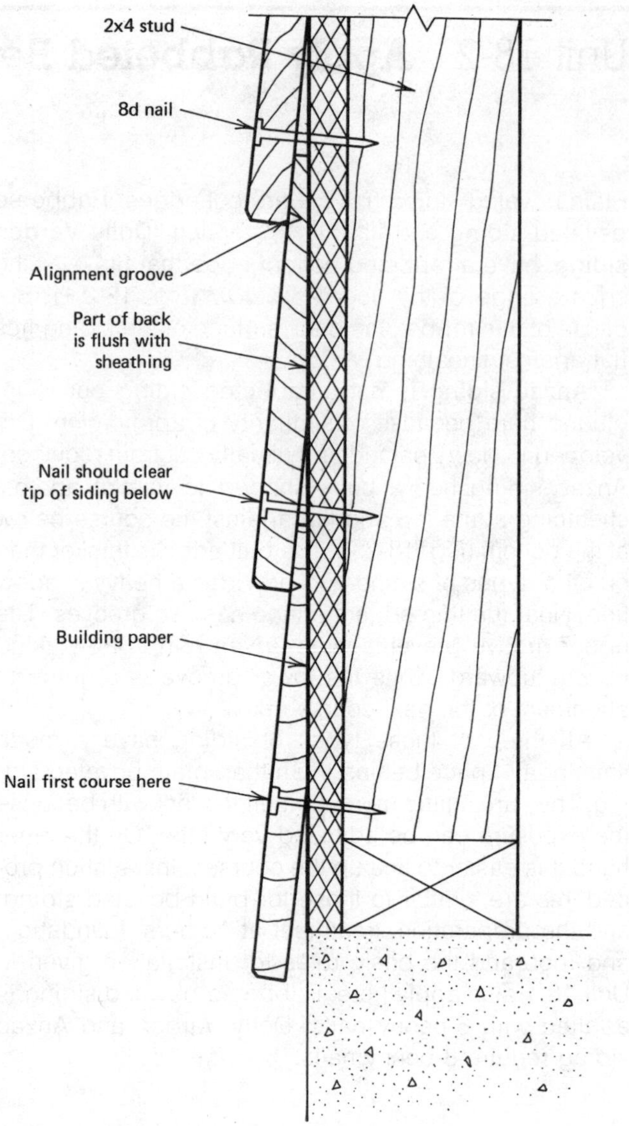

2x4 stud

8d nail

Alignment groove

Part of back
is flush with
sheathing

Nail should clear
tip of siding below

Building paper

Nail first course here

Fig. 18-2.4

6 Continue to apply courses upward as in step 4
and as described in Unit 18-1. **Note.** Except for
the first course, nails in Anzac siding should just
clear the thin edge of the course below, as in Fig.
18-2.4.

Questions

1 How does installation of rabbeted beveled siding differ from installation of
plain beveled siding at the foundation?

2 Give one reason why rabbeted beveled siding is easier to install than
plain beveled siding. and one reason why it is more difficult.

Unit 18-3 **Apply Lap Siding**

Not all siding is beveled. Exterior plywood, tempered hardboard, and particle board are manufactured in strips for use as siding, and their outside and inside surfaces are parallel. The strips come in lengths from 8 to 16' long and in widths that can be economically cut from a 48" wide sheet. All edges are cut square.

Some types of lap siding are prefinished on the exposed surface and primed on the back. Others may be factory primed on both surfaces or unfinished for application of paint or stain at the site.

The visual effect of lap siding is similar to that of beveled siding (Fig. 18-3.1), but the air space between courses is considerably larger. To assure a tight seal at butt joints in courses, the joints must be supported with wedges made from shingles or other tapered wood.

The first course of lap siding is nailed into a starter strip at the foundation wall. This strip should be $2\frac{1}{2}''$ high. So that the slope of all courses is the same, its thickness should be $\frac{3}{8}''$. Many builders, however, use a

Fig. 18-3.1 Lap siding looks like beveled siding. Here it is made of hardboard. (Courtesy of Georgia-Pacific Corporation.)

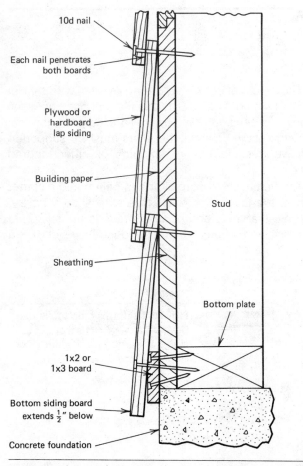

Fig. 18-3.2

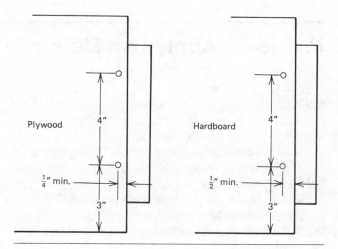

Fig. 18-3.3

primed 1 × 3 as a starter strip and let the bottom course project further from the foundation wall to minimize water problems.

Lap siding may be applied with threaded nails driven flush with the surface or with casing nails countersunk. Nails should be no smaller than 8-penny, and 10-penny nails are recommended.

Procedure

1 At a point 2″ above the foundation wall, snap a level chalk line across the sheathing.

2 Install a primed starter strip with its top on the chalk line. Nail it securely into the sill plate (Fig. 18-3.2).

3 At a point $\frac{1}{2}$″ below the starter strip, snap another chalk line on the foundation wall to mark point A.

4 Cut, prime, and install any corner boards.

5 Cut wedges 2″ wide and prime all surfaces.

6 Install wedges at corner boards and over studs at the end of each length of siding.

7 Prime all unprimed or newly cut edges of siding.

8 Attach the first course. Nail every 16″ into the starter strip about 3″ above the butt edge (Fig. 18-3.2). At ends nail into wedges every 4″ if siding is 12″ wide or less, and every 8″ into wider siding. **Precaution.** Nail $\frac{1}{4}$″ back from the ends of plywood siding, and $\frac{1}{2}$″ from the ends of hardboard or particle board siding (Fig. 18-3.3).

9 Use a spacer block to locate the second course.

10 Repeat steps 5 through 8 for the second and succeeding courses. Nail $\frac{1}{2}$″ above the butt edges through both courses (Fig. 18-3.2).

11 Follow procedures for beveled siding to complete the wall.

Questions

1 What are the recommended dimensions of a starter strip behind the first course of lap siding?

2 Give two reasons for using wedges at joints between lengths of siding.

3 From what three materials is lap siding usually cut?

Unit 18-4 **Apply Metal Siding**

Metal siding is made of either aluminum or steel, with aluminum by far the more common. The surface metal is very thin (about $\frac{1}{40}$" thick) and has a backing of insulating foam or fiberboard laminated to it. The metal comes prefinished, either with a baked-on enamel finish or a vinyl film and either smooth or embossed with various textures (Fig. 18-4.1).

Unlike wood, metal expands and contracts rapidly with changes in air temperature and under the heat of direct sunlight. Therefore the siding has nailing slots and interlocking edges that allow for this movement (Fig. 18-4.2). Exposure is fixed, however; most metal siding is designed for either a 4" or 8" exposure.

Note in Fig. 18-4.2 that the nailing flange on each strip of siding does not extend the full length of the piece. The flanges of adjacent lengths butt and estab-

lish the amount of overlap. If possible the overlap should face away from the prevailing wind. Overlaps should be staggered at least 24" between courses, and seams should not fall at the same vertical point more often than every fourth course.

Manufacturers of metal siding also produce a complete line of accessories to solve any construction problem (Fig. 18-4.3). The shapes shown are typical; they vary from manufacturer to manufacturer.

Metal siding is applied with nails of the same material as the siding. Cuts are marked with a utility knife and made by a power saw with a metal-cutting blade. The locations of joints must be planned in advance so that cut ends are hidden where lengths overlap. Siding should not be cut at corner posts or at openings; cuts should fall at overlaps in the middle of the wall. After

Fig. 18-4.1 Metal siding comes prefinished and ready to install. (All photos courtesy The Aluminum Association.)

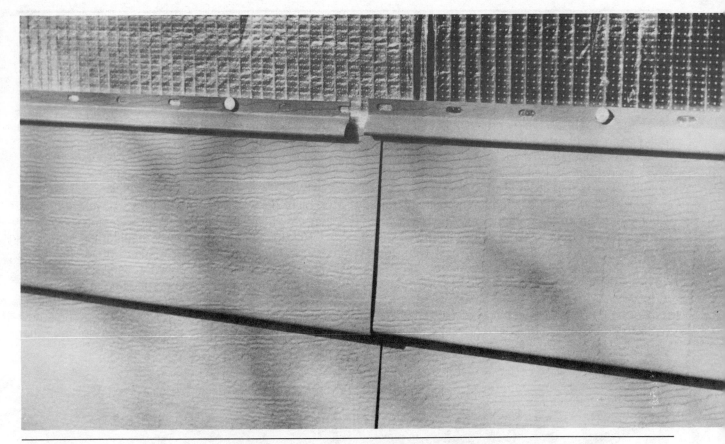

Fig. 18-4.2 Shape of typical interlocking edges of lengths of metal siding. Note that holes for nails are slotted.

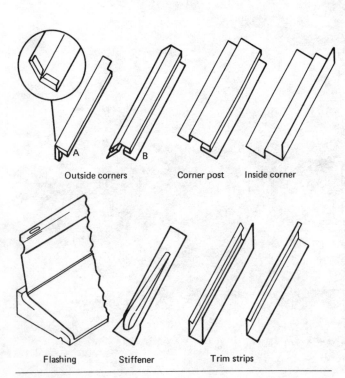

Outside corners   Corner post   Inside corner

Flashing   Stiffener   Trim strips

Fig. 18-4.3 Typical accessories for trimming walls of metal siding.

siding is cut, the raw edge should be touched up with a metal primer to forestall oxidation. If not, rain water may react with the raw edge and streak the surface.

To prevent metal siding from buckling and giving the wall a wavy appearance, it is important to leave a gap of $\frac{1}{8}$ to $\frac{1}{4}''$ where siding meets corner posts and trim. Gap also at joints between lengths of siding where the nailing flanges meet.

The tools required for a good job of installing metal siding are a chalk line, level, folding rule, combination square, metal shears, power saw, hammer, and caulking gun. To cut backing or furring strips where required you will also need a crosscut blade.

Procedure with Outside Corner Posts

1 Carefully read the manufacturer's instructions for application.

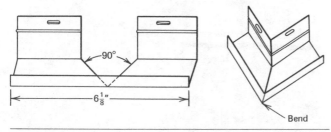

Fig. 18-4.4

Fig. 18-4.5

Fig. 18-4.6

Fig. 18-4.7

2 On the foundation wall snap a chalk line to mark the location of the bottom edge of the starter strip.

3 Install the starter strip on this line. **Precaution.** Do not drive nails far enough to bend the metal. The undersides of nail heads should just touch the starter strip.

4 Nail inside corner posts through the slots provided. The bottoms of the posts should be on the chalk line.

5 If recommended by the manufacturer, cut and fit wood strips at outside corners as a support for posts.

6 If corner posts do not have a flange at the bottom (B in Fig. 18-4.3), cut a section of window trim as shown in Fig. 18-4.4.

7 Install the trim on the chalk line at the bottom of each corner post. **Note.** This trim is not required with flanged posts (A in Fig. 18-4.3).

8 Fit outside corner posts on the chalk line and any closure, and nail through the wood backing.

9 Install trim pieces at the jambs of all doors and windows.

10 Install drip caps over all windows and doors.

11 Install trim pieces below all window sills.

12 Apply trim strips wherever walls meet a sloping roof. **Note.** The trim strips are always installed with the nailing slots away from the roof's surface.

13 Beginning at a corner post or trim piece at an opening, fit the first length of siding into the starter strip. **Note.** If the siding comes without a backing, fit stiffener strips (Fig. 18-4.5) in place behind the ends and every 32″ between.

14 Nail through the slots in the upper flange (Fig. 18-4.6).

15 Complete the first course, and fit succeeding courses on up the wall to window sills.

16 Precut siding to fit with a small gap around windows (Fig. 18-4.7), and prime the cut edges.

17 At heads of openings add furring strips of a thickness to support siding at the same slope as adjoining panels.

18 At eaves trim the top course to fit into the trim strip.

19 At rakes cut the siding as outlined for beveled siding in Unit 18-1.

20 Caulk joints where siding meets corner posts, at heads and jambs of openings, and where walls meet roofs.

21 Wash off the siding with a hose. Use mild detergent, if necessary, to remove fingerprints.

Procedure with Corner Caps

1 Complete steps 1 to 4 inclusive as just described for inside corner posts.

2 Complete steps 9 to 12 inclusive for installing trim at openings.

3 Beginning at an inside corner or at an opening, fit the first course of siding. Gap about $\frac{1}{2}$″ at outside corners (Fig. 18-4.8).

4 Bend each cap to fit the angle of the corner (Fig. 18-4.9).

5 Install the cap by slipping the bottom flanges under the butt edges of siding and pushing upward (Fig. 18-4.10). If necessary, use a putty knife slipped between panel locks to allow room for the flanges. Tap with a rubber mallet to fit the cap tight.

6 Secure the cap with nails long enough to penetrate $\frac{3}{4}$″ into the structure.

7 Complete corners of succeeding courses as described in steps 4, 5, and 6 of this procedure.

8 Complete application of siding by following steps 16 to 21 of the previous procedure.

Fig. 18-4.8

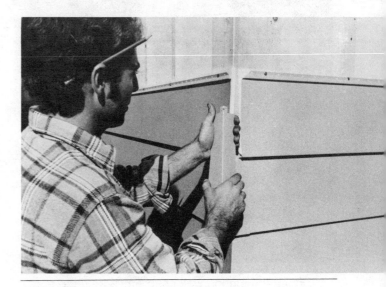

Fig. 18-4.9

Fig. 18-4.10

Questions

1 What are the most common exposures for metal siding?
2 How should siding be cut?
3 Where should the cuts occur?
4 How and why should cut edges be treated?
5 Where is caulking required?

Unit 18-5   **Apply Vinyl Siding**

In many ways rigid vinyl siding is similar to metal siding. The first course is applied over a starter strip, and each course has flanges at the top and bottom that provide a self-aligning lock with courses above and below (Fig. 18-5.1). Edges of adjacent lengths overlap. Manufacturers also make corner boards and trim pieces of the same material in matching colors. Vinyl may be cut with standard metal-working tools, and the siding is applied with rust-resistant nails. The siding comes in single courses with an 8″ exposure and in one-piece double courses with a 4″ exposure.

The two chief differences between metal and vinyl siding lie in the vinyl itself. The color is not applied to the surface but is part of the raw material molded into the finished shape. Therefore scratches don't show. Vinyl is also stiffer than metal and does not dent, although it can be shattered. From the standpoint of application procedures, the most important difference is that vinyl does not expand and contract any more than wood, and therefore it isn't necessary to gap at joints and corners. Cut edges do not require any special treatment.

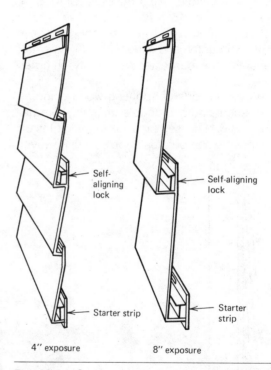

Fig. 18-5.1 Sections through typical shapes of vinyl siding and starter strips.

Procedure

1 Read the manufacturer's instructions.
2 Mark the location of the bottom edge of the starter strip with a chalk line, and install the strip. Make sure it is level. The bottom of the strip should overlap the foundation wall 1″.
3 Install corner posts as described in Unit 18-1 or 18-4.
4 Install vinyl trim around openings.
5 Slip the bottom edge of the first course into the slot in the starter strip, push upward until the joint is tight, then nail through the slots in the flange.
6 Install succeeding courses in the same manner.

Questions

1 Name three differences between metal and vinyl siding.

2 How much of a gap should be left between siding and trim at openings?

Unit 18-6   **Apply Wood Shingles**

The wood shingles used to finish exterior walls are the same as the wood shingles used on roofs (Unit 15-3). They come in lengths of 16″, 18″, and 24″ and vary in width from 3 up to about 14″. They are very popular in some parts of the country, and especially among small builders because one carpenter can apply shingles working alone.

There are two methods of applying wood shingles to walls: single-coursing and double-coursing. **Single-coursing** (Fig. 18-6.1) is used for applying shingles on a roof. The exposure can be greater on a wall, however

—by about 50%. With single-coursing all wall areas are covered by two thicknesses of shingle and by three thicknesses at butt edges. This is the traditional way of application. All nails are concealed.

With **double-coursing** (Fig. 18-6.2) each course consists of two layers of shingles laid with butt edges slightly offset. Shingles in the concealed course are a lower grade—No. 3 or undercoursing shingles. All wall areas are again covered with two thicknesses of shingle, but there are four thicknesses at butt edges. Shingles must be nailed at both top and bottom. Dou-

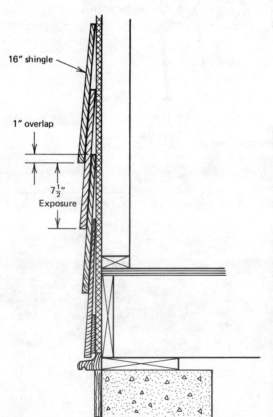

Fig. 18-6.1 Section through a single-course wall of wood shingles. Coverage is three shingles thick at butts, and the surface course overlaps the base course by a minimum of

1″ with maximum exposures. (All photos courtesy Red Cedar Shingle & Handsplit Shake Bureau.)

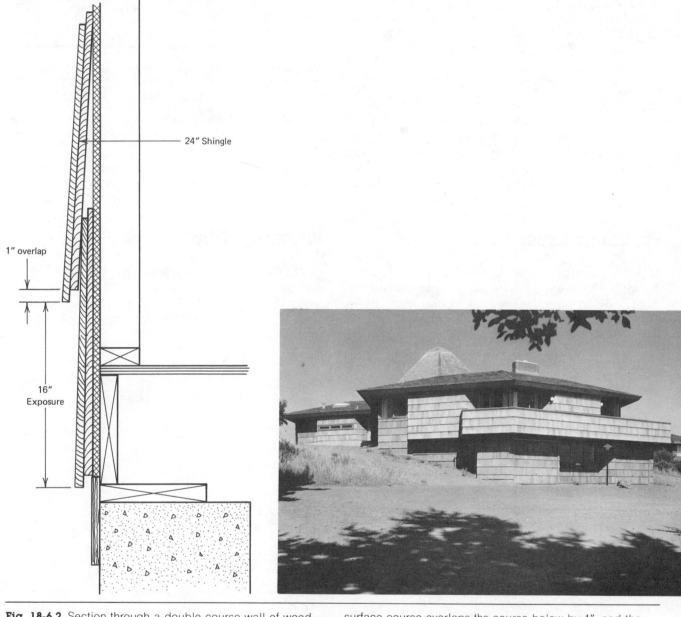

Fig. 18-6.2 Section through a double-course wall of wood shingles. Coverage is four shingles thick at butts. The surface course overlaps the course below by 1″, and the other two courses by at least 3″.

ble-coursing as a method resulted from the development of 24″ long shingles, which can have an exposure up to 16″. Shorter shingles may be applied double-coursed, but the economy in materials and application time lies with the longer units.

Preparation

Wood shingles may be applied directly to sheathing that has been treated or covered to keep out moisture.

Use rust-resistant siding nails if the sheathing is plywood or boards. With gypsum or insulating sheathing, ordinary nails may not hold; self-clinching nails or patented fasteners (Fig. 18-6.3) are recommended instead. An alternate method is to apply nailing strips horizontally over the sheathing at predetermined heights to accept shingle nails. The spacing of strips varies with shingle exposure and is equal to it.

On utility buildings such as garages and sheds, shingles may be applied to blocking either behind sheathing or, where ventilation is an advantage, in place of it. The blocks are 2 × 4s—either cutoffs from

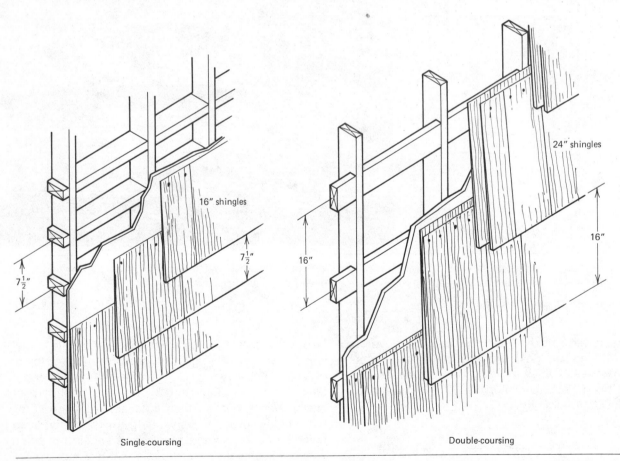

Fig. 18-6.3 Patented threaded nails are often used to secure shingles to nonwood sheathing.

other construction or utility grade lumber. With single-coursing nail the blocks flat at a spacing equal to shingle exposure (Fig. 18-6.4). With double-coursing nail the blocks upright so that butt edges fall 1½ to 2″ below the center line of the blocks. Again the spacing is the same as the shingle exposure.

Planning Exposure

The best exposure for wood shingles is calculated in the same way as exposure of siding. As with siding, the ideal spacing puts butts level with drip caps above windows and the bottom edges of window sills.

The best place to take measurements is at the most

prominent corner of the front wall. Here are the maximum allowable exposures.

Shingle length	Single-coursing	Double-coursing
16″	7½″	12″
18″	8½″	14″
24″	11½″	16″

In double-coursing, the exposure is figured from butt edge to butt edge of the outer layer.

Planning Trim

Any of the treatments shown in Figs. 18-1.2 and 18-1.3 may be used at the foundation line, and those shown in Figs. 18-1.4 and 18-1.5 may be used at the roof. Usually, however, the bottom course of shingles is doubled, as with a wood shingle roof at the eaves, and the

Single-coursing   Double-coursing

Fig. 18-6.4 Blocking is sometimes placed between studs as a nailing strip for wood shingles. Center-to-center spacing of blocks equals the exposure. Blocks must be set vertically to catch both rows of nails with double-coursing.

water table is not used. Even if a water table is specified, the bottom course should be doubled to make the slope of all courses uniform.

Outside corners may be finished in any of the ways shown in 18-1.8, they may be mitered, or they may be interwoven like a Boston hip in a wood shingle roof (see Fig. 15-3.9). At inside corners shingles may be mitered with a reverse angle of cut, or they may butt against a corner board. The total thickness of exterior material is greater at the butt edges of shingles than at the butts of siding. Therefore corner boards must be thicker—at least $1\frac{1}{2}''$ in actual thickness and possibly a full $2''$, depending on the exposure and method of coursing. All cut edges at corners must be primed before shingles are applied.

Wood shingles may be applied to a wall with nails or staples. Two nails are enough when shingles are no more than $8''$ wide. Add a center nail in line with the other two in wider shingles. Nails should be placed about $\frac{3}{4}''$ from the edge of the shingle and about $1''$ above the butt line of the covering course. Use galvanized or aluminum nails—3-penny nails with single-coursing and 5-penny nails with double-coursing.

Shingles in any course may be laid with **closed joints** (shingles fitted tightly together) or with **open joints** (gapped about $\frac{1}{8}''$). Closed joints provide a stronger horizontal shadow line but less texture to the wall than open joints. Prefinished or primed shingles may be laid in either pattern; unfinished shingles should be gapped so that their edges can be sealed with the tip of a paint brush. Vertical joints in lapping courses should be **broken**—that is, out of alignment—by at least $1\frac{1}{2}''$. If edges of adjoining shingles are not parallel, use a hatchet or power plane to shave them.

On most houses shingles are applied with their butts lined up. For a more rustic appearance the butts are sometimes staggered as much as $2''$ (Fig. 18-6.5). When staggered butts are specified, the shingles with the lower-hanging butts are lined up as described in the following procedure, and the height of the remaining shingles is established with a spacer block.

Procedure for Single-Coursing

1 Establish the exposure of individual courses.

2 Unless corners will be mitered, cut, prime, and install corner boards and beads.

3 Snap a level chalk line on the foundation wall about $\frac{1}{4}''$ below its top.

4 Lay an undercourse of shingles with their butts along the chalk line. Fasten this course 1 to $2''$ above the bottom edge.

5 Lay the bottom course of shingles over this undercourse, overlapping about $\frac{1}{2}''$ at the butt edge.

6 At several points on the bottom course, mark the location of the butt line of the next course.

7 At this exposure line tack a straight length of 2×4 as a guide for leveling the next course.

8 Continue courses up the wall in this manner, nailing as shown in Fig. 18-6.6.

9 Caulk all joints between shingles and trim at corners and around openings.

Fig. 18-6.5 Application of shingles with butts staggered takes an extra step, but isn't as complex as it looks. The method is specified when the architect or designer wants maximum texture and pattern.

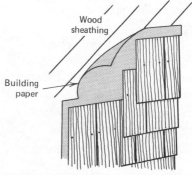

Fig. 18-6.6

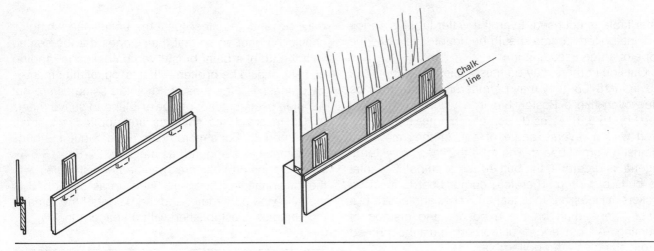

Fig. 18-6.7

Procedure for Double-Coursing

1 Establish the exposure of individual courses.

2 Unless corners will be mitered, cut, prime, and install corner boards and beads.

3 Snap a level chalk line on the foundation $\frac{1}{4}''$ below its top.

4 To the back side of a 10' length of siding, nail three undercourse shingles 3 to 5" wide (Fig. 18-6.7, left). **Precaution.** Treat this course with wood preservative before application if shingles will be within 12" of the finished grade.

5 Lay the board against the wall with its upper edge on the chalk line, and nail the three shingles to the wall (Fig. 18-6.7, right).

6 Using the top shiplapped edge as a guideline, attach the rest of the shingles in the undercourse (Fig. 18-6.8, top). **Note.** Leave the original three shingles in place, and butt adjacent shingles against them.

7 Again using the top shiplapped edge as a guideline, attach the second layer of undercourse shingles over the first layer (Fig. 18-6.8, center).

8 Using the bottom shiplapped edge as a guideline, attach the weather course of shingles (Fig. 18-6.8, bottom). Nail as shown in Fig. 18-6.9.

9 Carefully remove the length of shiplap, and set it aside for reuse.

10 Trim the exposed butts of the three original shingles with a sharp knife (Fig. 18-6.10).

11 Repeat steps 4 to 10 inclusive to complete the bottom course.

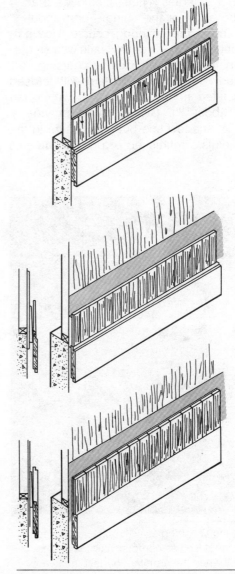

Fig. 18-6.8

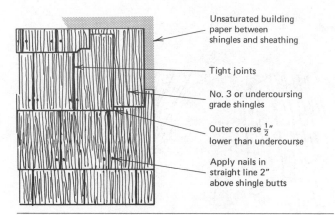

Unsaturated building paper between shingles and sheathing

Tight joints

No. 3 or undercoursing grade shingles

Outer course $\frac{1}{2}$" lower than undercourse

Apply nails in straight line 2" above shingle butts

Fig. 18-6.9

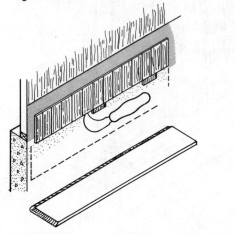

Fig. 18-6.10

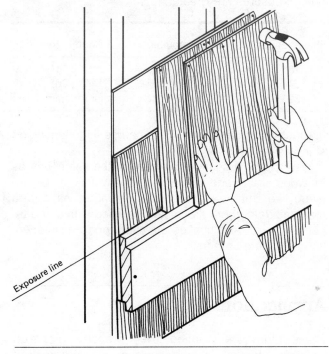

Exposure line

Fig. 18-6.11

12 Tack straight shiplapped boards at the exposure line as a leveling guide for laying the remaining double-courses (Fig. 18-6.11).

13 Fill nail holes left by guides.

14 Caulk all joints between shingles and trim at corners and around openings.

Questions

1 How many nails penetrate each shingle in single-coursing? How many in double-coursing?

2 If shingles are attached to nailing strips, on what is the spacing of the strips based?

3 Name three ways that outside corners may be finished in a shingle wall.

4 What is the minimum recommended distance between staggered shingle joints in adjacent courses?

5 Why do you think only one row of nails is necessary to attach single-course shingles, but two rows are recommended with double-coursing?

Unit 18-7 **Apply Cedar Shakes**

The same general procedures for applying cedar shakes to roofs should also be followed when applying them to walls. As with wood shingles, however, maximum exposures are greater.

Shake length	Single-coursing	Double-coursing
18″	$8\frac{1}{2}$″	14″
24″	$11\frac{1}{2}$″	20″
32″	15″	not recommended

In double-coursing, the undercourse is often wood shingles instead of shakes.

The best exposure is planned in the same way as for wood siding and shingles. Shake walls may be finished like shingle walls at the foundation, eave, and rake. Inside corners are usually finished with a square corner board, and shakes are overlapped or interwoven at outside corners.

Application

Shakes may be applied over properly prepared sheathing. In a quality job, however, an extra layer of building paper is applied under each course of shakes so that it overlaps the thin edge of the course below (Fig. 18-7.1).

The length of rust-resistant nails varies with the type of coursing and the exposure of the shakes. Usually 6-penny nails are long enough, but in some instances 8-penny nails are required. Because of the texture of the shakes, it is important to drive nailheads so they only touch but do not damage the shake's surface.

Procedure for Single-Coursing

1 Follow steps 1 to 3 inclusive for single-course application of wood shingles (Unit 18-6).

2 Apply a strip of building paper over the sheathing (Fig. 18-7.1). **Note.** This strip should extend from the chalk line to a point about 1″ above the top of the thin edge of the first course.

3 Follow steps 4 and 5 for application of wood shingles, using shakes for the bottom course.

4 Apply a strip of building paper over the bottom course as shown in Fig. 18-7.1.

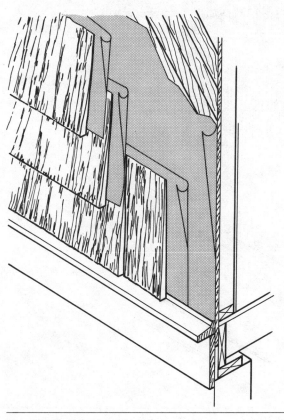

Fig. 18-7.1 A layer of building paper separates each course of shakes.

5 Follow steps 6 to 8 inclusive for wood shingles.

6 Caulk all joints between shakes, and trim at corners and around openings.

Procedure for Double-Coursing

1 Follow steps 1 to 3 inclusive for double-course application of wood shingles (Unit 18-6).

2 Apply a strip of building paper over sheathing as in step 2 for single-course application of shakes.

3 Follow steps 4 to 11 inclusive for double-course application of wood shingles.

4 Apply a strip of building paper over the bottom course as shown in Fig. 18-7.1.

5 Follow steps 12 and 13 for wood shingles.

6 Caulk all joints between shakes and trim at corners and around openings, and at interwoven corners.

Question

1 What is the chief difference between application of wood shingles and shakes on a wall?

Unit 18-8   Apply Asbestos-Cement Shingles

The asbestos-cement shingles manufactured for walls are larger than similar shingles for roofs, but are otherwise the same (Fig. 18-8.1). Some types are predrilled for nailing, while others are set into metal channels and are not themselves nailed.

Standard shingle widths are 24″, 32″, and 48″. Courses are applied with staggered joints. With 24″ and 32″ shingles start with a full shingle in every other course and a half-shingle in alternate courses. With wider shingles start the first course with a full shingle, the second with a two thirds-shingle, the third with a one third-shingle, the fourth with a full shingle again, and so on. Some manufacturers make partial shingles, but they can be made on the job by scoring the face and snapping the shingle along the scoring line. Sometimes extra nail holes must be drilled or punched in cut shingles.

Application

The typical application of asbestos-cement shingles has a tapered **cant strip** (Fig. 18-8.2) at the foundation line that gives the necessary slope (cant) to the bottom course. The cant strip should be installed before build-

Fig. 18-8.1 Asbestos-cement shingles look like most other shingles on a house except for their size. (Courtesy Supradur Manufacturing Corp.)

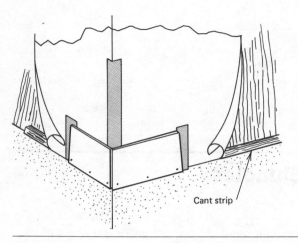

Fig. 18-8.2 The first course of shingles is applied over a cant strip.

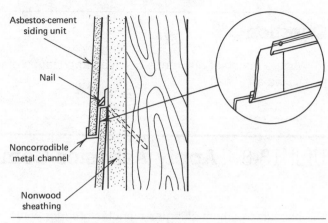

Fig. 18-8.3 Asbestos-cement siding shingles may be set in metal channels without nailing.

ing paper is applied over wood sheathing. If the paper is already in place, carefully unfasten the bottom edge and restaple over the cant strip.

At inside corners the shingles butt against square boards or metal interior corners made for the purpose. At outside corners they may butt against a corner board or be finished with metal corners. Asbestos-cement shingles can't be mitered.

To prevent breakage, asbestos-cement shingles must have a solid backing. Plywood or wood sheathing is adequate. If sheathing is nonwood, there are two methods of application. One is with metal channels supplied by the manufacturer (Fig. 18-8.3). The rust-proof channels fit on top of one course and are nailed into studs. The channels support the lower edge of the next course and hold it firmly in place. If the shingles are predrilled for nailing, they should be installed upside down so that the channel covers the holes.

The other method is to use **shingle backers** (Fig. 18-8.4), made by the shingle manufacturer in the same widths as the shingles themselves. Shingle backers act like an undercourse to keep out the rain water, protect shingles against breakage, and increase the weight of the shadow line. Joints in shingles and shingle backers must not fall at the same point, but should be staggered at least 12".

With all methods of application, strips of building paper must be applied behind each joint between shingles (see Fig. 18-8.4). The strips touch the backs of the shingles, regardless of how they are supported or protected.

The exposure of courses can be adjusted only slightly. The wise carpenter, therefore, measures wall heights at both corners to make sure the dimensions are the same. If not, the difference should be spread over all courses, not just the top course. To increase

the exposure using shingle backers, nail a little higher than recommended in the following procedures. To reduce the exposure, drill new holes near the tops of shingles in the course below. If channels are used, increase the exposure by nailing the channels with a slight gap above the shingle course below. To reduce the exposure, trim the top edge of shingles.

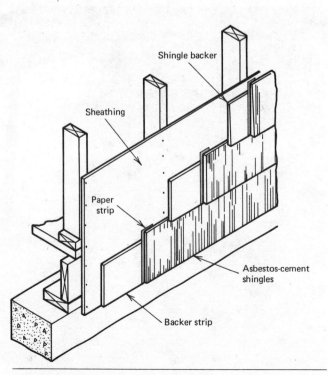

Fig. 18-8.4 Shingle backers provide the required support behind shingles, and strips of building paper provide moisture control behind joints between shingles.

Tools

To finish a wall with asbestos-cement siding shingles, you need a chalk line, carpenter's level, and hammer. To cut and drill holes in shingles, the best tool is a shingle-cutter.

Procedure for Installation over Wood Sheathing

1 Measure the wall height, and determine the amount of adjustment in exposure.

2 a. Cut, prime, and install any corner boards or beads.

2 b. If outside corners are to be finished with metal corners, flash them with a vertical strip of building paper.

3 Cut, prime, and nail a cant strip along the bottom edge of sheathing.

4 Install building paper over the cant strip.

5 Snap a chalk line at the level of the top of the first course and all other courses up the wall. **Precaution.** Don't rely on the level of the cant strip as a guide; it will seldom be level along its entire length.

6 Beginning at a corner, apply a full shingle with two nails. **Precaution.** Make sure the shingle is on the chalk line, plumb and level; if not, all other shingles will be off line.

7 Center a 4″ wide joint strip behind the unnailed end of the shingle so that it overlaps the cant strip (Fig. 18-8.5).

8 Drive the remaining nail in the first shingle. **Precaution.** Nails should be driven snug, but not so tight that they crack the shingle. Use 4-penny galvanized nails.

9 Repeat steps 6 to 8 for the remaining shingles in the first course. Butt shingles tight; no gap is required for expansion.

10 Cut the last shingle in the course to fit. **Precaution.** No shingle should be less than 6″ wide. If necessary to avoid this condition, cut the width of the next to last shingle also. Punch nail holes where necessary, using the pin in the shingle cutter for that purpose.

11 Begin the second course at the same corner with a half-shingle or a two thirds-shingle.

12 Snap horizontal chalk lines to mark the top of each course.

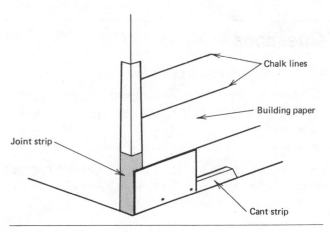

Fig. 18-8.5

13 Insert a nail in each nail hole, with the shank resting on the top of the shingle below. Then drive the nail.

14 Repeat the nailing and stripping procedure up the wall.

Procedure for Installation with Channels

1 Measure the walls to determine any adjustment in exposure.

2 Cut, prime, and install corner boards and beads, or flash corners with a second strip of building paper.

3 Snap a level chalk line along the foundation wall to mark the location of the bottom edge of the bottom course.

4 Nail a channel with its bottom flush with the chalk line. **Precaution.** Nails enter the structure at a 45° angle (see Fig. 18-8.3), so be sure that they penetrate the sill plate and don't bend on the foundation.

5 Set the first course of shingles in the channel.

6 Center a 4″ wide joint strip behind each joint so that it overlaps the top of the channel.

7 At the top of the first course install another channel, laying it atop the shingles.

8 Set the second course and joint strips in the channel.

9 Repeat steps 6, 7, and 8 on up the wall.

Questions

1 What is the purpose of a cant strip?

2 What treatment is required at vertical joints between shingles?

3 With each of the two methods of application, how do you increase shingle exposure?

4 With each method how do you reduce exposure?

5 What size would the first shingle in the fifth course be if the width of shingle used is 24″? If the width is 48″?

Unit 18-9   **Apply Drop Siding**

Drop siding is a classification for 1″ boards with a decorative pattern milled on the surface (Fig. 18-9.1). Some of the shapes are patented, and for this reason drop siding is sometimes called patent siding. The siding comes in more than one width in each shape.

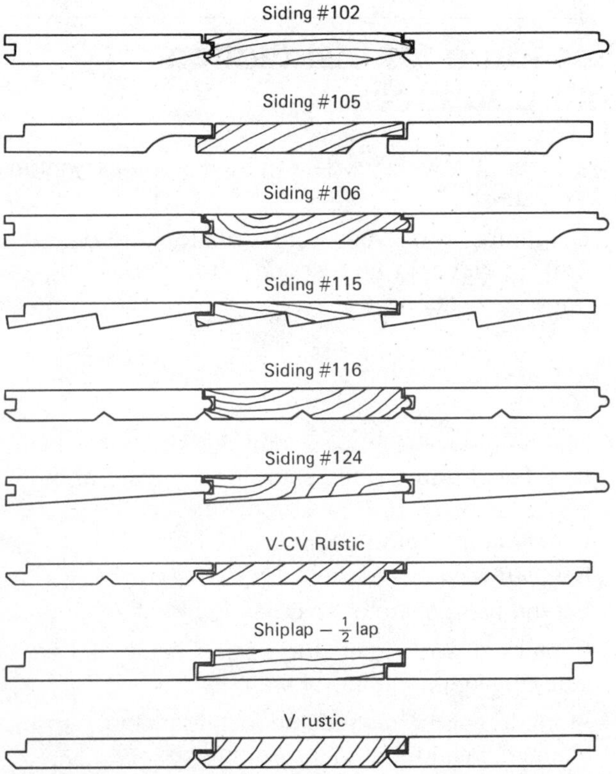

The use of drop siding has diminished a great deal in the last 50 years, partly because of the cost of the material and partly because of the time it takes to apply. It may be installed either horizontally or vertically. Horizontally, it may double as sheathing and be nailed directly to studs on such buildings as garages and sheds. Usually, however, it is applied in either direction over sheathing like other sidings. Drop siding is often used to finish gables when the exterior material on the main house is brick veneer or stone.

Except where broad overhangs protect the walls from heavy rains, special care must be taken to prevent water from seeping into vertical joints.

The procedure for applying drop siding horizontally is essentially the same as for any rabbeted siding. Vertical application begins at the center line of a wall or gable. Any of the corner treatments shown in this chapter may be used, or the corner may be overlaid with thin boards either mitered or butted. Many carpenters work out the spacing of the vertical siding to see where the

Siding #102

Siding #105

Siding #106

Siding #115

Siding #116

Siding #124

V-CV Rustic

Shiplap — $\frac{1}{2}$ lap

V rustic

Fig. 18-9.1 Here are some of the more than a dozen shapes for drop siding. Each shape is available in more than one width.

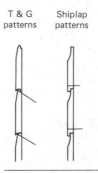

T & G patterns   Shiplap patterns

Fig. 18-9.2 Where to nail drop siding with shiplapped and tongue-and-groove edges.

edges of end pieces will fall, and then adapt a corner treatment to fit conditions, rather than rip boards to width.

Drop siding with shiplapped edges is applied with 8-penny siding nails driven about 1″ from the edge (Fig. 18-9.2). This is the proper nailing method whether siding is applied horizontally or vertically. If edges are tongue-and-groove, use 6-penny finishing nails driven into the tongue in horizontal application and 8-penny siding nails as with shiplapped edges in vertical application.

Procedure for Vertical Application

1 Find the midpoint of the length of the wall.

2 Divide the siding exposure into the dimension for half the wall's length.

3 a. If the remainder is less than half an exposed width of siding, snap a vertical chalk line at the midpoint of the wall.

3 b. If the remainder is more than half an exposed width, move the midpoint either direction a distance equal to half the exposure and snap a chalk line at that point.

4 Cut siding to length; the pieces should overhang the foundation wall by at least ½″.

5 Prime all surfaces and edges.

6 Snap a chalk line on the foundation wall to mark the bottom end of siding.

7 Lay the first piece against the two chalk lines and drive one nail as far up as you can reach.

8 Check the plumb before driving additional nails approximately 16″ o.c. as shown in Fig. 18-9.2.

9 Fit and nail adjoining lengths of siding in place. Joints should be as tight as possible.

10 At openings notch siding to fit, and prime cut edges before nailing.

11 a. If boards cover siding at corners, carry siding to the corner and let the two corner pieces overlap each other.

11 b. If siding butts against corner boards, install primed corner boards first, then trim the siding for a tight fit.

12 Caulk all joints at openings and corners.

Questions

1 What is the difference in nailing between horizontal and vertical application of tongue-and-groove drop siding?

2 Why do you think the first length of vertical siding is nailed at the center of a wall?

3 Why do you think greater care is required to install drop siding vertically than horizontally?

Unit 18-10 Apply Vertical Boards

Boards for vertical application to exterior walls may have smooth or rough-sawn surfaces, and are available clear or with knots for a rustic appearance (Fig. 18-10.1). The most common board size is 1 × 12. If the edges are square or shiplapped, they must be covered with **battens**—thin strips of wood (Fig. 18-10.2). Boards with tongue-and-groove edges may be applied without battens.

Vertical boards may be nailed directly into wood or plywood sheathing, and the procedure is similar to that for drop siding (see Unit 18-9). More often, however, insulating or gypsum sheathing is used, and the boards are applied to nailing strips spaced 24 to 30″ on centers. These nailing strips may be applied over the sheathing, or applied directly to studs between sheets of sheathing as described in the following procedure.

Board and Batten Patterns

A variety of patterns can be achieved with boards and battens (Fig. 18-10.3). In the standard pattern (A) the boards are applied first with gaps between, and the

Fig. 18-10.1 A board-and-batten exterior adds both texture and strong vertical shadow lines to a wall. (Courtesy American Plywood Association.)

gaps are covered with narrower battens. Both boards and battens are nailed at their midpoints. In the reverse board-and-batten pattern (B) the battens are applied first, spaced on centers the width of one board plus $\frac{1}{2}''$. In the board-on-board pattern (C) where boards are also used as battens, the spacing between boards is the width of one board less $1\frac{1}{2}''$, which is the standard amount of overlap. With the board-on-board pattern underboards and batten boards are not necessarily the same width. In both board-on-board and reverse board-and-batten patterns, the undercourse is applied with one nail per bearing, and the overlapping course with two. Note in Fig. 18-10.3 that 10-penny nails clear

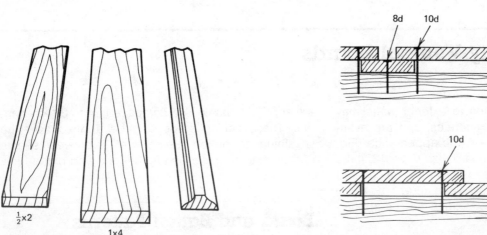

Fig. 18-10.2 Typical shapes of battens.

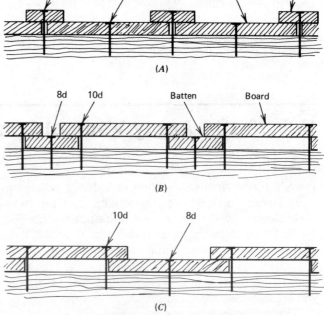

Fig. 18-10.3 Common board-and-batten patterns, and the accepted method for nailing. (A) Standard board and batten. (B) Reverse board and batten. (C) Board on board.

the underboards about $\frac{1}{4}''$ to allow for separate movement of the boards.

Walls of vertical boards may be trimmed at the top and bottom or finished with a horizontal strip of batten material. Corners may be mitered or finished with corner boards or battens. Openings are usually framed with battens.

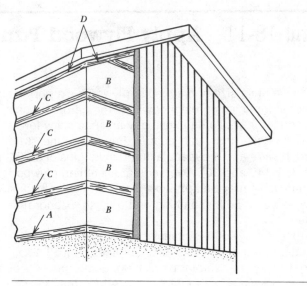

Fig. 18-10.4

Board Layout

The layout of the boards should be marked on the sheathing or building paper before the first board is applied. Battens should overlap boards by no less than $\frac{1}{2}''$. With standard $1\frac{1}{2}''$-wide battens, boards with square edges can be gapped $\frac{1}{4}$ to $\frac{1}{2}''$. Boards with shiplapped edges should gap no more than $\frac{1}{4}''$.

Procedure

1 Before studs are covered with sheathing, nail a 1×4 around the house just above the foundation wall (A in Fig. 18-10.4).

2 Install sheets of 24″ wide insulating or gypsum sheathing (B) above the 1×4.

3 Install a 1×6 nailing strip above the sheathing (C).

4 Repeat steps 2 and 3 on up the wall.

5 At eaves and rakes install a 1×4 (D).

6 Trim the top course of sheathing to fit the remaining openings between nailing strips.

7 Apply building paper over the entire wall with staples. **Note.** Staple close to sheathing nails, and use the staples as a guide for nailing the vertical boards.

8 If necessary, trim boards at corners to size.

9 Apply a prime and finish coat of paint or stain to the edges of all boards.

10 Nail the first board at a corner. **Precaution.** Make sure the corner away from the edge is absolutely plumb.

11 Cut succeeding boards to length and apply them at the established spacing. **Precaution.** Nail into structural members whenever possible. If insulation is not yet in place, clinch (bench over the ends of) nails that go only into nailing strips.

12 When all boards are in place, apply a prime coat of stain to the backs of battens, and let it dry.

13 Complete treatment at corners.

14 Install battens as trim around window and door openings.

15 Install batten trim at the top and bottom of the walls.

16 Cover all vertical joints with battens, nailing as shown in Fig. 18-10.3.

17 Caulk all joints around openings.

Questions

1 How much should battens overlap boards?

2 Name three common patterns using boards and battens.

3 What does it mean to clinch a nail?

4 Name two ways to install nailing strips for boards and battens.

Unit 18-11 **Apply Plywood Panels**

Because of its size and strength, plywood is a popular material for finishing exterior walls. The carpenter can cover a large wall area in one day and has many fewer joints to cover than with boards. Panels are the standard 4' wide and are manufactured in lengths of 8', 9', 10', 12', 14', and 16'. Over regular sheathing plywood as thin as $\frac{5}{16}''$ may be applied. Panels $\frac{5}{8}''$ thick or thicker may serve as both sheathing and a finish exterior material.

Exterior plywood is manufactured unfinished or finished and with plain or textured surfaces (Fig. 18-11.1). Edges may be shiplapped, tongue-and-groove, or square. Plywood with interlocking edges may be applied vertically or horizontally but, because the graining and texture run the long dimension of the sheets, it is usually applied vertically.

Panels are attached with 8-penny noncorrosive nails spaced 6" apart at the edges and 12" on centers in the field (the center of the panel). They must be fully supported at their edges, either by studs or blocking. No panel width should be less than 16". On a few occasions it may be necessary to butt two cut edges together. If joints will not be covered with battens, bevel the matching edges at opposite 45° angles (Fig. 18-11.2) and caulk not only in the joint but behind it.

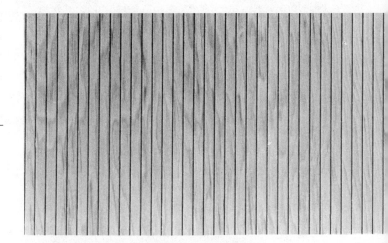

Procedure

1 Measure the wall and establish the locations of vertical joints. **Precaution.** Make sure there is a stud centered behind each joint. If not, add one.

2 If necessary, adjust the locations so that the first joints are equidistant from opposite corners of the wall.

3 On the sheathing mark joint locations.

4 If panel height is constant across the wall, precut panels to the correct length. Otherwise, cut panels one at a time to fit.

5 Prime edges of panels to be painted, or treat with water repellent the edges of panels to be stained.

6 If necessary, rip a corner panel to width.

7 Install the corner panel with the ripped edge at the corner.

8 Install full-width panels to the first window or door opening, with edges touching but not forced tight.

9 Measure the distance from the edge of the last full panel to the casing.

Fig. 18-11.1 Types of textured plywood for exteriors. (All photos courtesy American Plywood Association.)

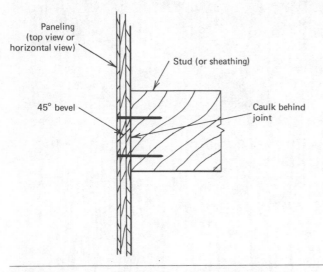

Fig. 18-11.2 How to seal plywood joints that do not interlock and are not protected by battens.

Fig. 18-11.3

10 Measure from the foundation line to the drip cap of windows and doors, and to the sill of windows.

11 Mark the panel for cutting, allowing for a $\frac{1}{8}''$ gap on all sides. **Precaution.** Mark the cuts on the back if you use a power saw, and on the front if you use a handsaw.

12 Recheck your measurements and marking.

13 Cut on the marking lines.

14 Install the cut panel (Fig. 18-11.3).

15 Measure from the edge of this panel to the opposite casing.

16 Using the measurements in steps, 10, 11, and 15, mark, cut, and install the opposite cut panel.

17 Repeat steps 8 through 16 as necessary toward the opposite corner.

18 Measure the distance from the last installed panel to the corner.

19 Rip a panel to this dimension. **Precaution.** The edge to be cut off is opposite the edge cut off in step 6.

20 Install the corner panel.

21 Install corner boards if required.

22 Install any battens on the centers specified.

23 Caulk all joints around openings.

24 Seal cut edges at corners with paint or water repellent.

Questions

1 What is the recommended spacing of nails for attaching exterior plywood?

2 How tight should panels fit against trim at openings?

3 Why do you think more nails are required at the edge of a panel than in the field?

Unit 18-12 Apply Asbestos-Cement Panels

Like plywood, asbestos-cement is also manufactured in large sheets ready to install. The surface, either smooth or striated, is light gray in color and can be left natural or painted. Like plywood, the edges of each panel must be fully supported by either studs or blocking.

The installation is quite different, however, largely because the panels have square edges that must be flashed at every joint. If vertical joints are not covered, the strips of building paper used for flashing lie behind the joint (Fig. 18-12.1, left). If either vertical or horizontal joints are covered with battens, the flashing goes behind the batten but over the panel edges (right). Horizontal joints may also be either lapped (Fig. 18-12.2) or fitted with noncorrosive metal moldings called **Z strips** (Fig. 18-12.3) that are similar to the channels often used with asbestos-cement shingles (Unit 18-8).

Wider flashing is also required at inside and outside corners. At inside corners two panels are butted together, and the joint is sealed with a bead of caulking (Fig. 18-12.4). Panels also butt at outside corners, but there the edges must be flashed again and covered with corner trim like corners of other exterior materials.

Manufacturer's directions for application vary somewhat, so study them carefully before beginning work. The panels, which are 4′ × 8′ in size, may be installed either horizontally or vertically. Joints between panels should be fitted tightly if they are not covered with battens. If covered, they should be gapped $\frac{1}{4}$″ to allow for nailing the battens.

Tools

Only one tool is needed for trimming asbestos-cement panels to size—a power saw with a masonry blade. To

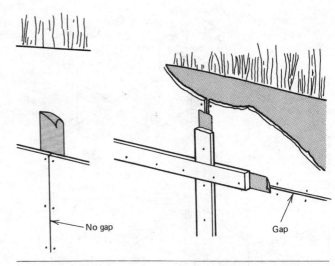

Fig. 18-12.1 Where flashing goes at vertical joints between panels of asbestos-cement.

avoid damage to panels, most carpenters lay the sheets out on four 2 × 4s, one on each side of the cutting line and one to support the outer edges of the cut sheet. The 2 × 4s support the carpenter's weight as he or she walks along making the cut, and also support the cut sheets so that they don't cause the saw to bind. Other tools needed for installation are a hammer, utility knife, power drill, and caulking gun.

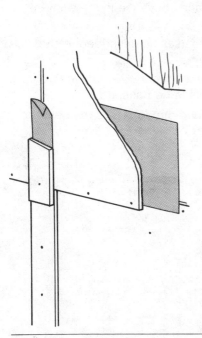

Fig. 18-12.2 Treatment when horizontal joints are lapped.

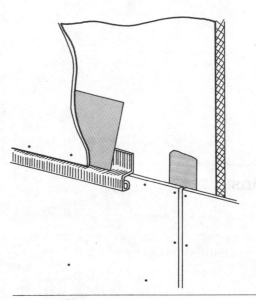

Fig. 18-12.3 Treatment when panels are applied with Z strips.

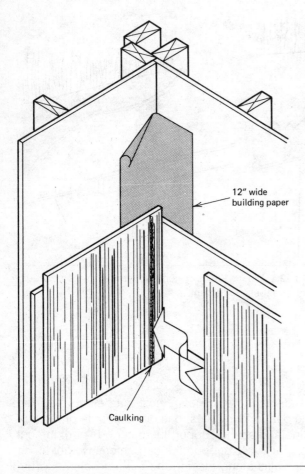

Fig. 18-12.4 How to seal an inside corner.

12″ wide
building paper

Caulking

Procedure

1 Read the manufacturer's instructions carefully.

2 Measure the wall and establish the locations of all vertical and horizontal joints.

3 Add studs and blocking, where necessary, behind these joints.

4 Cover outside corners with a continuous strip of building paper 12″ wide and centered on the corner.

5 Cut a corner panel to width.

6 Nail the panel in place at the corner and bottom.

7 If joints will not be covered with battens, slip strips of building paper behind the unnailed edges.

8 Complete the nailing of the first panel.

9 Repeat steps 6 to 8 inclusive to complete the first course.

10 At openings mark and cut panels for a tight fit.

11 Lay a bead of caulking compound around the trim at each opening.

12 Embed the panel in the caulking, and nail around all edges. **Precaution.** Predrill the panels for nails where necessary.

13 a. If panels overlap at horizontal joints, cover the top edge of the first course with a strip of building paper at least 6″ wide (see Fig. 18-12.2).

13 b. If panes butt at Z strips, install the moldings and cover the surface above the ledge with a 6″ strip of building paper (see Fig. 18-12.3).

14 Install remaining courses as described in steps 6 through 13.

15 With strips of building paper cover all vertical and horizontal joints to be battened.

16 Install battens, nailing through the gaps between panels into studs or blocking.

Questions

1 Where is caulking required to seal joints at edges of panels?

2 At what four points may flashing be required at horizontal joints, depending on the method of installation?

3 How many layers of flashing are required at outside corners?

19

Exterior Trim

"Trimming out" is the term many carpenters use to describe the many small but important tasks still to be completed outside the house. Finish roof and wall materials are in place, but trim must still be installed where the two come together. Gutters and downspouts are needed to carry potentially damaging rainwater away from the foundation. Windows and doors are in place, but may not be fully trimmed, particularly if the walls are masonry. Window decoration, such as shutters and flower boxes, are also not yet in place. Even some structural work remains, such as building decks, screening porches, and putting in exterior steps.

Large-volume builders often have one crew of carpenters installing exterior materials and another following them to apply trim. Small builders schedule trimming after all exterior materials are in place, or sometimes even finish one wall before going on to the next. The design of the house, particularly at the eaves, the size of crews, and the availability of materials all play a part in the schedule of work.

Trimming the edge of a roof is sometimes a complex process that takes a great deal of care and time to assure an attractive and tight-fitting result. Yet the modern finish carpenter has a much easier job then his fellow carpenters of the 18th and 19th centuries. In those earlier days, particularly during the Victorian era at the turn of the century, trim was so ornate that it was called "gingerbread". Many houses had towers with circular eaves, for example, and the trim had to fit the curve. Fascias were intricately carved, sometimes as much as 18″ wide, and made up of dozens of separate pieces. Porches, eaves, and gables were covered with trim specially carved or turned for the job. The trim on most houses today consists of relatively few, simple pieces. Its main purpose is to cover, not adorn.

Unit 19-1   **Build a Cornice**

The trim applied at the eaves and rakes of a roof is called a **cornice.** In its simplest form a cornice is a single molding that fits between the top course of wall material and the underside of roof shingles. On the other hand it may include four or five separate pieces that not only close the joints between wall and roof, but also improve the appearance of the house and provide ventilation at the eaves.

There are three basic types of cornices, and they relate to roof construction at the eaves:

- When rafters overhang the wall but are fully exposed, the joint is finished with an **open cornice.** The top course of siding or shingles is covered with a **frieze** (Fig. 19-1.1), as described in Unit 18-1, that must be notched to fit around the rafters. Between rafters the joint is covered with a bed molding. When the ends of rafters are covered with a fascia but the undersides remain exposed, (Fig. 19-1.2), the variation is called an **open soffit.** Blocking is sometimes needed, if the roof pitch is low, as a nailing surface for trim.

- When rafters overhang the wall but are enclosed, the construction is called a **box cornice** (Fig. 19-1.3). The ends of rafters are finished with a

fascia and shingle molding. A soffit hides the undersides of rafters or lookouts. Where the soffit meets the wall, the joint is trimmed with a frieze and molding. The soffit may be horizontal or follow the slope of the roof.

When the ends of rafters are cut flush with wall plates, the joint between roof and wall is finished with a **close cornice.** Wall sheathing extends to the top edge of rafters, and is covered with a frieze and molding. The treatment is similar to an open cornice except that no notching is required.

Many of the houses built along the east coast in colonial days had close cornices. Today they are rare because of one drawback: It is impossible to provide ventilation at the eaves. When an open cornice is used, openings are cut through the wall between rafters to let air through, and are then screened to keep out insects.

Ventilation through a box cornice may be provided in several ways: (1) holes are cut in the soffit for individual manufactured vents; (2) the soffit is installed in two pieces with a gap between them that allows air to circulate; (3) perforated metal is installed as a continuous soffit (Unit 19-3).

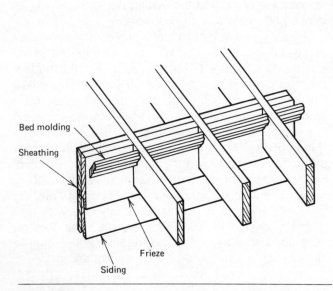

Fig. 19-1.1 An open cornice consists of a frieze and molding.

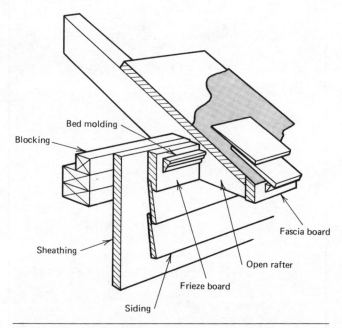

Fig. 19-1.2 An open soffit is an open cornice with a fascia at the ends of rafters.

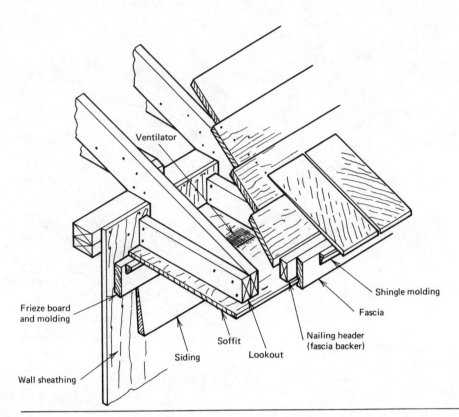

Ventilator

Frieze board
and molding

Wall sheathing

Siding

Soffit

Lookout

Nailing header
(fascia backer)

Fascia

Shingle molding

Fig. 19-1.3 A box cornice completely encloses the rafter tails. Its soffit may be horizontal or sloping.

Soffits

Three materials are commonly used for soffits: plywood, gypsum board, and hardboard. Plywood should be exterior grade, group 1,2,3, or 4. When supports—either rafters or lookouts—are spaced 16″ or 24″ on centers, $\frac{3}{8}$″ thick plywood is adequate. When supports are 48″ on centers, $\frac{5}{8}$″ plywood should be used.

Gypsum board for exterior use in areas protected from the weather has a blue surface paper and a water-resistant gypsum core. It is $\frac{1}{2}$″ thick, and comes in sheets 4′ wide and either 8′ or 12′ long. Its supports must be no more than 24″ apart. Joints between lengths of boards must be closed with joint treatment—a combination of special perforated tape and joint compound, as discussed in detail in Chapter 21. Gypsum board takes paint easily, but will not take stain.

Tempered hardboard used for soffits should either have a factory prime coat, or be primed on the job before application. Made in $\frac{3}{16}$″ and $\frac{1}{4}$″ thicknesses, it must be supported at all joints and edges. Panels are 4′ wide and come in standard lengths from 4 to 16′. Metal moldings are available that cover nailheads when soffits are not more than 24″ wide.

Fascias

With an open soffit the fascia board is attached only to the ends of rafters. When rafter spacing is 16″, the fascia board may be a clear 1″ board that is wide enough to cover the exposed edge of roof sheathing and the entire ends of rafters. Some builders use 2″ thick lumber for a fascia, regardless of rafter spacing, to minimize the danger of warping and poor exterior appearance. The top of the fascia should be beveled to the slope of the roof. The bottom edge is usually cut square and extends about $\frac{3}{8}$″ below the rafters to act as a drip edge.

With a box cornice a board is sufficiently stiff for a fascia, because the soffit provides the necessary stiffness along the lower edge. The soffit either fits into a groove cut $\frac{3}{8}$″ above the bottom edge of the fascia board, or against a 2 × 2 cleat attached to the back of the fascia.

There are two types of fascia construction. Some builders add a 1″ header across the ends of rafters after the roof is completely framed. This header, also called a **fascia backer,** acts as a guide for the installation of roof sheathing, and the sheathing is nailed into

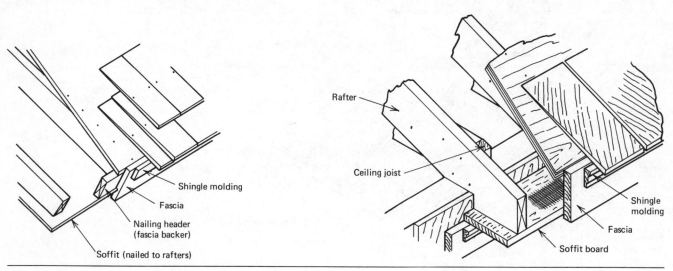

Fig. 19-1.4 The two possible locations of a fascia: beneath sheathing (left) and over the end of sheathing (right).

the header (Fig. 19-1.4, left). The fascia itself is installed during trimming out. This method is used only with a box cornice.

Other builders omit the nailing header, and bring roof sheathing flush with the trimmed ends of the rafters, as described in Unit 9-34. The fascia covers the end of the sheathing (Fig. 19-1.4, right). It is installed in an open cornice as soon as roof framing is completed and in a box cornice during trimming out.

Except on short walls, lengths of fascia must butt together. These joints must fall at the ends of rafters if there is no fascia backer across the ends of rafters. Seams can fall at any point along the shingle backer, but nails should be clinched to prevent the seam from opening up.

Lookouts

As discussed in Unit 9-15, lookouts are horizontal framing members that extend from the ends of rafters horizontally back to walls. They provide the necessary support for soffits. Usually lookouts are cut from 2 × 4 stock, but 1 × 4s may be used when soffits are narrow —that is, 12″ or less.

Lookouts may be installed at one of two times during construction. They are usually in the same plane as studs, and are sometimes toenailed into those studs as soon as rafters are in place. Installed this early in construction, however, they interfere with wall sheathing. Most builders prefer to install lookouts after the walls are sheathed but before any exterior materials are applied. In this way they can build the framework for the box cornice on the subfloor of the house, and raise the entire assembly into place.

At Corners

Details of cornices are usually included among the working drawings of a house. These drawings show what moldings to use, and also what happens at the corners of the house—the **cornice return.** On hip, mansard, shed, and flat roofs the cornice continues around side walls, and any soffit is mitered at a hip rafter, valley rafter, or corner rafter. On gable and gambrel roofs there are three treatments possible. If there is no overhang at the rake, the cornice is closed with a frieze board that follows the slope of the roof and the shape of the cornice. If the roof overhangs at the rake and the soffit follows the slope of the rafters, the corner is mitered as with a hip roof. If the house has a box cornice and an overhang at the rake, the cornice extends to the fly rafter, the rake fascia follows the slope of the roof and the shape of the cornice, and the intersection is boxed in. Construction of cornice returns is discussed in the next unit.

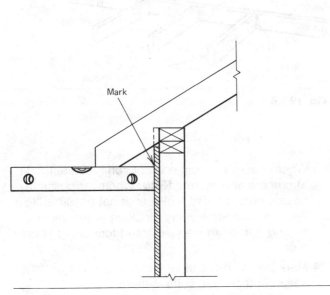

Mark

Fig. 19-1.5

Procedure for Framing a Box Cornice

1 Against the bottom edge of the two end rafters place a carpenter's level, and mark the point where the top of the level touches sheathing (Fig. 19-1.5).

2 Snap a chalk line between the two points, and check its level.

3 From the drawings determine the corner treatment.

4 Establish the overall length of the box cornice.

5 Above the chalk line tack 1 × 4 ledgers along the full length of the cornice.

6 Hold a straight board vertically against the side of each rafter, and mark along the edge of the board where it crosses the ledger (Fig. 19-1.6).

7 Indicate with an X on which side of the line the lookouts go.

8 Measure the distance from the face of the ledger to the ends of several rafters to determine the length of lookouts.

9 To allow for variation in the length of rafters and slight waviness of sheathing, reduce the length in step 8 by $\frac{1}{2}''$.

10 Cut the required number of lookouts to the length in step 9.

11 Determine the length and shape of corner lookouts, and cut them to size. **Note.** Unlike all others, corner lookouts fit against the bottoms of end rafters (Fig. 19-1.7) in gable and gambrel roofs, and against bottoms of hip rafters in hip and mansard roofs. Therefore they must be trimmed to the slope of the roof. In hip and mansard roofs corner lookouts get a double cheek cut at the inner end.

12 Remove the ledgers, and add a 1 × 4 cleat at each joint between lengths. **Precaution.** Be sure that cleats do not fall at Xs, and that the lengths are lined up straight before you attach them.

13 Cut the ledger to length. Recheck measurements to make sure it will fit properly.

14 Attach all except corner lookouts, using 8-penny coated nails driven through the back of the ledger (Fig. 19-1.8).

15 Raise the assembly and attach the ledger permanently in position above the chalk line. Nail through sheathing into studs wherever possible.

16 Nail corner lookouts in place.

Procedure for Installing a Fascia

1 From drawings determine the height, thickness, and length of the fascia.

2 Determine the thickness of any soffit material.

3 At a point $\frac{3}{4}''$ from the bottom edge of the fascia, cut a groove for any soffit material (Fig. 19-1.9).

4 If required, bevel the top edge of fascia boards.

5 Cut ends of individual fascia boards at 45° for a tight fit where they butt.

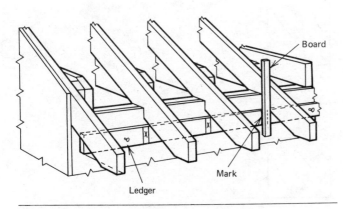

Board

Ledger

Mark

Fig. 19-1.6

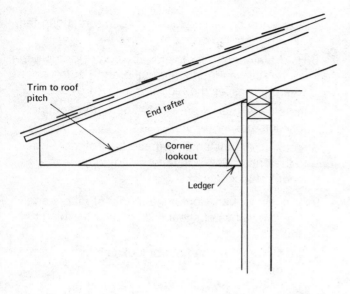

Trim to roof pitch

End rafter

Corner lookout

Ledger

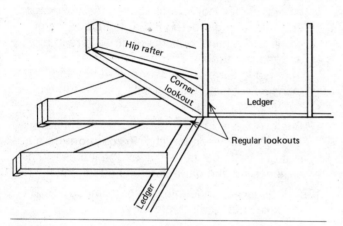

Hip rafter

Corner lookout

Ledger

Regular lookouts

Ledger

Fig. 19-1.7

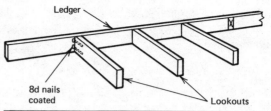

Ledger

8d nails coated

Lookouts

Fig. 19-1.8

3 Where necessary, cut ends of soffit material to fit at corners and seams. **Note.** Wherever possible, seams should fall at lookouts. If not possible, fit a 2 × 2 lookout to serve as blocking at the seam, nailing it through the fascia and toenailing at the ledger.

4 Mark the locations of ventilators, and cut or bore holes for them into soffit material.

5 If holes are screened, staple screening over the openings on the back sides of soffits. **Note.** If

6 Beginning at one end of the roof, fit fascia boards in place, with the top edge against roof sheathing, and the end overlapping the last rafter.

7 Nail fascia boards to the ends of rafters with finishing nails, and countersink them. **Precaution.** If necessary to keep the fascia absolutely straight, fit shims between the board and short rafter tails. At joints, nail through both boards.

8 Mold the joint between the fascia and finish roofing material.

Procedure for Installing a Soffit

1 Measure the distance from wall sheathing to the inside of the groove in the fascia.

2 Rip soffit material to the width in step 1.

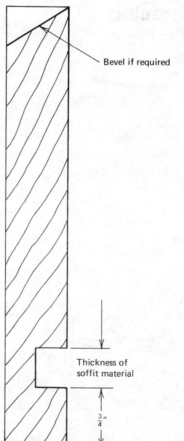

Bevel if required

Thickness of soffit material

$\frac{3}{4}''$

Fig. 19-1.9

ventilators are purchased, they are installed from outside the cornice after step 9.

6 Fit soffits into the groove in the fascia, then swing the lengths until they are tight against the ledger.

7 Attach soffits to the ledger and each lookout. With plywood use 4-penny nails spaced 6″ o.c. With gypsum board use annular ring wallboard nails or screws spaced 7″ o.c. With hardboard use 5-penny nails spaced 4″ o.c. around the perimeter and 6″ o.c. in the field. **Precaution.** With any soffit material, do not nail any closer than $\frac{3}{8}$″ to its edge.

8 Finish gypsum board as described in Unit 21-3.

9 If specified, add a thin batten or metal molding behind the fascia to help support the soffit.

2 Cut frieze members to length.

3 If the cornice is open, mark and notch the frieze to fit around rafters.

4 Miter ends as described in Unit 18-1.

5 If necessary, cut a groove in the bottom edge of the frieze to fit over the top course of exterior wall material.

6 Fit frieze members in position. They should fit against roof sheathing in an open or closed cornice, and against soffits in a box cornice.

7 Nail the frieze to wall sheathing, checking its level as you go.

8 Mold the joint between the frieze and sheathing or soffit.

Procedure for Installing a Frieze

1 Determine the type and size of frieze as described in Unit 18-1 and Fig. 18-1.4

Questions

1 What is the difference between an open soffit and an open cornice?

2 At what two points are moldings added in a box cornice?

3 Name three common soffit materials.

4 How can ventilation be provided with a box cornice? With an open cornice? With a close cornice?

5 What is another name for the diagonal lookout at the corner of a flat roof?

6 How is a soffit supported?

Unit 19-2   **Trim a Gable**

The method of trimming a gable depends partly on the amount of side overhang and partly on the type of cornice at the eave. A **close rake** has almost no overhang, and is similar to a close cornice (Fig. 19-2.1). It consists of a frieze board and a small molding nailed over wall material into the end rafter. Where a close rake meets a close cornice, the two friezes and their moldings are mitered for a tight fit. Where a close rake meets a box cornice, the gable frieze extends to the fascia and acts as a closure for the cornice.

When the roof overhangs the gable and eaves have an open cornice, the rake is usually open also. The molded friezes meet at the corner of the wall and are mitered as with a close rake and close cornice. With an open soffit, the rake fascia is nailed to the fly rafter and trimmed, and the fascia and trim are again mitered at the corner of the roof (Fig. 19-2.2).

Corner construction is a little more complex when the house has a box cornice at the eaves and the gable overhang is also boxed. Along the roof rake are a fas-

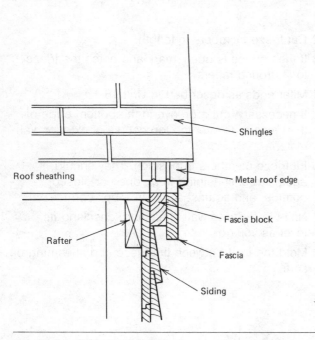

Fig. 19-2.1 A close rake is similar to a close cornice.

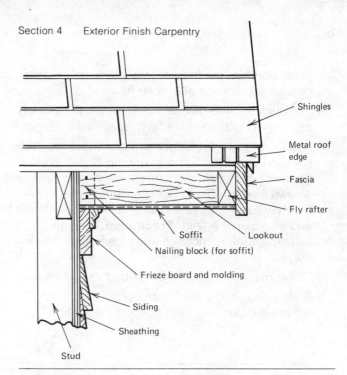

Fig. 19-2.2 The rake fascia is nailed to the fly rafter with an open rake.

cia and molding, a soffit, and either a frieze or molding where soffit and wall meet. The rake and eave soffits meet at a **cornice return** (Fig. 19-2.3), which must be cut and fitted after soffits are installed.

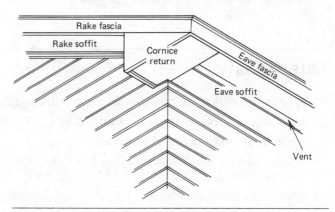

Fig. 19-2.3 A cornice return completes the junction of a box cornice and a gable overhang with a soffit.

Procedure for Building a Cornice Return

1 Determine the length of the fascia. It should match in height and thickness the fascia of the box cornice.

2 Cut a groove for soffit material.

3 Give both ends a plumb cut.

4 Bevel the lower end at 45° to fit tightly against the cornice fascia.

5 Nail the fascia board to fly rafters.

6 Mold the joint between fascia and finish roofing material.

7 Rip soffit material to width so that seams fall on ladders (see Unit 9-10, Fig. 9-10.1). **Note.** Soffits at rakes extend from the ridge to the end walls, not to the eaves (Fig. 19-2.4).

8 Fit soffits into the groove in the fascia, and nail them to ladders.

9 Cut the cornice closure to shape (A in Fig. 19-2.5). Study drawings to determine how far the closure goes around the side wall.

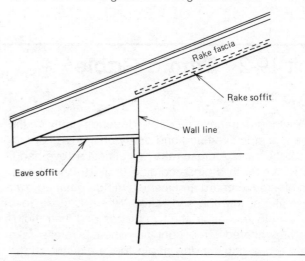

Fig. 19-2.4

10 Cut a groove for soffit material.

11 Nail the closure into the edge of the rake fascia and flush with it.

12 Cut and install a length of frieze material to support the soffit return (*B* in Fig. 19-2.5).

13 Cut the soffit return (*C*) and slide it into the grooves at the end and side and onto the frieze. Nail it securely.

14 Cut the end piece of the cornice return (*D*) to size, bevel the top edge and cut a groove near the bottom. **Note.** This board fits inside the cornice closure.

15 Nail the end piece into the closure and toenail to the rake soffit.

16 Add a frieze and molding from ridge to cornice return along the gable end.

17 Repeat this procedure at the other corners.

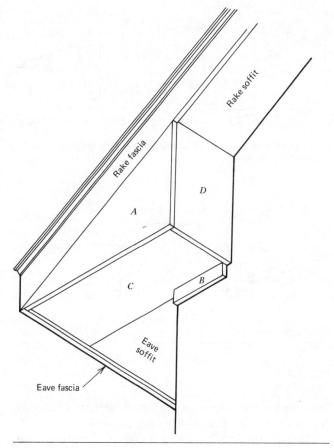

Fig. 19-2.5

Questions

1 At what points does a rake fascia end?
2 In what order are the following parts of a cornice return installed?
   a) soffit return
   b) cornice closure
   c) cornice end piece
   d) frieze support

Unit 19-3   **Install a Ventilating Metal Cornice**

Metal, usually aluminum, is often used in place of wood to trim eaves. Trim pieces are manufactured of thin sheets in various shapes (Fig. 19-3.1) to meet job requirements, and are designed to support metal soffits.

Soffit material is manufactured in widths ranging from 12 to 48″. Most types are made in sections that are crimped for stiffness and to form loose-locking seams. Others are made in coils up to 50′ long. Both types are made in sheets either solid or perforated for ventilation.

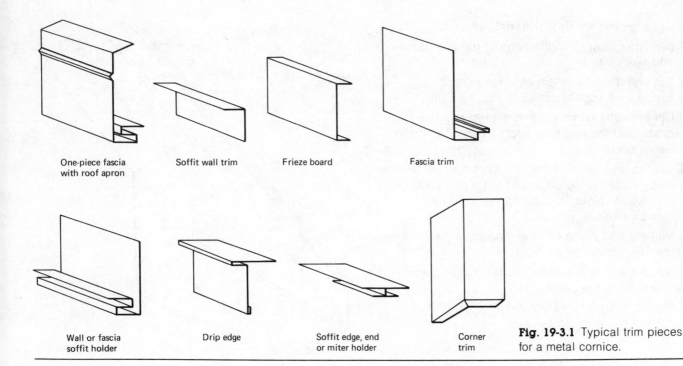

One-piece fascia
with roof apron

Soffit wall trim

Frieze board

Fascia trim

Wall or fascia
soffit holder

Drip edge

Soffit edge, end
or miter holder

Corner
trim

Fig. 19-3.1 Typical trim pieces
for a metal cornice.

They are prefinished, ready to install, and are fitted into the trim from the ends of the eaves.

To cut any trim piece to length, cut through the flanges with a hacksaw (Fig. 19-3.2). Then bend the remaining metal back and forth until it snaps off. Trim the rough edges with metal shears. You use a spline setter (Fig. 19-3.3) to insert polyethylene splines that prevent soffits from moving or rattling.

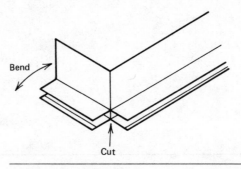

Bend

Cut

Fig. 19-3.2 Shorten metal trim in two steps: (1) cut the flanges with a hacksaw, and (2) bend the thinner metal until it breaks.

Procedure

1 Install a wood fascia board across the tails of rafters on one side of the house.

2 Against the bottom edge of the fascia board place a carpenter's level, and mark the point where the top of the level touches the sheathing. **Note.** If the wall is brick veneer, skip steps 2 and 3.

3 Snap a chalk line between the two points, and check its level.

4 Above the chalk line nail a continuous cleat. Use a 2 × 2 cleat when the soffit will be supported on wall trim (Fig. 19-3.4, top left), and a 1 × 2 when the soffit will rest on a frieze (top right). **Note.** If the exterior wall is brick veneer, set the cleat atop the upper course in a bed of mastic (lower left), or block as shown at lower right.

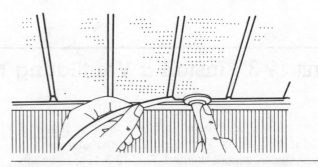

Fig. 19-3.3 A spline setter simplifies the task of inserting splines at metal soffits.

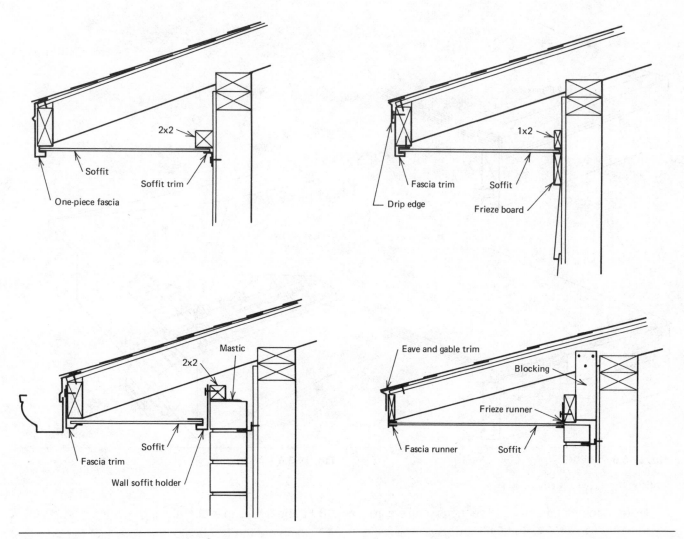

Fig. 19-3.4

5 a. If a one-piece metal fascia is used, attach the first length at a corner as shown at the top of Fig. 19-3.5, spacing nails no more than 24" apart.

5 b. If a two-piece fascia is used, attach the fascia trim piece first, then the metal drip edge (Fig. 19-3.5, bottom).

6 Install the remaining fascia pieces along the first eave, overlapping ½ to ¾" at seams. **Precaution.** Do not nail the metal at seams. It must be able to expand freely.

7 If a metal frieze board is specified, cut a board to fit inside it. Make the fit as tight as possible.

8 Determine the length of the wall support. With gable and gambrel roofs the support is the same length as the fascia. With hip, mansard, flat, and shed roofs it is the length of the wall.

9 Measure the height of the slot in the metal fascia.

10 a. If the wall support is a metal angle, attach it at the distance in step 9 below the cleat.

10 b. If the wall support is a frieze, fit the metal frieze over the backer, and attach them as a unit at the distance in step 9 below the cleat.

10 c. If the wall support is a wall soffit holder, snap a chalk line on the wall at the same level as the bottom of the fascia, and set the holder along the chalk line.

11 Measure the width of the soffit from inside of slot to inside of slot.

12 If necessary, trim soffit material to width. **Note.** The final width should be ⅜ to ⅞" less than the dimension in step 11.

13 Feed the soffit from one end of the cornice into the slots at the fascia and wall (Fig. 19-3.6).

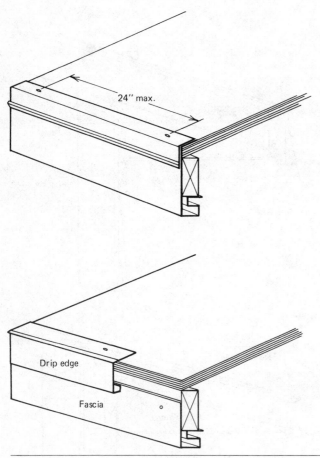

Fig. 19-3.5

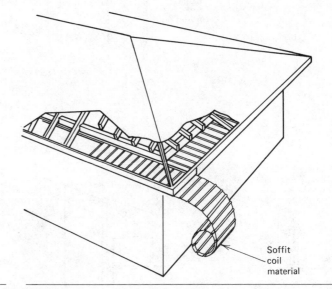

Fig. 19-3.6

Note. If the metal comes in sections, make sure the lengths overlap at crimps.

14 Close the open ends of a one-wall cornice with a short fascia board and metal frieze (*A* and *B* in Fig. 19-3.7). For other cornices, see step 16.

15 Enclose the corners with a piece of corner trim (*C*).

16 If the cornice continues around a corner, as under a hip roof, repeat steps 1 to 10 inclusive to points *X* and *Y* in Fig. 19-3.7.

17 Where side soffits butt against the front soffit, insert a soffit edge (*D*) to support the edge of the front soffit and the ends of side soffits.

18 Fit the soffits in place along the sides.

19 Complete the fascia along the back of the house (*E*).

20 Complete wall support along the back wall (*F*).

21 Insert soffit edges (*G*) and the last length of soffit.

22 Repeat step 15 at the third corner, and steps 14 and 15 at the last corner.

23 Insert a polyethylene spline underneath the soffit at all edges.

Questions

1 How are metal soffits installed?

2 Name three ways of supporting a soffit at the wall.

3 When two metal soffits meet at right angles, as under a flat roof, how are they joined?

4 What prevents soffits from rattling in a strong wind?

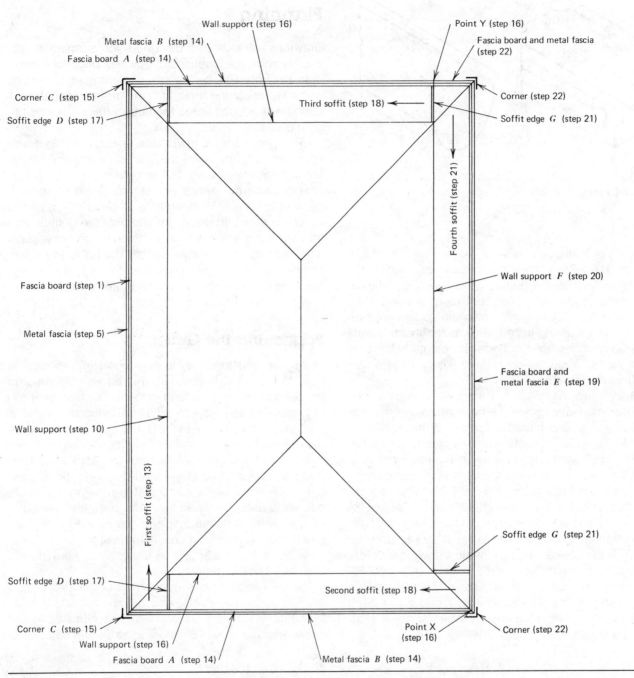

Fig. 19-3.7

Unit 19-4 **Install Gutters and Downspouts**

Gutters and downspouts together form a water collection system for the rain that falls on roofs and the snow that melts off them. Except in areas of the country with little rainfall or with highly absorptive sandy soil, water dripping or running off a roof can damage the foundation unless the runoff is carried safely away from the house.

Gutters, also called **eaves troughs,** collect the water, and **downspouts,** also called **leaders,** carry it to the ground and direct it away from the house. They may

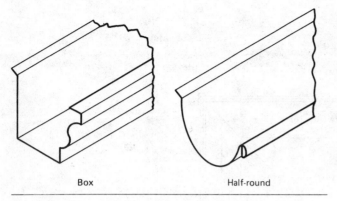

Fig. 19-4.1 The common shapes of residential gutters.

be made of fiberglass, vinyl, or metal—usually aluminum or galvanized steel, but sometimes copper and stainless steel. Wood gutters of redwood and fir are also made, primarily for use with roofs of wood shingles and shakes. Vinyl and aluminum gutters are prefinished and ready to install. Galvanized metal gutters are usually unfinished, and must be painted after they are in place. Wood gutters require two coats of water-repellent preservative.

Gutters are manufactured in several shapes. The most common shapes are the **half-round** and the **box** type, which is also called a **formed gutter, ogee,** or **Style K** (Fig. 19-4.1). Both types are made with a top dimension of 4 and 5″ for houses and in lengths of 10 to 40′. Manufacturers provide the necessary accessories for a complete installation (Fig. 19-4.2)—slip connectors, end sections, corners, outlets, and hangers.

Downspouts are made in two shapes: Round for use with half-round gutters and rectangular for use with box gutters (Fig. 19-4.3). Higher quality downspouts are corrugated for stiffness. They are made in 8 and 10′ lengths, and accessories include elbows of various angles and straps for attachment. One end of each piece is smaller than the other, and each piece is installed with the smaller end down.

Planning

Drawings of a house designed for a specific site usually show the locations of downspouts. Stock drawings do not. The builder or carpenter must decide where water can be brought to the ground so that it will run off without damaging the house, the site, or neighboring sites.

Two general rules govern the layout. First, no downspout should drain more than 40′ of gutter; where rainfall is frequently heavy, 30′ is better. Second, gutters should slope downward at a steady angle of about $\frac{1}{4}$″ per foot of length. With too little slope rain won't wash away the dust and debris that fall into the gutters, and they will clog and overflow. With too much slope water may overflow at the downspouts during heavy rains. The high point of a gutter with one downspout is the opposite end. The high point of a gutter with two downspouts is its midpoint.

Positioning the Gutter

The spike and ferrule type of support (upper right in Fig. 19-4.2) is attached at the eaves. Straps and brackets hangers are attached to the roof beneath roofing material, and may be adjusted to the roof's slope. The relationship between the gutter and the edge of the roofing material is important. A gutter should be centered under the edge of the roof to catch water that drips off the roof. Gutters should always be placed below the slope of the roof to catch water that runs off, but not to catch snow and ice that slide off. The exact height varies with the slope of the roof (Fig. 19-4.4), and is measured from a continuation of the slope to the outer lip of the gutter. These dimensions apply to the high point of the gutter.

Gutters should be supported on both sides of outlets, at ends and joints, and at intermediate points. Maximum spacing of supports is 48″ in climates free of snow and ice, and 18″ spacing is recommended in

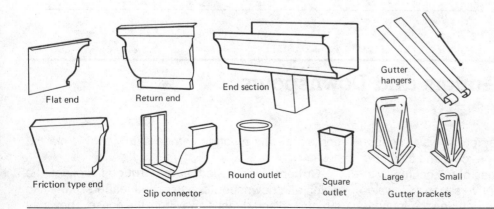

Flat end

Return end

End section

Gutter hangers

Friction type end

Slip connector

Round outlet

Square outlet

Large   Small

Gutter brackets

Fig. 19-4.2 Typical gutter accessories.

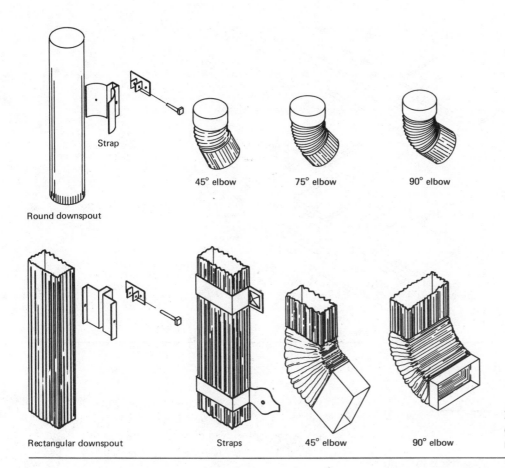

Strap

45° elbow

75° elbow

90° elbow

Round downspout

Rectangular downspout

Straps

45° elbow

90° elbow

Fig. 19-4.3 The common shapes of downspouts and elbows and methods of attachment.

areas where snow and ice are longlasting. Normal spacing is about 32".

Downspouts should be supported at least every 8' of length, and more often if there is any danger of ice forming inside them. The straps are nailed into siding and shingles. If the walls are masonry, holes must be drilled in mortar joints; attachment is made with lag bolts driven into expansion shields.

Estimating

Part of the job of planning is estimating the number of pieces of gutter, downspout, and each accessory. With

a hacksaw you can cut off lengths of gutter and downspout to suit the dimensions of the house, but all accessories should be used without adjustment.

Tools

The tools required to install a typical water collection system are a soldering iron or caulking gun, hacksaw or power saw with a metal-cutting blade, metal shears, hammer, steel tape or folding rule, carpenter's level, and a putty knife.

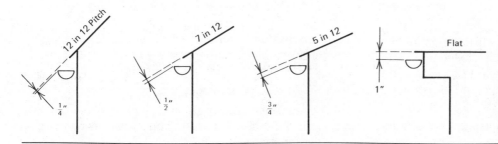

12 in 12 Pitch $\frac{1}{4}''$

7 in 12 $\frac{1}{2}''$

5 in 12 $\frac{3}{4}''$

Flat   1"

Fig. 19-4.4 Gutters should be centered under the edge of roofing and lie below the slope of the roof. The dimensions shown are for the high point of the gutter.

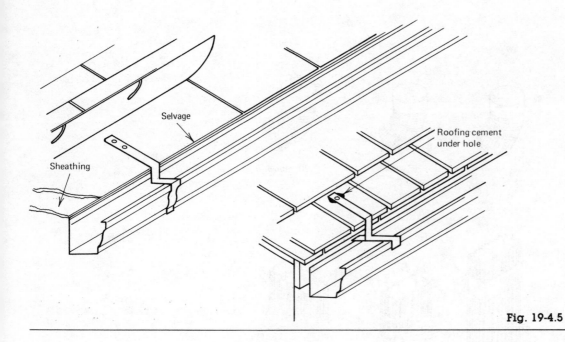

Selvage

Sheathing

Roofing cement
under hole

Fig. 19-4.5

Procedure

1 Establish the locations of gutters around the entire house.

2 Determine the overall length of each straight run of gutter.

3 Determine the locations of downspouts, and the best direction for water to flow from each of them.

4 List the lengths of gutters and downspouts required and all accessories for installation.

5 Assemble the longest gutter, using slipjoint connectors at joints.

6 Solder the joints in metal gutters, and seal joints in other types with caulking or with the sealant recommended by the manufacturer. **Precaution.** Wipe away any excess sealant so that it doesn't slow the flow of water.

7 Add and seal outlet sections and end caps.

8 Fit hangers around the gutter assembly.

9 On the fascia mark the high point of the gutter and its location below roof shingles.

10 With help from other people raise the gutter into position at the high point. **Note.** If the high point is in the middle, you need help to support both ends. If the high point is at the end of a gutter, you need a helper for every 20′ of length.

11 a. Carefully lift the roof shingles, and attach the highest hanger into sheathing with roofing nails

(Fig. 19-4.5, left). **Precaution.** Work on a warm day when shingles are soft enough so that they don't crack when raised.

11 b. When the roofing, such as wood shingles, can't be lifted, nail into the shingles through a dab of roofing cement to seal the hole (Fig. 19-4.5, right). Then cover nail heads with the same cement.

12 Working from the high point toward the end or ends, attach the remaining hangers. **Precaution.** At each hanger check the slope of the gutter and recheck its relationship with the edge of the roof.

13 When all gutters are in place, check the slope once more.

14 Fit elbows to outlet sections.

15 Determine the length of sloping sections of downspout, and cut them to length. **Precaution.** Remember to allow for the overlap at elbows at both ends.

16 Assemble the remainder of each downspout, soldering or otherwise sealing all joints. Each piece fits inside the piece below it.

17 Attach the vertical section with straps. **Precaution.** If holes must be drilled for anchors, support the bottom end of the downspout in the proper position until it is strapped in.

18 Seal the joint at the upper elbow.

19 Where specified, position concrete splash blocks below the downspouts to carry water away from the foundation.

Questions

1 Of what four materials are residential gutters most often made?

2 What is the shape of the downspout that should be used with an ogee gutter?

3 What is the ideal slope for a gutter?

4 How is the high point of a gutter determined?

5 How should joints between two lengths of gutter be connected and made watertight?

6 What is the recommended spacing of supports for gutters? For downspouts?

7 How should gutters be attached when the roof is asphalt shingles?

Unit 19-5   **Build a Deck**

Vacation houses and year-round homes built in climates that permit outdoor living more than eight months of the year frequently have decks. Working drawings seldom show much in the way of details. The proposed location of any deck is shown on the floor plan. Its exact size, shape, and construction, however, are usually determined on the site after a study of site conditions—the contours of the land, orientation of the deck, and the locations of nearby trees.

Decks must be built to the same code requirements as floor framing systems inside the house. Every deck has four parts: a platform, a floor frame, vertical supports, and guard rails.

The Platform

The deck itself is built of weather-resistant 2 × 4s or 2 × 6s laid flat. If the decking has vertical graining (the grain runs up and down when the lumber is flat), either side may be placed face up. If the decking has **flat** graining (the grain curves from side to side), each piece must be laid with the annular rings pointing down at the edges.

The deck may run parallel to the wall of the house, at right angles to it, or in a decorative pattern, such as a parquet or herringbone (Fig. 19-5.1). Whatever the pat-

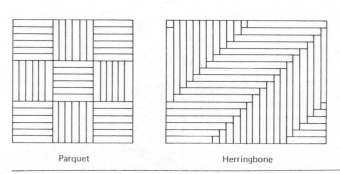

Parquet   Herringbone

Fig. 19-5.1 Decking is usually laid with long boards, but it may also be laid with short boards placed in decorative patterns. Both ends of every piece of decking must be supported. (All photos courtesy California Redwood Association.)

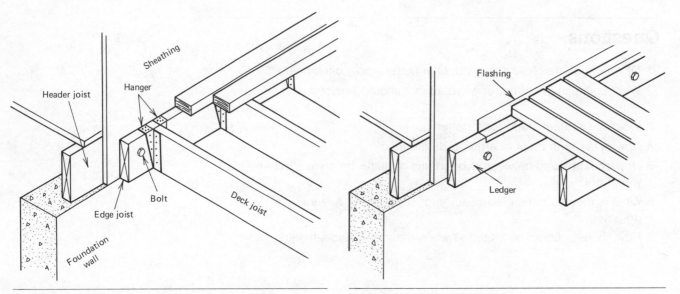

Fig. 19-5.2 Detail of deck support when decking runs parallel to the wall of the house.

Fig. 19-5.3 Detail of deck support when decking runs at right angles to the wall of the house.

tern, each piece of decking must be supported at both ends, and long lengths given intermediate support. Decking must run at right angles to framing beneath it.

Deck Frame

When a deck is supported around its perimeter on posts or piers, its frame is a simple box frame. When

decking runs parallel to the wall of the house, supporting joists are attached with joist hangers to an edge joist bolted through sheathing into a header joist (Fig. 19-5.2). At the opposite edge the joists may butt against a beam, as at the wall, or rest atop the beams. When decking runs at right angles to the house wall, the support at the wall is a ledger bolted to the header and flashed on the top (Fig. 19-5.3). When the deck is made up of short boards in a pattern, the frame is a lattice of joists cut to support the ends of individual pieces.

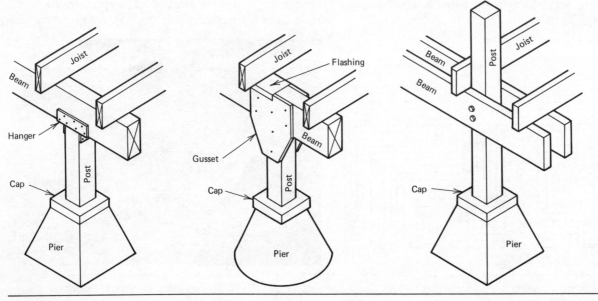

Fig. 19-5.4 Wood posts that support decks rest on caps anchored to concrete piers. The deck frame may be attached to posts in several ways.

Support

The outer edges of a deck frame are usually supported on 4 × 4 wood posts (Fig. 19-5.4). The posts rest on 2″ wood caps anchored to the tops of concrete piers. See Unit 6-5 for methods of anchoring. Caps may be omitted if posts are less than 12″ long. All piers are alike, and their tops must be at least 8″ above grade. If the site is not level, the lengths of posts must be varied so that their tops are at the same level to support the deck frame.

The beams that span from post to post may be 4″ timbers or a pair of 2″ joists (Fig. 19-5.4). The sizes of beams and the spacing of posts are related. Here are the maximum spans of typical beams for typical decks (Fig. 19-5.5):

Deck width	Post spacings with:		
	4 × 6 beam	4 × 8 beam	4 × 10 beam
6′	6′-9″	9′-0″	11′-3″
8′	6′-0″	8′-0″	10′-0″
10′	5′-3″	7′-0″	8′-9″
12′	4′-6″	6′-0″	7′-6″

It is usually more economical to use the larger beams and fewer posts.

Below are the maximum joist spans in the other direction.

Joist spacing	2 × 6 joist	2 × 8 joist	2 × 10 joist
16″ o.c.	6′-0″	9′-0″	13′-0″
24″ o.c.	5′-0″	7′-6″	10′-10″

The sizes of structural members must be determined well in advance of construction so that concrete piers can be poured at the required spacing.

Guard Rails

Any deck more than 24″ above grade must be surrounded with a railing. Posts supporting the railing must be part of the deck structure; they must not be simply toenailed to the deck. The posts may extend upward through the deck, as in Fig. 19-5.4, or else be bolted to joists below deck level. Spacing of supports every 4′ provides a sturdy railing; support on 6′ centers is acceptable.

The railing is usually a 2 × 6 set no less than 30″ nor more than 42″ above the deck, and centered over supports. The space between supports may be filled with a decorative metal grille, a wood latticework, or benches (Fig. 19-5.6).

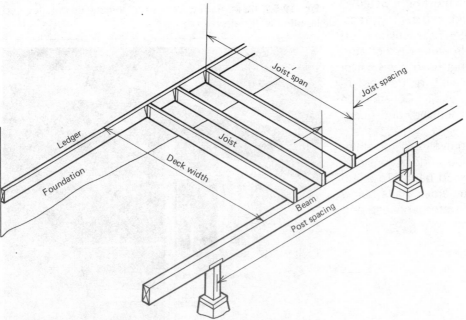

Fig. 19-5.5 Typical frame layout for determining the spans shown in the accompanying tables.

488

Cantilevered Decks

When a deck is 6' or more above grade, it may have no visible support at all. Instead it is **cantilevered**—that is, it projects from the foundation wall and is balanced on it (Fig. 19-5.7). A cantilevered deck is supported by joists that are a continuation of floor joists in the house. The header joist at the foundation is omitted, and later the space between joists is filled with blocking.

The construction of a cantilevered deck is usually detailed in working drawings. The slight difference in level between the finished interior floor and the deck may be overcome in one of three ways (Fig. 19-5.8). Floor joists may extend from the house under the deck at full height (top). At the entrance to the deck a threshold is set in mastic to prevent rain from running into the house. If the difference in level is slight, the joists may be trimmed to fit (center). With a greater difference in level, deck joists 2″ less in depth and twice as long as the amount of cantilever are bolted to interior floor joists (bottom).

As a general rule the amount of cantilever should never exceed one third the total length of the interior floor joist. Even then the deck may vibrate when people walk on it, and diagonal struts may be needed to stiffen the frame.

Materials

All lumber used to build any deck must be a weather-resistant species, such as redwood, cedar, or cypress. It should be preserved with water repellent applied under pressure at the mill; site treatment is not adequate except to preserve ends that have been cut or notched. Construction heart is the grade recommended for caps, posts, and beams. Construction common is the grade for joists, decking, and guard rails.

Only aluminum, hot-dipped galvanized, or stainless steel fasteners should be used in deck construction. Other types may stain the wood badly. To eliminate any chance of staining, countersink nails below the surface, brush the holes with water repellent, and fill them with an oilless wood filler.

Preparation

Careful planning is needed before construction starts on a deck. The first step is to determine the best relationship between deck level and the finished floor level

Fig. 19-5.6 The spaces between a railing, the deck, and supporting posts may be filled in many ways. Here all parts are redwood connected with stainless steel bolts.

in the house. The deck may be at the same level, or as much as 4″ lower without a step. The sizes and spans of beams and joists must be established. The direction or pattern of the decking must also be decided, because it affects the method of support.

To determine the exact length of each supporting post is the most difficult task. Use of an optical level is the best method. An alternate method of measurement takes two people. A helper places a joist upright with

Fig. 19-5.7 Long, narrow decks are sometimes cantilevered from the wall below.

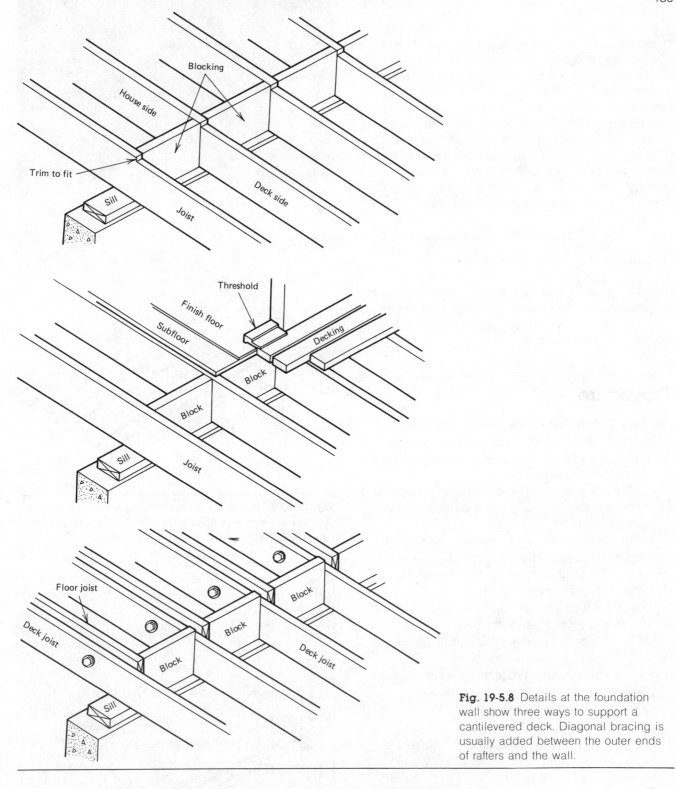

Fig. 19-5.8 Details at the foundation wall show three ways to support a cantilevered deck. Diagonal bracing is usually added between the outer ends of rafters and the wall.

one edge on the ledger at the wall and the other end held directly above a capped pier. The carpenter places the level at the open end, and the helper moves the joist up or down until it is level. Then the carpenter measures the distance between the underside of the joist and the cap. From this dimension he or she subtracts the depth of any supporting beam to arrive at the length of the post.

Application

Decking should always be laid with a space between the pieces to let rain drain off and to let the wood dry as quickly and evenly as possible. The ideal gap is $\frac{1}{8}$ to $\frac{3}{16}''$; when gaps are wider deck furniture gets caught in them, and things fall through. Decking is fastened with

16-penny nails, and many carpenters use a nail as a spacer (Fig. 19-5.9).

Pieces of decking should be nailed in a staggered pattern—one nail into a joist near one side of the piece, and one nail into the next joist near the other side (Fig. 19-5.10). Where two lengths butt, nail at opposite corners. Either predrill the holes, or blunt the ends of nails with a hammer to prevent splitting. Where long boards end at the edge of the deck, use two nails side by side. Again predrill the holes. Place nails about $\frac{3}{4}''$ in from edges and $\frac{3}{8}''$ in from ends.

Fig. 19-5.9 Carpenters use nails as spacers between widths of decking. (Photo Drew Leviton, Atlanta.)

Tools

The tools needed to build a deck are a brush for applying preservative, steel tape, chalk line, hammer, power drill, power saw, wrenches, carpenter's level, framing square, and metal shears.

Procedure

1. Snap a chalk line on the sheathing to mark the level of the top of the ledger or deck edge joist.
2. Cut the member to length, and liberally coat the cut ends with preservative.
3. Tack the ledger or edge joist in place, and drill holes in it and the header joist behind for bolts spaced 32" o.c.
4. Insert the bolts, thread on washers, and tighten.
5. Cut a strip of metal flashing 6" wide, and bend it so that it runs about 1" down the side of the ledger and $3\frac{1}{2}''$ up the sheathing.
6. Attach the flashing near the top edge of the ledger with nails of the same metal.
7. On the edge joist mark joist spacing, and install joist hangers at these points.
8. From 2" lumber cut all caps to cover the tops of piers, and treat them with water repellent.
9. Anchor the caps to the piers.
10. Determine the length of each post.
11. Cut, treat, and toenail posts to caps. Make sure each post is plumb, and that its top is level with the tops of adjacent posts.
12. Cut beams to length, treat them with repellent, and set them in position. Check their level, and adjust heights of posts by trimming or shimming.
13. Attach beams to posts by one of the methods shown in Fig. 19-5.4.
14. Cut, treat, and nail joists in place. Use a framing square to assure that they are at right angles to their supports.

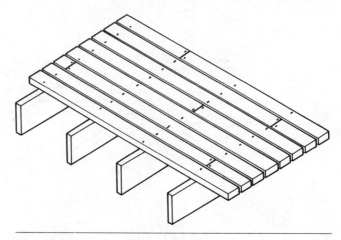

Fig. 19-5.10 The recommended nailing pattern for decking: one nail per joist in a staggered pattern, one nail at ends where lengths butt, two nails at an open end.

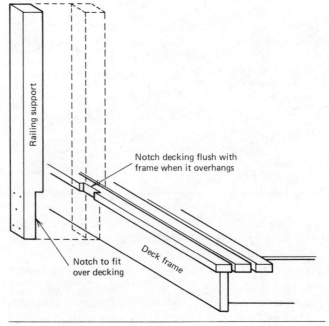

Railing support

Notch decking flush with frame when it overhangs

Notch to fit over decking

Deck frame

Fig. 19-5.11

15 Lay decking across the joists to determine the best spacing, and where to cut so that ends butt over joists.

16 Precut and preserve all lengths before installation.

17 Nail the decking in the pattern shown in Fig. 19-5.10. **Precaution.** If decking runs parallel to the wall, attach the first piece with a $\frac{1}{8}''$ gap at the wall.

18 Cut, treat, and attach railing supports. **Note.** If supports extend through the decking, they are attached before the last pieces of decking, which are notched to fit around the supports. If supports are bolted to beams, they are notched to fit over the completed deck (Fig. 19-5.11).

19 Cut, treat, and attach the railing.

20 Complete decorative treatment between railing supports.

Questions

1 What species and grades of lumber are most commonly used in building decks?

2 At what two points is flashing often required?

3 When must a deck have a railing around it?

4 From what materials should fasteners be made? Why?

5 How many nails are required in a piece of decking running the length of a 24′ deck supported by joists spaced 16″ o.c.?

6 When should holes be predrilled for nails?

7 When rail supports do not also support the deck frame, how should they be attached?

Unit 19-6   **Screen a Porch**

Porches are built with post and beam construction. The posts bear on a floor of concrete, stone, or wood decking laid without gaps. They are usually 4 × 4s or 6 × 6s finished with boards mitered at the corners (Fig. 19-6.1). Header beams laid across the posts support the upper wall structure and roof framing. The openings between floor and beam and from post to post are filled with individual screen units. The screening may be fixed or removable.

Wire screen mesh is manufactured in long rolls in widths up to 48″. Typical meshes have 18 wires per inch in one direction, and 14 or 16 wires per inch in the other. For residential screening the most popular materials are fiberglass and aluminum. Aluminum screening comes with both a mill finish (metallic) or anodized finish (coated electrically) in black or green.

The maximum width of any screen panel is 48″, and 36″ is recommended. Seven feet is a practical maximum height. To prevent damage, screening should have horizontal stiffeners. The stiffeners should be placed at a height where damage to the screen is most

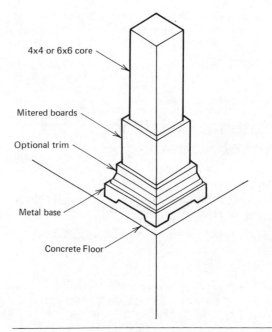

4x4 or 6x6 core

Mitered boards

Optional trim

Metal base

Concrete Floor

Fig. 19-6.1 Section through a typical porch post. Note that every post should be raised above the level of the floor.

likely to occur, but where they do not interfere with a view outdoors from seated eye level.

When screening is permanent, the opening between posts is fitted with a framework of 2"-thick lumber treated with water repellent. Screening is applied to the framework from the outside, and the edges covered with battens on all sides.

When panels are removable, the frames are made of select grade $\frac{5}{4}$ or $\frac{6}{4}$ millwork with high resistance to bending. Screening must be stretched tight, and has a tendency to cause the frame to bow inward. Framing members are connected at joints with glue and corrugated fasteners. Screening is applied from the outside, and edges are battened. The frames are inserted against stops on three sides, and are held tight to posts and beams with screws.

To prevent posts from rotting, they should be installed on pedestals, as shown in Unit 6-5, or on corrosion-resistant metal bases. Permanent base framing members should also be raised above floor level. Removable members should be designed to fit about $\frac{1}{4}$" above floor level. Some insects may crawl through this gap, but they are likely to crawl in anyway. The advantage of the gap, aside from protecting the frame from rot, is that the floor may be hosed down and the water will run off.

Tools

To install screens around a porch, the carpenter is likely to need a chalk line, power saw, paint brushes, trowel, nail gun, staple gun, hammer, screwdriver, level, steel tape, metal shears, sliding T-bevel, a brace and bits or power drill, and a jointer or circular saw for rabbeting joints.

Procedure for Installing Permanent Screening

1 Cut bottom and top framing members to length, and treat the cut ends with water repellent. Add a prime coat of paint or stain.

2 On the porch floor snap a chalk line to mark the location of the frame.

3 Between the chalk lines lay a bed of rich mortar about $\frac{3}{8}$" thick if the floor is masonry. **Note.** A manufactured gasket may be substituted for the mortar bed, or be used if the floor is wood.

4 Lay the bottom framing member in the mortar or on the gasket and attach it to the floor. **Note.** Use

power nails if the floor is concrete or masonry, and 16-penny nails if the floor is wood.

5 Nail the top framing member to the beam.

6 Cut vertical framing members to length, treat the cut ends, and add a prime coat.

7 On both horizontal framing members mark the spacing of vertical members.

8 Toenail the members in place on their marks.

9 Cut, treat, and prime horizontal dividers.

10 Mark the locations of dividers on vertical framing members.

11 Install the dividers.

12 Cut screening so that it overlaps each opening $\frac{1}{2}$" on each side. **Note.** Screening may be cut at dividers or continue over them. Because damage is most likely to occur in the bottom 3′ of a screen panel, replacement is easier and less expensive if screening is attached in two pieces instead of one.

13 Fit the screening over each opening from the outside, and staple across the top. Use coated staples with a $\frac{1}{2}$" leg and a $\frac{1}{4}$" crown, and space them about every 2".

14 Pull the screening taut across the frame, making sure it is centered in the opening.

15 Staple at the bottom, beginning in the center and working toward the corners.

16 Staple along one side.

17 Pull the screening taut and staple down the other side, beginning in the middle and working toward the corners. **Precaution.** Be sure to pull horizontally so that the screen wires are parallel to the floor, and not angling up or down.

18 Cut outer battens to length, and miter the corners.

19 Treat, prime, and install battens with box nails.

20 Cut, treat, prime, and install battens at dividers.

Procedure for Installing Removable Screening

1 Check the plumb of posts beside each opening to be filled.

2 Check the level at the floor and beam. **Precaution.** Many floors slope slightly for drainage, and beams above may follow the same slope. Joints of frames must be cut to fit the angle at each corner.

3 Cut vertical frame members to length. Their length should be the height of the opening less $\frac{3}{8}$".

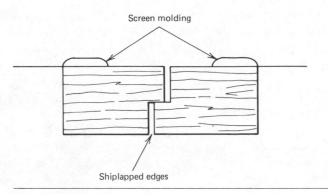

Screen molding

Shiplapped edges

Fig. 19-6.2

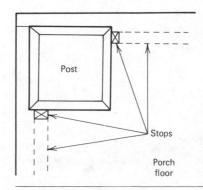

Post

Stops

Porch
floor

Fig. 19-6.3

4 If vertical framing members butt against each other in a wide opening, shiplap the edges (Fig. 19-6.2).

5 Cut and, if necessary, bevel bottom frame members.

6 Lay framing members on the floor in the opening to check the fit. They should gap about $\frac{1}{8}''$ at posts and about $\frac{1}{16}''$ where panels meet.

7 When the fit is good, cut and, if necessary, bevel top framing members.

8 Cut horizontal dividers to length with square ends.

9 Treat all members with water repellent and add a prime coat of paint or stain.

10 Assemble the frames, attaching at corners with glue and corrugated fasteners.

11 Cut, treat, and prime stops.

12 Install stops at the top and sides of all openings, $\frac{1}{2}''$ in from the outside edges of posts and beams (Fig. 19-6.3).

13 Insert screen panels in openings, and recheck the fit. **Note.** When the opening is wide, wedge a straight 2×4 between posts at floor level to serve as a temporary bottom stop.

14 Drill holes for screws through the frames into posts near the top, bottom, and a midpoint between.

15 Drill holes through frames into beams about 4″ from the corner of each frame.

16 Insert screws and tighten them. **Precaution.** Do not tighten the screws all the way; maintain the $\frac{1}{8}''$ gap at posts and beams.

17 Where two panels meet, attach hooks and eyes at the midpoints between the divider and the top and bottom members.

18 Move the temporary 2×4 stop to the inside of the porch, and wedge it between posts against the bottom of screen frames.

19 Attach screening as described in steps 12 to 17.

20 Cut perimeter screen molding to length, and miter the corners.

21 Treat, prime, and install the molding with brads.

22 Cut, prime, and install moldings at dividers. **Note.** Use two lengths of narrow molding if the screening is in two pieces, or one wide length if the screening is continuous.

23 Give screws another turn or two to stretch the screening taut and to secure the panels.

Questions

1 What is the maximum recommended size for a screen panel?

2 What other framing member in a house is installed like the bottom framing member in a permanently screened porch?

3 At what point is screening most subject to damage? How do you minimize the chance of damage?

4 How are the members of removable screen panels connected?

5 How are adjacent panels connected in wide openings?

5 Interior Finish Carpentry

Finishing the inside of a house is exacting work. As long as horizontals are level and verticals are plumb on the exterior of a house, a minor flaw doesn't show up readily. In the close-up view within the house, however, an error or poor workmanship is all too visible. At the same time that they are exercising the greatest care in finishing, carpenters must work in ever more confining quarters as finish wall and ceiling materials make rooms seem smaller and smaller. There is less space in which to move around; even the shortcuts between studs that allowed them to move from room to room are now closed off.

In addition to this confinement carpenters must often finish walls, ceilings, and floors among crews of plumbers, electricians, and sheet metal workers who are completing their installations. The work of all finishing trades must be carefully coordinated to avoid confusion, delays, and rework.

With rare exceptions, interior surface materials are applied from the top down. That is, ceilings are finished first, then walls, then floors. Interior trim may be installed in two ways. Some builders prefer to have their carpenters finish one room at a time so that painters can follow them immediately. Other builders prefer to install trim by type: All ceiling coves first, then window and door casings, then base moldings at the floor level. Cabinets are installed as soon as surface materials are in place.

Stairways are built at two points in the construction process. In a two-story or split-level house, and in any house with a basement, a rough stairway is put in as soon as possible to make it easier for workers to move themselves and materials from one level to the next with minimum delay and maximum safety. Finish stairways, both inside and outside the house, are normally not completed until the danger of damage from materials, tools, and workers is past.

20

Stairways

Several terms describe the means of walking from one level to another. The basic term is a **stair**, which is a single step. More than one step is called stairs, a flight of stairs, a staircase, or a **stairway.** A stairway may run straight, curve gently, turn continuously, or have right angles in it. It may be manufactured as a stock item, built to specific dimensions in a millwork shop. or constructed on the job. It may be made of metal or wood.

A **stairwell** is the framed opening in the floor through which the stairway rises. A stairway that has walls on both sides is called a **closed stairway.** An **open stairway** has either no wall on one side or no wall on either side.

Stairway Terms and Parts

Some terms that apply to stairways are also used in roof construction. The **run** of a stairway (Fig. 20-0.1) is its horizontal length from its start at one level to its end at another. The **rise** is its vertical height from **finished** floor to **finished** floor. A **flight** is the equivalent of a slope; it is not a measurement, but refers to continuous steps from one horizontal platform to another.

The term **tread** has two meanings. It is the horizontal dimension of a step from front to back (Fig. 20-0.1), and it is also the horizontal face of a step—the part you step on (Fig. 20-0.2). The term **riser** also has two meanings. It is the vertical dimension from tread to tread, and it is also the vertical face of a step. A **nosing** or **nose** is the part of a tread that overlaps a riser. A **nosing line** is an imaginary diagonal line along the nosings.

Treads and risers are supported on long diagonal supports called **stringers** or **carriages,** depending on the part of the country. Narrow stairways have two stringers; wide stairways have a third stringer at the center. A **landing** is a platform that separates flights of steps, and often supports one end of stringers. When a stairway turns without a landing, stair treads are not rectangular but wedge-shaped, and they are called **winders.**

Like most decks, most stairways require a railing, called a **handrail.** Handrails may be attached to walls, or supported on vertical members called **balusters** (Fig. 20-0.3). The combination of handrail and balusters is called a **balustrade.** The main support for the end of a balustrade is called a **newel** or **newel post.** A newel may be found at the upper end of a stairway, at the lower end, or at a landing. Its purpose is not only to support, but to stiffen the balustrade.

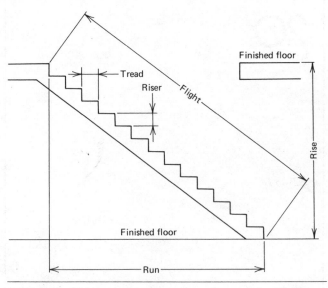

Fig. 20-0.1 Terms that describe the design of a stairway.

Stairway Design

About one third of all accidents in a home occur on stairs. To make stairways as safe and comfortable to use as possible, architects have developed rules for stairway design that are incorporated into building codes everywhere. Normally carpenters aren't concerned with stairway design. Sometimes, however, an alteration in the plans may cause a change in the stairway, and the carpenter may be called upon to make the adjustment. For that reason every carpenter should learn the six basic rules for stairway design. Here are the first three:

1 The product of the height of a riser (R) times the width of a tread (T) should equal about 75.

$$R \times T = 75$$

2 The height of a riser plus the width of a tread should equal between 17 and 18.

$$R + T = 17 \text{ to } 18$$

3 Twice the height of a riser plus the width of a tread should equal between 24 and 25.

$$2R + T = 24 \text{ to } 25$$

Take a moment to absorb these three rules. Building codes based on the Department of Housing and Urban Development (HUD) Minimum Property Standards allow a maximum riser height of $8\frac{1}{4}''$ and a minimum tread width of 9". Plug these dimensions into the three formulas, and the answers are $74\frac{1}{4}$, $17\frac{1}{4}$, and $25\frac{1}{2}$. The dimensions generally considered ideal for indoor stairways are $7\frac{1}{2}''$ risers and 10" treads, which give formula results of 75, $17\frac{1}{2}$, and 25. Outdoor stairways should be a little more gradual (Unit 20-7).

The rise and run of a stairway have an important bearing on the dimensions of risers and treads. A com-

mon rise for a stairway between two floors of a two-story house is $8'-11\frac{1}{4}''$ (Fig. 20-0.4) or $107\frac{1}{4}''$. If you divide this rise by the ideal $7\frac{1}{2}$, you need 14.3 risers. Since the number of risers must be a whole number, try both 14 and 15. With 15 risers, riser height is slightly more than $7\frac{1}{8}''$—a little low. With 14 risers, each is a little more than $7\frac{5}{8}''$, which is good.

With a $7\frac{5}{8}''$ riser, the ideal tread is $9\frac{7}{8}''$ wide. Since a stairway always has one fewer treads than risers (the

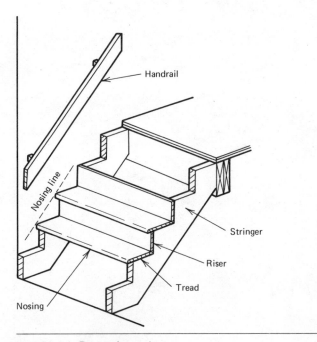

Fig. 20-0.2 Parts of a stairway.

top tread is the flooring at the upper level), to find the run you multiply tread width times the number of treads. In this example, the run would be $9\frac{7}{8}'' \times 13$, or $128\frac{3}{8}''$.

Before making a final decision on these dimensions, look at the other three basic rules for stairway design.

4 A main stairway should be no less than 36" wide from wall to wall. A width of 42" is better; it is difficult to move furniture on a narrow stairway. Basement stairways may be as narrow as 30" wide, but 36" is preferable.

5 Handrails should be 30 to 34" above the nosing line. In houses a handrail is required on only one side of a closed stairway, usually the right-hand side as you descend. With an open stairway without walls, a handrail is required on both sides.

6 Minimum headroom for a standard stairway is 6'-8", and 6'-4", for a basement stairway. Headroom is measured from the nosing line to the under side of the closest structure overhead. This point may be the ceiling of the first floor at the stairwell, an overhead beam, or the ceiling of the stairway itself. When walking up stairs, a person needs less headroom than when coming down stairs because he leans forward going up and is more erect coming down.

The stairways shown in drawings should not be revised unless the floor plan is changed. But their dimen-

Fig. 20-0.3 Parts of a stairway railing. (Courtesy Fine Hardwoods/American Walnut Association.)

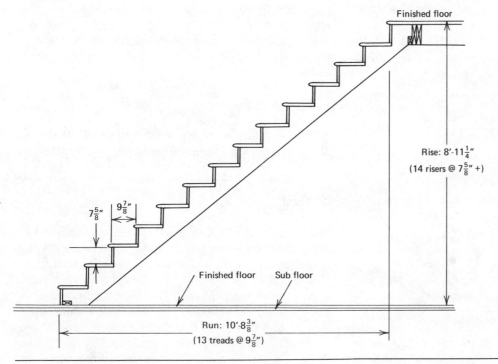

Fig. 20-0.4 Typical stairway dimensions.

sions should always be carefully checked. Knowledge of the rules of design gives the carpenter the needed background to lay out a stairway.

Questions

1 From what two points is the depth of a tread measured?

2 From what two points is the height of a riser measured?

3 List the three formulas that determine a good riser-tread ratio.

4 Between what two points is headroom measured?

5 What is the normal height of a handrail?

6 If a stairway has 14 steps with $7\frac{1}{2}''$ risers and $9\frac{1}{2}''$ treads, what are its run and rise?

Unit 20-1   **Lay Out a Stairway**

In the working drawings for most houses is a section through the main stairway (Fig. 20-1.1). It will give the number of risers and their height, and the number of treads and their depth. There will be details of any special condition, such as headroom, landings, and winders.

Even if complete stairway information is given, the carpenter should always check out the dimensions by using the formulas listed in this chapter. To check vertical dimensions, use whichever of the formulas below gives you the answer you are looking for:

$$\text{total rise} = \text{height of a riser} \times \text{number of risers}$$

$$\text{number of risers} = \frac{\text{total rise}}{\text{height of a riser}}$$

$$\text{height of a riser} = \frac{\text{total rise}}{\text{number of risers}}$$

To check horizontal dimensions use one of the following similar formulas:

$$\text{total run} = \text{width of a tread} \times \text{number of treads}$$

$$\text{number of treads} = \frac{\text{total run}}{\text{width of a tread}}$$

$$\text{width of a tread} = \frac{\text{total run}}{\text{number of treads}}$$

If dimensions don't check out, or if the future owner of the house wants a change in plan that also changes the stairway, the carpenter must lay out the stairway on paper to make sure it meets all code requirements. This should be done before any construction begins. Needed for this layout are a sheet of graph paper, a steel-edged ruler, and a sharp pencil.

If construction is already under way, redesign the stairway by working from existing job conditions. Follow steps 6, 10, 11, and 17 below to determine riser and tread dimensions. Then follow steps 15 and 16 to assure adequate clearance at the end of the stairs.

To determine the length of the stairwell opening, use this formula:

$$\text{well opening} = \frac{\text{tread width}}{\text{riser height}} \times R$$

where R equals required headroom plus total height of upper floor construction including finish materials.

Scale

The scale of the drawing should be as large as the sheet of graph paper will permit. Graph paper divided into $\frac{1}{8}''$ square is ideal, and each square can represent $1''$ of actual dimension. At this scale the sheet of paper should be at least 15" by 20". If the squares are $\frac{1}{4}''$, let each square represent 2". In both cases the scale is $\frac{1}{8}'' = 1''$, or $1\frac{1}{2}'' = 1'\text{-}0''$. If the largest paper available is a standard $8\frac{1}{2}'' \times 11''$, use a scale of $\frac{3}{4}'' = 1'\text{-}0''$. Then each $\frac{1}{8}''$ square equals 2" in actual dimensions, and each $\frac{1}{4}''$ square equals 4".

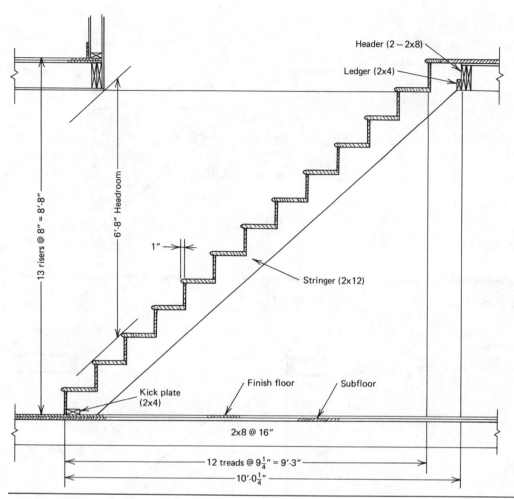

13 risers @ 8″ = 8′-8″

6′-8″ Headroom

1″

Header (2 – 2x8)

Ledger (2x4)

Stringer (2x12)

Finish floor

Subfloor

Kick plate
(2x4)

2x8 @ 16″

12 treads @ 9¼″ = 9′-3″

10′-0¼″

Fig. 20-1.1 Section through
a typical stairway.

Procedure

1 Decide on a scale for the drawing appropriate to
the size of the sheet of graph paper and the size
of the squares.

2 From drawings determine the length of the
stairwell opening.

3 Near the top of the paper draw in at scale the
construction around the stairwell (Fig. 20-1.1).
Note. Be sure to include the thicknesses of
subflooring and future finish ceiling and flooring
materials.

4 From details in drawings, determine the location
of the finished edge of the upper floor. Figure
20-1.2 shows the two most common conditions.

5 Mark this edge on the scale drawing.

6 From drawings determine the rise of the stairway.

7 Measure off this height on the drawing at scale,
and draw in the finished floor line at the lower
level. At this point your drawing should look like
Fig. 20-1.3.

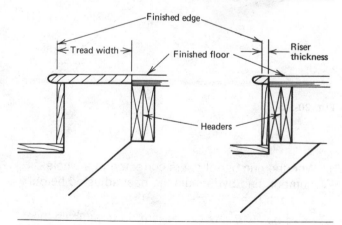

Finished edge

Tread width

Finished floor

Riser
thickness

Headers

Fig. 20-1.2

8 From the ceiling line (A in Fig. 20-1.4) measure
off 6′-8″ at scale and make a mark (B).

9 Draw a diagonal line from the top edge of the
stairwell (C) through point B to the lower floor.

10 Assuming for the moment that riser height is 7½″,
determine by formula the number of risers in the
stairway.

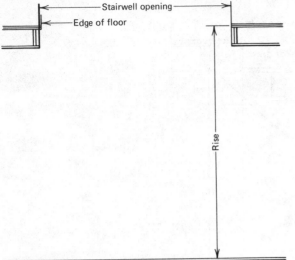

Fig. 20-1.3

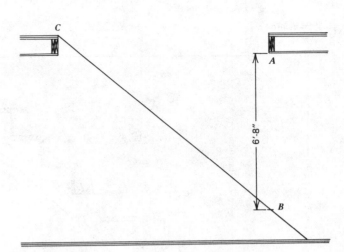

Fig. 20-1.4

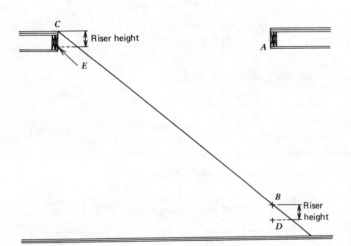

Fig. 20-1.5

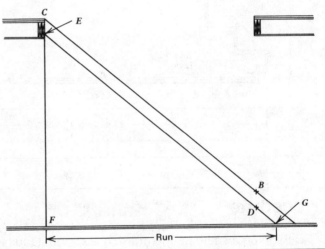

Fig. 20-1.6

11 With the number of risers corrected to a whole number, find by formula the best adjusted height for each riser.

12 From points *B* and *C* measure down at scale the height of a riser in step 11, and mark points *D* and *E* in Fig. 20-1.5.

13 Draw a line from *D* through *E* parallel to the nosing line *B-C* (Fig. 20-1.6).

14 Extend the line of the top riser (*C-E*) to the lower floor line.

15 Measure the distance from *F* to *G*. This is the run of the stairway.

16 Check the plan of the lower floor to make sure there is enough space for a stairway of this

length. **Note.** The distance from point *G* to the nearest wall should not be less than 36″.

17 By formula determine the width of a tread.

18 Plug the riser height in step 11 and the tread width in step 17 into the three formulas for stairway design.

19 a. If the results are within limits, go to step 20.

19 b. If the results are not within limits, revise the number of risers in step 10 up or down one riser, and repeat steps 11 to 18 until the results are within limits. **Note.** Line *D-E* will no longer be parallel to line *B-C*, but must be at a steeper angle.

19 c. Draw a new line from point *C* parallel to the new line *D-E*.

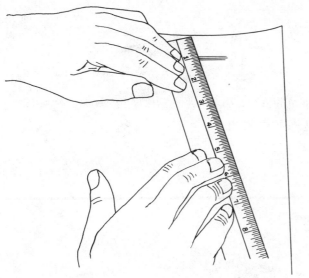

Fig. 20-1.7

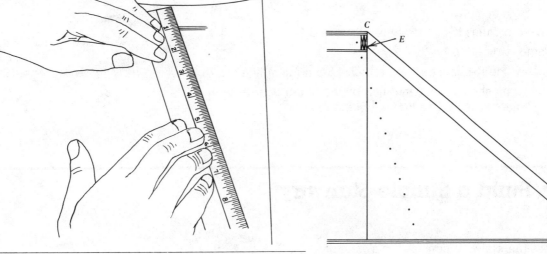

Fig. 20-1.8

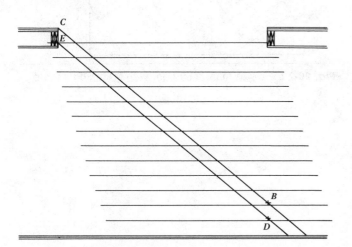

Fig. 20-1.9

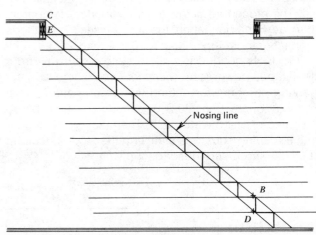

Fig. 20-1.10

20 Multiply the scale of the drawing times the number of risers. **Examples.** If the scale is $\frac{3}{4}''$ to the foot and there are 13 risers, the answer is $9\frac{3}{4}''$. If there are 14 risers, the answer is $10\frac{1}{2}''$.

21 Lay a ruler diagonally across the graph paper at one side of the scale drawing with the 1″ mark on the ruler at the upper floor level and, with the ruler touching the lower level at either the $10\frac{3}{4}''$ or $11\frac{1}{2}''$ mark, using the examples in step 20 (Fig. 20-1.7).

22 Place a dot at the scale distance (every $\frac{3}{4}''$ in the example) along the diagonal (Fig. 20-1.8).

23 Repeat steps 21 and 22 at the other side of the scale drawing.

24 Connect the pairs of dots with lightly drawn horizontal lines (Fig. 20-1.9). The vertical distance between each pair of lines should equal the height of a riser.

25 Wherever a horizontal line touches the nosing line, draw a vertical line downward to the tread below (Fig. 20-1.10). Each of these riser lines should touch the tread line below at the diagonal line D-E. The horizontal distance between each pair of lines should equal the width of a tread.

26 Strengthen all lines between the diagonals. The stairway layout is now complete.

Questions

1 What are the best formulas to use in steps 10, 11, and 17?
2 What is the name for line *B-C*?
3 Why must the new line *D-E* in step 19b lie below the original line *D-E*?
4 Why do you think you should place the ruler on the upper floor level in step 21 at the 1″ mark instead of at the end of the ruler?

Unit 20-2 **Build a Simple Stairway**

The simplest possible stairway consists of a pair of 2 × 8 stringers and a series of 2 × 10 treads with no risers. This type of stairway is often used temporarily during construction, but also makes an adequate permanent basement stairway with the addition of a handrail. The procedure below is for installing a temporary stairway. For a permanent installation, see the methods of attachment in Figs. 20-3.3, 20-3-4, and 20-3.5.

The tools needed to build a simple stairway are a steel tape, power saw, hammer, and some tool for marking stringers for cutting. Most carpenters use a framing square (Fig. 20-2.1), although a pitch board may also be used (Unit 20-3). The marking procedure with a framing square is similar to the step-off method of measuring rafters described in Unit 9-3.

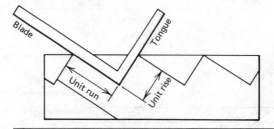

Fig. 20-2.1 How to mark a stringer with a framing square.

Procedure

1 Determine the rise of the stairway by actual measurement.

2 Determine the run of the stairway. **Note.** If the run is not restricted by walls or other obstructions, let the run equal $1\frac{1}{3}$ times the rise.

3 Using the Pythagorean theorem, determine the length of the stringers (Fig. 20-2.2).

4 Establish the safest riser height and tread width (Unit 20-1).

5 Using a pitch board or a framing square, mark the floor line across the end of a stringer (Fig. 20-2.3).

6 Working upward, draw lines on the stringer to mark the locations of the tops of treads, as shown in Figs. 20-2.1 and 20-2.3.

7 Mark on the stringer the point where it butts against the stair header.

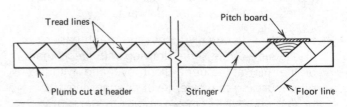

Fig. 20-2.3

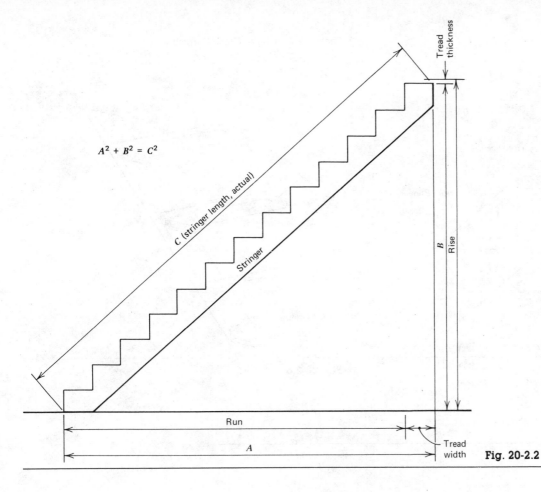

$A^2 + B^2 = C^2$

C (stringer length, actual)

Stringer

Tread thickness

B Rise

Run

A

Tread width

Fig. 20-2.2

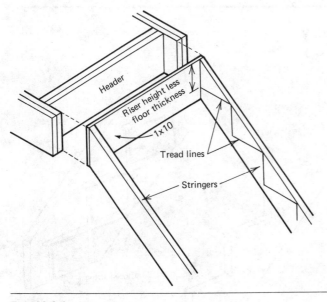

Header

Riser height less floor thickness

1 x 10

Tread lines

Stringers

Fig. 20-2.4

8 From this point draw a plumb cut line.

9 Cut the stringers along the floor line and plumb line.

10 Lay the stringer in position to check the fit at the floors.

11 Measure headroom to make sure it is adequate.

12 When the fit is good, mark and cut the second stringer.

13 Attach a 1 x 10 across the upper ends of the stringers (Fig. 20-2.4).

14 Nail the 1 x 10 securely into the floor header.

15 Cut treads to length.

16 Fit the treads just below the tread lines on the stringers, and nail through the stringers into the treads with at least three 8-penny nails on each side. **Note.** The edges of treads should be flush with the nosing line (Fig. 20-2.5).

17 For maximum tread strength, nail 1 x 2 cleats to the stringers beneath each tread (Fig. 20-2.6).

18 If possible nail the stringers to studs to prevent the lower end of the stairway from moving.

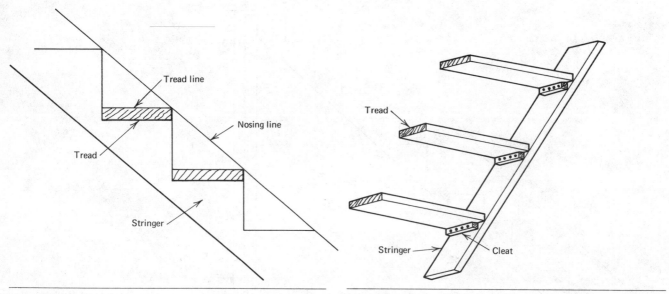

Fig. 20-2.5   **Fig. 20-2.6**

Questions

1 How is the minimum length of stringer stock determined?
2 What is the relationship of the cuts at opposite ends of stringers?

Unit 20-3   **Build a Straight Stairway**

Two types of stringers may be used to build finished stairways. An **open stringer**, also called a **cut stringer**, is notched to the outline of treads and risers (Fig. 20-3.1, left). A **closed stringer**, sometimes called a **housed stringer**, is routed out to receive treads, risers, and the wedges that hold them in place (Fig. 20-3.1, right). When a stairway has more than two stringers, the center stringer is always open. Open stringers are usually cut from 2 × 10 or 2 × 12 stock lumber, since they are hidden from view. Closed stringers are visible, and are shaped out of $\frac{5}{4}$ or $\frac{6}{4}$ millwork. They must be wider than open stringers because at least $1\frac{1}{2}''$ of the stringer must be visible above the nosing line.

Support

Finished stairways may be fixed in position in several ways. The bottoms of stringers are notched to fit over a

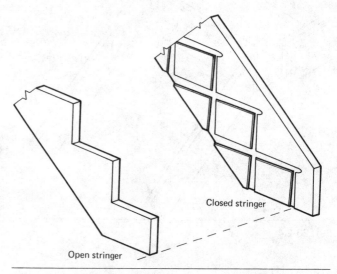

Fig. 20-3.1 An open stringer is cut to the shape of the stairway. A closed stringer is routed to receive treads and risers.

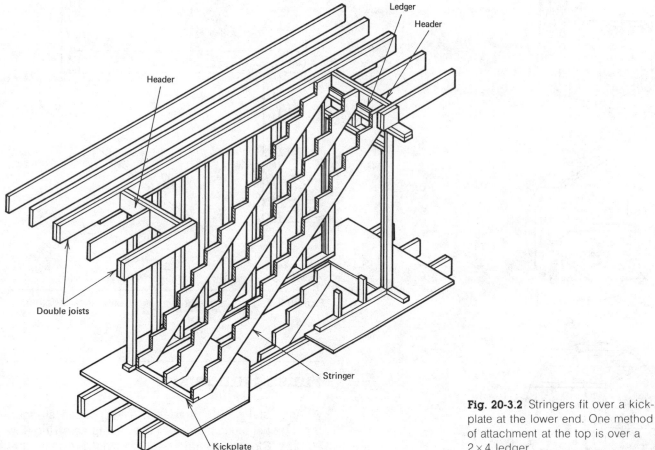

Fig. 20-3.2 Stringers fit over a kickplate at the lower end. One method of attachment at the top is over a 2×4 ledger.

kickplate (Fig. 20-3.2). The kickplate is laid flat and nailed into the subfloor. The tops of stringers are usually notched again, here to fit over a 2 × 4 ledger nailed upright to the headers that frame the stairwell. However, they may be attached to headers with metal joist hangers (Fig. 20-3.3), or be installed like a temporary stairway (Fig. 20-2.4).

When headroom is tight, vertical space can be saved in three ways. Stringers may butt against a wider piece of plywood, and the first tread is then located a full riser height below the edge of the stairwell (Fig. 20-3.4, top). Or they may be notched to fit over a ledger that is itself notched into studs (center). These two methods are practical only if there is a stud wall below the header. When there is no wall, the stringers butt against the header and are connected with straps to the backs of the headers (bottom).

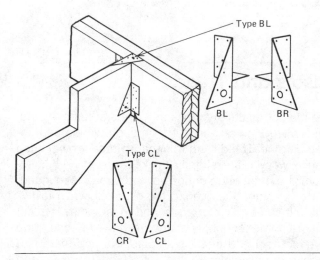

Fig. 20-3.3 The tops of stringers may also be supported by joist hangers.

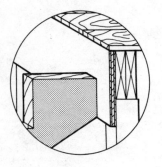

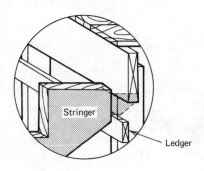

Stringer

Ledger

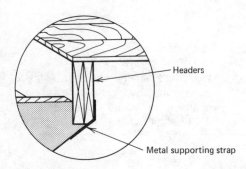

Headers

Metal supporting strap

Fig. 20-3.4 Three alternate methods of supporting the tops of stringers when headroom is limited.

Risers and Treads

Treads and risers are stock mill items at lumberyards. Risers are $\frac{7}{8}''$ thick and cut from such woods as Ponderosa pine. Treads, cut from oak, birch, or other such dense woods, are slightly thicker—usually $1\frac{1}{8}''$ minimum. The front edge or nose of each tread is rounded, and is designed to overlap the riser below it 1 to $1\frac{1}{2}''$. The other edges of treads and risers may be cut square. In better quality stairways, however one edge of each riser and tread is shiplapped, and there is a groove in one surface—the outer faces of risers and the undersides of treads. The shiplapped ends fit into the grooves (Fig. 20-3.5).

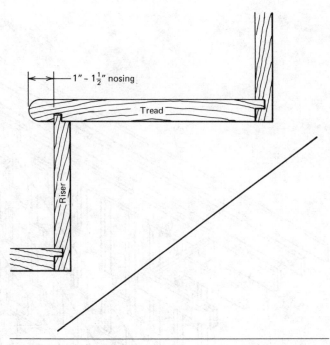

$1''$ – $1\frac{1}{2}''$ nosing

Tread

Riser

Fig. 20-3.5 How grooved risers and treads lock together.

Finish Stringers

When a stairway runs against a wall, open stringers butt against a finish stringer, often called a **string** (Fig. 20-3.6). Closed stringers serve a dual purpose; they not only act as trim at the wall, but also support treads and risers. When there is no wall, strings are cut like open stringers. Edges where the string and risers meet are mitered. Treads overlap the string the same distance they overlap risers.

Some builders prefer to finish walls in stairwells before having carpenters install strings. Others attach strings over continuous diagonal blocking (Fig. 20-3.7), and butt wall materials against the top of the strings. In either case the joint is concealed behind a cover molding.

Tools

Some carpenters always use a stair pole as a means of measuring vertical dimensions while building a stairway. A **stair pole** (Fig. 20-3.8) is simply a straight length of 2×4 long enough to extend from the first floor past second floor level. You stand it absolutely plumb in a corner of the stairwell, and mark on it the rough floor level and the height of ceiling framing. Then at

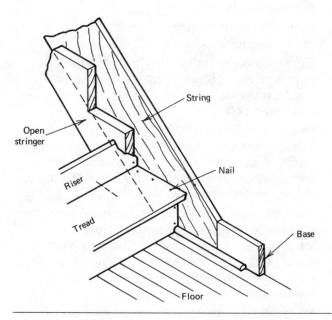

Fig. 20-3.6 At a wall an open stringer butts against a string.

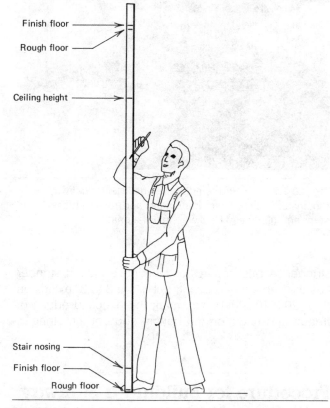

Fig. 20-3.8 A stair pole is a simple but useful tool for recording heights in a stairwell.

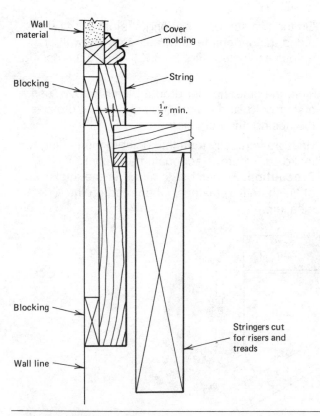

Fig. 20-3.7 When a finished stairway with strings is installed before walls are surfaced, the strings are nailed over diagonal blocking, and a cover molding finishes the joint.

both ends mark the future finished floor level, and add the finished ceiling level. You mark the nosing level during the procedure that follows.

No special tools are needed to install a stairway with open stringers. For marking and routing closed stringers a special template is made that makes the work quicker and easier (Fig. 20-3.9). The template can be adjusted to match the desired height of risers, the width of treads, the projection of nosings, and the size of wedges. Alternate marking tools are a framing square or a pitch board.

A pitch board for marking an open stringer consists of a triangle nailed down the center of a length of scrap 2 × 4 or 1 × 6 (Fig. 20-3.10, left). The length of one side equals riser height and the length of the other side equals tread width. You work along the top edge of the

Fig. 20-3.9 A routing template, adjustable to stairway dimensions, saves time in marking and routing closed stringers. (Courtesy Rockwell International.)

stringer. A pitch board for marking a closed stringer has the same triangle in it, but has a 2 to 3″ extension (Fig. 20-3.10, right). You mark the exposed edges of treads and risers on the bottom edge of the stringer.

Procedure for Building a Stairway with Open Stringers

1 Place a carpenter's level across the headers at each end of the stairwell to verify level from one side to the other.

2 With a long, straight 2×4 check the level at the sides of the stairwell.

3 If the level is off more than $\frac{1}{8}$″, shim under subflooring to establish true level.

4 From a corner of the stairway by the top step, drop a plumb bob, and mark the point where it hits the lower subfloor (Fig. 20-3.11). As an alternate, use a stair pole.

5 Measure the rise at this point, and check it against drawings. **Note.** If finish flooring is not yet installed, add the thickness of the upper flooring and subtract the thickness of the lower flooring to determine the actual rise.

6 From the mark in step 4 measure off the run of the stairway, and make another mark. **Note.** If the length of the run isn't given in drawings, there should be a note that reads, for instance, 13 treads @ $9\frac{3}{4}$″. Multiply the two figures, and use the product as the length of the run.

7 Measure the width of one tread beyond the run, and make another mark.

8 Set a scrap of upper level flooring material at the header to represent the edge of the step, and a scrap of lower level flooring material at the mark in step 7.

9 On the two scraps lay a long straight 2×4, and along the underside draw a line on the wall or across each stud (Fig. 20-3.12). This is the nosing line.

10 From the other header drop a plumb bob, or use a stair pole, and mark the point where it crosses the nosing line from step 9.

11 Measure vertically from this point to the ceiling to make sure there is adequate headroom. **Precaution.** Remember to allow for the thickness of finish ceiling material when reading the measurement.

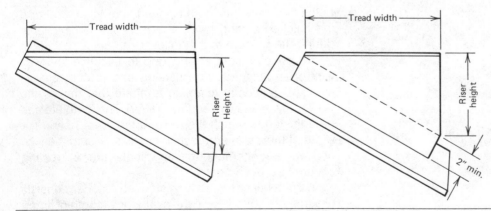

Fig. 20-3.10 A pitch board for marking open stringers is a simple triangle (left). Because tread and riser lines meet below the top edge of a closed stringer, the pitch board for marking them must have an extender added to the triangle (right).

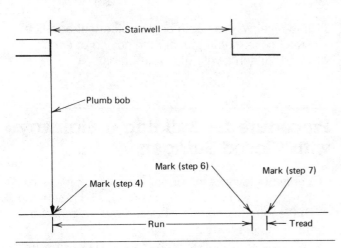

Fig. 20-3.11

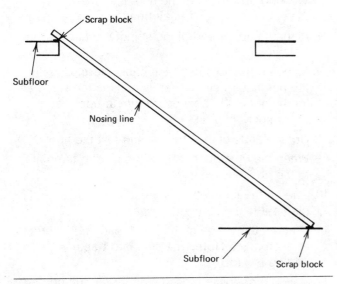

Fig. 20-3.12

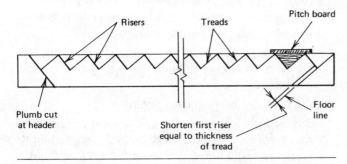

Fig. 20-3.13

12 Measure along the diagonal 2 × 4 to determine the minimum length of stringer required.

13 Using a framing square or pitch board, mark the floor line across the end of the stringer (Fig. 20-3.13).

14 At right angles to the floor line measure off the thickness of a tread. **Note.** If the lower floor is not yet finished, subtract the thickness of finish flooring from the tread thickness.

15 At this point draw a line parallel to the floor line.

16 Mark the notch for the kickplate.

17 Working upward from the floor line, mark the locations of all treads and risers. **Precaution.** Be as accurate as possible, and recheck your marks regularly.

18 Mark the upper end of the stringer as required for attachment (Figs. 20-2.4 and 20-3.2 to 20-3.4). **Precaution.** As in step 14, the point of marking depends on whether the upper level floor is finished.

19 Cut the stringer along the marks.

20 Test the fit.

21 When the fit is good, use the first stringer as a template for cutting the others.

22 If the stairway butts against a wall on either or both sides, cut the finish strings to length, giving them plumb cuts at both ends.

23 Nail strings into studs.

24 Nail stair stringers temporarily in place.

25 Check stringers for plumb, and set a carpenter's level on tread cuts to check the level across the stairway.

26 When the level and plumb are right, attach the stringers to the header.

27 Toenail the lower ends of stringers into the kickplate. **Precaution.** In nailing at both ends be sure that stringers are not twisted. Recheck level and plumb constantly.

28 Cut risers and treads to length.

29 Beginning at the bottom of the stairway, fit risers tightly against the vertical cuts in stringers, and nail them to stringers with 6-penny nails driven about 2″ above the bottom of each riser. **Note.** If stairs are to be carpeted, use 5-penny nails.

30 Spread glue liberally in the grooves of each set of risers and treads.

31 Fit the tread into the groove in the back riser, and press it firmly onto the shiplapped edge of the front riser.

32 Drive one 12-penny finishing nail through the

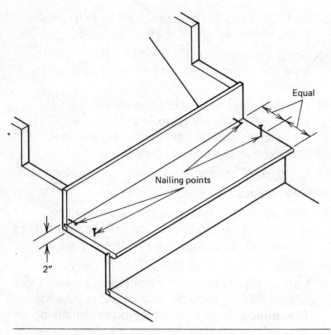

Fig. 20-3.14

tread into each stringer at about the midpoint of the tread (Fig. 20-3.14). **Note.** Nails simply hold treads in place until the glue sets. A good gluing job is the secret to a good, squeakless stairway.

33 Countersink all nails, and fill nail holes with compound.

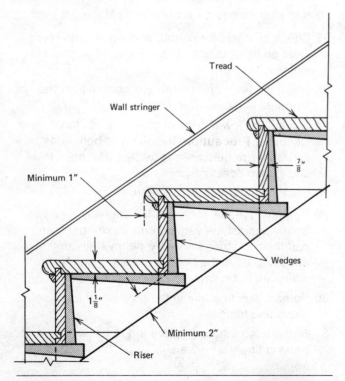

Fig. 20-3.15

34 If the stairway is not to be carpeted, cut and fit a small cove molding under the nosing of each tread. Attach it with glue and brads.

Procedure for Building a Stairway with Closed Stringers

1 Verify the level of the upper floor, check the rise, run, and headroom of the stairway, and determine the length of stringers as described in steps 1 to 12 of the previous procedure.

2 Mark the floor line as described in steps 13 to 15.

3 Mark the notch for the kickplate.

4 Mark the nosing line on the exposed face of the stringer.

5 Working upward from the floor line, mark the locations of the tops of all treads.

6 Mark the upper end of the stringer to fit over the header.

7 Mark the plumb cuts at the ends. **Note.** Usually the height of a stringer at its ends is the same as the height of the baseboard trim at upper and lower floors (Fig. 20-3.6).

8 Cut the ends of the stringer and test the fit.

9 When the fit is good, adjust the routing template to the desired dimensions.

10 Line up the template along each tread line, and clamp it to the stringer.

11 Rout out the shape of the tread, riser, and wedges at each mark. **Note.** The standard depth of the routing is $\frac{1}{2}$ to $\frac{5}{8}''$.

12 Mark the opposite closed stringer for cutting and routing. **Precaution.** Remember that the template must be used in reverse on the second stringer.

13 Nail the stringers temporarily in place, and check the cuts for plumb and level.

14 When level and plumb are right, attach the stringers to studs with spikes driven below the line of treads and risers.

15 Measure the distance from inside of routing to inside of routing to determine the length of treads and risers.

16 Cut the stair parts to this length.

17 Cut wedges to length. **Note.** Many experienced carpenters cut vertical wedges to the same length as risers excluding the shiplapped end, and cut horizontal wedges to the same length as treads. When installed, the wedges are then locked in place (Fig. 20-3.15).

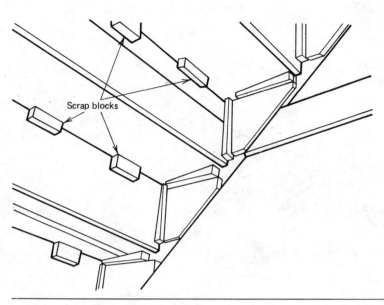

Scrap blocks

Fig. 20-3.16

18 Working from underneath the stairway, and from the top downward, slide each tread into the groove.

19 Fit each riser into the groove in the tread above and over the shiplapped end of the tread below.

20 Liberally coat three sides of a pair of short wedges with glue, and drive them upward to hold risers firmly in place.

21 Similarly coat a pair of long wedges and drive them horizontally to hold treads in place.

22 Tack the wedges to secure them until the glue dries.

23 After all wedges are installed, drive 5-penny nails through the backs of risers into treads about every 8".

24 Apply glue to the rabbeted edge of the bottom riser, and fit it into the first tread from the front of the stairway. Nail into the ends of the stringers.

25 About every 12" between stringers, glue scrap blocks where risers fit into treads (Fig. 20-3.16).

26 If the stairway is not to be carpeted, cut and fit a small cove molding under the nosing of each tread. Attach it with glue and brads.

Questions

1 Name four ways that stringers may be attached at stair headers.

2 When and where is a string used?

3 How much should the nose of a tread overlap the riser below?

4 Under what circumstances are treads installed before risers?

5 If finished flooring is installed before the stairway, how does this affect cutting and marking stringers?

6 In a stairway built with closed stringers, what nails are visible to anyone walking up the stairway?

Unit 20-4   Install a Balustrade

Handrails and all the parts of a balustrade are stock items. In a closed stairway handrails are almost always wood, and attached to metal brackets that in turn are attached through walls into studs with long screws. In an open stairway the parts may be all metal, all wood, or a combination of the two.

The lower point of anchorage for most balustrades is a newel. Newels are available in a variety of designs, or the carpenter can build one by mitering four pieces of $\frac{5}{4}$ stock (Fig. 20-4.1). The center line of the railing should meet the newel at its center line. In good quality construction the stringer, bottom tread, and bottom riser are all mortised into the newel. The newel in turn penetrates the subfloor and is securely bolted to the floor framing below.

In a stairwell with open stringers balusters may fit into treads in either circular or dovetailed joints (Fig. 20-4.2), or they may be centered on a low and sloping extension of the stair wall (Fig. 20-4.3). In a stairway with closed stringers, a subrail fits atop the string. A **subrail** is a molding that is grooved to receive balusters. Separating the balusters at a regular spacing are filler strips, called **fillets,** that also fit in the groove and project slightly above it.

Because the outside edges of treads are exposed in an open stairway, they are finished with a **return nosing** (Fig. 20-4.6). A return nosing matches the nosing on the front of the tread, but is usually attached as a separate piece. The nosing and return nosing meet at a mitered joint.

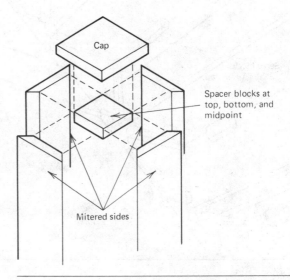

Fig. 20-4.1 How to build a newel with mitered corners.

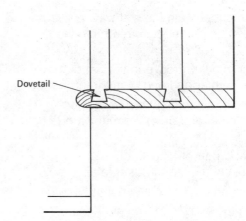

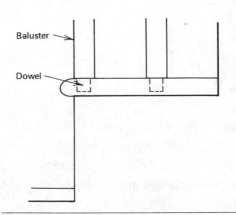

Fig. 20-4.2 Circular holes are bored or dovetailed joints are cut into treads and the treads finished with a return nosing.

Procedure for Installing a Newel

1 If subflooring is already installed, mark on it the location of the newel post.

2 Cut a hole through the subflooring for the newel. The hole must be large enough so that the end of the newel can move 2″ in the two directions away from the stairway.

3 Determine the height of the newel above the floor, and the length required below floor level for solid attachment to floor framing.

4 Cut the newel to length, and support it temporarily in position.

5 Mark the locations of grooves for the stringer, bottom tread, and bottom riser (Fig. 20-4.4).

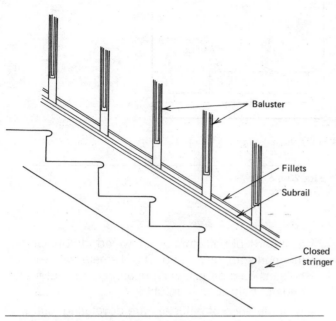

Fig. 20-4.3 How a balustrade is assembled atop a sloping wall or a closed stringer.

6 Rout the grooves. **Note.** The groove for the stringer should be $1\frac{1}{2}$ to 2″ deep.

7 Trim the bottom riser to length; it will be narrower than other risers.

8 Notch the bottom tread to fit around the newel.

9 Glue the bottom tread and riser together.

10 Fit the stringer into its groove in the newel.

11 Move the newel outward far enough so that the riser-tread assembly fits into the grooves in the newel and stringer.

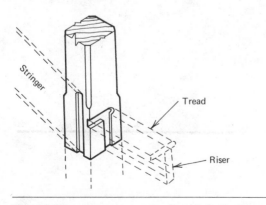

Fig. 20-4.4

12 Fit the riser-tread assembly into the opposite stringer.

13 With all parts tightly in their joints, check the newel for plumb.

14 Spike the newel permanently to the floor structure.

Procedure for Installing Balusters in a Subrail

1 Cut the subrail to length and bevel its ends as required.

2 Nail the subrail through its groove to the top of the stringer or to the cap trim on a sloping wall.

3 Determine the spacing of balusters; a standard horizontal dimension between centers is half a tread width.

4 Cut fillets to the required size and bevel their ends to the slope of the stairway. **Precaution.** Cut them for a tight fit.

5 If necessary, trim and bevel both ends of balusters to fit tightly into subrails.

6 Beginning at the newel, alternately glue fillets and fit balusters into the subrail.

7 Place another length of subrail beside the first length, and mark its ends and the locations of the edges of balusters.

8 Cut the upper subrail to length.

9 Wedge fillets between the marks.

10 Place the upper subrail over the balusters to test the fit.

11 When the fit is good, partly fill the gaps between fillets with glue, and set the subrail back on the balusters. Let the glue dry.

12 Center the railing over the subrail, and drill holes upward through fillets into the railing (Fig. 20-4.5).

13 Attach the railing with screws.

14 Attach the end of the railing to the newel with glue and screws.

Procedure for Installing Balusters in Treads

1 Before installing treads, mark the locations of balusters at the edge of each tread (Fig. 20-4.6). **Note.** Balusters are usually placed flush with the face of each riser and at the midpoint between risers.

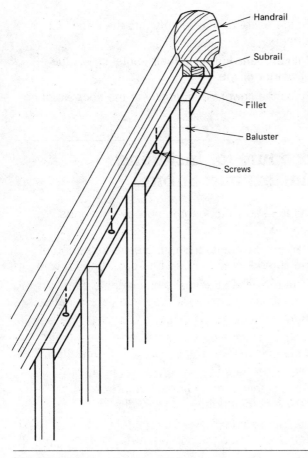

Fig. 20-4.5

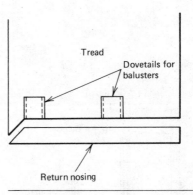

Fig. 20-4.6

4 a. If ends of balusters are doweled, drill holes on the marks about $\frac{3}{4}''$ deep. The diameter of the holes should be just wide enough to accept the ends of balusters in a tight fit.

4 b. If the ends of balusters are dovetailed, cut dovetails of the same dimensions in the edges of treads.

5 Check the fit of balusters into treads, then install the treads as described in Unit 20-2.

6 Fill holes or dovetails with glue, and insert the balusters.

7 Cut nosing returns to complete the edges of treads, and apply them with glue and enough nails to hold them while the glue dries.

8 a. Complete the balustrade as described in steps 8 to 14 of the previous procedure.

8 b. If the railing is grooved on the underside for balusters, fit the railing over the ends of balusters, and space them with glued fillets.

2 Determine the two lengths of balusters. **Note.** The balusters in the midpoints of treads will be half a riser longer than the balusters at the nosing of treads.

3 Cut all balusters to length, and slope their ends.

Questions

1 What is the difference between a baluster and a balustrade?

2 Under what circumstances does a tread receive a nosing return?

3 Under what conditions are balusters of different lengths?

4 Do you think it is possible to mortise a newel for a railing? Why?

Unit 20-5　**Build a Landing**

No flight of stairs should have more than 18 treads without a landing. Straight stairways long enough to require a landing are rare in houses, however. Stairways that turn a corner, regardless of their length, must have either a landing or winders (Unit 20-6).

A landing is nothing more than a small, lightweight

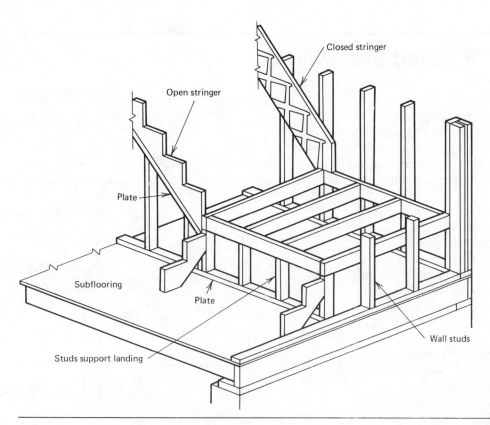

Fig. 20-5.1 Construction details of a landing.

floor frame. Where there is a wall beside it, the landing is supported on studs. Where there is no wall, it is supported on posts and studs nailed to a bottom plate (Fig. 20-5.1). Open stringers butt against the landing. Because most L-shaped stairways are open, the stringers are supported by studs capped with a diagonal plate. Closed stringers are notched to fit over a landing as they are at a floor. The floor of the landing is usually treated as an oversized tread.

The post that supports the inside corner of a landing is usually built up of two 2 × 4s. One is cut square and supports the landing itself. The other is beveled on top to support the plate for the upper stringer.

Procedure

1 From drawings determine the size of the landing and its height above the floor below it.

2 On the floor mark the location of the plate below the supporting post.

3 Cut and fit the plate.

4 Cut to length the studs supporting the landing, and nail them to the plate.

5 Build the framing for the platform.

6 On wall studs mark the height of the platform.

7 Nail the platform to wall studs and supporting studs.

8 Cut and install the diagonal plate that supports the open stringer.

9 Cut and install the beveled half of the built-up post.

10 Build the two flights of stairs as described in Units 20-3 and 20-4.

11 Cut and install the floor of the landing.

Questions

1 What is the shape of a landing in an L-shaped stairway? Why is it this shape?

2 Can you think of another shape for a landing in another shape of stairway?

Unit 20-6 **Build a Winding Stairway**

A stairway with winders can be very dangerous, and most designers avoid them if at all possible. Occasionally, however, when there is not enough floor space for a straight stairway and not enough headroom to turn with a landing, a winding stairway is the only alternate short of an elevator.

In most winding stairways the winders make a 90° turn between two short flights of straight stairs. Some older expensive homes have elliptical stairways, but they are rare today. An **elliptical stairway** is a graceful and continuously curving stairway that completes from a half to three-quarters of a circle. All winders are the same size and shape. A **spiral** or **circular stairway** completes a full circle. It is manufactured of steel, and all treads are attached to a vertical pole like spokes of a wheel. The various parts are shipped knocked down, and are assembled on the job with brackets. There are no risers.

In most winding stairways today, three winders are installed to turn 90°. Two rules govern their design (Fig. 20-6.1). First, the winders do not come to a point, but taper from a broad to a narrow tread. Second, the width of each tread at a point 15 to 18″ from the newel must be the same as the width of a tread in the straight flights. Stringers must be closed, and are mortised into the newel.

The only sure way to establish the correct dimensions and shape of the parts for winders is to lay them out full size on a large sheet of paper or a cleared area of the floor.

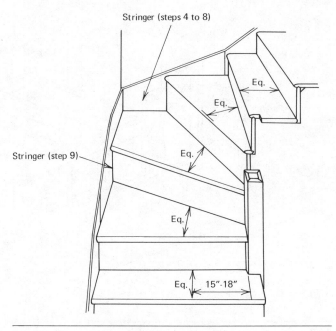

Fig. 20-6.1 The two safety rules governing the design of winders.

Procedure

1 Install the newel as described in Unit 20-4.

2 Install stringers for the flight above the winders.

3 Lay out the top winder at full size to determine the width of the tread at the wall.

4 Mark the shape of this tread and the riser and tread below on a short length of stringer.

5 Cut the stringer to length, with plumb cuts at both ends.

6 Cut or rout the stringer for treads and risers.

7 Cut and install strings if the stringer is open.

8 Tack the stringer in place.

9 Repeat steps 3 through 8 above for the lower winder and for the remainder of the corner winder.

10 Cut winders and risers to size, and test their fit.

11 When the fit is good, nail the stringers permanently.

12 Install stringers for the flight below the winders.

13 Install treads and risers for all steps, as described in Unit 20-3.

Questions

1 What two rules determine the shape of a winder?

2 When there are three winders, what size are the three angles against the wall?

3 What is the basis for the 15 to 18″ dimension in the second winder rule?

Unit 20-7   **Build Steps Outdoors**

The riser-to-tread ratio of exterior steps is not the same as for interior stairways. If space permits, a 6" riser and 12" tread are far safer for concrete steps, and a 7" maximum riser with a $10\frac{1}{2}$" minimum tread for wood steps. Furthermore, exterior treads should slope for drainage.

The basic procedures in Unit 20-2 for a simple stairway apply generally to wood steps from porches and decks to the ground. But there are several important differences. First, the stringers should be supported on a concrete base carried below the frost line (Fig. 20-7.1). The base must be large enough (about twice the width of a tread) so that it doubles as the bottom step. Second, the kickplate and ledger must be anchored into the base and porch or deck wall. Third, the treads of exterior stairways should slope forward for drainage about $\frac{1}{8}$". Therefore the cuts in open stringers for treads and risers will not be at right angles to each other.

Concrete steps are formed in the ground (Unit 5-3). Minimum thickness of concrete is 4" when the soil beneath is good, and 6" if the soil is poor. Therefore side forms should be either 2×10s or 2×12s or the plywood equivalent when there is soil beside the steps.

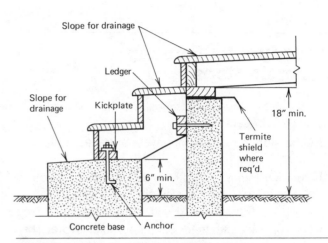

Fig. 20-7.1 A section through a short flight of exterior wood steps showing their construction and method of support.

Riser forms should be angled to give maximum toe space on treads (Fig. 20-7.2).

When the sides of steps butt against a wall, riser forms are held in place by angled supports nailed to a

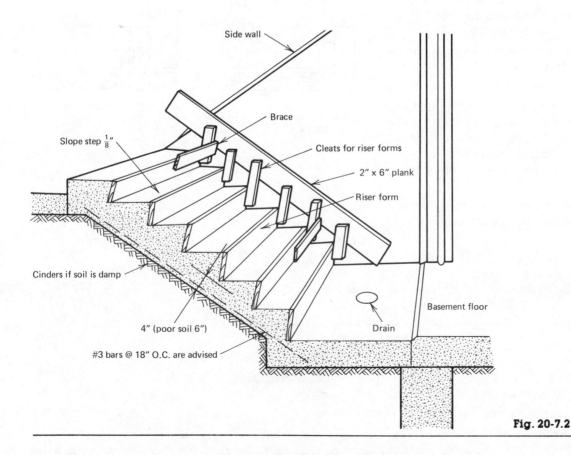

Fig. 20-7.2

2 × 6 plank that is braced against the wall. If possible the brace should be wedged against the wall. If not, it must be nailed, and the nail holes patched with concrete.

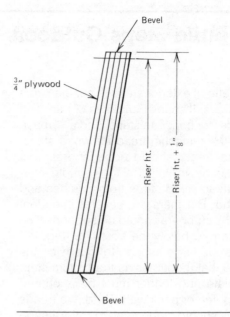

Fig. 20-7.3

Procedure for Forming Concrete Steps

1 Establish the width of treads and the vertical height of risers.

2 On a side form mark the layout of treads and risers.

3 Cut riser forms to length, and rip them to a width $\frac{1}{8}''$ greater than the riser height with slightly beveled edges (Fig. 20-7.3).

4 Cut two 1 × 2 cleats for each riser form. **Note.** The cleats should clear the bottom edge of the form by 1″, and extend about 6″ above the top edge.

5 Attach the cleats to the riser form, nailing through the back of the form into the edges of cleats (Fig. 20-7.4). **Note.** If the sides of the steps are open, nail cleats flush with the ends of the forms. If the sides are against a wall, nail 1″ away from the ends. See Fig. 20-7.2.

6 Stake the side form on an open side into the ground. **Precaution.** Make sure that tread lines have the required $\frac{1}{8}''$ slope forward.

7 Set the riser forms on the riser lines, and nail the cleats to the side form.

8 At a wall, wedge a 2 × 6 plank behind the cleats.

9 Beginning at the bottom of the steps, pound each riser form into position, and nail its cleat to the plank. Use a sliding T-bevel, folding rule, and carpenter's level at each form to check its position in relation to the opposite side of the steps.

10 When all forms are in place, tack braces between forms into the opposite sides of cleats (Fig. 20-7.2).

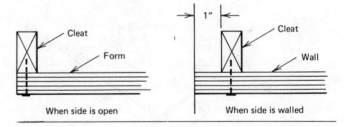

Fig. 20-7.4

Questions

1 When the rise of an exterior stairway is $3\frac{1}{2}'$, how many more concrete steps are required than wood ones?

2 How much should exterior steps slope, and in which direction?

3 Why should riser forms be set at an angle?

Unit 20-8 Install Disappearing Stairs

The attic area of many homes offers good and economical bulk storage space. Yet if this area under the roof is used only for storage, it seldom warrants the cost of a fixed stairway to reach it. A better answer is a movable stairway that can be pulled out of the ceiling when needed, but remains stored in the attic the rest of the time.

A **disappearing stairway** serves this purpose well. It is manufactured and comes ready to install, with complete instructions. There are two basic types, both built of wood. One type has three short sections of stairs that are hinged together (Fig. 20-8.1). When not in use, the sections are stacked on top of the door that closes the opening in the ceiling. When the stairway unfolds, a short handrail rises into place beside the top section. A short pull hangs from the door, and a gentle tug is enough to open the door and expose the stairway for unfolding.

The other type has only two sections, and the wood parts are a little thicker and heavier. The top section is hinged at the attic floor, as is the folding type, but the stair section slides into position (Fig. 20-8.2). When not in use, the sections are stacked, but the stair section, which is longer than the hinged section, extends beyond the opening. Sliding stairways require more attic headroom than folding stairs.

Both types are shipped already assembled in wood frames, and installation takes nothing more than a hammer. The carpenter who frames the ceiling must know the stairway's dimensions before installation, however, so that the rough opening in ceiling framing is the correct size and is properly located. When the stairway is extended, there should be at least 4' of floor space at the lower end so that boxes and other bulky items can be moved safely into the attic.

Fig. 20-8.1 A typical disappearing stairway that folds up. (Courtesy Bessler Disappearing Stairway.)

Procedure

1 From the manufacturer's literature or actual measurement of the manufactured stairway, determine the size of the rough opening in the ceiling.

2 Measure the opening to assure that actual and required dimensions are the same. Also check the opening for square.

3 Have a helper fit the stairway frame into the rough opening below, and hold it tight against the finished ceiling.

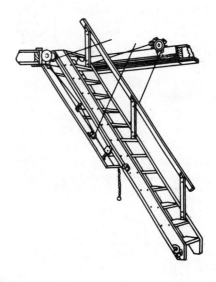

Fig. 20-8.2 A typical disappearing stairway that slides out.

4 Working in the attic, nail through the frame into headers and trimmers. If necessary for a tight fit, shim between the two frames on one side or end.

5 Test the stairway's operation.

Unit 20-9 **Install Basement Access Stairs**

Like disappearing stairways, basement access stairs are manufactured to serve a specific purpose, in this case to provide a way to reach basement storage space directly from outdoors. Like an attic, a dry basement provides good and economical bulk storage space, especially for furniture and sports equipment that are used outdoors at certain seasons. For basement storage space to be practical for outdoor gear, there should be direct access between the basement and outdoors. Concrete steps (Unit 20-7) are one answer. Manufactured basement access stairs are another.

Manufactured access stairs are designed to fit over and into a concrete areaway. The stairs themselves consist of a pair of steel stringers that support 2×10 treads (Fig. 20-9.1). There are no risers.

The weatherproof cover (Fig. 20-9.2) has a header, sill, two sides, and two doors. At the house the flange of the steel header fits against sheathing and is covered by finish wall material. The sides of the cover are triangular. Both sides and the front sill are bolted to the areaway. Completing the cover is a pair of hinged doors that open up and out, and may be locked from the inside.

Prior to installation, the concrete areaway should be formed but not poured. The inner form should be 1″ higher than the outer form so that the top of the areaway can be sloped to shed rain water. This top should be far enough above grade for good drainage—a minimum of 4″. Although the doors prevent rain and snow from entering, a drain in the floor of the areaway is recommended in case someone forgets to shut the doors. A standard basement door or a door cut from a sheet of plywood should be installed in cold climates at the base of the stairs to prevent heat loss and reduce condensation.

Fig. 20-9.1 Stringers in basement access stairways are steel set in a concrete areaway. Treads are usually wood. Photos courtesy The Bilco Company.

Procedure

1 Read the manufacturer's instructions for assembly and installation.

2 Form the areaway as shown in Fig. 20-9.3. Make sure the inner forms are level.

3 Assemble the header, sides, and sill.

Fig. 20-9.2 The frame for a basement access stairway fits against the foundation wall, and rests on an areaway. Doors are hinged to the frame.

4 Set the assembled frame in position on the inner form. **Note.** The bottom of the frame must be level and at the same height as the top of the areaway.

5 Fasten the header to sheathing with nails, or to masonry with $\frac{1}{4}'' \times 1\frac{1}{4}''$ flat head machine bolts and shields.

6 Set screws provided by the manufacturer in holes in the sides and sill, and add the nuts. **Note.** The screws and nuts hold the cover in the concrete when it is poured.

7 Install the two doors according to the manufacturer's instructions.

8 Check door alignment and operation, and recheck the unit's level. Then add braces to hold the cover in position while the areaway is poured.

9 After the concrete hardens, remove the forms and tighten all screws.

10 Caulk tightly around the perimeter of the areaway.

11 On the inner walls of the areaway mark the location and slope of the two stringers.

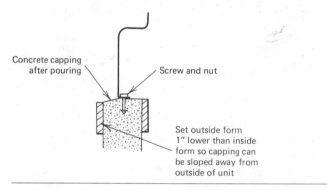

Concrete capping after pouring

Screw and nut

Set outside form 1" lower than inside form so capping can be sloped away from outside of unit

Fig. 20-9.3

12 Install the stringers along the lines with masonry nails.

13 Cut treads to length, and slide them into the slots provided for them in the stringers.

Question

1 Suppose the areaway around a basement access stairs is built of concrete block. What steps in this procedure would change, and how would you change them?

21

Drywall

Less than 50 years ago almost all interior walls and ceilings were finished with plaster. **Plaster** is a powder whose main ingredient is gypsum rock, a white, gray, or pink mineral found throughout the world. The rock is crushed, and all water driven off by a process known as **calcining.** When remixed with water, gypsum plaster can be spread easily with a trowel. When the mixture sets, it regains its original rocklike qualities. Thus gypsum is known as "the rock with a memory."

Plaster is applied over a base that is fastened to studs and joists. For many years that base was wood lath. But wood lath expanded and contracted with temperature changes, and plaster often cracked as a result. Development of lath made from steel mesh sharply reduced the cracking problem, and metal lath is still used today as a base for plaster in many commercial and institutional buildings.

Development of gypsum lath led to even better plastering work. Gypsum lath is processed gypsum rock formed between two sheets of paper, and the lath and plaster combine chemically to form a strong and durable surface. Gypsum lath is the usual plaster base today beneath flat plaster surfaces in residences. Metal lath is more often used for curved surfaces.

In spite of the improvement in plaster bases, plaster as a surfacing material has one major drawback. It is applied in two or three coats, and each coat must dry slowly and evenly. Even under the best drying conditions, it takes several weeks for plaster to dry, and the water remains in the air of the building for weeks after that. This delay is costly in residential construction, where short building time is important.

Just as the introduction of gypsum lath improved the quality of plastering, it also led to plaster's strongest competition. If plaster could be used in sheet form as a base for finish plaster, why couldn't it be used in place of plaster? In answer to this question, the drywall trade came into being.

Officially called **gypsum wallboard**, but also known as **plasterboard** and **Sheetrock** (which is a trademark), drywall is now used on more than 80 percent of the walls and ceilings of American houses. Gypsum wallboard consists of a core of processed gypsum rock poured as a **slurry** (watery mixture) between two sheets of paper and baked to rocklike hardness. It is manufactured in a continuous ribbon 4' wide. After the core has set, it is cut into sheets from 6 to 16' in length. Standard thicknesses are $\frac{1}{4}$, $\frac{3}{8}$, $\frac{1}{2}$, and $\frac{5}{8}$". The $\frac{1}{4}$" material is used primarily to resurface old walls, and must be fully backed. The $\frac{3}{8}$" material is used in low-cost construction when stud spacing is no greater than 16" o.c. It is also used in double-wall application to form a wall surface $\frac{3}{4}$" thick. For most residential work, the standard thicknesses are $\frac{1}{2}$ and $\frac{5}{8}$".

The paper surface on the back of most gypsum wallboard is a thin cardboard. On its exposed surface regular wallboard has an ivory-colored manila paper that takes paint or wallpaper easily. Wallboards for special uses, such as in bathrooms and soffits of eaves, have paper tinted in distinguishing colors, such as blue, gray, and green. Predecorated wallboard has a vinyl surface. The ends of all sheets of gypsum wallboard are cut square, and the core is exposed. The surface paper is wrapped around the long edges. The edges of prefinished wallboard are either beveled or slightly rounded (Fig. 21-0.1). Edges of wallboard that must be finished with joint treatment (Unit 21-5) are either tapered or eased.

With the differences in gypsum cores, surface papers, and edges, wallboard falls into six different categories: regular, insulating, water-resistant, predecorated, exterior, and fire-rated.

Regular

Regular wallboard is the product used on walls and ceilings under normal conditions. It is manufactured in thicknesses from $\frac{1}{4}$ to $\frac{5}{8}$", and in lengths from 6 to 16', but in only one width—4'. It comes with tapered or eased edges for finishing with joint treatment. It is also made on special order with beveled edges if joints are to be exposed or covered with battens.

Because of the fibers in the surface papers, wallboard is stronger along its length than across its width. If possible, then, it should be applied across ceiling joists and across studs. It may be applied with nails or screws (Unit 21-1).

Insulating

Except for the foil backing used in place of cardboard, insulating wallboard is just like regular wallboard. It is made in thicknesses from $\frac{3}{8}$ to $\frac{5}{8}$", in lengths from 8 to 16', and in a width of 4'. Edges are either tapered or

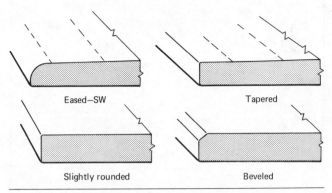

Fig. 21-0.1 Shapes of edges of wallboard panels.

eased. The foil side goes against structural members, where it acts as a vapor barrier and reflective insulation. To be most effective, however, there must be at least $\frac{3}{4}$" of air space behind the foil.

Water-Resistant

Made for use in rooms with high humidity, such as bathrooms and utility rooms, water-resistant wallboard has a core of gypsum mixed with asphalt. Both back and face papers are chemically treated to keep out moisture. It comes in thicknesses of $\frac{1}{2}$ and $\frac{5}{8}$", in lengths of 8 and 12', and with tapered edges (Unit 21-4).

Predecorated

Wallboard that requires no on-site finishing has an exposed surface consisting of a vinyl film on a fabric backing that is laminated to the gypsum core. The vinyl comes in solid colors and a variety of patterns, such as woodgrained, textile, and marblized. Panels are $\frac{3}{8}$ to $\frac{5}{8}$" thick, and 8 and 10' long for vertical application.

Square and beveled edges are left exposed. Panels with eased edges have an extra flap of vinyl 2" wide along one edge that overlaps the adjacent panel to form a solid surface uninterrupted by joints. Predecorated wallboard is usually fastened with nails that have heads painted to match the color of the board (Unit 21-6).

Exterior

Used primarily on soffits of cornices (Unit 19-1) and on ceilings of carports, exterior wallboard has a core and surface papers that are all water-resistant. It is made in sheets $\frac{1}{2}$ and $\frac{5}{8}$" thick, in lengths of 8 and 12', and with rounded edges. Joints may be finished with joint treatment, but they are generally covered with battens or fitted into H-shaped metal moldings.

Fire-Rated

Fire-rated wallboard is made especially to meet requirements of building codes for multifamily dwellings, such as apartment buildings, duplexes, and condominiums. It has a special core of gypsum mixed with other nonflammable materials such as glass fibers or vermiculite. It is $\frac{5}{8}$ thick and comes in lengths from 8 to 14'. Edges are eased or tapered. The surface paper may be regular or water-resistant, and the backing may be cardboard or aluminum foil.

A fire rating is based not on the surface material, but on the total construction of a wall or ceiling (Chapter 12). Depending on the fire rating and the sound transmission class that must be met, the wallboard may be applied directly to structural members or to resilient channels.

Attachment

Wallboard may be applied directly to studs and joists in three ways—with nails, screws, or adhesive.

Nails

When drywall was first used, a common problem was **nail pops**—that is, raised circles in the wallboard where nails were pushed out of framing members as they moved with settling or changes in temperature and moisture conditions. To prevent this problem the best nail to use is an annular ring nail (Fig. 21-0.2). It is made in two lengths—$1\frac{1}{4}''$ for wallboards up to $\frac{1}{2}''$ thick, and $1\frac{3}{8}''$ for $\frac{5}{8}''$ wallboard.

The best tool for driving nails is a wallboard hammer (Fig. 21-0.3). It has a slightly rounded head that is scored for accurate driving. Nails must be driven straight to a point just below the level of the wallboard's surface. This process is called **dimpling** (Fig. 21-0.4). The rounded head of the wallboard hammer dimples the wallboard without breaking its surface paper; other tools do not.

The nails used to attach prefinished wallboard are $1\frac{3}{8}''$ long, and are provided by the wallboard manufacturer—the only mechanical fasteners that are. They should be driven with a plastic-headed hammer (Fig. 21-0.5) that doesn't mar the paint matching the wallboard. These nails are not dimpled; the undersides of their heads should be just flush with the panel.

Screws

Many carpenters prefer screws to nails, because they have greater holding power and are less likely to break

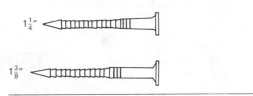

Fig. 21-0.2 Annular ring nails.

Fig. 21-0.3 Wallboard hammer. Note the waffle pattern on the head. (Courtesy United States Gypsum Company.)

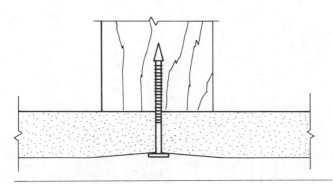

Fig. 21-0.4 In a correct application, annular ring nails must be driven at 90° to the surface of the wallboard and into a dimple made by the head of a wallboard hammer. The head of the nail must not break the surface paper.

Fig. 21-0.5 A plastic-headed hammer for nailing predecorated wallboard. (Courtesy United States Gypsum Company.)

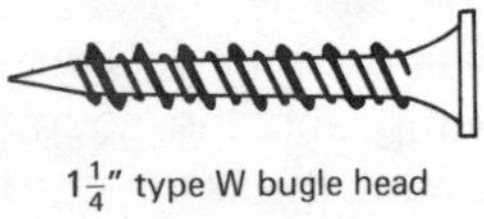

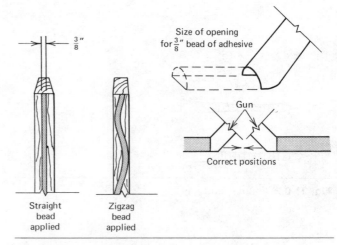

Fig. 21-0.6 A bugle-head screw.

Fig. 21-0.6 A bugle-head screw.

Fig. 21-0.8 How to lay a bead of adhesive.

the surface paper or crack the gypsum core. The best screw has a bugle head and alternately high and low threads (Fig. 21-0.6). There are three types of bugle-head screws: Type W for standard application to wood, type S for application to steel framing, and type G for application to gypsum base panels.

The tool used to drive screws is a power screwdriver with a magnetic Phillips bit. It also has adjust-

able depth control and a positive clutch. These last two features let the carpenter drive screws to the same depth as nails—just below the original surface of the wallboard—but not farther. When the desired depth is reached, the clutch disengages the bit so that it won't turn. Proper use of either a wallboard hammer or an electric screwdriver takes a little practice.

Fig. 21-0.7 Adhesive is applied to studs with a gun-type applicator. (Courtesy Gypsum Association.)

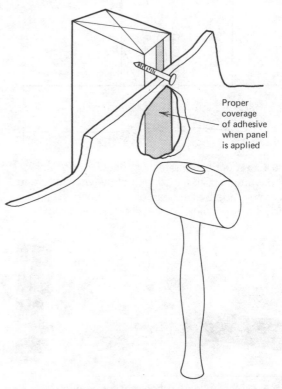

Fig. 21-0.9 Light blows by a rubber mallet spread adhesive and create a firm bond.

Adhesive

Specially formulated solvent-based adhesives may be used to attach wallboard directly to wood framing members (Fig. 21-0.7). The adhesive comes in 5-gallon cans and in $\frac{1}{4}$-gallon cartridges, and is spread with an applicator. When held at an angle of 45° to the stud or joist, the applicator produces a continuous bead of adhesive $\frac{3}{8}''$ wide. Where two panels of wallboard meet, the bead should be applied in a long zigzag pattern (Fig. 21-0.8); on other members a straight bead is adequate. Adhesive is not normally required in corners or along top and bottom plates.

To apply the wallboard, press the panel firmly in position, and nail immediately around the edges only with annular ring nails. To assure a good bond at all points, pound with a rubber mallet over each framing member to spread the bead of adhesive (Fig. 21-0.9).

Questions

1 In the table below, fill in the types of surface materials used on the backs and fronts of the various types of wallboard, and the types of edges.

Type of wallboard	Front surface	Back surface	Edges
Regular			
Insulating			
Predecorated			
Water-resistant			
Exterior			
Fire-rated			

2 Name the four methods of attachment and the tool used.

3 What is a dimple and how is it made?

Unit 21-1   Apply Wallboard to Ceiling

Gypsum wallboard is heavy. A 4 by 12′ panel weighs about 90 pounds. Because the longest possible panels should be used on a ceiling, installation overhead takes two people. Although wallboard is not easily damaged once it is applied, it may crack at a corner or even snap in two with rough handling.

Methods of Application

Ceilings may be finished with a single thickness of $\frac{1}{2}$ or $\frac{5}{8}''$ wallboard, or with a double layer of $\frac{3}{8}''$ wallboard. In the double-layer application the base layer is usually an economy wallboard called **backerboard** or **back**-ing board made only in 4 × 8′ sheets. The surface layer is regular wallboard; backer board should never be used for an exposed surface.

Single Layer

There are two methods of nailing—single nailing and double nailing. In the single-nailing method space the annular ring nails no more than 7″ apart and no closer than $\frac{3}{8}''$ to any edge of a panel (Fig. 21-1.1, left). In the double-nailing method, begin nailing in the middle of the panel 12″ o.c. Then drive the second set of nails 2 to $2\frac{1}{2}''$ from the first set after the panel is in place (Fig. 21-1.1, right). At the ends space nails 7″ o.c. Most manufacturers recommend that nails be omitted at the

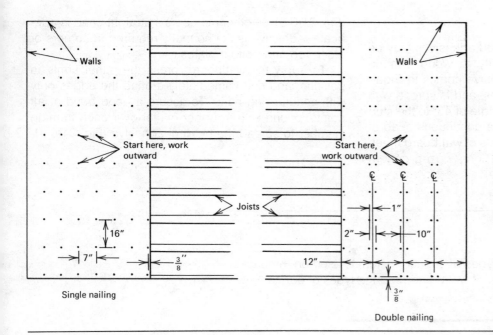

Fig. 21-1.1 The proper sequence and pattern for applying wall-board to walls with the single-nailing method (left) and with the double-nailing method (right).

edges along walls. Instead, nail about 8″ away (Fig. 21-1.2), and let the edges float; they are supported on wall panels anyway. This method reduces stresses in corners created by structural movement.

If screws are used instead of nails, drive them 12″ o.c. into all structural members (Fig. 21-1.3). Wallboard should not be attached to blocking or any member that is not part of the ceiling frame itself.

Double Layer

In double-layer application, the backer board is applied across joists. Nails should be spaced the same as with a single-layer installation. Screws should be placed 12″ o.c. when the joists are on 24″ centers, and every 16″ when joists are on 16″ centers. Backer board may also be applied with staples spaced 7″ o.c. The best staple is a 16-ga. staple with a $\frac{1}{2}$″ crown, $1\frac{1}{4}$″ leg, and divergent (spreading) points. The crown must run the same direction as the grain of the structural member into which the legs are driven, except at the edges that fall on these members (Fig. 21-1.4). There crowns should be parallel to the edge of the board.

The face layer in a double-layer installation may be applied with $1\frac{1}{2}$″ type G screws or 6-penny cement-coated gypsum nails with diamond points. This layer may be applied either at right angles to the base layer or in the same direction—whichever leaves the least total length of joint to finish. If both layers run in the same direction, however, the edges of panels must be offset at least 10″ (Fig. 21-1.5).

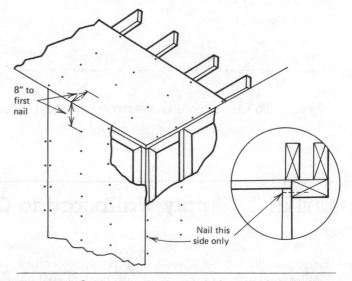

Fig. 21-1.2 Ceiling panels should not be nailed at edges along a wall. Where two walls meet at a corner (inset), nail only the second panel applied.

To Resilient Channels

To achieve certain fire ratings and sound transmission classes in multifamily dwellings, ceiling panels are often installed on resilient channels (Unit 12-1). The channels may be used with either single-layer or double-layer ceilings. With both methods the surface layer is attached to the channels.

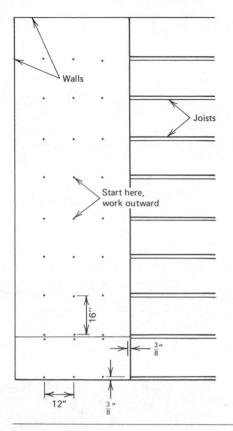

Fig. 21-1.3 The proper sequence and spacing of screws. Spacing is the same on both ceilings and walls.

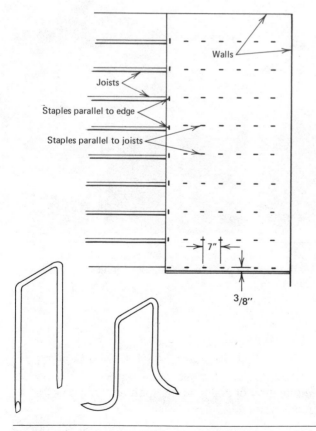

Fig. 21-1.4 Where to staple backer board and the proper staple to use.

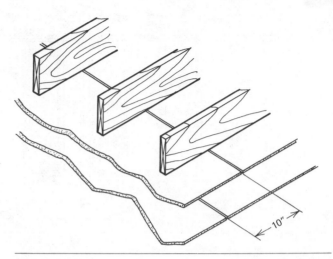

Fig. 21-1.5 When both layers of a double-layer ceiling or wall run in the same directions, edges must be offset at least 10″.

Problems and Solutions

Lifting a ceiling panel and positioning it require two pairs of hands. To get close enough to the ceiling carpenters may work from a portable scaffold or strap special metal stilts to their legs to raise them to the proper height. After one end of a panel is securely attached, the other end may be temporarily supported on a T-brace (Fig. 21-1.6). This site-built device substitutes for the helper, who is then free to prepare the next panel for installation.

At Openings

If a panel will extend part way over a framed opening, such as a stairwell, it may be cut to the shape of the opening with a utility saw after it is attached. Otherwise, the locations of holes and cutouts should be marked and cut before the panel is installed. Ceilings don't have many openings, but the working drawings should show the approximate locations of ceiling light fixtures, vent fans, access doors, and similar items for which cuts must be made.

Ridging

The most common flaw seen in wallboard ceilings is a little ridge that sometimes occurs when structural movement forces the butt ends of two panels together. This defect is called **ridging**, and it rarely occurs in

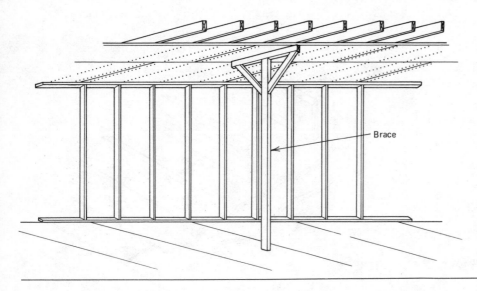

Fig. 21-1.6 A T-brace may be used to support the unfastened end of a ceiling panel.

double-layer ceilings. It can be prevented in single-layer ceilings with backblocking.

Backblocking is a method of fastening butt ends together with adhesive, and is frequently used when ceilings are supported on trusses. In a standard ceiling installation, the ends of panels butt together at joists or chords of trusses. With backblocking, the joints fall midway between supports. Behind them you fit pieces of scrap wallboard applied with adhesive (Fig. 21-1.7). When the adhesive dries, it draws the ends upward, leaving a tapered joint that is finished with joint treatment.

Planning the Ceiling Layout

The primary objective of any layout is to leave the fewest possible feet of joint to finish. A secondary objective is to have as few feet of butt joint as possible. Wallboard applied across joists provides the strongest ceiling, but this objective is only of third importance.

Manufacturers of wallboard recommend to the carpenter or drywall applicator the following rules for planning a layout and installing a ceiling:

- Install ceiling panels first, then wall panels.
- If possible, use panels long enough to extend from wall to wall without any end joints.
- If ends must butt, stagger the joints as far from the center of the ceiling as possible.
- *Never* place a butt end next to a tapered edge.
- Place tapered edges next to each other. If a panel must be cut lengthwise, place the cut edge at the wall.

Fig. 21-1.7 Backblocked joints fall between framing members. The blocks create a tapered joint that is easier to finish than a butt joint. (Courtesy United States Gypsum Company.)

- Butt panels loosely; do not force them together.
- Do not fasten panels to blocking or to the wide dimension of any lumber, such as the side of a joist.
- At ceiling openings, such as a stairwell, support all ends and edges on framing members.

Which Direction?

In order to determine the best direction to run ceiling panels, you may have to plan the layout both ways. Suppose that the room to be finished is a bedroom 12' wide and 14' long. If you use $3\frac{1}{2}$ panels 12' long, you have 36' of joint to finish (Fig. 21-1.8, top). If you use three 14' panels (bottom), you have only 28' of joint to finish. Obviously, the longer panels are better.

Now suppose the room is a living room 18' wide and 22' long. If panels run the long dimension of the room, you have 106' of joint—88' of tapered joint and 18' of butt joint. If panels run the short dimension of the room, you have 90' of tapered joint and 22' of butt joint for a total of 112'. Here again go with the fewest feet of joint.

But suppose that living room is 18' by 24'. Now you have 96' of tapered joint and 18' of butt joint in one direction for a total of 114'. In the other direction you have only 90' of tapered joint, but 24' of butt joint, which also totals 114'. In this case go with the layout that has the fewest feet of butt joint because it is harder to finish properly.

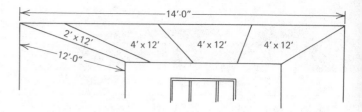
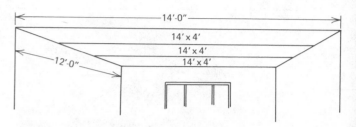

Fig. 21-1.8 The two ways to panel a ceiling. Which is better?

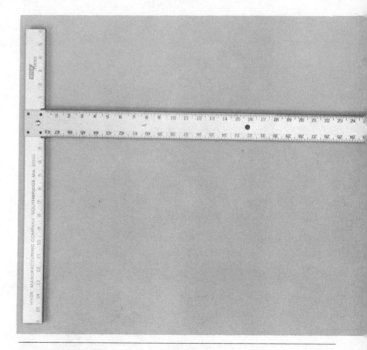

Fig. 21-1.9 A metal T-square is used for measuring, marking, and as a guide for cutting. (Courtesy Hyde Manufacturing Company.)

Tools

Several special tools are used only with drywall.

For measuring you need a steel tape and a **metal T-square** with a blade at least 48" long (Fig. 21-1.9). With a T-square you can not only measure for cuts, but also mark and make the cuts along the steel edge.

Cutting Panels

To make straight cuts the most useful tool is a **wallboard knife** (Fig. 21-1.10). To cut a panel to length, for example, lay the panel face up on a flat surface, such as a sheet of plywood laid across sawhorses. Using the T-square as a guide, draw the knife toward you (Fig. 21-1.11, top). Press hard enough to cut through the surface paper and score the gypsum core. Then, with a quick motion, snap along the cut line (top center); the panel should snap cleanly. Use the utility knife to cut the paper on the back of the panel (bottom center). Both edges should be straight and clean (bottom).

To cut the corner of a panel, make the shortest cut with a **drywall saw** (Fig. 21-1.12), then score and snap in the other direction. To cut curves and odd shapes, use a **keyhole saw**. To cut circles, use a **circle cutter**. When necessary, smooth rough edges with a rasp made commercially, or make one yourself by stapling a piece of metal lath or coarse sandpaper to a 2 × 4 block.

The process of cutting any gypsum product dulls tools quickly. Because of this, carpenters and drywall applicators often use one set of inexpensive tools only for this purpose, and replace them frequently.

Preparation

Any thin surface material such as wallboard tends to take on the shape of the structure to which it is applied.

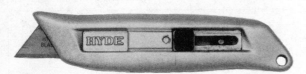

Fig. 21-1.10 A wallboard knife; the blade is retractable. (Courtesy Hyde Manufacturing Company.)

Fig. 21-1.11 The three steps in making a straight cut in wallboard are: score (top left), snap (top right), and cut (bottom left). Both edges should be straight and clean (bottom right). (Courtesy United State Gypsum Company.)

Fig. 21-1.12 Special tools for cutting wallboard and smoothing edges. Top to bottom they are: drywall saw, keyhole saw, circle cutter, and rasp. The other photograph shows how to hold a circle cutter. (Courtesy United States Gypsum Company.)

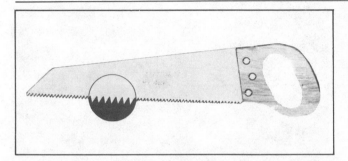

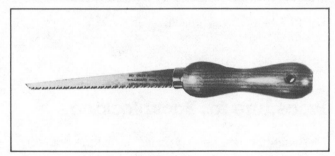

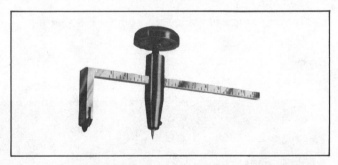

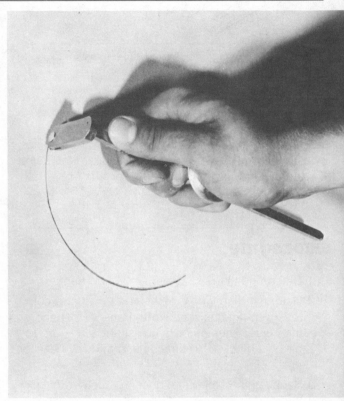

Therefore the bottom edges of ceiling joists must be level and in alignment with each other; if they are not level and aligned, the ceiling surface will be wavy, and the waves will show clearly in the finished job. Before applying any ceiling material, take three steps and make these corrections if necessary:

1 Check the alignment of joists. If they are not all in the same plane, align them with a **strongback**—a structural device consisting of a length of joist material attached at right angles to a 2×4 (Fig. 21-1.13). Nail the 2×4 across the joists, with the strongback upright. If necessary, push upward on joists with a plank to assure a tight fit between the 2×4 and each joist.

2 Look for cocked joists. Straighten then with solid bridging placed between the guilty joist and its neighbor.

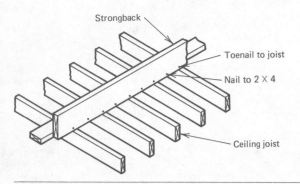

Fig. 21-1.13 A strongback brings ceiling joists into horizontal alignment.

3 Look for anything, such as bridging or blocking around ceiling light fixtures, that protrudes below the level of joists. Either renail or trim the obstructions.

Fig. 21-1.14 (Courtesy United States Gypsum Company.)

Procedure

1 Determine the best layout for ceiling panels.

2 Determine the length of the first panel and cut it to length if necessary. **Note.** Panels must end on the center lines of joists unless the joints are to be backblocked. Then they end midway between joists.

3 Because the first panel goes in a corner, check that corner for square. If the corner is much off square, scribe the panel to fit. **Precaution.** The first panel must be placed square with the ceiling structure, regardless of the condition of studs in the corner.

4 Raise the first panel in position, gapping about $\frac{1}{8}''$ at the wall plates.

5 Begin attachment with nails or screws at the center of the panel and work outward toward the edges and ends. **Precaution.** Press the panel firmly against joists, and make sure nails or screws are driven straight and into a dimple about $\frac{1}{32}''$ deep. Do not nail within 8" of walls.

6 If the first panel does not span from wall to wall, cut and fit the butting panel next, gapping slightly.

7 Complete the first row of panels before starting the next row.

8 Repeat the process until the ceiling is finished. **Precaution.** Remember to offset joints in adjoining rows.

Procedure for Backblocking

1 Attach the first ceiling panel so that the butt end falls midway between joists.

2 Cut four strips of wallboard about 12" square.

Fig. 21-1.15 (Courtesy United States Gypsum Company.)

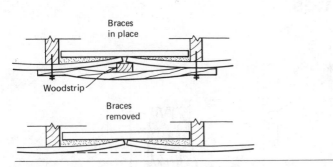

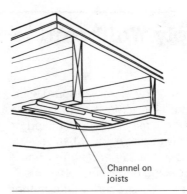

Fig. 21-1.16

Braces
in place

Woodstrip

Braces
removed

Channel on
joists

Fig. 21-1.17

3 Coat the back of each strip with joint compound, using a **mastic spreader** (Fig. 21-1.14). **Note.** The beads of cement should be about $\frac{1}{2}''$ high, $\frac{3}{8}''$ wide, and about $1\frac{1}{2}''$ apart.

4 Lay the backblocks adhesive side down along the entire butt end of the panel, so that they overlap about half their length (Fig. 21-1.15).

5 Install the next panel, butting the ends loosely.

6 Cover the seam with a strip of wood about $\frac{1}{2}'' \times 1\frac{1}{2}'' \times 47''$ long.

7 Brace the strip as shown in Fig. 21-1.16. Place braces about 6″ from the ends of the seam, and add one in the center if necessary to hold the strip flat.

8 After the adhesive has dried thoroughly, remove the braces and strip.

Procedure for Attachment to Resilient Channels

1 a. In a single-layer ceiling, attach resilient channels at right angles to ceiling joists with $1\frac{1}{4}''$ type W

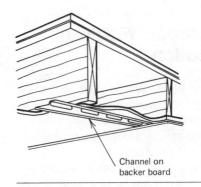

Channel on
backer board

Fig. 21-1.18

bugle-head screws, spacing the channels 24″ o.c. (Fig. 21-1.17). **Note.** The ends of ceiling panels must be supported on channels at the walls.

1 b. In a double-layer ceiling, install backerboard across joists, and attach resilient channels 24″ o.c. over the backerboard at right angles to joists, using $1\frac{1}{2}''$ type G screws (Fig. 21-1.18).

2 Apply the surface layer of wallboard to channels with $\frac{7}{8}''$ type S bugle-head screws.

Questions

1 What is the proper spacing of nails in a single-nailed ceiling installation?

2 What is the proper spacing of screws?

3 For what attachment purpose may staples be used?

4 What is the purpose of backblocking?

5 What is the fewest feet of joint possible in a ceiling above a room 11′-0″ × 17′-4″?

6 Describe in detail the steps for cutting a piece 12″ × 24″ out of a full sheet of wallboard.

Unit 21-2   Apply Wallboard to Studs

The same regular or insulating gypsum wallboard used on ceilings is also used on walls. It may be applied in a single layer or double layer. It may be single-nailed or double-nailed, or applied with screws or adhesive. Adhesive is applied to studs in the same patterns shown on joists in Fig. 21-0.8. Spacing between screws is still the standard 12″. Nails may be spaced a little farther apart, however—8″ instead of 7″. The rules for application are generally the same, but there are a few important differences.

Differences Between Wall and Ceiling Application

Predecorated wallboard is always applied vertically. All other types are applied horizontally across studs, with the panels that touch the ceiling installed first. If ends of panels butt, the joints should be toward the corners of the room.

At inside corners the panels overlap. Most drywall applicators cover the shorter walls first with single lengths, then finish longer walls that may have butt joints. At outside corners panels should be cut to meet flush, and are then covered with corner bead. **Corner bead** is a trimming accessory with a rounded nose and flanges of thin metal or metal mesh (Fig. 21-2.1). It is applied with screws, nails, or staples.

Walls have many more openings than ceilings. Cuts to fit around windows and doors should be marked and made on each panel before it is installed. The fits should be loose so that framing members can expand without putting any pressure on the wallboard.

Electrical outlets and switches should be in place before wall panels are installed. To mark their locations, hold the panel in place over the fixture, and tap with a rubber mallet on the face of the board to impress the shape of the fixture on the back of the panel. Cut along the impression with a keyhole saw, or punch out the opening with a special cutter.

Preparation

The same care must be taken to correct unevenness in wall framing that is taken to align ceiling joists, and the steps are similar:

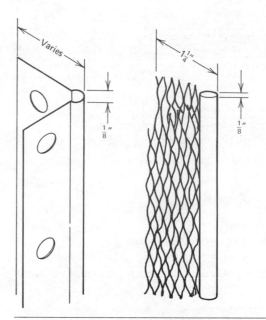

Fig. 21-2.1 Corner bead for trimming outside corners may have mesh or punched flanges and is installed with nails or staple. Fasteners must be driven. (Photos courtesy United States Gypsum Company.)

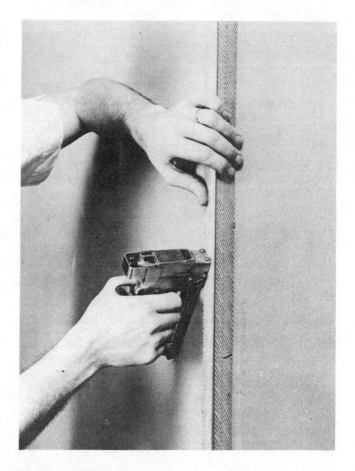

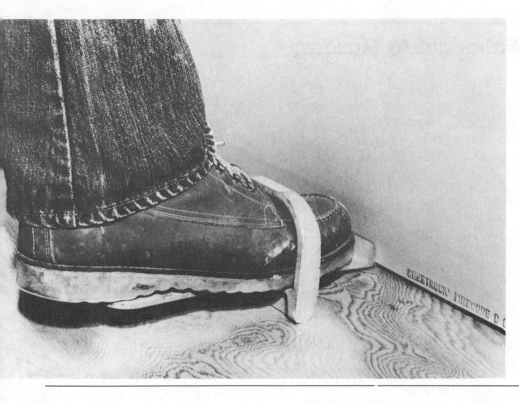

Fig. 21.2.2

1 Check the alignment of studs. With a hammer pound them back into line, and add nails to hold them in their new position. Interior studs must be aligned before wall materials are applied on either side.

2 Look for cocked or twisted studs. If possible, straighten them with blocking. If not, replace them.

3 Look for protrusions beyond the faces of studs, and eliminate them.

Procedure

1 Measure the shorter walls, and cut panels to length.

2 Check the corners at the ceiling and floor for square.

3 Raise the first panel and press it firmly against the ceiling panel.

4 Install the panel by the method of attachment selected. **Note.** Fasten the center of the panel first, then work upward and outward toward the upper corners of the panel. Work downward and outward from the center to complete attachment.

5 Set each bottom panel on a foot lifter (Fig. 21-2.2), and raise it until it lightly touches the top panel.

6 Install as in step 4.

7 Repeat steps 1 to 6 until the room is finished.

Questions

1 How does the spacing of nails and screws in a wall installation differ from the spacing in a ceiling installation?

2 What is the purpose of corner bead?

3 Give two reasons for applying regular wallboard horizontally instead of vertically.

Unit 21-3   **Apply Wallboard to Masonry**

Wallboard must not be applied directly to exterior walls of concrete, brick, block, or stone. Instead, panels are attached to furring strips. **Furring strips** are lengths of wood or metal that create an air space between the wall itself and the surface material. Either 1×2s or 2×2s may be used over masonry walls, or else special metal furring channels (Fig. 21-3.1).

Furring strips may be attached either vertically or horizontally, and the wallboard is applied at right angles across the strips. Applied vertically, strips should run from just above the floor to just below the ceiling on a 24″ spacing. Horizontal furring strips or channels should be placed within 4″ of the floor line, 4″ of the ceiling line, and at intervals between not exceeding 24″.

Openings for windows and exterior doors must also be furred, as must any outside corners. Even if other channels run vertically, channels at outside corners should be run horizontally and mitered. To miter a channel, cut the flanges and wings at 45° with a hacksaw, and bend the web 90°. Do not cut the web.

Channels are fastened every 24″ through alternate flanges. Use $\frac{5}{8}$″ concrete nails if the wall is concrete, and 2″ cut nails driven into mortar joints if the walls are masonry. Apply wood strips with power nails.

Wallboard may be attached to wood furring strips with nails or screws, but screws must be used with metal channels.

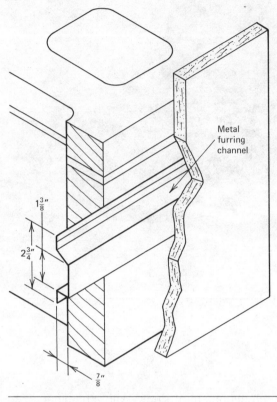

Fig. 21-3.1 Metal furring channel serves as a spacer between wallboard and exterior masonry walls.

Procedure

1 Determine the layout and direction of furring.
2 Install furring strips or channels around openings.
3 Install the main supporting strips.
4 Furr at any corners.
5 Apply the wallboard as described in Unit 21-2.

Questions

1 How do furring strips differ from resilient channels?
2 In what direction should wallboard be applied over a masonry wall?

Unit 21-4   **Install Water-Resistant Wallboard**

One of the most frequent uses of water-resistant wallboard is at a bathtub or shower enclosure. Here it can be painted, finished with a water-repellent wallpaper, or surfaced with tiles. Regardless of how it is finished, this type of wallboard is applied horizontally.

Before the wallboard can be applied, the bathtub or

shower base must be in place and attached to framing. Blocking must also be installed for soap dishes, towel racks, and similar accessories. Plumbing at water inlets must be stubbed in beyond the final surface of the wallboard or tile.

With water-resistant wallboard the manufacturer provides a sealant in one-pint cans. Although the surface paper and its core are resistant to moisture, any nicks or gouges and all cut edges must be thoroughly coated with sealant to keep water out of the core.

Procedure

1 Check the alignment of all framing in the walls to be covered.
2 Make sure all blocking and piping are in place.
3 Measure the length of the wall above the long edge of the tub, and cut a piece of wallboard to this length.
4 Brush sealant liberally on both ends of the piece.
5 Temporarily place a pair of spacers about $\frac{1}{4}''$ thick along the edge of the tub or shower base.
6 Rest one papered edge of the wallboard on the spacers, and attach the panel with nails or screws on standard spacing for wall application. **Note.** If

walls will be finished with ceramic tiles more than $\frac{5}{16}''$ thick, space nails 4″ o.c. and screws 8″ o.c. to carry the extra weight.

7 Cut end panels to length. **Note.** If the bathtub is standard size and the wallboard is 10′ long, it is usually possible to cut the side panel and both end panels from a single sheet.
8 Cut holes in end panels where necessary for plumbing.
9 Brush sealant liberally on both ends and at all holes.
10 Install end panels.
11 Cut upper panels to length and width, and seal the edges.
12 Install upper panels with the cut edges toward the ceiling.
13 Remove spacers and fill the joint around the tub or shower with joint caulking.
14 a. If the wallboard will be papered or painted, finish joints with joint treatment (Unit 21-5).
14 b. If the walls will be tiled, brush sealant over the heads of all fasteners. **Precaution.** Do not use joint treatment behind tiles.

Questions

1 Under what circumstances should the number of fasteners used to install water-resistant wallboard be increased?
2 How must cut edges be treated?

Unit 21-5   Finish Wallboard Joints

The process of finishing gypsum wallboard for painting or wall-papering is called **joint treatment**. It includes filling joints between panels to the level of the main surface, filling the dimples for fasteners, and repairing nicks or other damage to the surface.

Materials

Two materials are used in joint treatment—joint compound and joint tape. **Joint compound** is a chemical

mixture with a consistency similar to caulking. For many years joint compounds were powders with a casein base that were mixed with water on the job. Casein compounds are still popular because of uniform characteristics and years of reliable results. To eliminate mixing time, fast hardening powders were developed. To make spreading easier, ready-mix compounds were developed that can be used right out of the can. All types are commonly available, and have their champions.

Joint tape is a strong fibered tape with edges tapered to paper thinness. Hundreds of tiny pinholes

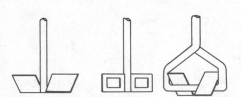

Fig. 21-5.1 Typical paddles for mixing powdered joint compound in large quantities.

Fig. 21-5.2 Compound is mixed in small quantities in a mud pan with a potato masher. (Courtesy Hyde Manufacturing Company, United States Gypsum Company.)

perforate the tape, and allow joint compound to seep through during embedding. The tape comes in rolls 2″ wide and from 60 to 500′ long. It has a slight crease down the center that is helpful in fitting it into corners.

Equipment for Mixing

Powdered joint compound in large quantities is usually mixed in a 5-gallon pail with power-operated paddles. There are three basic designs of paddles (Fig. 21-5.1). In small quantities the compound is mixed by hand in a mud pan with a potato masher (Fig. 21-5.2).

General mixing directions are to fill the container about half full with water, then sift in the powdered compound, mixing at the same time. Stir until the powder is uniformly damp—this takes less than three minutes with most types of powder. Let the mix sit and soak from 3 to 30 minutes, depending on the compound. Then stir again until the mixture is creamy smooth. The mixture can be thinned by adding more water, and thickened by adding more powder, but these additions should not be made until after the mix has soaked.

Tools for Application

Joint tape may be embedded by hand or with mechanical tools. Hand application takes a minimum of four tools, and on most jobs you will use six (Fig. 21-5.3). Joint compound is applied in three stages with flat-bladed knives—a 4″-wide knife for stage 1, a 6″ knife for stage 2, and a 10″ knife for stage 3. For finishing large areas you may need a trowel, and to clean off ex-

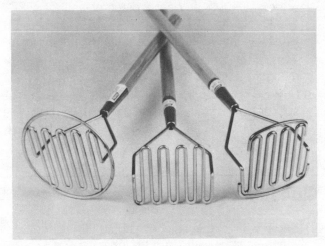

cess compound a broadknife is the best tool. To tuck tape into corners a tool with an L-shaped blade is ideal.

Professional drywall applicators who finish hundreds of feet of joint every working day use mechanical tools (Fig. 21-5.4). An automatic taper completes stage 1 in a single operation. A corner roller embeds tape in corners. Other mechanical tools are designed for specific finishing jobs in stages 2 and 3.

Although the procedures that follow are related to specific areas of finishing, it is standard practice to do all work in each stage at one time at all locations.

Procedure for Taping Tapered Edges

1 Press on each panel of wallboard along all rows of fasteners to make sure it lies tight against framing members.

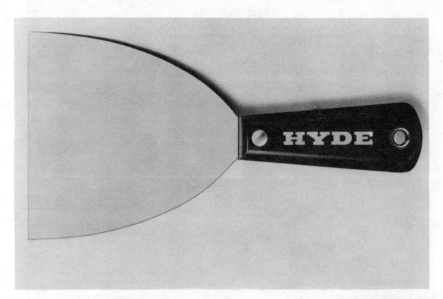

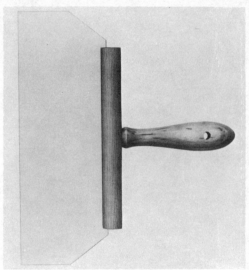

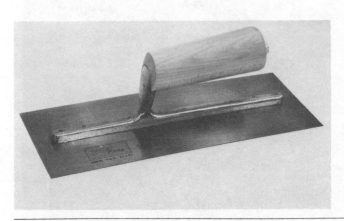

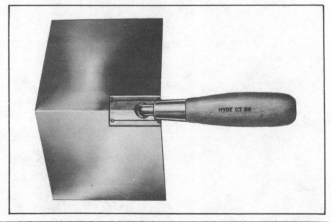

Fig. 21-5.3 Hand tools for finishing wallboard; top to bottom and left to right, 4″, 6″, and 10″ knives, trowel, and corner knife. (Courtesy Hyde Manufacturing Company.)

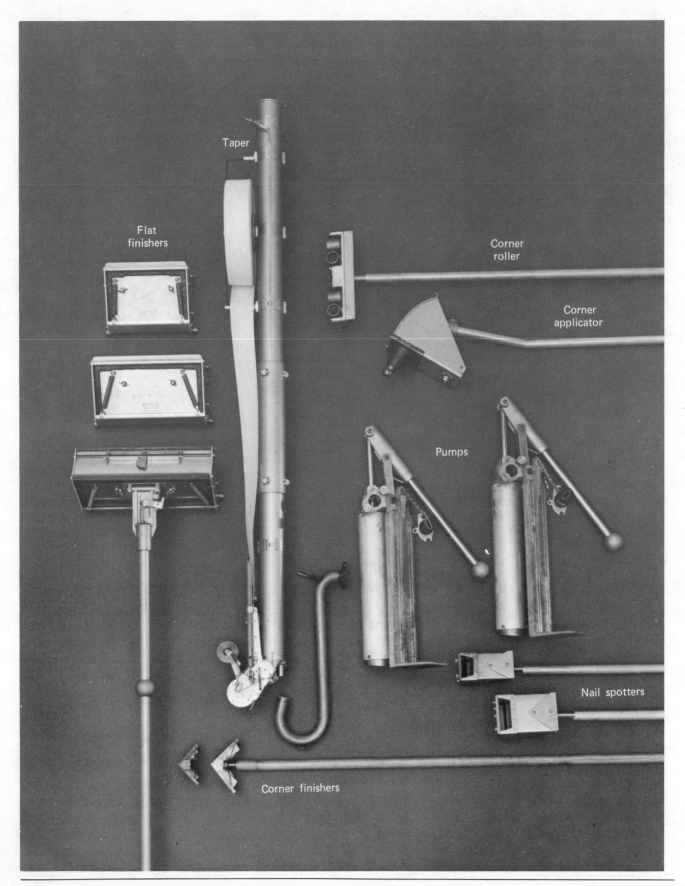

Fig. 21-5.4 Automatic tools for finishing wallboard.   (Courtesy Ames Taping Tool Systems.)

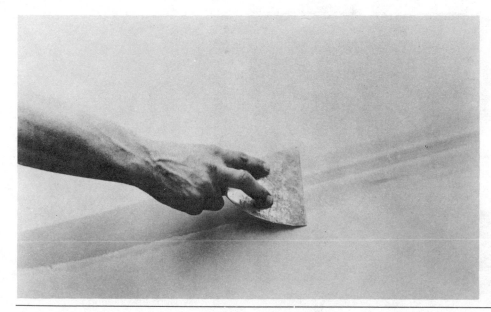

Fig. 21-5.5 (Photos United States Gypsum Company, Gypsum Association.)

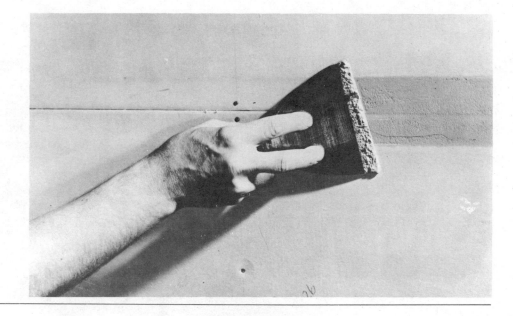

Fig. 21-5.6

2 Check for protruding nails or screws by running a finishing knife over dimples. If you hear metal hit metal, sink the fastener a little deeper.
Precaution. Be careful not to break the face paper.

3 Remove dust and grit from wallboard surfaces with a damp sponge.

4 Make sure adjoining panels are in the same plane; an invisible joint is almost impossible to make if one panel is higher than the other.

5 If the gap between any panels is more than $\frac{1}{4}''$, fill the joint with compound (Fig. 21-5.5), and let it dry for at least 24 hours.

6 With a 4″ knife butter the tapered joint between panels with joint compound (Fig. 21-5.6).
Precaution. Fill the shallow channel full and evenly but do not overfill it.

7 Center the joint tape over the seam, and press it into the fresh compound with a 4″ knife held at about a 45° angle to the wall, exerting enough

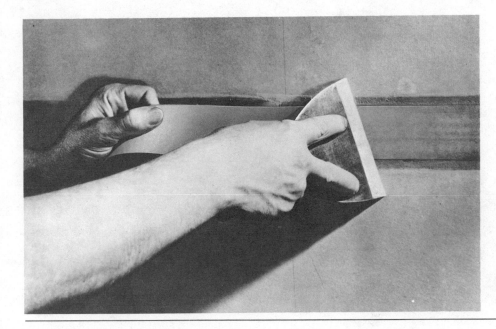

Fig. 21-5.7 (Photos courtesy United States Gypsum Company, Gypsum Association.)

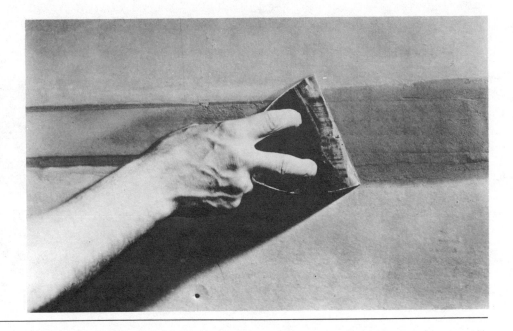

Fig. 21-5.8

pressure to remove excess compound (Fig. 21-5.7). **Note.** At the center of the joint the compound should be deep enough to create a good bond with the tape, but at the edges should be no more than $\frac{1}{32}$″ thick.

8 Apply a thin skim coat of compound over the joint (Fig. 21-5.8) to prevent the edge from wrinkling.

9 Let the first coat dry, using Table XV as a guide to the amount of drying time required.

10 Apply a second coat of compound with a 6″ knife,

and feather the edges approximately 2″ beyond the first coat on each side (Fig. 21-5.9).

11 Let the second coat dry.

12 Sand the dry second coat lightly to smooth the surface.

13 Apply a thin finishing coat of compound with a 10″ knife, feathering the edges another 2″ on each side of the second coat (Fig. 21-5.10).

14 Let the third and final coat dry.

15 Sand lightly.

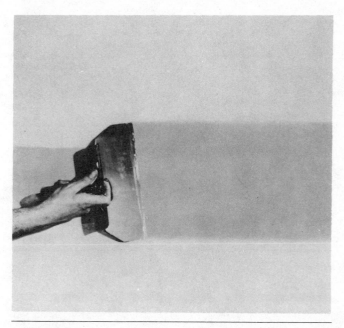

Fig. 21-5.9 (Photos courtesy United States Gypsum Company.)   **Fig. 21-5.10**

Table XV

Relative humidity	Temperature					
	50°F 10°C	60°F 16°C	70°F 21°C	80°F 27°C	90°F 32°C	100°F 38°C
10%	21H	14H	10H	7H	5H	$3\frac{1}{2}$H
20%	23H	16H	11H	8H	$5\frac{1}{2}$H	4H
30%	26H	18H	12H	9H	6H	$4\frac{1}{2}$H
40%	29H	20H	14H	10H	7H	5H
50%	36H	24H	17H	12H	9H	6H
60%	42H	29H	20H	14H	10H	8H
70%	$2\frac{1}{4}$D	38H	26H	19H	14H	10H
80%	$3\frac{1}{4}$D	$2\frac{1}{4}$D	38H	27H	19H	14H
85%	4D	3D	48H	34H	25H	18H
90%	6D	$4\frac{1}{4}$D	3D	49H	36H	26H
91%	7D	$4\frac{3}{4}$D	$3\frac{1}{4}$D	$2\frac{1}{4}$D	40H	29H
92%	8D	5D	$3\frac{1}{2}$D	$2\frac{1}{2}$D	44H	32H
93%	9D	6D	4D	$2\frac{3}{4}$D	48H	36H
94%	10D	7D	5D	$3\frac{1}{4}$D	$2\frac{1}{4}$D	41H
95%	12D	8D	6D	4D	$2\frac{3}{4}$D	48H
96%	14D	10D	7D	5D	$3\frac{1}{2}$D	$2\frac{1}{2}$D
97%	18D	12D	9D	6D	$4\frac{1}{2}$D	$3\frac{1}{4}$D
98%	26D	18D	12D	9D	6D	$4\frac{1}{2}$D

D = Days (24 hr)   H = Hours.

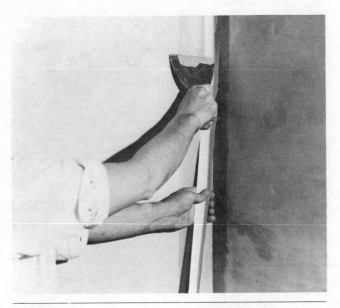

Fig. 21-5.11 (Photos courtesy United States Gypsum Company.)

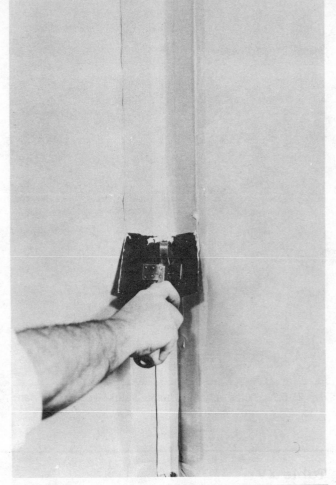

Fig. 21-5.12

Procedure for Taping Butt Ends

1 Repeat steps 1 to 4 for tapered edges.
2 Fill the gap between butt edges with joint compound, and let it dry for at least 24 hours. **Note.** Fill to slightly above the surface of the wallboard to allow for shrinkage.
3 Repeat steps 6 to 9 for tapered edges.
4 Apply the second coat with a 10″ knife, and feather the edges to a total width of about 10″.
5 After the second coat is dry, lightly sand the surface.
6 With a trowel apply a thin finishing coat of compound, feathering the edges to a total width of about 18″. **Note.** Wider feathering is necessary to avoid the appearance of a hump at a butt joint.
7 After the third coat is dry, sand lightly.

Procedure for Taping Inside Corners

1 Repeat steps 1 to 5 for tapered edges.

2 Fold the tape along its center crease.
3 With a 4″ knife or corner tool, butter the joint with compound.
4 a. Fit the creased tape into the corner, and press it into the compound with a 4″ knife, working first on one side of the corner and then on the other (Fig. 21-5.11).
4 b. Fit the creased tape into the corner, and press it into the compound with the corner tool in a single operation (Fig. 21-5.12). **Precaution.** As the tape is embedded, a ridge of compound builds up at the edges of the tape. Remove this excess with a knife or either blade of the corner tool.
5 Repeat steps 7 to 15 for tapered edges. **Precaution.** Use long vertical strokes to spread the compound evenly, but be careful not to tear the tape with the edge of the blade.

Procedure for Spotting Fastener Heads

1 Over fastener heads apply a first coat of compound level with the wall's surface (Fig.

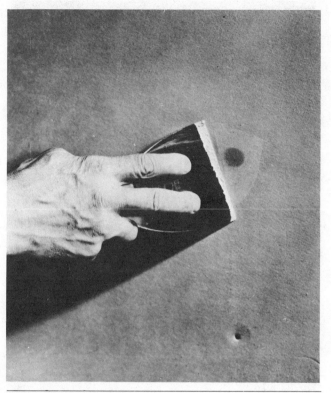

Fig. 21-5.13

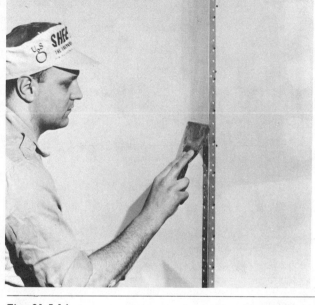

Fig. 21-5.14

21-5.13). **Note.** This coat is usually applied as the first step in finishing a wall or ceiling, just before the joints are taped.

2 Apply a thin second coat when the second coats are applied at joints.

3 After the second coat is dry, sand lightly.

4 Apply a thin finish coat after the other finish coats are applied.

5 After this coat is dry, sand lightly.

5 Apply a second coat, feathering the edges about 6″ on either side of the corner.

6 Let the second coat dry thoroughly, then sand it lightly.

7 With a trowel apply a thin finishing coat of compound, feathering the edges 8 to 10″ from the corner on each side.

8 After the third coat is dry, sand it lightly.

Procedure for Finishing Outside Corners

1 If necessary, cut strips of corner bead to length. **Note.** This metal corner reinforcement should gap about $\frac{1}{8}$″ at the floor and ceiling.

2 Nail or staple corner bead at all outside corners, making sure it is plumb.

3 With a 4″ knife butter the flanges on both sides of the rounded nose of the bead with joint compound (Fig. 21-5.14).

4 Let this coat dry thoroughly.

Procedure for Mechanical Finishing

1 Prepare the wall for taping as described in steps 1 to 4 for taping tapered edges.

2 With a special adapter on an Ames tool (Fig. 21-5.15), fill the joints between panels with compound. Let the compound dry.

3 With an automatic taper, embed tape in all flat joints (Fig. 21-5.16).

4 Touch up joints with a broadknife to remove excess compound.

Fig. 21-5.15 (Photos courtesy United States Gypsum Company.)

Fig. 21-5.16

5 Mechanically tape interior corners.

6 With a corner roller and corner finisher (Fig. 21-5.17), smooth the compound on both sides of the corner, and remove any excess with a broadknife.

7 Let the first coat dry.

8 With a hand finishing tool (Fig. 21-5.18) apply a second coat of compound on flat joints, and with a corner finisher apply a second coat at corners.

9 Using the hand finisher and a thicker mix of compound, apply a first coat at outside corners and spot fastener heads (Fig. 21-5.19).

10 Let the coats applied in steps 8 and 9 dry.

11 Apply the final coat at all points by hand, feathering at least 2″ beyond the previous coat.

12 After the final coat is dry, sand lightly.

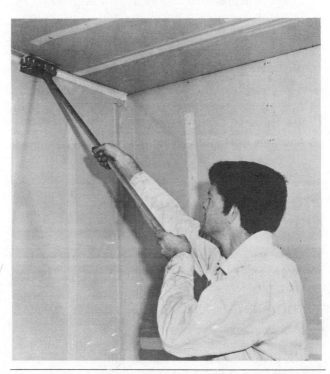

Fig. 21-5.17 (Courtesy United States Gypsum Company, Gypsum Association.)

Fig. 21-5.8

Fig. 21-5.19

Questions

1 What is the purpose of the holes in joint tape?
2 What is the recommended width of joint treatment at edges after each of the three stages?
3 What is the recommended final width of joint treatment at ends?
4 If humidity is 50% and temperature is 70°, how much time should elapse between the stages of joint treatment?
5 What is the test for improperly driven fasteners?

Unit 21-6   Apply Predecorated Wallboard

Predecorated wallboard is not a practical material for ceilings because butt joints are almost impossible to hide. On walls, however, panels go up quickly, and need no further treatment. The panels must be applied vertically, and can be attached directly to studs with adhesive or colored nails, or over backerboard.

Manufacturers suggest that vinyl wallboards with grained patterns or textures should be numbered in the order in which they will be put up. Then, for the most interesting and attractive effect, apply odd-numbered panels right side up and even-numbered panels upside down.

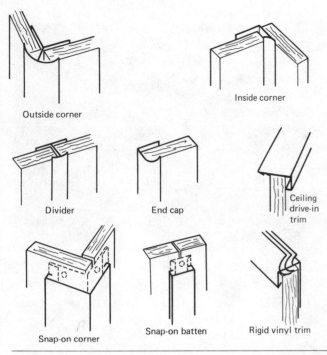

Fig. 21-6.1 Typical moldings for finishing predecorated wallboard.

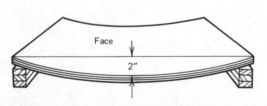

Fig. 21-6.2 How to prebow predecorated wallboard to eliminate edge fasteners.

Special care must be taken with cut panels. Before cutting any panel, cut through the vinyl surface with a utility knife, even if the core is cut with a saw. Cut edges should be placed in the corners of the room, or else covered with trim or battens. Holes for electrical outlets and switches should be cut from the back side

to prevent the vinyl from delaminating. Manufacturers make moldings in matching colors (Fig. 21-6.1) to finish raw edges and to conceal nails.

If both moldings and exposed fasteners are objectionable, there are two alternatives. One alternative is to prebow standard predecorated panels (Fig. 21-6.2). Lay the stack of panels face up, with their ends resting on three or four 2 × 4s and their centers touching the floor. Let the panels remain this way overnight, or until they take on a permanent 2" bow. Then apply the panels to studs or backer board with adhesive, and nail only at top and bottom where fasteners are concealed behind trim.

The other alternative is predecorated wallboard designed for seamless installation. Each panel has a flap of vinyl at one long edge that extends 2" beyond the edge of the board itself and hides fasteners. Procedures are not covered here, because they vary from manufacturer to manufacturer, and detailed instruction sheets are furnished with the material.

Procedure

1 Carefully read the manufacturer's instructions, especially the recommended method for cutting panels to size.
2 Determine the layout of the panels, and number them on the back. **Note.** In most rooms the best appearance is achieved by putting full panels in diagonally opposite corners, and butting any cut panels in the other two opposite corners.
3 Check the starting corner for square and plumb. The first panel must be applied absolutely plumb.
4 Install the first full panel at the corner.
5 Complete the first wall to the opposite corner.
6 Beginning at the same starting corner, complete the second wall.
7 Start at the diagonally opposite corner and repeat steps 3 to 6.
8 Install matching moldings if specified.

Questions

1 Which of the following installations is not acceptable with predecorated wallboard: double-layer, with adhesive, horizontal, single-layer, vertical, with screws?
2 Name three ways that exposed fasteners at edges of panels may be avoided.

22

Paneling

Although gypsum wallboard covers the walls of most rooms in homes today, four other materials are often used in some rooms for a change of pace and contrast. They are plywood paneling, hardboard paneling, tileboard, and wood boards.

Plywood

To give a feeling of warmth and richness to a room, few materials can equal wood. Of the wood or wood-based materials for residential use, predecorated plywood is by far the favorite of most homeowners.

Individual panels are 4' wide, come in lengths from 7 to 10', and in thicknesses from 4 millimeters (about $\frac{3}{32}''$ thick) up to $\frac{5}{16}''$. The most common thickness is $\frac{1}{4}''$. Although plain panels are available, most paneling has V-grooves cut in it at random intervals. So that panels may be split lengthwise to fit stud spacings,there is always a groove 16", 24" and 32" from an edge.

Face veneers range in tone from pale tans to dark browns, and the pattern of grain may be subdued or prominent, depending on the species of wood and the way the veneers are arranged on the panel. Most veneers are hardwood; birch, oak, pecan, maple, walnut, and mahogany (roughly from lightest to darkest) are the most popular.

Unfinished panels are also available. They are thicker than prefinished panels ($\frac{1}{4}$ to $\frac{1}{2}''$) and are usually stocked only in 4' × 8' size. Edges may be square, beveled, or shiplapped. The surface may be smooth or rough, and it may be painted or stained.

Hardboard

Hardboard for surfacing walls is 4' wide, 8 to 12' long, and is available in thicknesses from $\frac{1}{8}$ up to $\frac{1}{2}''$. Hardboard less than $\frac{1}{4}$" thick must have solid backing. Unfinished hardboard has a smooth tan or dark brown surface that takes paint easily. It may be plain or perforated. Prefinished hardboard comes with a wide variety of surface patterns that look like brick, marble, stucco, stone, leather, cork, and even wood. The best quality surfaces are embossed vinyl; others have the pattern printed on the hardboard surface.

Tileboard

Tileboard is tempered hardboard with a baked-on plastic finish that sheds water. Its primary use is on walls in bathrooms and kitchens. It comes in a range of soft colors and in patterns that look like ceramic tile or marble. The manufacturer of the tileboard also makes moldings to trim panels used in tub and shower enclosures.

Wood

Just as development of sheet materials for exterior use has reduced the popularity of drop siding, panels for interior use have largely taken the place of solid wood paneling. Yet solid wood is still specified in higher-priced homes and for special decorative purposes in smaller homes.

Solid wood paneling may be installed vertically or horizontally. The boards should be no wider than 8", and edges may be square, shiplapped, or tongue-and-groove. The surface may be plain or milled near the joint between boards. For informal rooms, knotty pine, wormy chestnut, and pecky cypress are popular; all three species have strong grain and defect patterns. For more formal rooms, walnut, redwood, mahogany, cherry, oak, and gum are usually considered more appropriate. Boards may be selected for uniformity of color or for the presence or absence of strong graining.

Unit 22-1 Install Prefinished Plywood

Panels of prefinished plywood may be installed in three ways. They may be applied directly to studs, over a base layer of gypsum backerboard, or to furring strips. They may be attached with nails, adhesive, or contact cement. The panels run vertically on walls; they are seldom used on ceilings.

To Studs

The most popular prefinished panels have V-grooves cut into them lengthwise. At the bottoms of the grooves the panels are quite thin and may deflect as much as $\frac{1}{2}$" when bumped hard. To prevent damage, builders that apply paneling directly to studs often add horizontal blocking at the heights where bumps are most likely to occur—about 36" up from the floor and 36" down from the ceiling.

A standard 8' panel obviously will not completely cover a wall that usually measures 8'-1" from subfloor to finished ceiling. Therefore you must gap at both ends. Since there is not much chance of a panel being bumped at a ceiling, allow the greater gap there— about $\frac{3}{4}$". Gap no more than $\frac{1}{2}$" at the floor so that you have a good nailing surface for base trim.

To Backerboard

A base layer of $\frac{3}{8}$" gypsum board makes an ideal backing for plywood paneling. It adds considerable strength to the wall, provides a plumb working surface, and reduces the passage of sound through the wall. Backerboard is applied horizontally across studs as in double layer wallboard construction (see Chapter 21). Joints do not require joint treatment.

To Furring Strips

Applied across studs, furring strips strengthen a paneled wall, eliminate problems with cocked studs, and reduce the chance that joints between panels will open up and expose raw studs. The best furring is strips 2" wide that are cut from sheets of $\frac{1}{4}$" plywood, either finished or unfinished. Long strips are nailed horizontally to top and bottom plates, and approximately every 16" between (Fig. 22-1.1). Shorter vertical strips are required over studs that support joints between panels. Framing around doors, windows, and other openings, such as for heating grilles, must also be covered with furring strips.

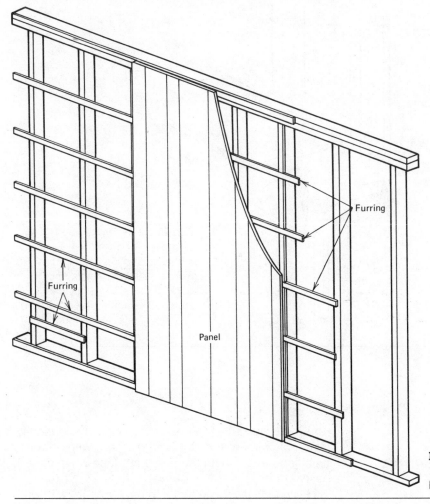

Furring

Furring

Panel

Fig. 22-1.1 Furring strips are needed every 16″ vertically, on studs behind seams between panels, and around openings.

Furring strips on masonry walls should be 2 × 2s. The surface must be given a coat of waterproofing before the strips are applied, and the space between strips must be insulated afterward. At the floor the bottom strip should clear by at least $\frac{1}{4}$″ to prevent moisture problems. At the ceiling the top strip is attached to ceiling joists or, if joists run parallel to the wall, to 6″ nailing blocks attached to the sill plate (Fig. 22-1.2). The usual method of attachment is with power nails into mortar joints.

wood panels. Use 4-penny nails into studs or furring strips and 5-penny nails through backerboard into studs.

Nails should be driven into grooves wherever possible because they are less visible there. To keep from splitting the panel, nail at an angle into the side of the groove (Fig. 22-1.3). When applying panels directly to studs, nail every 6″ along the edges and every 12″ in the field. When applying panels to furring strips, nail every 8″ along the edges and every 16″ in the field.

With Nails

If colored nails are used, they should be driven with a plastic-headed hammer flush with the surface of the panel. If finishing nails are used, they should be countersunk about $\frac{1}{32}$″, and the heads covered with **putty stick,** a wood putty that matches the color of the

With Adhesive

Various types of adhesives may be used to attach plywood to either studs or furring strips. The main difference between them is **open time** (drying time), which may range from half an hour for some types down to no time for contact adhesive that sticks immediately.

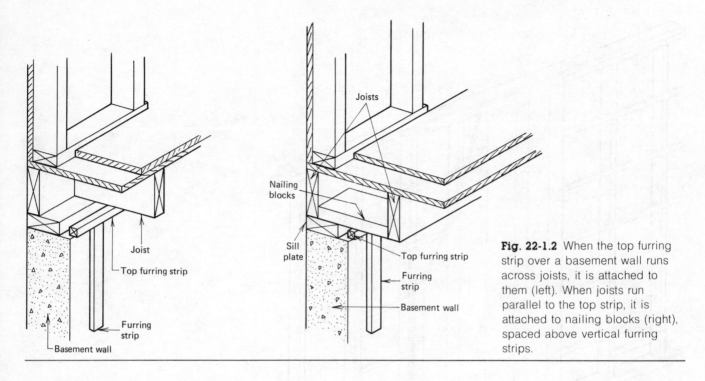

Fig. 22-1.2 When the top furring strip over a basement wall runs across joists, it is attached to them (left). When joists run parallel to the top strip, it is attached to nailing blocks (right), spaced above vertical furring strips.

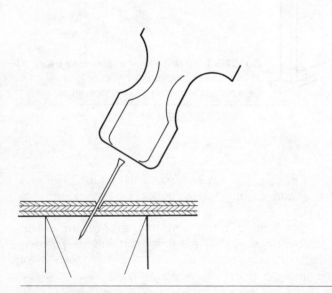

Fig. 22-1.3 How to nail into a V-groove in plywood panels.

Check the adhesive manufacturer's instructions carefully before beginning work.

Adhesive is applied with a gun in beads about $\frac{1}{8}''$ wide. It should be applied only to studs, not to any blocking or horizontal framing members. If the glue has a long open time, you can speed up the drying process by pressing each panel into the adhesive to spread it, then removing the panel at once. This transfers some adhesive to the panel and allows the solvent to evapo-

rate more quickly. Then reset the panel permanently. Nail only along the top and bottom with 4-penny finishing nails to hold the panel in place while the adhesive dries.

Plywood may be attached to backerboard with wallboard adhesive applied with a notched spreader. Nails are still needed at the top and bottom, however, to hold panels temporarily in place.

With Contact Cement

With adhesives you have some opportunity to adjust a panel that is not set quite right. With contact cement there is no chance to adjust; once the panel is applied it can't be moved without destroying it.

To prepare panels for contact cement, you must cut and fit each panel first. With the layout established and the panels cut, mark the center lines of studs on the backs of each panel. Then brush three coats of adhesive about $2\frac{1}{2}''$ wide along the center lines. Brush two or three coats on the faces of studs. Let the coats dry thoroughly.

To apply each panel, fit one edge in place either in a corner or against another panel (Fig. 22-1.4), then swing the panel into position. With a rubber mallet, pound along the entire length of each stud to assure a permanent bond.

Fig. 22-1.4 To attach panels with contact cement, carefully fit one edge of the panel in place, then swing the panel into the wall.

Planning the Job

The arrangement of panels is extremely important to the appearance of the finished room. To plan a proper layout, you must know the spacing of supports, treatment of seams at corners and between panels, and the locations of openings.

Treatment at Seams

When stud spacing is uniform and panels are applied directly to studs, joints between panels are usually left exposed, with a factory-finished edge of one panel touching the factory-finished edge of the next. Seams may be emphasized in several ways, however, for decorative effect (Fig. 22-1.5).

Application over horizontal furring strips or backerboard eliminates the need for vertical support every 48″. Now decorative strips may be inserted between panels (Fig. 22-1.6) to add richness to the layout.

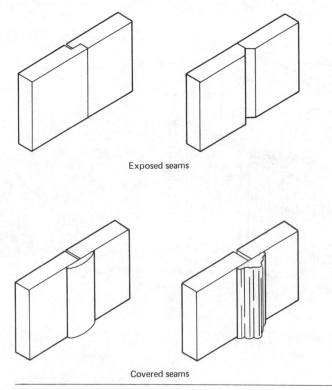

Exposed seams

Covered seams

Fig. 22-1.5 Vertical seams are usually left exposed, but may be finished with decorative battens.

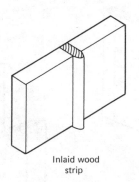

Inlaid wood strip

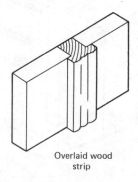

Overlaid wood strip

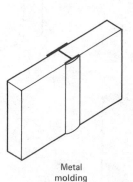

Metal molding

Fig. 22-1.6 Seams may be finished with decorative inserts when square-edged panels are supported on backerboard or furring strips.

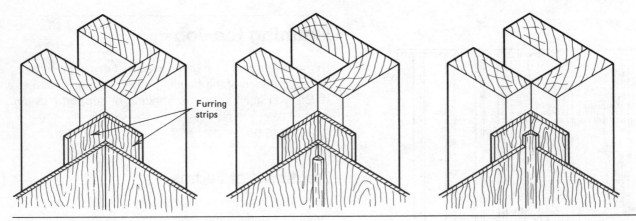

Fig. 22-1.7 Three ways to finish an inside corner.

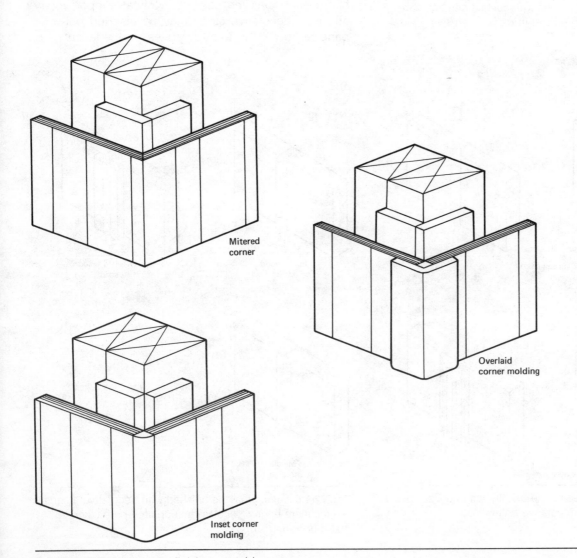

Mitered corner

Overlaid corner molding

Inset corner molding

Fig. 22-1.8 Three ways to finish an outside corner.

Treatment at Corners

At inside corners, panels may butt together if they will fit without trimming or scribing. When an exact butt fit isn't possible, gap the panels slightly and cover the gap with a quarter-round or small-cove molding (Fig. 22-1.7). Another solution is to fit a $\frac{3}{4}''$ molding into the corner, then butt the panels against it. This method puts the edges of panels on the center lines of studs.

Outside corners are more likely to be bumped than inside corners and are therefore more subject to damage. For this reason mitering is not recommended. Better answers are an inset corner molding or an overlaid corner molding (Fig. 22-1.8). Spacing of studs may be the most important factor in selecting the best solution.

Layout

Wherever possible, use full-width panels so that you are working with as many factory-finished edges as possible. If panels must be cut, use $\frac{1}{3}$, $\frac{1}{2}$, or $\frac{2}{3}$ panels and know in advance where the other cutoff piece can be economically used. At openings it is usually quicker, more economical, and more attractive to apply full panels, then rout off the excess. The offal makes good furring strips in other rooms.

There is always some difference in color and graining between panels. For the best appearance, then, stand each panel in place against the wall so that you can see the overall effect. When you have achieved the required beauty, number the panels on the back and also mark which end is up.

Handling

Being thin, plywood panels will take on a bow very quickly. To prevent this, store cartons of plywood flat in a dry place as soon as they arrive at the site. On the floor out of traffic is a good location. If panels must be raised to keep them dry, support them on spacers under the ends and every 24" between.

At least 24 hours before you intend to begin paneling, take the plywood from the cartons so that panels can adjust to the moisture content of the room.

Panels are packed between sheets of protective paper. Remove as many panels as there are between sheets at one time. Then lift each panel off the one below; don't ever slide them, or you will mar the finish. After you have numbered the panels, lay them flat again in the order in which you will apply them.

Procedure for Application Directly to Studs

1 Check specifications to see what special treatment, if any, is called for at joints between panels.
2 Establish the layout of panels for each wall.
3 Set the first panel in position in the most prominent corner of the room and check it for plumb. Scribe the edge if necessary.
4 Make sure the opposite edge is centered on a stud.
5 Precut any holes for switches or electrical outlets.
6 Install the first panel, making sure the outside edge is plumb.
7 Install the next panel, gapping about the thickness of the blade of a putty knife.
8 Precut panels to fit at openings if panels are applied with adhesive or contact cement. **Note.** If the offal is less than a reusable third of a panel, and panels are nailed, attach a full panel and nail at the opening. Then cut the panel to fit with a router or coping saw.
9 Repeat steps 4, 5, 7, and 8 to complete the first wall.
10 Repeat steps 1 to 9 to complete the room.

Procedure for Application over Backerboard

1 Determine any special treatment at joints.
2 Establish the layout of the panels on each wall.
3 Check the fit of the corner panel.
4 Install panels as described in steps 5 to 10 for stud application.

Procedure for Application to Furring Strips

1 Establish the layout of the panels on each wall.
2 Install a horizontal furring strip just above floor level.
3 Install a horizontal furring strip just below ceiling level.

4 Install vertical furring strips on studs that support seams.

5 Furr at openings.

6 Repeat steps 3 to 10 for stud application.

Questions

1 What must be done to protect plywood panels installed over a masonry wall?

2 Why should the gap between panels and floor be smaller than the gap between panels and ceiling?

3 Describe where and how plywood panels should be nailed.

4 Why should panels be stored flat?

5 What precautions should be observed in handling prefinished plywood?

6 How much of a gap should be left between panels? Why?

Unit 22-2   **Install Hardboard**

The general procedures for applying hardboard are quite similar to those for plywood paneling, but the materials themselves are somewhat different and must be treated differently.

Hardboard expands and contracts more in the presence of moisture. Therefore it must be protected by a vapor barrier—on the back of insulation, on the back of backerboard, or waterproofing over masonry walls. Hardboard is a little more brittle and must be tightly fastened at all edges; it is also more flexible and can be easily bent to follow a curve.

Decorative Hardboard

Decorative hardboard may be applied directly to studs, over backerboard, or to furring strips. It may be attached with colored nails, finishing nails, or adhesive.

With Nails

Nails are required every 4″ along all edges and 8″ apart at all other points of support. If colored 3-penny nails are used, drive them with a plastic-headed hammer flush with the hardboard surface. If finishing nails are used, predrill holes, countersink the heads, and fill the holes with matching putty stick.

With Adhesive

The strongest application, recommended by hardboard manufacturers, is over furring strips with adhesive. The strips are applied horizontally over studs behind the top and bottom of each panel and every 16″ between. Over masonry walls vertical strips are applied first to support all edges, then shorter horizontal strips are added on 16″ spacing. Panels must gap at least $\frac{1}{4}$″ at the floor.

Perforated Hardboard

Perforated hardboard is sometimes called Pegboard, which is a trade name. Its main use is as a hanging wall, although it is also practical for sliding doors to cabinets that need ventilation. The basic hardboard is punctured with small holes (either $\frac{3}{16}$″ or $\frac{9}{32}$″ in diameter, depending on the manufacturer). More than 60 different metal fixtures (Fig. 22-2.1) are made to fit into the holes for hanging anything from shelves to bottles and from coats to screwdrivers.

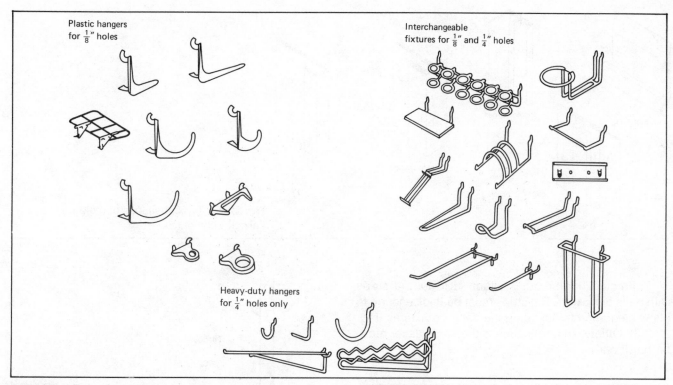

Fig. 22-2.1 These are just a few of the many fixtures for use with perforated hardboard walls.

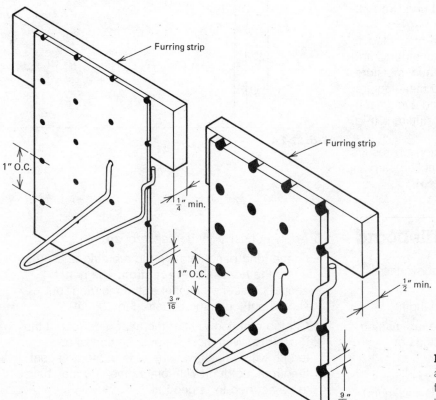

Fig. 22-2.2 Perforated hardboard used as a hanging wall must be applied over furring strips thick enough to permit insertion of hanging fixtures.

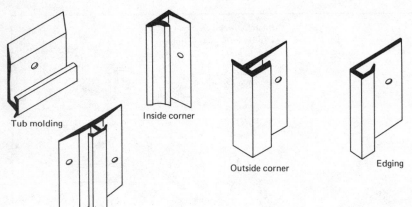

Tub molding

Inside corner

Outside corner

Edging

Divider

Fig. 22-2.3 Typical shapes of moldings for finishing tileboard.

Perforated hardboard is attached to furring strips with nails or screws. The strips must be thick enough to allow clearance for inserting the hanging fixtures (Fig. 22-2.2). Otherwise application is the same as for regular hardboard.

Tileboard

Tileboard is made in 4 × 8′ sheets, but it also comes in 7′ lengths that are wider—54″ and 60″—to match the sizes of standard bathtubs. Two sheets cover a tub recess—one full sheet on the long wall and two half-sheets on the shorter wall.

Although the surface of tileboard is enameled to shed water, edges must be protected by moldings and caulking to prevent damage by absorbed moisture. The moldings are made by the tileboard manufacturer of plastic or chrome-plated aluminum, and each manufacturer's molding shapes are slightly different (Fig. 22-2.3).

Tileboard may be nailed, but a better job results when it is applied with adhesive over backerboard.

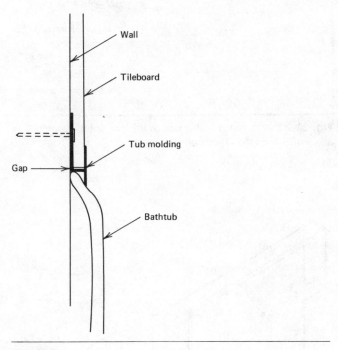

Wall

Tileboard

Tub molding

Gap

Bathtub

Fig. 22-2.4

Procedure for Installing Tileboard

1 Measure the length of the wall just above the longest edge of the bathtub.

2 Cut a tub molding or edging to this length.

3 Install the molding over the lip of the tub, nailing through all holes in the flange into studs or blocking.

4 Cut the panel of tileboard to width if necessary. **Precaution.** Allow for a slight gap ($\frac{1}{8}$″ maximum) at the corners for expansion.

5 Apply adhesive to the backerboard.

6 Fit the long panel into the tub molding and press it into the adhesive. **Precaution.** The panel should close the channel in the molding, but should not touch the bottom (Fig. 22-2.4).

7 Cut tub molding to fit at the head and foot of the tub. **Note.** With most shapes the end of one length will fit flush with the side of the adjacent length of molding. With others the end may have to be scribed for a good fit.

8 Install the short tub moldings.

9 Cut and nail the two inside corner moldings.

10 Determine the correct molding treatment at the outside edges of end panels. **Note.** These panels may end in an outside corner, a divider, or an edge molding (Fig. 22-2.5).

11 Allowing for the molding in step 10, cut end panels of tileboard to size.

12 Cut holes in one end panel to fit over inlet piping and valve controls, and check the fit. **Precaution.** Holes should be no more than $\frac{1}{8}''$ greater in diameter than the plumbing projections.

13 Apply adhesive to the backerboard.

14 a. If the outside edge of either panel fits into an edging or divider molding, install the molding. Then install the panel, fitting it between the vertical moldings and sliding it down into the tub molding.

14 b. If the outside edge of either panel fits into an outside corner molding, install the panel first, then add the molding.

15 When all panels are installed, caulk all joints thoroughly, following the recommendations of the tileboard manufacturer.

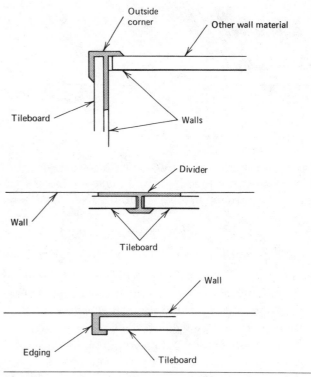

Fig. 22-2.5

Questions

1 Name three important differences between plywood and hardboard paneling.

2 What is the recommended method of application of decorative hardboard?

3 Why isn't this same method recommended for perforated hardboard?

4 Name four typical moldings around tileboard and state whether they are normally installed before or after the panel.

Unit 22-3   **Install Wood Paneling**

Solid wood paneling is much simpler in shape today than years ago when its use was much more common. Most lumber dealers carry two or three shapes in stock, of which those shown in Fig. 22-3.1 are typical. Boards may be applied horizontally, although vertical application is much more common.

work in that corner. The last board to be installed is the most difficult and, if possible, that board should be above a doorway or window where it is short, easiest to fit, and least noticeable.

Planning the Job

As experienced carpenters enter a room to be paneled, they look for a door adjacent to another wall and start

Preparation

Wood shrinks; it is easily dented; and it absorbs grease and moisture. Therefore all pieces should be fully seasoned when they are delivered, and they

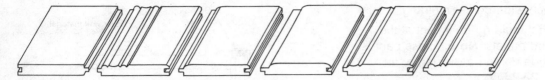

Fig. 22-3.1 Popular patterns for solid wood paneling.

should be stored under cover to protect them from damage and moisture. If the wood is to receive a stain or natural finish, all pieces should be of the same species.

When you are ready to begin paneling, take these precautions:

- If boards are to be applied to the inside of an exterior wall or to furring strips over a concrete or masonry wall, paint the back sides with a good primer-sealer or aluminum paint.

- Inspect each board before you put it in place, looking for rough spots. Hand-sand them if possible, sanding with the grain. If hand-sanding isn't enough, use a belt sander. A drum sander will leave scratch marks.

- If the boards have strong figuring or a lot of pattern, such as knots, sort them in advance so that the pattern is evenly spread around the wall or room, not concentrated in one place.

- If some boards have a variation in tone, such as both light and dark, when other boards are all light, set them aside for cutting into shorter pieces at openings.

- Work in a clean room with clean hands. Soil marks are almost impossible to remove and will show through stain.

- Keep pencil marks thin and light.

- Apply a first coat of stain or paint before you put up any piece of paneling that is to be finished. This helps to seal edges and prevent them from shrinking, and it protects the wood from soiling. If wood is to be left natural, use fine oils made specifically to seal and protect the surface.

When boards are applied vertically, you must provide horizontal support. One way is with 1×2 and 1×4 furring strips (see Fig. 22-1.1), but this method adds $1\frac{1}{2}''$ to total wall thickness and means deeper frames at window and door openings. The three methods shown in Fig. 22-3.2 are more common. Use 2×4 blocks (A) when both sides of the partition are to be paneled. This is the best method because the blocking also serves as a firestop. When panels go on only one side, you

can use 2×2 blocking in line (B), but it is difficult to install. Blocking is easier to install if you use 2×4s laid flat and offset $1\frac{1}{2}''$ (C) to maintain horizontal nailing lines 16'' o.c.

When boards are applied horizontally, they may be nailed directly to studs. Lengths must be cut so that they butt on the center lines of studs, however. Boards are applied from the bottom of the wall upward, and spacing should be worked out in advance so that the top board is the same width as the others. If any board is narrower, it should be the bottom board, which is the least noticeable. No blocking is required behind the boards, but in good construction a row of firestops is installed (see A in Fig. 22-3.2).

At Corners

At inside corners, vertical boards may butt without molding as long as the corner is square and the edges of both boards are plumb. The usual procedure, however, is to trim the tongue from board A, and let it overlap board B (Fig. 22-3.3, left). There should be a little gap between boards A and B to allow for expansion.

Another procedure is to install a $\frac{3}{4}'' \times 1''$ block in the corner (Fig. 22-3.3, right). Stop either vertically or horizontally applied boards short of the block and cover the gaps with a corner molding nailed into the block. With this method the boards can expand freely without any seams becoming visible.

At outside corners, edges of boards may be mitered or else gapped and covered with corner molding (Fig. 22-3.4). If you miter the edges, cut at slightly more than 45° so that there is contact only at the corner. Rounding off the corner not only helps appearance but reduces the possibility of damage or injury from bumping.

Attachment

The edges of all solid wood panels are designed for **blind-nailing** (concealed nailing). Drive 6-penny or 8-penny finishing nails at about a 45° angle at the inner edge of the tongue (Fig. 22-3.5), then countersink them below the surface with a nailset before the next board

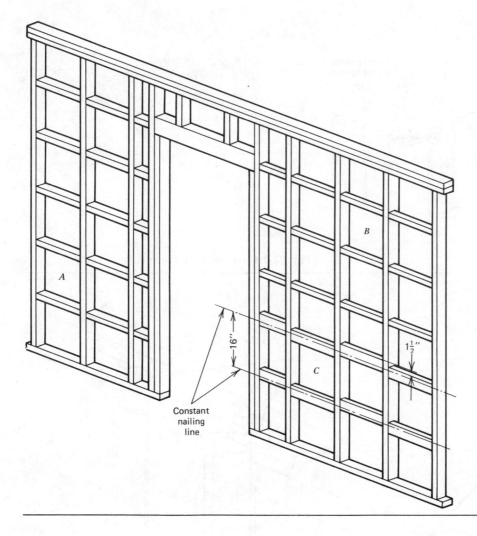

Fig. 22-3.2 Three ways to block behind vertical wood panels.

is installed. Face-nail at top and bottom behind future trim. If you must face-nail in the field to correct a slight warp, nail into the grain or near a knot where the fastener is least noticeable.

Expensive hardwood panels are sometimes applied with special clips that eliminate any possibility of

nails showing (Fig. 22-3.6). The clips fit over the sides of grooves in one panel, and the next panel slides in to conceal the clip and hold the tongue end firmly. To emphasize joints, clips may be used in pairs to support a fillet strip. To assure a smooth wall the clips should be nailed into S2S furring strips rather than blocking.

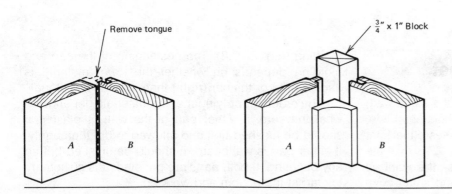

Fig. 22-3.3 Two ways to finish inside corners.

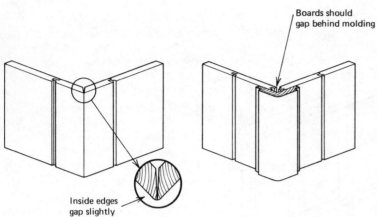

Fig. 22-3.4 Two ways to finish outside corners.

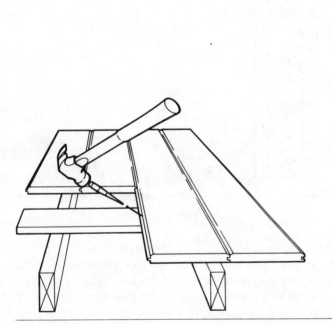

Fig. 22-3.6 Wood panels may be attached with concealed clips instead of nails. The decorative fillet strip is optional.

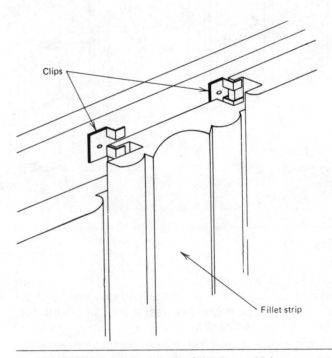

Fig. 22-3.7 When two materials of the same thickness meet, the joint may be covered, filled, or recessed.

Where Two Materials Meet

Many times, two materials are used on the same wall, with the materials usually meeting at a horizontal line. If the two materials are the same thickness, the joint may be covered, filled, or exposed (Fig. 22-3.7). If the materials are of different thicknesses, the joint is usually finished with a wainscot cap and small cove molding (Fig. 22-3.8). The positioning of the cap and molding depends on whether the lower material is thicker (left) or thinner (right) than the upper material.

The order of applying the two materials depends on what they are. If either part of the wall is plaster, it should be finished first and allowed to dry thoroughly. If either part is wallboard, it should be finished before any paneling. If both parts are paneled, it is customary to start at the bottom and work up.

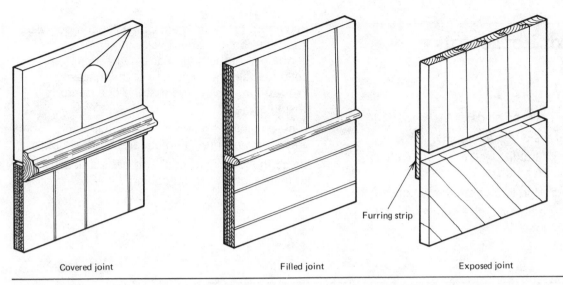

Covered joint   Filled joint   Exposed joint

Furring strip

Fig. 22-3.5 How to blind-nail solid wood panels.

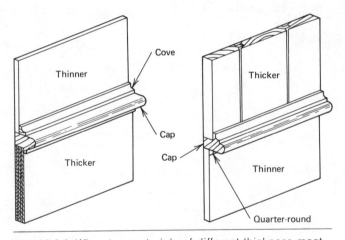

Thinner

Cove

Cap

Thicker

Thicker

Cap

Thinner

Quarter-round

Fig. 22-3.8 When two materials of different thickness meet, the best treatment is to fill the seam with a wainscot cap and add a cove against the thinner material.

Procedure for Vertical Application

1 Prime, sand, and sort boards as described in the introduction of this unit.

2 Install blocking between studs or furring strips over studs.

3 Install blocks at inside corners, if required.

4 Cut boards to length, if necessary, and prime the

cut ends. **Note.** Boards should gap about $\frac{1}{4}''$ at the floor. If walls are standard height and boards are 8′ long, gap $\frac{1}{2}''$ at the floor.

5 In the corner of the room selected as a starting point, install the first board with its groove facing the corner (board *B* in Fig. 22-3.3). Face-nail in the corner so that nails just clear the groove, and blind-nail at the tongue. **Precaution.** Check plumb at the tongued edge constantly as you nail.

6 Install full-width boards to the first opening, checking plumb frequently and blind-nailing. **Precaution.** Make sure that all joints are uniform in appearance—that is, all tongues fit into all grooves the same distances.

7 Lightly mark cuts for jambs, and make the cuts before installation.

8 Mark and cut holes for electrical boxes. Bore $\frac{1}{2}''$ holes at the ear locations of boxes and cut holes for the boxes with a saber saw. **Precaution.** Cut the holes $\frac{1}{8}''$ larger than the box to allow for installation of the boards.

9 Repeat steps 6, 7, and 8 to the next corner.

10 a. If the next corner is an inside corner, trim the tongue from the corner board for a slightly gapped fit, as shown in Fig. 22-3.3.

10 b. If the next corner is an outside corner, trim or miter the edge as shown in Fig. 22-3.4.

11 Repeat these procedures as required until the room is finished except for the last panel.

12 Cut the last board to length, then trim it to width with a saber saw, beveling the edge slightly for a tight wedge fit.

Procedure for Horizontal Application

1 Prime, sand, and sort boards.

2 Install firestops between studs.

3 Install blocks at inside corners.

4 Measure the height of the wall to determine the width of the bottom board.

5 Rip the grooved edge to the required width.

6 Cut boards to length so that their ends fall at the center lines of studs.

7 Beginning on the longest wall without a doorway, install the bottom board with its tongue up, gapping $\frac{1}{4}''$ at the floor. Face-nail $\frac{3}{8}''$ from the bottom edge and blind-nail at the tongue. **Precaution.** Make sure that the first board is absolutely level.

8 Trim remaining boards to fit at corners and openings, and precut holes for electrical boxes.

9 Install the top board so that it gaps at least $\frac{1}{2}''$ but no more than $1\frac{1}{4}''$ at the ceiling.

Questions

1 Why is it so important to work with clean hands when installing solid wood paneling?

2 Describe the possible methods for smoothing rough surfaces.

3 Name three ways of blocking for vertical installation and an advantage of each method.

4 Describe the procedure for mitering at an outside corner.

23

Acoustical Ceilings

Aside from finishing the overhead structure in a room, ceiling materials sometimes serve two other important purposes. They may control unwanted sound, and they may provide light.

A wallboard or plaster ceiling painted a light color does a good job of reflecting light from windows and lamps into the inner corners of rooms. Most gypsum surfaces, however, absorb little sound. Acoustical plaster does a good job of controlling noise, but it is more practical when applied on large ceiling areas of office and commercial buildings than in residences. Textured paints are sometimes applied to ceilings, but their effect is more decorative than acoustical.

To control sound in rooms that are the scene of noisy activity, such as game rooms or tool shops, acoustical ceiling tiles do the job well. There are two types—fiber tiles and mineral tiles. Fiber tiles are made of wood or vegetable fibers pressed into large sheets $\frac{1}{2}$ to $\frac{3}{4}''$ thick. Next, sheets are perforated with small holes or fissures and given a factory finish. Then they are cut into individual tiles. They come in squares of 12″, 16″, and 24″, and in rectangles 12 × 24″, 16 × 32″, and 24 × 48″. Their edges may be square, beveled, tongue-and-groove, or slotted.

Mineral tiles are made of rock that is heated to a molten state, then is sprayed into sheets almost an inch thick. In its raw state this material is a chocolate brown on the surface. Trimmed to final thickness, however, tiles are almost white, with fine holes or fissures in the surface. Unlike fiber tiles, mineral tiles do not burn.

Application

Ceiling tiles may be installed in four ways:

1 With staples to furring strips. If mineral tiles are used, the furring strips are nailed across ceiling

joists. If fiber tiles are used, furring strips must be applied over a wallboard or plaster ceiling that prevents the spread of fire.

2 With staples to a finished ceiling material, such as gypsum wallboard or backerboard.

3 With adhesive to a finished ceiling surface.

4 Laid in metal tees suspended from the ceiling structure. A suspended ceiling lowers the ceiling height of a room about 6″, and it is therefore more common in commercial buildings than in new homes.

The first method is the most common in new construction. The next two methods are more common in remodeling of newer homes, and the last method most common in remodeling of older homes with high ceilings.

Tile Patterns

On most ceilings the tiles form a series of squares, and the lines run straight from wall to wall. Each tile is just like its neighbors, and the effect is a uniform ceiling.

Specifications may call for patterns, however. Many types of mineral tiles, for example, have decorative surfaces, some with directional patterns. These are made with tongues and grooves in two positions, so that the tiles can be applied in alternate directions. Any type of square tile may also be laid in a pattern having offset joints with only a slight variation in standard procedures. Or square and rectangular tiles may be used together to create a pattern.

Questions

1 What are the six standard sizes of acoustical tiles?
2 What is the main difference between fiber and mineral tiles?
3 Name the four ways in which ceiling tiles are installed.

Unit 23-1   Apply Tiles to Furring Strips

Whether they are applied across ceiling joists or over a ceiling surface, the 1 × 3s used for furring strips must be flat and straight. If the structure to which they are nailed is uneven, the strips must be shimmed so that the finished acoustical ceiling is level. If the strips aren't straight, attachment with staples is as difficult as applying wallboard to warped studs.

Preparation

Before furring strips can be applied, the carpenter must make a layout of the ceiling. For the best-looking job, three important rules must be followed in making the layout:

1 Any cut tiles must go at the edge of the ceiling, where they are covered by trim.
2 No cut tile should be less than half a tile in width.

Small tiles don't look well and are difficult to install properly.

3 The cut tiles on opposite edges of the ceiling should be equal in width.

To illustrate how to apply these rules, assume that specifications call for 12″ square tiles on the ceiling of a room measuring 12′-2″ by 11′-6″. You could cover this area with 12 full tiles and 1 cut tile in one direction and 11 full tiles and 1 cut tile in the other direction. But the 13th tile in one direction would be less than half a tile, not permitted by rule 2. And while the 12th tile in the other direction would be half a tile, the 1st and 12th tiles would be different widths, not permitted by rule 3.

The answer is to use 11 full tiles and 2 cut tiles over the longer direction, with each of the cut tiles 7″ wide (Fig. 23-1.1). Then use 10 full tiles and two 9″ tiles over the shorter dimension. In both cases the cut tiles go at the edge of the ceiling, and the layout meets all three rules.

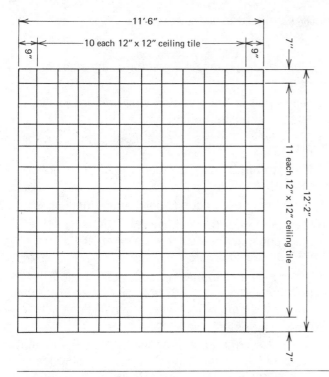

Fig. 23-1.1 A typical ceiling layout. Note that only edge tiles are not full-sized, and that tiles against opposite walls are the same width.

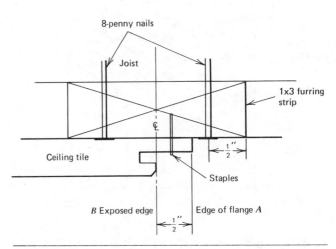

Fig. 23-1.2 The stapling flanges of tongue-and-groove tiles extend about ½″ beyond two edges of the exposed surface. Nails that attach furring strips should be placed near their edges away from the stapling line. Guide lines for aligning tiles are placed either at the edges of flanges (A) or the edges of the finished surface (B).

With the ceiling layout established, you then know where to place furring strips. They should be centered under the exposed seams between tiles (Fig. 23-1.2). Edge furring strips, which can be 1 × 2s, lie against finished walls.

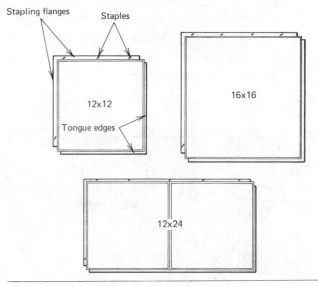

Fig. 23-1.3 The recommended locations for staples depend on the size of the tile. Staples may be driven parallel to the edges of flanges or at a slight angle.

Application

Install furring strips with two 8-penny nails at each joist. You can avoid hitting them when you staple if you nail near the edges of the strips—about ½″ in from each side (see Fig. 23-1.2).

Use tiles with tongue-and-groove edges and staple into the flanges according to the manufacturer's instructions. Use 3 staples in a 12″ flange, 4 staples in a 16″ flange, and 5 staples in a 24″ flange. Secure the other flange with a single staple (Fig. 23-1.3). It is easiest to drive the staples parallel to the edge of the tiles, but staples hold better when they are driven at a slight angle.

One type of stapling gun drives two staples at one time. The bottom staple spreads the top staple for a firm grip. Double-stapling is recommended when tiles are attached directly to a gypsum ceiling.

Cutting Tiles

Both fiber and mineral tiles are easy to damage, and they must be cut carefully and handled with clean hands. Mark cuts on the face of the tile with a sharp util-

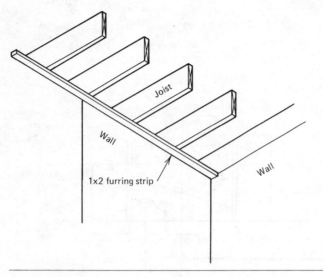

Fig. 23-1.4

ity knife and make the cuts with that knife or a coping saw with the tile face up.

Procedure for Installing Furring Strips

1 Determine the size and shape of the ceiling area to be covered by measuring accurately the length of each wall at ceiling level and checking corners for square.

2 Make a ceiling layout to establish the sizes of edge tiles.

3 Nail a pair of 1 × 2 furring strips across joists against opposite walls (Fig. 23-1.4).

4 From the walls, measure out the width of an edge tile established in step 2 and make marks. **Precaution.** Be sure to select the correct width of the two (A in Fig. 23-1.5).

5 Snap chalk lines across joists at the marks (B).

6 Measure the distance between chalk lines (C). It must come out in even feet (or be divisible by 16″ if tiles are that size).

7 When the measurement checks out, measure $1\frac{1}{4}″$ (half the width of a 1 × 3) from one chalk line toward the wall. Mark this location (D) on joists at opposite sides of the room.

8 Nail a 1 × 3 furring strip on the marks (D) so that it is centered on chalk line B. Check the strip for level and shim if necessary. **Note.** Ends of strips do not need to be supported; ceiling trim supports edge tiles.

9 With a steel tape, measure from the edge of the furring strip in step 8 every 12″ (or whatever the tile's width) and mark these locations on two joists (Fig. 23-1.6).

10 Install the remaining furring strips with their edges on these marks and check them for level. **Note.** The center of the last furring strip should fall on the center of the other chalk line snapped in step 5.

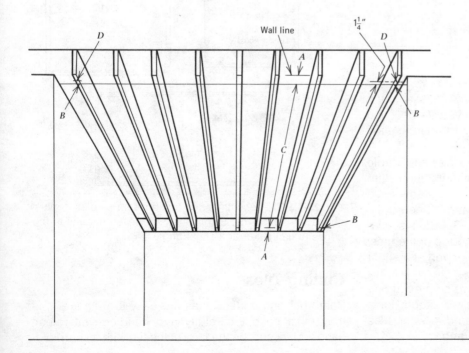

Fig. 23-1.5

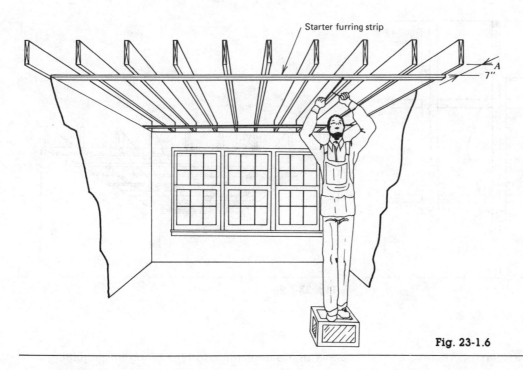

Starter furring strip

A

7"

Fig. 23-1.6

11 a. To mark the locations of the finished edges of tiles, snap chalk lines down the centers of furring strips.

11 b. To mark the locations of the edges of tile flanges, snap chalk lines $\frac{1}{2}$" off center away from the first wall (see Fig. 23-1.2).

12 From the remaining two walls, measure out the width of the other edge tiles established in step 2 and snap chalk lines (Fig. 23-1.7).

13 Measure the distance between chalk lines to make sure it comes out in even feet, and also measure the diagonals to make sure guide lines are at right angles to each other.

14 When the measurements check out, snap additional chalk lines across furring strips at distances equal to the width of a tile. **Note.** If chalk lines are to mark the edges of flanges, as in step 11b, snap a chalk line $\frac{1}{2}$" away from the chalk

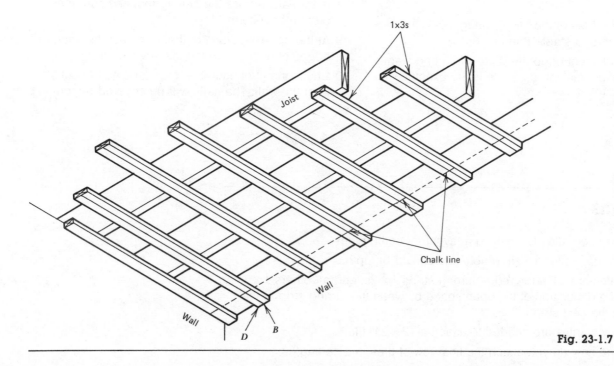

1x3s

Joist

Chalk line

Wall

Wall

D   B

Fig. 23-1.7

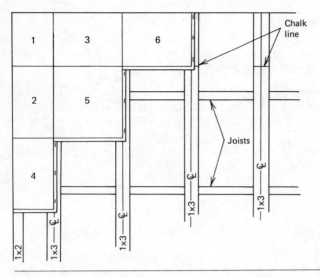

Fig. 23-1.8

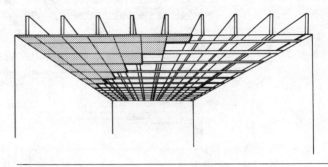

Fig. 23-1.9

line in step 12 before taking step 14. Then measure off the distances from the nearest chalk line.

Procedure for Stapling Tile to Furring Strips

1 Cut the first tile to fit in a corner. **Precaution.** Cut off the tongued sides, leaving the flanged side for stapling.

2 Align the edges of the tile (*1* in Fig. 23-1.8) on the chalk lines and staple it in place.

3 Cut the adjacent edge tile (*2*) to width and staple it in position, making sure it lines up exactly with the first tile.

4 Cut the adjacent edge tile along the adjoining wall (*3*) and staple it in position.

5 Repeat step 3 (tile *4* in Fig. 23-1.8).

6 Fit the first full tile in position (*5*). **Precaution.** Slide each pair of tongues into matching grooves gently but firmly for a snug fit. With too much pressure you can damage the tiles.

7 Repeat step 4 (tile *6*).

8 Work diagonally across the room, as in Fig. 23-1.9. **Precaution.** After stapling every two or three tiles, sight down the chalk lines and diagonally across the ceiling to make sure the pattern is uniform.

9 At the opposite sides of the room, cut the flanged edges of each tile in the last rows to fit.

10 Check the fit of each tile, then face nail about $\frac{1}{2}''$ from the wall. The nails will be covered by trim.

Questions

1 Name the three rules for making a layout for ceiling tiles.

2 When you cut a tile to size, which side should be up?

3 If edge tiles are $8\frac{3}{4}''$ wide, edge furring strips 1 × 2s, and all others 1 × 3s, what is the dimension of the open space between the furring strip at the wall and the next strip in?

4 How many staples are needed to attach a 12 × 24" tile?

5 Where and how do you attach the final row of tiles in a ceiling?

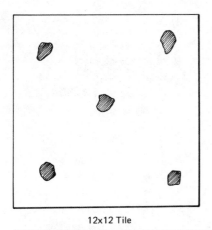

12x12 Tile

Fig. 23-2.1 Where to apply mastic to the back of a square ceiling tile. Use eight dabs on a rectangular tile by repeating the pattern.

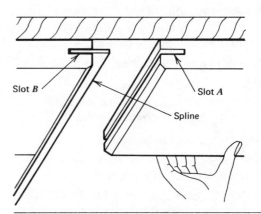

Slot *B*   Slot *A*

Spline

Fig. 23-2.2 Splines that hold slotted tiles at the same level are supple. They may be installed in the slots of a completed row and the next row fitted over them; or they may be threaded into pairs of slots from one end after both rows are completed except for the end tile.

Unit 23-2   **Apply Tiles to Surfaces**

Over an existing surface of plaster, wallboard, or plywood, flanged ceiling tiles may be installed in the same general manner as described in Unit 23-1. Tiles with square or beveled edges are applied with mastic. The rules for layout are the same, and tiles are cut as previously described.

Working with Mastic

Mastic, also called ceiling cement, is applied to the back of each tile with a brush or putty knife. Manufacturers recommend five small dabs of equal size and thickness, one about 2″ from each corner and one in the center in a pattern like the 5 on a dice cube (Fig. 23-2.1). When ceiling tiles are applied to furring strips, the ceiling will be level as long as the furring strips are level. When tiles are applied with mastic, you must constantly check the level of the tiles themselves. You can compensate for variations in the level of an existing surface by increasing or decreasing the size of the dabs of cement.

To set a tile, hold it close to its final location, with the mastic just touching the ceiling surface. Then slide it into position with a slight sideways motion, at the same time pressing against the ceiling surface to assure a bond. The edges of one tile should just touch the edges of adjacent tiles. Lay your carpenter's level across pairs of tiles in each direction to make sure your work is level. If a tile is too low, push straight upward to spread the mastic a little more and bring it into level. If a tile is too high, you still have time to remove it carefully and add a little more mastic.

Some bevel-edged tiles are slotted on opposite edges and are kept in alignment with plastic splines (Fig. 23-2.2). Most manufacturers recommend applying tiles in a straight line across a room, then setting the long splines in the slots before installing the next row of tiles with mastic. In the final corner of the room the easiest method of installation is to cut the last two edge tiles a little narrower than the measured dimension. Then preassemble the remaining four tiles with a spline (Fig. 23-2.3) and place the assembly in position in one operation. Nail at the wall.

Procedure

1 Determine the size and shape of the ceiling area to be covered by measuring accurately and checking corners for square.

2 Make a ceiling layout to establish the sizes of edge tiles.

3 From all four walls measure out the width of edge tiles from step 2 and make marks.

4 Snap chalk lines through the marks.

5 Measure the distance between pairs of chalk lines to make sure it will accept even tiles.

6 Decide in which corner you intend to start. Then, from the chalk line nearest the opposite short wall (*A* in Fig. 23-2.4), measure off a distance (*B*) equal to the difference in the number of full tiles in the two directions. **Example.** If the room requires 12 full tiles in one direction and 10 in the other, and tiles are 12″ square, measure off 24″ from the chalk line.

7 Snap a chalk line (*C*) at this point parallel to the other chalk lines. You now have a square outlined by chalk lines.

8 Snap diagonal chalk lines (*D*) through the square.

9 Install the first three edge tiles and the first full tile as described in steps 1 through 5 for stapling tiles to furring strips. **Precaution.** Check each tile for level in both directions before you dab mastic on the next one.

10 Work diagonally across the room. If you are accurate in your work, the corners of tiles should fall exactly on the diagonal chalk lines.

11 At the opposite sides of the room, cut each tile in the last rows to fit. **Note.** If splines are used, preassemble the last four tiles as shown in Fig. 23-2.3.

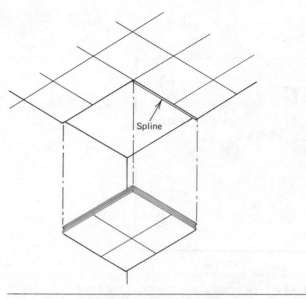

Fig. 23-2.3 Preassemble the last four tiles with a short spline, and apply them as a unit.

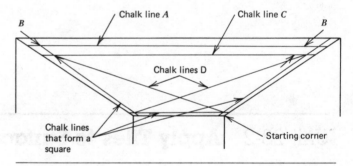

Fig. 23-2.4

Questions

1 What are the shapes of edges of tiles applied with mastic?

2 How many dabs of cement are needed on the back of a 12″ square tile?

3 Describe the correct action for setting a mastic-applied tile in position.

4 What is the main advantage to splines?

5 What is the difference in method of aligning tiles on furring strips and on a ceiling surface?

Unit 23-3 Hang a Suspended Ceiling

Whenever ceiling structure is high enough to permit it, a suspended ceiling offers several advantages. A suspended ceiling consists of a supporting grid of metal tees in which panels of ceiling material are laid. These panels may be the larger sizes of tiles, or they may be sheets of translucent plastic behind which fluorescent lights are installed. Thus a high ceiling can be lowered to give a room in an older house better proportions. And a suspended ceiling in either a new or old house can provide sound control and good overhead light at the same time.

The system of ceiling supports has relatively few parts (Fig. 23-3.1). Angles support the ceiling where it meets walls. Main tees, usually 12' long, are hung between angle supports on either 24" or 48" centers. Cross tees may be 12' long or shorter pieces just long enough to fit between main tees. They are installed at right angles. The method of interlocking main tees and cross tees varies considerably from one manufacturer to another. Some types snap in place, others slip into slots, and others lock with tension pins.

If the new ceiling is suspended from an existing surface, there must be at least 2" in height between the top of the supporting grid and an existing ceiling or exposed joists. You need that much clearance in order to insert ceiling panels in the grid.

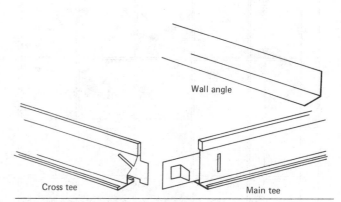

Fig. 23-3.1 Typical parts of a supporting grid for a suspended ceiling.

Preparation

Making a layout is the first order of business. Follow these general rules:

1 As in any acoustical ceiling, the side panels against opposite walls should be the same width and at least half a panel wide.

2 Run the main tees across joists. If the tees are to be attached over an existing ceiling, run them the long dimension of the room.

3 If the size of individual ceiling panels is specified, use their dimensions as a guide to layout. If the size is not specified, try layouts using available widths and lengths to determine which requires the fewest main tees. When a ceiling provides light, make sure the light falls on surfaces where it is needed most.

Application

Many grid members must be cut to length with either a hacksaw or metal snips. If a wall is longer than a length of angle, the two pieces of angle should overlap about an inch. Cross tees between main tees require no cutting; they may need to be shortened where they fit over wall angles, however.

Cutting main tees to length is the only operation that requires some thought. They are slotted or punched every 12" to accept cross tees, and the slots must be positioned according to the ceiling layout. If the cross tee nearest a short wall is 14" from it, for example, the main tee must be cut 14" from the *point of attachment*, not its end. When the main tee consists of two or more pieces, they fit together at a pair of slots and tabs that lock them together. Many carpenters cut one tee to length, then connect the pieces before making the other cut to make sure the length is correct.

Procedure for Laying out the Grid

1 From drawings, determine the height above the floor of the finished ceiling.

2 To this dimension, add the height of a wall angle (usually $\frac{3}{4}''$).

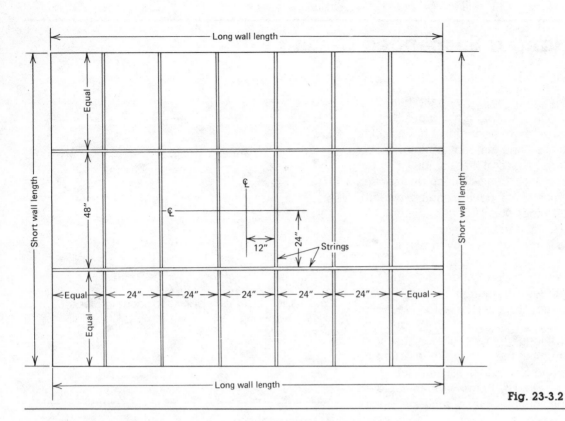

Fig. 23-3.2

Fig. 23-3.3 (Photos courtesy Armstrong Cork Company.)

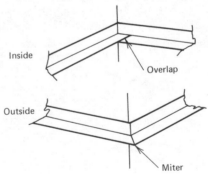

Fig. 23-3.4

3 Mark the height in step 2 on all wall surfaces near the corners of the room.

4 Snap chalk lines through all marks and check them carefully for level.

5 Accurately measure the lengths of all walls and check all corners for square.

6 Determine the size of individual ceiling panels.

7 Make a ceiling layout to establish the locations of main tees and cross tees (Fig. 23-3.2). **Note.** If walls are not exactly parallel, use the center line of the room as the fixed point from which to establish layout.

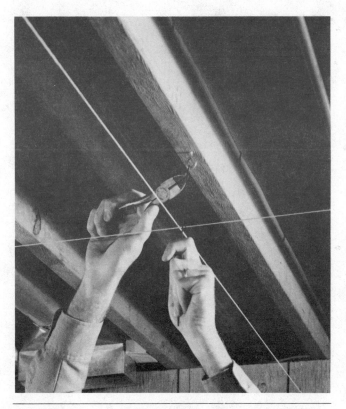

Fig. 23-3.5

Fig. 23-3.6

8 With the layout determined, nail wall angles with their tops on the chalk lines (Fig. 23-3.3).

9 Where wall angles meet, overlap at inside corners and miter at outside corners (Fig. 23-3.4).

10 On the wall angles along the two shorter walls, mark the center line of the room.

11 If the layout calls for main tees on both sides of the center line instead of on it, as in Fig. 23-3.2, measure off half the length of a panel from the center line in both directions and make new marks.

12 Between the marks in step 10 or one set of marks in step 11, stretch a string tight at ceiling level.

13 Repeat steps 10, 11, and 12 along the two longer walls.

14 Check to make sure the two strings cross exactly at right angles. If not, repeat steps 10 through 12 for both strings until they do.

Procedure for Installing the Grid

1 Directly above the longer string, snap a chalk line across the under sides of joists.

2 On the chalk line install eyes for wire into joists 4' o.c. (Fig. 23-3.5).

3 Cut one main tee to length.

4 Lay the tee on the wall angle centered directly over the string and fasten the uncut end to the eye with wire where the two strings cross (Fig. 23-3.6). **Note.** Run the wire through the hole in the tee directly below the eye and attach it temporarily so that the tee is at string level.

5 Tie the main tee at all other supports.

6 Recheck the level and alignment of the main tee.

7 When it is correctly positioned, wire it permanently. **Note.** If necessary to set the main tee level along its entire length, add other fastener wires at low points.

8 Repeat steps 1 to 7 inclusive for all other main tees. **Note.** Relocate the string (from step 12 in the layout procedure) under each new main tee.

9 Cut the first cross tee to length to fit under the second string (from step 13 in the layout procedure). **Note.** If cross tees are not continuous, but are short lengths, fit the first section between main tees, then cut lengths to fit onto the wall angles.

10 Install the first cross tee (Fig. 23-3.7) and recheck it for level and square with main tees.

11 Repeat steps 9 and 10 until the ceiling grid is complete.

Fig. 23-3.7

Fig. 23-3.8

Procedure for Installing Drop-In Panels

1 Measure and cut edge panels one at a time for a good fit. **Note.** If panels are acoustical tiles, cut them face up with a sharp utility knife or coping saw. If panels are translucent plastic, cut them on a circular saw with a combination blade.

2 Install all edge panels first, feeding them through their opening in the grid, and setting them gently on the flanges of the main tees and cross tees (Fig. 23-3.8).

3 Install center panels in the same manner.

Questions

1 How are ceiling panels attached in a suspended ceiling?

2 Why do you think the chalk line guide for attaching wall angles is at the top of the angle instead of the bottom?

3 To cut a main tee to length, from what point do you measure to mark the cut?

4 What is the purpose of stretching strings?

5 If the walls of a room are not parallel, what is the starting point for establishing the best layout?

24

Floor Finishing

Finish flooring materials fall into two general categories—those that are installed with nails, and those that are installed with adhesive, including mortar. Carpenters seldom lay floors of brick, stone, or ceramic tile, although they frequently must build a frame for a masonry floor (see Unit 6-15). The carpenter does install floors made of wood and various types of thin materials in sheet and tile form.

Wood Flooring

Wood flooring comes in three different forms: unfinished strips, prefinished strips, and prefinished blocks. The species used must have hard surfaces that resist wear. Oak meets requirements the best and constitutes about 80% of all wood flooring. Other hardwoods used are maple, birch, beech, walnut, pecan, and hickory. Softwood flooring is less common, but it is made from yellow pine, redwood, cedar, cypress, larch, Douglas fir, and western hemlock where these woods are native.

Unfinished Strips

Strip flooring comes in random lengths from as short as 15″ up to as long as 16′. Hardwoods for homes range in width from 1 to $3\frac{1}{2}''$ and in thickness from $\frac{3}{8}$ to $\frac{33}{32}''$. Softwoods range in width from $2\frac{3}{8}$ to $5\frac{3}{16}''$ and in thickness from $\frac{25}{32}$ to $1\frac{5}{16}''$. These are actual dimensions. The most common size of hardwood for new construction is $\frac{25}{32}''$ thick, with an exposed surface $2\frac{1}{4}''$ wide. Softwood is $\frac{25}{32} \times 4\frac{1}{4}''$. Thinner strips are used primarily for reflooring over existing surfaces.

Although square-sided and square-ended strips are made, the sides and ends of most strip flooring are tongue-and-groove for a tight fit. All edges are cut at right angles. The backs are **plowed** (hollowed out) to reduce cupping. The strips are attached with nails to a subfloor of plywood or wood, or to furring strips. See Unit 24-1.

Prefinished Strips

Although unfinished strips are still the most widely used wood flooring, prefinished strips have grown in popularity among builders because they take less time to install and are ready to use as soon as laid. They come in random lengths in thicknesses from $\frac{5}{16}$" up to $1\frac{1}{32}$". The most common width for standard strips is $2\frac{1}{4}$". Prefinished planks, a popular variation, are wider. Sides and ends may be tongue-and-groove or square. Unlike unfinished flooring, however, the long edges have a slight bevel. All types are applied with nails to a subfloor.

Prefinished Blocks

Of all wood flooring materials, prefinished blocks are the most expensive, but they also make the most interesting floors. Most blocks are 9×9", although other sizes of squares and some rectangular shapes are manufactured. Blocks range in thickness from $\frac{5}{16}$" up to $\frac{13}{16}$". The thinner blocks (less than $\frac{1}{2}$" thick) are laid in a bed of mastic over a subfloor of plywood or concrete. The thicker blocks are nailed like strip flooring. See Unit 24-4.

Resilient Flooring

The term **resilient flooring** applies to a wide variety of materials no more than $\frac{1}{8}$" thick. They come in two forms —in long rolls or individual tiles. Both types must be applied over an **underlayment**—a thin layer of smooth material in sheet form. See Unit 24-5.

Roll Goods

Roll goods is the general term for thin flooring made in rolls. For nearly 100 years the only resilient flooring available was linoleum. **Linoleum** is made of solidified linseed oil mixed with binders and fillers such as wood flour, gums, and ground cork. This jellylike mass is spread on a backing of canvas, felt, or burlap. Linoleum is available in three thicknesses: $\frac{1}{16}$", $\frac{3}{32}$", and $\frac{1}{8}$". Rolls are 6' wide and up to 90' long. The embossed patterns may look like anything from pebbles and tiles to bricks and carpeting. Surfaces receive several coats of protective lacquer.

Vinyl flooring is made in rolls 6' and also 12' wide. The thickness of the sheet varies with the manufacturer and type of backing, which may be asphalt, rubber, or vinyl foam. Vinyls, unlike linoleum, do not absorb grease or stains, and are thinner and easier to handle but not as durable. See Unit 24-7.

Resilient Tiles

Thin tiles for residential floors are made of linoleum, vinyl, vinyl-asbestos, asphalt, rubber, and cork. Most tiles are square; asphalt, linoleum, vinyl, and vinyl-asbestos tiles are all made 9×9", and 12×12". Rubber tiles range from 4×4" up to 18×18", while cork is available only in 6" and 12" squares. Some vinyls come in 4×36" planks that look like hardwood flooring, and rubber tiles are also made in rectangles 9" or 18" wide and 18" or 36" long. See Unit 24-6.

In addition all resilient flooring manufacturers make **feature strips**—long strips of the same basic material. The strips vary from $\frac{1}{4}$ up to 2" in width and usually come in solid colors. They are used to create unique floor patterns.

Questions

1 How do the edges of unfinished and prefinished wood strip flooring differ?

2 Of what species is most strip flooring made?

3 What is the most common size for floor tiles?

4 Name six materials from which resilient tiles are made.

Unit 24-1   **Lay Wood Strip Flooring**

Wood strip flooring, whether unfinished or prefinished, is nailed down. It may be nailed directly into a wood or plywood subfloor, or to furring strips called **sleepers** when the subfloor is concrete.

Before finish flooring is installed, all other work in the room must be completed except for interior trim. The subfloor must be fully prepared—all loose nails driven in tight, all hardened materials such as mortar scraped off, all loose materials swept away, and any high edges in the subfloor smoothed off with a power sander.

Although hardwood flooring is kiln-dried at the mill to a low moisture content (6 to 10%), it will warp, buckle, and crack if not properly cared for during storage at the lumber yard and site. It should be unloaded at the site in dry weather. If it will be applied in warm weather, it should be stored in a dry, well-ventilated room above ground level. If it will be applied in cold weather, the house should be heated to at least 65° F. (18°C) before delivery. Two full days before the beginning of installation, the flooring should be unbundled and spread out, so that it can adjust to temperature and moisture conditions in the house.

Preparation

When the subfloor is clean and level, it should be covered with a layer of building paper, with the strips overlapping 4″ at seams (Fig. 24-1.1). The paper keeps dust, moisture, and cold air away from the back sides of the floor where it can cause warping.

Planning the Job

The layout of the house and the relationship of rooms to be floored have a bearing on the best way to run the flooring and where to start work. The following rules will serve as a guide; they are listed in their order of

Fig. 24-1.1 A layer of building paper, overlapped at seams, is needed between subflooring and any wood flooring that is nailed in place. (Photos courtesy National Oak Flooring Manufacturers' Association, Inc.)

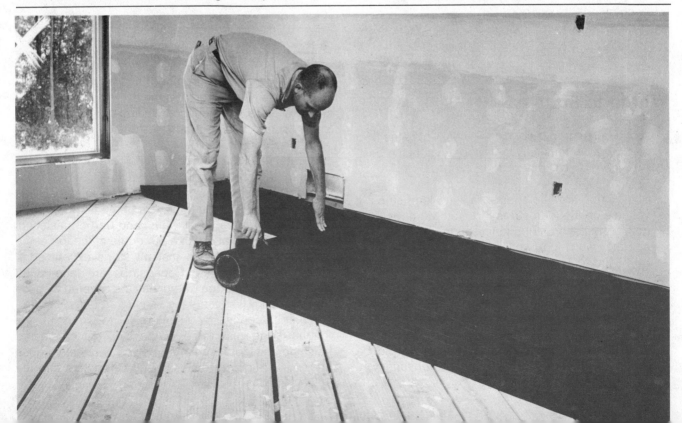

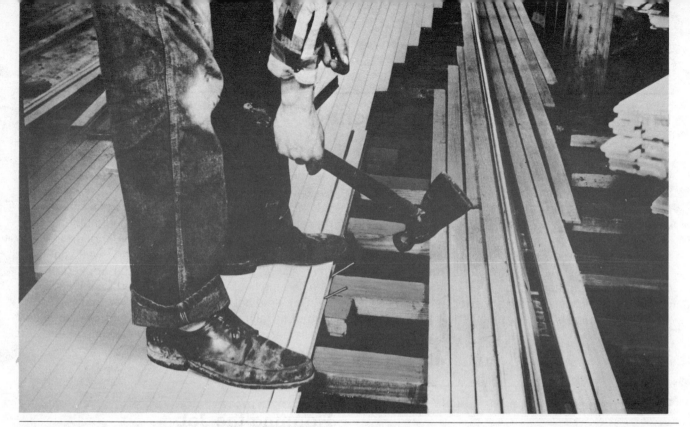

Fig. 24-1.2 For the best-looking floor carpenters sort out strips by length and color, and place a dozen rows in position to see how they look before installing the first row.

When carpenters work in pairs, one usually does the sorting and layout, while the other actually installs the strips.

importance if they conflict with each other, and frequently they will.

1 Run flooring parallel to the long walls of a room.
2 Do not change direction when rooms are open to each other.
3 Lay the strips at right angles to joists.

Where to Start

Most carpenters begin work in the room where the floor is most prominent, usually the living room. In the best-looking job:

- Short strips are interspersed among long strips. If you use up the long strips first, you will have nothing but short strips to finish the last room, and they won't look well.
- Use long pieces at doorways and in heavily traveled areas such as hallways.
- Use the less attractive strips, both long and short, in inconspicuous areas, such as closets.

What these recommendations indicate should be obvious: sort the flooring for length and uniformity of

color before you begin work (Fig. 24-1.2). Carpenters often lay out ten or twelve strips in advance to make sure that butt ends are staggered at least 12″ and that the strips blend well together.

Attachment

When the amount of strip flooring to be laid is limited, the tool used is either a hammer or a hatchet. Four types of nails are commonly hand-driven: cut nails, spiral screw-nails, barbed fasteners, and finishing nails (Fig. 24-1.3). It is best to follow the flooring manufacturer's recommendations. If none are given, use the nail lengths and spacings shown in Table XVI as a guide. The length of the nails and the spacing between them increase with the thickness of the flooring.

Cut nails, spiral screw-nails, and barbed fasteners should be driven at about a 50° angle at the top of the tongue (Fig. 24-1.4), so that all strips except edge strips are blind-nailed. Edge strips are face-nailed near the grooved edge where they are covered by base trim. Sometimes, if there isn't enough room to swing a hammer properly, face-nailing is required at the tongue edge also. Use barbed or cement-coated finishing nails for face-nailing. If you find that tongues are splitting, predrill holes. Many carpenters clip the

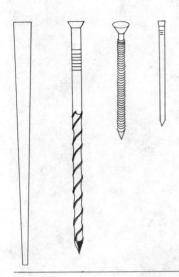

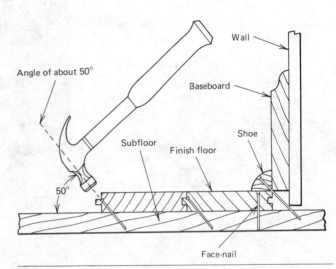

Fig. 24-1.3 The types of nails used for installing wood flooring by hand: (left to right) cut nail, spiral screw-nail, barbed fastener, and finishing nail. The drawing shows them full size.

Fig. 24-1.4 How to blind-nail strip flooring. Note the angle of the nail, and its position just above the shoulder of the tongue.

Table XVI   Nail usage

Flooring thickness	Type of edge	Nail size and type	Spacing
$\frac{25}{32}''$	T & G or square	7- or 8-penny cut or spiral nails; 6-penny barbed fastener; 8-penny coated nail	10–12"
$\frac{1}{2}''$	T & G or square	5-penny cut, spiral, or coated nail; 4-penny barbed fastener	8–10"
$\frac{3}{8}''$	T & G	4-penny cut, spiral, or coated nail; 3-penny barbed fastener	6–8"
$\frac{5}{16}''$	Square	1" cement-coated barbed brad	In pairs every 7"

This tool automatically sets special fasteners at the proper angle. You place the nailing machine in position and strike a plunger with a mallet. The plunger drives the nails. Plates on the bottom of the machine are adjustable to the thickness of the flooring.

Carpenters frequently work in pairs. While one nails the strips in place, the other lays out additional rows, and cuts strips to length.

Cutting

To cut a piece to a shorter length, turn it end for end so that the tongue end will be cut off. Then lay it as close as possible to its final position, and mark the cut (Fig. 24-1.6).

To fit around a door jamb or similar obstruction, place the strip flush against the problem to be cut around. Measure the gap between the finished floor and the loose piece, and measure off this distance on the face of the strip. Mark where the strip must be cut (the width of a doorway for example), and draw the finished shape on the top of the strip (Fig. 24-1.7). Where the strip will be covered by trim, gap about $\frac{1}{2}''$ to allow for expansion. Then cut along the lines. If the required cut is irregular, mark the shape with a scriber and cut with a coping saw along the scribed line.

head off a No. 6 finishing nail with wire cutters, then use the nail as a bit in a power drill. Holes should be bored $\frac{1}{2}''$ from the edge of the grooved side, and spaced to match the spacing of floor joists.

When a large floor area must be covered with strips, carpenters usually use a nailing machine (Fig. 24-1.5).

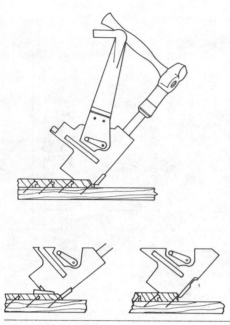

Fig. 24-1.5 A nailing machine is adjustable to the thickness of the flooring (bottom two sketches), and drives nails uniformly at the recommended angle. Note in the photo how the carpenter stands on the strips as he drives them to assure a good fit.

Sleepers

A level wood or plywood subfloor provides an adequate nailing surface for strip flooring. In top quality work sleepers are sometimes nailed directly into joists over the subfloor. The air space between sleepers and subfloor serves three purposes: it acts as an insulator to keep the floor warmer, it reduces the transmission into rooms below of the sound of people walking, and it serves as a **chase** (a small open space) for running electrical conduit.

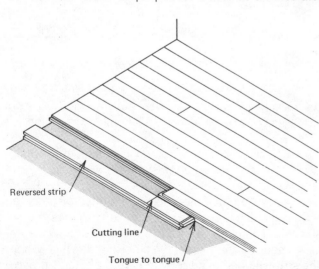

Reversed strip

Cutting line

Tongue to tongue

Fig. 24-1.6 To plan a cut the strip must be marked in reverse. Here the tongue end will be cut off.

Fig. 24-1.7 At a door jamb use a folding rule to mark off the gap between the last nailed strip and the next one. Then notch to fit around the jamb.

Fig. 24-1.8 Sleepers embedded in a coat of mastic over a concrete floor are 2 × 4s laid in a staggered pattern. Maximum center-to-center spacing is 16″. The mastic serves as a moisture barrier.

In Mastic

A wood floor laid over a concrete subfloor must be supported on sleepers, and be protected by a moisture barrier. There are three accepted methods of installing sleepers. One is with mastic, troweled to a thickness of about $\frac{1}{8}$″ (Fig. 24-1.8). The sleepers are 2 × 4s that must be flat even though they don't have to be straight. Short lengths of 18 to 30″ are best to correct any unevenness in the surface of the concrete.

Sleepers must be embedded in the mastic as soon as it has been spread, and strip flooring must be laid before the mastic can set. Otherwise the force of nailing can destroy the bond. Sleepers should be spaced no more than 16″ o.c., and laid in a staggered pattern with ends overlapping at least 4″. Leave a 1″ gap between edge sleepers and walls.

Over Felt

Mastic acts as a moisture barrier. So does a layer of 15-lb felt. With this method a thoroughly dry concrete floor must first be swept clean. Then it is coated with an asphalt primer allowed to dry overnight. The next day a coat of hot asphalt is mopped over the primer, and a layer of 15-lb felt is embedded in the asphalt and mopped in. A second layer of felt is then mopped over the first layer, with butt seams staggered. Sleepers are then power-nailed 12″ on centers over the felt base.

Over a Vapor Barrier

A simpler method that can be completed by the carpenter is to install a double thickness of sleepers with a vapor barrier between (Fig. 24-1.9). There are four steps.

On a clean, dry slab joist lay ribbons of adhesive with a gun 16″ o.c. at right angles to the finished floor (Fig. 24-1.10). Over the ribbons nail 1 × 2 sleepers treated with wood preservative, using $1\frac{1}{2}$″ concrete nails spaced 24″ apart. Cover the first course of sleepers with a polyethylene vapor barrier 0.004″ thick, lapping the edges at sleepers. Nail the second course of sleepers, which don't have to be treated, over the first course with 4-penny nails spaced 16″ o.c. If the sleepers must be leveled, shim under the second course.

Tools

To install sleepers you need either a trowel or adhesive gun and a hammer. To lay strip flooring you need a hammer, hatchet, or nailing machine, folding rule, nailset, portable power saw, string, scriber, crowbar or flatbar, and a broom.

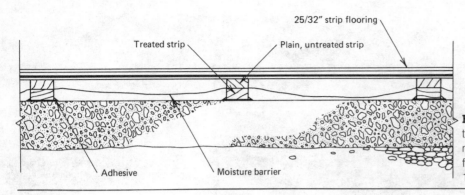

Fig. 24-1.9 Here 1 × 2 sleepers of treated wood are laid in strips of mastic, covered with a polyethylene film, then topped with another set of 1 × 2s, this set untreated.

Procedure for Installing Strip Flooring

1 Determine the direction that strip flooring will run.

2 Establish the best place to start work.

3 Clean the subfloor thoroughly.

4 a. If flooring is to be applied directly to the sub-floor, install a layer of building paper.

4 b. If flooring is to be nailed to sleepers, install the sleepers no more than 16″ o.c.

5 At a point equal to about three widths of flooring stretch a string from end wall to end wall parallel to the first long wall. **Note.** Because walls are seldom straight over their entire length, position

Fig. 24-1.10 The four steps to installing sleepers around a moisture barrier.

Fig. 24-1.11

the first few courses by measuring from the string instead of the wall. The string should be approximately level with the top of flooring strips (Fig. 24-1.11).

6 Place the first strip with its grooved edge toward the long wall and $\frac{1}{2}''$ to $\frac{3}{4}''$ from it, and with its grooved end $\frac{1}{2}''$ to $\frac{3}{4}''$ from the side wall.

7 Check the first piece for parallel by measuring from the string at one end, and face nail that end.

8 Repeat step 7 at every nailing point. **Precaution.** Except at ends against walls, do not nail within 5" of the end of a strip. This makes it easier to fit the next strip in a course onto the tongue.

9 Nail the first strip at the tongue as shown in Figs. 24-1.4 or 24-1.5. **Precaution.** Drive the nail only as far as possible without hitting the floor with your hammer. Most carpenters hold the strip in position by standing on it, and nail from a stooped position. Nails must be driven home with a nailset, laid flat as shown in Fig. 24-1.12.

10 Complete the first row of strips, cutting the tongue end to fit.

11 Fit and nail additional rows of strips, staggering

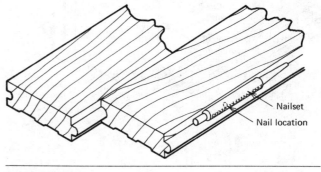

Fig. 24-1.12

joints at least 6". **Precaution.** Do not knock strips into place on tongues with a hammer. Line up the groove on the tongue, place a scrap block over the tongue of the length to be installed, and pound on the block to drive the strip tight against the previously laid strip (Fig. 24-1.13).

12 At the opposite wall, where space doesn't allow toenailing, fit the last two courses in position, gapping at the wall.

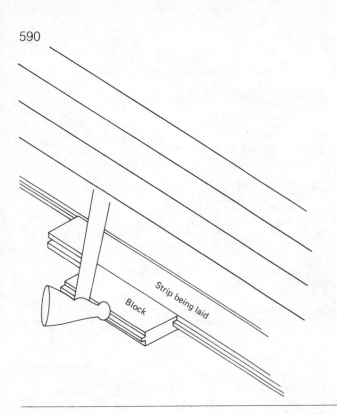

Fig. 24-1.13

Fig. 24-1.14

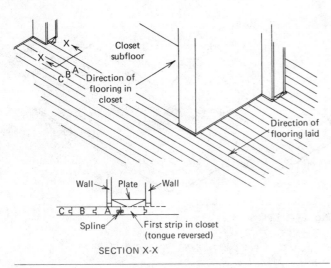

Fig. 24-1.15

13 Use a crowbar to force the last courses tight (Fig. 24-1.14). **Precaution.** Protect the wall by placing a piece of scrap lumber behind the crowbar. Except for the last course, use a scrap block to keep from damaging the tongue.

14 Face nail the final courses.

15 When a room layout requires doubling back, as at a closet or recess, insert a plastic spline in the groove of the last piece laid in the normal manner.

16 Fit the first piece in the recess or closet over the spline, so that the two pieces are groove to groove (Fig. 24-1.15).

17 Toenail in the standard method, and complete the floor as described in steps 12 to 14 inclusive.

18 When the last flooring strip is installed, sweep the entire floor surface to remove dust, bent nails, and all debris that might damage the surface.

Questions

1 What is the purpose of building paper between subflooring and strip flooring?

2 When are sleepers used and what is their purpose?

3 Name the six general rules for laying a good-looking strip floor.

4 What four fasteners may be used to install strip flooring?

5 When might you use a scriber during a flooring installation?

6 What is the best position for a carpenter to be in while nailing flooring?

7 Under what conditions is face-nailing permitted?

8 Under what circumstances are strips laid groove to groove?

Unit 24-2   **Lay Plank Flooring**

The procedures for installing unfinished and prefinished strip flooring are the same. For planks, however, they are often a little different. The type and width of planks determine the spacing of fasteners, and the specific method varies somewhat from manufacturer to manufacturer.

Attachment

Planks with tongue-and-groove edges are blind-nailed like strip flooring, but manufacturers sometimes recommend the use of adhesive at tongue-and-groove ends. Tongue-and-groove planks are almost always prefinished.

Planks with square edges, which may be finished or unfinished, must be face-nailed. Planks up to 4″ wide are usually attached every 24 to 32″ like decking, with the fasteners staggered. Spacing of fasteners into wider planks is 16″. The fasteners may be spiral screw-nails, finishing nails, or flat-head wood screws. It is customary to predrill the holes and countersink the fasteners.

The ends of some types of planks are also fastened with screws, two per wide plank and one in each narrow plank. The manufacturers predrill the holes for these fasteners, and also provide wood plugs to cover the screw heads (Fig. 24-2.1). The planks are almost always oak, and the plugs are a contrasting walnut. The finished floor looks as if it has been applied with dowels, which is the method used in Colonial days.

Procedure for Installing Square-Edge Planks

1 Follow steps 1 to 6 for attaching strip flooring.

2 Check the first piece for parallel by measuring from the string at one end, predrill the hole, and attach that end with a nail or screw.

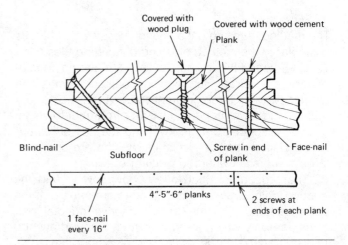

Fig. 24-2.1 Section and plan view showing method of attachment of plank flooring.

3 Continue attachment at every nailing point, working toward the unfastened end. **Note.** Some carpenters countersink the fasteners as they go, while others prefer to countersink all fasteners in a plank at one time.

4 At the end of the plank fit screws into the predrilled holes, and install them.

5 Repeat steps 2, 3, and 4 across the area to be floored.

6 After the floor is covered, fill the holes made by countersunk nails with wood cement.

7 Blow clean the holes made by screws, cover the heads with glue, and fit in the wood plugs. **Precaution.** Make sure the plugs are level with the surface of the glue, and wipe off any excess that oozes out.

8 Sweep the floor clean.

Questions

1 Where is adhesive sometimes used in installing plank flooring?
2 What fastener is frequently used with planks-that is not used with strip flooring?
3 If you were laying a plank floor, when would you countersink the fasteners? Why?

Unit 24-3   Lay Out a Floor Pattern

The method of laying out a floor of tiles or wood blocks is similar to the method of laying out ceiling tiles (Unit 23-1), provided the room is rectangular and no special pattern is required. But specifications sometimes call for tiles of two or three different colors (Fig. 24-3.1), or a pattern laid on a diagonal instead of parallel to walls, or even a special design involving the use of feature strips. Usually the details of the design are shown at large scale somewhere in working drawings.

The procedures that follow show how to find the starting point when all flooring units are square. If the design includes feature strips or other special patterns, the only way to begin is with a layout drawn either on paper at large scale, or on the floor itself.

Insets, such as initials or other irregular shaped designs, require different treatment (Fig. 24-3.2). They are usually cut from roll goods to exact size, then laid in place from the outside in. Drawings indicate the size and shape of the inset to scale. To bring the design up to full size, you make a template out of cardboard or thin plywood, using coordinates. **Coordinates** are pairs of crossing lines, as in graph paper. Figure 24-3.3 shows the general method of transferring a design from paper to a full-size template.

Procedure When Flooring Units Run Parallel to Walls

1 At floor level check all corners of the room for square.
2 Measure the overall dimensions of the room in inches. **Note.** If the room is not rectangular, measure the longest uninterrupted walls.
3 a. If the room is rectangular and corners are square, find the midpoints of opposite walls, and mark their locations with chalk on the underlayment.

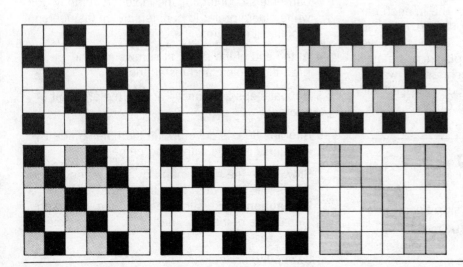

Fig. 24-3.1 Diagonal patterns can be created by laying tiles parallel to walls. These designs are typical.

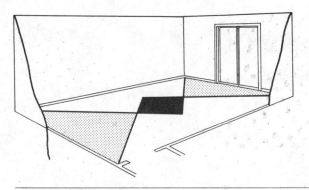

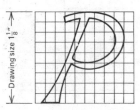

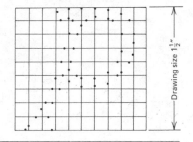

Fig. 24-3.2 Irregular patterns such as this must be laid out in advance either full size on the floor or on paper at a convenient scale.

Fig. 24-3.3 To transfer a design from drawing size to floor size, establish the final size of the design in feet or tiles. Then draw squares across the design at scale (this design is 10 units by 9 units), and lay out another series of squares at full size. Wherever a line in the design crosses a coordinate line in the drawing, mark that crossing on the larger coordinates at the same point. Eventually the shape of the design will show up in a series of marks, which is your cutting guide.

3 b. If the room is odd shaped, find the midpoints of straight walls, then measure off this same distance on opposite walls (Fig. 24-3.4).

4 Snap a chalk line between pairs of marks on opposite walls. Where they cross is the center of the room.

5 Check the accuracy of your work by stretching lines diagonally. They should pass directly over the center.

6 Divide the long dimension of the room in step 2 by twice the width of the flooring unit to be installed, and determine the remainder. **Example.** If the room is 186″ long and you are using 9″ square units, divide by 18, and the remainder is 6″.

7 a. If the remainder in step 6 is more than the width of a flooring unit, go to step 8.

7 b. If the remainder is less than the width of a flooring unit, measure off half the width of a tile from the long center line, and snap a parallel chalk line.

8 Repeat step 6 using the short dimension of the room.

9 a. If the remainder is more than the width of a flooring unit, you have established the location of the first unit.

9 b. If the remainder is less than the width of a flooring unit, measure off half that width from the short center line, and snap a parallel chalk line.

10 If you followed steps 7a and 9a, position the first tile or block as shown in *A* of Fig. 24-3.5. If you followed steps 7a and 9b, start as shown in *B*. If you followed steps 7b and 9a, start as shown in *C*. If you followed steps 7b and 9b, start as shown in *D*.

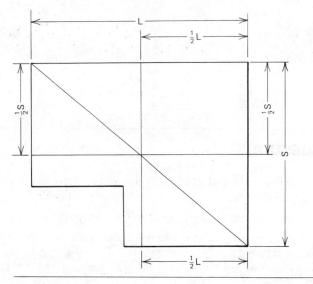

Fig. 24-3.4

Procedure When Flooring Units Run on the Diagonal

1 Find the center point of the room as described in steps 1 to 5 above.

2 Working both directions from the midpoint of the long wall, measure off half the length of the shorter wall, and make a mark on the floor (Fig. 24-3.6).

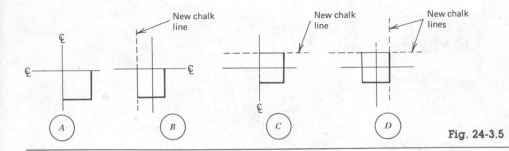

Fig. 24-3.5

3 Snap chalk lines across the room from these points so that they pass through the center.

4 Check the accuracy of your work. The diagonal chalk lines must cross at right angles to each other.

5 Position the first flooring unit where the 45° lines cross.

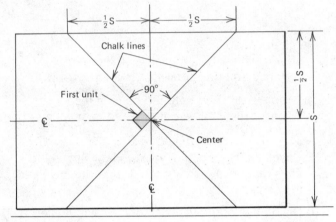

Fig. 24-3.6

Questions

1 How many possible starting points are there for the first tile in a floor laid parallel? How many in a diagonally laid floor?

2 What are coordinates, and how are they used?

3 In step 6, why do you use twice the width of a flooring unit in your calculations?

Unit 24-4 Lay Wood Blocks

Ever since it was first introduced in Europe some six centuries ago, the parquet floor has been the symbol of the ultimate in wood flooring. **Parquet** is a wood mosiac, a design in which the grain of boards or groups of boards runs at right angles to the grain of adjoining boards or groups. Originally, each board was laid individually. Today most parquet floors are made up of blocks, but they are still the most expensive of all wood floors.

There are two types of blocks. The more expensive blocks are made of individual pieces, the same thicknesses as strip flooring, glued together at their edges (Fig. 24-4.1). The blocks have tongues on two sides and grooves on the other two, and are blind-nailed like strip flooring. Tongues are already drilled for nailing.

The less expensive blocks are thinner, and consist of a hardwood veneer laminated under pressure to core plies of utility grade woods (Fig. 24-4.2). Each

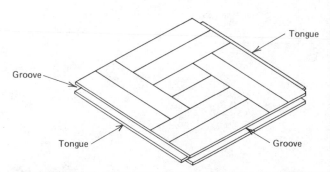

Fig. 24-4.1 Thick wood blocks have tongues and grooves on opposite sides. Note that the tongues extend only from the inside of one groove to the inside of the other groove.

The effect of this block floor is a series of short, interlaced strips. (Photos courtesy National Oak Flooring Manufacturers' Association, Inc.)

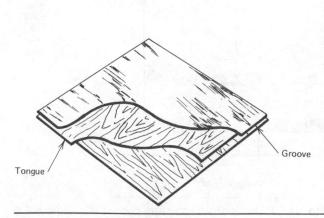

Fig. 24-4.2 Thin wood blocks are made like plywood, and there are no strip lines. The effect is a series of squares like a checkerboard.

block has a clear directional grain, and blocks are usually laid with the grain running in alternate directions. The blocks have predrilled tongues on opposite sides, however, to permit their installation in a parquet pattern. Some types have a groove on the back into which you insert a metal spline to prevent cupping.

Some thin wood blocks come with a self-sticking adhesive on the back, and are ready to lay as soon as you peel off the protective paper. Others are laid in a bed of thin mastic with a low moisture content. When mastic is used, the manufacturer's recommendations must be carefully followed.

Procedure

1 Clean the subfloor thoroughly.

2 Establish the starting point for the first block (see Unit 24-3).

3 a. If blocks are nailed, follow nailing procedures in Unit 24-1.

3 b. If blocks are laid in mastic, follow procedures in Unit 24-6, steps 3 to 6 inclusive.

4 At walls, cut blocks to gap $\frac{1}{2}''$ to $\frac{3}{4}''$, and face-nail at points that will be covered with base trim.

Questions

1 Describe the differences between the two types of wood blocks from the standpoints of thickness, appearance, and method of installation.

2 What is a parquet floor?

Unit 24-5 Apply Underlayment

Under any thin flooring material, such as linoleum or resilient tile, a layer of underlayment must be installed. Its purpose is to provide a smooth base without visible seams or ridges. The surface of resilient flooring has a high gloss, and any flaw in the base beneath it is telegraphed through the flooring.

The two most commonly used underlayments are $\frac{1}{4}''$ hardboard or $\frac{1}{4}''$ underlayment grade plywood. Underlayment may be safely installed directly over a plywood subfloor that has been properly laid. When the subfloor consists of boards, the boards should be covered with a layer of 15-lb felt, then underlayment. If resilient flooring is being used to resurface an uneven or badly cupped wood floor, it should be covered with a level coat of latex underlayment in place of the felt.

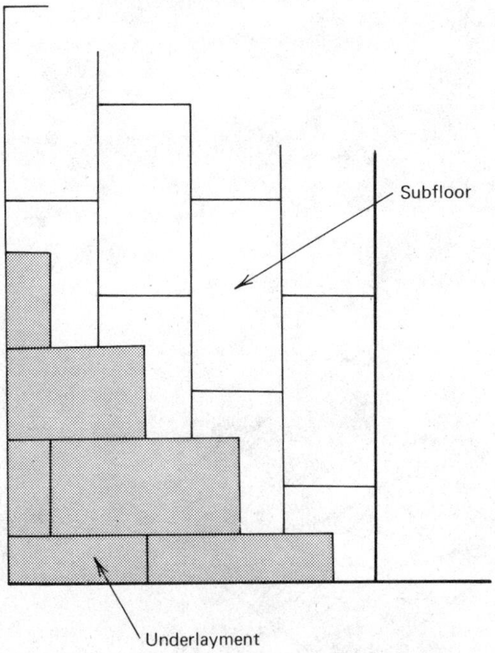

Application

Underlayment may be installed with staples or underlayment nails, spaced 3″ o.c. along the edges of each sheet and 6″ apart elsewhere. If the subfloor is ply-

Fig. 24-5.1 Underlayment must be placed so that no seam between sheets falls within 12″ of a seam in the subfloor below.

wood, stagger the seams in the underlayment so that they bridge the joints in the subfloor for the smoothest possible base (Fig. 24-5.1).

Procedure

1 Clean the subfloor thoroughly of all building debris.
2 Inspect the floor carefully for protruding nails and raised edges. Either drive loose nails deeper, or remove and replace them. Use a power sander to remove ridges and other high spots.
3 If felt is required, spread a layer of adhesive on the subfloor with a finely notched trowel, and embed the felt immediately with a roller.

Precaution. Butt the strips of felt; do not overlap them.

4 Lay the first sheet of underlayment in a corner, cutting it to size if necessary to avoid seams in the subfloor.
5 Fasten the sheet with nails or staples, beginning in the center of the sheet and working outward to the edges.
6 Install the remaining sheets of underlayment, working in a diagonal pattern across the subfloor. **Precaution.** Leave a gap the thickness of the cover of a matchbook between sheets.
7 When all underlayment is installed, check carefully for any raised fasteners, and drive them flush with the surface.
8 Sweep the underlayment clean.

Questions

1 What is the standard thickness of underlayment?
2 What may happen if underlayment is not used under thin flooring?
3 Under what conditions is a felt base required beneath underlayment?
4 At what point is the first fastener placed in a sheet of underlayment?

Unit 24-6   **Lay Resilient Floor Tiles**

The method of laying floor tiles is the same for all types of resilient tiles, regardless of their basic material. Some types will damage more easily than others, however. The tiles should be soft and flexible at the time of installation. Store them in a room no cooler than 70°F (21°C) for 24 hours before use, and keep room temperature at 65°F (18°C) or above during installation. The best place is in the last corner of the room to be floored.

Adhesive

Most manufacturers of floor tiles have their own brand of adhesive, formulated for their products. Some types are applied to underlayment with a fine-toothed trowel. Others are applied with a brush, and are similar in action to contact cement. Follow instructions on the container exactly. Be particularly careful to work in a well-

ventilated room, and keep the adhesive away from any open flame. Most types are highly flammable.

Application

Each tile must be carefully and accurately placed. Gently position the edge of one tile against the edge of a laid tile, then lower it into place (Fig. 24-6.1). Do not slide or twist the tiles; when you do so, you build up a ridge of adhesive between tiles that will show in the finished job. The seams between tiles should be almost invisible.

Tiles must always be cut to fit around pipes and similar vertical obstructions. They should also fit tightly at door jambs, and procedures for fitting are included in this unit. Fitting can be avoided, however, if door frames are trimmed slightly before they are installed. A

Fig. 24-6.1 The action of laying a tile is similar to the procedure for installing a wall panel in contact cement.

Position one edge against a laid tile, and gently lower it into position. (Courtesy Azrock Floor Products.)

tile laid under jambs and casings establishes the trimmed height for the door frame, which is installed with a gap for tiles. When the floor is laid, the tile slips into the gap and fills it for a tight fit.

When working over newly laid tiles, move slowly with your weight bearing straight down. Any quick or sideways movement can easily push tiles out of line. Stack tiles close enough so you can reach them with a minimum of movement. If any adhesive does get onto the face of a tile, remove it at once with a soft cloth dipped in cleaner listed by the tile manufacturer.

Fig. 24-6.2 Linoleum knife. (Courtesy Hyde Manufacturing Company.)

is applied to the material itself, not to the wall. Cove bases are very thin, and are applied with edges butting, even at inside corners. At outside corners it is usually necessary to trim the edges for a clean fit.

Cove Base

Manufacturers of resilient floor tiles also make base trim of the same material, usually in black or dark brown. The base is cut to fit just like tiles, but adhesive

Tools

To spread mastic you need either a trowel with fine teeth or a brush. To apply the tiles you need a folding

Fig. 24-6.3 (Courtesy Azrock Floor Products.)

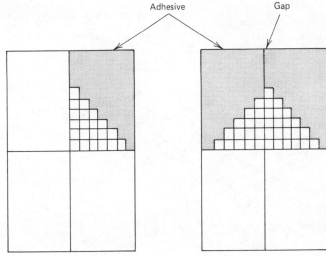

Adhesive | Gap

Fig. 24-6.4

rule, chalk line, and an awl, shears, or linoleum knife (Fig. 24-6.2).

Procedure

1 Prepare the base with underlayment as described in Unit 24-5.

2 Establish the location for the first tile as described in Unit 24-3.

3 Spread adhesive on half the underlayment in a small room, or one fourth the underlayment of a large room (Fig. 24-6.3). **Precaution.** You may cover most of the chalk guide lines, but leave the ends exposed at the wall and center of the room so that you still know where the lines are.

4 When the adhesive is **tacky** (no longer plastic but slightly sticky to the touch), snap new guide lines over the adhesive. **Note.** Open times of adhesive vary, but the average time between step 3 and step 4 will be about an hour.

5 Lay the first tile at the center of the room where chalk lines cross.

6 Work toward corners, laying the first tiles in diagonal rows on the chalk lines, then filling in. **Note.** The shape of the finished portion of floor is a triangle if you are covering one fourth the floor at one time, or a pyramid if you are covering half the floor (Fig. 24-6.4).

7 Install all full-size tiles in a section.

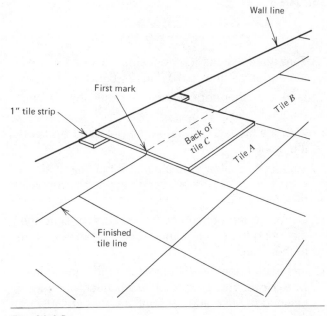

Fig. 24-6.5

8 To complete a rectangular pattern at a wall, cut a strip 1″ wide from a damaged tile, and lay it along the wall.

9 Place a full-size tile (C) upside down on the strip and finished floor, and line up the edges as shown in Fig. 24-6.5.

10 Where the loose tile crosses the finished tile line, mark the edge (first mark).

11 Move the strip to the right along the wall the width of a tile.

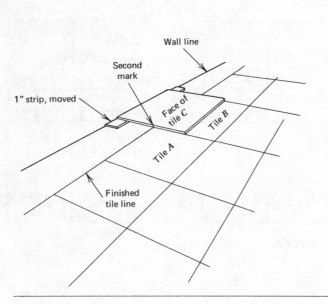

Fig. 24-6.6

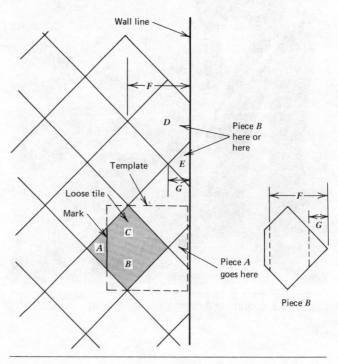

Fig. 24-6.7

12 Turn the loose tile (C) right side up, and line up the edges as in step 10 (Fig. 24-6.6).

13 Mark the other edge where it crosses the tile line (second mark).

14 With the tile face up, lay a straight edge between the marks.

15 Mark the cut along the straight edge, using a pencil on vinyl tiles or scoring asphalt or rubber tiles with an awl or linoleum cutter.

16 Cut along the pencil line with shears, or snap the tile along the scoring line, working from one edge to the other.

17 Lay each edge tile with the cut edge against the wall. **Note.** This procedure works even when the finished tile line and the wall are not parallel.

18 Roll the floor if manufacturer's instructions require it. **Note.** Rolling is usually recommended with thin tiles, such as vinyl.

Procedure for Finishing a Diagonal Pattern

1 Make a square template with sides equal in length to the diagonal dimension of a flooring unit. **Example.** The diagonal of a 9″ unit is $9 \times \sqrt{2}$ or 12¾″. The diagonal of a 12″ unit is 17″.

2 Place a full-size tile face up over the last full tile in a row (C in Fig. 24-6.7).

3 Lay the template on the floor with one edge against the wall, and two other edges just touching the corners of the loose tile.

4 Mark along the edge of the template where it crosses the loose tile.

5 Cut along this line. Piece A will fit against the wall as shown in Fig. 24-6.7. The remainder, piece B, will sometimes fit in the adjacent gap against the wall (D).

6 If piece B does not fit, it can be cut to fit at D or E by drawing a line parallel to the first cut at a distance of F or G from the uncut corner (inset).

7 Roll the floor if required.

Procedure for Fitting at a Door Jamb

1 With a scriber measure the uncovered distance (A in Fig. 24-6.8) between the last tile laid and the wall, measuring along a tile line if extended.

2 Mark this distance (point B) on the corresponding edge of a loose tile.

3 Lay the loose tile face up on the last tile in the row, with one edge against the jamb and the

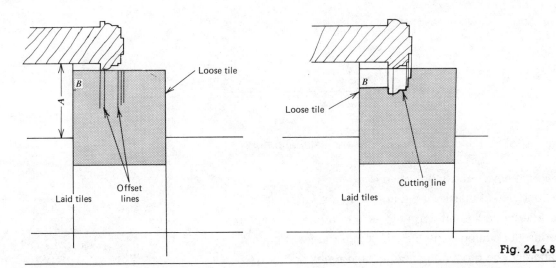

Fig. 24-6.8

adjoining edges lined up on the edges of the tile beneath.

4 On the loose tile draw with a pencil parallel lines showing the locations of offsets (left in Fig. 24-6.8).

5 Set the scriber so that the distance between legs is equal to the distance between the wall and the mark in step 2.

6 With the scriber mark the shape of the jamb on the tile (Fig. 24-6.8, right). **Precaution.** Make sure that the legs of the scriber are at right angles to the surface being scribed.

7 Cut along the marks. **Note.** To assure a tight and attractive fit, bevel the edge slightly away from the cut line, undercutting the tile.

8 Check the fit before laying the tile.

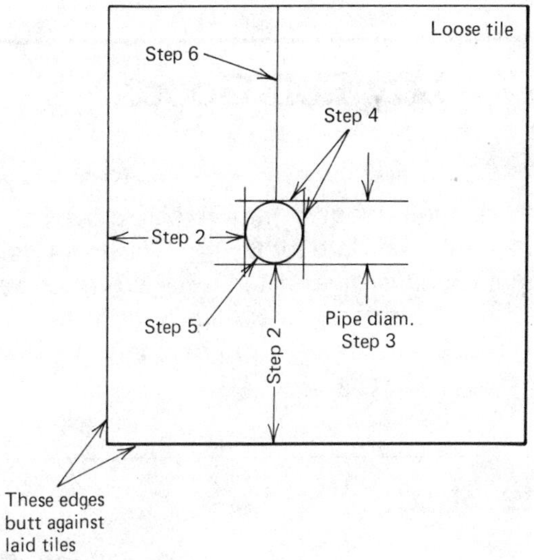

Fig. 24-6.9

Procedure for Fitting Around a Pipe

1 Measure from the edges of laid tiles to the pipe.

2 Transfer these measurements to a loose tile, drawing lines parallel to the edge.

3 Determine the diameter of the pipe.

4 Draw two more lines parallel to the first two to form a square with sides equal to the pipe diameter (Fig. 24-6.9).

5 With a compass draw a circle inside the square.

6 From the least visible edge of the tile when laid, make a straight cut to the circle, then cut the circle. **Note.** Make the straight cut with a sharp knife.

7 Fit the tile around the pipe and lay it in the adhesive. The cut line should be almost invisible.

Procedure for Installing a Cove Base

1 Cut pieces to length.
2 Spread adhesive on lengths of cove base. **Note.** Follow the adhesive manufacturer's instructions as

to whether to install the cove immediately, or to let the adhesive dry partially.

3 Install the base beginning at a corner. Butt all edges, even at corners.

Questions

1 What two room conditions are important to a good, safe installation of resilient tiles?
2 When and how do you remove excess adhesive?
3 What is the proper condition of mastic at the time tiles are laid in it?
4 Describe the procedure for cutting an edge tile to fit in a parallel pattern.
5 Describe the procedure for cutting edge tiles to fit in a diagonal pattern.
6 What is the main difference in procedures for installing tiles and cove base?

Unit 24-7   Lay Roll Goods

Like resilient tiles, roll goods is laid in mastic over an underlayment. The procedure for laying it, however, is a little different. Because the material is laid in long, wide strips, it must be carefully precut to fit at doorways, around pipes, and to the width of rooms.

Where to Start

The best starting point is along a long wall unbroken by doorways or jogs of any kind. Lay roll goods the long dimension of the room if possible; in this way you have the least amount of seam to fit. Work toward doorways; roll goods must be cut to fit into the opening anyway, and the narrower widths, which are likely to fall on the door side of a room, are easier to handle.

Tools

To lay roll goods you need a trowel for spreading mastic, a sharp linoleum knife, and a steel tape.

Procedure

1 Determine the best direction to run the flooring.
2 Unroll the roll goods onto the floor along the longest wall, with one end in the corner.
3 Check the fit at the corner, and trim if necessary. **Note.** Allow about a $\frac{1}{8}''$ gap for expansion at the end and side wall.
4 Reroll the flooring, and spread only enough mastic at the end of the room to hold the roll goods while you position the roll along the wall.
5 Press the roll in the adhesive at the corner, and unroll it far enough to make sure the roll is parallel to the long wall.
6 When the direction is right, spread mastic only to within a foot of the first obstacle.
7 As described in Unit 24-6, cut the roll goods to fit around the obstacle. **Precaution.** If the obstacle is a doorway, the roll must be cut to a narrower width along the walls so that a tongue of flooring fits through the doorway. Leave a tongue wider than the door opening (Fig. 24-7.1).

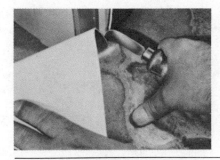
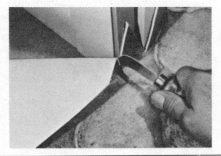

Fig. 24-7.1 (Courtesy Armstrong Cork Company.)

8 Apply mastic through the area of the cut, and press the roll goods in place.

9 Repeat steps 6 to 8 as necessary to reach the other end of the room.

10 Repeat the process for other strips until the room is floored.

Questions

1 When you start laying a floor of roll goods, how much mastic do you spread?

2 What is the ideal starting point for the first length of flooring?

3 What is the procedure at a doorway?

25

Interior Doors

As you learned in Chapter 17, an exterior door may serve from two to five purposes. Interior doors serve only two: to open up a passageway between two rooms or areas, or to close off that passageway, either for privacy or to screen off the view of whatever is beyond the door. There are two basic types of interior doors: those that swing on hinges, and those that slide on tracks.

Hinged Doors

The procedures for installing a frame for exterior hinged doors (Unit 17-1) and for hanging the door itself (Unit 17-2) apply also to interior doors. There are some minor differences, however. The doors will probably be flush hollow-core doors instead of panel types, and smaller in size. Therefore they are lighter in weight and easier to install. Most interior doors are 6'-8" tall, compared to 7'-0" for some exterior doors. They will be narrower; doors between major rooms are usually either 2'-8" or 2'-6" wide. Bathroom doors are often only 2'-4" wide. Hinged doors on broom and linen closets may be as narrow as 1'-8". Good closet design calls for doors as close to the width of the closet as possible.

The main difference between exterior and interior hinged doors lies in the hardware. Hollow-core doors often swing on one pair of hinges instead of $1\frac{1}{2}$ pairs. The exceptions are doors that are opened and closed many times during the day, and doors that are held shut with a dummy knob (see below).

In Unit 17-3 you learned about four of the six types of locking functions. The other two locks are used exclusively on interior doors. One is the **privacy lock** (Fig. 25-0.1), a variation of the button lock for bathroom and bedroom doors. When a button on the room side is pushed in or turned, the door locks; a turn of the knob or button releases the latch and unlocks the door. The

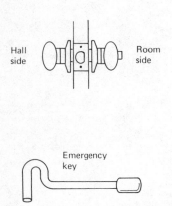

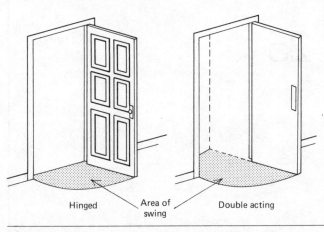

Fig. 25-0.1 A privacy lock has a locking button on the room side and a slot on the hall side for emergency unlocking.

Fig. 25-0.3 A double-acting door swings into either room and takes twice as much floor space as a hinged door in operation.

door can be unlocked from the hall side only with a small flat pin that fits into a hole in the knob.

The **knob latch** (Fig. 25-0.2) is for hinged doors to rooms where privacy is not a consideration. The most common type has knobs on both sides that are always free to turn, and either knob releases the latch. There are two variations for closet doors. One has a thumb

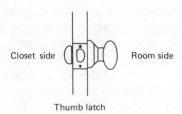

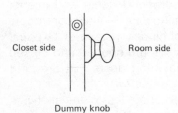

Fig. 25-0.2 Both knobs of a knob latch (top) turn freely, and the door cannot be locked. On the closet side of a door a thumb latch may substitute for a knob (center), or there may be no hardware at all (bottom). With a dummy knob a spring-operated release holds the door shut.

latch instead of a knob on the closet side, in case somebody is accidentally shut in the closet. The other has a dummy knob on the outside of the door only. It doesn't turn; a pull on the knob or a push on the door from the inside releases a spring-operated button or roller at the head or jamb that holds the door shut.

Double-Acting Doors

A hinged door opens in one direction only. A double-acting door opens into either of the rooms it separates (Fig. 25-0.3). There are two types. Full-size doors are used singly between such rooms as a kitchen and dining room where usage is heavy during meals but the door is shut most of the time to block the view of after-meal clutter. Shorter doors, called **cafe doors**, are hung in pairs to screen a view but permit passage of air and sound (Fig. 25-0.4).

Double-acting doors have push plates instead of knobs, and can't be locked. The doors are hung on special double-acting hinges with springs set in pairs of cylinders (Fig. 25-0.5). One type (left) is mounted on a hanging strip attached to the door jamb. The other type (right) is mounted on the jamb itself. Both types automatically return the door to a closed position, but they have catches so that the door can be left open temporarily.

Sliding Doors

Four types of interior doors slide instead of swing. They are a pocket door, a bypass door, a bifold door, and a folding door (Fig. 25-0.6).

A **pocket door** slides sideways from the door opening into a recess in the wall. It is not interchangeable with other types of doors, and wall space beside it must be free of any wiring, ducts, or other obstructions.

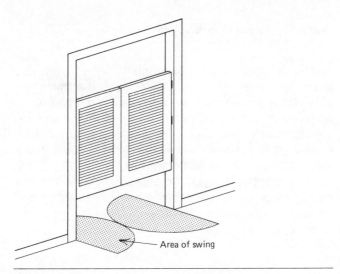

Fig. 25-0.4 Cafe doors are short double-acting doors whose main purpose is to block the view between rooms. They are hung in pairs and have no lock.

A pocket door is usually specified when the space to swing a hinged door is limited, or when two hinged doors would bump together. Its main advantage is as a space saver. Its main disadvantage is rather noisy operation compared to other types. See Unit 25-1.

Bypass doors slide past each other horizontally in tracks. They are installed on wide closets where floor space is limited—their main advantage. Their chief disadvantage is that only half the closet space is accessible when a door is open. See Unit 25-2.

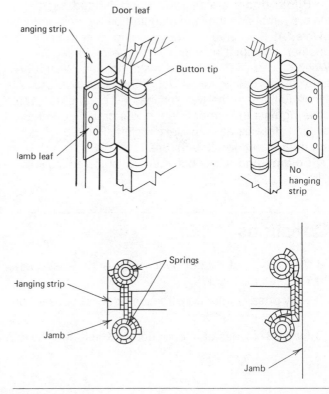

Fig. 25-0.5 The two types of double-acting hinges. They are mounted either on a hanging strip (left) or the jamb itself (right).

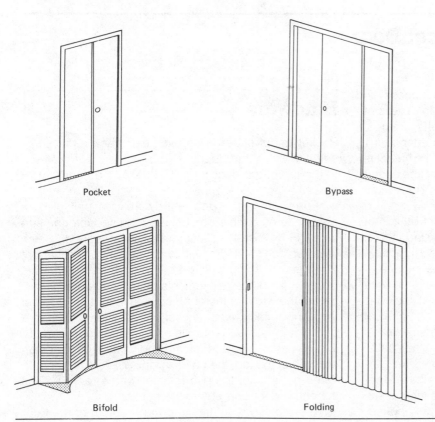

Fig. 25-0.6 The four common types of interior sliding doors and the amount of floor space their operation requires.

Bifold doors are pairs of doors hinged together. On the jamb side they are mounted on top and bottom pivots. At the center of the opening their support is another pivot that moves in a single overhead track. When closed, the doors form a straight line. When partially open, each pair of doors forms an outward V. When fully open the two sections lie parallel to each other against the jamb. Bifold doors provide good access to closet space, and take up little floor space during operation (Fig. 25-0.6). See Unit 25-3.

Folding doors, like bifold doors, are fixed at jambs and slide open or shut horizontally on an overhead track. The individual sections of which the doors are built, however, are narrow, and stack against one or both jambs without swinging into the room. Small folding doors are often used to screen off laundry equipment installed in an alcove off the kitchen. Wider doors are used to divide large floor areas into smaller units, such as to separate a pool or table tennis table from the cardplaying areas of a basement recreation room. Pairs of folding doors latch to each other; single doors latch at one jamb. See Unit 25-4.

Questions

1 Which type of interior door requires the least operating floor space? Which requires the most?

2 How does header construction above a pocket door differ from that over an ordinary interior door?

3 Of the six types of interior doors, which ones have no latching hardware?

Unit 25-1 **Install a Pocket Door**

Any standard-size door can be installed in a pocket as long as it is only $1\frac{3}{8}''$ thick—the standard thickness of interior doors. The opening is specially framed, however. Its width is twice the width of the door plus 1 to 2″, and headers span the entire opening. Its height is usually 6′-11$\frac{1}{2}$″ for a 6′-8″ door. Headers must be strong enough to carry the weight above without deflection, otherwise the door is likely to bind.

The pocket itself is made of two ladders built from 1 × 4s (Fig. 25-1.1). Thus the door clears the ladders on each side by about $\frac{1}{4}''$. The 1 × 4s must be straight lumber to prevent binding and damage to the surface of the door. Pocket doors are available complete with hardware and a lightweight steel frame that includes the track (Fig. 25-1.2).

The outside surfaces of the 1 × 4s lie flush with the sides of the header, and are surfaced like other walls. When finishing this section of wall, however, the carpenter must be very careful to use fasteners that are not so long that they go through the ladder into the pocket. The same care is required when moldings are applied.

Hardware

Pocket doors slide on an overhead track running the entire width of the opening. The hangers are screwed to the top of the door (Fig. 25-1.3). At the floor a steel spreader acts as a guide.

For opening the door, hardware is a pair of door pulls, one on each side, that are mortised into the surface (Fig. 25-1.4). Closing hardware varies. Some builders don't install any, but let the door project about $\frac{1}{2}''$ into the opening where it can be gripped by fingers to pull it out of the pocket. This has two disadvantages: part of the width of the opening is lost, and the edge of the door quickly becomes soiled. A better answer is a pull that is mortised into the edge of the door (Fig. 25-1.5). A latch pivots near the top of a recess. When pressed at the top, the bottom pops out as a finger pull. When not in use, the pull remains in the recess. Thus the entire door can fit into the pocket.

There are two solutions to the problem of the door

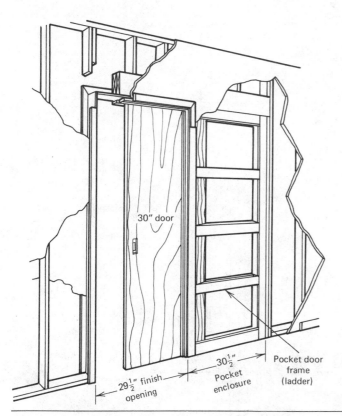

Fig. 25-1.1 Perspective and photo of a typical frame for a pocket door. Note that the length of the header is twice the width of the door itself. (Photos Drew Leviton, Atlanta.)

Fig. 25-1.2 Some pocket doors are available complete with steel pocket frames.

sliding too far into the pocket. The simplest hardware is a pair of bumpers or springs attached to the frame at the end of the pocket. The other is a pair of stabilizer arms attached to the inner edge of the door and the end of the pocket. They work like a pair of scissors.

Tools

To build the pocket walls you need the usual carpentry tools. To hang the door and install its hardware you need a screwdriver, folding rule, and tools for mortising.

Procedure

1 Check the size of the framed opening against requirements for the pocket door.

2 Install the casing on the solid door jamb. **Note.** There is no doorstop.

3 Build the two thin pocket wall frames.

4 On the floor and on the head and jamb of the opening mark the exact locations of the pocket

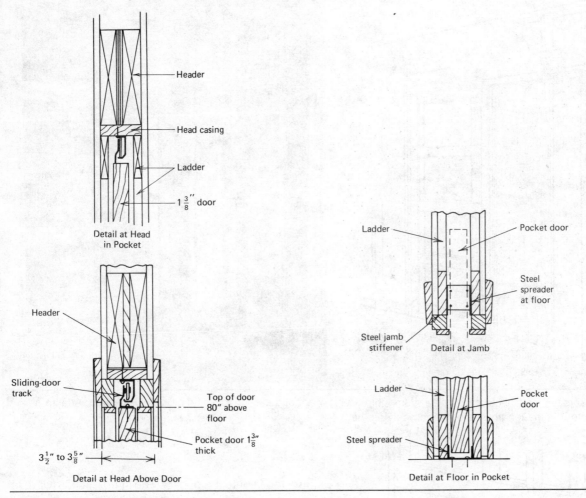

Fig. 25-1.3 Sections through a pocket door showing construction.

wall frames. **Note.** Mark the locations of both edges to make sure that the outsides line up with the headers and that the insides provide the required clearance.

5 Install the door guide spreader at the floor.

6 Install the frames on the lines and against the spreader, nailing with 8-penny nails through shims at the head and jamb. **Precaution.** Check plumb and level constantly, and make sure to stay within the guide lines.

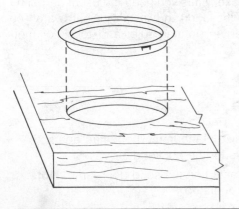

Fig. 25-1.4 Door pulls are recessed into the surfaces of the door. Some types have a set screw, while others stay in place with a friction fit.

Fig. 25-1.5 An edge pull fits into the edge of a pocket door. The pull pivots near the top, and swings out for gripping with a finger. (Courtesy Kwikset.)

7 Mark the center line of the overhead track.

8 Install the track on the center line.

9 Install any closing hardware at the exposed edge of the door.

10 Hang the door to test the fit; make any adjustments to framing at this time.

11 Install hardware as required at the end of the pocket to prevent the door from sliding too far in.

12 With the door in the pocket, tack 1 × 2 braces across the split jamb to hold the pocket walls apart at the proper spacing.

13 Install wall materials on the pocket frame to enclose the recess. **Precaution.** Use fasteners only long enough to do the job.

14 Remove the braces, and test the door's operation to make sure it clears all wall material fasteners.

15 Raise or lower the height of the door by adjusting the hangers. **Note.** Standard clearance at the floor is about $\frac{1}{2}$".

16 Install the split casing at the head and remaining jamb.

17 Retest the door and make final adjustments.

Questions

1 Of what two materials may a pocket be constructed?

2 What precaution must be taken in applying finish wall materials on a pocket frame?

3 How does the width of the door opening for a pocket door compare with the depth of the pocket itself? Why?

4 Why is no doorstop required with a pocket door?

Unit 25-2 **Install Bypass Doors**

The procedure for installing bypass doors is similar to that for patio doors (Unit 17-4) and pocket doors (Unit 25-1). They are installed in a standard door frame. The doors themselves may be hollow-core interior doors, louvered wood or metal doors, or may be cut from plywood or edge-laminated boards only $\frac{3}{4}$" thick. The meeting edges of thin doors should be stiffened with a metal channel or angle to prevent warping.

Bypass doors are not practical in openings less than 6'-0" wide. Openings up to 8'-0" should have one pair of doors. Wider openings, however, may have four doors, two per track. Rough opening height is 7'-0". Rough opening width is usually twice the width of one door, plus the thickness of two casings, less the dimension the doors overlap. Minimum overlap is 1", and it may be as much as the width of a stile on louvered doors.

eral types of each (Fig. 25-2.1). Some tracks come with a metal strip that hides the track and trims the opening across the head. With others the carpenter installs wood trim.

Usually there is no bottom track. Instead, a floor guide keeps doors in alignment. There are two types (Fig. 25-2.2). One type is attached to the floor with long screws where the two doors meet, and directs the path of both doors. The other consists of an angle guide attached to the floor and a crimped metal strip attached to the inside face of the door at the bottom; each door needs its own guide.

Bypass doors are operated by recessed pulls. A pair, one at the jamb edge of each door, is the minimum. In better-quality houses each door has two pulls so that either door may be at the left or right side of the opening.

Hardware

Bypass doors hang from overhead tracks on hangers attached to the backs of doors in pairs. There are sev-

Procedure

1 Check the frame opening for square.

2 Install the door casing.

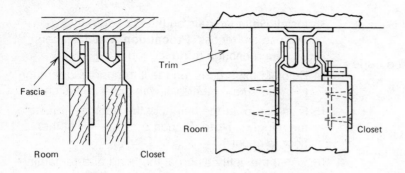

Tracks

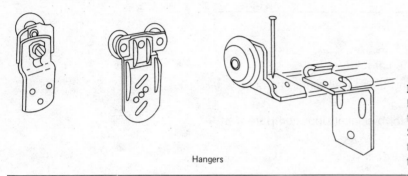

Hangers

Fig. 25-2.1 Typical tracks and hangers for bypass doors. Some tracks include a metal fascia (left), while other types must be trimmed. Some types of hangers are made for a specific thickness of door; others (right) are adjustable to any standard thickness.

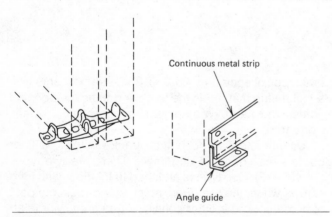

Fig. 25-2.2 Typical floor guides keep bypass doors from bumping against each other.

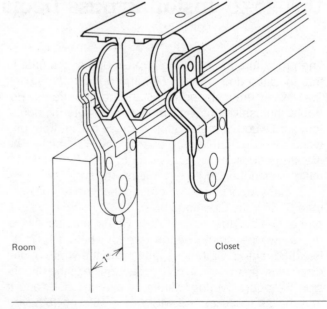

Fig. 25-2.3

3 Mark the location of the track assembly on the head, and install it according to manufacturer's instructions.

4 Install pairs of hangers on the backs of doors about 1″ from their vertical edges (Fig. 25-2.3).

5 Install door pulls.

6 Hang the doors and test their operation. **Note.** When closed, the doors should overlap about the width of a stile.

7 Adjust their height. **Note.** The bottoms of doors should clear the finished floor or carpeting by $\frac{3}{8}$″ to $\frac{1}{2}$″.

8 Push both doors to one side of the opening, and mark the location of the floor guide. Its center

should be in the center of the opening and
directly below the center line of the track.

9 Remove the doors and install the floor guide.

10 Retest the doors' operation.

Questions

1 Name the tools normally required to install bypass doors. There are five.

2 What is the main disadvantage of bypass doors?

3 When sheet materials are used for doors, what problem may occur and what is the solution?

Unit 25-3 **Install Bifold Doors**

Bifold doors are manufactured in three heights (6'-8", 7'-6", and 8'-0") and widths of 3', 4', 5', and 6' even. They come as four-door units, but a two-door unit may be installed in a narrower opening simply by cutting the single track in half. The doors may be made of wood, metal, or plastic. Often they are louvered to allow air circulation in the closed-off space. Metal and plastic doors come prefinished. Wood doors may be painted or stained.

The rough opening for bifold doors is framed in the same way as for a hinged door, although the finished opening size to which the doors are assembled varies from manufacturer to manufacturer.

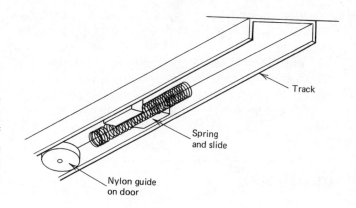

Hardware

Most bifold doors are shipped complete with all hardware except knobs, and with door hardware already attached. Pairs of doors are connected with $1\frac{1}{2}$ or 2 pairs of hinges mounted on the stiles with the pins on the inside. Thus hinges are completely concealed when doors are closed. On the top and bottom edges of the jamb-side door of each pair are spring-loaded pivots. The top pivots fit into sockets installed in an overhead track, and bottom pivots fit into similar sockets attached to the floor.

Where doors meet at the middle of an opening, a nylon guide holds the door in the track (Fig. 25-3.1, top). At the floor some sort of device is needed to keep doors aligned when closed. There are various devices, of which those in Fig. 25-3.1, bottom, are typical. Door aligners or butterflies bring the doors together and

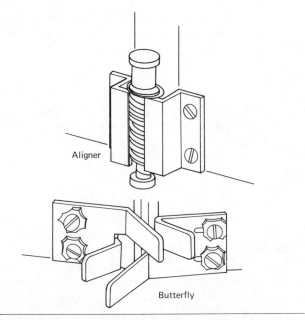

Fig. 25-3.1 A pivoting nylon guide (top) slides in the track and supports the center panels of bifold doors where pairs meet. At the bottom is typical hardware designed to keep bifold doors in line when closed.

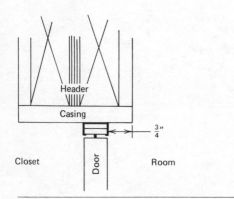

Fig. 25-3.2 The edge of the track is usually held back about $\frac{3}{4}''$ from the room side of the casing. If the track is trimmed with quarter-round, the distance may be increased to 1".

hold them plumb. The spring device slides loose in the overhead track, and exerts pressure on both doors to hold them shut.

The more or less standard opening hardware is a thin round knob mounted at standard knob height on the center doors only. The knobs are sometimes placed near the hinged edge, and sometimes are centered.

Installation

There is no set location for the track for a bifold door. Usually, however, it is placed about $\frac{3}{4}''$ back from the room edge of the head casing (Fig. 25-3.2). In this position the doors look well, and the closet has maximum depth for storage.

To install bifold doors, follow the manufacturer's instructions. The usual method is to fold the pairs of doors together, and insert the spring-loaded bottom pivot in the bottom socket. Then depress the top pivot, and let it snap into place in the top socket. The center door hangs from a nylon roller that snaps into the track. You adjust the height of the doors by turning the bottom pivot pin.

The only tools required to install preassembled bifold doors are a screwdriver, folding rule, plumb bob, and a small wrench.

Procedure

1 Check the frame opening for square.

2 Install the door casing.

3 If hinges were not installed at the factory, connect pairs of doors with their hinges.

4 Mark the location of the overhead track, and install it on the mark. **Note.** The track is usually a little shorter than the width of the opening, and should gap about $\frac{1}{8}''$ on each side.

5 With a plumb bob hung from the sockets in the track, mark the locations of the bottom sockets near the door jambs.

6 Install the bottom sockets. **Precaution.** The top and bottom sockets must line up exactly.

7 Fold one pair of doors together and fit the bottom pivot in the bottom socket.

8 Depress the top pivot and let it snap into place in the top socket.

9 Into the track snap the rollers that support the outer edges of the doors.

10 Repeat steps 7, 8, and 9 with the other pair of doors.

11 Test the doors for fit at the center, and for both vertical and horizontal alignment.

12 Adjust the bottom pivots as necessary to level the doors. **Note.** Doors should clear the floor or carpeting by $\frac{1}{2}$ to $\frac{3}{4}''$.

13 On the insides of the doors where they meet, mark the locations of closers.

14 Install the closers loosely to test the fit, then drive the screws home.

15 Install doorknobs.

Questions

1 What is the recommended location for a closed door track? Why?

2 Assuming that you are looking toward a closet at the left-hand pair of bifold doors, what hardware is required near each corner?

3 What do you think happens when bifold doors are installed without door aligners?

4 How much should the doors clear finish flooring?

Unit 25-4　**Install Folding Doors**

Folding doors consist of a series of narrow panels about equal in width to the depth of a door frame. Panels are hinged to each other on alternate sides. When closed, a folding door looks much like a paneled wall. When open, the panels stack up against each other at one door jamb, at both door jambs, or in a recess in a thick wall.

Accordion doors are a type of folding door consisting of a series of lightweight metal frames covered with a continuous sheet of fabric or vinyl on both sides. The name is taken from the musical instrument that opens and closes in a similar manner when being played. Folding doors come from the factory completely assembled and ready to install. Standard height is 6'-8". Standard widths range from 2'-4" up to 4'-0", but folding doors may be ordered to fill an opening of any width. They give clear access to about 90% of the storage space they shield.

Fig. 25-4.1 When folding doors are hung in pairs, they latch together. A single door latches to a catch fitted to a jamb molding. (Courtesy Modernfold.)

Hardware

Except for handles at the edge of each door, all hardware is concealed. Door sections move on rollers in an overhead track that is hidden behind trim that matches the doors. There is seldom a bottom track in residential installations; in hotels and offices, where doors are often as high as the ceiling and as wide as a room, bottom tracks are required. Hinging hardware is hidden behind moldings at the edges of folding doors sections, and under the surface of accordion doors.

When a pair of folding doors fills an opening, the edges of the doors have interlocking latches that hold them shut. When a single door fills the opening, a latch in the end post of the door closes on a catch attached to a full-height jamb molding (Fig. 25-4.1).

Installation

Doors that stack up on jambs are installed in a standard door frame. If they are recessed, the pocket must be at least 3" wider than the stacked thickness of the doors (Fig. 25-4.2). The recess may be left open or be covered with a small door that completely conceals the folding door. The molding at the opening serves as both trim and a stop for the sliding jamb panel. Working drawings should indicate the details of any required recess.

Procedure

1. Check the frame opening for square.
2. Install any door casing specified. **Note.** Sometimes folding doors are attached directly to walls or door openings finished with plaster or wallboard.
3. Read the manufacturer's instructions for installation. They may vary somewhat from these procedures.

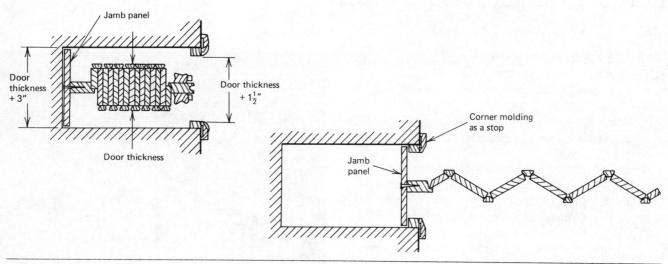

Fig. 25-4.2 Detail of a jamb pocket for a folding door. The depth of the pocket is determined by the width of the door when open. The width of the pocket should be about 3″

greater than the thickness of the door. A pocket may be left open or be covered with its own narrow door.

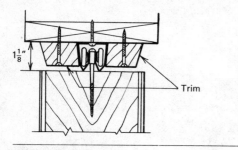

Fig. 25-4.3

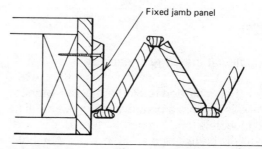

Fig. 25-4.4

4 On the head casing mark the location of the overhead track. It is usually centered.

5 Install the track on its mark.

6 Install the trim pieces that conceal the track (Fig. 25-4.3).

7 Mark the location of jamb panels. **Precaution.** Study the manufacturer's installation drawings to determine the relationship between the center of the overhead track and the edge of jamb panels.

8 Hang the door on its track, with the jamb panel toward the door jamb.

9 With screws fasten the jamb panel through the casing into studs (Fig. 25-4.4).

10 a. If two folding doors meet at the center of the opening, repeat steps 8 and 9 for the second door.

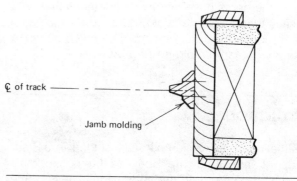

Fig. 25-4.5

10 b. If a single door fills the opening, locate and
install the jamb molding centered directly
beneath the center line of the track (Fig. 25-4.5).

11 Test the door's operation, and make sure it
latches properly.

Questions

1 How does an accordion door differ from a regular folding door?

2 When is a bottom track required?

3 Under what circumstances might a door track be attached directly to the
ceiling of a room?

26

Cabinetmaking

One of the reasons that Americans have become members of a throwaway society is that their homes often don't have enough storage space. Early in this century few houses were built without an attic and basement for bulk storage. Kitchens had few cabinets, but many houses had a whole room, called a **pantry**, for storage of dishes, glassware, silverware, and dry foods. Bedrooms sometimes had no closets, but clothes were stored in large pieces of furniture—chiffoniers, highboys, and wardrobes. Bathrooms had little storage space, but most houses had a large closet nearby for towels, sheets, blankets, and the like. And almost every living room had a fireplace with a mantel and some bookshelves.

The last 50 years have seen great changes in residential design. Houses have become smaller and more efficient. New designs for heating equipment and the development of automatic laundry equipment eliminated the absolute need for a basement. Better insulation and the popularity of low roof pitches cut the height of attic space in half. Therefore little bulk storage space exists. Moving the garage up to the house and enlarging it provided some. Items used in the kitchen are now stored in the kitchen, and the pantry has disappeared. Items used in the bathroom are stored in the bathroom, and storage space for linens and bedding is frequently limited. Except for flat clothing, furniture is seldom used as a major source of storage space.

Today the storage units that a carpenter is called on to build consist primarily of cabinets for kitchen and bathroom, shelving and hanging rods in clothes closets, storage walls for general purposes, and built-ins designed for special storage purposes.

Good storage space has three elements of equal importance:

- *Good design.* All too often designers of homes do not think through a family's needs for storage. The space they provide is the space left over when the room arrangement is laid out on paper. Good storage must be of the proper size and shape for the items to be put away. It must be at the most convenient height commensurate with the number of times the space is used. It must be as close as possible to the point of use.

- *Good materials.* Most storage units are exposed to view as furniture, for which they are an alternate. Therefore the materials must be attractive and capable of taking a smooth finish. Storage units get a lot of use; therefore their materials must be durable. Some units carry heavy weights over long periods of time; therefore their materials must be strong and strongly supported.

- *Good craftsmanship.* Of all the work a finish carpenter does, building cabinets and other storage elements requires the most care and talent. To do a proper job, the carpenter must have available and know how to use a wide selection of tools, ranging from large table saws down to chisels and awls. All tools must be sharp and in top working condition. The joints a carpenter makes should be tight and virtually invisible. Surfaces should be smooth and unmarred by tools, bumps, or scrapes. Fasteners should be completely concealed. All cuts should be square and true. And the finished storage unit should be as attractive as a piece of fine furniture.

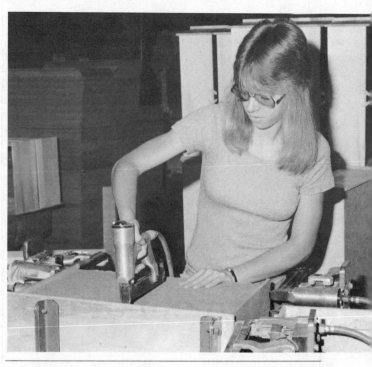

Fig. 26-0.1 Many cabinets are manufactured off site where compressed air for staple guns is readily available. (Courtesy Senco Products, Inc.)

Competition

The storage cabinets in many houses are built in a factory or millwork shop (Fig. 26-0.1). Manufacturers of factory-built cabinets offer a complete line of units for kitchens and bathrooms. They are available either knocked down or preassembled. They include **base cabinets**—those that are attached to the floor—and **overhead** or **wall cabinets**—those that are attached to the wall. Base cabinets are made in one height and one depth—usually $35\frac{1}{4}''$ high and 24" deep. Overhead cabinets are made in two or three heights from 15 to 33", and one depth of 12 to 14".

The widths of manufactured base and wall cabinets vary from 12" up to 60". Three inches is the established module of width. Therefore cabinets are made in widths of 12", 15", 18", and so on up to 33". Wider cabinets are usually 42", 48", and 60". If kitchen plans call for a cabinet 51" long, it can be composed of several narrower cabinets, such as two 18" wide and one 15"

wide, or two of 15" and one of 21"—whichever combination looks best. The quality of factory-built cabinets ranges from excellent to average, depending on attention to the three elements of good storage.

For any specific cabinet many optional features are available. A 24"-wide base cabinet, for example, may have two or three shelves; one wide door or two smaller ones; one drawer and two doors, or four drawers; sliding shelves, turning shelves, bins, etc. Because the parts from which the cabinets are made are stock parts, uniformity of appearance, flexibility of usage, and smoothness of operation are major selling points that the on-site carpenter must compete against.

For specialized storage, such as bookshelves, music centers, and workshops, builders may use their own carpenters or have the units built at a local millwork shop. Here the design of the storage unit is established by working drawings, and the same materials are available to both the on-site carpenter and the millwork carpenter. Now the competition is in craftsmanship alone.

The procedures that follow in this chapter will help carpenters on the site compete successfully with their counterparts in the factory and local shop.

Questions

1 What is the module for the width of cabinets?
2 What are the three elements that determine good storage space?
3 If a wall space is 57″ wide, and the cabinet arrangement must be symmetrical, what is the fewest number of cabinets that you can install in that space, and how many size combinations are there? Assume that the narrowest cabinet is 12″, the widest 33″, and all sizes on a 3″ module between those extremes are available.

Unit 26-1   Assemble Base Cabinets

A base cabinet not only provides planned storage space, but also serves as a support for a countertop. Therefore you will find base cabinets in working drawings whenever the use of a room requires counter space. Kitchens always have them and, in today's houses, they are usually planned also in bathrooms and utility rooms, and sometimes in basements and garages.

Cabinet Dimensions

As mentioned earlier in this chapter, cabinet widths are almost always divisible by 3″. If wall space is not divisible by this 3″ module, the difference is closed with a filler strip (Fig. 26-1.1) or by a wider stile on the face of the cabinet. Only the face is wider, however; the width of the cabinet itself remains modular.

The 24″ standard depth of kitchen and laundry base cabinets is measured from the face of the cabinet to the wall against which it rests. This is the maximum depth of the cabinet from front to back, excluding door and drawer fronts (Fig. 26-1.2). At the floor the cabinet is undercut to provide toe space. The depth of toe space, called a **toe kick,** varies from a minimum of 2″ to a maximum of 4″, with $2\frac{3}{4}$″ a common dimension. Height of a toe kick is usually 4″.

Standard height is 36″ for kitchen and utility cabinets. This is the overall height from the floor, including the thickness of a counter. Unfortunately, people are not made as standard as cabinets, and a counter at standard height may be quite uncomfortable for a very tall or very short person. Ideally, the best height for a kitchen counter is 3″ below the elbow of the person using the kitchen when the person's arm is extended as

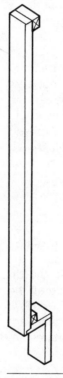

Fig. 26-1.1 A filler strip looks like a section through a base cabinet and is the method used by cabinet manufacturers to compensate for odd dimensions. This problem is solved in site-built cabinets with a wider stile.

though shaking hands (Fig. 26-1.3). Thus, for a small person only 5′-2″ tall, a 34″ counter height is the best, while for a person 5′-10″ tall a $38\frac{1}{2}$″ height is likely to be much more comfortable.

Base Cabinet Parts and Materials

The simplest base cabinet consists of a base, bottom shelf, lightweight frame, sides, back, and front. Its interior parts vary with the usage of each cabinet.

The Base

The base is a rectangular frame 4″ high (Fig. 26-1.4). It is as wide as the cabinet at the front, and this same width less the thickness of side materials at the back. It is as deep as the cabinet less toe space and the thickness of back material. Concealed parts may be cut from any sound 1″ boards or $\frac{3}{4}$″ plywood. The toe kick may be made of prefinished plywood, or unfinished material of good enough quality to take stain or paint. Cabinets more than 24″ wide usually have a center support.

The Bottom Shelf

The bottom shelf may be made of plywood or particleboard at least $\frac{1}{2}$″ thick. Thinner materials are not adequate. The shelf is the same width as the base, but projects beyond it at the front of the cabinet. Its depth depends on the type of cabinet. If the shelf is exposed and used for storage, it ends in a dado in the face material (Fig. 26-1.5). If drawers fill the cabinet and the shelf is hidden, it butts against the face material.

The Frame

A frame of good quality is made of 1 × 2s, although 1 × 1s are sometimes used in inexpensive cabinets. The length of horizontal members varies with the dimensions of the cabinet. Vertical members are all the same length. A vertical stiffener is usually added in cabinets more than 24″ wide. Horizontal stiffeners are added only as supports for shelves or drawer hardware (Fig. 26-1.6).

Sides

The primary purpose of sides is to enclose the cabinet as protection against dust and dirt. Concealed sides, such as those of cabinets flanked by other cabinets, are made of $\frac{1}{4}$″ plywood or hardboard. Exposed sides of end cabinets match the front in appearance, but are usually thinner. The sides cover both the frame and the base, and are notched for the toe space.

Back

A back on a base cabinet is not a structural necessity; the bottom shelf and countertop provide adequate

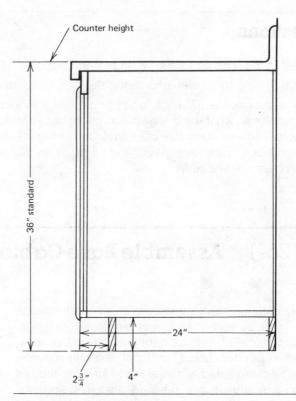

Fig. 26-1.2 Where the depth of a base cabinet is measured.

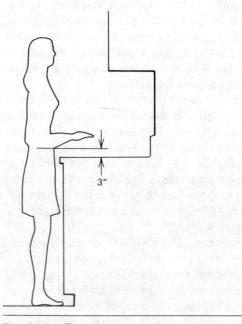

Fig. 26-1.3 The ideal height for a counter is 3″ below the elbow of the person using the counter the most. Although 36″ is a standard, more comfortable heights range from 34″ up to 39″.

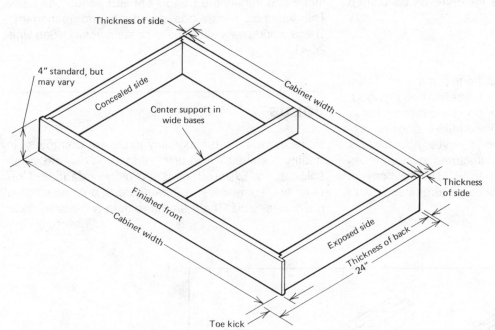

Fig. 26-1.4 The base of a base cabinet. Only the front needs to be cut from finish materials.

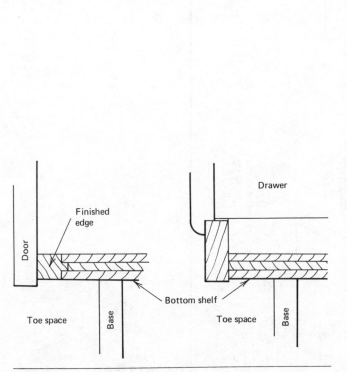

Fig. 26-1.5 The edge of an exposed bottom shelf must be finished. When concealed, the edge may be left raw.

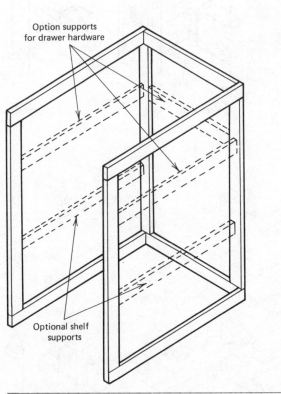

Fig. 26-1.6 Additional 1 × 2s are required to stiffen large frames and to support drawer hardware or shelves.

strength. But a back does keep out insects that somehow find a home in walls, and keeps out dust and moisture, especially in cabinets below or near the sink. Backs are cut from the same materials as concealed sides.

Front

The fronts of cabinets are cut from $\frac{3}{4}''$ material. Three types are frequently used. One is prefinished plywood, which looks well when cabinet doors are closed, but is difficult to finish properly and sometimes does not accept hardware easily. Another is vinyl-wrapped stiles and rails. These parts are manufactured in long pieces of standard widths from $1\frac{1}{2}''$ up to $5\frac{1}{2}''$. The vinyl covers three sides, and provides a finished appearance with doors closed or open.

Most fronts, however, are cut from four woods known for their strength, light weight, workability, and fine appearance. The common softwoods used in cabinetry and joinery are Douglas fir and ponderosa pine. The common hardwoods are birch and mahogany. These woods take either paint or stain easily. (See Unit 26-4).

Joints

In inexpensive cabinets many joints are butt joints. In quality cabinets, few are. Hidden joints are often shiplapped. Exposed joints are usually dadoed. Corners are sometimes spline-mitered, but more often haunch-mitered. There are more than two-dozen ways

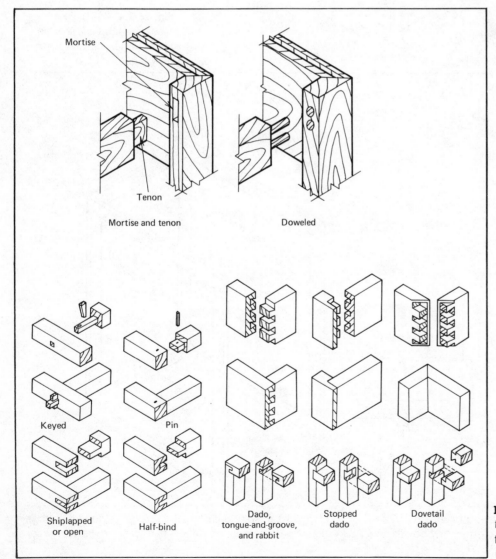

Mortise

Tenon

Mortise and tenon

Doweled

Keyed

Pin

Shiplapped or open

Half-bind

Dado, tongue-and-groove, and rabbit

Stopped dado

Dovetail dado

Fig. 26-1.7 There are many ways to join two pieces of wood together. These are among the more common methods.

to join two pieces of wood. Figure 26-1.7 shows just a few of the more common methods.

Fasteners

Using a good glue and applying enough of it is the secret to good cabinet joints. The fastener then serves only a temporary purpose—to hold the joint tight until the glue dries. Finishing nails are still the most commonly used fastener. Where compressed air is available, however, the use of blind pins driven with a nailer is popular because it is fast.

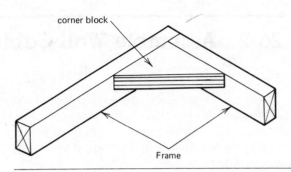

Fig. 26-1.8

Procedure

1 Determine the overall dimensions of the base cabinet to be built. **Precaution.** The height should be without countertop.

2 Rip base material to 4″ width.

3 Allowing for the thickness of back material and for shiplap joints, cut side base members to length.

4 Allowing for the thickness of side materials and base members, cut the back base member to length.

5 From finish stock cut the front base member to the full width of the cabinet.

6 Rabbet ends of base members where required (see Fig. 26-1.4). Miter the front base member where it meets any exposed side.

7 Assemble the base with glue and fasteners. **Precaution.** Be certain that the base is absolutely square at all four corners.

8 Cut the bottom shelf to size, allowing for the thickness of side materials and treatment at the front edge (see Fig. 26-1.5).

9 Nail the shelf on the base, countersink the nails, and fill holes with wood putty.

10 Cut 1 × 2 frame members to length, allowing for butt joints.

11 Cut side panels to dimension, and notch them for the toe kick. **Note.** Side panels usually overlap back panels.

12 Cut the back panel to size.

13 Attach framing members to sides and back by fastening through the panels. **Precaution.** Measure to make sure that the dimension from the bottom framing member to the bottom of the panels is the same as the distance from the bottom of the base to the surface of the shelf (usually $4\frac{1}{2}″$).

14 Attach side panels to the base.

15 Attach the back panel to the base and side panels.

16 Check all corners for square and plumb.

17 If necessary to maintain square, cut triangular blocks of 1″ material and glue them in the corners as shown in Fig. 26-1.8.

Questions

1 What are the most common height and depth for a toe kick?

2 How is the best height for a kitchen counter determined?

3 Name the five parts of a base cabinet, and a common material used for each.

4 What joint would you use where:

a) Two pieces of base come together?

b) Two framing members meet?

Unit 26-2 Assemble Wall Cabinets

Wall cabinets serve only one purpose: storage on shelves. They are found in plans in the same rooms as base cabinets, with the exception of the bathroom. Their construction is similar, but there are differences, particularly in dimensions.

Cabinet Dimensions

Like base cabinets, wall cabinets have modular widths divisible by 3″. Most designers try to plan cabinets so that the joints between base cabinets and the wall cabinets above line up vertically.

Most wall cabinets are 12 to 13″ deep, depending on the material used for the bottom shelf. If the shelf is wood, the width of a 1 × 12 is the governing dimension. If the shelf is a sheet material, the sheet is ripped into four widths that, allowing for the saw kerf, measure about $11\frac{7}{8}$″. Add the thicknesses of front and back materials to arrive at overall cabinet depth.

The height of wall cabinets varies according to conditions and accepted design standards. The most important standard is that the top shelf of any cabinet should not be more than 72″ above the finished floor. Otherwise it is difficult and dangerous to reach the contents on the shelf. Therefore ceilings are usually furred down above wall cabinets (see Unit 27-3) to a height of about 7′-0″, or 84″.

Minimum clearance between the underside of a wall cabinet and a countertop is 15″, and 18″ is better. With a counter 36″ above the floor, then, maximum cabinet height is 33″ when the ceiling is furred; 30″ is a more common height (Fig. 26-2.1). Above a range building codes require a 30″ clearance unless the underside of the cabinet is protected against fire. Carpenters must check local building codes before building or installing any cabinet above a range. A cabinet 15″ high meets requirements even under a furred ceiling.

At one time a 9″ clearance above a refrigerator was standard to allow exhausted heat to escape. Modern refrigerators exhaust the heat of cooling through the front, however, and only enough clearance is needed to move the refrigerator into position. At the same time refrigerators have become taller, so the rule is still about what it was: 69″ from the floor to the underside of a cabinet above a refrigerator.

Cabinets are seldom installed above a sink except in apartments where space is limited. The rule, however, is 22″ clearance to the underside of the cabinet.

If all these rules were observed, the bottom line of overhead cabinets would look like the teeth in a Halloween pumpkin. Most kitchens, then, have only two heights of wall cabinets: one approximately 30″ and one approximately 15″.

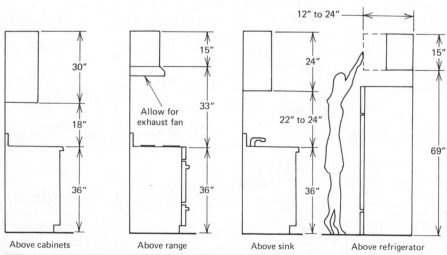

Fig. 26-2.1 Many of the rules for positioning wall cabinets are established by building codes. The clearances shown here are generally accepted industry standards.

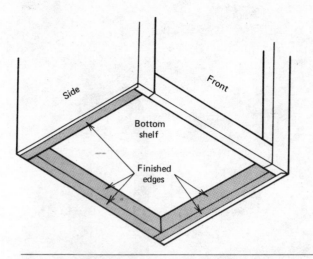

Fig. 26-2.2 The framing below the bottom shelf of a wall cabinet is made of finish material, with any prefinished sides exposed. This view is looking up at the bottom of the cabinet.

Wall Cabinet Parts and Materials

Because they have no floor to support them, wall cabinets must be more strongly built than base cabinets. They have the same parts, however, and the materials are the same—with one exception. The base that supports the bottom shelf is part of the lightweight frame and is exposed to view, especially in short cabinets. When prefinished materials are used, they must be as-

sembled with the unfinished face against the wall or adjacent cabinet (Fig. 26-2.2).

Procedure

1 Determine the overall dimensions of the wall cabinet to be built.
2 Cut side panels to dimension. **Note.** If either side is exposed, it must be cut from finished stock.
3 Cut the back panel to size so that it fits between end panels.
4 Cut 1 × 2 members to length.
5 Attach framing members to panels.
6 Attach the back panel to side panels.
7 Cut bottom shelf supports to length from finished stock (Fig. 26-2.2).
8 Install the supports on side and back panels so that they are flush at the bottom.
9 Cut the bottom shelf to size. **Note.** Side-to-side dimension is the distance between side panels. Front-to-back dimension is the distance from the back panel to the front edge of the front support.
10 Notch the shelf to fit around vertical framing at the four corners.
11 Install the shelf.
12 Check all corners for square and plumb.

Questions

1 What are the minimum clearances between a wall cabinet and the following?
   a) Range top
   b) Counter
   c) Sink
2 What is the maximum height above the floor for the upper shelf in a wall cabinet?
3 What governs the depth of a site-built wall cabinet?

Unit 26-3   Assemble a Vanitory

The base cabinet in a bathroom usually has a lavatory built into it. Such a unit is called a **vanitory** (Fig. 26-3.1) because it combines the use of a lavatory with the counter space of a vanity, which is a dressing table. A

vanitory is similar in construction to a sink cabinet, except that frequently the back and one side lie against walls while the second side is exposed to view.

Fig. 26-3.1 The name vanitory was coined by combining the beginning of vanity and the ending of lavatory. A vanitory combines a lavatory in a vanity. (Courtesy Eljer Plumbingware.)

Cabinet Dimensions

The standard depth for a vanitory is 21", although depths from 24" down to 18" are not uncommon. The width and depth generally vary with the size and shape of the bathroom itself.

Heights range from 30" up to 33". Because bathrooms are used regularly by more than one family member, one counter height can't suit everyone. In a bathroom used primarily by adults, the 33" height is the most common. In a child's bathroom the 30" dimension is better. When people of all heights use the same bathroom, it is better to stick with the 30" height.

Plumbing

In a vanitory or sink cabinet it is not uncommon for plumbing connections to extend as much as 18" below the countertop. The depth of the lavatory itself is 8 to 10", and the drain pipe and P-trap below block out the rest of the space. Therefore a vanitory has a bottom shelf like any other base cabinet, but no other full-width shelf. Any intermediate shelf must either be cut to allow for drain piping, or framing must be provided for smaller shelves.

Similarly, in a sink cabinet the space beneath the sink must be left clear for the sink itself, a continuous waste, and a garbage disposal unit. Sink cabinets usually have only a bottom shelf, and the rest of the cabinet is empty.

Procedure

1 Determine the overall dimensions of the vanitory to be built. **Precaution.** The height should be without countertop.

2 Build the base as described in Unit 26-1, steps 2 to 7 inclusive.

3 Cut and install the bottom shelf.

4 Cut 1 × 2 framing members to length.

5 a. Cut concealed side panels to dimension, and notch them for the toe kick.

5 b. Cut exposed side panels to length, notch them for the toe kick, and miter the front edges.

6 Cut the back panel to size.

7 On the back panel mark the locations of holes for water lines and drain lines. **Note.** At the time cabinets are being built, all plumbing is usually stubbed through the walls. If cabinets are built in advance, locations of center lines of piping are shown in installation drawings for the lavatory or sink (Fig. 26-3.2).

8 Drill holes for piping.

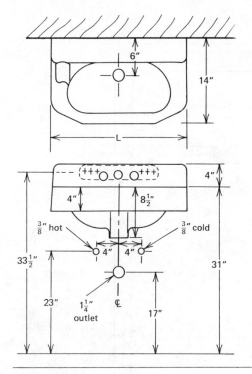

Fig. 26-3.2

9 Assemble the cabinet as described in Unit 26-1, steps 13 to 17.

10 Install supports as required for intermediate shelves.

Questions

1 How do the dimensions of a vanitory differ from those of a standard kitchen base cabinet?

2 About how far below the counter does plumbing extend for a lavatory or sink?

3 How does the carpenter know where to cut holes for water and drain pipes?

Unit 26-4 **Build Cabinet Fronts**

Cabinet fronts are sometimes cut from a single piece of plywood. Holes are carefully routed for doors and drawer, and the routings saved for use as flush fronts for those doors and drawers. A better, more economical, and simpler method is to cut and assemble individual stiles and rails, then cut door and drawer fronts to fit.

Top, bottom, and intermediate rails are 1 × 2s or their plywood equivalent (Fig. 26-4.1). Mullions between doors are usually the same size. The width of

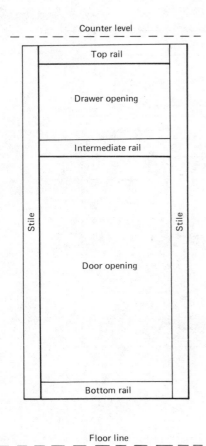

Fig. 26-4.1 In a typical cabinet front, rails are $1\frac{1}{2}''$ wide. Stiles range in width from $1\frac{1}{2}''$ up to $3\frac{1}{2}''$ in $\frac{1}{2}''$ increments.

stiles, however, varies with the width of the cabinet and the size of door or drawer front, and their construction. They may be overlaid, flush, or lipped.

Types of Doors and Drawer Fronts

Overlaid fronts, also called **overlapping fronts** (Fig. 26-4.2), are the simplest to build, but use the most material. Doors and drawers overlap the cabinet front on all sides, and the front acts as a stop to prevent them from closing too far. Overlaid doors often have beveled edges, which act as finger grips and eliminate the need for door and drawer pulls. This is an advantage, because it is easy to bruise knees and thighs on exposed hardware on base cabinets in a kitchen.

Sometimes door and drawer fronts are cut square at the top and bottom and rounded at the vertical edges. This so-called **waterfall** edge is common on cabinets in lower-cost housing. Door pulls are required.

In cabinets with **flush fronts** the fronts of doors and drawers lie in the same plane as the frame (Fig. 26-

4.3). Fronts are cut $\frac{1}{8}$ to $\frac{3}{16}''$ smaller than the opening in which they fit to allow for expansion and smooth operation. When all pieces are cut for a close, attractive fit, they sometimes bind in operation. When the gap is large enough to assure smooth operation, appearance suffers. Pulls are required.

The most common construction combines the advantages of overlaid fronts and flush fronts, and avoids their major disadvantages. The **lipped front,** also called an **inset front** (Fig. 26-4.4), fits into the opening with enough clearance to assure smooth operation. Yet its edges, which are rounded, cover the opening to assure a neat, clean appearance. A $\frac{3}{8}''$ dado allows $\frac{1}{8}''$ clearance and a $\frac{1}{4}''$ lap on all sides.

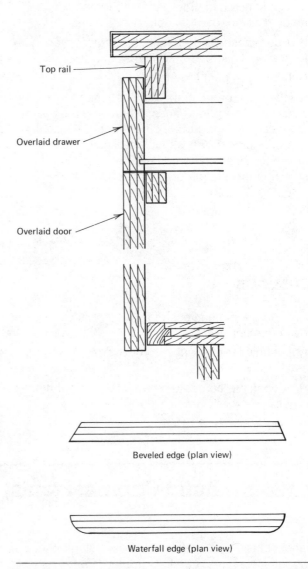

Fig. 26-4.2 An overlaid front overlaps the opening on all sides. The edges of doors and drawers may be square, beveled inward on all sides, or square at top and bottom and rounded at edges.

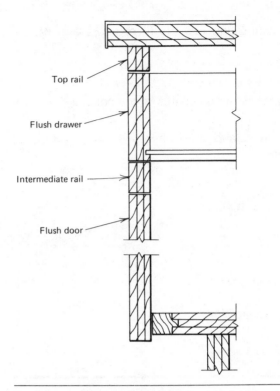

Fig. 26-4.3 In a flush front, the faces of doors and drawers are flush with the front of the cabinet.

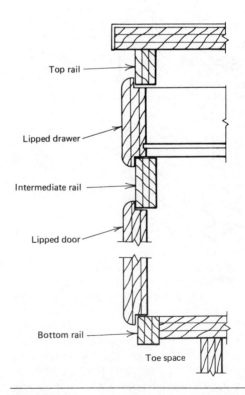

Fig. 26-4.4 In a lipped front, doors and drawers overlap the openings but also fit part way into them.

Materials

Economical usage of materials requires a good understanding of cabinet design. As in a closet, the wider the door opening, the more useful is the storage space behind it. Usually a cabinet less than 12″ wide, which has an opening of only 9″, is useless except for special storage, such as for card tables. Cabinets up to 21″ wide have single doors. A 24″ cabinet may have one door of any style, or a pair of flush or overlaid doors. Lipped doors meet at an intermediate stile, which interferes with effective use of the storage space. Cabinets more than 24″ wide have two or three doors.

When doors are cut from sheet materials such as plywood, their dimensions should be planned to minimize waste. A 4′ × 8′ sheet will yield a dozen doors $15\frac{7}{8}″$ wide and $23\frac{7}{8}″$ high—the standard height in a base cabinet with one drawer. Similarly, six doors $23\frac{7}{8}″$ wide and $29\frac{7}{8}″$ high plus two drawer fronts $23\frac{7}{8}″ × 5\frac{7}{8}″$ can be cut from one sheet with no waste.

Suppose that the cabinet to be built is 18″ wide with lipped fronts. The $15\frac{7}{8}″$ door is ideal; it fits over a 15″ opening between two $1\frac{1}{2}″$ stiles. But suppose the cabinet is 21″ wide. Rather than cut a door $18\frac{7}{8}″$ wide, which is wasteful, use a $15\frac{7}{8}″$ door and widen the stiles to 3″ to make up the difference. This is just one example of the planning required to build economical cabinets if drawings do not dictate specific dimensions.

Joints

The stiles and rails of a cabinet front must be tightly fastened together in joints as invisible as possible. The two types of joints that do the best job are mortise-and-tenon or doweled construction (Fig. 26-1.7).

Procedure

1 Determine the length and width of stiles.
2 From materials of the required width, cut stiles to length.
3 Determine the length and quantity of rails.
   Precaution. Allow for the type and depth of joint.
4 Cut rails to length, and shape their ends for fitting into mortises.
5 On stiles mark the size and location of mortises for rails.

6 Mortise the stiles.

7 Fill mortises with glue and insert rails.

8 Hold the assembly in clamps until the glue dries.

9 Cut doors to size.

10 If required, shape the edges of doors.

11 Cut any drawer fronts to size.

12 If required, shape their edges.

13 When the glue applied in step 7 has dried, remove the clamps and test the fit of doors and drawer fronts.

14 With finishing nails attach the front to the cabinet frame built in Units 26-1, 26-2, or 26-3.

15 Countersink the nails and fill the holes with wood putty.

16 Set drawers and drawer fronts aside in a protected location.

Questions

1 What is the common size for rails in cabinet fronts?

2 If a cabinet is 27″ wide, and has one door, what is the most economical width of door and stiles?

3 If the cabinet has two doors, what size would you make doors and stiles?

Unit 26-5   **Build a Drawer**

The sides and back of a drawer are stock items at most lumber yards. They are $\frac{1}{2}$″ boards of various widths, often edge laminated, and already planed smooth. Alternate materials are 1″ boards and $\frac{3}{8}$″ or $\frac{1}{2}$″ plywood. Bottoms are usually $\frac{1}{2}$″ plywood, although hardboard is adequate if drawers are small and the items likely to be stored are lightweight.

A good drawer is rather difficult to build with hand tools, but easy if you use a router with a dovetail attachment. The dovetail joint is by far the best for building sturdy drawers. The sides and back meet in continuous dovetail joints (Fig. 26-5.1). Fronts may be attached to sides in several ways. Most common is in a dovetail dado (Fig. 26-5.2. top). extending from the bottom of the front either to the top or just below it. A lapped joint or tongued lapped joint (center) is also used. If the drawer front is special, the drawer may be built like a box with its own front dovetailed into sides, and the exposed front attached from inside with screws (bottom). All four sides of a drawer should be routed for the bottom.

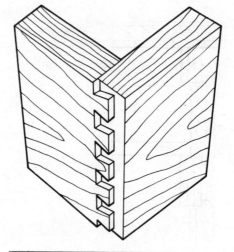

Fig. 26-5.1 The back and sides of good drawers are locked in a continuous dovetail joint.

Drawer Slides and Guides

There are many types of guides to keep drawers in line and operating smoothly. The simplest guide is a smooth runner of wood or plywood securely attached to the front and back of the cabinet frame (Fig. 26-5.3). Either the back of the drawer is notched to ride on the guide, or a block on the bottom of the drawer is grooved to fit over the runner. Nylon glides attached to the cabinet front keep the door in line and sliding smoothly.

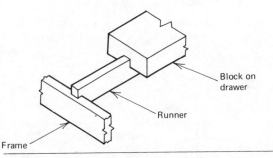

Fig. 26-5.3 The simplest drawer slide is a wood runner that fits into a notch or block on the drawer itself.

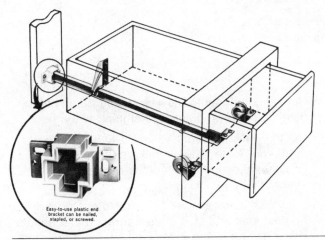

Fig. 26-5.4 Typical metal drawer slides. (Courtesy Amerock Corporation.)

In better quality construction drawers slide on nylon rollers in metal tracks (Fig. 26-5.4). The tracks may be under the drawer, above it, or on both sides. The hardware includes stops that prevent the drawer from being pulled all the way out of the opening.

The type and location of slides and guides must be decided in advance of construction. There must be a place on the cabinet frame for attachment of tracks, and clearance around the drawer for the hardware.

Procedure

1 Determine the type of drawer guide or slide to be used.

2 Determine the exterior dimensions of the drawer, excluding its front. **Note.** It is standard practice to allow a minimum clearance of $\frac{3}{4}''$ between the drawer and the frame of the cabinet.

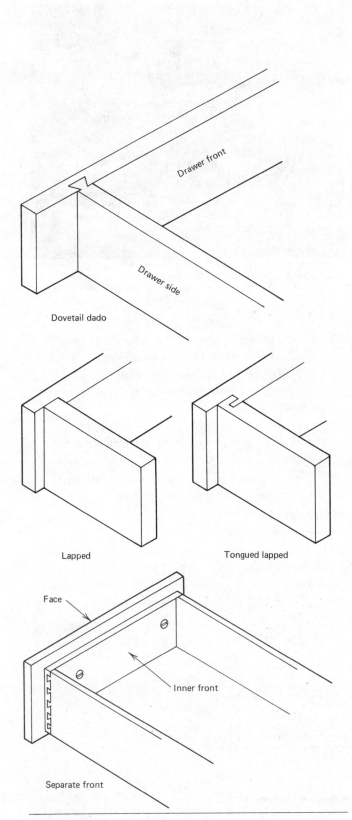

Fig. 26-5.2 Sides and fronts may be connected in a dovetail dado (top), lapped joint (left center), or tongued lapped joint (right center). Some drawers have double fronts that are screwed together (bottom).

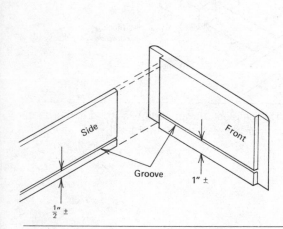

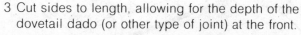

Fig. 26-5.5

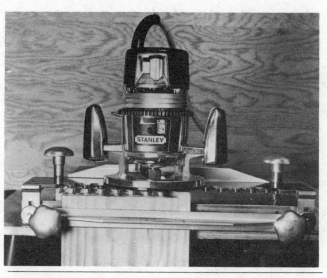

Fig. 26-5.6 (Courtesy Stanley Tools.)

3 Cut sides to length, allowing for the depth of the dovetail dado (or other type of joint) at the front.

4 Cut the back to length, allowing for dovetails at each end. **Note.** The overall length of the back is usually the overall width of the drawer less $\frac{1}{4}''$.

5 On the sides and back mark the location and thickness of the drawer bottom. **Note.** Depending on the types and size of drawer guide, the bottom of the drawer is $\frac{1}{2}$ to $\frac{3}{4}''$ above the bottom edge of the sides and back.

6 On the front mark the location of the drawer bottom. **Precaution.** The groove in the front is seldom at the same point as in the sides. Usually it is about $\frac{1}{2}''$ higher, and stops at the edge of the drawer (Fig. 26-5.5).

7 Rout the groove for the bottom where marked in steps 5 and 6.

8 Mark the location of the two dovetail dadoes in the front, and cut them to the height required.

9 Rout the ends of drawer sides to fit into the dovetail joint.

10 Test the fit, and measure the distance between sides to verify the width of the drawer back.

11 Using a dovetail template and router (Fig. 26-5.6), cut dovetails in the other ends of sides and in the back at both ends.

12 Check the fit. If the fit is too tight or too loose, correct it by turning the template adjusting nut and reworking the joint.

13 Cut the drawer bottom to size, measuring from inside of groove to inside of groove.

14 Spread glue in the dovetail dadoes, and slide the sides into the cuts in the front.

15 Insert glue in the grooves for the bottom in sides and front and slide the bottom into position.

16 Spread glue on the dovetails and in the groove in the back, and fit it into position.

17 Check all corners for square.

18 Hold the drawer in clamps until the glue dries.

Questions

1 What type of joint is most commonly used to attach the back of a drawer to its sides?

2 Describe two ways of attaching drawer sides to a front.

3 A cabinet is 15″ wide and 24″ deep, and has $1\frac{1}{2}''$ stiles. What are the approximate overall dimensions of a drawer with a lipped front?

Unit 26-6 **Build Cabinet Shelves**

To offer the homeowner the maximum use of storage space, shelves should be adjustable. Then they can be raised or lowered in cabinets to suit the height of the glassware, dinnerware, utensils, or foods that will be stored there.

Shelves up to 24″ wide must be supported at all four corners. Wider shelves require extra support at midpoints, or must be stiffened to prevent them from sagging under loads, or must be built of thicker materials. There are no hard and fast rules for shelf thickness. A shelf that holds only small cans of spices doesn't need to be as strong as a shelf carrying 2-liter bottles of soft drinks.

As a guide to shelf thickness, however, use the following table:

Area of shelf in square feet	Minimum thickness	Materials
1 or less	$\frac{1}{4}$″	Plywood or tempered hardboard
1 to 2	$\frac{1}{2}$″	Plywood or particleboard
More than 2	$\frac{3}{4}$″	Plywood

Shelf Dimensions

Fixed shelves should be cut for a tight fit, and supported at the ends on 1 × 1 wood cleats, with intermediate supports if necessary. Adjustable shelves should be cut for a loose fit. They may be supported on clips that fit into holes drilled into the cabinet frame, or on **standards**—slotted strips of metal that accept metal shelf supports.

The shelves in overhead cabinets run the full depth of the cabinet, with minimum clearance at cabinet doors. Intermediate shelves in base cabinets should not be more than half the depth of the cabinet. Otherwise items stored at the back of the bottom shelf can't be removed without unloading half the shelf.

Shelf Edges

Because the front edges of cabinet shelves are exposed to view when doors are open, they should be fin-

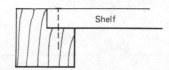

Fig. 26-6.1 A rabbeted 1 × 1 both stiffens and finishes the edge of a thin shelf.

ished. There are several methods. Thin ($\frac{1}{4}$″) shelves are finished with a strip of 1″ wood rabbeted to accept and stiffen the edge of the shelf (Fig. 26-6.1). Thicker materials may be finished simply with a batten nailed and glued along the edge. For the best appearance to shelves $\frac{1}{2}$″ thick or thicker, add a stock band. There are three shapes: T, V, and Y (Fig. 26-6.2). With each band the edge of the plywood must be routed to the shape of the band.

Procedure

1 Determine the overall dimensions of the shelf.
2 Cut the shelf to these dimensions, allowing for a slight gap at sides and back, and a $\frac{1}{2}$″ gap at the front edge.
3 Notch the corners to fit around the cabinet frame where necessary.
4 Check the fit.
5 Rout the front edge of the shelf to fit the shape of the band selected.
6 Insert the band at one end of the shelf to check the fit, and smooth any rough places.
7 Spread glue in the routed edge, and fit the band tightly in place.

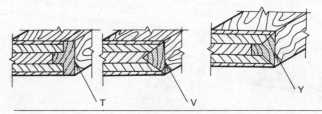

Fig. 26-6.2 Three ways to finish the edge of a plywood shelf $\frac{1}{2}$″ to $\frac{3}{4}$″ thick.

8 Add small finishing nails at the ends and middle
   to hold the band in place.
9 Countersink the nails and fill the holes with wood
   putty.
10 When the glue has dried, sand the joints between
   shelf and band, and ease the edges slightly.

Questions

1 Describe four ways of finishing the exposed edge of a cabinet shelf.
2 Why should an intermediate shelf in a base cabinet be only half as deep
   as the bottom shelf?

Unit 26-7   Build Countertops

Like cabinets, countertops may be made to order in a
shop or built on site. They consist of a base of smooth
$\frac{3}{4}''$ plywood or particleboard surfaced with a thin layer
of **plastic laminate.** Front edges and exposed sides
are finished in various ways. Where a countertop butts
against walls it usually turns upward into a **backsplash**
made of the same materials (Fig. 26-7.1).

The Base

The best countertop is seamless, and good kitchens
are designed so that no joints are necessary. Even
L-shaped sections are made in one piece. When a
seam in the base is necessary, make a butt joint lo-

Fig. 26-7.1 A countertop has a
working surface, a finished edge,
and usually an extension that
goes up the wall, called a backsplash. (Courtesy H. J.
Scheirich Co.)

cated between frames of a small base cabinet, which can provide maximum support. Hold the joint tight by backblocking with a strip of base material laid in mastic. Make sure that the backblock clears the framework at the front and rear of the cabinet.

Laminate

Sheets of laminate are only $\frac{1}{16}''$ or $\frac{1}{20}''$ thick. They are made in lengths of 5', 6', 7', 8', 10', and 12'. Standard widths are a nominal 24", 30", 36", and 48". Most manufacturers make sheets wider than nominal, however. Two tops 25" wide, for example, can usually be ripped from a single 48"-wide sheet.

Cutting

Laminate should always be cut $\frac{1}{8}$ to $\frac{1}{4}''$ oversized. The material chips easily, so the edge being cut must be supported as close to the cutting line as possible. The best cutting tool is a router and you work with the decorative side of the laminate face *down*. You may also cut with a fine-toothed handsaw or metal shears, working with the decorative side *up*. Do not use a utility knife; you are likely to ruin both the laminate and the knife blade.

Joints in laminate should be avoided. When necessary make them at one of two places: either at a 45° angle in a corner or at the edge of a countertop range. Seams in laminate should never lie over seams in a base. A seam at a sink is not a good idea, either It is scarcely visible but, unless it is watertight, moisture can seep into the base and cause delamination.

To cut a hole for a sink, lavatory, or countertop range, wait until the laminate is fully bonded to its base. Then mark the hole with a template, which allows the necessary $\frac{1}{16}''$ clearance on all sides. Make the cut with a saber saw.

Laminate is quite brittle until bonded to its base. Then it can withstand impacts, boiling water, and temperatures up to 275 °F (135°C) without damage. To do so, however, the base must be absolutely smooth and the bond tight over the entire surface. Particleboard makes an excellent base.

Adhesive

Contact cement is the adhesive used by most carpenters to bond plastic laminate to countertop bases. Slow setting glues may be used, but the laminate must be

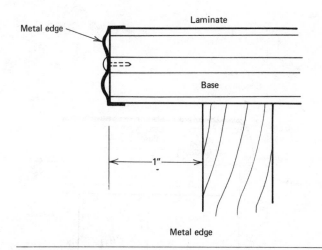

Fig. 26-7.2 A metal molding conceals the raw edge of the countertop base. The edge is the same thickness as the countertop itself.

clamped to the base until the glue dries. Use of contact cement is quicker and provides an excellent bond, but the result is only as good as the workmanship in preparing the base and accurately applying laminate.

Adhesive is applied with a brush, roller, or notched spreader to both the base and the laminate. A roller is best on broad surfaces and a brush on narrow strips. Both coats must be dry before you begin application, but the job must be completed within 2 hours under most conditions. Read instructions on the label of the can of contact cement for information on open time.

Edges

Edges and exposed sides of countertops may be finished with a metal molding attached with screws into holes predrilled in the edge of the countertop (Fig. 26-7.2). The molding may fit flush at the top, or form a lip to prevent liquids from running off the counter onto the floor. Another method is to fit a plastic T molding into a slot cut into the edge of the countertop (Fig. 26-7.3). Use a rubber mallet to drive the molding into the slot, and tack it on the underside with brads. Both plastic and metal moldings must be wide enough to cover the edge of the counter completely, including the laminate.

The most attractive countertops have laminate not only on the top but also on edges. Edge laminate is mounted on a strip of base material previously attached with nails and glue to the edge of the counter (Fig. 26-7.4). The width of the countertop base must be reduced by the thickness of the edge base. Height of

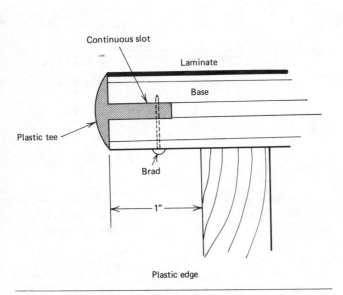

Fig. 26-7.3 A plastic T wedged into a slot in the counter-top base is an inexpensive method of finishing an edge.

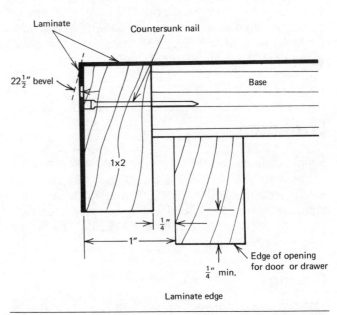

Fig. 26-7.4 The most attractive edge on a countertop is a selfedge. It consists of a strip of laminate on its own base. The width of this type of edge may be from two to three times as great as the thickness of the countertop itself.

the edge varies. It is usually $1\frac{1}{2}''$ in small kitchens, and as much as 2″ in large kitchens. Regardless of its width, the bottom edge must clear the tops of doors and drawers in the base cabinet by at least $\frac{1}{4}''$.

Countertops overhang base cabinets about 1″ at the front and exposed sides. Allow $\frac{1}{2}''$ clearance beside a refrigerator or freezer. Allow only $\frac{1}{8}''$ clearance beside a freestanding range or other appliance that is the same height as the counter.

Backsplash

The backsplash, also called a **splashboard**, may be built as part of the countertop or as a separate piece. Unless walls are absolutely straight and corners are square, it is usually easier to install a separate unit, and appearance will be better, too.

A backsplash rises 4 to $5\frac{1}{2}''$ up the wall above the countertop. In a one-piece unit the base of the back-splash extends to the bottom of the countertop base, and all laminate is applied before the unit is installed (Fig. 26-7.5, left). A separate backsplash rests on the countertop (right), and the base is attached to the wall before laminate is added. The joint between counter

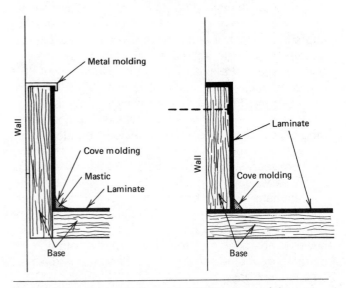

Fig. 26-7.5 A backsplash may be built as part of the countertop (left) or as a separate unit (right). Its top may be trimmed with laminate or metal. A small plastic cove set in mastic is required to seal the joint at the countertop.

and backsplash is sealed with a small plastic cove molding set in mastic. Where two lengths meet at a corner, miter.

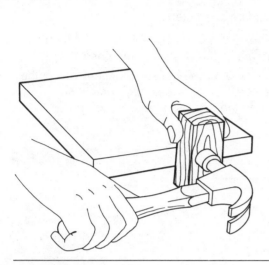

Fig. 26-7.6

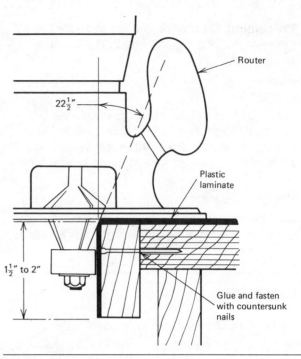

Fig. 26-7.7

Procedure for Laminating

1 Determine the dimensions of each section of base, cut each piece to size, and check the fit, especially at corners.

2 Study the surface for flaws, and fill all defects with crack filler.

3 Sand the surface lightly until it is smooth, and remove all dust. **Precaution.** The surface of the base must be smooth, clean, dry, and free of any foreign material such as grease.

4 Repeat steps 1 to 3 for any strips of base for edges or backsplashes.

5 Glue and nail base strips to the countertop base, countersink the nails, and fill the holes.

6 Cut pieces of laminate to size. **Precaution.** Allow for an overlap of the base on all sides.

7 If the countertop has a laminated edge, apply two coats of contact cement to the edge's base and another coat to the strip of laminate. With any other type of edge, go to step 13.

8 Let the two coats dry until they are slightly tacky to the touch.

9 Cautiously position the strip of laminate flush with the bottom of the edge base, and press it in place. **Precaution.** Be accurate in placement; once the two coats of adhesive touch, they can't be separated without damage to both base and laminate.

10 Assure a good bond by tapping a wood block with a hammer along the entire edge (Fig. 26-7.6).

11 With a router trim the excess laminate so that the edge is flush with the top of the countertop base.

12 File the edge smooth. **Precaution.** File away from you, never toward you, so that the file won't pull the edge loose.

13 Spread adhesive on the countertop and the back of wide sheets of laminate.

14 Let both coats dry.

15 Lay pieces of brown wrapping paper on the base, overlapping where they meet. Cover the entire surface to be laminated.

16 Position the laminate on the paper so that it butts against the backsplash or rear edge. Let it overlap at the front edge.

17 Gently pull out the first piece of paper a few inches at a time, at the same time pressing the laminate into the exposed contact cement.

Precaution. Be very careful not to knock the laminate out of position while contact is first being made.

18 Repeat step 17 down the entire length of the counter.

19 With a 3″ hand roller press heavily on the laminate, beginning in the center and working toward the edges. **Precaution.** Make sure to roll every square inch of the surface to assure a good bond.

20 Wipe off any excess adhesive.

21 a. If the countertop has a laminated edge, trim excess laminate at the front edge using a router

with bevel trimming bit, set at $22\frac{1}{2}°$, to bevel the two edges where they meet (Fig. 26-7.7). As an alternate use a block plane and hand file, filing only on the downward stroke.

21 b. If the countertop has a metal or plastic edge, trim laminate as in step 11.

22 Install any metal or plastic edging.

23 Trim the rear edge flush with the base.

24 After the base for the backsplash is nailed to the wall (see Unit 27-4), repeat the laminating process on the front and top edge of the backsplash.

Questions

1 What material makes the best use for a countertop? Why?

2 If a countertop is 25″ from front to back, has a laminated edge and a separate backsplash, what is the depth of the base?

3 What is the method suggested for preventing laminate from bonding to the base before you are ready for that to happen?

4 How do you finish an edge where two pieces of laminate meet?

5 Describe the two recommended ways to cut laminate.

6 If the finish dimensions of a countertop are 48″ × 25″, what size should you cut the laminate?

Unit 26-8   Build Simple Bookshelves

In many homes bookshelves are built into walls, either as a separate unit or part of a complete storage wall. The cabinet in Fig. 26-8.1 is a typical example. Its base is similar to a kitchen cabinet. It has storage space behind hinged doors, and long, flat, open shelves for such items as magazines and books too tall to stand upright. Above are more shelves for display of books and objets d'art—a fancy name for vases, bowls, and heirlooms that the family is proud enough of to place on public view. Sides and shelves more than $11\frac{1}{2}″$ wide are made either of hardwood plywood or edge-laminated boards. Narrower parts may be cut from the same materials or from S4S boards.

This type of cabinet is beyond the skill and experience of most young carpenters. But simple bookshelves are not. Almost any room can benefit by a bookshelf or two—in the kitchen for cookbooks, in a

shop for how-to books, in bedrooms for a mystery novel or two, and in the living room or den for reference books.

There are two ways to put up shelves. One is to attach standards to studs behind wall surfaces, add brackets, and fit the shelves onto the brackets (Fig. 26-8.2). The quantity, length, and location of the shelves vary with wall space available and the number of books to be stored.

The other type of bookshelf is a simple box with well-made joints (Fig. 26-8.3). One box provides two shelves. A taller box may have adjustable shelves inside. Or three boxes properly spaced can provide six shelves in an interesting design.

These shelves may be attached to a wall in several ways. Here are a few:

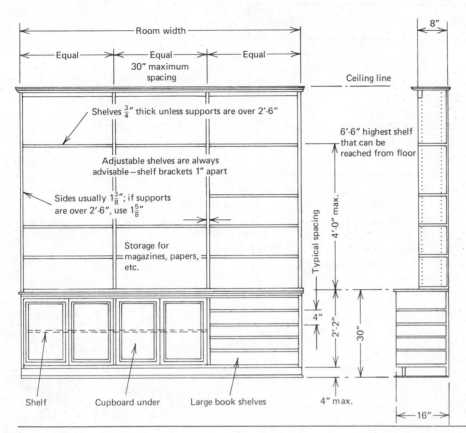

Fig. 26-8.1 Details of a good storage wall for a living room or family room.

- Hung on standards.
- Hung on shelf brackets attached to perforated hardboard.
- Nailed or screwed to studs through a back of plywood at least $\frac{3}{8}$" thick.
- Attached through cleats into studs, like wall cabinets.

Shelf Spacing

The best distance between shelves depends on the type of books to be stored. A spacing of 9" is adequate for paperbacks. A spacing of $11\frac{1}{2}$" is enough for storing cookbooks, novels, and most textbooks. For large reference books, such as atlases and some texts, 14" is the best spacing.

Materials

Bookshelf units $11\frac{1}{2}$" deep will hold most books. The best materials, then, are 1 × 12 boards of No. 1 clear

Fig. 26-8.2 Loose shelves rest on brackets locked into standards to provide convenient storage space for books in a paneled family room. (Photo: Leviton, Atlanta.)

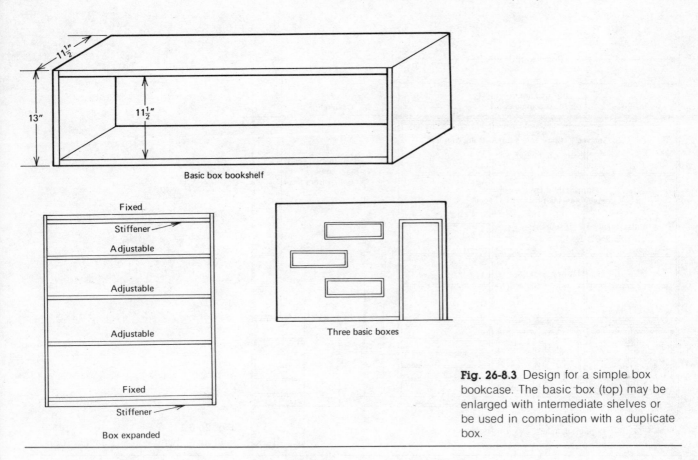

Basic box bookshelf

Fixed
Stiffener
Adjustable

Adjustable

Adjustable

Fixed
Stiffener

Box expanded

Three basic boxes

Fig. 26-8.3 Design for a simple box bookcase. The basic box (top) may be enlarged with intermediate shelves or be used in combination with a duplicate box.

ponderosa pine or white pine. Shelves up to 4' long do not require stiffening. Long shelves need a 1×2 strongback (see Fig. 26-8.3) to prevent them from deflecting under the constant load of heavy books.

Procedure

1 Determine the spacing between shelves. **Note.** The spacing must be measured from the top of one shelf to the underside of the shelf above, or to the bottom edge of any stiffener.

2 Cut the sides, top, and bottom of the box.

3 Cut continuous dovetail joints in the ends of all pieces, as described in Unit 26-5.

4 Assemble the unit with glue and finishing nails.

5 Cut and install any back or hanging cleats.

6 Add any stiffener if required.

7 Cut any intermediate shelves.

8 Add short standards in pairs on the sides of units to support intermediate shelves. **Note.** Screw holes in the standards must be equidistant above the bottom shelf. Otherwise intermediate shelves will not be level.

Questions

1 Name four ways to support bookshelves on a wall.

2 Can you think of any other ways?

3 How do you determine where to attach standards for intermediate shelves?

27

Cabinet Installation

Installation of cabinets is the next to the last operation in finishing the interior of a house: only trimming remains. The ceiling is finished and painted. and recessed light fixtures are usually in place. Walls are fully surfaced and painted. Flooring is in place and carried all the way to walls under base cabinets. Although it may never be seen. flooring should extend under cabinets to protect the subflooring from possible leaks and spills around the sink and dishwasher. In other words, rooms in which cabinets are to be set are as complete as if they had no cabinets at all.

Builders furr down the ceiling above wall cabinets at one of two stages of construction. Some prefer to frame the cabinet soffits as soon as ceiling joists are in place. and before walls are surfaced. This method saves materials and assures a good structural relationship between furring and the ceiling and wall framing. Others prefer to surface the walls and ceilings first. then furr the ceiling. Although this method takes a little more material. it takes less labor because much cutting and fitting is eliminated. The procedures in this chapter call for finishing wall and ceiling surfaces before furring begins.

Wall cabinets are usually installed first. They must be held in place by one carpenter while another attaches them to the wall and ceiling, and this job is easier when base cabinets are not in the way. Base cabinets come next, and countertops last.

When installing cabinets, the carpenter should keep two goals in mind. One is to make sure that all cabinets are firmly attached and capable of holding the loads that will be placed in them. The other is appearance. In many ways a kitchen is a laboratory, and a look of neatness and efficiency is of primary importance. All cabinets must be level and plumb, with all edges flush with each other and hardware carefully aligned.

Unit 27-1   **Build Soffits for Wall Cabinets**

Furring is a light framework that forms a soffit to which most wall cabinets are attached and a fascia to complete the wall above them. The cabinet soffit is sometimes called a **bulkhead.** It may be installed piece by piece, or be preassembled and installed as a unit. It may be installed before or after walls and ceilings are surfaced.

A cabinet soffit may run the entire length of a wall, even though cabinets don't cover the entire wall. It provides an excellent place for a recessed light over the sink, for example. Or it may be only as long as cabinets, and exposed on one end.

The framework is made of 2 × 2s and/or 2 × 4s (Fig. 27-1.1). When the furring is attached to exposed structural members, 2 × 2s are usually adequate. When the wall and ceiling are already surfaced, 2 × 4s are recommended because of their broader surface for nailing.

To assure a solid base for attachment, blocking is required between studs at the end of a bulkhead (*A* in Fig. 27-1.2). It is advisable to add a block (*B*) between studs to support an open-end soffit. A similar block is recommended for the same purpose between ceiling joists at right angles to the cabinet. When ceiling joists run parallel to the face of the cabinet, a ladder of blocking (*C*) is required as a nailing surface for the corner member of the soffit.

The faces of wall cabinets are usually recessed $1\frac{1}{2}$ to 2″ from the fascia and the joint finished with a small cove molding (Fig. 27-1.3, left). Manufactured cabinets are made with their top rails flush with the top of the cabinet, and are almost always installed this way. Cabinets built on site may be built with a wider top rail, however, and installed flush (Fig. 27-1.3, right). The framework is the same except that the 2 × 4 edge member is installed flat with recessed cabinets and upright with flush cabinets.

Procedure

1 Determine from drawings whether wall cabinets are to be installed flush or recessed.

2 Write down the overall dimensions of the furring. **Precaution.** Be sure to allow for the thickness of finish materials on the fascia and soffit.

3 Cut long horizontal members to length. They should all be the same length (see Fig. 27-1.1).

4 Cut horizontal cross members to length. They are also equal in length.

5 Cut vertical members to length. Except for a 2 × 2 to frame the end of a short bulkhead (shown dotted in Fig. 27-1.1), these are also equal in length.

6 Assemble the vertical framework.

7 Assemble the horizontal framework.

8 Connect the two frames by cross-nailing.

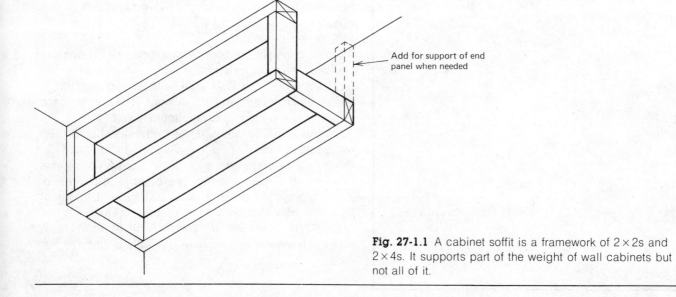

Add for support of end panel when needed

Fig. 27-1.1 A cabinet soffit is a framework of 2 × 2s and 2 × 4s. It supports part of the weight of wall cabinets but not all of it.

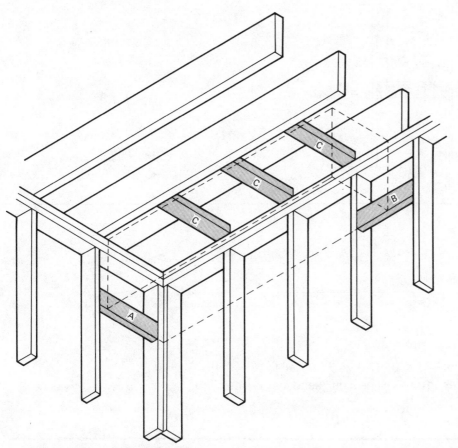

Fig. 27-1.2 Where blocking is required between studs and joists to support the furring.

9 On the surface of the wall snap a chalk line to mark the height of the bottom of the frame. **Note.** The usual height is 84" above the floor plus the thickness of finish soffit material.

10 Install the framework on the lines, nailing into studs, ceiling joists, and blocking. **Precaution.** Check level and plumb before driving nails home.

11 To strengthen the soffit, add 1 × 2 braces every 24" as shown in Fig. 27-1.4.

12 Surface the soffit with wallboard as described in Units 21-2 and 21-5.

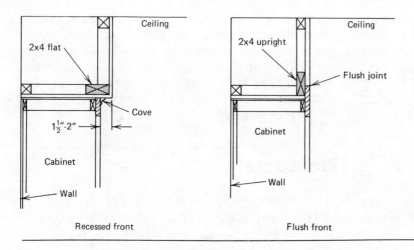

Fig. 27-1.3 Most wall cabinets are inset from the fascia, and trimmed with a cove molding (left). Occassionally, however, cabinets have a wider top rail that lies flush with the fascia (right).

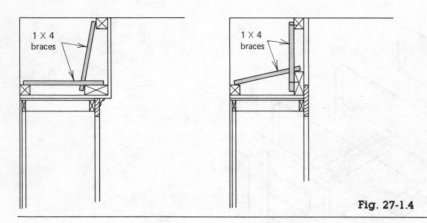

Fig. 27-1.4

Questions

1 What framing member in a cabinet soffit is always a 2 × 4?

2 What determines whether this member is placed flat or upright?

3 Name an advantage of each of the two times when a bulkhead may be installed.

Unit 27-2   **Install Wall Cabinets**

The most prominent edges of wall cabinets are the bottom line and exposed corners. In setting cabinets then. these are the two edges that must be level and plumb. respectively. Any unevenness at the soffit or in corners can be hidden behind trim.

Because of the weight of wall cabinets and the need for accuracy of installation. setting them is a two-person job. It is easier to install wall cabinets first before base cabinets get in the way. But there is one disadvantage. In most kitchens there is little leeway to adjust the locations of base cabinets. while wall cabinets can be adjusted so that they line up with the cabinets below.

Attachment

Cabinets are attached with screws at three points (Fig. 27-2.1). Drive one set horizontally through the top frame at the back of the cabinet into blocking in the walls. Drive a second set vertically through the framework just inside the top rail into the soffit. Drive the third set horizontally through the support for the bottom shelf into another set of blocking. If walls do not contain blocking. cabinets may be attached with wall anchors to the finish wall material.

Holes must be predrilled for screws. The best locations are about 3" from the ends of the cabinet and approximately every 16" between. and centered in cabinet framing members. Drill holes slightly upward for screws that go into walls. so that there is room to operate a ratchet screwdriver.

Begin the installation in a corner and work outward. If two cabinets meet in any corner. start there. When two or more wall cabinets are installed in a row. most carpenters place them together on the floor. align them carefully. then drill holes and drive screws through stiles (Fig. 27-2.2). Then the cabinets are hung as a unit. and good alignment is assured.

Procedure

1 Measure the height of the cabinet at its back.

2 Measuring down from the soffit. mark this height lightly along the wall with a chalk line. **Note.** If

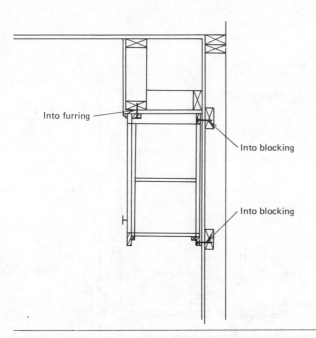

Fig. 27-2.1 The three points of attachment of wall cabinets.

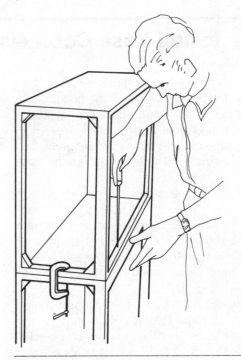

Fig. 27-2.2 When more than one cabinet covers a wall, it is easier to assemble them on the floor than to hang them individually. Clamp the cabinets together to assure alignment at stiles and bottom rail, and connect them with screws through holes predrilled in stiles.

cabinets of more than one height are to be installed on a wall. mark the bottom edges of all cabinets.

3 Check the line for level.

4 Measure the widths of all cabinets.

5 Mark the locations of cabinet sides along the height line. **Note.** Unless the total width of cabinets and the length of wall are exactly the same, allow a slight gap at corners.

6 If the cabinet frame is not already predrilled, drill holes for screws.

7 Raise the cabinet or cabinet assembly. into position along the line. and with its corner on the vertical marks made in step 5.

8 While a helper holds the cabinet on its mark and firmly against the wall and soffit. drive home screws in the upper back of the cabinet.
Precaution. Use flat-head wood screws long enough to penetrate 1″ into blocking.

9 Recheck the cabinet's location for plumb and level.

10 Drive screws upward into the soffit. Shim if necessary to assure a tight installation.

11 Complete attachment by driving in screws below the bottom shelf.

12 Test the installation by tugging on the cabinet.

13 Install remaining wall cabinets.

14 Fill exposed screw holes with wood putty or matching putty stick.

Questions

1 Describe the points of attachment of wall cabinets and the angle of the fasteners.

2 Where may it be necessary to shim a wall cabinet?

Unit 27-3 **Install Base Cabinets**

When hanging wall cabinets. you keep the bottom level and shim at the top. When setting base cabinets, you keep the tops level and shim at the bottom. Tops must be exactly level if the countertop is to fit properly, and fronts must be vertical and line up flush with each other.

Like wall cabinets. base cabinets are attached with screws at three points (Fig. 27-3.1). Drive one set upward at about a 30° angle through the upper frame into blocking. Drive a second set downward at about a 30° angle through the bottom frame into either blocking or studs. Drive the third set at about a 60° angle through the base into the subfloor.

Because base cabinets rest on the floor and can be moved without much difficulty. lay out the cabinets on the floor with chalk lines. By doing so, you can quickly see how good the fit is. and where adjustments must be made. Aim for close fits. especially next to freestanding appliances. Leave just enough room to move appliances into position and to remove them if necessary for repair or future replacement. A gap gathers dust, dirt, and spilled food, and is very difficult to clean.

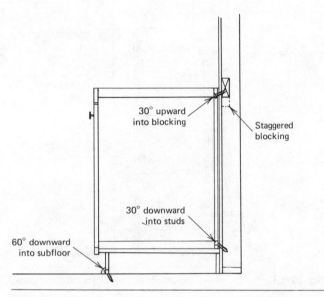

Fig. 27-3.1 The three points of attachment of base cabinets.

Procedure

1 Measure the depths of all cabinets at the base to make sure they are identical in dimension.

2 If the cabinet frames are not already predrilled. drill holes for screws.

3 On the floor snap a chalk line at the distance in step 1 from walls.

4 Move the cabinets into position along the chalk line.

5 Check the fit at corners and at openings for appliances. **Note.** If appliances are not on hand to measure. check the manufacturer's literature for dimensions.

6 If the cabinet has a sink in it. make sure it lines up properly with water supply and drain connections.

7 Check the level of each cabinet from front to back. and the level across the tops of all aligned cabinets.

8 If the top line is uneven, mark the top of the tallest cabinet on the wall. and snap a level chalk line through that point.

9 Remove the cabinets. then reposition them. at the same time leveling them with shims.

10 When all fronts are aligned and all tops are level. install the first cabinet in the corner with screws driven into blocking as shown in Fig. 27-3.1.

11 Align. level, and install all remaining base cabinets.

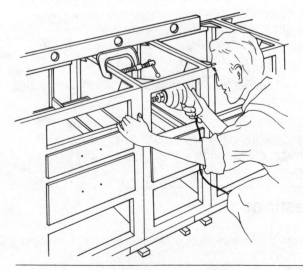

Fig. 27-3.2

12 Fasten cabinets together with wood screws driven into holes predrilled in the stiles. **Precaution.** To assure that cabinets don't move during attachment. hold them together with C clamps (Fig. 27-3.2).

13 Recheck the installation for level and plumb.

14 Fasten cabinets through the base into subflooring.

Questions

1 Describe the points of attachment of base cabinets. and the angle of the fasteners.

2 If necessary for leveling, where do you shim base cabinets?

3 Why is it advisable to attach wall cabinets to each other before installation. but not base cabinets?

Unit 27-4   **Install an Island Cabinet**

Some large kitchens have a cabinet or two in the middle of the room and not close to any wall (Fig. 27-4.1). Often these units. called **island cabinets,** are designed to serve a special purpose. such as a cooking center or food preparation center. They are likely to differ from standard base cabinets in three ways: all sides are exposed. the top may be higher or lower than other counters, and they may require separate water. fuel. power. or drain lines.

At the time the cabinet is ready to install. utility lines should be stubbed in. and extend far enough above the floor and bottom shelf of the cabinet for later finish plumbing or wiring. The carpenter must work around these utility lines.

Fig. 27-4.1 Island cabinets are base cabinets attached only to the floor in the middle of the kitchen. Frequently they are specially designed and built. (Courtesy H. J. Scheirich Co.)

Procedure

1 From the floor plan of the house establish the exact location of the island cabinet.

2 Mark the location of the base on the floor with chalk lines. **Note.** Island base cabinets often have toe space on more than one side. so allow for this in marking the floor.

3 Measure the base of the cabinet to verify the fit.

4 Measure from the chalk lines to the centers of any stubbed-in plumbing or electrical lines.

5 Accurately transfer these measurements to the bottom shelf.

6 Cut holes in the shelf for connections.

7 Carry the cabinet into position, and set it over the stubbed-in utility lines, with the utilities rising through the bottom shelf.

8 Check the fit of the base on the chalk lines.

9 When the fit is right, screw the base to the floor as described in Unit 27-3 and shown in Fig. 27-3.1.

Questions

1 In what three ways is an island cabinet likely to differ from a regular base cabinet?

2 Why must an island cabinet usually be lifted rather than slid into place?

Unit 27-5 Install Countertops and Backsplashes

When walls are straight and corners are square, it is easiest to build a countertop and backsplash as a unit, laminate them at the same time, and install them in one piece. But when walls are uneven and side walls are not at right angles, the carpenter has several alternatives.

One method is to cut the base for an exact fit, install it, then apply the laminate. Another—and this works well if the unevenness is not great—is to install a plastic or rubber cove along edges of the counter that meet the wall (Fig. 27-5.1, left). The cove compresses like a gasket, and hides the unevenness. This is a good solution if there is no backsplash.

When there is a backsplash, the countertop doesn't need to fit exactly except at the front edge. The backsplash hides minor unevenness. Any gap may be concealed by a small cove (Fig. 27-5.1, right), or the base of the backsplash may be made of a thinner material—$\frac{1}{2}''$ instead of $\frac{3}{4}''$—that is supple enough to bend.

In such a case the backsplash must be built separately, and be laminated after the base is installed.

Whatever the method of countertop construction, the procedures for installing it are approximately the same. One person can usually install a countertop up to 4' long alone. Longer tops require four hands.

Occasionally a countertop will not have cabinets under its entire length. Typical examples are above the kneehole in a planning desk, and in a corner above a lazy Susan revolving storage unit. At such points a 1×2, carefully set level with adjoining cabinet tops, is fastened along the wall into studs (Fig. 27-5.2).

Procedure

1 At all four corners of each base cabinet unit, predrill holes for screws. **Note.** If corners are blocked (see Fig. 26-1.8), drill through the blocks.

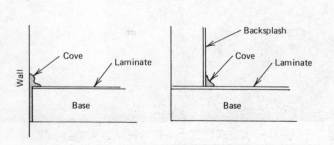

Fig. 27-5.1 Plastic coves seal seams between countertop and wall, backsplash and wall, and countertop and separate backsplash.

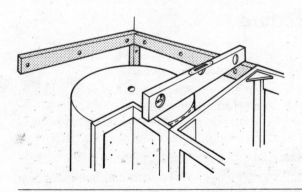

Fig. 27-5.2 Countertops must be fully supported along the wall when there is no cabinet framework beneath them.

2 If the countertop has no backsplash, glue a cove molding to the back of the base (Fig. 27-5.1, left).

3 Place the countertop on the cabinets about 6" from the back wall. **Precaution.** *Set* it down at this point; avoid dragging it across the cabinet frames as much as possible.

4 Gently but firmly shove the countertop against the wall. If it touches more than one wall, make sure it is as tight as possible against all of them.

5 With the top held tight in position. drive screws upward through all holes into the base. **Precaution.** Screws should penetrate no more

than half way into base material. Make sure the screws draw the base tight to the cabinets.

6 a. If the countertop has a separate backsplash, nail its base to the wall, and finish it with laminate as described in Unit 26-7.

6 b. Cover the joint between countertop and backsplash with a small plastic cove set in mastic (Fig. 27-5.1, right).

7 Seal the seam between backsplashes and walls with a bead of waterproof sealant spread with a putty knife. Wipe off all excess sealant.

Questions

1 At what points may you need to install a cove with a flange on it?

2 At what points may you need a small cove without a flange?

3 At how many points is a countertop attached to each base cabinet? Where are these points?

4 If attachment is made through boards in the corners of cabinets into a countertop base of $\frac{3}{4}$" particleboard, how long should the screws be?

Unit 27-6   **Cut Holes in Laminate**

Holes for a sink or lavatory are cut in a countertop after it is installed. In this way the carpenter is assured of cutting in the exact location necessary for attachment of plumbing in the cabinet below.

Most sinks and some lavatories have a separate stainless steel rim about $\frac{11}{16}$" wide that covers the joint between the countertop and the flange of the plumbingware (Fig. 27-6.1). Other types are called **self-rimming**—that is. the rim is part of the lavatory and fits over the hole in the countertop instead of into it (Fig. 27-6.2).

The required hole may be circular. oval. or rectangular with rounded corners. It is important to cut the shape accurately. and at the location shown in working drawings. Usually connections for water lines lie within the cutout.

Bowls are fastened to countertops with metal clips (Fig. 27-6.3). The edge of the hole must clear the cabinet frame by at least 2" to allow room for tightening the clips.

Fig. 27-6.1 Some sinks and lavatories have a separate metal rim that covers the edge at the countertop. (Photos courtesy Eljer Plumbingware.)

Fig. 27-6.2 Some sinks and lavatories have an extended lip that covers the edge of the hole. They have no separate rim.

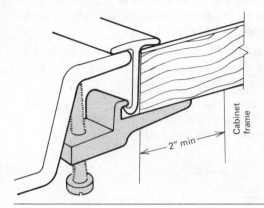

Fig. 27-6.3 Metal clips hold lavatories and sinks in place in countertops. They are attached to the rims of rimmed bowls or the concealed flanges of self-rimming bowls.

Procedure

1 From drawings determine the location of the fixture in the countertop. **Note.** If the precise location is not shown in details in the working drawings. use the center line of a window or medicine cabinet as one coordinate. For the right-angle coordinate use the midpoint between the edge of the counter and the front edge of the backsplash.

2 Mark these center lines with a pencil on the countertop. The length of the lines should be greater than the dimensions of the plumbing fixture.

3 Lay a self-rimming fixture or a separate rim upside down on the counter so that its outer edges are equidistant from the point where the center lines cross (Fig. 27-6.4).

4 Trace around the edge of the fixture or rim.

5 Determine the distance from the outline to the point of cutting. **Note.** When the sink or lavatory is self-rimming, this dimension is usually $\frac{3}{8}''$. When the rim is separate, the dimension is about $\frac{1}{4}''$.

6 Remove the fixture or separate rim and. with a scriber set at the dimension in step 5. draw a parallel outline inside the outline in step 4.

7 Within this outline bore a hole for the blade of a saber saw.

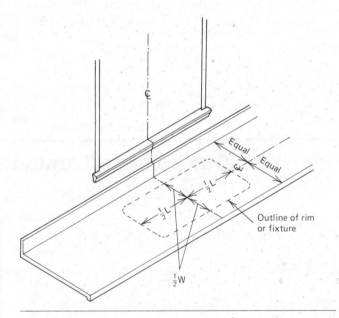

Fig. 27-6.4

8 With the saber saw cut along the inner outline until the hole is about two-thirds cut.

9 Tack a 1 × 2 temporarily under the cut area to support the cutout until you complete the cut.

10 Finish the cut and remove the cutout.

11 Check the fit. and adjust where necessary. The sink or lavatory should fit with $\frac{1}{16}''$ to $\frac{1}{8}''$ clearance on all sides.

12 If the sink is self-rimming. apply a bead of plumber's putty to its edge. and set the unit in the hole. **Note.** Omit this step if the rim is separate.

13 Clamp or otherwise secure the plumbingware in place.

14 Coat the underside of a separate rim with plumber's putty. and press the rim firmly over the seam.

15 Remove excess putty with a putty knife.

Questions

1 Describe the differences between the two types of lavatories designed for countertop installation.

2 How are sink and lavatory units usually attached to the countertop?

3 How is the seam between countertop and plumbingware sealed?

4 What tools do you use to cut holes in a countertop?

Unit 27-7   **Install Cabinet Doors**

Most cabinet doors are hinged. Occasionally they slide, usually when there is not enough room for a door to swing. Sliding doors are always installed after cabinets are in place. although their tracks are usually installed after cabinets are assembled. Hinged doors may be attached before or after the cabinet is installed. They are easier to position properly and hang accurately when the cabinets can be laid face up on the floor. But doors often get in the way of cabinet installation by swinging open at inopportune moments.

Hinged Doors

There are four types of cabinet hinges: surface, pin, semiconcealed. and invisible. Designs vary to meet decorative requirements and the type of cabinet door to be hung.

Surface Hinges

As their name implies, surface hinges are mounted on the face of the door and the stile of the cabinet. They are known by their shape (Fig. 27-7.1). Their purpose is not only to support the door but to add a decorative touch to the cabinet. Often the hinges are hammered brass with a flat black coating baked on. Flush doors require flush hinges; lipped and overlaid doors require offset hinges.

Pin Hinges

Pin hinges are like standard hinges for interior doors except for their size. They are mounted on the back side of the door and stile. and only the pins are visible from outside the cabinet. Pin hinges are made of several metals. but are usually finished in brass. bronze. or chromium. all either polished or dull.

Semiconcealed Hinges

Semiconcealed hinges have a vertical leaf and a horizontal leaf (Fig. 27-7.2). The vertical leaf is attached to the back and edge of the cabinet stile. and the horizontal leaf to the back of the door. The two leaves turn on a pivot that fits in a narrow slot cut into the edges of the door. This type works well with overlaid doors.

Invisible Hinges

Invisible hinges are designed for use with flush doors. One type is mortised into the edge of the door and the edge of the stile (Fig. 27-7.3, left). Another type (right) requires no mortise; it is mounted on the back of the door and stile.

Other Hardware

Many cabinet hinges are **spring loaded.** They have a small spring in them which automatically pushes the door shut. When doors don't shut by themselves. hardware is required to hold them closed. The most

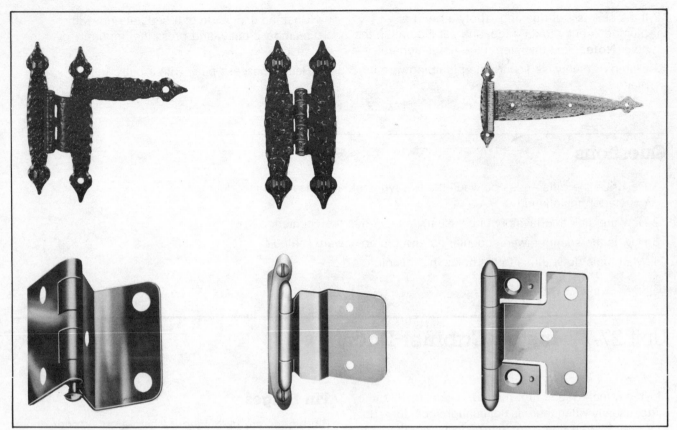

Fig. 27-7.1 Typical shapes of surface-mounted hinges. Flush hinges are identifiable by their shapes: L, H, and strap, for example (top row, left to right). Offset hinges (bottom row) are swaged. (Photos courtesy Amerock Corporation.)

common device is a two-piece magnetic catch (Fig. 27-7.4). One piece is a flat metal plate attached to the inside surface of the cabinet door near the edge and at midheight. The other is a small metal box with mag-netic leaves in it. It may be mounted on the stile at mid-height, or on the under side of a center shelf.

Unless doors are the overlaid type and edges are beveled for finger gripping, a knob or pull is needed to

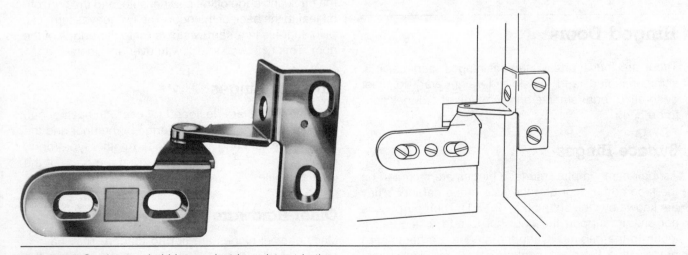

Fig. 27-7.2 Semiconcealed hinges pivot in a slot cut in the cabinet door.

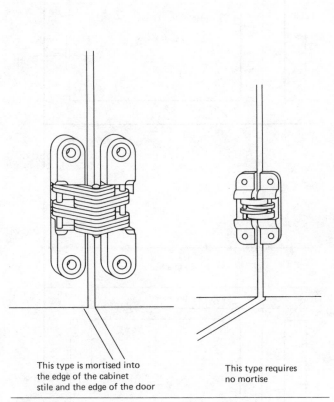

This type is mortised into
the edge of the cabinet
stile and the edge of the door

This type requires
no mortise

Fig. 27-7.3 Invisible hinges are designed either to fit into
mortises (left) or for flush mounting on the inside of the
cabinet and door (right).

open each door. Hundreds of designs and styles are
available in metal, wood, plastic, or ceramic. Knobs
and hinges usually have the same finish. Almost all
types are attached with a screw inside the door.

Fig. 27-7.4 A magnetic catch fits on the bottom of a shelf
or inside a stile and draws to it a plate mounted on the
inside of the cabinet door. This model holds a pair of
doors shut.

Sliding Doors

Sliding cabinet doors are made of $\frac{1}{4}''$ sheet material—
plywood, plain hardboard, or perforated hardboard.
They slide in a pair of top and bottom tracks. Manufactured tracks are made of nylon, plastic, or metal in
strips that you cut to length. Carpenter-made tracks are
grooves cut into thin strips of clear, straight hardwood
board glued inside the door openings (Fig. 27-7.5).

Bottom tracks must be shallower than top tracks,
and the height of the door should be about $\frac{1}{4}''$ greater
than the vertical opening between tracks (Fig. 27-7.6).
This arrangement permits doors to be inserted and
removed when necessary. Doors overlap about $\frac{1}{2}''$ at the
center of the opening.

Sliding cabinet doors are operated either by a recessed pull or a hole 1″ in diameter cut through the
door as a finger grip.

Hardware Locations

In a room with many cabinets, it is important to attach
visible hardware at uniform locations. There is no set
standard for hinges, but a common location is centered
3″ from the top and bottom edges of the door, regardless of its type. Thus the upper hinges of all wall cabinets and the bottom hinges of all base cabinets will
line up around the room. The heights of the second
hinges will vary, depending on the height of upper
cabinets and whether lower cabinets have a drawer.
But the heights will vary uniformly.

Knobs are centered on lines drawn at 45° from the
lower corner of doors on wall cabinets and the upper

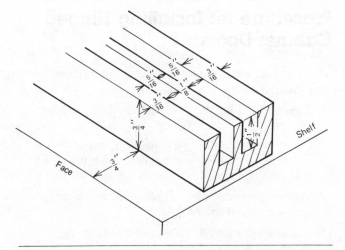

Fig. 27-7.5 Tracks may be made by cutting grooves
in hardwood boards.

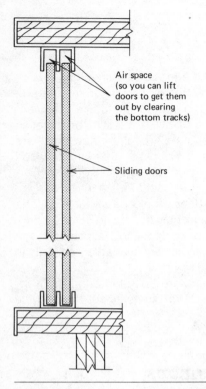

Air space (so you can lift doors to get them out by clearing the bottom tracks)

Sliding doors

Fig. 27-7.6 Tracks are placed about $\frac{3}{4}''$ in from the edge of cabinets. Note that the top track is deeper than the bottom track.

corner of doors on base cabinets. A common location is between $1\frac{3}{4}''$ and $2''$ along the diagonal, which is about $1\frac{1}{4}$ to $1\frac{3}{8}''$ from each edge (Fig. 27-7.7). Pulls are usually installed vertically on doors and horizontally on drawers. One good location is with the screw nearest the counter at the same location as the screw for a knob.

Procedure for Installing Hinged Cabinet Doors

1 From the package of hinges pull out the template provided by the manufacturer.

2 Establish the locations of hinges; they must be uniform.

3 Using the template. mark the locations of the hinge leaves and screw holes on both the door and stile.

4 Make starter holes for screws with an awl or screw starter (Fig. 27-7.8).

5 Cut any slots for semiconcealed hinges or mortises for concealed hinges.

6 Install the hinges on the door first.

7 Attach the hinges to the stile with a single screw.

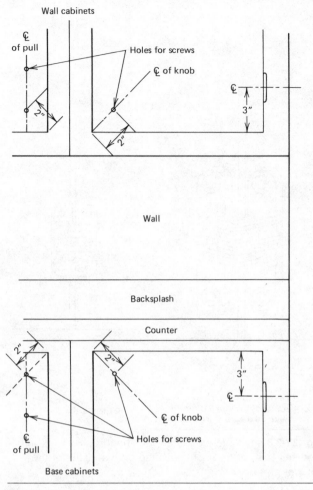

Wall cabinets

$\frac{\mathbb{C}}{}$ of pull

Holes for screws

$\mathbb{C}$ of knob

2"

2"

3"

Wall

Backsplash

Counter

2"

2"

3"

$\mathbb{C}$ of knob

Holes for screws

$\frac{\mathbb{C}}{}$ of pull

Base cabinets

Fig. 27-7.7 Typical locations for cabinet door hardware.

8 Test the fit.

9 When the fit is good. install the remaining screws.

10 Establish the locations of knobs or pulls; they must be uniform.

11 Install knobs.

12 Locate magnetic boxes on shelves or stiles. using the manufacturer's template for positioning the box at the proper distance from the inner face of the door.

Fig. 27-7.8 (Courtesy The Irwin Auger Bit Company.)

13 Install the box. **Note.** Do not sink screws at this time.

14 Mark the location of the metal plate on the back of the door.

15 Install the plate on the marks.

16 Test the effectiveness of the catch by opening and closing the door.

17 Drive screws home that were left loose in step 13.

Procedure for Installing Sliding Cabinet Doors

1 Determine the length of track required.

2 Cut both upper and lower tracks to this length.

3 Mark the location of the outer edges of tracks on the supporting shelves. **Note.** Tracks are usually inset $\frac{1}{2}$ to $\frac{3}{4}''$ from the edge of the shelf.

4 Install the deeper upper track on the line. **Precaution.** Make sure the edge of the track is parallel with the edge of the shelf.

5 Install the shallower lower track on the line. again making sure it is parallel to the edge.

6 Cut doors to size.

7 Cut or install door pulls.

8 Push the inner door all the way to the top of the upper track. then let it drop into its lower track.

9 Test the smoothness of operation. **Note.** If the door tends to drag. coat the bottom edge with soap or paraffin.

10 Repeat steps 8 and 9 with the outer door.

Questions

1 Name the four types of cabinet hinges. and their points of attachment on doors and cabinets.

2 What is the purpose of a spring-loaded hinge?

3 What is the alternative piece of hardware?

4 Along what line are cabinet door knobs installed?

5 If the opening for a pair of sliding doors is $22\frac{1}{2}''$ wide and 18" between tracks. what size should the finished doors be?

Unit 27-8   **Install Drawers**

A cabinet door that doesn't work smoothly or latch properly is an annoyance to the homeowner. A drawer that doesn't work properly is pure frustration. The fault for poor drawers lies either with the sliding hardware selected. with the carpenter. or both.

Good drawer hardware is readily available, and it is no place to cut costs. When installing drawers, the carpenter should do the best possible job of installation with the hardware available. and the hardware that is used should operate smoothly when installed according to directions. If it doesn't. the carpenter should discuss with the supervisor or the builder the need for providing better hardware.

Drawers up to 24" wide receive a single knob or pull. Wider drawers need two. Single knobs or pulls are centered between the sides of the drawer. Pairs are lo-

cated about 6" from each side. Both knobs and pulls are usually centered between top and bottom of fronts of drawers up to 8" deep. When the bottom drawer in a cabinet is very deep. such as a bread drawer. the knob or pull is usually moved up from the center so that the user of the drawer doesn't have to stoop so far to open it (Fig. 27-8.1).

Procedure

1 Read the manufacturer's installation instructions and study the diagrams provided with the drawer slides.

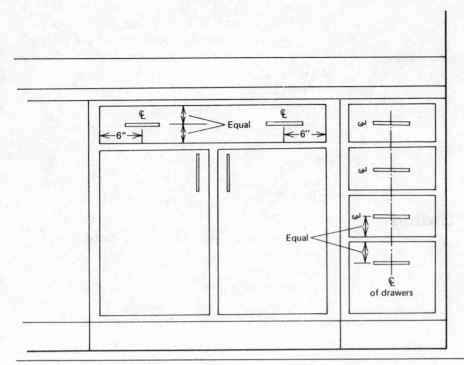

Fig. 27-8.1 Typical locations for drawer hardware.

2 Make sure that the cabinet has framing members in the locations needed for attachment of slides. If not. cut and install them.

3 Install drawer slides and guides in the cabinet. **Precaution.** Recheck all fasteners to make sure they are tight. and that slides and guides will not move out of place during years of use.

4 Install matching parts on the drawer itself, again doublechecking their tightness.

5 Fit the drawer into the opening, and test its operation when empty. **Note.** Move the drawer in and out to make sure it moves easily and does not bind. Attempt to move the drawer from side to side at various degrees of openness to make sure the drawer remains in line. Attempt to pull the drawer all the way out to make sure the stop prevents it.

6 Wrap one or two bricks in cloth (or something of equivalent weight) and repeat step 5.

7 When the drawer passes all tests, install the knob or pull.

Questions

1 If a drawer front is 30" wide. how far apart should the knobs be?

2 Name three problems with drawers that can be avoided with proper installation and good hardware.

Unit 27-9   Install Medicine Cabinets

The standard medicine cabinet is designed to fit into a wall between studs. Heights vary with the manufacturer. but widths are either $14\frac{1}{2}$" or $30\frac{1}{2}$". Some makes are just $3\frac{1}{2}$" deep, and only the door and door frame protrude from the wall. Others are slightly deeper and extend a little farther into the bathroom. The opening is framed like any other wall opening with a header and sill, and walls are already finished to the edge of the opening at the time of installation. Cabinets have pre-drilled holes in the sides for attachment with screws.

A relative newcomer in bathroom storage is the cabinet designed to hang on the wall. Often called a **cosmetic box,** it is a short but wide box, usually made of metal, that extends about 6″ out from the wall. Attachment is made with screws into studs or blocking in the wall.

Medicine cabinets come with several shallow shelves that are either attached to fixed brackets or rest on supports that permit random locations. Cosmetic boxes rarely have shelves. Medicine cabinets may have hinged or sliding doors, usually mirrored. Doors to cosmetic boxes always slide.

Procedure for Installing a Medicine Cabinet

1 Compare the dimensions of the rough opening with the width and height of the cabinet. The cabinet should fit with little tolerance.
2 Check the opening for square.
3 Remove sliding doors. **Note.** Do not remove a hinged door.

4 Fit the cabinet into the opening. with the flange of the cabinet flush against the wall on all sides. **Precaution.** Make sure it is both level and plumb. Shim as required.
5 Attach the cabinet to studs with screws and washers:
6 Replace sliding doors.
7 Clean the mirror in the door.

Procedure for Installing a Cosmetic Box

1 From drawings determine the location of the box. **Note.** It is usually centered above the lavatory.
2 On the wall surface mark the bottom line of the box, and check the line for level.
3 Remove the sliding doors. and set the box in position on the line.
4 Install the box with screws driven into studs or blocking. or with wall anchors into finished surface material.
5 Replace and clean the sliding doors.

Questions

1 What is the primary difference in installation between a medicine cabinet and a cosmetic box?
2 What do you think is the reason for removing sliding doors before a bathroom cabinet is installed?

28

Interior Trim

Trim is any material that finishes off another material, or covers a seam between two materials. A molding is trim stock shaped to serve a specific purpose. Figure 28-0.1 shows typical locations for interior trim and the names of the moldings required in each location.

Moldings have been used to finish residential interiors for centuries. Before the development of wood carving machinery, moldings were handmade by the finest artisans. They were usually thick and rather heavy in scale. Over the years moldings have gradually become smaller and thinner. In remodeling a house only 50 years old, for example, it is impossible to match the original moldings without having them made to order.

Trim Materials

Wood is still the basic trim material. Most moldings are shaped from such softwoods as ponderosa pine and Idaho white pine, but gum, cedar, and local woods are also used for moldings that will be painted. Moldings to be stained are usually milled from hardwoods, such as ash, birch, oak, and poplar, and such richly decorative woods as walnut and mahogany. The grade of woods used for trim is always #1 clear and almost completely free of defects of any kind.

Recently developed substitutes for wood that are growing in popularity among builders are hardboard, vinyl, and plastic reinforced with glass fibers. All three are more dimensionally stable than wood, and nonwood moldings are less likely to break or split. Moldings are also made of aluminum, rubber, and ceramic tile for specific trimming purposes. They are commonly specified in places where moisture would cause wood to deteriorate quickly — such as around bathtubs, shower stalls, countertops beside lavatories and sinks, and along floors that are likely to be washed.

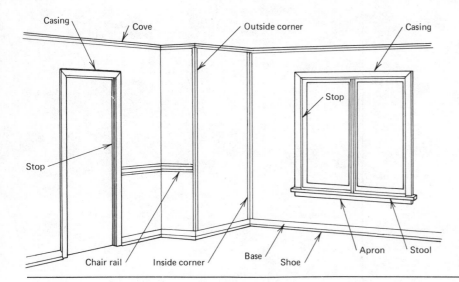

Fig. 28-0.1 The names of common moldings and their locations.

How Trim Comes

Trim may be bought in three ways:

- In random lengths from 7' on up. The longest pieces are used along walls and the shorter pieces around openings.
- In standard lengths of 10', 12', 14', 16' and occasionally 20'. Some lumber dealers buy odd-foot lengths, such as 7', 9', and 11'.
- By the package. For standard-sized windows and doors complete trim packages are available, sometimes already primed. All the carpenter has to do is open the package or unwrap the bundle and begin installation.

Care

Trim warps and breaks easily. Therefore it should be stored out of traffic, flat on the floor, and safely away from any moisture. When trim gets damp, the moisture raises the grain, and the surface must then be carefully sanded before it can be stained or painted. Ideally, trim should be completely dry when installed. Then, as it takes on moisture from the air, joints tend to close tighter. It is a good idea to prime the hidden sides of all trim, and the cut ends of pieces that will be painted but not stained.

Installation

The standard order of trim installation is to finish off exterior and interior doors, then windows (Units 28-1 and 28-2). Next is to mold each room beginning at the top and working down (Units 28-3 to 28-6). Last on the list are finishing closets (Unit 28-7) and installing special items such as mantels and shutters (Units 28-8 and 28-9).

All trim is light in weight, and one carpenter alone can install most trim at the rate of about one room per day. Long pieces are quite flexible and hard to hold still, however. A second pair of hands is very useful when the piece to be installed is more than 8' long.

At the beginning of this section we emphasized the importance of quality in interior finish. In no part of a carpenter's work is quality more important than in installation of interior trim. A poor fit or a bad joint shows up like a beacon. So work slowly and carefully to do the job right the first time. And if it isn't right, don't leave it. Do it over.

Questions

1 In what three ways is trim available for purchase?

2 Why should trim be kept as dry as possible?

3 Name four woods used in milling moldings.

4 Of what other materials are moldings made?

Unit 28-1   **Install Door Casings**

The molding used to finish interior doors is called a casing, and the process of trimming a door is called **casing the opening.** It is a simple procedure with few steps.

Casings vary in width from as narrow as 2″ to as wide as 4″. Standard thickness is $\frac{3}{4}″$, although some wide casings are thicker. Figure 28-1.1 shows six typical shapes. The simplest shapes are designed for houses with a contemporary flavor, and the more ornate designs for more traditional houses, such as colonial.

Door trim consists of two jamb casings and a head casing. They may be fitted together in three ways (Fig. 28-1.2). The standard method (left) is to miter the corners, a procedure that can be used with any shape of casing. The second method is easier, but the result does not look as finished. It is to cut the ends of casings square (center), then cover the edges with a backband (Fig. 28-1.3). The third method (right) is to cut the ends square and fit square blocks in the corners. This was a common method in fine homes a couple of generations ago. The blocks are thicker than the casings and other moldings, and act as a neat stop. They are

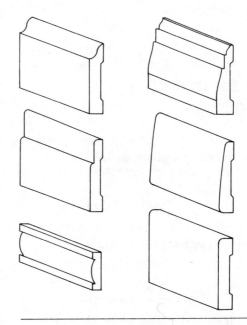

Fig. 28-1.1 Typical casings for doors and windows. Note that they are usually plowed out on the back so that one edge rests flat on the door frame and the other flat against the wall, even when the two surfaces are not exactly in the same plane.

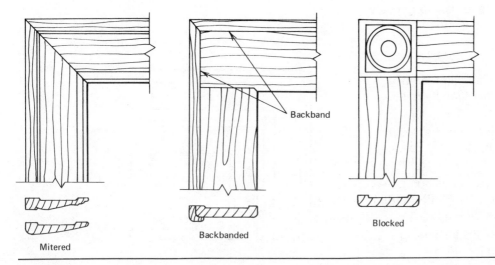

Mitered

Backbanded

Backband

Blocked

Fig. 28-1.2 Two lengths of casing may meet at a corner in a mitered joint (left), a butt joint covered by backbands (center), or at a butt joint against a decorative corner block.

usually the same width as the casing, and may be plain or have a design cut into their faces.

Casings are always installed with a slight setback from the edge of the door frame. The standard setback is $\frac{1}{4}''$, but may be as little as $\frac{1}{8}''$ if wall space around the casing is severely limited.

Except when blocks called **plinth blocks** are specified at floor level, jamb casings extend all the way to the finish floor, and base trim butts against them.

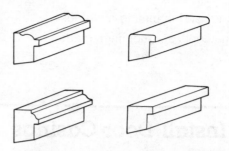

Fig. 28-1.3 Typical backbands. The lip fits over the edge of the door casing.

Procedure for Installing Mitered Casings

1 Measure off the setback on all three sides of the opening, and mark the distance at the top, bottom, and midpoint of both jambs, and at the corners of the head (Fig. 28-1.4).

2 Square the lower end of a piece of jamb casing.

3 Set the casing on the finish floor and line it up on the three setback marks on the jambs.

4 On the inside edge mark the location of the start of the miter cut.

5 Make the cut in a miter box with a backsaw.

6 Tack the casing temporarily in place. **Note.** Use 4- or 5-penny nails at the thinner jamb edge of shaped casings, and 7- or 8-penny nails at thicker edges.

7 Miter one end of the head casing, and test the fit. **Precaution.** The fit will rarely be exact the first time, but one or two light passes with a block plane will usually create a tight fit.

8 Tack the mitered end of the head casing in place with a single nail, and mark the location of the second miter cut.

9 Make the second cut.

10 Repeat steps 2 through 6 for the other jamb casing. **Precaution.** Remember that the miter is reversed on the other jamb casing.

11 With both jamb casings tacked in place, test the fit of the head casing, and adjust the second miter as necessary.

12 When the fit is flawless, nail jamb casings in place, spacing nails about every 16". **Precaution.** If casings are hardwood, predrill the holes.

13 Cross-nail at the corners (Fig. 28-1.5), using 6-penny nails and predrilling the holes.

14 Sink all nails with a nailset, and fill the holes.

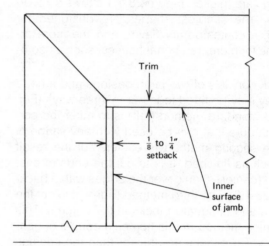

Fig. 28-1.4

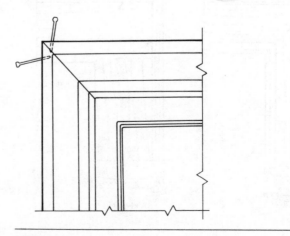

Fig. 28-1.5

Procedure for Installing Squared Casings

1 Follow steps 1 to 3 of the previous procedure for both jamb casings.

2 Measure the distance from the floor to the head setback line along both jambs. The measurements should be equal.

3 Cut both jamb casings to the length in step 2.

4 Tack jamb casings in place on the setback lines.

5 Mark the length of the head casing. **Note.** It extends from outside edge to outside edge of jamb casings, and should be flush with the ends of both casings and the head setback line.

6 Tack the head casing and check the fit. Scribe and shape the bottom edge at jambs if necessary.

7 When the fit is good, nail all casings in place.

8 Nail into drilled holes through the head casing into the jamb casing at both corners (Fig. 28-1.6).

9 Square one end of backbands for jambs.

10 Fit the backbands in place, and mark the miter cuts.

11 Make the cuts.

12 Tack one jamb backband in place, and mark the miter cut for the head backband.

13 Make the cut, fit the backband in place, and mark the second miter cut.

14 Make the second cut, and test the fit at the second jamb backband.

15 When the fits are tight, nail the backbands to the casings. **Note.** At the miters cross-nail as shown in Fig. 28-1.7.

16 Countersink all nails and fill the holes.

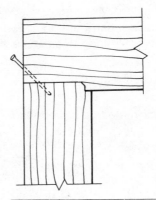

Fig. 28-1.6

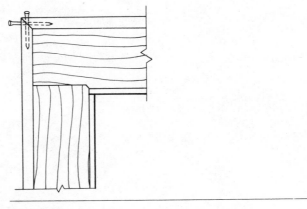

Fig. 28-1.7

Procedure for Installing Casings with Corner Blocks

1 Measure off and mark the three setback lines. **Note.** Carry lines onto wall surfaces (Fig. 28-1.8).

2 Install plinth blocks at the floor on the setback lines (Fig. 28-1.9).

3 Install corner blocks flush with the pairs of setback lines where they cross.

4 Measure the distance between blocks to determine the lengths of casings.

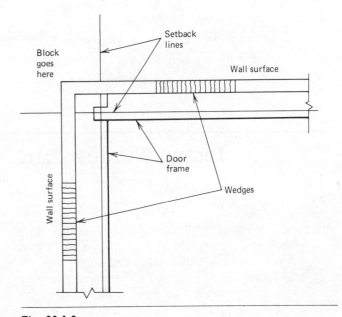

Fig. 28-1.8

5 Square one end of each casing, and cut it to
length. **Precaution.** Cut about $\frac{1}{16}''$ long to assure a
tight fit.

6 Fit the casings in position on the setback lines,
and nail them in place.

7 At the floor toenail downward into holes predrilled
through the jamb casing into the plinth blocks
(Fig. 28-1.9).

8 At upper corners nail through the blocks into
jamb and head casings.

9 Countersink all nails, and fill the holes.

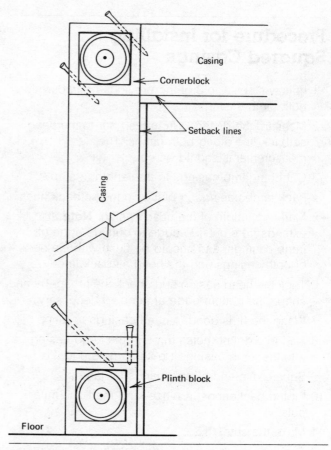

Fig. 28-1.9

Questions

1 What is the purpose of a backband?

2 What do you think is the reason for a setback at a door or window casing?

3 Describe the three common ways of casing an opening.

4 What step was omitted from these procedures?

Unit 28-2   **Install Window Trim**

Like doors, windows are trimmed at the head and
jambs with casings, which are installed in the same
way as door casings. At the sill they may be finished in
two ways—with a projecting **stool** and **apron** (Fig. 28-
2.1, left), or with a flush stool and casing (right).

In some installations the carpenter must also install
window stops. If the windows come from the manufacturer already hung in their frames, the stops are in
place.

Stools

The stool is installed first. It has a flat, smooth top surface, and a shaped outer edge (Fig. 28-2.2). The inner
edge is usually cut square, but if provided by the window manufacturer may have a tongue that fits into a
matching groove in the window frame (Fig. 28-2.3). The
underside is either flat or cut to the slope of the sill

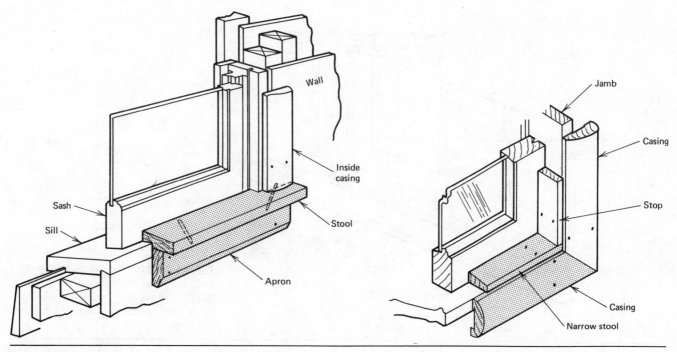

Fig. 28-2.1 Most windows are trimmed at the sill with a stool and apron (left). Sometimes small windows or banks of windows have only a narrow stool (right), and the opening is cased on all four sides.

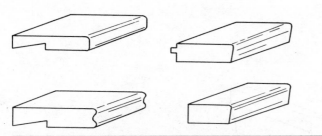

Fig. 28-2.2 Typical shapes of window stools. The two at the left are rabbeted to fit over the sloping window sill common with doublehung windows. The two at right are cut to fit tightly against framing members of manufactured windows such as casements and sliding windows.

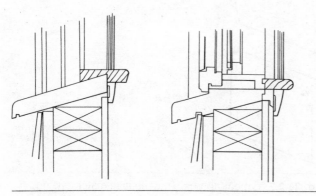

Fig. 28-2.3 How stools fit against frames.

member of the window frame. Stools are a stock item at lumberyards.

Each stool must be notched to fit into the window opening (Fig. 28-2.4). During the marking and fitting process, the stool must be kept absolutely level. One common method is to tack to the sill a pair of thin plywood strips long enough to support the stool against the wall (Fig. 28-2.5). If the apron is beveled on the under side, cut off a short piece before marking it. and test the fit against the angle of the window sill. The apron should be exactly level as it projects into the room.

The stool must also be trimmed to length. Its length is the width of the window opening. plus the width of the two jamb casings. plus two returns (Fig. 28-2.6). A **return** is the extension beyond the casings at the sides of the window. A standard return is about 1″. Returns are rounded to match the front edge of the stool.

Aprons

Aprons (Fig. 28-2.7) are similar in shape to casings. and in most houses they match them in design. The only difference is that casings are plowed out on the back, and aprons are not. The length of the apron is equal to the dimension from outside of casing to outside of casing—in other words. the length of the stool

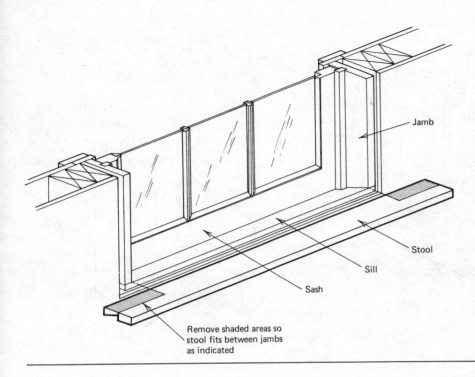

Fig. 28-2.4 Stools must be notched to fit between jambs. The depth of the notches is the distance between the surface of the wall and the face of the sash.

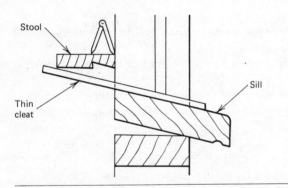

Fig. 28-2.5 To hold a stool level during marking, support it on a pair of thin cleats nailed to the sill.

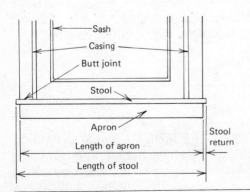

Fig. 28-2.6 How the lengths of a stool and apron compare and where they are measured.

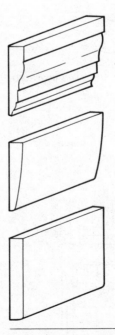

Fig. 28-2.7 Typical aprons. They usually match casings in shape, but their backs are flat instead of plowed.

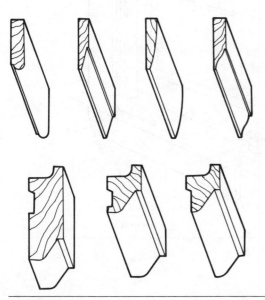

Fig. 28-2.8 Typical shapes of stops. Those at the top may be used with either doors or windows. Those at the bottom are for use with windows that swing outward.

without returns. Often the ends are shaped with a coping saw to match the contours of the apron's face.

Stops

Window stops are similar in shape to door stops, but are often a little smaller. There are two basic types (Fig. 28-2.8). One type is thin, and is designed for use with sashes that slide. The other type is thicker, and is routed on the inside to prevent hinged sashes from binding as they open and close.

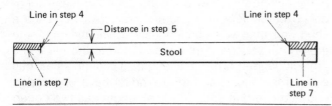

Fig. 28-2.9

Procedure for Installing a Stool

1 Tack plywood cleats to the window sill as shown in Fig. 28-2.5.

2 Set the stool against the wall. approximately centered on the window opening. and hold it level.

3 Check to see how flat the edge of the stool is against the wall.

4 Across the top of the stool mark the width of the window opening into which it must fit. **Note.** If stops are in the way, remove them. The stool fits under the stops.

5 With a scriber measure the distance between the inner edge of the stool and the inner face of the sash.

6 Mark off the distance in step 5 on the lines drawn in step 4.

7 a. If the stool is flat against the wall in step 3, draw lines from the marks to the ends of the stool parallel to the wall (Fig. 28-2.9). The lines in steps 4 and 7a mark the shape of the notches to be cut in the stool.

7 b. If the check in step 3 shows the wall is uneven, use the scriber to mark the contours of the wall from the marks to the end of the stool. The lines in step 4 and 7b mark the shape of the notches to be cut in the stool.

8 Cut the notches with a saber saw if lines are straight, or a coping saw if lines are irregular.

9 Cut the stool to length, and round the edges.

10 Test the fit, again holding the stool absolutely level on the cleats. The stool must fit exactly at the walls and jambs, and with only enough tolerance at the sash (about $\frac{1}{32}$″) to allow for coats of paint.

11 When the fit is tight, remove the cleats and recheck the level. **Precaution.** If the stool doesn't sit exactly level, plane the rabbet lightly.

12 Prime all hidden surfaces to keep moisture out of the stool.

13 Predrill holes, and attach the stool to the sill with 8-penny casing or finishing nails. Blind-nail at the jambs where the nails will be covered by casings.

Procedure for Installing an Apron

1 Install casings around the window as described in Unit 28-1. **Precaution.** Be careful not to knock the stool out of position.

2 Determine the length of the apron by measuring the distance between outside edges of casings.

3 On the wall below the stool lightly mark the locations of apron ends.

4 Cut the apron to length.

5 Shape the ends as required.

6 Predrill holes at the locations shown in Fig. 28-2.10. and start nails into the holes.

7 Set the apron between the marks and push it flush with the underside of the stool.

8 Nail the apron firmly to the sill and studs.

9 At the midpoint of the stool nail into the sill with an 8-penny casing nail. and into the apron with a 3-penny nail (see Fig. 28-2.10). **Precaution.** This last nail maintains a tight joint. Be sure to drive it straight so that it doesn't come through the face of the apron.

10 Countersink all exposed nails, and fill the holes.

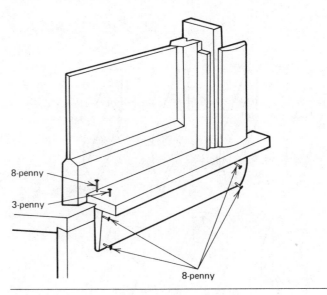

8-penny

3-penny

8-penny

Fig. 28-2.10

Questions

1 In general terms. what is the difference between the length of a stool and the length of its apron?

2 Why are casings installed after the stool but before the apron?

3 Why are some window stops routed out on the hidden side?

4 At what point do window stops end?

5 What do you do if the wall beside the window is uneven?

Unit 28-3   Install Ceiling Trim

The trim specified where the ceiling meets walls may vary from none to a three-piece cornice. even in the same house. When walls and ceilings are either plaster or wallboard. moldings are not often specified in small bedrooms. the kitchen. and bathrooms. Where trim is required. it is usually a crown-and-bed molding that is in scale with the size of the room (Fig. 28-3.1). Only in traditional houses are multipiece moldings called for (Fig. 28-3.2). The main purpose of ceiling moldings. called **cornices,** is to relieve the bare look of a room with nothing but flat surfaces.

Of all trim. ceiling moldings are the most difficult to install properly. and the work must be carefully planned in advance.

Planning the Job

All experienced trim carpenters develop their own sequence of work, and they may use a different sequence at the ceiling than they do at the floor. For less experienced carpenters it is usually easiest to begin on a short wall, then work continuously around the room—to the right if you are right-handed, and to the left if you are left-handed. The first length of molding has butt ends. The second and third lengths have one coped end and one butt end. The final length has two coped ends. and must be sprung into place. A long piece is easier to spring than a short one.

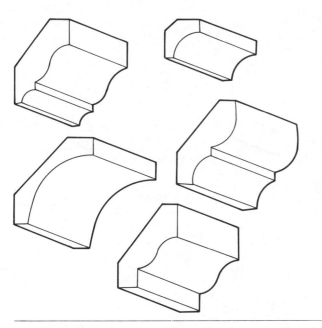

Fig. 28-3.1 Typical cove moldings. The three with irregular shapes are crown-and-bed moldings.

Coping

Making a coped joint is a three-step process. The first step is to set the molding upside down in a miter box. and make a 45° cut with a saw having 10 to 12 teeth per inch (Fig. 28-3.3). Next, mark a 90° cutting line along the bottom edge of the molding. Finally, with a coping saw started on the 90° line, cut by following the surface shape of the molding. The coped end should fit exactly on the contours of another piece of molding running at 90° to the cut piece.

Coping takes practice. and it is worth the time to make several practice cuts with short lengths of molding. In a good coped joint the moldings should fit exactly at their faces. even though they gap behind the joint. Adjustments along the thin edge may be made with sandpaper or a wood file.

Beveling

In many rooms two walls will be longer than the longest length of molding available. When two pieces are needed to trim one length of wall. the ends of the two pieces are bevelled for a tight sprung fit (Fig. 28-3.4). The joint. called a **lap miter joint**, should fall at a stud so that the trim can be nailed into something more solid than wall surface material.

Mitering

Ceiling trim is seldom mitered; coping provides a better fit. The only exception is at an outside corner. Here the ends may be mitered at 45°. and backcut to assure a good fit. Some carpenters cut their miters at about 50° so that only the edges of the molding touch each other. After the pieces are installed. a light pass with a piece of sandpaper smoothes the edges and provides an almost invisible joint.

Priming

Before any trim is installed. the backs and all cut edges should be primed to prevent absorption of mois-

Fig. 28-3.2 This three-piece molding graces the ceiling in the entry, living room, and dining room of a traditional house less than 12 years old. (Photo Jay Leviton, Atlanta.)

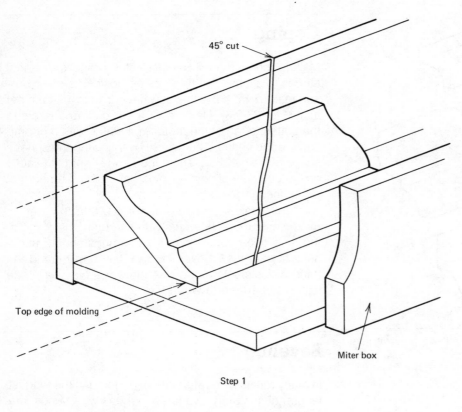

45° cut

Top edge of molding

Miter box

Step 1

90° angle of coping saw

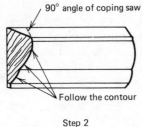

Follow the contour

Step 2

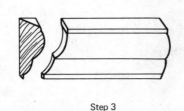

Step 3

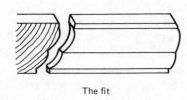

The fit

Fig. 28-3.3 Coping is done in three steps. Step 1 is to make a 45° miter cut with the molding upside down in the miter box. Step 2 is to mark a 90° line on the top of the molding. Step 3 is to cut with a coping saw along the 90° line, following the contours of the molding. Undercutting slightly at step 3 assures an exact fit.

ture. This step is not included in the procedures in Units 28-3. 28-4. and 28-5 because it would be repeated too often. But priming should become as automatic a step as nailing.

Nailing

Ceiling moldings should be nailed every 8 to 10" over their entire length to prevent them from sagging and pulling away from the ceiling. The length of nail and the point of nailing vary with the molding. The best place to nail a **convex** molding—one that curves outward—is at the outermost point on the curve, where the molding is least likely to be damaged by hammermarks (Fig. 28-3.5). For the same reason the best place to nail a **concave** molding—one that bends inward like a cave—is both horizontally and vertically near the edges. Holes should be predrilled in hardwood trim to prevent splitting. All moldings should be cross-nailed at outside corners.

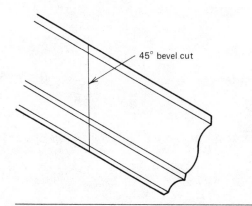

Fig. 28-3.4 In a lap miter joint, butting ends are mitered at 45°. One cut is made with the molding upside down in the miter box; the other is made with the molding right side up.

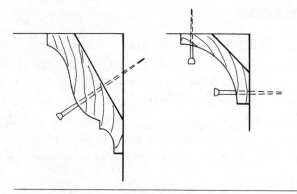

Fig. 28-3.5 Where to nail coves to avoid damaging the wood. Drive nails about halfway in, then finish driving with a nailset.

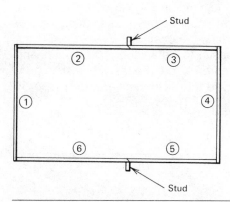

Fig. 28-3.6 The sequence of application of ceiling trim described in the procedure. This is a view looking up at the ceiling.

In the procedures that follow, assume that the room to be molded is 21'-4" long and 11'-6" wide (Fig. 28-3.6), and that all pieces of trim on hand are 12' long.

Procedure

1 With a steel tape measure accurately the length of one short wall at ceiling level.

2 Cut a length of molding (1 in Fig. 28-3.6) for an exact fit, with both ends square.

3 Fit the molding in position. **Precaution.** Make sure that both edges are flush with wall and ceiling surfaces.

4 Tack the molding in place, nailing first about 4" from each end, then every 8 to 10" between.

5 Cope the end of piece 2.

6 Since two lengths must be fitted along the long second wall, determine the length of piece 2 so that it ends at a stud.

7 Bevel the end, with the shorter side against the wall.

8 Tack this piece in place, but do not nail within 12" of the open end. Nail from the corner outward.

9 Measure the remaining unmolded length of wall. **Precaution.** Allow for the bevel fit against piece 2.

10 Cut piece 3 with one end beveled and the other square.

11 Spring piece 3 into position, and check the fit at the bevel.

12 When the fit is good, lightly nail, attaching the ends first.

13 Cut piece 4 to fit along the second short wall, with one end coped and the other cut square.

14 Tack this piece in place.

15 Determine the lengths of the two remaining pieces to trim the second long wall.

16 Cut the shorter piece (5) first, with one end coped and the other beveled as in steps 5 and 7.

17 Install piece 5 as in step 8.

18 Cut the final piece (6) with one end beveled and the other coped. **Precaution.** Cut this piece $\frac{1}{16}$" to $\frac{1}{8}$" long to assure a tight fit.

19 Install the final piece.

20 Check the fit around the entire room.

21 When the fit is tight, drive nails home, countersink them, and fill the holes.

Questions

1 Why is a coped joint better than a mitered joint?

2 In a room that can be trimmed with four lengths of cove, how many ends must be coped if the procedures above are followed?

3 What is the best way to join two pieces of trim on a long wall?

4 How are two pieces of trim joined and fastened at an outside corner?

5 When two pieces of trim meet at a lap miter joint, which should be installed first, and in which direction is the miter made?

Unit 28-4   **Install Base Trim**

The largest piece of trim at the floor is called a **base** or **baseboard** (Fig. 28-4.1). If a room is carpeted, a base is often the only molding installed. With any other finish flooring, a **base shoe** is also required. In some traditional houses a third piece, called a **base cap,** is specified. All pieces are cut and installed in the same way.

and butts against them nicely (Fig. 28-4.2). The ends of base shoe, however, must be beveled so that a narrow end also butts against the casings.

As a rule it is easier to install short pieces first, and leave the longer, more flexible pieces until last. In general, the fewer coped joints, the easier and quicker is the installation.

Planning the Job

Planning the sequence of installation of base trim is similar to planning for ceiling trim, except that now there are door openings to think about. The base is almost always thinner than the edges of door casings,

Joints

At an outside corner, all pieces of base must be mitered and cross-nailed to assure a good looking fit with no grainy edge exposed (Fig. 28-4.3). Treatment at in-

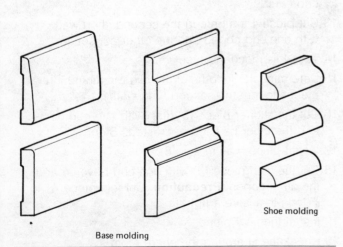

Fig. 28-4.1 Typical shapes of bases and base shoes.

Base molding

Shoe molding

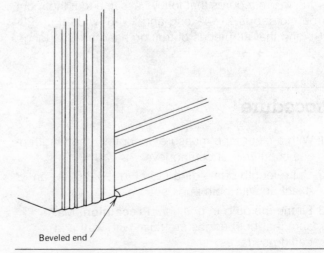

Fig. 28-4.2 Ends of base shoe are usually beveled where they extend beyond a door casing.

Beveled end

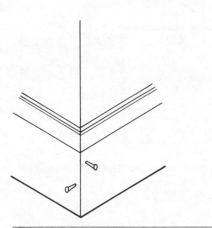

Fig. 28-4.3 Where to cross-nail a base at an outside corner.

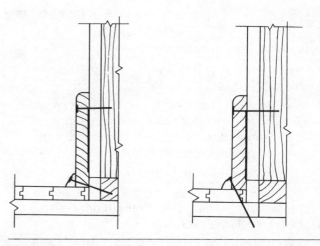

Fig. 28-4.5 Lengths of base are nailed near the top through a flat surface. Base shoe may be nailed either to the base and bottom plate (left) or to the finished floor and subfloor (right).

side corners varies (Fig. 28-4.4). Square-edged bases may be butted or rabbeted. When the base has an irregular shape, ends may be mitered or coped. Walls seldom meet at exactly 90°, and are often slightly off plumb. Under such conditions, coping assures the best joint.

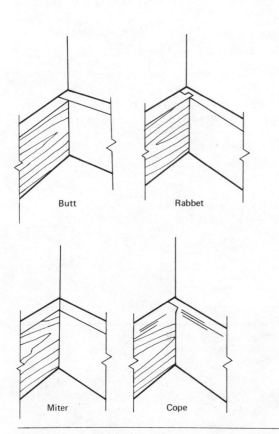

Fig. 28-4.4 Four ways to join lengths of base at an inside corner.

Nailing

Bases are nailed through their flat faces as near the top edge as practicable. Spacing is every 16″ into studs; the bottom wall plate is too low to reach. If the base is applied without a shoe, a second set of nails is driven into the bottom plate about every 8 to 10″.

The base shoe, which is usually nothing more than a quarterround, may be attached in two ways. One way is at about a 30° angle to the horizontal through the base and into the bottom plate (Fig. 28-4.5, left). With this method the shoe moves with the base, and any cracks from settling open up at the floor. The other way (right) is at a 60° angle to the horizontal into the subfloor. With this method the shoe moves with the floor, and any cracks occur between the base and the shoe. Usually the first method is preferable for two reasons: first, a crack at the floor is less noticeable than a crack at the base, and second, the base and shoe are usually painted or stained alike and should move together.

In the procedure that follows, assume that the room is the same as in the previous procedure for installing ceiling trim, with a door in one long wall (Fig. 28-4.6).

Procedure

1 With a steel tape, measure along the floor line from the door casing to one corner.

2 Cut a length of base (1 in Fig. 28-4.6) for an exact fit, with both ends cut square.

3 Tack the base in position.

4 Repeat steps 1 to 3 for piece 2 on the opposite side of the door.

5 Measure and cut pieces 3 and 4 for the two short walls, cutting one end square and coping the other. **Precaution.** Remember that the cuts are reversed for the two pieces.

6 Tack these two pieces in position.

7 Determine the lengths of pieces 5 and 6 needed to trim the remaining wall. **Precaution.** They should meet at a stud.

8 Cut the shorter piece (5) first, with one end coped and the other beveled inward.

9 Tack this piece in position.

10 Cut piece 6 about $\frac{1}{16}''$ long, with one end coped and the other bevelled outward.

11 Spring the final piece into position.

12 Check the fit around the entire room.

13 When the fit is good, drive nails home and countersink them.

14 Beginning on the long wall, determine the lengths of shoe needed. **Note.** Install the shorter piece in the opposite corner from the shorter piece of base, so that the seams don't fall together.

15 Give each piece of shoe a square cut in the corner and a bevel cut at the open end.

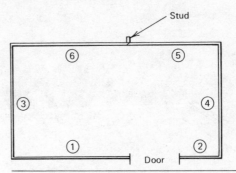

Fig. 28-4.6 The sequence of applying base trim described in the procedure. This is a plan view.

16 Install the short piece, then spring the longer piece into position.

17 Cut pieces of shoe for the end walls with one end square and the other coped.

18 Install these two pieces.

19 Cope the end of a piece of base shoe to fit between the corner and door casing.

20 Fit the piece in the corner, then mark the location of the bevel at the casing.

21 Make the bevel cut, and install the piece.

22 Repeat steps 19 to 21 for the remaining piece of shoe.

23 Check the fit.

24 Countersink all nails, and fill all nail holes in both base and shoe.

Questions

1 What cut must be made in base shoe that is not required in either base or ceiling cove?

2 Name an advantage and a disadvantage to each of the two methods of nailing base shoe.

3 Why is it easier to fit long pieces in place last?

4 What is the advantage to cutting the last piece against a wall a little long?

Unit 28-5 Install Corner Trim

Corner trim is rarely specified in houses except when walls are paneled. Then the panels that fit in corners often have cut edges that must be covered for a clean appearance.

Corner moldings (Fig. 28-5.1) are small and thin. They may be made of wood, hardboard, plastic, or vinyl. They fit into an inside corner and over an outside corner, and are installed after ceiling and floor

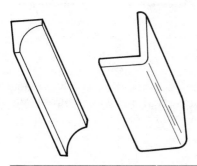

Fig. 28-5.1 Typical inside and outside corner moldings. They are used primarily on paneled walls.

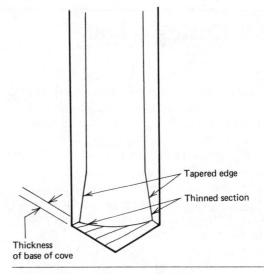

Fig. 28-5.3

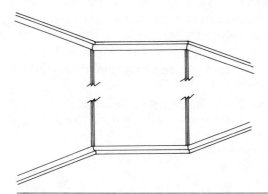

Fig. 28-5.2 Ceiling and floor trim normally runs continuously around corners, and any corner moldings butt against the bottom edge of coves and the top edge of bases.

trim is in place. Rooms look best when top and bottom trim is continuous around walls, and not broken by corner trim (Fig. 28-5.2). Therefore lengths of vertical trim are usually cut to fit between bands of horizontal trim. Thick corner moldings are sometimes carried to the floor like door casings. At the ceiling, however, they stop at the lower edge of cornice moldings, and are tapered for a level butt fit.

Procedure

1 Measure the distance between cornice and base in each inside corner.
2 Check the corners for square at the top. bottom. and midpoint.
3 Cut inside corner molding to the lengths in step 1.
4 Test each piece for fit. If necessary. plane the backs for a flush fit against both walls along their entire length.
5 Check at top and bottom for face lap. If corner trim is thicker than base or cornice trim. put a gradual taper on the outer edges for a flush fit (Fig. 28-5.3).
6 Install moldings, nailing with brads or colored nails in pairs about every 12″ into wall materials.
7 Repeat steps 1 to 6 at outside corners.

Questions

1 Why does ceiling and floor trim run continuously around a room, unbroken by corner trim?
2 When corner trim is thicker than a cove or base, what steps are necessary to achieve an attractive fit?

Unit 28-6   Install a Rail

In most houses today one material covers walls from floor to ceiling. When the carpenter installs base trim and any moldings at the ceiling and in corners, work in that room is done.

Occasionally, however, working drawings call for horizontal moldings, called **rails,** across wall surfaces (Fig. 28-6.1). A **chair rail** is installed about 32 to 36″ above the floor. Its original purpose was to prevent the backs of chairs from damaging walls in a dining room. Today its main purpose is to mold the top of a **wainscot**—the lower part of a wall covered with two different materials. A **picture rail** is a molding installed about 7′ above the floor as a means of hanging pictures. Today a high rail is uncommon except to cover a horizontal joint in paneling in a room with a high ceiling.

When the wall surface is flat, a rail is installed like any other horizontal molding with coped corners. When two wall materials have different thicknesses, as dis-

cussed in Unit 22-3, the usual molding is a two-piece wainscot cap. A one-piece chair rail with a rabbet may also be used, however (Fig. 28-6.2).

Procedure

1 Cut a small piece of chair rail, and check the fit at all points in the room. **Note.** An insert should fit with no more than $\frac{1}{8}$″ gap at any point. An overlaid rail should fit with no gap.

2 Cut the molding to length. Cope the ends in corners, cut ends square that butt against door or window casings, and bevel ends that meet other pieces in long walls.

3 Insert the rail with the flat side up when the upper wall material is thinner than the lower wall

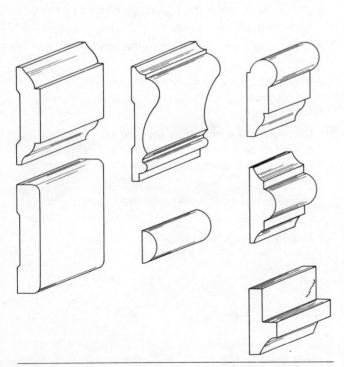

Fig. 28-6.1 Many moldings are available for use as chair rails and picture rails. These are typical; they may be used throughout a house applied either horizontally or vertically.

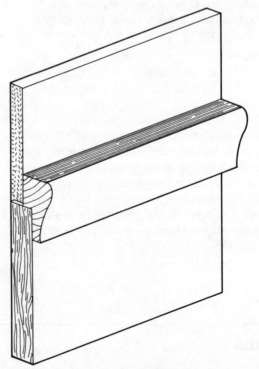

Fig. 28-6.2 A rabbeted chair rail may be used at a seam between two materials of different thickness.

material, and with the flat side down when the upper wall is thicker (see Fig. 28-6.2).

4 Toenail through predrilled holes every 16″ into studs at points just outside the thinner wall surface.

5 Use a nailset to countersink the nails.

Questions

1 What is one of the main uses of a picture rail in today's houses?

2 In what two ways can a seam be molded between two wall materials of different thicknesses?

Unit 28-7 **Install a Rod and Shelf**

In a bedroom closet the minimum storage requirement is a hanging rod and a single shelf. A second shelf is useful for storage of large and little-used items, such as boxes of out-of-season clothing and small traveling bags. Illustrated in Fig. 28-7.1 are the specific rules for locating these storage elements that are important to maximum practical use of closet space:

• Standard height of the rod above the floor is 66″ for men, measured to its center line. Add or subtract ½″ for every 1″ of a man's height more or less than 6′-0″. For example, the best height for a man 6′-4″ tall is 68″. At this height bathrobes and trousers hung by the cuffs clear the floor easily.

• Standard height of the rod for women is 62″. Add or subtract 1″ for every 1″ of her height above or below 5′-6″. At this height full-length gowns won't touch the floor, yet all clothes are within easy reach.

• Fit rods in children's closets on adjustable supports that can be raised as the children grow taller.

• The center of the rod should be no less than 11″ from any wall or door, and 12″ is much better.

• The clearance between the top of the rod and the bottom of the shelf should be 3″ to allow for placing and removing clothes hangers.

• The shelf should be no less than 12″ nor more than 14″ in depth. A narrow shelf isn't much good for storage, and a deeper shelf gets in the way of hanging clothes.

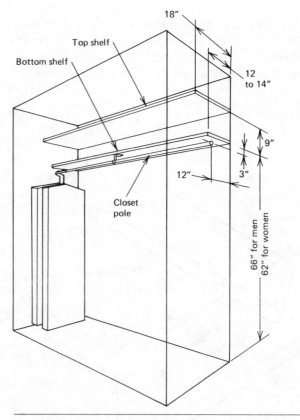

Fig. 28-7.1 Good closet storage meets the dimensional requirements and clearances illustrated in this drawing.

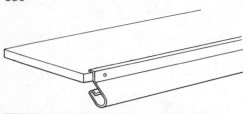

Fig. 28-7.2 A metal bar used for hanging clothes in a closet also stiffens the leading edge of the closet shelf.

- Allow at least 9″ clearance between the bottom shelf and any upper shelf.
- The upper shelf should be the same depth as the lower shelf except in walk-in closets or closets with full-width, floor-to-ceiling doors. Then the upper shelf may be as much as 18″ deep.

The Rod

The most practical rod for hanging clothes is a $\frac{3}{4}$″ steel pipe. It won't bend even under heavy loads of clothing, and is small enough to accept hangers easily. It is not a thing of beauty, however. More attractive is a $1\frac{1}{4}$″ or $1\frac{3}{8}$″ wood dowel. Dowels will deflect under loads over the years, however, and the extra diameter makes

hanging a little more difficult. An alternative solution is a metal hanging bar (Fig. 28-7.2) that is attached to and strengthens the front edge of the bottom shelf.

At its ends the rod is best supported in a slot cut at an angle in the shelf support (Fig. 28-7.3). A wood rod more than 48″ long also requires a center support. Figure 28-7.4 shows four examples. A metal rod less than 72″ long needs no center support.

The Shelf

The most attractive shelves are made of $\frac{3}{4}$″ edge-laminated boards. They require no edge finishing, although the front edge may be shaped in various ways (Fig. 28-7.5) as well as cut square. Plywood makes strong closet shelving; the exposed edge should be finished with a molding strip. Minimum thickness is $\frac{1}{2}$″; thicknesses of $\frac{5}{8}$″ or $\frac{3}{4}$″ are much stronger. Each shelf must be supported at the walls by no less than a 1 × 2 cleat.

Procedure

1 Cut a pair of clear 1 × 6 rod supports to length. Their length should be at least 3″ greater than the depth of the bottom shelf.

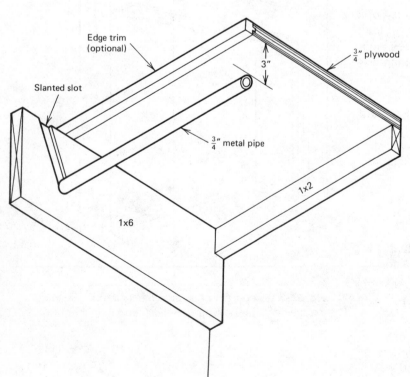

Fig. 28-7.3 The surest end supports for a closet rod are slots cut into cleats that are nailed into closet walls.

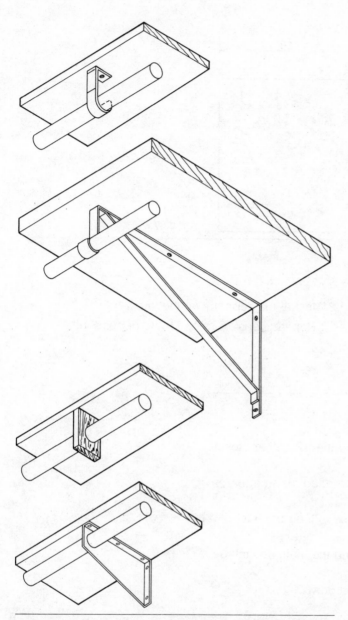

Fig. 28-7.4 Four ways to support the center of a long closet rod. Support is needed for a wood rod more than 4' long and for a metal rod more than 6' long.

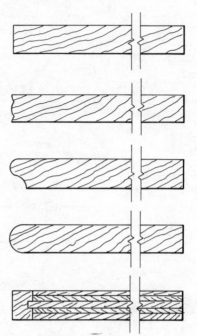

Fig. 28-7.5 The edges of closet shelves may be finished in many ways. Here are just a few.

5 From this mark draw a line tangent to the circle.

6 Draw a line parallel to the line in step 5 tangent to the circle on the other side.

7 Following the lines, rout out a channel ⅜" deep.

8 Notch the corner for the 1 × 2.

9 Round or otherwise finish off the leading edge of the 1 × 6 as required.

10 On all walls mark the height of the underside of the bottom shelf, following the rules in this unit as a guide to the correct height.

11 Install the two 1 × 6 supports with nails driven into studs or wall anchors set into surface materials. **Note.** If the house was carefully built, there should be blocking between studs at the ends of all closet shelves for nailing.

12 Cut and install the 1 × 2 support for the rear edge of the shelf.

13 Cut the rod to length. **Precaution.** It should fit into the slots with no more than ¹⁄₁₆" clearance at each end.

14 Install the rod.

15 Cut the bottom shelf to size, and test the fit. It should fit at the sides with no more than ¹⁄₁₆" tolerance.

16 Finish the leading edge of the shelf as required.

17 Set the shelf in place; do not nail.

2 On the tops mark the location of the edge of the shelf and the notch for the 1 × 2 shelf support (Fig. 28-7.6).

3 Mark the location of the rod, using the rules in this unit as a guide to placement. Draw a circle ⅛" greater in diameter than the rod.

4 Measure out 1" from the edge of the shelf along the top of the support and make a mark.

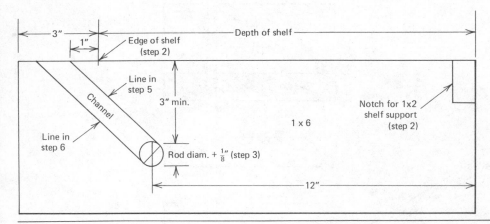

Fig. 28-7.6

18 Install any center shelf support.

19 Mark the location of any upper shelf.

20 Cut 1 × 2 shelf supports for three edges.

21 Install the supports on the lines.

22 Repeat steps 15 to 17 for the upper shelf.

Questions

1 For a woman 5'-2" tall, how far above the floor should the rod in her closet be?

2 What are the maximum and minimum depths recommended for an upper shelf?

3 How should rods in children's closets differ from those in their parents' closets?

4 Name the three common clothes hanger supports, and the main advantage of each.

5 Why do you think closet shelves should not be nailed down?

Unit 28-8   Install a Mantel

The wood mantel surrounding a fireplace was at one time the focal point of decoration in living areas, and sometimes in dining rooms and bedrooms. In many contemporary homes today the mantel has been eliminated, and the face material of brick, stone, or marble provides the decorative effect.

A mantel is still desirable in some homes of traditional design. Then either it is designed to fit the fireplace and is built in a local millwork shop to specifications; or the fireplace is designed to accept a stock mantel. Installation procedures are generally the same.

Mantels usually come assembled and ready to install.

There are two types of mantels. One has a flat back for use when the face of the fireplace and flanking walls are in the same plane (Fig. 28-8.1). This type is attached directly to blocking provided for it behind the wall surface. When the face of the fireplace extends beyond the wall, the mantel has three sides—a face and returns that fill the gap between the face and the two surfaces (Fig. 28-8.2). This type is fastened to blocking attached to the wall on both sides of and above the fireplace. At the opening a slip molding fills the gap.

Fig. 28-8.1 The face of this fireplace and the surrounding wall are in the same plane, and the mantel is flat on the back. It serves as a decorative molding over the seam between brick and wallboard. (Photo Jay Leviton, Atlanta.)

Procedure

1 Study the manufacturer's installation details carefully.

2 Install surface blocking where required by the manufacturer. **Precaution.** Make sure all blocking is plumb and level.

3 If the mantel has a flat back, mark its centerline on the face above the opening.

4 Find the center line of the fireplace, and mark its location on the wall.

5 Place the mantel against the wall, either fitting it against the blocking installed in step 2 or lining up the center lines in steps 3 and 4.

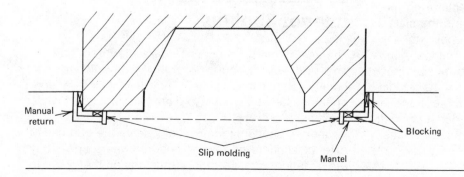

Manual return

Slip molding

Mantel

Blocking

Fig. 28-8.2 The mantel around this fireplace is three-sided, because the fireplace projects beyond the plane of the wall. This is a plan view.

6 Carefully nail the mantel in place, driving nails only far enough to hold.

7 Recheck the positioning.

8 When the position is correct, finish driving the nails, using a nailset to countersink them. **Note.** Nail through flat surfaces; do not nail through curves.

9 Fill all nail holes with wood putty.

10 If required, fit the slip moldings in place, and nail them to the sides of the mantel's face (Fig. 28-8.2).

Questions

1 Name the two basic types of mantels and the method of attachment.

2 What is a slip molding?

Unit 28-9   Install Shutters

In houses of traditional design shutters are often specified for both the exterior and interior. However, their purposes are entirely different.

Exterior Shutters

Shutters were used in Europe long before the first settlers came to America, and they brought the idea with them. As you learned in Chapter 16, the original meaning of the term *window* was an opening in a wall. To close the opening against the weather, shutters were installed in pairs. They latched together to provide some security and a little privacy. Slats inside the frames pivoted, so that air and light could enter rooms even when shutters were closed. It is still a common sight today to see shutters closed against the midday sun throughout southern European countries.

Operating shutters are as tall as the window opening and half as wide. Exterior shutters on most homes today, however, do not close, and the slats in them are fixed. They serve only one purpose: to make windows appear wider, and give a traditional house more pleasing scale. From an architectural standpoint shutters should still be half as wide as the window opening (see Fig. 28-9.4). They seldom are, however, and many times the too-narrow shutters detract from instead of improving exterior appearance. Shutters should look as if they would cover the window completely, even though they may be fixed.

Installation

Most exterior shutters today are bought as a manufactured item, and installed permanently. Even though shutters don't move, they should be removable. Because of the thinness of the parts and the many joints between slats and frame that are exposed to weather, the life of shutters is far less than the life of the house. They should be resealed and repainted at least once every five years.

Fixed exterior shutters are often attached with nails, screws, or clips directly to wall materials. When this is done, however, rain water tends to linger behind the shutters and cause rotting. Also, the tight spaces offer a welcome place for bees, wasps, and flying insects to nest. Both problems can be reduced by attaching shutters to blocking behind the four corners. If shutters flank a door opening, add blocking also at the midpoint.

Interior Shutters

The use of interior shutters also traveled from Europe, and they were popular in colonial days in the hot, humid southeastern states. They were used in door openings between rooms to permit air circulation while preserving privacy.

Interior shutters today usually serve the purpose of exterior shutters of several hundred years ago. They are installed across window openings on the inside to

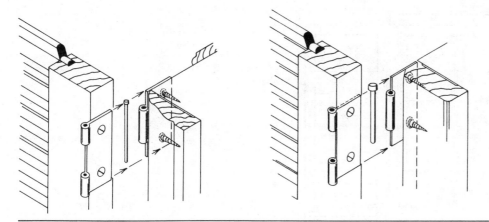

Fig. 28-9.1 Flush hinges are used when the jamb surface is broad enough for attachment of the leaf, and shutters overlay the opening.

control sunlight, direct air flow, and create privacy. They are similar in operation to bifold doors, except that there are no tracks. Shutters are hinged at window jambs, and latch at the center. They always come in pairs. Interior shutters are narrow, and you may need eight pairs of shutters to cover an opening.

Installation

The carpenter may face any one of six conditions when installing hinges for interior shutters. Loose pin butt hinges are used when the jamb casing is thick enough and flat enough to accept hinges (Fig. 28-9.1, left), or when the hinges are attached to a broad jamb (right). Either offset or flush door hinges may be used to mount shutters on a hinge strip (Fig. 28-9.2). At double-hung windows the hinge strip replaces the stop. Offset hinges are used when attachment is made to the insides of jambs (Fig. 28-9.3). When hinges are applied in reverse (right) shutters fit outside window trim and cover the casing. Only when offset hinges are installed in the normal way (left) do shutters fit completely inside the window opening.

Hinges are usually centered about $1\frac{1}{2}''$ from the top and bottom edges of shutters. Use one pair at the jamb on shutters up to 36″ tall, and $1\frac{1}{2}$ pairs on taller shutters. If the shutters fit inside the window opening, they should clear the head and sill by at least $\frac{1}{8}''$ but not more than $\frac{1}{4}''$.

Procedure for Installing Fixed Exterior Shutters

1 Measure the height of the window or door opening inside the casing, and compare it with

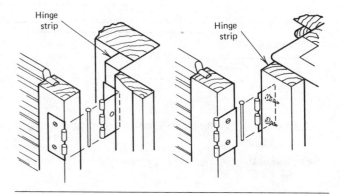

Fig. 28-9.2 A hinge strip may be added as an attachment surface at either of two points. Hinges must be flush or offset to match.

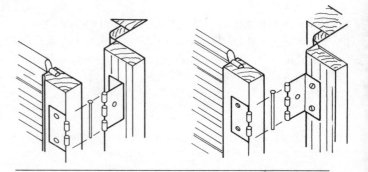

Fig. 28-9.3 Offset hinges may be attached to the insides of jambs. When installed as shown at right, shutters fit inside the opening. This application is common on shutter doors because the hinges permit them to swing 180°.

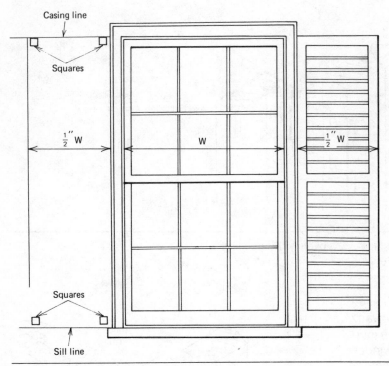

Casing line

Squares

$\frac{1}{2}'' W$ W $\frac{1}{2}'' W$

Squares

Sill line

Fig. 28-9.4

the height of shutters. The measurements should be identical.

2 On the wall surface mark an extension of the top of the sill.

3 Similarly mark the line of the inside of the head casing (Fig. 28-9.4).

4 Cut four 1″ squares of $\frac{1}{4}''$ plywood, and give them two coats of paint. The paint should be the same color as the wall.

5 Measure out the width of the shutter from the jamb casing, and mark this dimension on the lines drawn in steps 2 and 3.

6 About $\frac{1}{2}''$ in from each corner, nail a plywood square.

7 Fit the shutter in place, flush against the jamb casing and on the horizontal extension lines.

8 Attach the shutters at each corner with screws or nails driven into the squares.

Procedure for Installing Interior Shutters

1 From reading specifications or studying jamb conditions, determine the best method of attachment.

2 If required, install hinge strips at the jambs.

3 Determine and mark the locations of hinges on the jambs.

4 Install jamb leaves of hinges with a single screw in each hinge.

5 Locate and install the door leaves of hinges on the shutters.

6 Hang the shutters and test their operation.

7 Adjust hinge locations at jambs as necessary.

8 When operation is smooth, insert the second screw in each jamb hinge and drive both screws home.

Questions

1 What was the original purpose of exterior shutters at windows, and what is the usual purpose today?

2 If a window is $2\frac{1}{2}'$ wide, what is the ideal dimension from the outside edge of one flanking shutter to the outside edge of the other? Why?

3 On double-hung windows, what does the hinge strip for interior shutters replace? Why is this necessary?

4 Why should exterior shutters be blocked away from the wall surface?

Glossary of Building Terms

Adhesive Any substance that holds the surfaces of two materials together. Common adhesives in building are cements, glues, and mastics.

Air-drying A method of removing excess moisture from lumber by natural circulation of air. Compare kiln-drying.

Anchor Bolt A device for connecting wood members to concrete or masonry.

Annular Ring The two-color layer of wood grown in a tree in one year, as seen in a section through the trunk. The colors indicate the growth of springwood and summer-wood.

Apron (1) The piece of window trim applied vertically beneath the sill.

Apron (2) A hard-surfaced area in front of a building, such as the concrete apron of a garage.

Arbor The axle on which a cutting tool, such as the blade of a circular saw, is mounted.

Asbestos A fibrous mineral that is mixed with cement to form various building materials that will not burn.

Asphalt A thick, blackish substance used in the manufacture of various waterproofing materials and coatings. Asphalt occurs in nature on the earth's surface as pitch, but it is also made from waste left over from the manufacture of petroleum products.

Awning Window A window that is hinged at the top and swings outward. Compare hopper window.

Back To bevel the top edge of a member, usually a hip rafter. Same as *chamfer*.

Backband The outer molding sometimes used to cover the edge of a window or door casing.

Backblocking A method of applying gypsum wallboard to a ceiling with joints falling between framing members.

Backerboard A low-cost material applied for strength behind thin surface materials, such as metal siding and wallboard.

Backsplash A short extension of a countertop up a wall. Also called a *splashboard*.

Balloon Framing A method of construction in which vertical structural members are continuous pieces running the entire height of the wall or partition between plates, and floors are supported by studs. Studs in bearing partitions butt against only one joist where joists overlap at a girder. Compare braced framing.

Baluster A vertical support for a handrail in a stairway. Also called a *banister*.

Balustrade A complete handrail assembly, including rails and balusters, subrails and fillets.

Band An edge joist; the term is most often used in farm and utility buildings. Also called a *band joist*.

Base, Baseboard The molding covering the joint between a finished wall and a floor.

Base Cap A small molding sometimes added above a flat-topped base.

Base Shoe A molding added at the bottom of a baseboard to cover the edge of finish flooring or carpeting.

Batt A type of flexible insulation cut to length during manufacture.

Batten A thin, narrow, strip of wood or plastic used to cover joints between boards.

Batter Board One of several simple structures used to mark temporarily the corners of a building and the height of foundation walls early in construction.

Bay Window A structure that projects from an exterior wall containing at least one window, and usually several windows at an angle to each other. Compare bow window.

Beam A horizontal framing member of wood or steel. A wood beam is no less than 5″ thick and at least 2″ wider than it is thick.

Bearing Block A wood block attached to a wood post to help support a girder. Used in pairs.

Bearing Partition Any interior divider that supports the weight of the structure above it. Compare nonbearing partition.

Benchmark A point from which other measurements are made. Also, the device marking that point.

Bevel Any angle not at 90°. Also, a tool for marking such an angle, and the method of creating the angle.

Bifold Door An interior door that is hinged in the middle and folds against the jamb. Usually installed in pairs.

Bird's Mouth The notch in a rafter that rests on the rafter plate. Also called a *seat cut* or *plate cut*.

Blade The longer of the two legs of a framing or rafter square.

Blanket A form of flexible insulation manufactured in long rolls.

Blind-nailing Fastening with nails that can't be seen when the joint is finished.

Blind Stop A small rectangular molding in a window frame against which screens and storm windows fit.

Blocking Any nonstructural member in a floor, wall, or ceiling that serves primarily as a point for fastening finish materials or accessories.

Board Any piece of lumber more than 1″ wide but less than 2″ thick.

Board and Batten A finished wall surface consisting of boards applied vertically with gaps between them, and with battens covering the gaps.

Board Foot (B.F.) A unit of lumber measure equal to 144 cubic inches. The basic unit is 1″ in nominal thickness by 12″ square.

Body The length of a rafter between its points of support.

Bow Warp upward along the length of a piece of lumber laid flat. Compare crook.

Bow Window A structure that projects from an exterior wall containing a series of window panes that form a continuous curve. Compare bay window.

Box Sill The exterior frame for the floor structure in the platform framing system.

Braced Framing A method of construction in which corner posts run the entire height of the wall in a two-story house, but other vertical framing members are only one story in length. Studs in bearing partitions fit between floor joists where joists overlap at a girder. Compare balloon framing.

Bridging One of three methods of bracing between floor joists to stiffen floor structure and distribute live loads. Also, the braces themselves.

BTU The abbreviation for British thermal unit; also written *Btu*. It is a measure of heat and heat loss.

Buck A frame assembly that forms the rough opening for a door or window.

Building Line The line nearest the street along which a building may be built. Also called the *front setback line*.

Building Paper A general term for thin sheet materials impregnated with asphalt for use in weatherproofing. Also called *roofing paper*.

Built-up Roof A surface for a flat or nearly flat roof that consists of several layers of roofing felt, each hot-mopped with asphalt, and a topping of gravel or crushed stone. Also called a *tar-and-gravel roof*.

Butt (1) A type of joint in which materials meet edge to edge. Also the action of forming such a joint.

Butt (2) A hinge.

Bypass Door One of two or more doors that slide from side to side in parallel tracks, thus bypassing each other.

Caneboard A rough-textured insulating material manufactured in sheets from long vegetable fibers.

Cantilever Any part of a structure that projects beyond its main support and is balanced on it. Also the act of projecting and balancing the structure.

Cant Strip A length of lumber with a triangular cross-section used to change the direction of the material covering it. It is most often used where a roof meets a wall at a right angle, and where siding or shingles overlap a foundation wall.

Carriage The diagonal support for treads in a stairway. Also called a *stringer, string,* or *rough horse*.

Carrying Joist A trimmer.

Casement A type of window or door hinged at the side. Casement windows are usually installed in pairs; casement doors are always installed in pairs.

Casing A piece of wood or metal trim that finishes off the frame of a door or window.

Cement (1) Any substance that connects two materials by drying and hardening.

Cement (2) The basic ingredient in concrete, also known as *Portland cement*.

Chamfer A bevel cut on the edge of a piece of lumber. Also, the act of cutting such a bevel.

Check A lengthwise crack across the grain in a piece of lumber.

Chipboard A type of particleboard with treated wood chips the basic ingredient.

Chord Any principal member of a truss. In a roof truss the top chord replaces a rafter, and the bottom chord replaces a ceiling joist.

Clapboard Beveled wood siding.

Cleat Any strip of material attached to the surface of another material to strengthen, support, or secure a third material.

Clerestory A short (in height) exterior wall between two sections of roof that slope in different directions. The term also describes the window frequently used in such walls.

Collar Beam A horizontal board that connects pairs of rafters in opposite roof slopes.

Common Rafter Any of several identical structural members of a roof that run at right angles to walls and end at right angles to main roof framing members.

Concrete A mixture of cement, sand, gravel, and water.

Condensation The process by which moisture in the air becomes droplets of water or ice on a surface whose temperature is colder than the air's temperature.

Conduction Transfer of heat from one material to another through direct contact.

Convection Transfer of heat through movement of a liquid or gas.

Core The center ply or layer of a sheet of plywood.

Corner Bead (1) A one-piece wood molding for finishing corners of some exterior walls.

Corner Bead (2) A metal accessory for finishing and strengthening exposed corners of walls finished with wallboard or plaster.

Corner Post The structural assembly installed wherever two exterior walls meet in wood frame construction.

Cornice (1) The horizontal projection of a roof overhang at the eaves, usually consisting of fascia, soffit, and moldings.

Cornice (2) An old-fashioned molding attached to walls just below ceiling level.

Countersink To bore a recess in wood for the head of a nail or screw. Also, to drive the fastener into the recess below the wood's surface.

Course The exposed layer of an exterior finishing material, such as bricks, shingles, or siding.

Cove A wood molding for covering the joint between a ceiling and walls.

Crawl Space The area bounded by foundation walls, first floor joists, and the ground in a basementless house.

Cricket A small structure built on a roof to divert water, usually away from a chimney. Also called a *saddle*.

Crook Warp upward along the length of a piece of lumber set on edge. Compare bow.

Crossband A layer of wood between the core and face of a sheet of plywood.

Cross-bridging Diagonal wood braces that form an X between floor joists.

Cross-nailing A method of attaching two pieces of wood at right angles by nailing through each piece into the other.

Cross Partition A partition that runs across, usually at right angles to, the floor and ceiling structure. Compare parallel partition.

Crown (1) The high point of a piece of lumber with crook in it.

Crown (2) The horizontal part of a staple between its two legs.

Cup Warp across the grain.

Dado A rectangular groove cut into a board across the grain. (In architecture a dado has two other definitions.)

Dead Load The weight of all structure in place.

Deadman A stake used to hold a brace in position.

Dew Point The temperature at which moisture in air begins to form condensation.

Dimpling The process of setting the heads of fasteners below the surface of gypsum wallboard.

Dormer A structure that projects from a sloping roof, with at least one vertical wall large enough for a window or ventilator.

Double-cheek Cut A two-sided cut that forms a V at the end of some rafters, especially in hip and gambrel roofs.

Double-coursing A method of applying shingles to walls that permits maximum exposure of each course.

Double-glazing A method of insulating glass areas by fitting two pieces of glass into a sealed frame with a vacuum between them.

Double-hung Window A type of window with two sashes that slide vertically past each other.

Doubler The uppermost plate of a wall. Also called a *cap plate* or *rafter plate*.

Downspout A pipe for carrying water from gutters to the ground. Also called a *leader*.

Drip Cap A molding above the frame of a window or exterior door that directs water beyond the surface of the frame.

Drip Edge A strip of metal used to protect the edges of roof structure from water damage.

Drop To lower the top of a hip rafter. Compare back.

Drop Siding Boards with tight-fitting edges and shaped surfaces that are applied as an exterior wall covering, usually vertically.

Drywall A method of covering wall and ceiling surfaces with dry materials, as compared to wet materials such as plaster. The term drywall today refers primarily to the application of gypsum wallboard, which itself is also called drywall.

Dub Off To notch a hip rafter for a loose fit at the rafter plate.

Eave The part of a roof that projects beyond its supporting walls.

Eaves Trough A gutter.

Edge-grained Lumber that has been cut from the log at 45° to 90° to its center, or approximately across the annular rings. Also called *quarter-sawed*.

Edge Joist The outer joist of a floor or ceiling system that runs parallel to other joists. Compare header joist.

Elevation The side of a building as seen straight on. In drawings, a side drawn in two dimensions without foreshortening (perspective).

Escutcheon A decorative plate behind the knob and lock of an exterior door.

Expansion Joint A strip of flexible material, usually asphalt or treated lumber, inserted between sections of a large concrete slab to permit expansion and prevent cracking.

Exposure The distance between the edges of adjacent courses of exterior materials that are applied in layers, such as shingles and siding.

Face-nailing Exposed fastening at right angles to the surface. Compare blind-nailing and toe-nailing.

Fascia A vertical board attached across the lower ends of rafters that forms part of a cornice.

Fiberboard Any of several sheet materials made of various fibers, usually wood.

Field The area of a sheet of building material away from its edges.

Fillet A short length of trim separating balusters.

Finger-jointing A factory method of cutting, fitting, and gluing short lengths of lumber together to form longer lengths.

Fire Cut The angled cut given the end of any wood joist that rests on a masonry wall.

Fire-rated Given a specific F.R.R. after laboratory tests of identical construction. The term applies to finished structural systems.

Firestop A type of blocking between structural members in walls and floors that prevents the spread of flames and heat.

Flakeboard Particleboard made from wood shavings.

Flange A projecting surface, as of an I-beam. Compare web.

Flashing Any water repellent sheet material, such as building paper and some metals, used to cover joints between exterior materials to make them watertight.

Flight An unbroken series of stairs between two floors or a floor and a landing.

Flush In the same plane or at the same level.

Fly Rafter The end rafter of a gable roof with a side overhang.

Footing A base, usually of reinforced concrete, for any vertical part of below-ground construction, such as a foundation wall, column, post, or chimney,

Form A mold of metal or wood used to shape concrete until it has set.

Foundation The part of a building that rests on a footing and supports all of the structure above it.

Frame (1) The woodwork that surrounds a window or door opening.

Frame (2) The wood skeleton of a building. Also called *framing*.

Framing Anchor A device for connecting wood framing members that meet at right angles.

Frieze The part of a cornice that covers the joint between the soffit and exterior wall material.

Frost Line The level below the earth's surface to which the ground freezes in winter.

F.R.R. Abbreviation for *fire resistance rating*—a measure of the ability of a finished structural system to withstand fire.

Furring A light framework for lowering the ceiling in part of a room, such as above cabinets.

Furring Strips Thin wood, plywood, or hardboard strips fastened across studs or joists to form a level or plumb surface for finish wall or ceiling materials.

Gable The part of an end wall between the eaves and ridge of a house with a gable or gambrel roof.

Gable Roof A roof shape characterized by two sections of roof of constant slope that meet at a ridge.

Gain A mortise, most commonly for a hinged door.

Galvanic Action The corrosion that results when two dissimilar metals touch each other in the presence of water.

Gambrel A roof shape similar to a gable roof but with two sections of roof on each side of the ridge, the lower section being steeper than the upper.

Gauge A measure of thickness, usually of metals. Also spelled *gage*.

Girder A main horizontal beam of steel or wood.

Glue An adhesive made from animal substances.

Glue-laminating A factory method of building beams and wide boards by gluing lengths of like-sized lumber together either side to side or edge to edge.

Grade (1) A designation of quality, especially of lumber and plywood.

Grade (2) Ground level. Also the slope of the ground on a building site.

Grade Stamp A mark on lumber or plywood giving its grade and other information about the material.

Grain The direction of fibers in wood.

Gravel Stop A strip of metal at the edge of a built-up roof that is shaped to retain the gravel.

Gusset A plate of metal or plywood used to strengthen the joints of a roof truss.

Gutter A slightly sloping horizontal trough for catching water off a roof. Also called an *eaves trough*.

Gypsum A chalklike rock found throughout the world. It is the basic ingredient of plaster and wallboard.

Hand The side to which a door or window swings.

Hardboard A type of vegetable fiberboard pressed into thin, dense sheets.

Hardwood Any tree or wood from a tree that sheds its leaves. Compare softwood. The two terms do not in any way refer to the hardness of the wood. An oak is a hardwood with hard wood. A poplar is a hardwood with soft wood.

Header Any structural wood member used across the ends of an opening to support the cut ends of shortened framing members in a floor, wall, or roof.

Header Joist The outer joist of a floor or ceiling system that runs across other joists. Compare edge joist.

Heartwood The oldest and densest wood at the center of a tree.

Hip The outside angle where two adjacent sections of roof meet at a diagonal. The opposite of a valley.

Hip Rafter The diagonal rafter that forms a hip.

Hip Roof A roof shape characterized by four or more sections of constant slope, all of which run from a uniform eave height to the ridge.

Hopper A type of window that is hinged at the bottom and opens inward. Compare awning window.

I.I.C. The abbreviation for *impact insulation class*—a rating of the ability of various structural systems to absorb structure-borne sound.

Intermediate Stud A stud that serves no other purpose except to provide vertical support.

Isolation Joint A divider made of a flexible water-repellent material placed at the edges of a concrete slab floor to allow it to move independent of adjacent walls. Also, a similar divider used where foundation walls and basement partitions meet.

Jack Rafter A short rafter, usually running between an eave and a hip rafter or between a ridge and a valley rafter.

Jack Stud A shortened stud supporting the header above a door or window. Also called a *trimmer* or *jamb stud*.

Jalousie A type of window that consists of a series of

glass slats, set into a frame, that open and close together. Similar units are made for exterior doors.

Jamb The side of a window or door opening. Also, the side member of a door or window frame.

Joint The line along which two pieces of material meet.

Joint Compound A chemical mixture in powder or paste form for finishing joints, filling dimples, and covering nicks in gypsum wallboard.

Joint Tape A perforated tape that is embedded in joint compound to reduce the chance of cracking at joints in gypsum wallboard.

Joint Treatment The process of finishing the surface of gypsum wallboard.

Joist A horizontal structural member that, together with other similar members, supports a floor or ceiling system.

Joist Hanger One of a number of variously shaped framing anchors for holding joists in position against a girder.

Kerf The cut made by a saw blade.

Keyway A groove formed in the top of a continuous footing to provide a good mechanical bond between the footing and a poured concrete foundation wall.

Kiln-drying A method of removing excess moisture from lumber by heat in a special oven. Compare air-drying.

Knee Wall A short wall under a slope, usually in attic space.

Knocked Down Not assembled. Abbreviated *K.D.*

Knot A dark, dense, oval defect in lumber caused by the growth of a branch at that point.

Ladder (1) A series of blocks between two studs or joists for attachment of an intersecting partition.

Ladder (2) A series of short pieces of rafter lumber supported by a gable plate for attachment of a fly rafter.

Ladder (3) A device for climbing up or down between two levels.

Lally Column A circular steel pipe for supporting girders and beams.

Landing A platform between flights of stairs.

Leader A downspout.

Leaf The flat part of a hinge that is attached to the edge of a door or window, or to the frame around the opening.

Ledger A strip of lumber attached to the side of a girder near its bottom edge to support notched joists. Also, any similar supporting strip.

Let-in Brace A board set into notches cut diagonally across studs at the corner of a house built with wood frame construction.

Level (1) Horizontal.

Level (2) One of two basic types of optical tools for establishing a series of points at the same level. The two common types are a *builder's level,* also called a *dumpy level,* and a *transit level.*

Level (3) A hand tool for checking the level or plumb of any piece.

Light A pane of glass.

Live Load The total variable weight on a structural member or system. It includes the weights of people, furnishings, snow, and wind. Compare dead load.

Lockset The hardware for locking a door that includes a lock, knobs, roses, latch, and strike plate.

Lookout A horizontal framing member between stud walls and the ends of rafters at the eave.

Loose Fill Pellets of insulation that are poured or blown between joists above a ceiling.

Lug The extension of a jamb member beyond the header or sill of a window or door frame. Also a similar extension of the stiles of a door.

Lumber Wood cut at a sawmill into usable form.

Mansard Roof A type of roof with two slopes on each of four sides, the lower slope steeper than the upper and ending at a constant eave height.

Mantel The facing around a fireplace, including the shelf, which is also called a mantel.

Mastic A gummy mineral substance used as an adhesive for applying some interior materials.

Miter The joint formed by two pieces of material meeting at an angle. Also, the process of cutting at that angle.

Module A common and repetitive unit of measure, such as 16" on centers.

Moisture Content The amount of water remaining in wood after drying, stated as a percentage of the water the wood could hold.

Molding A strip of decorative trim made of wood, wood fibers, plastic, or metal.

Mortise A recess cut into wood.

Mud Sill The lowest plate in a frame wall. Better known as a sill or sill plate.

Mullion A structural divider, usually vertical, between two windows.

Muntin A nonstructural divider between lights in a window or door.

Newel The post supporting a handrail at the bottom, top, or a turn of a stairway.

Nonbearing Partition A dividing wall that supports none of the structure above it.

Nose The rounded front edge of a stair tread. Also called a *nosing.*

Notch A hollow cut made in the edge of a piece of lumber. The sizes and locations of notches in structural members are limited by building codes.

O.C. Abbreviation for *on centers*—a measurement from one center line to the next, usually of structural members.

Offal The part cut off; usually the term refers to the piece that falls off when a panel is cut to size.

Open Time The length of time between application of an adhesive and the moment when it is too dry to stick.

Ordinance A local law.

Overhang The part of a roof that extends beyond supporting walls.

Panel (1) A sheet of building material.

Panel (2) A rectangular piece set into a frame, as in a panel door.

Parallel Partition A partition that runs in the same direction as the structural members in the floor or ceiling. Compare cross partition.

Parquet A type of wood block flooring laid with the grain of alternate blocks at 90° to each other. Also, the pattern itself.

Particleboard A relatively thick fiberboard manufactured from wood chips or shavings.

Parting Strip or **Stop** A small, shaped length of wood that separates upper and lower sashes at the jambs and head of double-hung windows.

Partition An interior wall between rooms or areas.

Party Wall A fire-rated wall between two adjoining living units, such as apartments, in a multifamily dwelling.

Pedestal A separate base, usually concrete, that raises the bottoms of columns or posts above floor level.

Penny A designation, written as *d,* for the length of a nail at least 1″ long.

P.E.T. The abbreviation for *precision end trimmed*—cut to a specified length at the mill.

Pitch (1) The slope of a roof.

Pitch (2) A tarlike substance for sealing seams watertight.

Pitch Pocket A defect in lumber created by a hole in the wood that has filled with sap or pitch.

Plain-sawed Lumber that has been cut from the log by sawing tangent to the annular rings. Also called *flatsawed.* Compare edge-grained.

Plan A horizontal section through a building or part of a building, usually taken at a point about 5′ above floor level.

Plane (1) A flat surface or its imaginary continuation.

Plane (2) A tool for producing a smooth surface.

Plank A piece of lumber with nominal dimensions of at least 2″ in thickness and 6″ in width that is installed flat.

Plaster A powder that is mixed with sand and water and then applied over a plaster base to form a hard finish surface on walls and ceilings. Also, the finished product itself.

Plasterboard Wallboard made of a core of gypsum plaster sandwiched between surface coatings, usually paper.

Plastic Laminate A very thin, durable material applied as a finish surface over a thicker base material.

Plate A horizontal framing member laid flat.

> **Bottom Plate** The lowest plate in a wall in the platform framing system, resting on the subfloor. Also called a *sole plate.*
>
> **Cap Plate** The uppermost plate in a frame wall that has two plates at the top. Also called a *doubler* or *rafter plate.*
>
> **Gable Plate** The top plate of a gable wall.
>
> **Rafter Plate** A cap plate on which rafters bear.
>
> **Sill Plate** The structural member attached to the top of the foundation that supports the floor structure. Also called a *sill.*
>
> **Sole Plate** A bottom plate.

Top Plate The framing member nailed across the upper ends of studs and beneath any cap plate.

Plate Cut The bird's mouth cut into a rafter to fit over the rafter plate.

Platform Framing A method of construction in which wall framing is built on and attached to a finished box sill. Joists and studs are not fastened together as in balloon and braced framing.

Plot Plan A drawing that shows the size and shape of a site and the location of all buildings on it.

Plumb Vertical.

Plumb Cut Any cut in a piece of lumber, such as at the upper end of a common rafter, that will be plumb when the piece is in its final position.

Ply A layer.

Plywood A building material manufactured by laminating thin layers of wood to each other, with the grain at right angles in alternate plies.

Pocket Door A sliding interior door that fits into a recess in a partition.

Polyethylene A plastic material manufactured in thin sheets for use as a vapor barrier.

Post A timber set on end as a vertical support.

Post and Beam A method of construction that requires fewer but heavier structural members than other methods. Also called *plank and beam.*

Purlin (1) The horizontal framing member in a gambrel roof between upper and lower rafters.

Purlin (2) A horizontal member that supports long rafters in other types of roofs.

Putty Stick A wood putty in various wood tones used to touch up nicks and scratches in prefinished panels.

Pythagorean Theorem The square of the hypotenuse of a right triangle is equal to the sum of the squares of the other two sides.

Quarter-sawed Same as edge-grained.

R Factor A number that expresses the resistance of a material, such as insulation, to the transmission of heat.

Rabbet A groove cut in or near the edge of a piece of lumber to receive the edge of another piece.

Radiation Transfer of heat directly from its source to the surface of another material without heating the air between them.

Rafter The main structural member in a roof.

Rafter Seat The notch in a rafter that fits onto the rafter plate. Also called a *bird's mouth.*

Rail (1) A horizontal member between panels of a door, and the main member, either horizontal or vertical, of a window.

Rail (2) A horizontal molding, functional as well as decorative, applied across certain walls. Typical examples are a chair rail and a picture rail.

Railing A protective barrier or guard beside a stairway or between posts, such as around a deck. Also called a *handrail.*

Rake The sloping edge of a roof. Compare eave.

Rebar Short for *reinforcing bar*—a metal rod used to strengthen concrete footings and walls.

Regular Rafter A common rafter.

Return A short piece for finishing the side of a projection, such as a stair tread or mantel. Also, a dimension equal to the depth of that projection.

Reveal The depth of a window or door opening, usually measured from the surface of the exterior wall to the face of the door or window.

Ribband A board set on edge into notches in studs to support joists in a multistory house built with balloon framing. Also called a *ribbon*.

Ridge The uppermost line of a roof where two opposite slopes meet.

Ridgeboard The horizontal board at the ridge to which the top ends of rafters are attached. Also called a *ridge beam* or *ridge pole*.

Rise (1) In a roof, the vertical distance between the top of the rafter plate and the point where a line, drawn through the edge of the rafter plate and parallel to the roof's slope, intersects the center line of the ridgeboard. See Fig. 9-0.4.

Rise (2) In a stairway, the vertical height of the entire stairway measured from floor to floor.

Riser The vertical board between two treads in a stairway. Also, the vertical distance between treads.

Roll Roofing A finish roofing material manufactured in rolls that is composed of fibers saturated with asphalt and surfaced with mineral chips or an asphalt coating.

Roofing Felt Roll roofing without a coating, used in built-up roofs.

Roofing Paper A general term for building paper or roofing felt, used as a base under finish roofing materials.

Rose The part of a lockset that fits around the knob and holds it in place.

Rout To cut out by gouging.

Run (1) In a roof with a ridge, the horizontal distance between the edge of the rafter plate and the center line of the ridgeboard.

Run (2) In a stairway, the horizontal distance between the top and bottom risers plus the width of one tread.

Saddle A two-slope structure, shaped like a saddle, that fits between a chimney and any roof sloping toward it, designed to divert water away from the chimney. Also called a *cricket*.

Sash The operating unit of a window.

Scaffold Any temporary working platform and the structure to support it.

Scale (1) Proportion. In a drawing it is the proportion between the size of an object as drawn and that object full size, such as $\frac{1}{4}'' = 1' - 0''$.

Scale (2) A type of ruler marked with various scales for making scale drawings.

Scarf A lapped joint made by notching the ends of two pieces of lumber for a tight fit.

Schedule A table or listing in a set of working drawings, such as a window schedule, door schedule, or finish schedule.

Schematic A sketch, usually not drawn to scale, that shows in rough detail how a utility system, such as a drainage system, goes together.

Screed A device for achieving the desired thickness of concrete or plaster.

Scribe To mark with a pointed instrument, such as a scriber, for an irregular cut.

Seat Cut The notch in a rafter that fits over the rafter plate. Also called a *bird's mouth*.

Section A drawing that shows a slice through a house or any of its components. A plan is a horizontal section.

Selvage A starter strip of finish material placed at the edge of a surface, such as a roof, as a base for the final layer.

Set The spread given to saw teeth.

Setback The distance between a property line and the nearest point at which a building can be built, as established by local ordinance.

Shake (1) A lengthwise separation along annular rings in a piece of lumber.

Shake (2) A thick wood shingle split by hand.

Sheathing A structural material, usually in sheet form, applied across exterior studs and rafters for strength and protection.

Shed Dormer A dormer with a shed roof.

Shed Roof A roof that slopes in only one direction

Shim A wedge used to move a member or component into position and hold it there.

Shingles Thin pieces of material applied in overlapping courses as a finish on walls and roofs.

Shiplap An L-shaped edge cut into boards and some sheet materials to form an overlapping joint with adjacent pieces of the same material. Also, a board with a shiplapped edge, and the action of cutting such an edge.

Shortening Line The cutting line for the upper end of a rafter that is parallel to the plumb cut but half the thickness of the ridgeboard away from it.

Shutters Originally, pairs of louvered covers, hinged at the sides of windows and designed to close and shield openings from light, weather, and enemy attack. Today, exterior shutters may be louvered or solid panels, are usually fixed and decorative only. Interior shutters come closer to the original purpose; they are hinged and louvered, providing privacy and controlling the passage of light and air.

Sidelight A glass panel or series of panes in a frame beside an entrance door.

Siding In general, any exterior material for finishing a wall. More commonly, an exterior material applied horizontally with lapped horizontal joints.

Sill (1) A sill plate.

Sill (2) The structural member forming the bottom of a rough opening for a door or window. Also the bottom member of a door or window frame.

Single-cheek Cut A bevel cut at the end of a rafter, especially in hip and gambrel roofs.

Single-coursing A method of applying shingles to walls that provides the maximum total thickness of material.

Sleepers Strips of lumber laid over a concrete slab floor as a nailing base for wood flooring.

Slope (1) The variation from straight in the grain of wood.

Slope (2) The pitch of a roof, expressed as inches of rise per 12″ of run.

Soffit The underside of a projection, such as a cornice.

Softwood Any tree or wood from a tree that has needles instead of leaves. Compare hardwood.

Soil Stack The main vertical pipe of a waste disposal system. Also called a *soil pipe*.

Solid Bridging Blocking between joists cut from the same material as the joists themselves and used to stiffen the floor.

Spaced Beam A built-up beam with its main members separated by spacers.

Spacer Any piece of material used to maintain a permanent space between two members. Compare spreader.

Span The distance between structural supports, measured horizontally. In a typical roof the span is twice the run.

Splashboard A short extension of a countertop up a wall. Also called a *backsplash*.

Splice A joint at which two pieces are joined to each other. Also, the action of connecting the two pieces.

Splice Plate A flat piece of perforated metal or plywood used to connect members at splices.

Spline A long, flexible strip of wood, plastic, or metal that fits into slots in adjoining edges of material to hold those edges in line.

Split A separation along the grain that runs completely through a piece of lumber.

Spreader Any piece of material used temporarily to maintain a space, such as between form boards. Compare spacer.

Square (1) At 90° or a right angle. Also, the process of cutting at a right angle.

Square (2) Any of several tools for marking at right angles and for laying out structural members for cutting or positioning.

Square (3) A measure of roofing and some siding materials equal to 100 square feet of coverage.

Stack Wall The partition containing the soil stack.

Stair A single step. The term *stairs* refers to a series of steps.

Stairway A flight of stairs. Also called a *staircase*.

Stairwell The opening in a floor for a stairway.

Standards Metal strips with slots for inserting shelf supports.

Station Mark The starting point for establishing level.

S.T.C. Abbreviation for *sound transmission class*—a rating of the ability of various structural systems to shut out airborne sound.

Stickers Narrow strips of scrap wood laid between layers of lumber or plywood to let air circulate around them.

Stile A vertical member of a panel door. Compare rail.

Stool The flat shelf that trims the bottom of a window frame on the inside of a wall.

Stop In general, any device that prevents movement.

More specifically, a strip of wood against which a hinged door or window closes.

Strike Plate The part of a lockset attached to the door frame into which the latch fits to hold a door closed.

Stringer In its broadest usage, any horizontal framing member that ties other members together. More specifically in stairway construction, the diagonal member that supports treads. Also called a *string, carriage,* or *rough horse*.

Strongback A long structural framing member, usually the size of a joist, that is attached upright across the tops of ceiling joists as a stiffener.

Stud The main vertical framing member in a wall or partition.

Subfloor The rough floor laid across floor joists and under finish flooring.

Subflooring The material, usually plywood, used for a subfloor.

Subrail A molding on the underside of a handrail grooved to receive balusters.

Suspended Ceiling A false ceiling, hung below ceiling joists, that contains acoustical or lighting panels laid into a lightweight metal grid.

Swaged Shaped to fit, such as the leaves of a hinge that are bent at the knuckle.

T & G Abbreviation for *tongue-and-groove* and also for *tar-and-gravel*.

Tab A flap, as on an asphalt shingle.

Tail The part of a rafter, if any, between the rafter plate and an overhanging eave.

Tail Joist A shortened joist that butts against a header.

Tar-and-gravel Roof A built-up roof.

Template A full-sized pattern.

Thickened-edge Slab A concrete floor slab with thick edges on all sides that rest directly on the ground instead of a foundation wall. Also called a *turned-down slab*.

Threshold A beveled piece of wood, metal, stone, or concrete, set at the sill of a hinged exterior door and sometimes under interior doors, to cover the joint between different finish flooring materials.

Timber A wood framing member with nominal dimensions of at least 4″ by 6″.

Toekick A recess under the front of a base cabinet at the floor line.

Toenail To nail at an angle to the surface.

Tongue The shorter of the two legs of a framing or rafter square. Compare blade.

Tongue-and-groove A type of tight-fitting joint cut at the edges of many kinds of sheet materials, finishing materials, and some lumber. A tongue of material in the center of the edge of one piece fits into a groove in the center of the edge of the adjoining piece.

Tread The horizontal platform of a step. Also, the horizontal distance between risers.

Trestle A low working platform supported on legs.

Trim Any material, usually decorative, used to finish off corners between surfaces and around openings.

Trimmer A short joist or rafter, fastened to another joist or rafter between headers at the sides of an opening, to narrow or stiffen the opening. Also a short stud serving a similar purpose at a door opening.

Truss An assembly for bridging a broad span, most commonly used in roof construction.

Turned-down Slab A thickened-edge slab.

Twist Warp in a board in two directions.

Underlayment A thin, smooth sheet material, usually hardboard or plywood, laid over subflooring as a smooth base for application of thin finish flooring materials.

Unit Length In roof or stairway construction, the hypotenuse of a triangle whose other two sides are the unit rise and the unit run.

Unit Rise The amount of rise, stated in inches, for every foot of run in a roof.

Unit Run Always 12″.

Valley The inside angle where two adjacent sections of a roof meet at a diagonal. The opposite of a hip.

Valley Rafter The diagonal rafter that forms a valley.

Vapor Barrier A very thin sheet material or coating that prevents the passage of moisture or water vapor. Polyethylene is the most common sheet material, and some paints also act as vapor barriers.

Veneer A thin layer, usually the exposed surface layer. Most commonly applied to plywood and to masonry walls one brick or stone thick.

Vertical-grained Same as edge-grained.

Wainscot The surface material on the lower part of an interior wall, if different from the material on the upper part.

Walers Lengths of heavy lumber or steel used horizontally as braces, usually to lock concrete forms together. Also called *whalers*.

Walking Steps Cleats temporarily attached to the surface of steep roofs as a safety device to help carpenters climb around the roof while finishing it.

Wallboard A material for finishing the surfaces of interior walls and ceilings that is manufactured in large sheets out of gypsum or wood or mineral fibers.

Wane A defect in lumber caused by sawing too close to the outside edge of the log at the mill and leaving an edge either incomplete or covered with bark.

Warp Any variation from straight in a piece of lumber. See bow, cup, crook, and twist.

Water Closet A toilet assembly.

Water Table (1) The level below the surface of the ground at which water is present.

Water Table (2) A two-piece trim for finishing the bottom of an exterior wall surfaced with wood siding or shingles.

Web The center section of an I-beam. Compare flange.

Western Framing Platform framing.

Winder A nonrectangular stair tread used at changes in direction of a stairway without a landing.

Zoning A division of populated communities into smaller districts in order to limit the type of construction allowed within each district.

Table XI Square Root Table

No.	Sq. Root	No.	Sq. Root	No.	Sq. Root	No.	Sq. Root	No.	Sq. Root	No.	Sq. Root	No.	Sq. Root	No.	Sq. Root	No.	Sq. Root	No.	Sq. Root	No.	Sq. Root	No.	Sq. Root
1	1.000	51	7.141	101	10.050	151	12.288	201	14.177	251	15.843	301	17.349	351	18.735	401	20.025	451	21.237	501	22.383	551	23.473
2	1.414	52	7.211	102	10.100	152	12.329	202	14.213	252	15.875	302	17.378	352	18.762	402	20.050	452	21.260	502	22.405	552	23.495
3	1.732	53	7.280	103	10.149	153	12.369	203	14.248	253	15.906	303	17.407	353	18.788	403	20.075	453	21.284	503	22.428	553	23.516
4	2.000	54	7.348	104	10.198	154	12.410	204	14.283	254	15.937	304	17.436	354	18.815	404	20.100	454	21.307	504	22.450	554	23.537
5	2.236	55	7.416	105	10.247	155	12.450	205	14.318	255	15.969	305	17.464	355	18.841	405	20.125	455	21.331	505	22.472	555	23.558
6	2.449	56	7.483	106	10.296	156	12.490	206	14.353	256	16.000	306	17.493	356	18.868	406	20.149	456	21.354	506	22.494	556	23.580
7	2.646	57	7.550	107	10.344	157	12.530	207	14.388	257	16.031	307	17.521	357	18.894	407	20.174	457	21.378	507	22.517	557	23.601
8	2.828	58	7.616	108	10.392	158	12.570	208	14.422	258	16.062	308	17.550	358	18.921	408	20.199	458	21.401	508	22.539	558	23.622
9	3.000	59	7.681	109	10.440	159	12.610	209	14.457	259	16.094	309	17.578	359	18.947	409	20.224	459	21.424	509	22.561	559	23.643
10	3.162	60	7.746	110	10.488	160	12.649	210	14.491	260	16.125	310	17.607	360	18.974	410	20.249	460	21.448	510	22.583	560	23.664
11	3.317	61	7.810	111	10.536	161	12.689	211	14.526	261	16.156	311	17.635	361	19.000	411	20.273	461	21.471	511	22.605	561	23.685
12	3.464	62	7.874	112	10.583	162	12.728	212	14.560	262	16.186	312	17.664	362	19.026	412	20.298	462	21.494	512	22.627	562	23.707
13	3.606	63	7.937	113	10.630	163	12.767	213	14.595	263	16.217	313	17.692	363	19.053	413	20.322	463	21.517	513	22.650	563	23.728
14	3.742	64	8.000	114	10.677	164	12.806	214	14.629	264	16.248	314	17.720	364	19.079	414	20.347	464	21.541	514	22.672	564	23.749
15	3.873	65	8.062	115	10.724	165	12.845	215	14.663	265	16.279	315	17.748	365	19.105	415	20.372	465	21.564	515	22.694	565	23.770
16	4.000	66	8.124	116	10.770	166	12.884	216	14.697	266	16.310	316	17.776	366	19.131	416	20.396	466	21.587	516	22.716	566	23.791
17	4.123	67	8.185	117	10.817	167	12.923	217	14.731	267	16.340	317	17.805	367	19.157	417	20.421	467	21.610	517	22.738	567	23.812
18	4.243	68	8.246	118	10.863	168	12.962	218	14.765	268	16.371	318	17.833	368	19.183	418	20.445	468	21.633	518	22.760	568	23.833
19	4.359	69	8.307	119	10.909	169	13.000	219	14.799	269	16.401	319	17.861	369	19.209	419	20.470	469	21.656	519	22.782	569	23.854
20	4.472	70	8.367	120	10.955	170	13.038	220	14.832	270	16.432	320	17.889	370	19.235	420	20.494	470	21.680	520	22.804	570	23.875
21	4.583	71	8.426	121	11.000	171	13.077	221	14.866	271	16.462	321	17.917	371	19.261	421	20.518	471	21.703	521	22.825	571	23.896
22	4.690	72	8.485	122	11.045	172	13.115	222	14.900	272	16.492	322	17.944	372	19.287	422	20.543	472	21.726	522	22.847	572	23.917
23	4.796	73	8.544	123	11.091	173	13.153	223	14.933	273	16.523	323	17.972	373	19.313	423	20.567	473	21.749	523	22.869	573	23.937
24	4.899	74	8.602	124	11.136	174	13.191	224	14.967	274	16.553	324	18.000	374	19.339	424	20.591	474	21.772	524	22.891	574	23.958
25	5.000	75	8.660	125	11.180	175	13.229	225	15.000	275	16.583	325	18.028	375	19.365	425	20.616	475	21.795	525	22.913	575	23.979
26	5.099	76	8.718	126	11.225	176	13.267	226	15.033	276	16.613	326	18.056	376	19.391	426	20.640	476	21.817	526	22.935	576	24.000
27	5.196	77	8.775	127	11.269	177	13.304	227	15.067	277	16.643	327	18.083	377	19.417	427	20.664	477	21.840	527	2.2957	577	24.021
28	5.291	78	8.832	128	11.314	178	13.342	228	15.100	278	16.673	328	18.111	378	19.442	428	20.688	478	21.863	528	22.978	578	24.042
29	5.385	79	8.888	129	11.358	179	13.379	229	15.133	279	16.703	329	18.138	379	19.468	429	20.712	479	21.886	529	23.000	579	24.062
30	5.477	80	8.944	130	11.402	180	13.416	230	15.166	280	16.733	330	18.166	380	19.494	430	20.736	480	21.909	530	23.022	580	24.083
31	5.568	81	9.000	131	11.446	181	13.454	231	15.199	281	16.763	331	18.193	381	19.519	431	20.761	481	21.932	531	23.043	581	24.104
32	5.657	82	9.055	132	11.489	182	13.491	232	15.232	282	16.793	332	18.221	382	19.545	432	20.785	482	21.955	532	23.065	582	24.125
33	5.745	83	9.110	133	11.533	183	13.528	233	15.264	283	16.823	333	18.248	383	19.570	433	20.809	483	21.977	533	23.087	583	24.145
34	5.831	84	9.165	134	11.576	184	13.565	234	15.297	284	16.852	334	18.276	384	19.596	434	20.833	484	22.000	534	23.108	584	24.166
35	5.916	85	9.220	135	11.619	185	13.602	235	15.330	285	16.882	335	18.303	385	19.621	435	20.857	485	22.023	535	23.130	585	24.187
36	6.000	86	9.274	136	11.662	186	13.638	236	15.362	286	16.912	336	18.330	386	19.647	436	20.881	486	22.045	536	23.152	586	24.207
37	6.083	87	9.327	137	11.705	187	13.675	237	15.395	287	16.941	337	18.358	387	19.672	437	20.905	487	22.068	537	23.173	587	24.228
38	6.164	88	9.381	138	11.747	188	13.711	238	15.427	288	16.971	338	18.385	388	19.698	438	20.928	488	22.091	538	23.195	588	24.249
39	6.245	89	9.434	139	11.790	189	13.748	239	15.460	289	17.000	339	18.412	389	19.723	439	20.952	489	22.113	539	23.216	589	24.269
40	6.325	90	9.487	140	11.832	190	13.784	240	15.492	290	17.029	340	18.439	390	19.748	440	20.976	490	22.136	540	23.238	590	24.290
41	6.403	91	9.539	141	11.874	191	13.820	241	15.524	291	17.059	341	18.466	391	19.774	441	21.000	491	22.159	541	23.259	591	24.311
42	6.481	92	9.592	142	11.916	192	13.856	242	15.556	292	17.088	342	18.493	392	19.799	442	21.024	492	22.181	542	23.281	592	24.331
43	6.557	93	9.644	143	11.958	193	13.892	243	15.589	293	17.117	343	18.520	393	19.824	443	21.048	493	22.204	543	23.302	593	24.352
44	6.633	94	9.695	144	12.000	194	13.928	244	15.621	294	17.146	344	18.547	394	19.849	444	21.071	494	22.226	544	23.324	594	24.372
45	6.708	95	9.747	145	12.042	195	13.964	245	15.653	295	17.176	345	18.574	395	19.875	445	21.095	495	22.249	545	23.345	595	24.393
46	6.782	96	9.798	146	12.083	196	14.000	246	15.684	296	17.205	346	18.601	396	19.900	446	21.119	496	22.271	546	23.367	596	24.413
47	6.856	97	9.849	147	12.124	197	14.036	247	15.716	297	17.234	347	18.628	397	19.925	447	21.142	497	22.294	547	23.388	597	24.434
48	6.928	98	9.899	148	12.166	198	14.071	248	15.748	298	17.263	348	18.656	398	19.950	448	21.166	498	22.316	548	23.409	598	24.454
49	7.000	99	9.950	149	12.207	199	14.107	249	15.780	299	17.292	349	18.682	399	19.975	449	21.190	499	22.338	549	23.431	599	24.475
50	7.071	100	10.000	150	12.247	200	14.142	250	15.811	300	17.321	350	18.708	400	20.000	450	21.213	500	22.361	550	23.452	600	24.495

Index